D1760822

Biogeochemistry of Estuaries

Biogeochemistry of Estuaries

Thomas S. Bianchi

Department of Oceanography, Texas A&M University

UNIVERSITY PRESS

2007

OXFORD
UNIVERSITY PRESS

Oxford University Press, Inc., publishes works that further
Oxford University's objective of excellence in research, scholarship,
and education.

Oxford   New York
Auckland   Cape Town   Dar es Salaam   Hong Kong   Karachi
Kuala Lumpur   Madrid   Melbourne   Mexico City   Nairobi
New Delhi   Shanghai   Taipei   Toronto

With offices in
Argentina   Austria   Brazil   Chile   Czech Republic   France   Greece
Guatemala   Hungary   Italy   Japan   Poland   Portugal   Singapore
South Korea   Switzerland   Thailand   Turkey   Ukraine   Vietnam

Copyright © 2007 by Oxford University Press, Inc.

Published by Oxford University Press, Inc.
198 Madison Avenue, New York, New York 10016
www.oup.com

Oxford is a registered trademark of Oxford University Press

Library of Congress Cataloging-in-Publication Data
Bianchi, Thomas S.
Biogeochemistry of estuaries / Thomas S. Bianchi.
p. cm.
Includes bibliographical references and index.

ISBN-13: 978-0-19-5160826
1. Biogeochemical cycles.   2. Estuarine ecology.   I. Title.
QH344.B53 2006
577.7'86—dc22   2005033998

Printed in the United States of America
on acid-free paper

To Jo Ann and Christopher
for their unending support and patience

As we progress further into truly coupled analyses of ecosystems, it is becoming essential for a greater percentage of us to pool our expertise. This professional altruism is increasingly critical to our discipline.

Robert G. Wetzel

Preface

Over the past decade there has been a rapid increase in human population growth along coastal regions of the world. Consequently, many estuarine systems have been affected by the serious environmental impacts of this encroachment. A greater knowledge of the biogeochemical cycling in estuaries, which involves the transformation, fate, and transport of chemical substances, is critical in understanding the effects of these environmental alterations—from a regional and global context. For example, the impact of eutrophication, which is widespread in many estuaries around the world, can only be fully understood in the context of the physical dynamics of each system. Approaching estuarine science from a biogeochemical perspective requires a fundamental background in the subdisciplines of chemistry, biology, geology, and in many cases atmospheric science. My motivation for writing this book is that many, if not all, of the books on estuarine biogeochemistry to date are edited volumes too diffuse and difficult to use as textbooks for advanced undergraduate and graduate courses. This book is focused on biogeochemical cycles and attempts to comprehensively examine the physical, geochemical, and ecosystem properties of estuaries in the context of global and environmental issues. Following an introductory chapter, Estuarine Science and Biogeochemical Cycles, the book is divided into the following parts: I. Physical Dynamics of Estuaries; II. Chemistry of Estuarine Waters; III. Properties of Estuarine Sediments; IV. Organic Matter Sources and Transformation; V. Nutrient and Trace Metal Cycling; VI. Anthropogenic Inputs to Estuaries; VII. Global Impact of Estuaries.

In chapter 1, the reader is provided with a general background on the historical importance of estuaries during the advance of human civilizations (via trade, transportation, food resources, etc.); it is no coincidence that most of the largest cities in the world are situated on estuaries. This chapter also introduces the general concepts of biogeochemical cycling as they relate to estuaries. Part I of the book begins with the description of how and when different estuaries were formed and how they are currently classified from a geomorphological/physical mixing perspective; this part is critical as it sets the framework from which biogeochemical cycling will be controlled across different systems.

Part II examines the molecular properties of estuarine waters and the effects of mixing freshwater and seawater, as well as dissolved gases. Many estuaries and rivers have been shown to be net sources of carbon dioxide to the atmosphere; this part includes a discussion on the controls of carbon dioxide cycling in estuaries and the ramifications that estuarine carbon dioxide fluxes (and other greenhouse gases) may have in global warming issues. In part III, I begin my discussion on estuarine sediments, the repository for historical changes in estuarine watershed and water column processes. Interactions between sediments and the water column in estuaries are particularly important due to shallow water columns commonly found in these systems. Primary production and decomposition will be discussed in the context of system level dynamics in part IV; this part will cover the dynamics of such processes as they relate to the previously described parts on element cycling. I will also introduce many of the important bulk and chemical biomarker techniques used to "trace" organic matter inputs to these highly dynamic systems. In part V, an overview is provided on nutrient and trace metal cycling in estuaries with a focus on the major natural and anthropogenic sources of nutrients. General cycles are first discussed, followed by more detailed case studies for specific nutrients and metals. Part VI covers the dominant organic and inorganic contaminants in estuaries with an emphasis on the role of partitioning and binding coefficients in controlling the availability of contaminants to estuarine organisms. The natural cycling of dissolved organic carbon as well as the mineralogical properties of suspended particles in a system will have significant effects on these exchange processes. This part also provides some classic case studies on contaminant cycling as well as some new insights on the management of these systems. Finally, part VII provides an introduction to river-dominated margins, the zones where the major rivers/estuaries of the world meet the sea. These zones are different from the traditional estuarine/coastal zones in that the processing and residence times of dissolved and particulate materials is usually shorter, with a greater potential for the transport of terrigenous materials to the deep ocean. Recent work has also shown that fluvial inputs from rivers and estuaries are not the only important inputs from the continents to the coastal ocean. Groundwater inputs may serve as another transport route for materials processed on the continents to make it to the ocean; this part discusses the recent work on groundwater inputs as well as the ramifications of these inputs to estuarine flux models.

This book is designed for graduate level classes in estuarine biogeochemistry and/or ecosystem dynamics. Prerequisites for such a course may include introductory courses in inorganic and organic chemistry, environmental and/or ecosystem ecology, and some basic knowledge of calculus. This book should also prove to be a valuable resource for researchers in marine and environmental sciences because of the diversity of illustrations and tabular data used in the examples of estuarine case studies, as well as the exhaustive bibliographic sources. The basic organization of the book is derived from classes in global and estuarine biogeochemistry that I have taught for several years.

Acknowledgments

Over the 4 years it has taken to write this book, other people have helped along the way, and I am eternally grateful for their input. I want to especially thank Rebecca Green (Tulane University) and Jo Ann Bianchi for carefully reading all the chapters and providing helpful comments. One or more chapters were reviewed by the following scientists: Mark Baskaran (Wayne State University); Elizabeth Canuel (Virginia Institute of Marine Science [VIMS]—College of William and Mary); Daniel. L. Childers (Florida International University); Dan Conley (National Environmental Research Institute, Denmark); John M. Jaeger (University of Florida); Ronald Kiene (Dauphin Island Sea Lab [DISL]—University of Southern Alabama); Rodney Powell (Louisiana Universities Marine Consortium [LUMCON]); Peter A. Raymond (Yale University); Sybil P. Seitzinger (Rutgers University); Christopher K. Sommerfield (University of Delaware); and William J. Wiseman (National Science Foundation—Polar Programs). While all these people have helped greatly to improve the book, I am responsible for any errors that remain. I also thank my colleagues in the department of Earth and Environmental Sciences at Tulane University for the many discussions we have had in recent years on coastal environments, particularly Mead Allison, Brent McKee, George Flowers, and Franco Marcantonio, as well as Mike Dagg (LUMCON) for stimulating conversations in our many car trips to LUMCON. Michel Meybeck (Université de Pierre et Marie Curie), Sid Mitra (S.U.N.Y., Binghamton), Scott Nixon (University of Rhode Island), and Hans Paerl (University of North Carolina) provided useful references for certain chapters. Jeffrey S. Levinton and Robert G. Wetzel, with their extensive experience in book writing, provided invaluable advice and wisdom on how to cope with all the stresses of such an endeavor.

I owe special thanks to Charlsie Dillon for typing all the tables in the book and coordinating all the correspondence involved with getting permission rights for select illustrations—Jeremy Williams also assisted in the early stage of this as well. Cathy B. Smith did a superb job of redrawing all the illustrations, of which there were many, in the book. All the library research was conducted at LUMCON, where I am especially grateful to librarians John Conover and Shanna Duhon for their tireless assistance and hospitality. Special thanks to Michael Guiffre who helped design the book cover.

I would like to thank my family, Jo Ann, Christopher, and Grandmaster Chester, for their support and patience, and my parents for their continued inspiration over the years. Finally, I would like to thank Lyle More for taking the time to inspire a young fledgling in need of guidance at a very critical age.

Contents

Biogeochemistry of
Estuaries

Chapter 1

Estuarine Science and Biogeochemical Cycles

Importance of Estuaries

Estuaries are commonly described as semi-enclosed bodies of water, situated at the interface between land and ocean, where seawater is measurably diluted by the inflow of freshwater (Hobbie, 2000). The term "estuary," derived from the Latin word *aestuarium*, means marsh or channel (Merriam-Webster, 1979). These dynamic ecosystems have some of the highest biotic diversity and production in the world. Not only do they provide a direct resource for commercially important estuarine species of fishes and shellfish, but they also provide shelter and food resources for commercially important shelf species that spend some of their juvenile stages in estuarine marshes. For example, high fish and shellfish production in the northern Gulf of Mexico is strongly linked with discharge from the Mississippi and Atchafalaya rivers and their associated estuarine wetlands (Chesney and Baltz, 2001). Commercial fishing in this region typically brings in 769 million kg of seafood with a value of $575 million. Fisheries production and coastal nutrient enrichment, via rivers and estuaries, are positively correlated within many coastal systems around the world (Nixon et al., 1986; Caddy, 1993; Houde and Rutherford, 1993). The coupling of physics and biogeochemistry occurs at many spatial scales in estuaries (figure 1.1; Geyer et al., 2000). Estuarine circulation, river and groundwater discharge, tidal flooding, resuspension events, and exchange flow with adjacent marsh systems (Leonard and Luther, 1995) all constitute important physical variables that exert some level of control on estuarine biogeochemical cycles.

Description of Estuarine Science

There has been considerable debate about the definition of an estuary because of the divergent properties found within and among estuaries from different regions of the world.

3

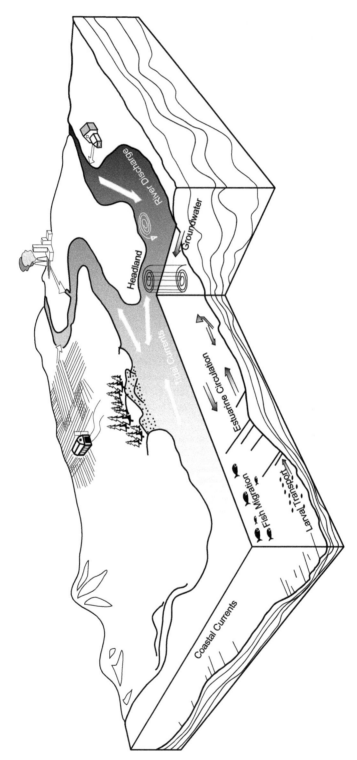

Figure 1.1 Schematic showing important linkages between physical (e.g., tidal currents, river discharge, and groundwater) and biological (e.g., fish migrations, larval transport) processes in estuaries. (Modified from Geyer et al., 2000.)

Consequently, there have been numerous attempts to develop a comprehensive and universally accepted definition. Pritchard (1967, p. 1) first defined estuaries based on salinity as "semi-enclosed coastal bodies of water that have a free connection with the open sea and within which sea water is measurably diluted with fresh water derived from land drainage." A general schematic representation of an estuary, as defined by Pritchard (1967), and further modified by Dalrymple et al. (1992) to include more physical and geomorphological processes, is shown in figure 1.2. In this diagram, we see a wide range of salinities (0.1–32), wave processes that dominate at the mouth of the estuary, tidal processes that occur in the middle region, and river or fluvial processes at the head of the estuary. The relative importance of physical forcing from each of these regions can vary seasonally (e.g., coastal wave energy versus river discharge), and ultimately determine the mixing dynamics of both water and sediments in estuaries. More recently, Perillo (1995, p. 4) provided an even more comprehensive definition of an estuary as "a semi-enclosed coastal body of water that extends to the effective limit of tidal influence, within which sea water entering from one or more free connections with the open sea, or any other saline coastal body of water, is significantly diluted with fresh water derived from land drainage, and can sustain euryhaline biological species from either, part or the whole of their life cycle." This definition provides the basis from which to work from in this book. Others studies have also shown the utility of using a geomorphological classification in determining characteristic "signatures" of estuaries in distinct regions in North America, such as the west coast

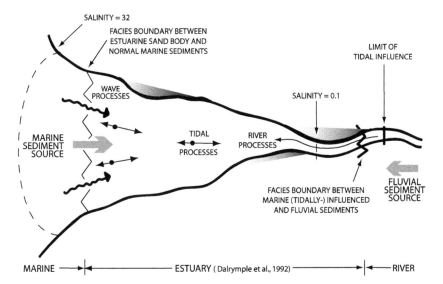

Figure 1.2 Classic estuarine zonation depicted from the head region, where fluvial processes dominate, to the mid- and mouth regions where tidal and wave processes are the dominant controlling physical forces, respectively. Differences in the intensity and sources of physical forcing throughout the estuary also result in the formation of distinct sediment facies. (Modified from Dalrymple et al., 1992, with permission.)

(Emmett et al., 2000) and northeastern (Roman et al., 2000) and southeastern (Dame et al., 2000) Atlantic coasts. Further details on the origin and complexity of estuarine geomorphology are provided in chapter 2.

The field of estuarine science has in essence, suffered from an identity crisis from its early inception (1950s and 1960s) to the present—as reflected in the extensive list of ambiguous definitions of an estuary found in the literature (Elliot and McLusky, 2002). Some of this perceived "identity crisis" has resulted from the diversity of vernacular terms (e.g., bay, sound, harbor, bight) commonly used in place of "estuary." The inconspicuous usage of the term "estuary" has in part, impeded the development of "Estuarine Science" as a discipline distinct from that described as unique boundary environments in the fields of "Oceanography" and "Limnology." In fact, due to the unique complexity and dynamic nature of estuaries there has been a recent call for a "synthesis approach" in defining estuaries on a broader categorical scale (Hobbie, 2000). The proposed synthesis would allow for a greater ability to describe changes in estuarine ecosystems as they relate to physical changes and to provide more statistical and mathematical correlations between biological and environmental parameters. Moreover, it was suggested that mathematical simulation models could serve to combine process descriptions with correlation determinations, along with a number of other interrelated processes in a single system or region. There is clearly a need for improved classification of estuaries if we are to effectively contend with the growing legislative, administrative, and socio-economic demands in coastal management which requires more unambiguous terminology than currently exists in estuarine science (Elliot and McLusky, 2002). Fortunately, these changes have already begun in places such as Australia and Europe with classification schemes like "The Australian-Environmental Indicators: Estuaries and Sea" (Ward et al., 1998); and "The European Union Water Framework Directive" (European Union, 2000), respectively.

Human Impact on Estuaries and Management Issues

Recent estimates indicate that 61% of the world population lives along the coastal margin (Alongi, 1998). These impacts of demographic changes in human populations have clearly had detrimental effects on the overall biogeochemical cycling in estuaries. Nutrient enrichment is perhaps the most widespread problem in estuaries around the world (Howarth et al., 2000, 2002). For example, 44 estuaries along the entire U.S. coastline have been diagnosed as having nutrient overenrichment (figure 1.3; Bricker et al., 1999). From a broader perspective, Hobbie (2000) recently summarized the findings of an earlier U.S. National Research Council report focused on the major effects of human population growth on estuaries. A slightly modified version of this summary is as follows: (1) nutrients, especially nitrogen, have increased in rivers and estuaries resulting in harmful algal blooms and a reduction in water column oxygen levels; (2) coastal marshes and other intertidal habitats have been severely modified by dredging and filling operations; (3) changes in watershed hydrology, water diversions, and damming of rivers have altered the magnitude and temporal patterns of freshwater flow and sediment discharge to estuaries; (4) many of the commercially important species of fishes and shellfish have been overexploited; (5) extensive growth and industrialization has resulted in high concentrations of both organic [polycyclic aromatic hydrocarbon (PAHs) and polychlorinated

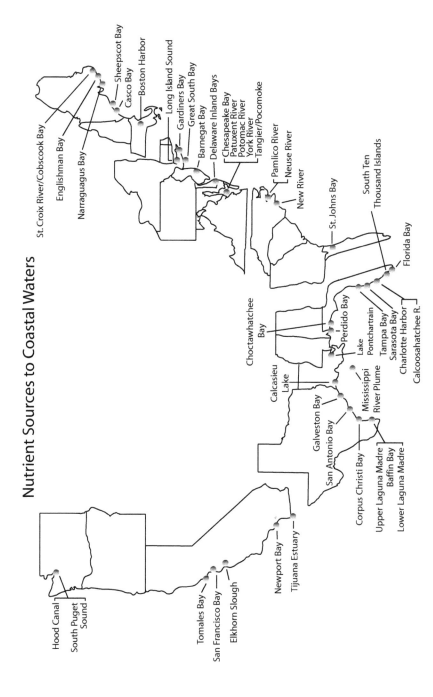

Figure 1.3 Forty-four estuaries along the U.S. coastline that have been diagnosed as having nutrient overenrichment. (From Bricker et al., 1999, with permission.)

biphenyls (PCBs)] and inorganic contaminants (heavy metals) in estuarine sediments and waters; and (6) introduced species have resulted in alterations in habitats, loss of native species, and a reduction in commercially important species. Model predictions indicate that, by 2050, an estimated 8.5 billion people will be living in *exoreic* watersheds (e.g., watersheds draining the ocean); this increase is greater than 70% based on numbers from 1990 (Kroeze and Seitzinger, 1998).

An understanding of the role that biogeochemical and physical processes play in regulating the chemistry and biology of estuaries is fundamental to evaluating complex management issues (Bianchi et al., 1999a; Hobbie, 2000). Biogeochemistry links processes that control the fate of sediments, nutrients, and organic matter, as well as trace metal and organic contaminants. Thus, the discipline requires an integrated perspective on estuarine dynamics associated with the input, transport, and either accumulation or export of materials that largely control primary productivity. The metabolism of in situ primary production and the utilization of *allochthonous* organic matter are also linked to patterns of secondary productivity and fishery yields in estuaries. As humans alter regional watersheds and local landscapes of estuaries, our ability to detect abiotic and biotic signals reflective of biogeochemical change in these systems will be critical in determining how we manage these unique coastal ecosystems. While there remain opposing views concerning how to effectively manage these diverse systems, a synthetic perspective [as described earlier (Hobbie, 2000)] will clearly lead to a more comprehensive approach.

Fortunately, we are beginning to actually detect measurable improvements in the water quality of some estuaries due to scientifically based nutrient reductions and extensive long-term monitoring studies. For example, the Patuxent River in Maryland (USA), a tributary of the Chesapeake Bay, experienced extensive eutrophication from sewage inputs and non-point sources over four decades (1960–2000; D'Elia et al., 2003).

During the late 1970s, scientists began to develop a dialogue with policy makers in this region, as well as local and national funding agencies, which resulted in stable funding for well-structured monitoring programs. In fact, the results from these studies led to some of the first documented nutrient control standards for an estuarine basin in the United States. Furthermore, based on preliminary indications it appears that the proposed nitrogen (N) removal strategy has succeeded in improving water quality in the Patuxent River estuary (D'Elia et al., 2003). This further corroborates that long-term and scientifically based monitoring programs can result in effective remediation strategies for environmental problems in estuaries (D'Elia et al., 1992). However, it should be noted that gaining general acceptance of nutrient overenrichment in the larger Chesapeake Bay system, along with consensus for nutrient controls, was a long and arduous task for scientists in the region (Malone et al., 1993). Nevertheless, similar reductions in nutrient loading have also resulted in significant water quality improvement in other regions of the world, such as the Baltic Sea (Elmgren and Larsson, 2001). Factors leading to the general success of estuarine management programs have recently been summarized as: having key individuals, a lead agency, an institutional structure, long-term scientific data, widespread public perception of the problem, and an ecosystem-level perspective (Boesch et al., 2000; Boesch, 2002).

New approaches using integrated ecological and economic modeling are providing new tools for adaptive management of estuaries (Constanza and Voinov, 2000). For example, a Patuxent Landscape Model (PLM) was developed to serve as a tool for the analysis of physical and biological variables, based on socioeconomic changes in the region. This adaptive approach allows for optimizing models, based on changes in the resolution of the model over time, along with "consensus building" where scientists and policy makers are frequently involved in continuously changing stages of model development. The PLM has been very effective in addressing land-use changes in the basin and how these changes control hydrologic flow and ultimately nutrient delivery into the Patuxent River estuary (Costanza and Voinov, 2000). The use of General Ecosystem Models (GEMs) in estuaries, like the PLM, has grown from prior applications in the Coastal Ecosystem Landscape Spatial Simulation (CELSS) model (Costanza et al., 1990) and wetland systems like the Florida Everglades.

Biogeochemical Cycles in Estuaries

Some of the first applications of the integrative field of "Biogeochemistry" are derived from organic geochemical studies where organisms and their molecular biochemistry were used as an initial framework for interpreting sources of sedimentary organic matter (Abelson and Hoering, 1960; Eglinton and Calvin, 1967). Biogeochemical cycles involve the interaction of biological, chemical, and geological processes that determine sources, sinks, and fluxes of elements through different reservoirs within ecosystems. Much of this book will use this basic box-model approach to understand the cycling of elements in estuarine systems. Therefore, we need to first define some of the basic terms before we can understand how fluxes and reservoirs interact to determine chemical budgets in a biogeochemical box model (figure 1.4). For example, a *reservoir* is the amount of material (M), as defined by its chemical, physical, and/or biological properties. The units used to quantify material in a reservoir, in the box or compartment of a box model, are typically of mass or moles. *Flux* (F) is defined as the amount of material that is transported from one reservoir to another over a particular time period (mass/time or mass/area/time). A *source* (S_i) is defined as the flux of material *into* a reservoir, while a *sink* (S_o) is the flux of material *out* of the reservoir (many times proportional to the size of the reservoir). The *turnover time* is required to remove all the materials in a reservoir, or the average time spent by elements in a reservoir. Finally, a *budget* is essentially a "checks and balances" of all the sources and sinks as they relate to the material turnover in reservoirs. For example, if the sources and sinks are the same, and do not change over time, the reservoir is considered to be in a *steady state*. The term *cycle* refers to when there are two or more connecting reservoirs, whereby materials are cycled through the system—generally with a predictable pattern of cyclic flow.

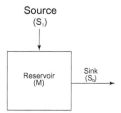

Figure 1.4 Schematic of box model commonly used in biogeochemical cycling work, showing reservoirs (M), sinks (S_o), and sources (S_i).

The spatial and temporal scales of biogeochemical cycles vary considerably depending on the reservoirs considered. In the case of estuaries, most biogeochemical cycles are based on regional rather than global scales. However, with an increasing awareness of the importance of atmospheric fluxes of biogases (e.g., CO_2, CH_4, N_2O) in estuaries and their impact on global budgets (Seitzinger, 2000; Frankignoulle and Middelburg, 2002), some budgets will involve both regional and global scales.

Summary

1. The coupling of physics and biogeochemistry occurs at many spatial scales in estuaries. Some of the dominant physical forcing variables that impact biogeochemical cycles are estuarine circulation, river and groundwater discharge, tidal flooding, exchange flow with adjacent marsh systems, and resuspension events.
2. According to Perillo (1995, p. 4), an estuary is defined as "a semi-enclosed coastal body of water that extends to the effective limit of tidal influence, within which sea water entering from one or more free connections with the open sea, or any other saline coastal body of water, is significantly diluted with fresh water derived from land drainage, and can sustain euryhaline biological species from either part or the whole of their life cycle."
3. An improved classification scheme for estuaries is needed if we are to effectively contend with the growing legislative, administrative, and socioeconomic demands in coastal management which requires more unambiguous terminology than currently exists.
4. Approximately 61% of the world population lives within and around the estuarine watersheds (Alongi, 1998). Demographic changes in human populations have clearly had detrimental effects on the overall biogeochemical cycling in estuaries, with nutrient enrichment being the most widespread global problem.
5. There have recently been successful nutrient abatement programs in estuarine systems around the world. Factors leading to the overall success of these management programs are having key individuals, a lead agency, an institutional structure, long-term scientific data, widespread public perception of the problem, and an ecosystem-level perspective.
6. New approaches using GEMs in estuarine management have allowed for optimization of models based on changes in the resolution of the model over time, along with "consensus building"—where scientists and policy makers are frequently involved in continuously changing stages of model development.
7. Box models are commonly used biogeochemical studies that involve the interaction of biological, chemical, and geological processes that determine sources, sinks, and fluxes of elements through different reservoirs within ecosystems.

Part I

Physical Dynamics of Estuaries

Chapter 2

Origin and Geomorphology

Age, Formation, and Classification

Geologically speaking, estuaries are ephemeral features of the coasts. Upon formation, most begin to fill in with sediments and, in the absence of sea level changes, would have life spans of only a few thousand to tens of thousands of years (Emery and Uchupi, 1972; Schubel, 1972; Schubel and Hirschberg, 1978). Estuaries have been part of the geologic record for at least the past 200 million years (My) BP (before present; Williams, 1960; Clauzon, 1973). However, modern estuaries are recent features that only formed over the past 5000 to 6000 years during the stable interglacial period of the middle to late Holocene epoch (0–10,000 y BP), which followed an extensive rise in sea level at the end of the Pleistocene epoch (1.8 My to 10,000 y BP; Nichols and Biggs, 1985). There is general agreement that four major glaciation to interglacial periods occurred during the Pleistocene. It has been suggested that sea level was reduced from a maximum of about 80 m above sea level during the Aftoninan interglacial to 100 m below sea level during the Wisconsin, some 15,000 to 18,000 y BP (figure 2.1; Fairbridge, 1961). This lowest sea level phase is referred to as low stand and is usually determined by uncovering the oldest drowned shorelines along continental margins (Davis, 1985, 1996); conversely, the highest sea level phase is referred to as high stand. It is generally accepted that low-stand depth is between 130 and 150 m below present sea level and that sea level rose at a fairly constant rate until about 6000 to 7000 y BP (Belknap and Kraft, 1977). A sea level rise of approximately 10 mm y^{-1} during this period resulted in many coastal plains being inundated with water and a displacement of the shoreline. The phenomenon of rising (*transgression*) and falling (*regression*) sea level over time is referred to as *eustacy* (Suess, 1906). When examining a simplified sea level curve (figure 2.2), we find that the rate of change during the Holocene is fairly representative of the Gulf of Mexico and much of the U.S. Atlantic coastline (Curray, 1965). While there is considerable debate about the controls of current sea level changes around the world we can generally conclude that tectonic conditions, regional subsidence rates, and regional climatic changes account for much of this variation. When factoring in these regional differences, rates of sea level

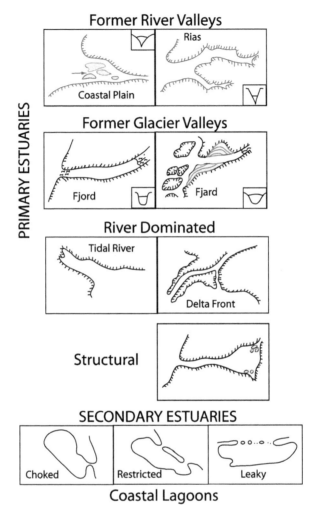

Figure 2.5 Morphogenetic classification of estuaries. (From Perillo, 1995, with permission.)

characteristics. The four categories within primary estuaries are former fluvial valleys, former glacial valleys, river dominated, and structural. The formal definitions of these primary estuaries and their respective subcategories are as follows:

1. *Former fluvial valleys* are formed by flooding of river valleys in the Pleistocene–Holocene during the last interglacial transgression. The two subcategories are as follows: (a) *coastal plain estuaries* occur on low relief coasts due to high sedimentation and in-filling of a river valley; (b) *rias* are former river valleys in regions of high relief (mountains and cliffs).

2. *Former glacial valleys* are formed by flooding of river valleys in the Pleistocene–Holocene during the last interglacial transgression. The two subcategories are as follows: (a) *fjords* are high relief systems formed by glacial scouring; (b) *fjards* are low relief systems formed by glacial scouring.
3. *River-dominated estuaries* are formed in high river discharge regions where the valley is presently not drowned by the sea. The two subcategories are as follows: (a) *tidal river estuaries* are associated with large rivers systems that are influenced by tidal action with the salt front usually not well developed at the mouth; (b) *delta-front* estuaries are found in sections of deltas that are affected by tidal action and/or salt intrusion.
4. *Structural estuaries* are formed by *neotectonic* (within the Quaternary period or past 1.8 My BP) processes such as faulting, volcanism, postglacial rebound, and isostacy that have occurred since the Pleistocene.

Secondary estuaries have been modified more by marine than river discharge processes, since the time that sea level reached its current position. Coastal lagoons are inland bodies of water that run parallel to the coast and are isolated from the sea by a barrier island where one or more small inlets allow for connection with the ocean. Subcategories of lagoons are as follows: (a) *choked* are lagoons that have only one long narrow inlet; (b) *restricted* are lagoons with very few inlets or having a wide mouth; (c) *leaky* are lagoons with many inlets separated by small barrier islands.

As discussed in chapter 1, there remains considerable disagreement on how to best classify estuarine systems because of their highly divergent characteristics. Thus, while the details provided in the aforementioned definitions may seem tedious and highly specialized, it is important to establish such a framework on morphogenetic differences prior to discussing the central focus of biogeochemical cycles throughout the book. Moreover, there is a need and/or urgency for establishing a unified set of definitions of estuaries because of increasing legislative and socioeconomic concerns involving estuaries worldwide—all of which requires an established terminology (Elliott and McLusky, 2002).

Distribution and Sedimentary Processes within Estuarine Types

Coastal plain estuaries, which occupy former river valleys and occur on low relief coastlines, are the most well-studied systems in the world (figure 2.5; Perillo, 1995). A few examples of studies involving sediment dynamics of coastal plain estuaries around the world are Chesapeake Bay (USA) (Nichols, 1974; Biggs and Howell, 1984), Delaware Bay (USA) (Schubel and Meade, 1977; Kraft et al., 1979), Thames (UK) (Langhorne, 1977), and Yang-Tze (China) (McKee et al., 2004). Globally, coastal plains cover an area of about 5.7 million km^2 (Colquhoun, 1968). These are low relief regions with unconsolidated sediments that in most cases have one or more rivers traversing them. These coastal plain deposits are formed by either fluvial inputs of sediments derived from higher mountain regions or from marine deposition during transgression periods (Bokuniewicz, 1995). As mentioned earlier, from about 17,000 to 6000 y BP these estuarine coastal plains were inundated by the sea. Since that time sea level has been

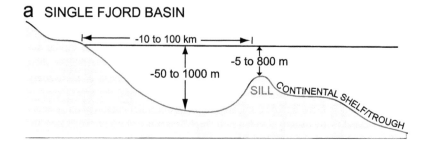

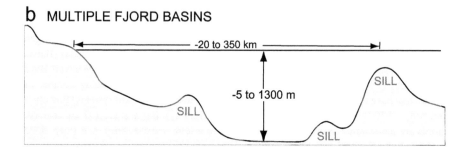

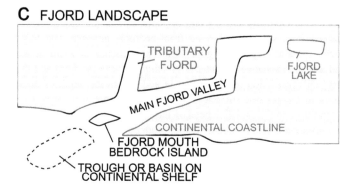

Figure 2.8 Schematic showing fjords characterized by one or more sills at the mouth of each system. (From Syvitski and Shaw, 1995, with permission.)

tidal classification along coastlines are *microtidal* (tidal range of <2 m), *mesotidal* (tidal range between 2 and 4 m), and *macrotidal* (tidal range >4 m) (Hayes, 1975). To make matters even more confusing, an estuarine system such as the Gironde Estuary (France) is essentially a coastal plain and tidal-dominated estuary. So, in some cases these estuarine types do overlap; details concerning the complications and rationale for not separating

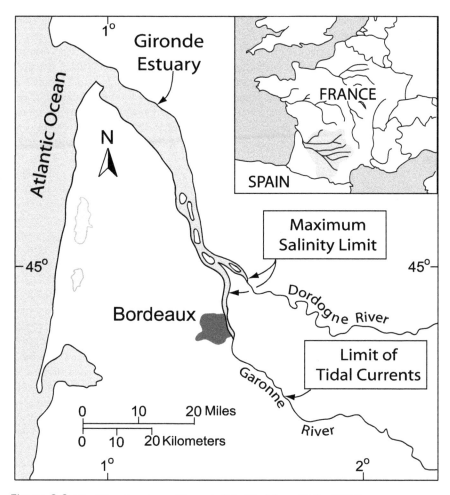

Figure 2.9 The Gironde estuary (France). (Modified from Wells, 1995.)

these tidal-dominated systems can be found in Wells (1995). In general, tidal-dominated estuaries are funnel-shaped with tidal currents that largely control the transport of river-borne sediments (Wells, 1995). The large opening at the mouth of the estuary, such as the Gironde, enhances the tidal wave-producing macrotidal ranges and strong tidal currents which allows for bedload transport of sand waves (figure 2.9). Some of the more well-studied examples of tidal-dominated estuaries are the Bay of Fundy (Canada) (Dalrymple et al., 1992), Gironde Estuary (France) (Ruch et al., 1993), Ord River Estuary (Australia) (Wright et al., 1975), and the Severn Estuary (UK) (Harris and Collins, 1985). Many of these estuaries are characterized by subtidal sand ridges, sand wave migration, and bordering intertidal mudflats and wetlands (marshes and mangroves). Channel sands are the dominant sediment facies near the central and mouth regions of tidal-dominated estuaries with finer sediments in tidal flat facies occurring in the low-energy margins of the funnel and the narrow sinuous head of the estuary (Nicholls et al., 1991). These distinct sedimentary facies should reflect differences in physical forcing, carbon loading, remineralization efficiency rates, and burial. For example, Aller (2001) recently used six sediment facies to describe the *diagenetic* subsystems that typically exist in tidal river and delta-front estuaries found along river-dominated margins (or RiOMars) as recently described by McKee et al. (2004). More details on the processing, transport, and exchange of particulate and dissolved materials in these environments will be discussed in chapters 6, 8, and 16.

Tidal river estuaries are essentially a subset of tidal-dominated estuaries, having similar morphometric and sedimentological features yet being distinct in their association with rivers that have high discharge. Two examples of tidal river estuaries are the lower Amazon River (Brazil) and Rio de La Plata (border of Uruguay and Argentina) (Wells, 1995). Both of these tidal river estuaries have limited or no seawater entering the mouth of the river—a characteristic feature of these estuaries. Consequently, the trapping of suspended sediments at the seawater concentration limit is significantly displaced seaward from the mouth of the Rio de La Plata (Urien, 1972) and Amazon River (Meade et al., 1985) which results in the formation of subaqueous deltas. In fact, the Amazon shelf has received considerable attention in past years, one such interdisciplinary study was "A Multidisciplinary Amazon Shelf SEDiment Study" (AMASSEDS) (Nittrouer et al., 1991, 1995; Trowbridge and Kineke, 1994; Cacchione et al., 1995; Geyer and Beardsley, 1995; Kineke et al., 1996). The combination of a high tidal range (4–8 m at the mouth) and high river discharge results in significant mixing deposition on the shelf. In fact, it has been estimated that approximately half of the sediments discharged from the Amazon River are deposited on the shelf (Kuehl et al., 1986; Nittrourer et al., 1986). The highest sediment accumulation rates are generally found in the outer *topset* and *foreset* regions of the shelf (figure 2.10). This high-energy shelf environment maintains high concentrations of suspended sediment in the tidal boundary layer, and the presence of *fluid muds* has a significant impact on a number of physical, chemical, and sedimentological processes occurring on the shelf (Kineke et al., 1996).

Delta-front estuaries are found in regions of deltas that are affected by tidal action and/or salt intrusion (figure 2.5; Perillo, 1995). The formation of a delta (shoreline protuberance) results because river-derived sediments accumulate faster in a coastal/river water body than dispersal due to redistribution processes (e.g., waves, coastal and tidal currents). More specifically, Wright (1977, p. 859) has defined a delta as "coastal accumulations, both subaqueous and subaerial, of river-derived sediments adjacent to, or in close proximity to, the source stream, including the deposits that have been secondarily molded by various marine agents, such as waves, currents, or tides." The tectonic history of a margin (*active* or *passive*) is very important in determining the development of a delta (Elliott, 1978). Passive or trailing-edge margins are more conducive to the formation of deltas than active or leading-edge margins because of the extensive drainage or receiving basins that typically form along low-relief passive margins. Certain characteristics of the receiving basin such as depositional slope, subsidence rate, size and shape, and tidal dynamics (e.g., macro- versus microtidal) strongly influence the *progradation* of a delta (Hart et al., 1992).

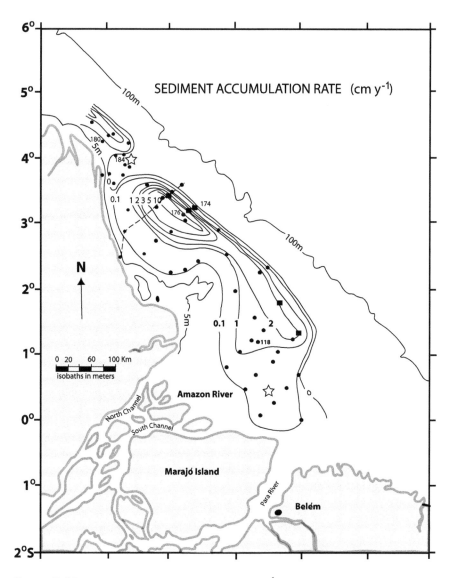

Figure 2.10 Sediments accumulation rates (cm y^{-1}) on Amazon River shelf. (From Kuehl et al., 1986, with permission.)

Deltas are generally divided into the following physiographic zones: *alluvial feeders*; *delta plain*; *delta front*; and *prodelta/delta slope* (figure 2.11; Coleman and Wright, 1975). The alluvial feeder is a valley within the drainage basin that supplies the water and sediment to the delta. Some of the alluvial feeders consist of a single channel or multiple channel (*line-source*) to the coast (Hart, 1995). The *upper* delta plain is an older section of

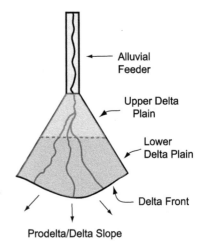

Figure 2.11 Physiographic zones of a delta complex. (From Coleman and Wright, 1975, modified by Hart, 1995, with permission.)

the delta that is not currently affected by tidal processes. The *lower* delta plain, consisting of subaerial and intertidal zones, is dominated by channels [*distributary* (deliver fluvial discharge seaward) and *tidal* (drain tidally inundated regions)] and their deposits. The distribution of sediments which empty out of the distributary channels in deltas, as both suspended materials and bedload, are strongly influenced by marine processes (e.g., coastal and tidal currents and waves). Many of the fine suspended sediments are transported greater distances away from the mouth than coarse sediments, which generally remain closer to the channel mouth. The seaward edge of the delta plain merges with the subtidal or subdelta region of the delta called the delta front, where tidal processes and the mixing between fresh and saline waters form delta-front estuaries (Hart, 1995). Most coarse-grained sediments from the river are deposited within the delta front. The prodelta is an area seaward of the delta front where most of the fine-grained sediments are deposited along a steep gradient within the delta slope.

Numerous studies have examined temporal changes in the geomorphology and/or sedi-mentary dynamics of major delta-front estuaries [Coleman, 1969; Allison, 1998 (Ganges–Brahmaputra Rivers); Lofty and Frihy, 1993 (Nile River); Xue, 1993 (Yellow River); Milliman, 1980 (Fraser River); Saucier, 1963; Coleman and Gagliano, 1964 (Mississippi River)]. Much of the early literature on deltas focused on the Mississippi River as a classic delta model (Trowbridge, 1930; Russell, 1936; Fisk, 1944, 1955, 1960; Coleman and Gagliano, 1964; Frazier, 1967). However, it was soon realized that a single model delta was not adequate to describe the unique and complex character of different delta sys-tems around the world, so different "schemes" of interacting forces (e.g., fluvial, wave, and tidal processes) were used to characterize deltaic systems (Galloway, 1975). Two of the major historical components of deltas involves the constructional phase of the delta where the system *progrades* and the destructional or *abandonment phase* caused by *avulsion* (channel switching within the delta plain) of the distributory channel (Elliott, 1978). Channel switching or avulsion results in a cut-off of the supply of sediment to the active delta which precludes further progradation. For example, the Holocene

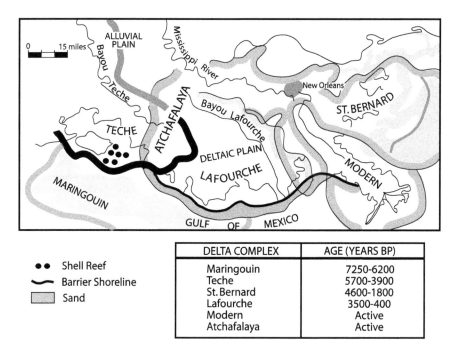

DELTA COMPLEX	AGE (YEARS BP)
Maringouin	7250-6200
Teche	5700-3900
St. Bernard	4600-1800
Lafourche	3500-400
Modern	Active
Atchafalaya	Active

•• Shell Reef
◝◜ Barrier Shoreline
▨ Sand

Figure 2.12 The four major pre-modern complexes and the modern Mississippi Delta that have resulted from various stages of abandonment. (Modified from Boyd and Penland, 1988.)

portion of the Mississippi delta is approximately 5000 to 6000 years old, and contains 16 recognizable lobes (with four major pre-modern complexes) (figure 2.12) that have resulted from various stages of abandonment (Boyd and Penland, 1988). The present lobe began to form about 700 years ago and consists of the main delta of the Mississippi River and the Atchafalaya delta (Swenson and Sasser, 1992; Roberts, 1997; McManus, 2002). The Atchafalaya delta is a site of deposition and land building in a coastal zone that is currently undergoing loss of land at a rate of 155 km^2 y^{-1} (Turner, 1990). Prior to the construction of the current levee system on the Mississippi River, natural levees were developed adjacent to distributary channels, from sedimentary accumulations during flooding events. The classic "birds-foot" delta shape of the current lobe has resulted from *crevasses* (breaches or cuts in the natural levee) which allowed sediments to be further distributed away from the main channel in fan-shaped deposits called splays (figure 2.13) (Coleman and Gagliano, 1964). During progradational growth the crevasses' development leads to the growth of the subdelta. The fact that the total volume of sediments in a subdelta continues to increase despite the loss of subaerial land is direct evidence of subsidence. If we assume an average subsidence of 1.5 cm y^{-1} and an average life of 150 years, then 2.25 m of subsidence will occur during the existence of a subdelta (Wells, 1996). The subsequent construction and consequences of the man-made levee system on the sedimentary and biogeochemical dynamics of the Mississippi River

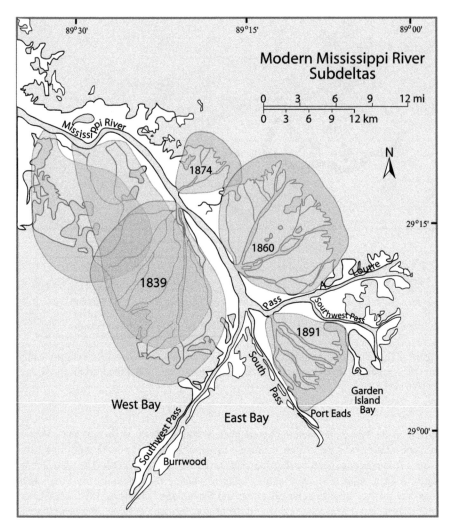

Figure 2.13 The classic "birds-foot" delta shape of the current Mississippi River lobe. (Modified from Coleman and Gagliano, 1964.)

delta-front estuary, and the adjacent marsh systems, will be further discussed in chapters 6 and 16.

Structural estuaries are formed by neotectonic processes such as tectonism [faults and diastrophic processes (larger scale folding and other deformations)], vulcanism (formed by volcanic activity), postglacial rebound, and isostacy—which have occurred since the Pleistocene (figure 2.5; Perillo, 1995; Quivira, 1995). Earlier attempts to characterize these unique estuaries based on their tectonic features (Schubel, 1972; Fairbridge, 1980) differed significantly from all other descriptors used to characterize estuaries. A more recent classification of these estuaries divides them into the following two categories: (1) those formed by tectonism; and (2) vulcanism (Hume and Herdendorf, 1988). Faulted estuaries, such as San Francisco Bay (USA) and Hawke and Tasman Bays (New Zealand), are typically elongate in shape, and are dominated by tidal currents and small inlets. The San Francisco Bay estuary was formed by vertical and lateral crustal movements along a number of faults (Atwater et al., 1977). Tectonics in the San Francisco Bay region is predominantly controlled by the San Andreas *strike-slip fault* system, which occurs along a major boundary between the Pacific and North American plates. In addition to dominant tectonic forces that created this estuary, Holocene sea level rise drowned the basin creating San Francisco Bay as we know it today (Atwater, 1979). Most of the estuaries along the west coast of North America are geologically younger than those along the Atlantic coast, yet typically have surface bedforms composed of Miocene (approximately 24 My BP) deposits derived from uplifting of ancient marine sediments (McKee, 1972; Emmett et al., 2000). Finally, volcanic estuarine systems are formed within the caldera of a crater which has been breached by sea level; some examples are the Panmure Crater and Lyttelton Harbour (New Zealand) (Quivira, 1995).

Coastal lagoon estuaries are inland shallow bodies of water that usually run parallel to the coast and are isolated from the sea by a barrier island where one or more small inlets allow for connection with the ocean. These lagoons have also been modified more significantly by marine than river discharge processes (figure 2.5; Kjerfve and Magill, 1989; Kjerfve, 1994; Perillo, 1995; Isla, 1995). These systems have also been commonly referred to as *bar-built* estuaries (Fairbridge, 1980). The three subcategories of coastal lagoons used by Perillo (1995) were previously described by Kjverfve and Magill (1989) as (1) choked, where diffusive transport dominates due to limited flushing via one inlet; (2) leaky, where multiple inlets allow for flushing that is dominated by advective processes; and (3) restricted, which are transitional between leaky and choked. Further details on the importance of these flushing characteristics (along with residence time, precipitation, and rainfall) on salinities in lagoons will be discussed in chapter 3. There is great variability among these shallow water environments, largely controlled by climatic and tidal conditions. Similarly, there is great variability in the habitats associated with the aforementioned physical conditions in lagoons, tidal inlets, tidal deltas, barriers, tidal flats, marshes, and mangroves. In microtidal lagoons, unusually high salinities may develop due to limited exchange with the open sea from minimal exchange through tidal inlets. For example, Laguna Madre and Baffin Bay (Texas) are hypersaline lagoons where local runoff and precipitation are lower than evaporation. Salinities as high as 50 to 100 have been recorded in the upper reaches of these lagoons (Collier and Hedgepeth, 1950). Salinities in mesotidal systems are generally more normal due to greater exchange through tidal inlets. Due to limited exchange through inlets in the adjacent barrier islands of lagoons, storm surges commonly produce *washover fans* consisting of sand deposits that extend into the lagoon. For example, numerous washover fans and channels can be seen in the Matagorda Bay lagoon system (Texas), particularly after Hurricane Carla in 1961, as shown in the lower map of figure 2.14 (McGowen and Scott, 1975). There has been considerable debate about the origin and development of barrier island–coastal lagoon systems that has focused on the importance of emergence versus submergence in their evolution (Isla, 1995).

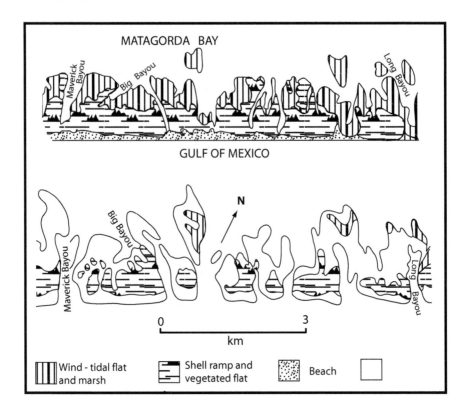

Figure 2.14 Washover fans and channels in the Matagorda Bay lagoon system (Texas, USA), after Hurricane Carla in 1961. (Modified from McGowen and Scott, 1975.)

Summary

1. Modern estuaries are recent features that only formed over the past 5000 to 6000 years during the stable interglacial period of the middle-to-late Holocene epoch (0–10,000 y BP), which followed an extensive rise in sea level at the end of the Pleistocene epoch (1.8 My to 10,000 y BP).

2. Sea level changes around the world are regionally affected by tectonic conditions, regional subsidence rates, and regional climatic changes. When factoring in these regional differences, rates of sea level change are referred to as relative sea level (RSL) rise or fall. The Intergovernmental Panel on Climate Change (IPCC) has projected that from 1990 to 2100 there will be a 48 cm rise in sea level.

3. Primary estuaries are formed from terrestrial and/or tectonic processes with minimal changes from the sea that have essentially preserved their original characteristics. The four categories within primary estuaries are former fluvial valleys, former glacial valleys, river dominated, and structural.

4. Secondary estuaries have been modified more by marine than river discharge processes, since the time that sea level reached its current position. The subcategories are choked, restricted, and leaky lagoons.

5. The geomorphology in coastal plain estuaries is derived from their ancestral rivers. Consequently, many of these system are at different stages of infilling with sediments since sea level has stabilized.

6. Rias are estuaries which occupy former river valleys and occur on high relief coastlines; the major difference in these systems is that in most rias the fluvial discharge is considerably weaker than that found in coastal plain systems.

7. Fjords and fjards are high and low relief systems, respectively, formed by glacial scouring during sea level changes during the Quaternary period. The presence of a sill at the mouth of these systems significantly reduces the exchange of water circulation between the fjord/fjard and the adjacent coastal waters.

8. Tidal river estuaries are associated with large rivers systems, influenced by tidal action with a salt front that is usually not well developed at the mouth.

9. Delta-front estuaries are found in regions of deltas that are affected by tidal action and/or salt intrusion. Wright (1977, p. 859) defined a delta as "coastal accumulations, both subaqueous and subaerial, of river-derived sediments adjacent to, or in close proximity to, the source stream, including the deposits that have been secondarily molded by various marine agents, such as waves, currents, or tides." Deltas are generally divided into the following physiographic zones: alluvial feeders, delta plain, delta front, and prodelta/delta slope.

10. Structural estuaries are formed by neotectonic processes such as tectonism, vulcanism, postglacial rebound, and isostacy—which have occurred since the Pleistocene.

11. Coastal lagoon estuaries are inland shallow bodies of water that usually run parallel to the coast and are isolated from the sea by a barrier island where one or more small inlets allow for connection with the ocean. These lagoons have also been modified more significantly by marine than river discharge processes. The three subcategories of coastal lagoons are (1) choked, where diffusive transport dominates due to limited flushing via one inlet; (2) leaky, where multiple inlets allow for flushing that is dominated by advective processes; and (3) restricted, which are transitional between leaky and choked.

Chapter 3

Hydrodynamics

The hydrologic cycle has received considerable attention in recent years with particular interest in the dynamics of land–atmosphere exchanges as it relates to global climate change and the need for more accurate numbers in global circulation models (GCMs). Recent advance in remote sensing and operational weather forecasts have significantly improved the ability to monitor the hydrologic cycle over broad regions (Vörösmarty and Peterson, 2000). The application of hydrologic models in understanding interactions between the watersheds and estuaries is critical when examining seasonal changes in the biogeochemical cycles of estuaries.

Hydrologic Cycle

Water is the most abundant substance on the Earth's surface with liquid water covering approximately 70% of the Earth. Most of the water (96%) in the reservoir on the Earth's surface is in the global ocean (figure 3.1). The remaining water, predominantly stored in the form of ice in polar regions, is distributed throughout the continents and atmosphere— estuaries represent a very small fraction of this total reservoir as a subcomponent of rivers. Water is moving continuously through these reservoirs. For example, there is a greater amount of evaporation than precipitation over the oceans; this imbalance is compensated by inputs from continental runoff. The most prolific surface runoff to the oceans is from rivers which discharge approximately 37,500 km^3 y^{-1} (Shiklomanov and Sokolov, 1983). The 10 most significant rivers, in rank of water discharge, account for approximately 30% of the total discharge to the oceans (Milliman and Meade, 1983; Meade, 1996). The most significant source of evaporation to the global hydrologic cycle occurs over the oceans; this occurs nonuniformly and is well correlated with latitudinal gradients of incident radiation and temperature. The flow of water from the atmosphere to the ocean and continents occurs in the form of rain, snow, and ice. Average turnover times of water in these reservoirs can range from 2640 y in the oceans to 8.2 d (days) in the atmosphere (Henshaw et al., 2000; table 3.1). The aqueous constituents of organic materials, such as overall biomass,

34

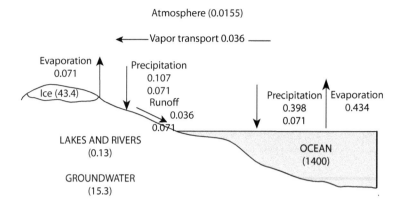

Figure 3.1 The hydrologic cycle. Arrows indicate fluxes (10^{18} kg y^{-1}) and inventories are in parentheses (10^{18} kg). (Modified from Berner and Berner, 1996.)

Table 3.1 Average turnover times of water in different reservoirs on Earth.

Reservoir	Volume (km^3)	Avg. turnover time
Oceans	1.338×10^9	2640 y
Cryosphere	24.1×10^6	8900 y
Groundwater/permafrost	23.7×10^6	515 y
Lakes/rivers	189,990	4.3 y
Soil moisture	16,500	52 d
Atmosphere	12,900	8.2 d
Biomass	1120	5.6 d

From Henshaw et al. (2000), with permission.

have an even shorter turnover time (5.3 d). These differences in turnover rate are critical in controlling rates of biogeochemical processes in aquatic systems.

As mentioned earlier, a significant amount of evaporation from the oceans is balanced by surface runoff from land. Runoff can be divided into the following two general categories: surface flow (overland runoff and rivers) and subsurface flow (groundwater). Both forms of runoff interface significantly with biogeochemical cycling in estuaries. The following section provides a general description of some of the basic hydrological principles associated with each of these types of runoff.

Water balance models have frequently been used to examine the surface runoff from watersheds. Some of these models, focused more on climate change, are called Soil–Vegetation–Atmosphere Transfer Schemes (SVATs) (Vörösmarty and Peterson, 2000). These model simulations use different parameters such as vegetation cover, *soil texture* (different sizes of mineral particles), water-holding capacity of soils, *surface roughness*, and *albedo* (the fraction of light reflected by a body or surface), to make predictions on soil moisture, evapotranspiration, and runoff. Unfortunately, the simulation of runoff has not been very effective as input to water budgets or for constructing hydrographs for rivers (Abramopoulos et al., 1988; Henderson-Sellers, 1996). When examining the pathways of water transport on land in an idealized SVAT model, we see that the movement of water is complex and difficult to measure on land (figure 3.2). For example, evaporation over land is strongly affected by vegetation. In fact, the term *evapotranspiration* is a term commonly used to describe the evaporation of water from the land surface as well as that lost through *transpiration* in plants. Plants may also reduce the impact energy of raindrops on soils (via leaf drip, figure 3.2) allowing for greater infiltration of water into soils than on barren land where surface flow develops very rapidly (Bach et al., 1986). The amount of water

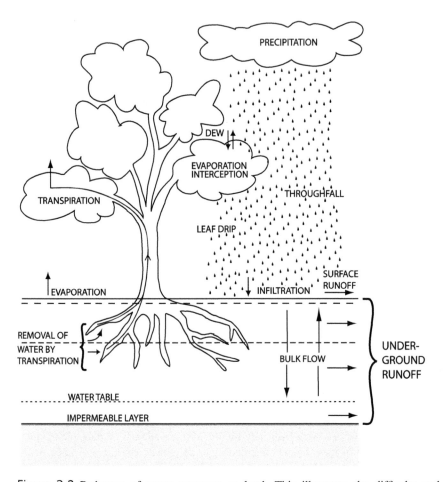

Figure 3.2 Pathways of water transport on land. This illustrates the difficulty and complexity of measuring the movement of water on land. (Modified from Vörösmarty et al., 2000.)

intercepted by dense forests can range from 8 to 35% of the total annual precipitation in a particular region (Dunne and Leopold, 1978).

Rainfall that reaches the land surface can infiltrate permeable soils, with each soil having a different, but limited, capacity to absorb water. The *infiltration capacity* (or rate) will vary depending on the soils' current moisture content (Horton, 1933, 1940). Surfaces of dry soil particles will develop a capillary action as they come into contact with water, resulting in a higher infiltration rate. As moisture content increases the soil swells and the infiltration capacity decreases eventually reaching an equilibrium value (Fetter, 1988; figure 3.3). The infiltration capacity curve can be described using the following equation from Horton (1933, 1940):

$$f_p = f_c + (f_o - f_c)\,e^{-kt} \tag{3.1}$$

where: f_p = infiltration capacity (or rate) at time t (m s^{-1}); f_c = equilibrium infiltration capacity (or rate) (m s^{-1}); f_o = initial infiltration capacity (or rate) (m s^{-1}); k = constant representing the rate of decreased infiltration capacity (s^{-1}); and t = time from the beginning of infiltration (s).

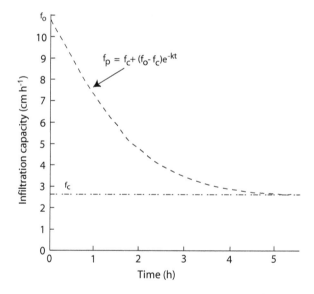

Figure 3.3 Relationship between infiltration capacity (cm h^{-1}) and time (h); this illustrates how capacity decreases over time and will eventually reach an equilibrium value. The infiltration capacity curve can be described using the following equation: $f_p = f_c + (f_o - f_c)e^{-kt}$, where: f_p = infiltration capacity (or rate) at time t (m s^{-1}); f_c = equilibrium infiltration capacity (or rate) (m s^{-1}); f_o = initial infiltration capacity (or rate) (m s^{-1}); k = constant representing the rate of decreased infiltration capacity (1/s); and t = time from the beginning of infiltration (s). (Modified from Fetter, 2001.)

According to Fetter (1988), the following three scenarios can be used to describe the relationship between precipitation and infiltration rates: (1) when the total precipitation rate is lower than the equilibrium infiltration capacity all of the precipitation reaching the land surface should infiltrate; (2) when the precipitation rate is higher than the equilibrium capacity and less than the initial infiltration capacity (at the start of the precipitation event) all of the precipitation will infiltrate initially; and (3) when the precipitation rate is greater than the initial infiltration capacity there is an immediate accumulation of water on land. In general, when the precipitation rate exceeds infiltration capacity, overland flow of water occurs, commonly referred to as *Horton overland flow* (HOF) (Horton, 1933, 1940).

The physical properties of soils and sediments are particularly important in determining infiltration rates in soils as well as groundwater flow velocities. One feature of particular importance is the void space or *porosity* of soils, rocks, and sediments. Water is capable of moving from one void space to another in these materials thereby allowing the flow of water. Total porosity is defined mathematically by the equation:

$$n = 100\left[1 - \left(\rho_b/\rho_d\right)\right] \tag{3.2}$$

where: n = total porosity as a percentage; ρ_b = bulk density of aquifer material (g cm^{-3}); and ρ_d = particle density of the aquifer material (g cm^{-3}).

The typical density of rock and soil materials is approximately 2.65 g cm^{-3}—the density of quartz. The high *hydraulic conductivity* of sandy sediments is ideal for producing groundwater. Groundwater flow velocities have been found to be in accordance with *Darcy's law*:

$$\upsilon = K/n\,(\mathrm{d}h/\mathrm{d}L) \tag{3.3}$$

where: υ = velocity; K = coefficient of hydraulic conductivity or permeability (units of length/time); n = total porosity; and $\mathrm{d}h/\mathrm{d}L$ = hydraulic gradient.

Hydraulic conductivities range from 10^{-12} to $3\,\mathrm{cm\,s}^{-1}$ for unfractured igneous/metamorphic rocks and porous limestone/gravel, respectively (Henshaw et al., 2000). If sediments are relatively homogeneous in their hydraulic conductivity, infiltrated water will generally move in a vertical direction. However, if the hydraulic conductivity decreases vertically in less permeable soils the flow will change to a horizontal direction; this is called *interflow* and may represent a substantial amount of the total runoff. The reservoir of groundwater is balanced by newly infiltrated waters and the *baseflow*, which discharges into a stream (Freeze and Cherry, 1979; Fetter, 1988). Although baseflow may show some annual variability, it is generally more constant than surface runoff. Therefore, most of the variability in total discharge of a stream is due to episodic changes in precipitation events that alter the contribution of overland flow, *interflow* (lateral flow of water through soil pores), and *direct precipitation* (direct input of rainfall to the water body—assuming equal distribution). When examining a hypothetical storm hydrograph for a period of 2 d (with evenly distributed precipitation) we see that changes in overland flow are responsive to precipitation and correlate well with changes in total stream flow; conversely, groundwater inputs (baseflow) remain essentially constant during this period (Fetter, 1988; figure 3.4).

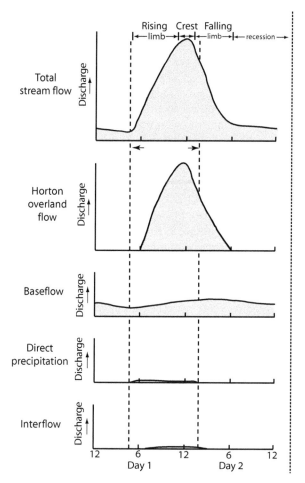

Figure 3.4 Hypothetical storm hydrograph for a period of 2 days (with evenly distributed precipitation) showing how changes in overland flow are responsive to precipitation and correlate well with changes in total stream flow. Conversely, groundwater inputs (baseflow) remain essentially constant during this period. (Modified from Fetter, 2001.)

Although the majority of runoff from land to the oceans is dominated by riverine inputs, with minimal inputs directly from groundwater runoff, recent work has shown that submarine groundwater discharge (SGD) into estuaries and coastal waters is more significant than previously thought and may have significant impacts on the cycling of nutrients and contaminants (Spiker and Rubin, 1975; Freeze and Cherry, 1979; Valiela and D'Elia, 1990; Moore, 1996, 1999; Burnett et al., 2001; Kelly and Moran, 2002; Burnett et al., 2003). Burnett et al. (2003, p. 6) recently defined SGD as "any and all flow of water on continental margins from the seabed to the coastal ocean, regardless of fluid composition or driving force." Assuming a mean river flow of 37,500 km^3 y^{-1},

SGD represents approximately 0.3 to 16% of the global river discharge (Burnett et al., 2003). Determining groundwater discharge into estuarine and coastal systems remains a difficult task because of the diffuse flow and overall heterogeneous character of groundwater flow. The chemical composition of emerging SGD in coastal waters is generally distinct between river and marine end-members; interactions between major ions' solid phases during transport are likely responsible for these differences (Burnett et al., 2003). Recent work has suggested that the chemical interactions between SGD and solid phases at the coastal boundary are similar to what occurs in estuarine systems, and that these coastal aquifers are serving as "subterranean estuaries"—in a biogeochemical sense (Moore, 1999).

Seep meters and hydraulic models have commonly been used to make estimates of diffusive groundwater inputs (Freeze and Cherry, 1979; Cable et al., 1996, 1997; Moore, 1999). Radium (Ra) isotopes have also been effective in tracing groundwater inputs to the coastal zone (Cable et al., 1996; Moore, 1996; Kelly and Moran, 2002; Krest and Harvey, 2003). In general, this method is based on the balancing of ^{226}Ra inputs from groundwater with the loss or removal of excess ^{226}Ra as a result of tidal flushing. The overall conclusions drawn from this Ra-tracer work suggests that groundwater inputs are spatially and temporally variable and can represent a significant source of dissolved constituents to the coastal margin. For example, Kelly and Moran (2002) found that the seasonal range of groundwater flux to the Pettaquamscutt estuary, Rhode Island, ranged from 1.5 to 22 L m^{-2} d^{-1}. Similarly, SGD was found to represent greater than 50% of the NO$_3$$^-$ inputs into Great South Bay estuary (USA) (Capone and Bautista, 1985; Capone and Slater, 1990). The relative importance of groundwater flux and its associated nutrients in porous carbonate systems, like Florida Bay (USA), are likely to be even more pronounced because of the permeability of *karst* systems (Corbett et al., 1999). Further details on the impact of groundwater inputs on the biogeochemistry of estuaries and coastal waters, and the radionuclide tracers used to track such inputs, will be covered in chapters 16 and 7, respectively.

Watershed analysis is critical if we are to understand the processes controlling hydrology and land–water interactions at the coastal margin. To date, the most effective approach has been modeling efforts that have examined process-level questions of water transport for small *catchment* and *hill-slope* scales (McDonnell and Kendall, 1994). Because of the strong linkage that exists between biogeochemical cycling in terrestrial ecosystems and inputs of dissolved and particulate constituents (e.g., nutrient and contaminant loading) into estuarine systems, an understanding of the correct water balances on land is critical in understanding estuarine biogeochemistry. For example, nutrient inputs to the catchment and streams/rivers have shown good correlations on both small regional (figure 3.5) and larger continental scales (figure 3.6). In one case, NO$_3$$^-$ concentrations in streams were found to be directly related to fertilizer applications in croplands in the mid-Atlantic region of North America (Jordan et al., 1997). Similarly, wet and dry deposition of oxidized nitrogen in the larger North Atlantic Basin was found to be a good predictor of river nitrogen export to the North Atlantic Ocean (Howarth et al., 1996). Therefore, understanding the surface runoff and groundwater processes that control the fate and transport of dissolved and particulate constituents in the watershed is critical in making better estimates of *point* and *non-point source* inputs to estuaries.

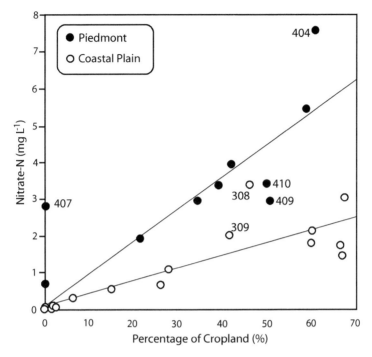

Figure 3.5 Correlation between nutrient inputs to the catchment and streams/rivers on a small regional scale. Nitrate concentrations in streams are shown to be directly related to fertilizer applications in croplands in the mid-Atlantic region of North America. (Modified from Jordan et al., 1997.)

General Circulation, Mixing Patterns, and Salt Balance

Estuarine circulation is one of the most important characteristics of an estuary because it determines the salt flux and horizontal dispersion and is a key variable that affects stratification. It has recently been suggested that the magnitude of stress between bottom inflow and surface outflow are not large enough to significantly affect mixing dynamics (Geyer et al., 2000). Just as estuaries were classified based on their geomorphometic features in chapter 2, they have also been traditionally classified based on the circulation patterns (Pritchard, 1952, 1954, 1956; Stommel and Farmer, 1952; Dyer, 1973, 1979; Officer, 1976; Bowden, 1967, 1980; Officer and Lynch, 1981). More specifically, these early theories can be divided into two general approaches: hydraulic control in two-layer flow (Stommel and Farmer, 1952, 1953) and spatial mixing as a viscous–advective–diffusive (VAD) balance in stratified layers (Pritchard 1952, 1954, 1956; Rattray and Hansen, 1962; Hansen and Rattrey, 1965; Fisher, 1972). Differences between the density and elevation of these layers are largely responsible for *gravitational circulation* in estuaries. While many of these classic and recent circulation models invoke a rectangular channel cross section, transverse depth variations are more realistic in natural systems (Uncles, 2002).

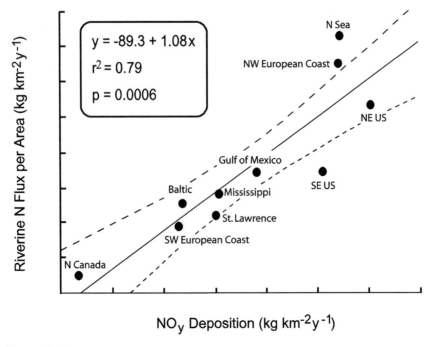

NO$_y$ Deposition (kg km^{-2}y^{-1})

Figure 3.6 Wet versus dry deposition of oxidized nitrogen in the larger North Atlantic Basin; this was found to be a good predictor of river nitrogen export to the North Atlantic Ocean. (Modified from Howarth et al., 1996.)

Recent advances using the Acoustic Doppler Current Profiler (ADCP), ocean surface current radar (OSCR), and satellite technology, such as the advanced very high-resolution radiometer (AVHRR) and sea-viewing wide field-of-view sensor (SeaWiFS), have led to more effective approaches in quantifying transverse variability (Uncles, 2002). These techniques, particularly the ADCP have also been useful in tracking frontal structures associated with propagation of density and tidal currents up and down estuaries (O'Donnell et al., 1998; Marmorino and Trump, 2000). Flow convergence, which is largely responsible for front formation, can occur in longitudinal and tranverse forms. In fact, transverse structure has been shown to be linked with the formation of flow convergence in near-surface convergence zones (Nunes and Simpson, 1985; O'Donnell, 1993; Valle-Levinson and O'Donnell, 1996).

All estuarine systems are influenced by rotational effects (e.g., Coriolis forces) that serve as important forcing functions and lead to complicated circulation patterns in estuaries (Angel and Fasham, 1983). For example, the St. Lawrence estuary is large enough to contain *mesoscale eddies* and cross-shore currents (Ingram and El-Sabh, 1990). The consequences of these circulation patterns can have significant effects on biogeochemical cycling; the downward transport of oxygen and the upward transport of nutrients are examples of such cycling (Boicourt, 1990; Kuo et al., 1991). The general categories of estuarine circulation have been identified as: (a) *well-mixed estuaries*, where there is minimal vertical stratification in salinity; (b) *partially mixed estuaries*, where the vertical mixing is inhibited to some degree; (c) *highly stratified estuaries*, with lower freshwater discharge than the salt wedge system; and (d) *salt wedge estuaries* and many fjords (Bowden, 1980; figure 3.7). Vertical transport and gravitational circulation in estuaries are influenced by both stratification and *turbulent mixing*, which results from the destabilizing forces of tides and/or winds (Hansen and Rattray, 1966; Pritchard, 1989; Etemad-Shahidi and Imberger, 2002). Under highly stratified conditions, two-layered flow is largely determined by the interaction of *baroclinic* and *barotropic* forcing (Chuang and Wiseman, 1983; Schroeder and Wiseman, 1986, 1999; Stacey et al., 2001).

The general mixing properties of an estuary have been examined using the following approaches (Bowden, 1980): (1) simple steady-state one-dimensional models (longitudinal properties considered but cross-section just averaged); (2) two-dimensional models (averaging vertical and transverse directions); and (3) three-dimensional (3-D) models where the vertical and transverse variations are too dissimilar to model. The mixing of saltwater and freshwater in estuaries occurs through the processes of mixing and diffusion. Turbulent mixing is essential for the basic patterns of estuarine circulation (Peters, 1999; Peters and Bokhorst, 2000). The interpretation of turbulence measurements in estuaries remains extremely controversial but should improve as more data continue to be gathered on different systems (Simpson et al., 1996; Stacey et al., 1999; Geyer et al., 2000). The biogeochemical consequences of these mixing processes are very important from the perspective of fluid dynamics as they relate to the downward flux of oxygen from the surface and the upward transport of nutrients from bottom waters (Kuo et al., 1991). In fact, the geometrical complexity of estuaries has raised many questions that concern the relative importance of hydraulic control points and overall distribution of mixing patterns (localized or very broad) (Seim and Gregg, 1997). Using a modeling approach, recent work has shown that tides alone can cause peak circulation in Charlotte Harbor estuary; in general, this is attributed to turbulent entrainment across boundary layers (Weisberg and Zheng, 2003). The basic premise here is that turbulent entrainment results in net transport of seawater into the upper freshwater lens; tides can further enhance this entrainment resulting in a horizontal density gradient thereby allowing circulation to occur. Continued mixing and entrainment can result in transitions from a salt-wedge to a partially mixed estuary. Such models emphasize the need for further work that examines the interaction of frictional forces on circulation patterns at the micro- and macroscale in estuaries (Weisberg and Zheng, 2003).

If we assume that mixing in estuaries can be treated as a simple mixing chamber, saltwater and freshwater will be mixed on the incoming flood tide, which results in an average estuarine salinity (S) that will be further mixed and discharged from the estuary during ebb tide (Solis and Powell, 1999). Thus, the salinity for a stationary system with complete mixing can be defined with the following equation:

$$S = \frac{\sigma \times V_t}{V_t + V_{fr}} \tag{3.4}$$

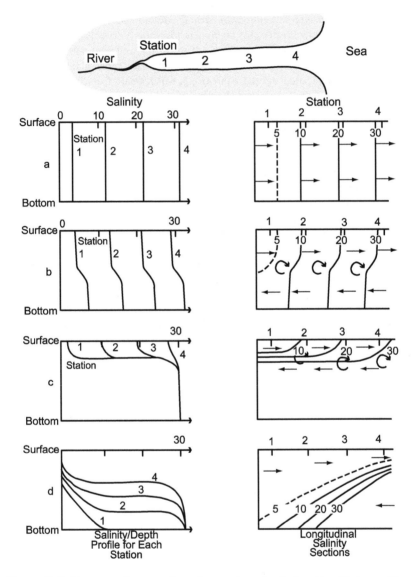

Figure 3.7 The general categories of estuarine circulation identified as: Type A, well-mixed estuaries, where there is minimal vertical stratification in salinity; Type B, partially mixed estuaries, where the vertical mixing is inhibited to some degree; Type C, highly stratified with lower freshwater discharge than the salt wedge system; and Type D, salt wedge estuary and many fjords. (Modified from Bowden, 1980.)

where: S = average salinity of mixed water in estuary; V_t = average *tidal prism* volume that enters during flood tide; V_{fr} = total volume of inflowing freshwater (rivers, ground-water, precipitation); and σ = salinity of saltwater influx (standard seawater salinity) through the tidal prism.

Solis and Powell (1999) further modified this equation to reflect a normalized average salinity of the entire estuary as follows:

$$\frac{S}{\sigma} = \frac{1}{1 + (V_{fr}/V_t)} = \frac{1}{1 + (V_r + V_p - V_e)/V_t} \tag{3.5}$$

where: V_r = average volume of river and groundwater inflow water; V_p = average precipitation volume falling on the estuary; and V_e = average volume lost to evaporation.

This equation essentially defines the ratio of total freshwater inflow volume relative to the tidal prism volume. It should be noted that there are a number of assumptions in these models that are not stated here; for further details on this please refer to Solis and Powell (1999). More specifically, the (V_{fr}/V_t) ratio alone is most critical in determining estuarine salinity. This ratio has been used as a general descriptor of the physical mixing characteristics of estuaries by Bowden (1980), with ratios of 1.0, 0.1, and 0.01 indicative of salt-wedge, partially mixed, and well-mixed estuaries, respectively. The above equation unrealistically assumes that the saltwater introduced is completely mixed during one tidal cycle. In fact, it has been well accepted for many years that the inflowing water of flood tide are not homogenously mixed throughout the entire estuary and that the waters in an ebb tide may not reflect any changes from mixing within the estuary proper. For example, the assumption of fractional entrainment of tidal waters into estuarine waters has been utilized for quite some time (Dyer and Taylor, 1973; van de Kreeke, 1988). Assuming imperfect mixing of inflowing tidal waters, Solis and Powell (1999) recently modified the equation from van de Kreeke (1988) for a bulk mixing term (e) to be:

$$e = \frac{V_{fr}}{V_t} \times \frac{S/\sigma}{1 - S/\sigma} \tag{3.6}$$

Using a data set from the National Oceanographic Atmospheric Administration (NOAA), Solis and Powell (1999) estimated that bulk mixing efficiencies for estuaries along the Gulf of Mexico ranged from 0.03 to 0.58 (figures 3.8a and 3.8b). As pointed out by Solis and Powell (1999) other factors that are likely to affect mixing, not considered in this mixing equation, are the geomorphology, general circulation patterns, and climatological forcing in estuaries. This wide range in mixing efficiencies along the Texas coast are consistent with the extreme gradient in rainfall patterns; there are higher rainfall amounts along the north Texas coast (Sabine Estuary) compared to estuaries in the south (Corpus Christi and Aransas Bay) (figures 3.8a and 3.8b). As will be discussed later in this chapter, there is a significant inverse relationship between mixing efficiency and residence times of these Texas estuaries—this can have serious consequences on biogeochemical processes.

Hansen and Rattray (1966) first introduced the idea of using stratification–circulation diagrams to describe a spectrum of circulation and geomorphometric types of estuaries that can be defined by stratification (figure 3.9). The basic classification parameters are as follows: the stratification is defined by $\delta S/S_0$, where δS is the difference in the salinity between surface and bottom water and S_0 is the mean-depth salinity, both averaged over a tidal cycle; and U_s/U_f, where U_S is the surface velocity (averaged over a tidal cycle) and U_f is the vertically averaged net outflow. The subdivisions "a" and "b" in figure 3.9 represent values where $\delta S/S_0 < 0.1$ and $\delta S/S_0 > 0.1$, respectively; subscripts "h" and "l" refer to high and low river flow. The curved line at the top represents the limit of surface freshwater outflow. In this general classification scheme proposed by Hansen and Rattray (1966) the following estuarine types are described. *Type 1 estuaries*: well-mixed estuaries with mean flow in the seaward direction and the salt balance being maintained by diffusive processes—via tidal transport; *Type 2 estuaries*: partially mixed estuaries where the net flow reverses at depth and the salt flux is maintained by both diffusive and advective processes; *Type 3 estuaries*: these estuaries include fjords with two distinct layers and advection accounting for the majority of salt flux; *Type 4 estuaries*, these are salt-wedge estuaries where freshwater flows over a stable more dense bottom layer (figure 3.9).

Other parameters used to characterize the types of circulation, mixing, and stratification in estuaries, principally derived from the method of Hansen and Rattray (1966), are the *densiometric Froude* (F_m) and *estuarine Richardson* (Ri_E) numbers (both defined later in this section) (figure 3.10; Fisher, 1976). Fisher (1976) demonstrated that there is a relationship between $\delta S/S_0$ and Ri_E; Jay et al. (2000) interpreted this ratio as the tendency

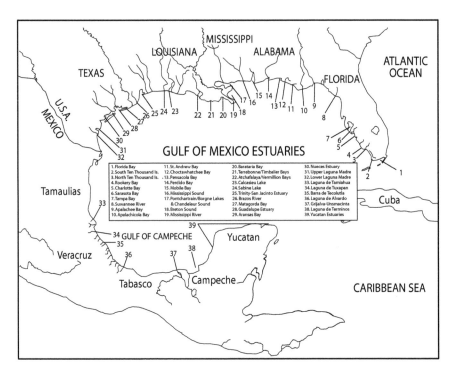

Figure 3.8a Map of estuaries in the Gulf of Mexico. (From Solis and Powell, 1999, with permission.)

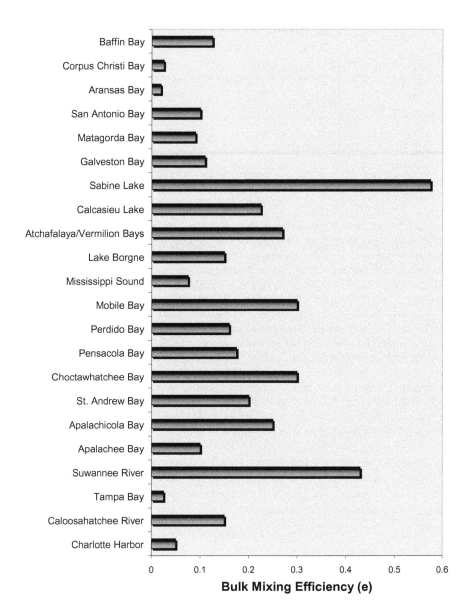

Figure 3.8b Bulk mixing efficiency (*e*) in estuaries from the Gulf of Mexico. (From Solis and Powell, 1999, with permission.)

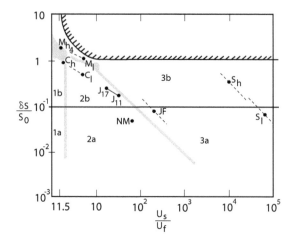

Figure 3.9 Stratification–circulation diagrams used to describe a spectrum of circulation and geomorphometric types of estuaries that can be defined by stratification. Estuarine types are as follows: Type 1 estuaries are those without upstream flow requiring tidal transport for salt balance; Type 2 estuaries are partially mixed (e.g., Marrows of the Mersey (NM) (UK), James River (J) (USA), Columbia River estuary (C) (USA); Type 3 estuaries are representative of fjords [e.g., Siver Bay (S), Strait of Juan de Fuca (JF) (USA)]; and Type 4 estuaries indicative of salt wedge estuaries [e.g., Mississippi River (M) (USA)]. The basic classification parameters are as follows: the stratification is defined by $\delta S/S_0$, where δS is the difference in the salinity between surface and bottom water and S_0 is the mean-depth salinity, both averaged over a tidal cycle; and U_s/U_f, where U_s is the surface velocity (averaged over a tidal cycle) and U_f is the vertically averaged net outflow. The subdivisions "a" and "b" represent values where $\delta S/S_0 < 0.1$ and $\delta S/S_0 > 0.1$, respectively; subscripts "h" and "l" refer to high and low river flow. The curved line at the top represents the limit of surface freshwater outflow. (From Hansen and Rattray, 1966, as modified by Jay et al., 2000, with permission.)

for freshwater flow to become stratified in a system as it relates to the tidal forces that tend to disrupt the stratification. This is further related to F_m, which Jay et al. (2002) described as the ratio of freshwater flow speed to the internal wave speed. The densiometric Froude number is defined by the following equation:

$$F_m = U_f/(gh\Delta\rho/\rho)^{1/2} \tag{3.7}$$

where: g = gravitational acceleration; $\Delta\rho$ = density difference between water layers; ρ = water density; and h = water depth.

Fischer (1972) defined the estuarine Richardson number (Ri_E) by the following equation:

$$Ri_E = \frac{g(\Delta\rho/\rho)(U_f/b)}{U_t^3} \tag{3.8}$$

where: b = total mean width of estuary; and U_t = root mean square of tidal velocity.

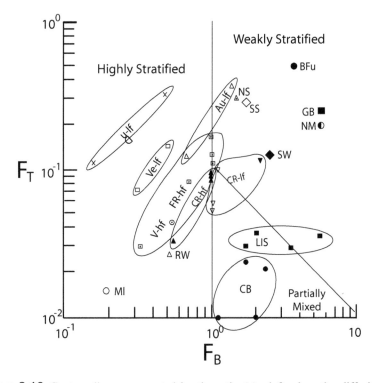

Figure 3.10 Contour lines represented by the ratio (v), defined as the diffusive salt transport (tidal driven) to the total landward salt flux as related to contours of Froude [$F_m = U_f/(gh\Delta\rho/\rho)^{1/2}$] Richardson numbers [$Ri_E = (g[\Delta\rho/\rho][U_f/b])/U_t^3$]. A clear decrease is found when moving from Type 1 to Type 3 estuaries (From Fisher, 1976, as modified by Jay et al., 2000, with permission.)

The contour lines represented by the ratio (v_s), defined as the diffusive salt transport (tidal driven) to the total landward salt flux, show a clear decrease when moving from Type 1 to Type 3 estuaries (figure 3.10). As discussed by Jay et al. (2000), this predictive capability in understanding the maintenance of salt balance in different estuarine types can be very important in comparing biogeochemical and ecological cycles. However, the assumption of a steady salt balance in the classification scheme by Hansen and Rattray (1966) can lead to erroneous information concerning the balance between tidal forcing and river flow (Jay and Smith, 1988). Improved methods that better constrain estuarine geometry and circulatory forcing functions, as they relate to residence time, may increase the applicability of these classification schemes to a broader diversity of estuarine types (Jay et al., 2000).

Stommel and Farmer (1952) showed that the partitioning of flow at the mouths of estuaries results in higher salinities during flood than ebb flow; this will produce a net salt flux in the landward direction. *Tidal pumping* generally results from a correlation between salinity and velocity variations on a tidal time scale. Another concept, referred to as *tidal trapping*, occurs when parcels of water are advected into lateral "traps" during one phase of the tidal cycle and then moved out in another resulting in an overall upstream transport (Okubo, 1973). Hughes and Rattray (1980) found that tidal pumping was the dominant mechanism for the landward flux of salt in a section of the Columbia River during high discharge. Tidal pumping has been shown to have considerable variability in magnitude compared with other mechanisms of salt flux (Geyer and Nepf, 1996). For example, tidal pumping in the Hudson River estuary was found to have significant variability between different tidal periods, with the greatest effects on salt flux during the periods of intense stratification (Geyer and Nepf, 1996). Other work in the Hudson River estuary has shown that the largest vertical salt flux occurred during spring ebb tides (about 30% of the total fortnightly salt flux), with flood tides providing the remaining flux (Peters, 1999). More recently it was shown that both VAD balance and hydraulics were important along the *thalweg* of the Hudson River (USA), which further supports the inclusion of both processes in estuarine models (Peters, 2003).

The effects of wind on subtidal current circulation in estuaries have been shown to be significant in many systems (Pollak, 1960; Weisburg and Sturges, 1976; Kjerfve et al., 1978; Smith, 1978; Wang and Elliott, 1978; Wong and Valle-Levinson, 2002). Wind effects can occur from both local and remote forcing. Winds on the continental shelves, adjacent to estuaries, can generate sea level changes at the mouth of estuaries as well as propagating free waves (Nobel and Butman, 1979). Conversely, local wind effects operate directly on the surface of the estuary, affecting estuarine circulation. For example, local wind effects have been shown to be important in controlling the bidirectional flow of currents in the Delaware Bay estuary (Wong and Moses-Hall, 1998). More recently, it was shown that local winds have significant effects on the bidirectional flow and subtidal exchange between the Chesapeake Bay and adjacent shelf (Wong and Valle-Levinson, 2002).

Residence Times

Residence time is defined as the ratio of the mass of a *scalar* (quantity in mathematics consisting of a single real number used to measure magnitude) (e.g., salinity) in a reservoir to the rate of renewal of the scalar, under steady-state conditions (Geyer et al., 2000). Numerous other definitions of residence time exist, including the time it takes to replace equivalent freshwater in an estuary with freshwater inputs (Bowden, 1967; Officer, 1976). The physical factors that have been used to estimate residence times are flushing and vertical mixing/entrainment (Jay et al., 2000). The actual methods for calculating residence times can be categorized as follows: (1) the freshwater fraction (or flushing time) method (Bowden, 1980); (2) the tidal prism method (Officer, 1976); (3) box-model methods dependent on age and removal of particles (Officer, 1980; Zimmerman, 1988; Miller and McPherson, 1991); and (4) numerical hydrodynamic models (Geyer and Signell, 1992; Sheng et al., 1993; Oliveira and Baptista, 1997). Drifter-tracking technology (Hitchcock et al., 1996) and the application of chemical tracers (SF_6) (Clark et al., 1996) should hold promise for future Lagrangian-type experiments. Unfortunately, many of these different methods can at times yield very different estimates of residence time. The complexity of mixing, circulation, and evaporation/precipitation associated with estuarine systems can make interpreting residence times a very difficult task. This has been particularly troubling for the more recent efforts that utilize numerical modeling techniques. Thus, the more simplistic freshwater fraction and tidal prism methods remain quite common today. Staying within the biogeochemical scope of this book only the freshwater fraction method will be covered. In the most simple of terms freshwater is used as a tracer in this method, where it is assumed that the freshwater being removed is being replaced at the same time by inflowing river water (Bowden, 1980). The freshwater content of an estuary is a function of the estuary volume and the equivalent freshwater fraction (f), where f is defined as follows:

$$f = \frac{\sigma - S}{\sigma} \qquad (3.9)$$

where: σ = salinity of saltwater influx (standard seawater salinity) through the tidal prism; and S = average salinity of mixed water in estuary.

This equation can then be used to calculate residence time (t) with the following equation:

$$t = \frac{V_f}{Q} \qquad (3.10)$$

where: V_f = volume of freshwater in the estuary; and Q = freshwater inflow rate

Solis and Powell (1999) used a NOAA data set to estimate residence times for estuaries in the Gulf of Mexico (figure 3.11). The range of residence times for these estuaries is very broad, from less than 5 to greater than 300 d (Solis and Powell, 1999).

Residence times in estuaries can have important consequences for elemental cycling. For example, the low retention of ^{7}Be and ^{210}Pb radionuclides in surface sediments of the Sabine Estuary (figure 3.8a) has been attributed to the short residence time (figure 3.11) of this system (Baskaran et al., 1997).

Residence times are commonly used to estimate exchange rates of biogeochemical processes such as nutrient fluxes (Nixon et al.,1996), chlorophyll concentrations (Monbet, 1992), primary production (Jørgensen and Richardson, 1996), and benthic faunal production (Josefson and Ramussen, 2000). While there are a limited number of studies that have effectively made such linkages between water exchange rates and biogeochemical processes, more studies are needed to better represent the diversity of estuaries. There has been considerable development of numerical hydrodynamic models in estuaries over the past decade (Dyke, 2001). For example, recent applications of two commonly used estuarine 3-D models, the Princeton Ocean Model (POM) and Hamburg Ocean Primitive Equation Model (HOPE), have been used effectively to determine the effects of external forcing on circulation patterns in estuaries (Walters, 1997; Valle-Levinson and Wilson, 1994a,b). Unfortunately, many of the complex input variables needed for hydrodynamic models are not available in many estuarine systems. Recent work has shown that simple morphological models, formulated using a constant volume flow per unit entrance width, were effective in estimating water exchange for shallow Danish estuaries in the absence of more sophisticated modeling (Rasmussen and Josefson, 2002).

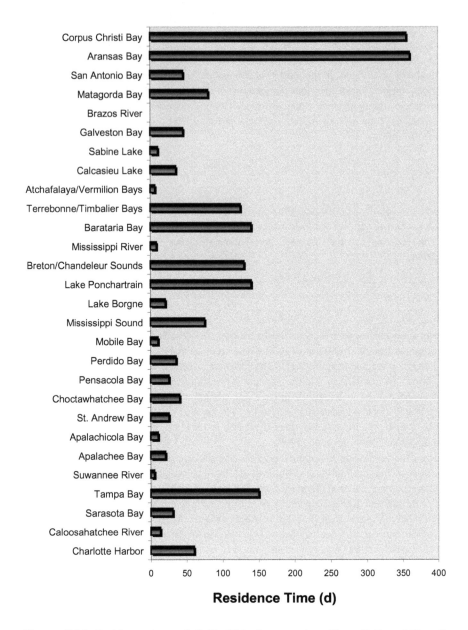

Figure 3.11 Residence times of Gulf of Mexico estuaries. (From Solis and Powell, 1999, with permission.)

Summary

1. Water is the most abundant substance on the Earth's surface with liquid water covering approximately 70% of the Earth with most of the water (96%) in the Earth's surface global ocean reservoir.
2. The 10 most significant rivers, in rank of water discharge, account for approximately 30% of the total discharge to the oceans.
3. SVATs are simulated models that use parameters such as vegetation cover, soil texture, water-holding capacity of soils, surface roughness, and albedo, to make predictions on soil moisture, evapotranspiration, and runoff.
4. Evapotranspiration is the evaporation of water from the land surface as well as that lost through transpiration in plants. Plants may also reduce the impact energy of raindrops on soils.
5. Rainfall that reaches the land surface can infiltrate permeable soils, with each soil having a different, but limited, capacity to absorb water. The infiltration capacity (or rate) will vary, depending on the soils' current moisture content.
6. When precipitation rate exceeds infiltration capacity, overland flow of water occurs, commonly referred to as Horton overland flow.
7. Hydraulic conductivity decreases in less permeable soils resulting in flow that is horizontal in direction; this is called interflow and may represent a substantial amount of the total runoff. The reservoir of groundwater is balanced by newly infiltrated waters and the baseflow, which discharges into a stream.
8. The majority of runoff from land to the oceans is dominated by riverine inputs, with minimal inputs directly from groundwater runoff. However, recent work has shown that SGD into estuaries and coastal waters is more significant than previously thought and may have significant impacts on the cycling of nutrients and contaminants.
9. The general categories of estuarine circulation have been identified as: highly stratified as in a salt-wedge estuary and many fjords; partially-mixed estuaries, where the vertical mixing is inhibited to some degree; and well-mixed estuaries, where there is minimal vertical stratification in salinity.
10. Vertical transport and gravitational circulation in estuaries are influenced by both stratification and turbulent mixing, which result from the destabilizing forces of tides and/or wind processes. Under highly stratified conditions, two-layered flow is largely determined by the interaction of baroclinic and barotropic forcing.
11. Hansen and Rattray (1966) introduced a general classification scheme for estuaries based on stratification/circulation that is divided into the following four estuarine types: Type 1 estuaries: well-mixed estuaries with mean flow in the seaward direction and the salt balance being maintained by diffusive processes—via tidal transport; Type 2 estuaries: partially mixed estuaries where the net flow reverses at depth and the salt flux is maintained by both diffusive and advective processes; Type 3 estuaries: these estuaries include fjords with two distinct layers and advection accounting for the majority of the salt flux; Type 4 estuaries: these are salt-wedge estuaries where freshwater flows out over a stable more dense bottom layer.

12. Two parameters used to characterize the types of circulation, mixing, and stratification in estuaries are the densiometric Froude (F_m) and estuarine Richardson (Ri_E) numbers.

13. Residence time is defined as the ratio of the mass of a scalar to the rate of renewal of the scalar under steady-state conditions. Residence times are commonly used to better understand variability in biogeochemical processes such as nutrient fluxes, chlorophyll concentrations, primary production, and benthic faunal production.

Part II

Chemistry of Estuarine Waters

Chapter 4

Physical Properties and Gradients

Before discussing the chemical dynamics of estuarine systems it is important to briefly review some of the basic principles of *thermodynamic* or *equilibrium models* and *kinetics* that are relevant to upcoming discussions in aquatic chemistry. Similarly, the fundamental properties of freshwater and seawater are discussed because of the importance of salinity gradients and their effects on estuarine chemistry.

Thermodynamic Equilibrium Models and Kinetics

Stumm and Morgan (1996) described how different components of laboratory- and field-based measurements in aquatic chemistry are integrated. Basically, observations from laboratory experiments are made under well-controlled conditions (focused on a natural system of interest), which can then be used to make predictions and models, which are ultimately used to interpret complex patterns in the natural environment (figure 4.1). Due to the complexity of natural systems, equilibrium models can tell you something about how chemical constituents (gases, dissolved species, solids) under well-constrained conditions (no change over time, fixed temperature and pressure, and homogeneous distribution of constituents). Equilibrium models will tell you something about the chemistry of the system at equilibrium but will not tell you anything about the kinetics with which the system reached equilibrium state. The laws of thermodynamics are the foundation for chemical systems at equilibrium. The basic objectives in using equilibrium models in estuarine/aquatic chemistry is to calculate equilibrium compositions in natural waters, to determine the amount of energy needed to make certain reactions occur, and to ascertain how far a system may be from equilibrium (Stumm and Morgan, 1996).

The first law of thermodynamics states that energy cannot be created or destroyed (i.e., the total energy of a system is always constant). This means that if the internal energy of a reaction increases then there must be a concomitant uptake of energy usually in the form of heat. *Enthalpy (H)* is a parameter used to describe the energy of a system as heat

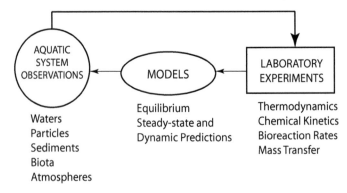

Figure 4.1 Schematic showing progression of how observations from laboratory experiments can then be used to make predictions and models, which are ultimately used to interpret complex patterns in the natural environment. (From Stumm and Morgan, 1996, with permission.)

flows at a constant pressure; it is defined by the following equation:

$$H = E + PV \tag{4.1}$$

where: E = internal energy; P = pressure; and V = volume.

The second law (at least one form of it) states that a spontaneous reaction is when there is a decrease in the system's free energy and an increase in the entropy (S). The *Gibbs free energy* (G) of a system is related to enthalpy and *entropy* by the following equation:

$$G = H - TS \tag{4.2}$$

where: S = entropy; and T = absolute temperature in kelvin [$T = t$ (°C) + 273.15]

Under standard conditions G, H, and S are represented as ΔG^0, ΔH^0, and ΔS^0, and are referred to as the changes in *standard free energy of formation, enthalpy of formation,* and entropy, respectively, at standard temperature and pressure (STP): 0°C or 273.15 K; one atmosphere = 101.325 kPa (1 Pa = 1 pascal = 1 newton per square meter).

The chemical reaction for this, when the temperature of reactants and products are equal, is as follows:

$$\Delta G^0 = \Delta H^0 - T\Delta S^0 \tag{4.3}$$

The more negative ΔG^0, the more spontaneous the system is, making it have a greater potential to react. For a given reaction the standard free energy ΔG^0 can be calculated by the following equation:

$$\Delta G^0 = \sum \Delta G^0(\text{products}) - \sum \Delta G^0(\text{reactants}) \tag{4.4}$$

The more negative ΔH^0 (a measure of bond strength in reactants and products) is the more spontaneous the reaction. Finally, ΔS^0, which is a measure of the disorder of reactants and products, will be more positive when the reaction is more spontaneous. In conclusion, chemical thermodynamics provide a way to predict how spontaneous a reaction will proceed. An example of a basic equilibrium model for species AB would be as follows:

$$A + B = AB, \qquad\qquad K = \frac{\{AB\}}{\{A\}\{B\}} \qquad (4.5)$$

where: $K = $ *equilibrium* or *stability* constant for the reaction.

In this example, one key objective would be to determine the concentrations of [AB], [A], and [B] that fit with the equilibrium constant. The values of ion activities, not shown (discussed later), will also be dependent on these concentrations at equilibrium. It is generally assumed that while most natural systems are far from equilibrium conditions, if the reactions between reactant and product states are rapid, equilibrium can be applied (Butcher and Anthony, 2000). For example, in aquatic systems, NH_4^+ and $NH_3(aq)$ are considered to be in equilibrium, as shown below, because the proton exchange reaction is so rapid (Quinn et al., 1988):

$$NH_3 + H_2O \leftrightarrow NH_4^+ + OH^- \qquad (4.6)$$

More details of equilibrium models commonly used for gas phase equilibria, mixed phases, isotope effects, oxidation/reduction, electron activity, and pH stability diagrams can be found in Stumm and Morgan (1996) and Butcher and Anthony (2000).

Chemical kinetic models provide information on reaction rates that cannot otherwise be obtained in chemical thermodynamics. However, in many situations information such as kinetic rate constants, needed for such models, are not available. The basic premise in kinetics is to relate the rate of a process to the concentration of reactants. For example, we can examine the formation and dissociation of species AB as it relates to reactants A and B in the following reactions:

$$A + B \xrightarrow{k_a} AB \quad \text{and} \quad AB \xrightarrow{k_b} A + B \qquad (4.7)$$

where: $k_a = $ rate constant of formation; and $k_b = $ rate constant of dissociation.

Butcher and Anthony (2000) used the following common equation in kinetics to determine the relationship between rate and concentrations:

$$\frac{dA}{dt} = -kA^m B^n C^p \qquad (4.8)$$

where: $A = $ reactant; $dA/dt = $ rate of change of A; $k = $ *rate constant*; and m, n, and $p = $ exponents that determine the *order of reaction* (e.g., first order)

The units of the rate constants (e.g., seconds, days) will depend on the units of concentration as well as the exponents. Temperature is another important factor that is critical in affecting rate constants. It is well established that temperature increases chemical reaction rates and biological processes—particularly important in estuarine biogeochemical cycles are the effects on microbial reactions (discussed in part IV). The following well-known equation used to describe these effects is referred to as the *Arrhenius equation*:

$$k = Ae^{-Ea/RT} \qquad (4.9)$$

where: k = rate constant; $A = frequency\,factor$ (number of significant collisions producing a reaction); $E_a = activation\,energy$ (amount of energy required to start a reaction, in Joules); $R = universal\,gas\,constant = 0.082057$ (dm^3 atm mol^{-1} K^{-1}); and $T = absolute$ *temperature* (K).

The rates of chemical processes usually increase in the range 1.5–3.0, and biological processes by a factor of 2.0, for a 10°C increase, respectively (Brezonik, 1994). The most common way, not necessarily the best, of dealing with the temperature dependence of biological processes began with a study of fermentation rates (Berthelot, 1862), where it was suggested that since rate increases with temperature, k at $T + 1$ is greater than k at $T - 1$ by a fixed proportion. It has become common to use a ratio over a 10° difference, or what is called the Q_{10}, defined as follows:

$$Q_{10} = k(T + 10)/k(T) \qquad (4.10)$$

Because log k does not go up linearly with temperature, Q_{10} will be dependent on the temperatures of comparison. The larger Q_{10} the greater the effect temperature has on the reaction. A value of $Q_{10} = 1$ implies that temperature has no effect on the reaction. Consequently, many of the global biogeochemical cycles mediated by biological processes are highly dependent on temperature, perhaps most notably trace gases (e.g., CO_2 and CH_4) cycles—as they relate to global warming. Other factors involved in controlling the chemical kinetics of reactions might include catalytic reactions, kinetic isotope effects, and enzyme-catalyzed reactions (Butcher and Anthony, 2000).

Physical Properties of Water and Solubility of Salts

As discussed in chapter 3, water occurs on the Earth's surface in the forms of atmospheric water vapor, ice and liquid phases in estuaries, lakes, rivers, oceans, and groundwater, as well as being bound in mineral structures. The unique structural properties of water, which are critical for life on the Earth, still make this one of the least understood molecules. For example, classic thermodynamic theory does apply to ideal gases but not to water (Degens, 1989). In the water molecule, both hydrogen atoms are located on the same side opposite the oxygen atom, with their bonds separated by 104.5° from the oxygen atom (figure 4.2). The oxygen atom carries a net negative charge while the hydrogen atoms have a net positive charge. These net charges are weaker than that typically found associated with ions in ionic bonds and are represented by symbols $\delta+$ and $\delta-$ (figure 4.3). This opposing positive and negative charge distribution creates a strong *dipolar molecule*. Some of the unique properties of water that stem from this dipolar character are as follows: (1) an excellent *solvent* (dissolving power) for a variety of salts and other polar compounds; (2) *thermal expansion*, where the maximum density of pure liquid water is approximately 1 g cm^{-3} at 4°C, a temperature greater than its freezing point (0°C); (3) high *surface*

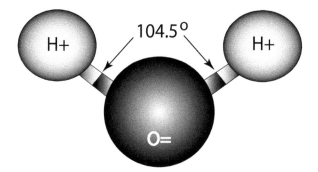

Figure 4.2 Chemical structure of water molecule.

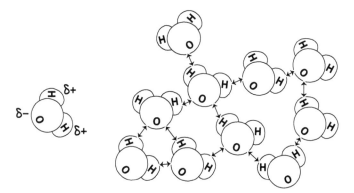

Figure 4.3 Schematic showing dipole–dipole electrostatic interactions between the negatively and positively charged ends of different water molecules which form the basis of hydrogen bonding. (From Henshaw et al., 2000, with permission.)

tension (strength of a liquid surface) and *viscosity* (resistance to distortion or flow of a fluid); (4) high *dielectric constant*; (5) high *specific heat* (amount of heat required to raise the temperature of a unit mass of substance 1°C; and (6) high *latent heats* (amount of heat required to melt a unit mass of substance at the melting point) of *fusion* and *evaporation*. More details on theses anomalous characteristics of water can be found in table 4.1.

Dipole–dipole electrostatic interactions between the negatively and positively charged ends of different water molecules form the basis of *hydrogen bonding* (figure 4.3). When ice forms, the maximum number of hydrogen bonds are formed—four per molecule; these four hydrogen bonds are arranged in a tetrahedral fashion (figure 4.4). This open lattice tetrahedral structure of ice results in it being less dense than liquid water causing ice to float. If this unique property between these two phases of water did not exist ice would sink, resulting in the displacement and loss of organisms from aquatic systems as we know them on Earth. Conversely, when pure ice melts the open structure is lost

Table 4.1 Unique physical properties of liquid water.

Property	Comparison with other substances
Specific heat ($= 4.18 \times 10^3 \mathrm{J\ kg^{-1}\ °C^{-1}}$)	Highest of all solids and liquids except liquid NH_3
Latent heat of fusion ($= 3.33 \times 10^5 \mathrm{\ J\ kg^{-1}\ °C^{-1}}$)	Highest except NH_3
Latent heat of evaporation ($= 2.23 \times 10^6 \mathrm{\ J\ kg^{-1}}$)	Highest of all substances
Thermal expansion	Temperature of maximum density decreases with increasing salinity; for pure water it is at 4°C
Surface tension ($= 7.2 \times 10^9 \mathrm{\ N\ m^{-1}}$)[a]	Highest of all liquids
Dissolving power	In general dissolves more substances and in greater quantities than any other liquid
Dielectric constant[b] ($= 87$ at 0°C, at 20°C)	Pure water has the highest of all liquids except H_2O_2 and HCN
Electrolytic dissociation	Very small
Transparency	Relatively great
Conduction of heat	Highest of all liquids
Molecular viscosity ($= 10^{-3} \mathrm{\ N\ s\ m^{-2}}$)[a]	Less than most other liquids at comparable temperature

[a]N = newton = unit of force in kg m s^{-2}.
[b]Measure of the ability to keep oppositely charged ions in solution apart from one another.

and there is a decrease in the volume of water until it reaches a maximum density at approximately 4.0°C (figure 4.5). This high-density phenomena is thought to be related to a certain fraction of water molecules that are arranged into ice-like structures, which are believed to be transient features that change in statistical frequency of occurrence as a function of temperature and pressure; they have also been referred to as *flickering clusters* (figure 4.4; Frank and Wen, 1957).

A similar type of clustering of water molecules can also have consequences on the *hydration* of ions in water. That is to say, when salts (e.g., NaCl) are added to pure water, ionic bonds of NaCl are broken due to a *primary hydration sphere* that develops around each ion (hydration) (figure 4.6); the free ions [Na^+ (aq) and Cl^- (aq)] are now considered to be dissolved or *hydrated*. The dissolution of salts in water results in an interference in the overall arrangement of water molecules, thereby affecting the physical properties of water. For example, the freezing points and maximum density temperatures for pure water and seawater (salinity of 35) are 0.0°C, −1.91°C and 3.98°C, −3.52°C, respectively (Horne, 1969). The shift occurs because salts inhibit the tendency of water molecules to form ordered groups, as described above, thereby making the density more controlled by thermal expansion. It should also be noted that, while there is an increase in the density of water with increasing dissolution of salts, the actual volume of solution is

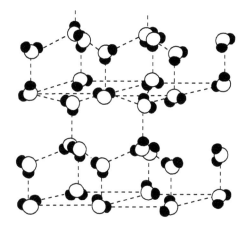

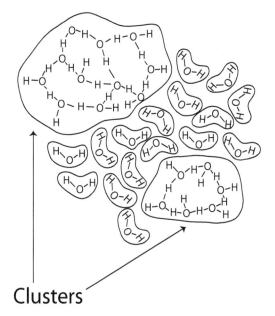

Clusters

Figure 4.4 Lattice structure of ice created when the maximum number of hydrogen bonds are formed—four per molecule; these four hydrogen bonds are arranged in a tetrahedral fashion. This open lattice tetrahedral structure of ice results in ice being less dense than liquid water causing ice to float. (From Stumm and Morgan, 1996, with permission.)

decreased due to a phenomenon known as *electrostriction*. In this process water molecules aggregate in higher densities than predicted near salt ions (e.g., Na^+); these "pockets" of water molecules have higher densities than the bulk water, resulting in a compression or reduction of the solvent. This is important because this may affect the mobility of ions across different salinity gradients.

The chemical weathering and solubility of different mineral salts in drainage basins can be quite different and are important controlling variables on salt inputs to fresh waters. In complex solutions such as estuarine waters, where the proportion of ions in solution is

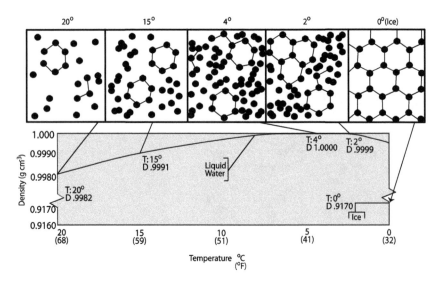

Figure 4.5 Open-lattice tetrahedral structure of ice which results in ice being less dense than liquid water causing ice to float. When pure ice melts the open structure is lost and there is a decrease in the volume of water until it reaches a maximum density at approximately 4.0°C. T = temperature and D = density. (Modified from Thurman, 1985.)

commonly not the same as that in the solids from which they are derived, it is necessary to use the *solubility product constant* defined as:

$$K_{sp} = [A^+][B]   \quad (4.11)$$

where: K_{sp} = solubility product constant; and $[A^+]$ $[B^-]$ = concentrations of cations and anions in a saturated solution in equilibrium with the solid phase.

To determine the degree to which a solution is supersaturated or undersaturated, the *ion activity product* (IAP) can be compared to the K_{sp}. For example, if the IAP is greater than the K_{sp}, the solution is considered to be supersaturated with respect to the mineral salt in question and precipitation will proceed spontaneously. On the other hand, if the IAP is less than the K_{sp}, it is considered to be undersaturated and dissolution will proceed spontaneously. If we consider the dissolution of a simple mineral salt like NaCl where the ions are efficiently hydrated by water molecules, the ion interactions between Na^+ and Cl^- are essentially insignificant, as well as with H^+ and OH^-. Moreover, as NaCl dissociates into ions there can be an infinite combination of ion activities that satisfy the equilibrium equation—only the IAP is required to equal the equilibrium constant. More details will be provided later in the chapter on ion activity coefficients. Thus, excluding any ion activity effects, the solubility calculations for NaCl are straightforward. Conversely, the solubility of other mineral salts may be enhanced with the formation of ion pairs, a *salting-in effect*, requiring the inclusion of ion speciation effects (Pankow, 1991). In contrast, there is commonly a *salting-out effect* of dissolved constituents across a salinity gradient; this can be particularly important when examining more hydrophobic organic compounds (HOCs),

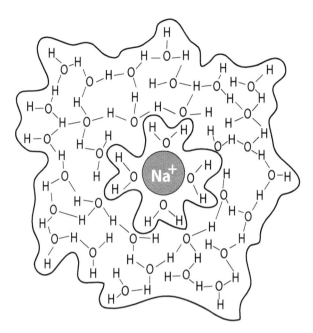

Figure 4.6 Clustering of water molecules which has significant consequences on the hydration of ions in water. When salts (e.g., NaCl) are added to pure water, ionic bonds of NaCl are broken due to a primary hydration sphere that develops around each ion (hydration)—as shown here for Na^+. (From Degens, 1989, with permission.)

such as aromatic hydrocarbons in estuaries (Means, 1995). In particular, the sorption of HOCs to suspended sediments is enhanced with increasing salinity in estuaries because the solubility of HOCs is inversely proportional to salinity (Schwarzenbach et al., 1993). Certainly, with any calculation under equilibrium conditions, the solubility of these mineral salts is a function of temperature, pressure, and ionic strength. Thus, effects of changing salinity gradients in estuarine environments will have a dramatic effect on the solubility of minerals.

Before discussing more details on the controls of ion interactions and speciation in fresh and marine waters, the following two sections will provide a brief description of what is generally known about sources of salts in river and estuarine waters and the general concept of salinity.

Sources and Mixing of Dissolved Salts in Estuaries

Prior to discussing the factors that control concentrations of the major dissolved components in rivers, estuaries, and oceans, it is important to discuss the operationally defined size spectrum for different phases (dissolved, colloidal, and particulate) of an element. The conventional definition for dissolved materials is the fraction of total material that

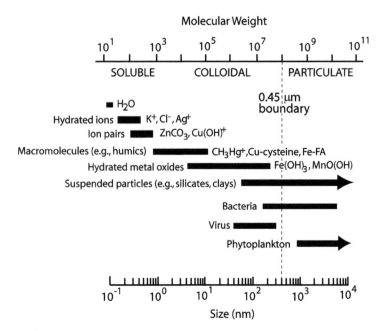

Figure 4.7 Conventional definition for dissolved materials shown as the fraction of total material that passes through a membrane filter with a nominal pore size of 0.45 μm. (From Wen et al., 1999, with permission.)

passes through a membrane filter with a nominal pore size of 0.45 μm (figure 4.7; Wen et al., 1999). Although ultrafilters are available for isolation of colloidal materials, this fraction is traditionally included within the dissolved fraction. However, *colloids* are not considered to be truly dissolved even though some of them pass through a 0.45 μm filter. More details on colloidal collection techniques and the importance of partitioning of C, N, P, S, trace metals, and organic contaminants between dissolved and particulate phases in estuaries will be discussed in chapters 10–15.

All natural waters in the world have a certain amount of salts dissolved in them. From a chemical perspective, estuarine environments are places where seawater is measurably diluted by freshwater inputs from the surrounding drainage basin. As discussed in chapter 3, the mixing of river water and seawater in estuarine basins is highly variable and typically characterized by sharp concentration gradients. In simple terms, estuaries contain a broad spectrum of mixing regimes between two dominant end-members—rivers and oceans. Rivers have highly variable amounts of salts in them, typically in the range of a few hundreds of milligrams per liter, while the oceans have more stable concentrations in the range of grams per liter. In the following section, a comparison of the chemical differences in both the particulate and dissolved constituents of rivers and the ocean will be made—as they relate to sources of salts in estuaries.

The sources of salts in rivers are primarily derived from the weathering of rocks in the drainage basin of rivers and estuaries, in addition to human activities (e.g., agriculture)

(Livingstone, 1963; Burton and Liss, 1976; Meybeck, 1979; Berner and Berner, 1996). When examining the relationship between drainage basin area and total sediment discharge in major rivers of the world it becomes clear that factors other than basin area are important (table 4.2). In addition to basin area other factors typically include the following: relief

Table 4.2 Sediment discharge and rank, water discharge and rank, and drainage basin area of the dominant rivers in the world.

River	Sediment discharge (10^6 t y^{-1})	Sediment discharge rank	Water discharge (10^9 m^3 y^{-1})	Water discharge rank	Drainage basin area (10^6 km^2)
Amazon, Brazil	1150	1	6300	1	6.15
Zaire, Zaire	43	22	1250	2	3.82
Orinoco, Venezuela	150	11	1200	3	0.99
Ganges–Brahmaputra, Bangladesh	1050	3	970	4	1.48
Yangtze (Changjiang), China	480	4	900	5	1.94
Yenisey, Russia	5		630	6	2.58
Mississippi, USA	210	7	530	7	3.27
Lena, Russia	11		510	8	2.49
Mekong, Vietnam	160	9	470	9	0.79
Parana/Uruguay, Brazil	100	14	470	10	2.83
St. Lawrence, Canada	3		450	11	1.03
Irrawaddy, Burma	260	5	430	12	0.43
Ob, Russia	16		400	13	2.99
Amur, Russia	52	20	325	14	1.86
Mackenzie, Canada	100	13	310	15	1.81
Pearl (Xi Jiang), China	80	16	300	16	0.44
Salween, Burma	100	15	300	17	0.28
Columbia, USA	8		250	18	0.67
Indus, Pakistan	50	21	240	19	0.97
Magdalena, Colombia	220	6	240	20	0.24
Zambezi, Mozambique	20		220	21	1.2
Danube, Romania	40	24	210	22	0.81
Yukon, USA	60	19	195	23	0.84
Niger, Africa	40	25	190	24	1.21
Purari/Fly, New Guinea	110	12	150	25	0.09
Yellow (Hwanghe), China	1100	2	49		0.77
Godavari, India	170	8	92		0.31
Red (Hunghe); Vietnam	160	10	120		0.12
Copper, USA	70	17	39		0.06
Choshui, Taiwan	66	18			0.003
Liao He, China	41	23	6		0.17

Data from Milliman and Meade (1983) and Meade (1996).

(elevation) of the basin, amount of water discharge, influence of lakes/dams (e.g., storage) along the river, geology of the basin, and climate (Milliman, 1980; Milliman and Syvitski, 1992). For example, the sediment yield in the Yellow (Huanghe) River is quite large despite its small basin size; this is attributed to the high erodability of heavily farmed soils in this region (Milliman et al., 1987). Consequently, the composition of suspended materials in rivers is largely a function of the soil composition of the drainage basin. However, significant differences exist between the chemical composition of suspended materials in rivers and the parent rock material (table 4.3). This is due to differences in the solubility of different elements in parent rock materials. For example, elements like Fe and Al are less soluble than Na and Cl, making them less abundant in the dissolved materials and more abundant in the suspended load of rivers, respectively (Berner and Berner, 1996). This enrichment of Fe and Al is further supported by the element weight ratio, which if greater than 1 indicates elemental enrichment (table 4.4). In some cases, rivers may receive the majority of salt inputs through precipitation and evaporation processes. This relationship was established when Gibbs (1970) plotted the total dissolved solids (TDS) concentration in rivers versus the compositional indices of $Na^+/(Na^+ + Ca^{2+})$ and $Cl^-/(Cl^- + HCO_3^-)$. A modification of the Gibbs (1970) plot shows that evaporation-controlled rivers are in arid regions, rock-controlled rivers are in intermediate rainfall areas, and atmosphere-controlled rivers are in high-rainfall areas (figure 4.8; Berner and Berner, 1996). Generally speaking, the most dominant ions found in river water are Ca^{2+} and HCO_3^-, which are principally derived from limestone weathering (Meybeck, 1979). Therefore, rivers like the Mississippi plot in a region that has relatively low $Na^+/(Na^+ + Ca^{2+})$ and/or $Cl^-/(Cl^- + HCO_3^-)$, indicative of limestone weathering in the drainage basin. Conversely, in arid regions where evaporation rates may be high, Ca^{2+} and HCO_3^- can be lost through precipitation of $CaCO_3$, while other ions like Na^+ and Cl^- are concentrated. However, the dissolved constituents in most major rivers around the world are controlled by rock weathering—as reflected by the high concentrations of Ca^{2+} and HCO_3^- (table 4.4). Long residence times and evaporative processes in the ocean basins result in a dominance of ions such as Na^+ and Cl^- in seawater.

A historical account of measurements of major dissolved components of seawater indicate that the most abundant elements, in order of decreasing abundance are Cl^-, Na^+, Mg^{2+}, SO_4^{2-}, Ca^{2+}, and K^+ (table 4.5) (Millero, 1996). In contrast to rivers, the major constituents of seawater are found in relatively constant proportions in the oceans, indicating that the residence times of these elements are long (thousand to millions of years)—highly indicative of nonreactive behavior (Millero, 1996). This relative constancy of major (and many minor) elements in seawater is referred to as the *rule of constant proportions or Marcet's principle*. More specifically, these elements are considered to be *conservative elements*, whereby changes in their concentrations reflect the addition or loss of water through physical processes. While these elements may be involved in other chemical or biological reactions, concentration changes from these processes are too small to change the constancy of the elemental ratios (Wangersky, 1965; Libes, 1992). The remaining elements in seawater are termed *nonconservative* because they do not remain in constant proportion due to biological (e.g., uptake via photosynthesis) or chemical (e.g., hydrothermal vent inputs) processes. In estuaries, as well as other oceanic environments (e.g., anoxic basins, hydrothermal vents, and evaporated basins), the major components of

Table 4.3 Concentrations of major elements in continental rocks and soils and in riverine dissolved and particulate matter.

| | Continents | | | Rivers | | | | Element Weight Ratio | | |
| | Surficial rock concentration (mg g^{-1}) | Soil concentration (mg g^{-1}) | Particulate concentration (mg g^{-1}) | Dissolved concentration (mg L^{-1}) | Particulate load (10^6 tons y^{-1}) | Dissolved load (10^6 tons y^{-1}) | | River particulate/ rock | Particulate/ (particulate+ dissolved) |
Element									
Al	69.3	71.0	94.0	0.05	1457	2		1.35	0.999
Ca	45.0	35.0	21.5	13.40	333	501		0.48	0.40
Fe	35.9	40.0	48.0	0.04	744	1.5		1.33	0.998
K	24.4	14.0	20.0	1.30	310	49		0.82	0.86
Mg	16.4	5.0	11.8	3.35	183	125		0.72	0.59
Na	14.2	5.0	7.1	5.15	110	193		0.50	0.36
Si	275.0	330.0	285.0	4.85	4418	181		1.04	0.96
P	0.61	0.8	1.15	0.025	18	1.0		1.89	0.96

Elements with no gaseous phase only. Particulate and dissolved loads based, respectively, on the total loads, 15.5×10^9 tons solids y^{-1} and 37,400 km^3 water y^{-1}. Data sources: Martin and Meybeck (1979); Martin and Whitfield (1981); Meybeck (1979, 1982). Modified from Berner and Berner (1996).

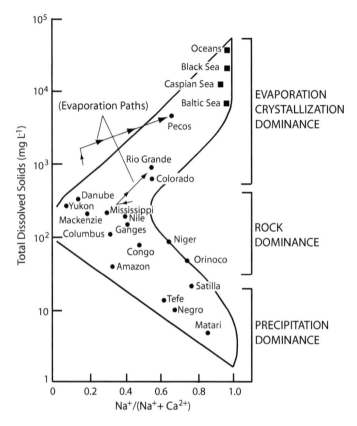

Figure 4.8 A modification of the Gibbs (1970) plot shows that evaporation-controlled rivers are in arid regions, rock-controlled rivers are in intermediate rainfall areas, and atmospheric-controlled rivers are in high rainfall areas. (Modified from Gibbs, 1970.)

seawater can be altered quite dramatically due to numerous processes (e.g., precipitation, evaporation, freezing, dissolution, and oxidation).

Concepts and Measurement of Salinity

In its early inception, the concept of salinity was simply a measure of the total amount of salts in a given mass of seawater. However, the required temperatures for the effective drying of these salts are coupled with problems of thermal decomposition of certain seawater components (e.g., bicarbonates and carbonates; see review by Millero, 1996). *Salinity* was first rigorously defined by Knudsen (1902, p. 28) as "the weight in grams of the dissolved inorganic matter in one kilogram of seawater after all bromide and iodide have been replaced by the equivalent amount of chloride and all carbonate converted to oxide." The relative constancy of the major ions in seawater is constant enough that

Table 4.5 Relative composition of major components of seawater ($pH_{sws} = 8.1$, $S = 35$, and 25°C).

Solute	gi/CI(‰)			
	A	B	C	D
Na^+	0.5556	0.5555	0.5567	0.55661
Mg^{2+}	0.06695	0.06692	0.06667	0.06626
Ca^{2+}	0.02106	0.02126	0.02128	0.02127
K^+	0.0200	0.0206	0.0206	0.02060
Sr^{2+}	0.00070	0.00040	0.00042	0.00041
Cl^-	0.99894			0.99891
SO_4^{2-}	0.1394		0.1400	0.14000
HCO_3^-	0.00735			0.00552
Br^-	0.00340		0.003473	0.00347
CO_3^{2-}				0.00083
$B(OH)_4^-$				0.000415
F^-				0.000067
$B(OH)_3^-$	0.00137			0.001002
$\sum =$	1.81484			1.81540

Column A: from Lyman and Flemming (1940); column B: from Culkin and Cox (1966); column C: from Riley and Tongadai (1967) and Morris and Riley (1966); column D: update of earlier calculations of Millero (1982) using new dissociation constants for carbonic (Roy et al., 1993) and boric acids (Dickson, 1992, 1993). Atomic weights (appendix 1). The values of TA/Cl(‰) = 123.88 mol kg^{-1} (Millero, 1995) and B/Cl(‰) = 0.000232 (Uppström, 1974) were also used to determine total carbonate and borate. Modified from Millero (1996).

determination of one major component could be used to determine the other components in a sample. Due to the accuracy and reproducibility of the measurement, chloride was chosen as the ion of choice. Libes (1992, p. 54) defines *chlorinity* as "the mass in grams of halides (expressed as chloride ions) that can be precipitated from 1000 g of seawater by Ag^+." This is referred to as the Mohr titration, where silver nitrate is used to titrate seawater with potassium chromate as an indicator. Chlorinity can also be estimated using density and conductivity measurements (Cox et al., 1967). Thus, salinity is now commonly measured using an inductive *salinometer*, where the conductivity of water is measured; in essence the electrical current is controlled by the movement and abundance of ions, the more dissolved salts, the greater the conductivity. Since much of the earlier work was presented in terms of salinity and chlorinity, the two units are related, by definition, according to the following equation:

$$S(‰) = 1.80655 \ Cl(‰) \qquad (4.12)$$

In 1978, the Joint Panel for Oceanographic Tables and Standards (JPOTS) decided that a new definition was needed for salinity that was based more on a salinity/conductivity ratio.

This new relationship was termed the *practical salinity scale* and is based on a background paper by Lewis (1978). The practical salinity of a water sample is defined in terms of the conductivity ratio, K_{15}, which is defined as follows:

$$K_{15} = \frac{\text{conductivity of water sample}}{\text{conductivity of standard KC1 solution}} \tag{4.13}$$

More specifically, practical salinity is related to the ratio K_{15} by the following equation:

$$S = 0.0080 - 0.1692\,(K_{15})^{1/2} + 25.3851\,K_{15} + 14.0941\,(K_{15})^{3/2}$$

$$- 7.0261\,(K_{15})^2 + 2.7081\,(K_{15})^{5/2} \tag{4.14}$$

Therefore, a standard seawater sample with a salinity (S) of 35 (with no ‰ units needed) has a conductivity ratio of 1 at 15°C and 1 atmosphere, using a standard KCl solution of 32.4356 g in a 1 kg mass of solution. Finally, recent applications of microwave remote sensing have been used to determine surface water of gradients of salinity in coastal regions, particularly in river plume regions (Goodberlet et al., 1997).

Reactivity of Dissolved Constituents

As discussed in chapter 3, the mixing of river water and seawater can be quite varied in different estuarine systems, resulting in a water column that can be highly or weakly stratified/mixed. These intense mixing and ionic strength gradients can significantly affect concentrations of both dissolved and particulate constituents in the water column through processes such as *sorption/desorption* and *flocculation* (discussed later in this chapter), as well as biological processes. The reactivity of a particular estuarine constituent has been traditionally interpreted by plotting its concentration across a conservative salinity gradient. As shown by Wen et al. (1999), the simplest distribution pattern, in a one-dimensional, two end-member, steady-state system, would be for a *conservative constituent* to change linearly with salinity (figure 4.9). For a *nonconservative constituent*, when there is net loss or gain in concentration across a salinity gradient, extrapolation from high salinities can yield an "effective" river concentration (C^*). This "effective" concentration can be used to infer reactivity of a constituent and can be used to determine total flux of the constituent to the ocean. For example, when $C^* = C_0$, the constituent is behaving conservatively, when $C^* > C_0$, there is removal of the constituent (nonconservative behavior) within the estuary, and when $C^* < C_0$, the constituent is being added (nonconservative behavior) within the estuary. River flux (F_{riv}) to the estuary and ultimately to the ocean are commonly estimated using this simplified mixing model. The flux of material into an estuary is $F_{riv} = RC_0$, where R is the river water flux. Similarly, flux from the estuary to the ocean (F_{ocean}) can be estimated by $F_{ocean} = RC^*$. Finally, the overall net internal flux (F_{int}) from input or removal can be estimated by $F_{int} = R(C^* - C_0)$.

Despite widespread application of the standard mixing model in estuarine systems there are numerous problems when invoking these simple steady-state mixing assumptions. Early work has shown that even the nonreactive constituents will show nonconservative

Estuarine Mixing Index

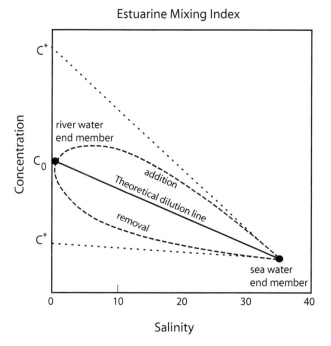

Figure 4.9 Illustration of the simplest distribution pattern, in a one-dimensional, two end-member, steady-state system for a conservative constituent to change linearly with salinity. (From Wen et al., 1999, with permission.)

mixing plots due to time-scale differences in river concentrations of the constituent relative to estuarine mixing time-scales (Officer and Lynch, 1981). Similarly, work in the Mississippi River salt-wedge estuary showed that concentrations of trace metals, based on freshwater and salinity mixing diagrams, were erroneous due to the "extended" estuary effect of metals having a longer residence time in brackish waters on the shelf (Shiller and Boyle, 1987). Fluxes of trace elements such as Cd and Zn to the Louisiana shelf region have been shown to be derived from upwelling events on the shelf-edge break, similar in magnitude to river fluxes (Shiller, 1996). As mentioned in chapter 3, the relative importance of fluxes of dissolved constituents from groundwater inputs to estuarine and shelf waters is yet another transport mechanism that adds complexity to the estuarine mixing index. Groundwater inputs of nutrients to estuaries can be quite considerable (Kelly and Moran, 2002) enhancing the nonconservative behavior of nutrients.

Ion Activity, Speciation, and Equilibrium Models

Chemical species may not always "behave" as expected based on their concentrations. In fact, ions commonly appear to be more or less concentrated than they really are because of differences in their *activity* $\{i\}$ *or* $\{a_i\}$. It is commonly accepted that the reactivity between different ions is largely a function of their activity and not their concentration (Pankow, 1991). This discrepancy between ion activity and concentration is referred to as *nonideal* behavior. To resolve this problem, a correction factor called an *activity coefficient* (γ_i), is used to scale for such differences in activity and concentration. Activity $\{i\}$ is defined as follows:

$$\{i\} = a_i = \gamma_i([i]/[i]^0) \qquad (4.15)$$

where: γ_i = activity coefficient (dimensionless) of ion i; $[i]$ = *molal* concentration (molality = moles of solute kg^{-1} solvent water); and $[i]^0$ = standard concentration has numerical value of 1.0 with same dimensions as $[i]$.

In thermodynamics it is common to make certain parameters (such as $[i]$) dimensionless, simply for convenience. For example, in the equation above we accomplish this through the division of $[i]$ by $[i]^0$, which now makes $\{i\}$ dimensionless. When an ion is behaving *less* concentrated or *more* concentrated than it really is, the activity coefficients are <1.0 and >1.0, respectively. As we will see, activity coefficients are commonly less than 1.0 and are highly dependent on the ion in question as well as the matrix. This can be important in estuaries where steep gradients in salinity exist. When an ion's activity is equal to its concentration it is considered to be an *ideal solution*. Another factor that has perhaps the most significant effect on the activity coefficient is *ionic strength*. According to Stumm and Morgan (1996, p. 101), ionic strength (I) is defined as "a measure of interionic interactions that are primarily derived from electrical attractions and repulsions," as shown in the following equation:

$$I = \tfrac{1}{2} \sum {}_i m_i z_i^2 \qquad (4.16)$$

where: m_i = molal concentration of each ion; and z_i = charge of ion.

All cations and anions are included in the summation term of this equation. The typical I for river water is 0.0021 m compared to 0.7 m in seawater (Libes, 1992).

The ability to calculate successfully an ion's activity coefficient is increased in dilute solutions. This occurs because as the ionic strength of a solution decreases so does the interionic effects on charged ions. Ions that are well hydrated, such as strong electrolytes like Na^+ and Cl^-, tend not to interact with each other, but can have some effects depending on the ionic strength and composition of the water. Some ions are so well hydrated that they do not interact (except for nonspecific interactions) and are termed *free ions*. Other ions that are not well hydrated can get close enough to other ions to allow for electrostatic interactions; this is referred to as *ion pairing*. Millero (1996) described the four types of ion pairs as follows: (1) *complexes* (ions held by covalent bonds); (2) *contact* ion pairs (no covalent bonds, linked electrostatically); (3) *solvent-shared* ion pairs (linked electrostatically, separated by a water molecule); and (4) *solvent-separated* ion pairs (linked electrostatically, separated by more than one water molecule). The major ions in natural waters are shown in table 4.6. From this we can see that many of the dominant ions are complexed in natural waters. In the case of complex pairing, the reactions are between metal cations (M^+) and species like HCO_3^-, CO_3^{2-}, OH^-, Cl^-, SO_4^{2-}, and NH_3, known as *ligands* (L). These interactions between ligands and metal ions are

Table 4.6 Major ion composition of average river and seawater.

Ion	Average river[a] (mM)	Average seawater[b] (mM)
HCO_3^-	0.86	2.38
SO_4^{2-}	0.069	28.2
Cl^-	0.16	545.0
Ca^{2+}	0.33	10.2
Mg^{2+}	0.15	53.2
Na^+	0.23	468.0
K^+	0.03	10.2

[a]Data from Berner and Berner (1987). Note that the reported concentrations exclude pollution.
[b]Data from Holland (1978).
Modified from Morel and Hering (1993) and Stumm and Morgan (1996).

Table 4.7 Concentration range of some ligands in natural waters (log M).

	Fresh water	Seawater
HCO_3^-	−4 to −2.3	−2.6
CO_3^{2-}	−6 to −4	−4.5
Cl^-	−5 to −3	−0.26
SO_4^{2-}	−5 to −3	−1.55
F^-	−6 to −4	−4.2
HS^-/S^{2-} (anoxic conditions)	−6 to −3	
Amino acids	−7 to −5	−7 to −6
Organic acids	−6 to −4	−6 to −5
Particle surface groups	−8 to −4	−9 to −6

From Stumm and Morgan (1996), with permission.

commonly referred to as *coordination chemistry*, whereby a *coordination number* is used to identify the number of nearest ligand atoms of a particular atom. Ligands are electron donors and are usually negatively charged. The dominant ligands found in natural waters are shown in table 4.7. The formation of ion pairs and complexes with the stepwise addition of ligands is represented by the following equilibrium equation:

$$M^+(\text{soln}) + L_{n-1}^-(\text{soln}) = M^+L_{n-1}^-(\text{soln}) \tag{4.17}$$

where: M^+ = metal ion; and L^- = ligand.

As in equation 4.5, we can write an equilibrium constant using the following equation:

$$K_{eq} = \{M^+L_{n-1}^-\}(\text{soln})/\{M^+\}(\text{soln})\{L_{n-1}^-\}(\text{soln}) \tag{4.18}$$

Since equilibrium constants are defined by ion activities, which are defined by their concentrations and coefficients (see equation 4.15), they do not include ion pairing or complexation effects. In a multi-ion and multiligand solution, where ion pairing is common, it is necessary to use thermodynamic equilibrium constants to convert the ion-pair concentrations to concentrations of free ions. This equilibrium constant (K_c) is defined by concentrations, making it useful to compute ion speciation. The thermodynamic equilibrium constant (K_{eq}^*) used in calculating K_c is based on the following conditions: $I = 0$ m, $25°C$ and 1 atm. Thus, K_c is defined by the following equation:

$$K_c = [(\gamma M^+)(\gamma L^-)/\gamma M^+ L^-]K_{eq}^* \tag{4.19}$$

There are a number of computer packages that calculate these equilibrium constants through a series of iterations, if K_{eq}^0 is known. Dissolved organic matter (DOM) (such as organic acids), which typically reaches high concentrations in estuaries, has been shown to be a significant complexing agent of metals (Santschi et al., 1999). More details on metal complexing ligands are provided in chapter 14.

The significant oxygen gradients commonly found in estuarine waters and surface sediments is largely controlled by stratification, which is a function of tidal and wind mixing, and organic matter loading (Officer et al., 1984; Borsuk et al., 2001). Consequently, redox and acid–base reactions are also important in determining the state of an ion in estuarine waters. The half-reactions typically used to describe reduction/oxidation (*redox*) are as follows:

$$OX_1 + ne^- = RED_1 \text{ (half-reaction 1)} \tag{4.20}$$

where: OX = oxidized form; RED = reduced form; and n = number of electrons (e^-) transferred.

$$RED_2 = OX_2 + ne^- \text{ (half-reaction 2)} \tag{4.21}$$

When combined, the half-reactions and the final redox equation are as follows:

$$OX_1 + RED_2 = RED_1 + OX_2 \tag{4.22}$$

This equation represents the exchange of electrons between OX_1 and RED_2 to produce RED_1 and OX_2. In this, OX_1 is considered to be the *oxidant* because it oxidizes RED_2 to OX_2 and RED is the *reductant* because it reduces OX_1 to RED_1. A typical redox half-reaction that occurs in many estuarine waters is the reduction of Fe^{3+} to Fe^{2+}:

$$Fe^{3+} + e^- = Fe^{2+} \tag{4.23}$$

The application of equilibrium speciation models in aquatic systems works best when the oxidation state remains relatively constant and when complexes formed from solution or absorption are reversible (Tipping et al., 1998). For example, the application of such models to changing oxidation states, dissolution and/or formation of oxide precipitates, or formation of organometallic complexes will not prove useful because many of these

Table 4.8 Equations for activity coefficient γ_i where the activity $\alpha_I = \gamma_i \mu_i$.

Name	Equation	Applicable ionic strength range
Debye-Hückel (D-H)	$\log \gamma_i = -Az_i^2(I)^{1/2}$	$I < 10^{-2.3}$
Extended Debye-Hückel	$\log \gamma_i = -Az_i^2[(I)^{1/2}/(1 + Ba(I)^{1/2})]$ a is size parameter for ion i; it should not be confused with the activity	$I < 10^{-1.0}$
Guntleberg	$\log \gamma_i = -Az_i^2[(I)^{1/2}/(1 + (I)^{1/2})]$ Equivalent to extended D-H, with an average value of $a = 3$	$I < 10^{-1.0}$
Davies	$\log \gamma_i = -Az_i^2[(I)^{1/2}/(1+(I)^{1/2}))-0.2I$	$I < 0.5$

Ionic strength ranges are applicable for the equations $\xi_i \sim \gamma_i \sim y_i$ where ξ_i and y_i are the activity coefficients on the mole fraction and molarity concentration/activity scales, respectively. The parameter A depends on $T(K)$ according to the equation $A = 1.92 \times 10^6 \, (\varepsilon T)^{-3/2}$ where ε is the temperature-dependent dielectric constant of water; $B = 50.3 \, (\varepsilon T)^{-1/2}$. For water at 298 K (25°C), $A = 0.51$ and $B = 0.33$. Applicable ionic strength range obtained from Stumm and Morgan (1981). Modified from Stumm and Morgan (1996).

processes are governed by biological and kinetic processes (Brezonik, 1994). The standard set of equations commonly used in equilibrium models to predict activity coefficients of individual ions are shown in table 4.8 (Stumm and Morgan, 1996). In general, the Debye–Hückel equation works best for dilute solution while the Davies equation is more effective for higher concentrations (Turner, 1995; Stumm and Morgan, 1996). More recently, the Windermere Humic–Aqueous Model (WHAM) has been used to calculate equilibrium chemical speciation of alkaline Earth cations, trace metals, radionuclides in surface waters of rivers and estuaries, groundwaters, sediments, and soils (Tipping et al., 1991, 1995a,b, 1998; Tipping, 1993, 1994). This model is particularly well suited to work in aquatic systems that have high concentrations of DOM and humic substances. For example, recent work applied the WHAM model to estimate the chemical speciation of six divalent trace metals (Co, Ni, Cu, Zn, Cd, and Pb) in the Humber rivers, estuary, and coastal zone (Humber system) (UK) (Tipping et al., 1998). Interaction between metals and inorganic ligands (OH^-, HCO_3^-, CO_3^{2-}, SO_4^{2-}, Cl^-) and humic substances are included in the computations of the WHAM. Comparisons were made between speciation calculations of the WHAM and simple ion-binding model for humic matter made earlier by Mantoura et al. (1978) (table 4.9). In general, it appears that the WHAM showed more complexation of trace metals to humic substances than did the model used by Mantoura et al. (1978). This was particularly evident for Cu in the seawater end-member, which may have been expected, considering the Mantoura et al. (1978) model showed stronger binding between trace metals and Ca^{2+} and Mg^{2+} than other more recent modeling approaches (Tipping, 1993; Benedetti et al., 1995). More complexation with HCO_3^- and Co and Ni was also found in the WHAM compared to the Mantoura et al. (1978) model in the low salinity waters (table 4.9).

Table 4.9 Calculated species distributions of dissolved metals.

| | Low salinity (4) | | | | Seawater | | | |
| | M^{2+} | | M-FA | | M^{2+} | | M-FA | |
	Mant	WHAM	Mant	WHAM	Mant	WHAM	Mant	WHAM
Co	0.61	0.25	0.00	0.00	0.32	0.43	0.00	0.00
Ni	0.51	0.16	0.00	0.06	0.21	0.34	0.00	0.01
Cu	0.01	0.00	0.85	0.90	0.01	0.02	0.10	0.58
Zn	0.81	0.29	0.00	0.31	0.43	0.51	0.00	0.02
Cd	0.28	0.23	0.00	0.03	0.02	0.04	0.00	0.00

Comparisons are made between the more recent data of Tipping et al. (1998) and those of Mantoura et al. (1978). Mant = Mantoura et al. (1978); WHAM = Windermere Humic-Aqueous Model; M-FA = metal–fulvic acid.
Modified from Tipping et al. (1998).

Effects of Suspended Particulates and Chemical Interactions

Particulates in estuarine systems are composed of both *seston* (discrete biological parti-cles) and inorganic lithogenic components. An operational cutoff definition of 0.45 μm is used in this section to discriminate particulates—more details on the role of colloidal par-ticles are discussed in chapters 8, 14, and 15. The highly dynamic character of estuarine systems (e.g., tides, wind, resuspension) can result in considerable variability in particle concentration over diurnal time intervals (Fain et al., 2001). Moreover, the reactivity of these particles can change over short spatial intervals due to rapid changes in salinity, pH, and redox conditions (Herman and Heip, 1999; Turner and Millward, 2002).

Water column particulates in estuaries, primarily derived from rivers, adjacent wetland systems, and resuspension events, are important in controlling the fate and transport of chemicals in estuaries (Burton and Liss, 1976; Baskaran and Santschi, 1993; Leppard et al., 1998; Turner and Millward, 2002). A recent review by Turner and Millward (2002), with particular emphasis on metals and hydrophobic organic micropollutants (HOMs), showed that processes such as ion exchange, adsorption–desorption, absorption, and precipitation–dissolution were critical in controlling the partitioning of chemical species in estuaries (figure 4.10). Biological processing of particulates by both pelagic and benthic micro- and macroheterotrophs is also critical in estuaries.

Lithogenic particles are derived from weathering of crustal materials and mostly con-sist of the primary minerals quartz and feldspar, secondary silicate minerals such as clays, and hydrogenous components (Fe and Mn oxides, sulfides, and humic aggregates) formed in situ by chemical processes (Turner and Millward, 2002). Mineral surfaces on these particles have been shown to be important in binding organic molecules, gels, and microag-gregates (Oades, 1989; Mayer, 1994a,b; Aufdenkampe et al., 2001). The fate of organic molecules in estuarine systems largely depends on whether or not they are sorbed to min-eral surfaces (Keil et al., 1994a,b; Baldock and Skjemstad, 2000). For example, selective partitioning of basic amino acids (positively charged) to clay particles (negatively charged)

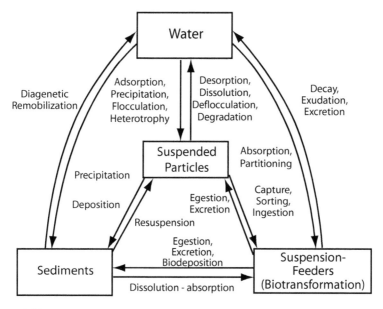

Figure 4.10 Processes critical in controlling the partitioning of chemical species in estuaries with particular emphasis on metals and hydrophobic organic micropollutants (HOMs). (Modified from Turner and Millward, 2002.)

has significant effects on the composition of dissolved amino acids in river waters of the Amazon basin (Aufdenkampe et al., 2001). Similarly, concentrations of many trace metals in estuarine waters are also influenced by sorption–desorption interactions with suspended particles (Santschi et al., 1999).

Biogenic particulates derived from fecal pellets and planktonic and terrestrial detrital materials are also important in controlling chemical interactions. Other suspended particulates composed of complex aggregates of biogenic and lithogenic materials have similar effects. Many of these biogenic particles are degraded and converted to DOM which can then be sorbed to lithogenic particles, providing an organic coating. These coatings have been shown to be important in controlling the surface chemistry of particulates in aquatic environments (Loder and Liss, 1985; Wang and Lee, 1993). Organic detrital particulates have also been shown to affect the adsorption of amines (Wang and Lee, 1990) and ammonium (Mackin and Aller, 1984) in sediments.

Summary

1. The laws of thermodynamics are the foundation for chemical systems at equilibrium.
2. Under standard conditions G, H, and S are represented as ΔG^0 and ΔH^0, and ΔS^0 and are referred to as the standard free energy of formation, enthalpy of

formation, and entropy, respectively. The chemical reaction for this, when the temperature of reactants and products are equal, is $\Delta G^0 = \Delta H^0 - T \Delta S^0$. The more negative ΔG^0 the more spontaneous the system is, making it have a greater potential to react. The more negative ΔH^0 (a measure of bond strength in reactants and products) is, the more spontaneous the reaction. Finally, ΔS^0 which is a measure of the disorder of reactants and products, will be more positive when the reaction is more spontaneous.

3. It is well established that temperature increases chemical reaction rates and biological processes. Particularly important in estuarine biogeochemical cycles are the effects on microbial reactions commonly described by the well-known Arrhenius equation: $k = Ae^{-Ea/RT}$.

4. Some of the unique properties of water that stem from it's dipolar character are as follows: (1) an excellent solvent; (2) thermal expansion; (3) high surface tension and viscosity; (4) high dielectric constant; (5) high specific heat; and (6) high latent heats of fusion and evaporation.

5. Dipole-dipole electrostatic interactions between the negatively and positively charged ends of different water molecules form the basis of hydrogen bonding. The open lattice tetrahedral structure of ice results in it being less dense than liquid water, causing ice to float.

6. In complex solutions such as estuarine waters, where the proportion of ions in solution is commonly not the same as that in the solids from which they are derived, it is necessary to use the solubility product constant defined as $K_{sp} = $ [A$^+$] [B$^-$]. To determine the degree to which a solution is supersaturated or undersaturated, the ion activity product (IAP) can be compared to the Ksp.

7. The solubility of mineral salts may be enhanced with the formation of ion pairs, a salting-in effect, requiring the inclusion of ion speciation effects. Conversely, there is commonly a salting-out effect of dissolved constituents across a salinity gradient. This can be particularly important when examining more hydrophobic organic compounds (HOC), such as aromatic hydrocarbons in estuaries.

8. The conventional definition for dissolved and particulate materials is the fraction of total material that passes through or is retained on a membrane filter with a nominal pore size of 0.45 μm, respectively.

9. The sources of salts in rivers are primarily derived from the weathering of rocks in the drainage basin of rivers and estuaries, in addition to other human activities (e.g., agriculture).

10. The relative constancy of major (and many minor) elements in seawater is referred to as the rule of constant proportions or Marcet's principle. These elements are considered to be conservative elements, whereby changes in their concentrations reflect the addition or loss of water through physical processes. The remaining elements in seawater are termed nonconservative because they remain in constant proportion due to biological or chemical processes.

11. Salinity was first rigorously defined by Knudsen (1902, p. 28) as "the weight in grams of the dissolved inorganic matter in one kilogram of seawater after all bromide and iodide have been replaced by the equivalent amount of chloride and all carbonate converted to oxide." In 1978, the JPOTS decided that a new definition was needed for salinity that was based more on a salinity/conductivity ratio and was termed the practical salinity scale.

12. The reactivity of a particular estuarine constituent has been traditionally interpreted by plotting its concentration across a conservative salinity gradient. The simplest distribution pattern, in a one-dimensional, two end-member, steady-state system, would be for a conservative constituent to change linearly with salinity. For a nonconservative constituent, there is net loss or gain in concentration across a salinity gradient.

13. Ions commonly appear to be more or less concentrated than they really are because of differences in their activity $\{i\}$ or $\{a_i\}$. It is commonly accepted that the reactivity between different ions is largely a function of their activity and not their concentration.

14. Some ions are so well hydrated that they do not interact and are termed free ions. Other ions that are not well hydrated can get close enough to other ions to allow for electrostatic interactions—this is referred to as ion pairing.

15. Equilibrium models provide information about the chemistry of the system at equilibrium but will not tell you anything about the kinetics with which the system reached equilibrium state. The basic objectives in using equilibrium models in estuarine/aquatic chemistry is to calculate equilibrium compositions in natural waters, to determine the amount of energy needed to make certain reactions occur, and to ascertain how far a system may be from equilibrium.

16. Estuarine water column particulates, primarily derived from rivers, adjacent wetland systems, and resuspension events, have been shown to be important in controlling the fate and transport of chemicals in estuaries. In particular, mineral surfaces on these particles have been shown to be important in binding organic molecules, gels, and microaggregates.

Chapter 5

Dissolved Gases in Water

Dissolved gases are critically important in many of the biogeochemical cycles of estuaries and coastal waters. However, only recently have there been large-scale collaborative efforts addressing the importance of coupling between estuaries and the atmosphere. For example, the Biogas Transfer in Estuaries (BIOGEST) project, which began in 1996, was focused on determining the distribution of biogases [CO_2, CH_4, CO, non-methane hydrocarbons, N_2O, dimethyl sulfide (DMS), carbonyl sulfide (COS), volatile halogenated organic compounds, and some biogenic volatile metals] in European estuaries and their impact on global budgets (Frankignoulle and Middelburg, 2002). The role of the estuaries and other coastal ocean environments as global sources and/or sinks of key greenhouse gases, like CO_2, have also been a subject of intense interest in recent years (Frankignoulle et al., 1996; Cai and Wang, 1998; Raymond et al., 1997, 2000; Cai, 2003; Wang and Cai, 2004). Similarly, O_2 transfer across the air–water interface is critical for the survival of most aquatic organisms. Unfortunately, many estuaries around the world are currently undergoing eutrophication, which commonly results in low O_2 concentrations (or *hypoxic* ≤ 2 mg L^{-1}), due to excessive nutrient loading in these systems (Rabalais and Turner, 2001; Rabalais and Nixon, 2002).

To understand how gases are transferred across the air–water boundary we will first examine the dominant atmospheric gases and physical parameters that control their transport and solubility in natural waters. The atmosphere is also composed of *aerosols*, which are defined as condensed phases of solid or liquid particles, suspended in state, that have stability to gravitational separation over a period of observation (Charlson, 2000). Chemical composition and speciation in atmospheric aerosols is important to understanding their behavior after deposition, and is strongly linked with the dominant sources of aerosols (e.g., windblown dust, seasalt, combustion). The importance of aerosol deposition to estuaries and coastal waters, via precipitation (rain and snow) and/or dry particle deposition, has received considerable attention in recent years. For example, *dry and wet deposition* of nutrients (Paerl et al., 2002; Pollman et al., 2002) and metal contaminants (Siefert et al., 1998; Guentzel et al., 2001) has proven to be significant in biogeochemical budgets in wetlands and estuaries. Further details on these inputs are provided in parts V and VI.

Composition of the Atmosphere

The composition of dry air at sea level is primarily composed of N_2 (78%) and O_2 (21%) (table 5.1). Temporal and spatial variability in concentrations of atmospheric gases will largely depend on how reactive they are, source and sink processes (e.g., links to anthropogenic sources), and source strength. Another approach in analyzing this variability is through comparisons between stability and *residence time* of atmospheric gases (Junge, 1974). Using the equation for residence time (t) defined in chapter 3 for freshwater exchange in estuaries, the freshwater terms can be replaced in the equation with the volume and turnover of the dissolved gas of interest to obtain t for an atmospheric gas (figure 5.1). We can see that gases with short residence times (Rn and H_2O) are highly variable, while those with long residence times (O_2 and N_2O) have less variability.

Total atmospheric pressure (P_t) is the sum of all *partial pressures* (P_i) exerted by each of the gases in the entire mixture of air—this is referred to as *Dalton's law of partial pressures*. The partial pressure of each gas is assumed to follow the ideal gas law,

Table 5.1 Composition of the atmosphere.

Constituent	Formula	Abundance by volume
Nitrogen	N_2	$78.084 \pm 0.004\%$
Oxygen	O_2	$20.948 \pm 0.002\%$
Argon	Ar	$0.934 \pm 0.001\%$
Water vapor	H_2O	Variable (% to ppm)
Carbon dioxide	CO_2	348 ppm[a]
Neon	Ne	18 ppm
Helium	He	5 ppm
Krypton	Kr	1 ppm
Xenon	Xe	0.08 ppm
Methane	CH_4	2 ppm
Hydrogen	H_2	0.5 ppm
Nitrous oxide	N_2O	0.3 ppm
Carbon monoxide	CO	0.05 to 0.02 ppm
Ozone	O_3	Variable (0.02 to 10 ppm)
Ammonia	NH_3	4 ppb
Nitrogen dioxide	NO_2	1 ppb
Sulfur dioxide	SO_2	1 ppb
Hydrogen sulfide	H_2S	0.05 ppb

[a]As of 1987.
Data from the Handbook of Environmental Chemistry (1986), U.S. Standard Atmosphere NOAA/NASA/U.S. (1976), and Walker (1977).
Modified from Stumm and Morgan (1996).

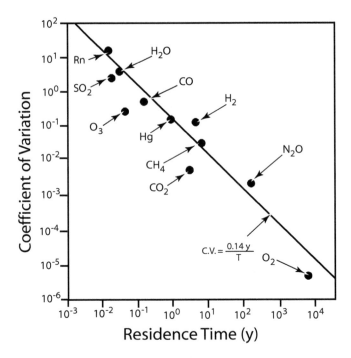

Figure 5.1 Relationship between residence time and coefficient of variation of concentrations of gas in the atmosphere. Gases with short residence times (Rn and H$_2$O) are highly variable while those with long residence times (O$_2$ and N$_2$O) have less variability. (Modified from Junge, 1974.)

as defined by the following equation:

$$P_i = n_i RT/V \tag{5.1}$$

where: n_i = number of moles of gas i; R = 8.314 liter kPa K^{-1} (Pa = pascal = one newton per square meter); T = absolute temperature; and V = volume of gas.

Therefore, *total atmospheric pressure* (P_t) would be defined by the following equation:

$$P_t = \sum P_i = P_{N_2} + P_{O_2} + P_{H_2O} \cdots \tag{5.2}$$

where: P_i = partial pressures of major gases in the atmosphere.

Since many of the atmospheric gases deviate from the "ideal" gas law, nonideal gas behavior is estimated using the *van der Waals* equation of state, which is defined as follows:

$$(P_i + n_i^2 a/V^2)(V - n_i b) = n_i RT \tag{5.3}$$

where: a and b = the van der Waals constants at standard temperature and pressure (STP).

The van der Waals constants were introduced to account for mutual attractions of the molecules and the space occupied by each molecule.

Atmosphere–Water Exchange

The direction in exchange of gases across the air–water interface will change accordingly as atmospheric and aqueous concentrations of respective gases change over time. When rates of exchange for a particular gas are equal across the air–water interface the gas is considered to be in *equilibrium*; this is when the concentration of the gas in both the aqueous (P_A) and gas phases (P_i) are equal. The equilibrium concentration of a gas in the aqueous phase is directly proportional to the pressure of that gas—this is referred to as *Henry's law of equilibrium distribution*, as defined by the following equation:

$$P_i = K_H P_A \qquad (5.4)$$

where: K_H = Henry's law constant; and P_A = concentration of gas in the aqueous phase (expressed as mol kg^{-1}).

The solubility of gases is influenced by their molecular weight. In general, the heavier the molecule the greater its solubility, excluding cases where there are molecules which interact more strongly with water. For example, CO_2 and NH_3 represent a weak acid and base, respectively, that partially dissociate in water enhancing their solubility. As temperature decreases the solubility of gases increases; this means greater gas solubility for estuaries in high latitudes. Salinity is another factor affecting solubility of gases in estuaries—the higher the abundance of dissolved ions in estuarine and coastal waters, the lower the solubility. While there are generally small variations in the partial pressure of gases over estuaries, it is more important to consider the effects of salinity and temperature since both parameters can be highly variable in estuaries. Weiss (1974) used a combination of the *Setchenow salting-out and van't Hoff equations* to describe the effects of temperature and salinity on the solubility of gases in sea water. Using a combination of the Setchenow equation, Weiss (1974) produced:

$$\ln C = B_1 + B_2 S \qquad (5.5)$$

where: C = solubility of gases (mol kg^{-1}); B_1 and B_2 = constants for a particular gas in water of a particular salinity (S) at a given temperature; and S = salinity and the van't Hoff equation:

$$\ln C = A_1 + A_2/T + A_3 \ln T + A_4 T \qquad (5.6)$$

where: A_1, A_2, A_3, and A_4 = constants for solubility of gas in water at various temperatures.

Weiss (1974) produced the following equation to express concentration of a particular gas in seawater:

$$\ln C = A_1 + A_2(100/T) + A_3 \ln(T/100) + A_4(T/100)$$
$$+ S[B_1 + B_2(T/100) + B_3(T/100^2] \qquad (5.7)$$

Table 5.2 Solubilities of nitrogen, oxygen, argon, neon, and helium in seawater at a salinity of 35.

	μmol kg^{-1}			nmol kg^{-1}	
$t(°C)$	N$_2$	O$_2$	Ar	Ne	He
0	616.4	349.5	16.98	7.88	1.77
5	549.6	308.1	15.01	7.55	1.73
10	495.6	274.8	13.42	7.26	1.70
15	451.3	247.7	12.11	7.00	1.68
20	414.4	225.2	11.03	6.77	1.66
25	382.4	206.3	10.11	6.56	1.65
30	356.8	190.3	9.33	6.36	1.64

Data from Kester (1975).

Applying this equation over a range of temperatures we can see what the solubilities of N$_2$, O$_2$, Ar, Ne, and He are in seawater at salinity 35, at 1 atmosphere, and at 100% humidity (table 5.2). It becomes apparent that many of the changes observed in the solubility of these gases might occur over different seasons, particularly with N$_2$ and O$_2$. These *normal atmospheric equilibrium concentrations* (NAECs) are based on expected equilibrium conditions, between water and atmosphere, at a particular pressure, temperature, salinity, and humidity. However, there are many physical and biological factors that cause many gases to behave in a nonconservative manner resulting in deviations from predicted NAEC. For example, phytoplankton can rapidly alter O$_2$ concentrations via photosynthetic production of O$_2$. Similarly, bacteria can alter N$_2$ concentrations through denitrification and N$_2$ fixation processes, as well as CO$_2$ production from decomposition. Gas concentrations can also be altered by nonbiological processes such as Rn production via radioactive decay.

In situations where equilibrium conditions are not applicable, rates of gas exchange across the atmosphere–water boundary can be calculated using kinetic models (Broecker and Peng, 1974; Kester, 1975). The most common kinetic model used is the *Stagnant Film Model* (figure 5.2). This model essentially has the following three regions of importance: (1) a well-mixed *turbulent atmospheric zone*; (2) a *well-mixed thin-film* liquid zone; and (3) a *laminar zone* separating the two turbulent regions. In the model, the thin film is considered permanent with a thickness defined as z. The average thickness of this film for the ocean has been estimated to be 17 μm (Murray, 2000), using isotope measurements (Broecker and Peng, 1974; Peng et al., 1979). There are uniform partial pressures of all gases in the turbulent zones, while liquid motion in the laminar zone flows parallel to the atmosphere–water interface. Movement of gases through the laminar zone is assumed to occur by molecular diffusion processes—this is the rate limiting step. It should also be noted that the Stagnant Film Model is only for soluble and slightly soluble gases. For nonsoluble gases the exchange of gases between the atmosphere-water interface is governed by their transport through the atmospheric boundary—making the dynamics very different.

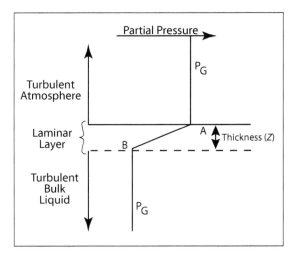

Figure 5.2 The most common kinetic model used to estimate rates of gas exchange across the atmosphere–water boundary is the Stagnant Film Model. This model essentially has the following three regions of importance: (1) a well-mixed turbulent atmospheric zone (PG); (2) a well-mixed thin-film liquid zone (PG); and (3) a laminar zone (A–B) separating the two turbulent regions. The thin-film is considered permanent with a thickness defined as z. (From Broecker and Peng, 1974, with permission.)

By combining Henry's law (described earlier) with *Fick's first law* described by the following equation:

$$dC_i/dt = D_i[dC_i/dz] \qquad (5.8)$$

where: C_i = concentration of species i; t = time; D_i = diffusion coefficient; and dC_i/dz = concentration gradient between the top and bottom of the thin film as represented by a vertical depth of z, the following equation can be used to describe the rate of gas flux through the atmosphere–water interface during disequilibrium:

$$dC_i/dt = (AD_i/zK_H)[P_i(gas) - P_A(soln)] \qquad (5.9)$$

where: A = interfacial area; and K_H = Henry's law constant.

Implicit in this model is the assumption that molecular diffusivity and Henry's Law constant are directly and inversely proportional, respectively, to the gas flux across the atmosphere–water interface. Molecular diffusion coefficients typically range from 1×10^{-5} to 4×10^{-5} cm^2 s^{-1} and typically increase with temperature and decreasing molecular weight (table 5.3). Other factors such as thickness of the thin layer and wind also have important effects on gas flux. For example, wind creates shear that results in a decrease in the thickness of the thin layer. The sea *surface microlayer* has been shown to consist of films 50–100 μm in thickness (Libes, 1992). Other work has referred to this layer as the mass boundary layer (MBL) where a similar range of film thicknesses has been

Table 5.3 Molecular diffusivity coefficients of various gases in seawater.

Gas	Molecular weight (g mol^{-1})	Diffusion coefficient ($\times 10^{-5}$cm^{2}s^{-1})	
		0°C	24°C
H$_2$	2	2	4.9
He	4	3	5.8
Ne	20	1.4	2.8
N$_2$	28	1.1	2.1
O$_2$	32	1.2	2.3
Ar	40	0.8	1.5
CO$_2$	44	1	1.9
Rn	222	0.7	1.4

Modified from Broecker and Peng (1974) and Broecker and Peng (1982).

found in estuarine and river systems (Zappa et al., 2003). Enhanced gas dissolution from *submersed air bubbles* during intense mixing events (e.g., storms) can also be a factor affecting gas concentrations in estuaries and oceanic waters (Aston, 1980).

A more simplified equation used to examine the flux (*F*) of gas transfer across the atmosphere-water interface is as follows:

$$F = k\alpha([P_i K_H] - P_A) \tag{5.10}$$

where: $F = dC_i/dt$ (mol cm^{-2} s^{-1}) $k = $ *gas transfer velocity* (cm s^{-1}), k is proportional to D_i/z (see more details on mathematical derivation in Millero (1996); $\alpha = $ *coefficient of chemical enhancement*, for gases (e.g., CO$_2$ and NH$_3$) that chemically react with water—as described earlier

In addition to being called the gas transfer velocity, k is also commonly referred to as the piston velocity, gas exchange coefficient, permeability coefficient, mass transfer coefficient, absorption coefficient, and exit coefficient (Millero, 1996). This equation has been commonly used to estimate gas fluxes across the atmosphere-water interface in estuaries and freshwater systems (Cai and Wang, 1998; Raymond et al., 2000; Crusius and Wanninkhof, 2003). Moreover, choosing a reliable k has proven critical in making comparisons between flux estimates from different estuarine systems (Raymond and Cole, 2001). In the remaining sections of this chapter, I will focus exclusively on flux measurements and cycling of some key nonconservative gases in estuaries.

Water-to-Air Fluxes of Carbon Dioxide and Other Dissolved Gases in Estuaries

As nonconservative gases, carbon dioxide and oxygen are closely coupled to the organic carbon pool through *autotrophic* (e.g., *photosynthesis*) and *heterotrophic* (e.g., *respiration*) processes. The dominant *primary producers* in estuaries and the coastal ocean are *benthic* and *pelagic microalgae (phytoplankton)*. The average atomic C-to-N-to-P ratio found in marine phytoplankton was found to be 106:16:1 (Redfield et al., 1963), which is commonly referred to as the *Redfield ratio*—more details on this in chapter 8. Thus, the basic stoichiometric reaction involving the reduction or fixation of CO_2 (photosynthesis) and oxidation of phytoplankton-derived organic matter by O_2 (aerobic respiration) is as follows:

$$\leftarrow \text{oxidation}$$
$$106\,CO_2 + 16\,HNO_3 + H_3PO_4 + 122\,H_2O \leftrightarrow (CH_2O)_{106}\,(NH_3)_{16}H_3PO_4 + 138\,O_2$$
$$\text{photosynthesis} \rightarrow$$

$$(5.11)$$

There is a critical balance between heterotrophic and autotrophic processes that control CO_2 and O_2 concentrations in estuarine waters. For example, high respiration rates in European estuarine waters resulted in high production of CO_2, where partial pressures exceeded 1000 ppm, generating high CO_2 fluxes to the atmosphere (Frankignoulle and Middelburg, 2002). In fact, in the polluted Scheldt estuary (The Netherlands) the partial pressure of CO_2 (pCO_2) can be as high as 9000 ppm, which is 25 times the current atmospheric equilibrium value ($\sim$370 ppm) (Frankignoulle et al., 1998). Similarly, these high rates of bacterial metabolism and pCO_2 during significant *detrital* decomposition can also result in significant decreases in O_2 concentrations.

Vertical O_2 exchange in estuaries is considered to be dominated by diffusive rather than advective processes (Officer, 1976). Recent air–water gas exchange coefficients for O_2 in the Waquoit Bay estuary (USA), measured using a floating chamber method, indicated that an atmospheric correction is needed for diel changes in O_2 (Kremer et al., 2003a) which are commonly used as measurements for ecosystem metabolism (Murphy and Kremer, 1983). In shallow estuaries, O_2 patterns are governed by shorter time scales, making O_2 modeling attempts very difficult (Stanley and Nixon, 1992). Hence, many of the O_2 modeling efforts have been in deeper estuaries with seasonal tidal mixing (Kemp and Boynton, 1980; Kemp et al., 1992). Perhaps the best approach in predicting O_2 concentration is determined by the last mixing event and water temperature (Borsuk et al., 2001). More details on O_2 cycling and hypoxia events will be discussed in chapters 10 through 13—as it relates to nutrient cycling.

Carbon dioxide water-to-air fluxes have been shown to be significant in estuaries (Frankignoulle and Borges, 2001; Wang and Cai, 2004). Estuaries and their associated marshes along the southeastern United States have received considerable attention on this topic (Cai and Wang, 1998; Cai et al., 2003; Wang and Cai, 2004). Calculated values of pCO_2, based on dissolved inorganic carbon (DIC) and pH data, showed the highest pCO_2 (1000 to >6000 µatm) at the lowest salinities (<10) in estuarine waters of the Satilla and Altamaha Rivers (USA) (figure 5.3; Cai and Wang, 1998). Corresponding CO_2 water-to-air fluxes across this gradient of low salinities ranged from 20 to >250 mol m^{-2} y^{-1}. The high pCO_2 and CO_2 water-to-air fluxes in the Satilla and Altamaha Rivers were attributed to inputs from organic carbon respiration in the tidally flooded salt marshes and groundwater (Cai and Wang, 1998). Recent work has shown the pCO_2 of groundwater entering the South Atlantic Bight (SAB) (southeastern coast of United States) to be high (0.05 to 0.12 atm) (Cai et al., 2003). This is consistent with other work, which suggests that the high pCO_2 in rivers is, in part, due to groundwater

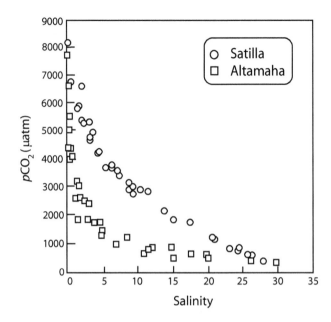

Figure 5.3 Calculated values of partial pressure of CO_2 (pCO_2), based on dissolved inorganic carbon (DIC) and pH data, versus salinity in estuarine waters of the Satilla and Altamaha Rivers (USA). (Modified from Cai and Wang, 1998.)

inputs (Kempe et al., 1991; Mook and Tan, 1991). More recently, export of DIC from marshes was found to rival that of riverine export to coastal waters of the SAB (Wang and Cai, 2004).

This work further proposed that the pathway of CO_2 being fixed by marsh grasses and then exported to coastal waters in the form of organic and inorganic carbon can be described as a "marsh CO_2 pump." Other studies have shown that marsh-influenced estuaries are important sources of DIC to adjacent coastal waters (Raymond et al., 2000; Neubauer and Anderson, 2003). For example, high pCO_2 and CO_2 water-to-air fluxes have been found in the Scheldt Estuary (Frankignoulle et al., 1996) and Rhine River (Kempe, 1982), where microbial consumption of O_2 is high due to excessive inputs of municipal wastes. Recent work has also documented the marsh pump in mangrove systems (Bouillon et al., 2003). Average pCO_2 values for other estuarine systems in Europe and the United States indicate that estuaries are net sources of CO_2 to the atmosphere (table 5.4). Highly dynamic regions in estuaries, such as the estuarine turbidity maximum (ETM), have been shown to be particularly important as net sources of CO_2 (Abril et al., 1999, 2003, 2004). However, fluxes of CO_2 across the air–water interface in estuarine plumes are poorly understood because of high spatial and temporal variability (Hoppema, 1991; Reimer et al., 1999; Brasse et al., 2002). While recent work has shown that estuarine plumes can be net sinks of CO_2, particularly in the outer-plume region (Frankignoulle and Borges, 2001), the overall estuarine system is generally a net source (Borges and Frankignoulle, 2002).

Table 5.4 Average pCO_2 ranges for various U.S. and European estuaries.

Estuary	Number of transects	Average pCO_2 range (ppmv)
Altamaha (USA)[a]	1	380–7800
Scheld (Belgium/The Netherlands)[b]	10	496–6653
Sada (Portugal)[b]	1	575–5700
Satilla (USA)[a]	2	420–5475
Thames (UK)[b]	2	485–4900
Ems (Germany/The Netherlands)[b]	1	560–3755
Gironde (France)[b]	5	499–3536
Douro (Portugal)[b]	1	1330–2200
York (USA)[c]	12	352–1896
Tamar (UK)[b]	2	390–1825
Hudson (NY, USA)[d]	6	517–1795
Rhine (The Netherlands)[b]	3	563–1763
Rappahannock (USA)[c]	9	474–1613
James (USA)[c]	10	284–1361
Elbe (Germany)[b]	1	580–1100
Columbia (USA)[e]	1	590–950
Potomac (USA)[c]	12	646–878
	Average	531–3129

The average range was obtained by averaging the low and high concentrations for each transect, and the estuaries are ranked by the high average range.
[a]From Cai and Wang (1998) and Cai et al. (1999).
[b]From Frankignoulle et al. (1998).
[c]From Raymond et al. (2000).
[d]From Raymond et al. (1997).
[e]From Park et al. (1969).
Modified from Raymond et al. (2000).

More direct measurements of pCO_2 are needed in estuaries and rivers because using pCO_2 derived from DIC and alkalinity may have some problems. Others studies have argued that the accuracy of pH measurements is not sufficient to allow for conversion to pCO_2 (Herczeg and Hesslein, 1984; Stauffer, 1990). The low cost, simplicity, and directness of floating chambers to estimate air–water exchange is perhaps the most reasonable technique available to researchers (Kremer et al., 2003b). More recently, a floating platform catamaran, that reduced any air and water side flow distortion, was found to work well when measuring pCO_2 in the Plume Island Sound estuary (USA). (Zappa et al., 2003). A regression of direct measurements of pCO_2 versus calculated pCO_2 in the Hudson River estuary reveals that calculated pCO_2 values underestimated true values by approximately 15% over the range of observed pCO_2—so most direct methods are the preferred choice over calculated values (figure 5.4). However, if accurate pH measurements can be obtained (which can be difficult), the general consensus is that pCO_2 derived from DIC and alkalinity is acceptable.

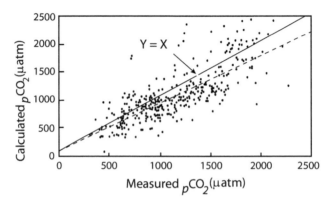

Figure 5.4 A regression of direct measurements of pCO_2 versus calculated pCO_2 in the Hudson River estuary. (From Raymond and Cole, 2001, with permission.)

It has recently been suggested that more direct measurements of the gas transfer velocity (k) are needed in estuaries and rivers, because of the high spatial and temporal variability in these systems (Raymond and Cole, 2001). Another concern, due to the lack of k measurements in estuaries, is what k value should be used in equation 5.10 when making comparisons between estuarine systems. When using this equation across a range of pH values in estuaries, CO_2 flux should be primarily governed by k and the concentration gradient of CO_2 between air and water (Raymond and Cole, 2001). To make comparisons across estuaries, k has commonly been reported as k_{600}, which is the k for CO_2 at 20°C at a Schmidt number of 600 (Carini et al., 1996). The *Schmidt number* is a dimensionless number, characteristic of each gas, which varies strongly with temperature and weakly with salinity, and is used to account for viscosity effects on the diffusion of gases. Schmidt numbers for CO_2 can be calculated for different salinities and temperatures using relationships found in Wanninkhof (1992). Recent work has concluded, based on purposeful gas tracer experiments, average wind speeds, tidal velocities, and estuary depth (table 5.5), that k_{600} should range from 3 to 7 cm h^{-1} (Raymond and Cole, 2001). The effects of wind on gas transfer velocities becomes insignificant at low wind speeds $< \sim 3.7$ m s^{-1}; other mixing properties in the water column are likely to be more important at this stage (Crusius and Wanninkhof, 2003). However, a simple generic relationship of gas transfer velocity as a function of wind speed in estuaries can lead to erroneous flux estimates, because the site-specific nature of k will vary as a function of wind speed, tidal currents, and *fetch* (Borges et al., 2004). In general, wind forcing and boundary friction generate turbulence which significantly alters transfer velocities in estuaries (Zappa et al., 2003).

Methane is an important greenhouse gas in the atmosphere that has a mean concentration of 1.7 ppm. While this is considerably lower than that of CO_2 (350 ppm) it has a greater *radiative forcing* capability (Cicerone and Oremland, 1988). Despite the smaller

Table 5.5 Average gas transfer velocities corrected to a Schmidt number of 600 (k_{600}) for rivers and estuaries.

System and study	Type of study	K_{600} (cm h^{-1}) Range	K_{600} (cm h^{-1}) Average
Hudson River (USA); Clark et al. (1994)	Purposeful gas tracer	1.5–9.0	4.8
Parker River, (USA); Carini et al. (1996)	Purposeful gas tracer	1.4–6.1	3.8[a]
South San Francisco Bay (USA); Hammond and Fuller (1979)	Natural gas tracer (^{222}Rn)	1.0–6.7	4.3
South San Francisco Bay (USA); Hartman and Hammond (1984)	Floating dome	1.1–12.8	5.7
Amazon and Tributaries; Devol et al. (1987)	Floating dome	2.4–9.9	6.0
Narragansett Bay (USA); Roques (1985)	Floating dome	4.5–11.0	7.4
Hudson River (USA); Marino and Howarth (1993)	Floating dome	3.3–26.0	11.6
Pee Dee River (USA); Elsinger and Moore (1983)	Natural gas tracer (^{222}Rn)	10.2–30	12.6
Hudson River (USA); Clark et al. (1992)	Natural gas tracer (CFC)	2.0–4.0	3.0
Raymond and Cole (2001)	Predictive equations	3.0–7.0	

Estuarine/river systems were limited to depths greater than 1 m. Reported k values were converted for O_2 and radon studies to k_{600} based on equations in Wanninkhof (1992). For the Pee Dee River and South San Francisco Bay natural gas tracer experiments, the equation in Elsinger and Moore (1983) was used to estimate the diffusivity of radon at different temperatures.
[a] Data from rain events are excluded.
Modified from Raymond and Cole (2001).

global surface area of estuaries relative to the global ocean, the contribution to total global CH_4 emission from estuaries ($\sim$7.4%) is remarkably within the range found for oceanic environments (1–10%) (Bange et al., 1994). Recent water-to-air flux estimates indicate that estuaries are contributing 1.1–3.0 Tg CH_4 y^{-1} to the global budget (Middelburg et al., 2002). Tidal creeks and marshes are the dominant source of CH_4 in estuaries (Middelburg et al., 2002). Groundwater also tends to be highly enriched in CH_4 (Bugna et al., 1996) and is likely responsible for the general increase in riverine CH_4 with increasing river size (Wassmann et al., 1992; Jones and Mulholland, 1998). Methane concentrations in rivers are typically one to two orders of magnitude higher than in open ocean waters (Scranton and McShane, 1991; Jones and Amador, 1993; Middelburg et al., 2002). The typical concentration range of CH_4 in rivers and some upper estuaries around the world are listed in table 5.6. There exists a wide range of spatial and temporal variability in CH_4 concentrations in rivers and estuaries (De Angelis and Lilley, 1987; De Angelis and Scranton, 1993; Bianchi et al., 1996).

Methane oxidation can be an important sink in estuaries as well, and is highly dependent on temperature and salinity—with lower oxidation rates at higher salinities (De Angelis and Scranton, 1993; Pulliam, 1993). In fact, turnover of the dissolved CH_4 pool in the upper Husdon River estuary (USA) (De Angelis and Scranton, 1993) and Ogeechee River (USA) (Pulliam, 1993) can occur as fast as 1.4^{-9} d to <2 h^{-1} d, respectively, depending on seasonal temperature. In conclusion, there is considerable temporal and spatial variability

Table 5.6 Concentrations of methane in rivers (and some upper estuaries).

River	Concentration (μM)	Reference
Mississippi River (USA)	0.1–0.37	Swinnerton and Lamontagne (1974)
York River (USA)	0.03–0.04	Lamontagne et al. (1973)
Potomac River (USA)	1.7	Lamontagne et al. (1973)
Pacific Coast Range rivers (USA)	0.02–1.7	De Angelis and Lilley (1987)
Cascade Range rivers (USA)	0.005–0.08	De Angelis and Lilley (1987)
Willamette Valley rivers (USA)	0.5–1.1	De Angelis and Lilley (1987)
Amazon River (Brazil)	0.053 + 0.091	Richey et al. (1988)
Saale River (Germany)	0.33–0.56	Berger and Heyer (1989)
Hudson River (USA)	0.02–0.94	De Angelis and Scranton (1993)
Upper Scheldt estuary (The Netherlands)	0.4–0.6	Scranton and McShane (1991)
Elbe River (Germany)	0.06–0.12	Wernecke et al. (1994)
Rivers of the Pantanal wetland (Brazil)	0.03–8	Hamilton et al. (1995)
Rivers in Florida (USA)	0.04–0.69	Bugna et al. (1996)
Creeks near Tomales Bay (USA)	0.14–0.95	Sansone et al. (1998)
Upper Elbe estuary (Germany)	0.111	Rehder et al. (1998)
Kaneohe stream (USA)	0.033	Sansone et al. (1999)
Rivers in the Great Bay (USA)	0.58–2.44	Sansone et al. (1999)

Modified from Middelburg et al. (2002).

in the sources and sinks of CH_4, water-to-air fluxes, as well as mechanisms of transport (e.g., *ebullition, diffusion, plant mediated*) in estuarine systems; more details on these topics will be provided in chapter 13.

Estuaries have been shown to be active sites of nitrous oxide (N_2O) production (Bange et al., 1998; Seitzinger and Kroeze, 1998; de Wilde and De Bie, 2000; Usui et al., 2001; Bauza et al., 2002). Nitrous oxide is one of the major greenhouse gases in the Earth's atmosphere (Wang et al., 1976; Khalil and Rasmussen, 1992). In fact, the Global Warming Potential of N_2O (310) is considerably higher than that of CO_2 (1) (Houghton et al., 1995). Furthermore, N_2O also plays a role in stratospheric ozone depletion (Hahn and Crutzen, 1982). Enhanced nutrient loading to estuaries stimulates microbial processes that include N_2O production (Seitzinger et al., 1983; Seitzinger and Nixon, 1985; Seitzinger, 1988; Seitzinger and Kroeze, 1998; Usui et al., 2001). Chemoautotrophic nitrification and denitrification are the primary processes that produce N_2O in natural systems (Yoshinari, 1976)—more details on the linkage of microbial cycles with estuarine nitrogen cycling are provided in chapter 10. Estuarine sediments are characterized by high spatial and temporal variability in N_2O production (Seitzinger et al., 1983, 1984; Middelburg et al., 1995; Usui et al., 2001). Efflux of N_2O from estuarine sediments is generally positively correlated with rates of denitrification in sediments (Seitzinger and Nixon, 1985; Jensen et al., 1994). However, recent work in the Tama estuary (Japan) has shown an increase in N_2O production with increases in both denitrification and nitrification (Usui et al., 2001). This work proposed that, in addition to activities of nitrification and denitrification, the change in metabolism of N_2O during denitrification via a balance between the total requirements for electron acceptor and sources of NO_3^- and NO_2^- is important in regulating N_2O production. Typical values for the evolution of N_2O from sediments in this estuary were within the range found in other coastal systems. Ratios of N_2O to N_2 in sediments generally range from 0.1 to 0.5%, with greater sediment yields (up to 6%) in estuaries with high anthropogenic loading (Seitzinger, 1988). Global estimates of N_2O emissions from rivers, estuaries, and continental shelves are 55, 11, and 33%, respectively (Seitzinger and Kroeze, 1998). Dissolved inorganic nitrogen (DIN) loading to sediments was positively correlated with N_2O fluxes, indicating that eutrophication may be linked with increases in global emissions of N_2O (figure 5.5; Seitzinger and Nixon, 1985). In fact, a model of DIN transport by world rivers and estuaries indicates that 90% of the denitrification in rivers and estuaries occurs in the northern hemisphere (Seitzinger, 2000).

Although there have only been a few studies to date, it has been suggested that coastal plumes (Turner et al., 1996; Simo et al., 1997) and estuaries (Iverson et al., 1989; Cerqueira and Pio, 1999) may be important atmospheric sources of DMS. DMS, a compound produced by certain phytoplankton, has been shown to have possible implications for climate control once released into the atmosphere (Charlson et al., 1987). DMS is formed by cleavage of dimethylsulfoniopropionate (DMSP) (Kiene, 1990). In fact, DMSP, shown to be correlated with bacterial activity, may provide as much as 100% of the sulfur and 3.4% of the carbon required for bacterial growth in oceanic waters (Kiene and Linn, 2000). Other sulfur compounds such as COS and carbon disulfide (CS_2) have also been shown to be possible sources of S in estuaries. For example, significant concentrations of COS and CS_2 were found in four European estuaries, 220 ± 150 and 25 ± 6 pM (Sciare et al., 2002). COS is the most abundant sulfur compound in the

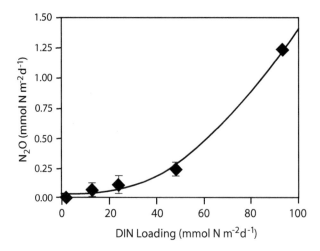

Figure 5.5 Nitrous oxide emission rates (mmol m^{-2} d^{-1}) versus dissolved inorganic nitrogen (DIN) loading (N m^{-2} d^{-1}) in mesocosm experiments at the Marine Ecosystem Research Laboratory. (Modified from Seitzinger and Nixon, 1985.)

atmosphere, and both COS and CS$_2$ may play an important role in the global radiation budget (Zepp et al., 1995). Recent estimates on CS$_2$ water-to air fluxes from estuaries indicate that they may be comparable to open ocean fluxes, with both representing approximately 30% of the global budget (Watts, 2000). However, it should be cautioned that these estimates are based on very few measurements and that further work is clearly needed in this area (Sciare et al., 2002). More details on sulfur cycling are provided in chapter 12.

The first attempt to include emissions of biogases in a full transient, one-dimensional reactive transport model for estuaries was recently developed for the Scheldt estuary (The Netherlands) (Regnier et al., 1997, 1998). The CONTRASTE (Coupled, Networked, Transport–Reaction Algorithm for Strong Tidal Estuaries) model is designed to provide a full description of estuarine residual circulation—daily freshwater discharge and tidal oscillations. It is also designed to describe physical/chemical and biological processes that are mediated by kinetically controlled and equilibrium reactions (Vanderborght et al., 2002). The four biogases (N$_2$O, O$_2$, CO$_2$, NH$_3$) and the biogeochemical processes affecting their flux rates across the air–water interface are shown in figure 5.6. Due to the short- and long-term changes in physical parameters controlling these flux rates in estuaries, along with the aforementioned difficulties associated with direct flux measurements, such models should have widespread appeal. In short, the CONTRASTE model revealed that water-to-air fluxes of N$_2$O and CO$_2$ from the Scheldt estuary were significant (Vanderborght et al., 2002). This work provides further evidence that highly polluted estuaries with strong tidal currents can produce significant contributions of biogases to the atmosphere.

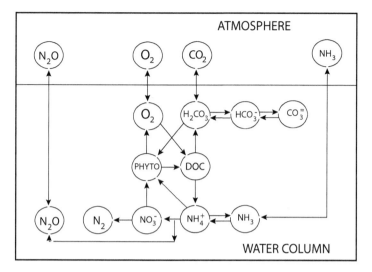

Figure 5.6 Biogeochemical processes affecting their flux rates across the air–water interface of the four biogases (N_2O, O_2, CO_2, NH_3) analyzed in a full transient, one-dimensional reactive transport CONTRASTE model (Coupled, networked, transport-reaction algorithm for strong tidal estuaries) for the Scheldt estuary (The Netherlands) (Modified from Vanderborght et al., 2002.)

Summary

1. Despite the critical role dissolved gases have in many of the biogeochemical cycles of estuaries and coastal waters, only recently have there been large-scale collaborative efforts (e.g., BIOGEST) addressing the importance of coupling between estuaries and the atmosphere.
2. The importance of aerosol deposition, via precipitation and/or dry particle deposition, has received considerable attention in recent years, showing that dry and wet deposition of nutrients and metal contaminants are significant in biogeochemical budgets in wetlands and estuaries.
3. The composition of dry air at sea level is primarily composed of N_2 (78%) and O_2, (21%). Temporal and spatial variability in concentrations of atmospheric gases will largely depend on gas stability, source and sink processes, source strength, and residence time.
4. Total atmospheric pressure (P_t) is the sum of all partial pressures (P_i) exerted by each of the gases in the entire mixture of air and is referred to as Dalton's law of partial pressures.
5. The equilibrium concentration of a gas in the aqueous phase is directly proportional to the pressure of that gas—this is referred to as Henry's law of equilibrium distribution, defined as $P_i = K_H\,P_A$.
6. In situations where equilibrium conditions are not applicable, rates of gas exchange across the atmosphere–water boundary can be made using kinetic models such as the Stagnant Film Model.

7. The best approach in predicting O_2 concentration is determined by the last mixing event and water temperature.

8. Carbon dioxide water-to-air fluxes have been shown to be significant in estuaries. In estuaries with extensive marsh systems, the pathway of CO_2 being fixed by marsh grasses and then exported to coastal waters in the form of organic and inorganic carbon can be described as a "marsh CO_2 pump."

9. More direct measurements of the gas transfer velocity (k) (e.g., floating chambers, floating-platform catamaran) are needed in estuaries and rivers because using pCO_2 derived from DIC and alkalinity may have some problems. Wind forcing and boundary friction generate turbulence which significantly alters transfer velocities in estuaries.

10. Methane is an important greenhouse gas in the atmosphere that has a mean concentration of 1.7 ppm and has a radiative forcing capability greater than that of CO_2. The contribution to total global CH_4 emission from estuaries is $\sim$7.4%.

11. Methane oxidation can be an important sink in estuaries as well, and is highly dependent on temperature and salinity, with lower oxidation rates at higher salinities. Considerable temporal and spatial variability exists in the sources and sinks of methane, water-to-air fluxes, as well as mechanisms of transport (e.g., ebullition, diffusion, plant mediated) in estuarine systems.

12. Estuaries have been shown to be active sites of N_2O production. DIN loading to sediments was positively correlated with N_2O fluxes, indicating that eutrophication may be linked with increases in global emissions of N_2O. In fact, a model of DIN transport by world rivers and estuaries indicates that 90% of the denitrification in rivers and estuaries occurs in the northern hemisphere.

13. Although there have only been a few studies to date, it has been suggested that coastal plumes and estuaries may be important atmospheric sources of DMS. DMS, a compound produced by certain phytoplankton, has been shown to have possible implications for climate control once released into the atmosphere.

14. The first attempt to include emissions of biogases in a full transient, one-dimensional reactive transport model for estuaries was recently developed for the Scheldt estuary (The Netherlands). The CONTRASTE model is designed to provide a full description of estuarine residual circulation—daily freshwater discharge and tidal oscillations.

Part III

Properties of Estuarine Sediments

Chapter 6

Sources and Distribution of Sediments

Weathering Processes

The uplift of rocks above sea level on the Earth's surface over geological time, produces rock material that can be altered into soils and sediments by *weathering* processes. Over geological time, a fraction of sediments can be sequestered for storage in the ocean basins—with most of it stored in the coastal margin. However, much of this material is modified via processing in large river estuarine systems which can ultimately affect the long-term fate of these terrigenous materials. Sediments produced from weathering of igneous, metamorphic, and sedimentary rocks are principally transported to the oceans through river systems of the world. The major routes of sediment transport from land to the open ocean can simply be illustrated through the following sequence: streams, rivers, estuaries, shallow coastal waters, canyons, and the abyssal ocean (figure 6.1). It should be noted that significant and long-term storage occurs in river valleys and floodplains (Meade, 1996). Submarine canyons are also thought to be temporary storage sites for land-derived sediments; however, episodic events such as turbidity currents and mud slides can move these sediments from canyons to the abyssal ocean (more details on coastal margin transport to the deep ocean are provided in chapter 16). The annual sediment flux from rivers to the global ocean is estimated to range from 18 to 24×10^9 metric tonnes (Milliman and Syvitski, 1992). Conversely, estuaries will eventually fill-in with fluvial inputs of sediments over time, and ultimately reach an equilibrium whereby export and import of sediment supply are balanced (Meade, 1969). For example, recent studies have shown that sediment accumulation in the Hudson River estuary, both short (Olsen et al., 1978) and long term (Peteet and Wong, 2000), is in equilibrium with sea level rise. More specifically, it is believed that river flow controls the direction of sediment flux in the Hudson, while variations in spring-neap tidal amplitude control the magnitude (Geyer et al., 2001).

Weathering is typically separated into two categories: physical and chemical. *Physical weathering* involves the fragmentation of parent rock materials and minerals through processes such as *freezing*, *thawing*, *heating*, *cooling*, and *bioturbation*

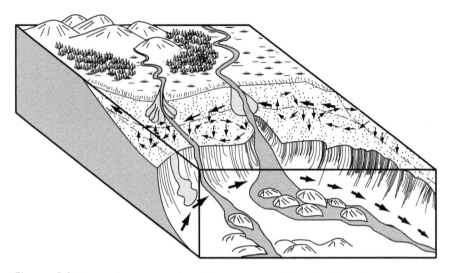

Figure 6.1 The major routes of sediment transport from land to the open ocean can simply be illustrated through the following sequence: streams, rivers, estuaries, shallow coastal waters, canyons, and finally the abyssal ocean. Arrows indicate the effects of coastal currents and resuspension events on the transport and distribution of particles delivered to the coastal zone. There can be substantial storage in each of these environments. (From Degens, 1989, with permission.)

(e.g., endolithic algae, fungi, plant roots, and earthworms). Conversely, *chemical weathering* involves the chemical alteration of minerals and is enhanced as the surface area of exposed rocks increases. The six dominant processes involved in the chemical weathering of rocks are: *dissolution* (e.g., minerals dissolve into component ions in water); *hydration* (e.g., incorporation of water–mineral structure); *acidolysis* (i.e., weathering via H^+); *chelation* (e.g., organic acids act as chelators and break down minerals); and *oxidation* or *reduction* (e.g., minerals weather via the transfer of electrons) (Montgomery et al., 2000). In essence, the interactive effects of both chemical and physical weathering processes are not easily separated in nature since they typically occur at the same time.

There are approximately 3000 different minerals found in rock formations; however, much of the Earth's crust consists of only 50 dominant mineral forms (Degens, 1989). The primary and secondary minerals commonly found in soils are listed in table 6.1. As mentioned above, chemical weathering processes in soils are important in the transformation of *primary* to *secondary* minerals. For example, when examining the weathering of feldspars (this most abundant group of minerals in the Earth's crust), K-feldspar is transformed into kaolinite through the following reaction:

$$KAlSi_3O_8 + H_2CO_3 + H_2O \rightarrow Al_2SiO_5(OH)_4 + 2K^+ + 4H_4SiO_4 + HCO_3^-$$
(K-feldspar)   (kaolinite)

$$(6.1)$$

Table 6.1 Primary and secondary minerals commonly found in soils.

Primary minerals	Approximate composition	Weatherability
Quartz	SiO_2	−
K-feldspar	$KAlSi_3O_8$	+
Ca,Na-plagioclase	$CaAl_2Si_2O_8$	+ to (+)
Muscovite	$KAlSi_3O_{10}(OH)_2$	+(+)
Amphibole	$Ca_2Al_2Mg_2Ge_3Si_6O_{22}(OH)_2$	+(+)
Biotite	$KAl(Mg,Fe)_3Si_3O_{10}(OH)_2$	++
Pyroxene	$Ca_2(Al,Fe)_4(Mg,Fe)_4Si_6O_{24}$	++
Apatite	$[3Ca_2(PO_4)_2]CaO$	++
Volcanic glass	Variable	++
Calcite	$CaCO_3$	+++
Dolomite	$(Ca,Mg)CO_3$	+++
Gypsum	$CaSO_4 \cdot 2H_2O$	+++
Secondary minerals	Approximate composition	Type
Kaolinite	$Al_2Si_2O_5(OH)_4$	1:1 Layer-silicate
Vermiculite	$(Al_{1.7}Mg_{.3})Si_{3.9}Al_{0.4}O_{10}(OH)_2$	2:1 Layer-silicate
Montmorillonite	$(Al_{1.7}Mg_{.3})Si_{3.9}Al_{0.1}O_{10}(OH)_2$	2:1 Layer-silicate
Chlorite	$(Mg_{2.6}Fe_{.4})Si_{2.5}(Al,Fe)_{1.5}O_{10}(OH)_2$	2:1 Layer-silicate
Allophane	$(SiO_2)_{1-2}Al_2O_5 \cdot 2.5-3(H_2O)$	Pseudocrystalline, spherical
Imogolite	$SiO_2Al_2O_3 \cdot 2.5H_2O$	Pseudocrystalline, strands
Hallyosite	$Al_2Si_2O_5(OH)_4 \cdot 2H_2O$	Pseudocrystalline, tubular
Gibbsite	$Al(OH)_3$	Hydroxide
Goethite	$FeOOH$	Oxyhydroxide
Hematite	Fe_2O_3	Oxide
Ferrihydrite	$5Fe_2O_5 \cdot 9H_2O$	Oxide

Modified from Degens (1989).

During the formation of kaolinite, a K^+ ion is also released from the soil, while Al is retained in the solid phase. Similarly, the weathering of goethite by H^+ can release Fe^{3+} in the following reaction:

$$FeOOH + 3H^+ \rightarrow Fe^{3+} + H_2O$$
(goethite)

(6.2)

Weathering of goethite under oxic conditions results in the release of Fe^{3+}, while, under anoxic soil conditions, Fe^{2+} is released (Birkeland, 1999). Although Fe is released in this reaction, both Al and Fe tend to accumulate in weathered aerobic soils as oxides [e.g., goethite, gibbsite, and hematite (table 6.1; Montgomery et al., 2000)]. For example, primary minerals containing Fe (e.g., pyroxenes, amphiboles, *mafic* rocks) are weathered, releasing Fe in solution, which is then precipitated and accumulated in oxides. This is clearly illustrated when examining the composition and percentage of clay and free oxides in Hawaiian soils, underlain by mafic volcanic rocks (figure 6.2; Sherman, 1952). Basically, basalt and clay minerals disappear with increasing rainfall as they are

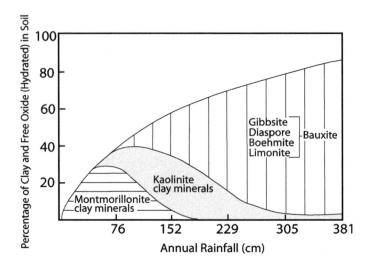

Figure 6.2 Primary minerals containing Fe (e.g., pyroxenes and amphiboles in basaltic volcanic rocks) are weathered, releasing Fe into solution, which is then precipitated and accumulated in oxides as illustrated when examining the composition and percentage of clay and free oxides in Hawaiian soils. (Modified from Sherman, 1952.)

replaced by Al, Fe, and Ti oxides and hydroxides. When considering particle–particle and particle–dissolved constituent interactions in aquatic systems, clay particles are particularly important because of their small size and high surface area. These interactions have broad-reaching effects on biogeochemical cycling in estuaries because of the strong linkages between the transformation and composition of clay minerals in drainage basin soils and estuarine sediments.

Erosion, Transport, and Sedimentation

Sediments in estuaries are derived from a diverse group of sources such as atmospheric inputs, fluvial, continental shelf, biological activity, and erosion of estuarine margins. Sources can vary considerably in the upper and lower regions of the estuary with biological inputs usually being more important in the higher salinity regions than the upper lower salinity region where terrigenous inputs dominate. However, certain blackwater river systems (e.g., St. John's River, USA) can have higher productivity in the upper reaches of the estuary (personal communication, John M. Jaeger, University of Florida, USA). Gradients in estuarine processes are primarily driven by changes in river inflow, tidal currents, waves, and meteorological forces.

The four dominant processes controlling sediment dynamics in estuaries are as follows: (1) *erosion of the bed*; (2) *transportation*; (3) *deposition*; and (4) *consolidation of deposited sediments* (Nichols and Biggs, 1985). *Erosion* will strongly depend on the cohesiveness of sediment; *cohesion* is defined as the result of interparticle surface attraction between clay minerals as well as the binding forces from organic mucous layers. The overall erodibility of estuarine sediments largely depends on the grain size and shear stress of the bed. Sediments can move whenever the shear imparted on the bed by fluid flow reaches *critical shear stress*, which is essentially the same as the *yield strength*—or the force required to break any cohesive bonds (Migniot, 1968). As sediments become more consolidated, surface "fluffy" (high water content) deposits become more condensed (low water content) via dewatering with depth. As these sediments become more compact, aggregates within these deposits form and become less erodible (Krone, 1962). The erodibility of sediment in estuaries can vary considerably temporally and spatially due to changes in resuspension frequency, which leads to higher water content in phytodetrital inputs to surface sediments, pelletization, and bioturbation of sediments from the benthos (Rhoads, 1974; McCall and Tevesz, 1982; Rice et al., 1986). More specifically, pelletization and bioturbation from benthic macrofauna (e.g., deposit feeders) can increase the potential for bulk flow through sediments (Aller, 2001) and average porosity (McCall and Fisher, 1980), affecting the overall erodibility of sediments. The potential effects of macrofauna on the texture of the sediment surface will largely be controlled by salinity gradients in estuaries; many of the late successional stage benthic communities (more bioturbation) occur at higher salinities (more details on animal–sediment relations are provided in chapter 8).

Microbial mats and *biofilms*, defined as surface layers of microbes entrained in a matrix of *extracellular polymeric substances* (EPS) (Characklis and Marshall, 1989), are also important in changing the surface texture and erodibility of sediments in estuaries (de Beer and Kühl, 2001). The EPS are primarily composed of cellular-derived polysaccharides, polyuronic acids, proteins, nucleic acids, and lipids (Decho and Lopez, 1993; Schmidt and Ahring, 1994). The EPS can serve as a cementing agent for surface sediment particles, thereby affecting the erodibility of sediments as well as the flux of dissolved constituents across the sediment–water interface (de Beer and Kühl, 2001).

The sedimentation of particles in the water column represents an important mechanism for sediment transport and is most fundamentally controlled by gravitational setting. Gravitational forces, which act in a downward direction, are also opposed by the viscous resistance of water in an upward direction. The sinking speed for spherical, slow sinking, single particles is described by *Stoke's law* (Allen, 1985):

$$w_{si} = [(\rho_{si} - \rho_w)g d_i^2)]/18\mu \tag{6.3}$$

where: w = sinking speed (cm s^{-1}); ρ_{si} = sediment density of particle in class i, (2.650 g m^{-3} for siliciclastic sediments); ρ_w = water density (1.007 g m^{-3} at 25°C); d = disaggregated particle diameter (μm); μ = dynamic viscosity (measure of the resistance to flow of a fluid under an applied force) (8.91 $\times$ 10^{-4} N s^{-1} m^{-2} at 25°C); and g = gravitational constant.

It should be noted that single particles are not often found in estuaries, and that most particles generally sink as flocs (see following section), nevertheless, Stokes law as a general background on the basic concept of particle settlement has been included.

Settling of particles less than 0.5 μm is slowed by *Brownian motion* (random motion of small particles from thermal effects) in the water. Conversely, large sand-sized particles are not affected by viscous forces and typically generate a frontal pressure or wake as they sink. Thus, Stokes law can only apply to particles with *Reynolds numbers* (Re) that are less than unity. The *particle* Reynolds number according to Allen (1985) is defined as follows:

$$Re = \rho w_{si} d_i / \mu \qquad (6.4)$$

For particles typically greater than 0.1 to 0.2 mm in diameter settling velocity varies according to the square root of the diameter — as described by the *impact law* (Krumbein and Sloss, 1963; figure 6.3). These larger particles generate a *turbulent wake* that after rapid acceleration slows the particle settling velocity; conversely, small particles have a fluid flow that is *laminar* in nature (Degens, 1989).

Estuaries typically have elevated concentrations of suspended fine sediments that are highly *cohesive* and readily *flocculate*. Coagulation of fine particles leads to the formation of the following composite particles in estuaries: (1) *agglomerates* (organic and inorganic matter bound by weak surface tension forces); (2) *aggregates* (inorganic particles bound by strong inter- and intramolecular forces); and (3) *floccules* (nonliving biogenic material bound by electrochemical forces) (Schubel, 1971, 1972). Neglecting the effects of aggregation in fine-sediment environments can seriously underestimate settling velocities (Kineke and Sternberg, 1989; Blake et al., 2001). Settling velocities estimated for flocs

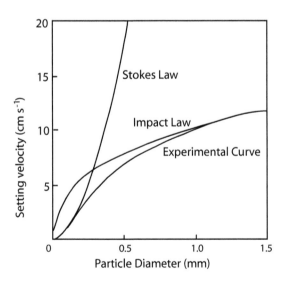

Figure 6.3 The sinking rates for particles (<100 μm) that follow Stokes law and for particles typically greater than 0.1 to 0.2 mm in diameter settling velocity varies according to the square root of the diameter—as described by the impact law. (Modified from Krumbein and Sloss, 1963.)

have been shown to range from 0.8 to 17.9 mm s^{-1} (Milligan et al., 2001) which agrees well with settling velocities estimated from image analysis of flocs (8.8 ± 1.85 mm s^{-1}) (Sternberg et al., 1999). High-resolution video cameras attached to sediment traps have allowed for new and improved time-lapsed images of aggregates settling through the water column (Sternberg et al., 1999). The cohesion of particles is primarily controlled by the repulsive electrochemical forces and the attractive *van der Waals forces*. As described earlier, clay particles have a layered lattice structure with a high negative charge. In freshwater, these negative forces dominate, allowing repulsion to dominate, preventing flocculation of particles. However, as salinity begins to increase within the mixing zone of estuaries repulsive forces are destabilized by an increasing abundance of cations (Ca^{2+}, Mg^{2+}, and Na^+); coagulation occurs at salinities as low as 0.1 (Stumm and Morgan, 1996). As repulsive forces decrease, van der Waals forces will dominate and flocculation will proceed. The strong salinity and tidal gradients in estuaries make them ideal environments for coagulation processes compared to other natural systems and actually rival those of wastewater systems (figure 6.4). It should also be noted that the collision efficiency (α), in highly contrasting environments such as the deep sea, is elevated because of the high ionic strength of the water, allowing for higher α. This occurs in the open ocean despite having a lower particle concentration (ϕ) and velocity gradient (G), relative to estuarine or wastewater systems (Stumm and Morgan, 1996).

Coagulation processes in estuaries are affected by other factors such as clay composition, particle size, and concentration of dissolved organic matter, to mention a few. For example, early work has shown that metal hydroxides can flocculate from dissolved/colloidal organic matter during the mixing of river-derived iron and seawater in the mixing zone of estuaries (Sholkovitz, 1976, 1978; Boyle et al., 1977; Mayer, 1982) (more details are provided on metal colloidal interactions in chapter 14). Surface sediments in

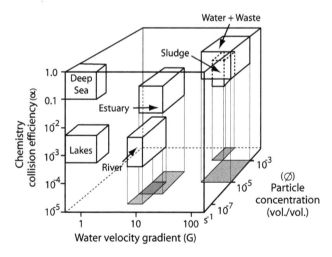

Figure 6.4 Comparison of key variables that control coagulation in natural and artificial systems; G = mean water velocity gradients (s^{-1}), Φ = particle concentration, and α = collision efficiency. (From Stumm and Morgan, 1996, with permission.)

the Pamlico River estuary (USA) indicate that illite flocculates in the high-salinity regions, while kaolinite dominates near the head of the estuary (Edzwald et al., 1974; Edzwald and O'Melia, 1975). Formation of fluid muds (discussed in the following section) requires flocculation of similarly sized particles with similar behavior (Postma, 1980). For example, montmorillinite (with a grain size of approximately 6 μm) was shown to double in concentration in fluid muds of the Loire estuary (France), showing a distinct preference over minerals such as illite, kaolinite, and chlorite (Gallenne, 1974). However, as discussed later, fluid muds can also form solely from physical resuspension (McKee et al., 2004). The abundance of bacterioplankton and phytoplankton can also affect rates of flocculation by providing mucilage to which particles can adhere (Ernissee and Abbott, 1975). Aggregation by microorganisms, commonly know as *bioflocculation* (in wastewater treatment), has been suggested as a potentially important process in natural systems (Busch and Stumm, 1968; Stumm and Morgan, 1996). More specifically, the actual cell surfaces of microorganisms and/or the polymers secreted by these organisms (e.g., mucopolysaccharides) may affect oxide stability as well as the initial adhesion phases of colloids (van Loosdrecht et al., 1990).

Suspended sediment in estuaries occurs in a broad spectrum of sizes. Since individual clay particles are too small for measurements in the field, *microflocs* (<150 μm in diameter) and *macroflocs* (comprising of microflocs) represent the population of particulates commonly studied in estuaries (Dyer and Manning, 1999). Microflocs tend to be more spherical, higher in density, and more robust, while macroflocs are easily disrupted and more amorphous (Eisma, 1986). Flocculation results from collisions induced, in part from Brownian motion, which commonly results in a fraction of microflocs being converted into macroflocs. The frequency of collisions is positively correlated with particle concentration. However, with high amounts of shear and particles, turbulence can disrupt particle formation, resulting in smaller floc sizes (van Leussen, 1988). Thus, settling rates of fine-grained particulates in estuaries are largely determined by the size distribution of aggregate flocs, the concentration of suspended particles, and velocity shear (Hill et al., 2000). While laboratory studies have been useful in studying flocculation processes, other work has suggested that the significantly smaller floc sizes typically generated under laboratory conditions are the result of inadequate time intervals for particle collision (Winterwerp, 1998). If true, this has major implications for our understanding of flocculation processes in rivers and estuaries over different discharge periods with changing water residence times.

High levels of flocculation typically result in high settling velocities and fluxes to the bottom. Differences in the settling behavior of flocs in a suspended load can be characterized by the *Rouse parameter*, as described by Middleton and Southward (1984):

$$Ro = \frac{w_s}{\beta \kappa u^*} \tag{6.5}$$

where: w_s = floc settling velocity; β = *proportionality coefficient* between eddy viscosity and diffusivity; κ = *von Karman's constant* (0.41); and u^* = *frictional velocity* (calculated using the *quadratic stress law*—see Blake et al., 2001).

Rouse values significantly less than 1 indicate that particles will be distributed throughout the water column—if less than 1, particles are near the bottom; if greater than 2.5,

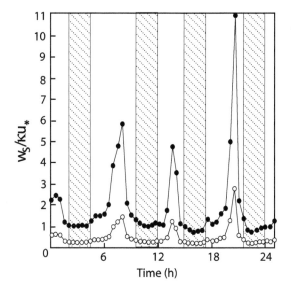

Figure 6.5 Distinct differences in transport behavior between pools of surface and bottom flocs over several tidal cycles, as determined by R_0 values, in the ACE Basin (USA). Hatched areas are times of maximum current speed. w_S = sediment settling velocity, β = proportionality coefficient between eddy viscosity and diffusivity, κ = von Karman's constant, and u_* = frictional velocity. (From Milligan et al., 2001, with permission.)

particles are part of the sea bed (Middleton and Southward, 1984). Distinct differences in pools of surface and bottom flocs, as determined by Ro values, can be seen in the Ashepoo, Combahee, and Edisto Rivers (ACE) Basin (USA) (figure 6.5; Milligan et al., 2001). These results indicate that, during periods of high discharge, fine particulates can become trapped in high-salinity regions, where floc size will increase along with the settling velocity of flocs.

In general, sediments are transported as *suspended load* or *bed load* and are directly discharged into open oceans, coastal shelves, or stored in estuaries (McKee and Baskaran, 1999). The majority of sediment that is transported to the coast is usually suspended load (90–99%), with relatively minor amounts as bed load (Syvitski et al., 2000). Bed load will generally move slower than the mean flow of water and occurs near the bed through *saltational* ("jumping" of particles along bottom) interparticle high collision movement. Suspended load will move more closely with water flow and is largely independent of bed load. The faster the settling of particles in the water column the more energy will be needed to maintain particles as a suspended load.

Estuarine Turbidity Maximum, Benthic Boundary Layer, and Fluid Muds

The *estuarine turbidity maximum* (ETM) is defined as a region where the suspended particulate matter (SPM) concentrations are considerably higher (10–100 times) than in adjacent river or coastal end-members in estuaries (Schubel, 1968; Dyer, 1986). Some of the most extensive early work on the ETM, conducted in Chesapeake Bay, suggested that the primary mechanisms of particle trapping in ETMs were due to simple convergence at the limit of salt intrusion, in addition to slow particle settling (Schubel, 1968;

Schubel and Biggs, 1969; Schubel and Kana, 1972; Nichols, 1974). Later studies suggested that the mechanisms controlling particle density in ETMs were more complex, involving gravitationally induced residual currents (Festa and Hansen, 1978), tidal asymmetry (Uncles et al., 1994), and tidal straining of particles on the ebb tide (Geyer, 1993). For example, asymmetrical tidal resuspension and tidal transport are responsible for the formation of settling aggregates in the ETM of the Chesapeake Bay (Sanford et al., 2001). A conceptual diagram illustrates that, while the ETM roughly tracks the limit of salt, it is often decoupled from the salt front (defined here as the isohaline with a salinity of 1 due to a lag between ETM sediment resuspension/transport and rapid meteorologically driven movement of the salt front (figure 6.6). Recent studies focused on sediment transport dynamics in the ETM have typically used acoustic Doppler profiles (ADPs) of

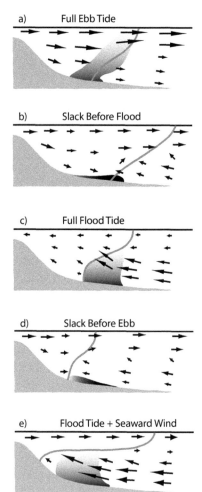

Figure 6.6 A conceptual diagram showing that while the ETM roughly tracks the limit of salt, it is often decoupled from the salt front (defined here at the isohaline with a salinity of 1), due to a lag of ETM sediment resuspension/transport from rapid meteorologically driven movement of the salt front. (From Sanford et al., 2001, with permission.)

velocity, acoustic backscatter (ABS), and optical back-scatterrance sensors (OBS) (Fain et al., 2001), in addition to remote sensing tools such as the sea-viewing wide field-of-view sensor (SeaWiFS) (Uncles et al., 2001).

The location of the ETM is generally considered to be controlled by tidal amplitude, volume of river flow, and channel bathymetry (Brenon and Le Hir, 1999; Rolinski, 1999; Kistner and Pettigrew, 2001). Recent work has focused on the development of a secondary turbidity maximum (STM) downstream of the salt limit in the Hudson River estuary (USA) (Geyer, 1993; Geyer et al., 1998) and York River estuary (USA) (Lin and Kuo, 2001). The Hudson River estuary ETM is formed by interactions between topography and flow (Geyer, 1993; Geyer et al., 1998). For example, suppression of turbulent mixing through density stratification enhances particle trapping efficiency downstream as the salt-wedge front migrates in and out (figure 6.7) (Geyer, 1993). As illustrated here, the "mud reach" is an area with fine sediments that is easily eroded, where during flood tide much of the suspended particulates near the sea bed are from resuspended materials. However, during

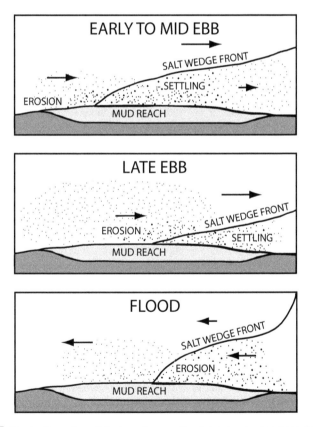

Figure 6.7 Illustration of particle trapping as it relates to resuspension in the "mud reach" and the location of the salt wedge over a tidal cycle. (Modified from Geyer, 1993.)

ebb tide there is a seaward flow in the lower layer allowing materials to settle out in this highly stratified particle trapping zone, which are then redistributed back during flood tide in the landward direction (figure 6.7). The location of particle trapping is a function of the positioning of the salt wedge as it moves across the mud reach during a tidal cycle. Similarly, increases in freshwater runoff/river discharge move the ETM of the Elbe and Weser estuaries (northern Germany) further downstream (Kappenberg and Grabemann, 2001). Other estuaries where river discharge is considered to be a controlling factor on ETM location are the ACE Basin (USA), Scheldt estuary (The Netherlands) (Fettweis et al., 1998), Humber–Ouse estuary (UK) (Uncles et al., 1998), and Gironde estuary (France) (Allen et al., 1980). The ETM has also been shown to be a highly productive region, important to heterotrophic communities (Boynton et al., 1997). High microbial biomass/turnover and recycling of nutrients (Baross et al., 1994; Ragueneau et al., 2002) in ETMs may also be important to estuarine food webs. The ETM has been shown to be an important region for fish larvae (Dodson et al., 1989; North and Houde, 2001). High retention of fish larvae in this region is likely due to the following: (1) high phytoplankton and zooplankton biomass which serve as vital food resources (Boynton et al., 1997), (2) reduced predation due to high turbidity conditions (Chesney, 1989), and (3) isolation from the osmotically stressful conditions of high-salinity waters (Winger and Lasier, 1994).

Due to rapid and high sedimentation rates in the ETM, the accumulation of particles in the *benthic boundary layer* (BBL) can result in the formation of *mobile* and *fluid muds*. The BBL is defined by Boudreau and Jørgensen (2001 p. 1) as "those portions of sediment and water columns that are affected directly in the distribution of their properties and processes by the presence of the sediment–water interface." An idealized diagram of the BBL illustrates the important components and their relative scale within the BBL (figure 6.8). Although not relevant to coastal or estuarine environments, the thickness of the BBL is defined by an Eckman scale in deep ocean environments, where, $u*$ is the friction velocity and f is the Coriolis parameter (McCave, 1976; Boudreau and Jørgensen, 2001). The meter-to-millimeter scale thickness is relevant to BBLs found in coastal margins. In particular, BBL research has been focused on shelf regions (Hinga et al., 1979; Smith, 1987; Nittrouer et al., 1995; Shimeta et al., 2001) and in some cases estuaries that are deep and stratified (Santschi et al., 1999; Mitra et al., 2000a). The viscous sublayer and diffusive boundary layer (DBL) are particularly important in controlling molecular diffusive rates across the sediment–water interface (more discussion on these processes is in chapter 8). Other BBL-related phenomena, such as mobile muds, defined as high-porosity BBL–upper seabed layers where diagenetic transformation processes are enhanced, commonly occur in river-dominated ocean margins (RiOMars), within the lower river of the estuary (salt-wedge region), and on the adjacent shelf (Aller, 1998; Chen et al, 2003a, b; McKee et al., 2003). The remineralization rates of organic matter in these mobile muds have been found to be highly efficient, as in the Amazon and Fly delta (Papna New Guinea) (Aller, 1998), compared to other shelf environments. Mobile muds found in the lower Mississippi River estuary and adjacent shelf, are sites of active organic matter remineralization (Chen et al., 2003a, b; Sutula et al., 2004) and are important regions for the transport of river-derived terrestrial organic matter (Corbett et al., 2003; Wysocki et al., 2006).

Similarly, fluid muds occur in estuarine regions where there is a large source of floc-culated particles (e.g., the ETM), a boundary in the water column that reduces upward

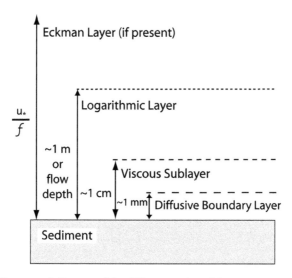

Figure 6.8 Conceptual diagram of the different scales of the components of the benthic boundary layer (BBL). In bottom water above the sediment–water interface where the Eckman layer occurs as flow is affected by the rotation of the Earth and bottom friction, where $u_* = $ friction velocity and $f = $ Coriolis parameter; the logarithmic layer predominates when the velocity profile is well described using a logarithmic function; a viscous sublayer is formed by molecular viscosity; a diffusive boundary layer forms, whereby solute transport is controlled by molecular diffusion. (Modified from Boudreau and Jørgensen, 2001.)

mixing of particles, and sufficient shear to overcome the reduced settling of particles to the bed (Trowbridge and Kineke, 1994; Allison et al., 2000). Fluid muds are defined as a high concentration of sediment suspended in mass concentrations of greater than 10 g L^{-1} (Kemp, 1986). Fluid muds have been found in estuarine and shelf environments (Gallene, 1974; Allen et al., 1980; Wells, 1983; Nichols, 1984; Wolanski et al., 1988; Wright et al., 1990; Odd et al., 1993; Kineke and Sternberg, 1995; Allison et al., 1995, 2000). Most of the classic work on fluid muds has been along the Amazon delta where extensive fluid muds are found — annually representing sediment masses of approximately 31×10^9 tons on the delta topset (Keuhl et al., 1986; Allison et al., 1995; Jaeger and Nittrouer, 1995; Kineke and Sternberg, 1995; Aller et al., 1996). As shown in figure 6.9, fluid muds on the Amazon shelf are formed primarily during neap tide when the water column is stratified, due to salinity, and sedimenting particles are trapped at the frontal zone (Kineke et al., 1996). During spring tide, stratification is disrupted, and the water column and fluid muds are resuspended forming "mixed'' fluid muds or mobile muds. On the Amazon shelf, fluid muds are important in reducing boundary shear stresses, affecting fluxes across the sediment–water interface, and promoting the growth of a subaqueous delta (Kineke et al., 1996). Fluid muds have also been shown to be important in transporting river-derived organic materials from Atchafalaya Bay estuary to the Louisiana shelf at certain times

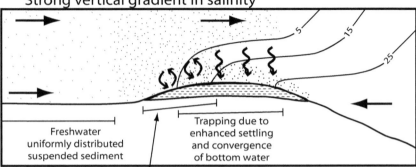

(a) Neap Tide - stratified fluid mud
   Strong vertical gradient in salinity

Freshwater
uniformly distributed
suspended sediment

Trapping due to
enhanced settling
and convergence
of bottom water

Transient resuspension due to
migration of front, tidal variation
in bed stress, and locus of trapping

(b) Spring Tide - mixed fluid mud
   Strong horizontal gradient in salinity

Freshwater

Trapping due to
lateral convergence
of bottom water

Resuspension of stratified fluid mud
and seabed to form mixed fluid mud or
moderate concentration suspension

0.01   0.01-0.1   0.1-1   1-10   10-100   100's   mud bed

Suspended Sediment Concentration (g L^{-1})

Figure 6.9 Theoretical diagram of fluid mud formation and disruption at the frontal zone showing: (a) neap tide stratified fluid mud formation—with strong vertical salinity gradient; and (b) spring tide mixed or mobile fluid mud formation—with strong horizontal salinity gradient. (Modified from Kineke et al., 1996.)

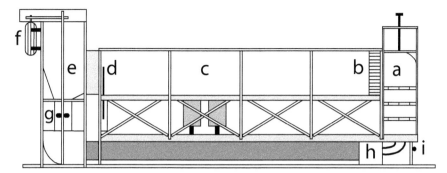

Figure 6.10 An example of a recirculating straight open-channel design flume. The components are: a = head box, b = collimator, c = open channel, d = weir, e = tail box, f = motor, g = axial pump, h = return pipe and cooling coil, and i = electric jacks. (Modified from Khalili et al., 2001.)

of the year (Allison et al., 2001; Gordon et al., 2001; Gordon and Goni, 2003). Similar to mobile muds, recent studies have shown that fluid muds are active for biogeochemical transformation reactions (Abril et al., 2000; de Resseguier, 2000; Tseng et al., 2001; Schafer et al., 2002). More work is needed to further understand the effects of these bottom transport processes on the biogeochemical character of particles and water masses at the estuarine/shelf boundary, particularly in RiOMar regions.

Detailed in situ investigations of bottom transport processes, such as mobile and fluid muds and the BBL, are in many cases not possible with the conventional field equipment [e.g., tripods equipped with submersible pumps, conductivity/temperature/depth (CTD), OBS, and ADP], as they are too coarse in their sampling resolution (e.g., one measurement every 25 cm above the bed). Consequently, flumes have been a useful experimental tool for understanding detailed near-bed benthic processes in an environment that simulates near-bottom flow conditions. The basic flume designs commonly used for boundary layer transport studies are straight, racetrack, and annular channels (Khalili et al., 2001). An example of a recirculating, straight, open-channel design is shown in figure 6.10; other examples of flume design and operation can be found in Williams (1971), Vogel (1981), Nowell and Jumars (1987), Trowbridge et al. (1989), and Huettel and Gust (1992). Flume studies have provided important information on the effects of boundary layer (BL) flow on larval settlement (Butman and Grassle, 1992), benthic worm tube assemblage interactions with BL flow (Eckman et al., 1981), and effects of BL flow on transport of dissolved (e.g., metals, nutrients, oxygen) and particulate (fine sediments) constituents (Sleath, 1984; Huettel et al., 1996).

Summary

1. The uplift of rocks above sea level on the Earth's surface over geological time produces rock material that can be altered into soils and sediments by

weathering processes. Weathering is typically separated into two categories: physical weathering, which involves the fragmentation of parent rock materials and minerals, and chemical weathering, which involves the chemical alteration of minerals.

2. The four dominant processes controlling movement of sediments in estuaries are as follows: (1) erosion of the bed; (2) transportation; (3) deposition; and (4) consolidation of deposited sediments.

3. Microbial mats and biofilms, defined as surface layers of microbes entrained in a matrix of EPS, are also important in changing the surface texture and erodibility of sediments in estuaries.

4. The sinking speed for spherical, slow sinking, single particles is described by Stokes' law. However, single particles are not representative of the floc settlement that typically occurs in estuaries.

5. Settling particles less than 0.5 μm are slowed down by Brownian motion in the water, while large sand-sized particles are not affected by viscous forces and typically generate a frontal pressure or wake as they sink.

6. Estuaries typically have elevated concentrations of suspended fine sediments that are highly cohesive and readily flocculate. Coagulation of fine particles leads to the formation of the following composite particles in estuaries: (1) agglomerates; (2) aggregates; and (3) floccules.

7. As repulsive forces decrease with increasing salinity, van der Waals forces will dominate and flocculation will proceed. Strong salinity gradients in estuaries make them ideal environments for coagulation processes.

8. Sediments in rivers are transported as suspended load or bed load and are directly discharged into open oceans, coastal shelves, or remain stored in estuaries and rivers.

9. The ETM is defined as a region where the SPM concentrations are considerably higher (10–100 times) than adjacent river or coastal end-members in estuaries. The location of the ETM is generally considered to be controlled by tidal amplitude, volume of river flow, and channel bathymetry.

10. Due to rapid and high sedimentation rates in the ETM, the accumulation of particles in the BBL can result in the formation of mobile and fluid muds. Fluid muds are defined as sediments suspended in concentrations of greater than 10 g L^{-1}.

Chapter 7

Isotope Geochemistry

Basic Principles of Radioactivity

There is a broad spectrum (approximately 1700) of *radioactive isotopes* (or *radionu-clides*) that are useful tools for measuring rates of processes on Earth. The term *nuclide* is commonly used interchangeably with *atom*. The major sources of radionuclides are: (1) *primordial* (e.g., ^{238}U, ^{235}U, and ^{234}Th-series radionuclides); (2) *anthropogenic* or *transient* (e.g., ^{137}Cs, ^{90}Sr, ^{239}Pu); and (3) *cosmogenic* (e.g., ^{7}Be, ^{14}C, ^{32}P). These isotopes can be further divided into two general groups, the *particle-reactive* and *non-particle-reactive* radionuclides. Transport pathways of non-particle-reactive radionuclides in aquatic systems are more simplistic and primarily controlled by water masses. Conversely, particle-reactive radionuclides adsorb onto particles, making their fate inextricably linked with the particle. Consequently, these particle-bound radionuclides are very useful in determining sedimentation and mixing rates, as well as the overall fate of important elements in estuarine and coastal biogeochemical cycles.

Radioactivity is defined as the spontaneous adjustment of nuclei of unstable nuclides to a more stable state. Radiation (e.g., *alpha*, *beta*, and *gamma rays*) is released in different forms as a direct result of changes in the nuclei of these nuclides. The general composition of an atom can simply be divided into the *atomic number*, which is the number of protons (Z) in a nucleus. The mass number (A) is the number of neutrons (N) plus protons in a nucleus (A = Z + N). *Isotopes* are different forms of an element that have the same Z value but a different N. Instability in nuclei is generally caused by having an inappropriate number of neutrons relative to the number of protons. Some of the pathways by which a nucleus can spontaneously transform are as follows: (1) *alpha decay*, or loss of an alpha particle (nucleus of a ^{4}He atom) from the nucleus, which results in a decrease in the atomic number by two (two protons) and the mass number by four units (two protons and two neutrons); (2) *beta (negatron) decay*, which occurs when a neutron changes to a proton and a *negatron* (negatively charged electron) is emitted, thereby increasing the atomic number by one unit; (3) emission of a *positron* (positively charged electron) which results in a proton becoming a neutron and a decrease in the atomic number by one unit;

and (4) *electron capture*, where a proton is changed to a neutron after combining with the captured extranuclear electron (from the K shell)—the atomic number is decreased by one unit.

The experimentally measured rates of decay of radioactive atoms shows that their decay is *first order*, where the number of atoms decomposing in a unit of time is proportional to the number present—this can be expressed in the following equation (Faure, 1986):

$$dN/dt = -\lambda N \tag{7.1}$$

where: N = number of unchanged atoms at time t; and λ = decay constant.

If we rearrange and integrate equation 7.1, from $t = 0$ to t and from N_0 to N we obtain:

$$-\int dN/N = \lambda \int dt$$

$$-\ln N = \lambda t + C \tag{7.2}$$

where: $\ln N$ = is the logarithm to the base e of N; and C = constant of integration.

When $N = N_0$ and $t = 0$, equation 7.2 can be modified to:

$$C = -\ln N_0 \tag{7.3}$$

We can then substitute into equation 7.2 to obtain the following equation:

$$-\ln N = \lambda t - \ln N_0 \tag{7.4}$$

$$\ln N - \ln N_0 = -\lambda t$$

$$\ln N/N_0 = -\lambda t$$

$$N/N_0 = e^{-\lambda t}$$

$$N = N_0 e^{-\lambda t} \tag{7.5}$$

In addition to λ, another term used for characterizing rate of decay is *half-life* ($t_{1/2}$), the time required for half of the initial number of atoms to decay. If we substitute $t = t_{1/2}$ and $N = N_0/2$ into equation 7.5 we obtain the following equation:

$$N_0/2 = N_0 e^{-\lambda t_{1/2}}$$

$$\ln \tfrac{1}{2} = -\lambda t_{1/2}$$

$$\ln 2 = \lambda t_{1/2}$$

$$t_{1/2} = \ln 2/\lambda = 0.693/\lambda \tag{7.6}$$

Finally, the term used to examine the average life expectancy, or *mean life* (τ_m), of a radioactive atom is expressed as follows:

$$\tau_m = 1/\lambda \tag{7.7}$$

On the assumption that the radioactive parent atom produces a stable radiogenic daughter (D^*), that is zero at $t = 0$, we can describe the number of daughter atoms at any time with the following equation:

$$D^* = N_0 - N \tag{7.8}$$

This assumes no addition or loss of daughter atoms and that any loss of the parent atoms is due to radioactive decay. If we then substitute equation 7.5 into 7.8 we obtain the following:

$$D^* = N_0 - N_0 e^{-\lambda t}$$

$$D^* = N_0(1 - e^{-\lambda t}) \tag{7.9}$$

With further substitution:

$$D^* = Ne^{\lambda t} - N = N(e^{\lambda t} - 1) \tag{7.10}$$

Figure 7.1 illustrates the hypothetical decay of a radionuclide (N) to a stable radiogenic daughter (D^*); the successive loss of atoms of N from radioactive decay is followed by a proportional increase in the daughter atoms. Assuming that the total number of daughter atoms is formed radiogenically plus those initially present at $t = 0$, the total number of daughter atoms present in a system (D) is:

$$D = D_0 + D^* \tag{7.11}$$

with further substitution from equation 7.10:

$$D^* = N(e^{\lambda t} - 1)$$

$$D = D_0 = N(e^{\lambda t} - 1) \tag{7.12}$$

Using equation 7.12, rocks and minerals can be dated based on the radiogenic formation of a stable daughter atom. In a decay series (e.g., ^{238}U), where the half-lives of the parent atoms are considerably longer than their respective daughter atoms, the parent atoms remain essentially constant over time relative to the half-lives of the daughter atoms (Faure, 1986). In this case, isotopes are considered to be in *secular equilibrium*, where the rate of decay of the parent is equal to that of its daughter. This is expressed as follows:

$$N_1 \lambda_1 = N_2 \lambda_2 = N_i \lambda_i \tag{7.13}$$

where: $N_1 \lambda_1$ represents the parent atom and $N_2 \lambda_2$ and/or $N_i \lambda_i$ represents the series of daughter half-lives.

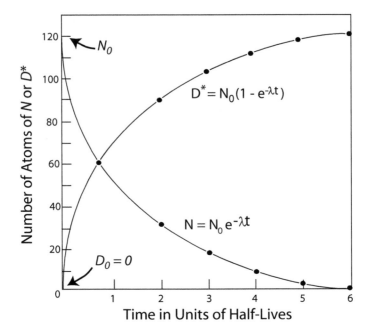

Figure 7.1 Hypothetical decay of a radionuclide (N) to a stable radiogenic daughter (D^*); the successive loss of atoms of N from radioactive decay are followed by proportional increase in the daughter atoms. (Modified from Faure, 1986.)

The radioactivity of a sample is measured as the number of disintegrations per minute (dpm). Since it is not practical to describe radioactivity by the number of atoms remaining (N) in aforementioned equations, the observed *specific activity* (A) of a sample is usually described by the following equation:

$$A = \lambda N \qquad (7.14)$$

The specific activity (A) of a radionuclide represents the observed counting rate in a sample. Radionuclides are useful in evaluating processes over time scales that are four to five times their half-lives, where beyond that period there is typically only 1% or less of the nuclide remaining and cannot be measured effectively (Nittrouer et al., 1984). For example, ^{210}Pb has a half-life of 22.3 y and a useful life of approximately 112 y; thus, dating sediment or rocks older that about 140 y is beyond the range of the nuclide.

Radionuclides in Estuarine Research

Radionuclides have been used for studying a very broad range of biogeochemical processes in estuaries (table 7.1). In this section more details are provided on the cycling

Table 7.1 Radionuclides used to investigate biogeochemical processes in estuaries and adjacent coastal systems.

Nuclide	Half-life	Utility	Reference
[7]Be	53 d	Residence time of particles and particle-sorbed pollutants, particle mixing, and short-term sedimentation rates	Aaboe et al. (1981); Baskaran and Santschi (1993); Baskaran et al. (1997)
[137]Cs	30 y	Watershed erosion, sedimentation, and mixing	Ritchie et al. (1974); Baskaran and Naidu (1995)
[210]Pb	22.1 y	Residence times of particles and particle-sorbed pollutants, particle mixing, and sedimentation rates	Rama et al. (1961); Baskaran and Santschi (1993); Baskaran et al. (1997)
[224]Ra	3.66 d	Biogeochemical processes in salt marsh	Bollinger and Moore (1993)
[234]Th	24.1 d	Rates of removal of Th and particle-reactive pollutants from solution, short-term particle mixing, and sedimentation rates in estuaries	Aller et al. (1980); McKee et al. (1986)
[228]Th	1.9 y	Rates of removal of Th and particle-reactive pollutants from solution, short-term sediment accumulation	Kaufman et al. (1981); Minagawa and Tsunogai (1980)
[238]U	4.5×10^9 y	Removal of metals from low-salinity estuarine waters	Borole et al. (1982)
[239,240]Pu	2.4×10^4 y	Sedimentation and mixing	Benninger et al. (1979); Ravichandran et al. (1995a)

Modified from Baskaran (1999).

of both particle-reactive (^{234}U/^{238}U, ^{234}Th,^{210}Pb,^{7}Be,^{137}Cs) and non-particle-reactive (^{222}Rn, ^{226}Ra) radionuclides commonly used in estuaries. More emphasis will be placed on the particle-reactive radionuclides because of their widespread applications to biogeochemical rates and processes. The primary processes affecting the distribution of particle-reactive radionuclides in estuaries are as follows (Baskaran, 1999): (1) removal or *scavenging* of nuclides from the water column by precipitation, ion exchange, and hydrophobic interactions with particle surfaces (Gearing et al., 1980); (2) complexation with organic materials that are adsorbed on particle surfaces (Santschi et al., 1979, 1980, 1999); (3) flocculation of radionuclides in colloidal material (Edzwald et al., 1974; Sholkovitz, 1976); (4) sorption on Fe and Mn oxide coatings that undergo coprecipitation (Saxby, 1969; Boyle et al., 1977; Swarzenski et al., 1999); (5) direct removal by biological processes (Santschi et al., 1999); (6) desorption from suspended sediments (particularly from rivers) (Duinker, 1980); and (7) release from sediments via physical or biological mixing/uptake (Li et al., 1977; Gontier et al., 1991; Baskaran and Naidu, 1995). Before providing details and case studies on each of the radionuclides commonly used in estuaries a few of the general equations used in making estimates of some of the aforementioned processes will be described more broadly.

Scavenging/Removal Rates, Residence Times, Inventories, and Resuspension Rates

To understand fully the cycling of many of the dissolved and particulate constituents in estuaries, it is necessary to determine the residence time and removal rates in the system. One of the most common means of obtaining such information has been through the use of the U/Th decay series. More details on the U/Th disequilibrium are provided later in this chapter. Generally, two approaches are used here: (1) making use of the parent/daughter nuclide disequilibrium; and (2) making precise measurements of fluxes and concentrations of atmospherically derived radionuclides. As mentioned earlier, particles in estuaries play an important role in the removal of particle-reactive radionuclides as well as many contaminants in these systems (Baskaran and Santschi, 1993). At steady state, the *residence time* (τ) of an element is defined as the ratio of the element's standing stock to the removal rate or supply. When assuming steady-state conditions, the residence time for irreversible removal of a radionuclide such as ^{234}Th, on to particles (τ_s) or from the water column (τ_r), is defined by the following equation (Baskaran and Santschi, 1993):

$$\tau = \tau_m \times R/(1 - R) \tag{7.15}$$

where: τ_m = mean life ($1/\lambda$) of a radionuclide (in the case of ^{234}Th, τ_m = 34.8 d); and R = activity ratio in the filter passing (for dissolved residence time) or unfiltered water (for total residence time).

The total and dissolved residence times of ^{234}Th with respect to removal on to particles in estuaries along the Texas coast indicate that there are distinct seasonal patterns with respect to removal processes in these estuaries (table 7.2)—more details on Th cycling are provided later in this chapter. The removal of particle-reactive radionuclides by particles with a pore size of 0.4 μm or greater can be made using an empirically determined *scavenging rate constant* (λ_s) which equals 1/residence time (see reviews by Baskaran and Santschi, 1993, and Baskaran, 1999). Further details on the kinetics of particle dynamics can be obtained by examining the distribution coefficient, K_d, which was discussed in chapter 4 and is a measure of the partitioning of a radionuclide between the filter-passing and filter-retained particulate phases, defined as follows:

$$K_d = A_p/A_w \tag{7.16}$$

where: A_p = activity of radionuclide in suspended particulates; and A_w = activity of radionuclide dissolved in water.

The equation for the residence time of radionuclides with atmospheric inputs (e.g., ^{210}Pb and ^{7}Be; more details on the cycling of these radionuclides are provided

Table 7.2 Total and dissolved residence time of ^{234}Th with respect to removal onto particles in Texas estuaries.[a]

Sample code	Total ^{234}Th residence time (d)			Dissolved ^{234}Th residence time (d)		
	Spring	Summer	Winter	Spring	Summer	Winter
Copano Bay	3.9	NM	0.7	2	NM	0.4
San Antonio Bay	5.3	NM	1.7	1.5	NM	0.9
Aransas Bay	6.1	1.1	1.2	4.9	0.16	0.3
Baffin Bay	7.8	1	1.1	3.9	0.28	0.7
Corpus Christi Bay	0.9	1.6	1.5	0.4	0.08	0.9
Laguna Madre	0.9	1.3	1.3	0.5	0.09	1.1
Aransas Bay	NM	NM	NM	NM	NM	NM
Cedar Pass	NM	NM	NM	NM	NM	NM
Galveston Bay	NM	1.1	NM	NM	0.1	NM

[a] ^{238}U concentrations used in the calculation of residence time were estimated by assuming riverine end-member ^{238}U concentrations to be 0.5 dpm L^{-1} (Chen et al., 1986) at salinity of 35. For water samples with salinities greater than 35, a linear relationship between ^{238}U concentration and salinity was assumed.

NM = not measured.

Modified from Baskaran and Santschi (1993).

later in this chapter) can be slightly modified, in the case of ^{210}Pb, to be (Baskaran and Santschi, 1993):

$$\tau_r = \ln 2 \times A_{Pb} \times h/I_{Pb} \qquad (7.17)$$

where: A_{Pb} = total activity of excess ^{210}Pb (dpm m^{-3}); I_{Pb} = atmospheric input rate of ^{210}Pb (dpm m^{-2} d^{-1}); and h = mean depth of the estuary (m).

The total residence times and distribution coefficients of ^{7}Be and ^{210}Pb in different estuaries illustrate the importance of the concentration of dissolved and particulate constituents in controlling the dynamics of these radionulides (table 7.3).

To examine further the fate of particle-reactive radionuclides we will use a simple one-dimensional model that, in the case of ^{234}Th, would supply ^{234}Th to the seabed from a source of ^{238}U in the water column. To test this model, a sedimentary *inventory* (*I*) of ^{234}Th would have to be calculated based on the excess ^{234}Th profiles in cores using the following equation described by McKee et al. (1984):

$$I = \sum \rho_s X_i (1 - \varphi_i) A_i \qquad (7.18)$$

where: I = inventory of excess activity of radionuclide (dpm cm^{-2}); ρ_s = particle density (g cm^{-3}); X_i = depth in sediment; i = sampling interval (cm); φ = porosity of sediments; and A = average excess activity (dpm g^{-1}) for each interval.

Table 7.3 Total residence times and distribution coefficients of [7]Be and ^{210}Pb in different estuarine systems in the United States.

Name of estuary	Water depth (m)	DOC (mg L⁻¹)	Suspended particle conc. (mg L⁻¹)[a]	Hydraulic residence time (d)[b]	[7]Be residence time (d)[c]	[210]Pb residence time (d)[c]	[7]Be—K_d (10⁴ cm³ g⁻¹)
Copano Bay[d]	1.1	NM	30.4	NM	1.9	NM	0.71
San Antonio Bay[d]	1.4	4.0–5.8	43.5	39	0.2	NM	2.9
Aransas Bay[d]	2.4	NM	16.4	36	3.7	NM	2.0
Baffin Bay[d]	2.4	NM	206	NM	9.1	NM	0.11
Corpus Christi Bay[d]	3.2	6.7–7.6	17	356	5.7	NM	2.2
Laguna Madre[d]	1.4	NM	21.2	NM	2.8	NM	8.2
Cedar Pass[d]	0.5	NM	7.6	41	1.8	NM	NM
Galveston Bay[d]	2.0	5.0–5.8	11	—	0.87	29–117	31–113
Upper Chesapeake Bay[e]	8	NM	NM	—	NM	2.3–20	NM
James River[e]	3.5	NM	NM	—	NM	0.76–16	NM
Hudson River[e]	6	NM	NM	—	NM	1.3–7.5	NM
Raritan Bay[e]	6	NM	NM	—	NM	3.9–16	NM
Savine–Neches Estuary[f]	1.8	4.8–21.0	34.7	9	3.5–27	0.15–8.7	0.26–3.7

[a]The suspended particle concentrations is the average value of two to three seasons wherever data are available (Baskaran and Santshi 1983).
[b]The hydraulic residence time is taken from Solis and Powell (1999). [c]The Texas estuaries were sampled during winter and summer months, and the residence time is the average during these two seasons; the Chesapeake and Raritan Bays and the James and Hudson Rivers were sampled during summer, and the Sabine–Neches estuary was mainly sampled during spring and fall. NM = not measured. [d]Baskaran and Santschi (1993). [e]Olsen et al. (1986). [f]Baskaran et al. (1997).
Modified from Baskaran (1999).

Information from these observed inventories (I_0) can be compared to predicted inventories (I_p) based on production in the overlying water column to create a relative inventory (I_r), which equals I_0/I_p. If this ratio is greater than unity it means that there is more excess radionuclide than is predicted from complete scavenging and deposition of the radionuclide on the seabed from production in the overlying water column (Aller et al., 1980; McKee et al., 1984). This ratio can provide a simple evaluation of where the one-dimensional model breaks down in estuarine systems.

In shallow estuaries, resuspension of particles from the seabed can be an important factor in determining the overall fate of particle-reactive radionuclides by removing dissolved activity. The resuspension rate can be determined using a box model of mass balance for particle-reactive radionuclides (Baskaran et al., 1997; Baskaran, 1999). In the case of a box model for the inputs and removal of excess particulates ($^{210}\text{Pb}_{xs}$) in sediments and particles of a shallow estuary (figure 7.2), the inputs are as follows: (1) input from parent production in the particulate phase, ^{222}Rn ($\lambda_{Pb} A_{Rn}$); (2) input of particles sorbed from riverine sources (I_R^{op}); (3) removal of total dissolved ^{210}Pb [total dissolved ^{210}Pb (A_{Pb}^{Td}) = riverine dissolved input + direct atmospheric input + production from dissolved ^{222}Rn] onto suspended particles ($\psi_c^0 A_{Pb}^{Td}$), where ψ^0 is the composite first-order rate constants for removal from particles (y^{-1}); and (4) resuspension of bottom sediments ($S A_{Pb}^S/H$), where S is the sediment resuspension rate (g cm^{-2} y^{-1}), A_{Pb}^S is the ^{210}Pb in surficial sediments that is resuspended, and H is the mean depth of the estuary (m). Similarly, the losses of particulate ^{210}Pb are from the following: (1) decay of ^{210}Pb in the particle reservoir ($\lambda_{Pb} A_{Pb}^p$, which is the total particulate concentration in the particulate pool); (2) settlement of particles containing ^{210}Pb ($S A_{Pb}^R/H$), where A_{Pb}^R is the ^{210}Pb concentration in suspended particles; and (3) desorption of ^{210}Pb from particles

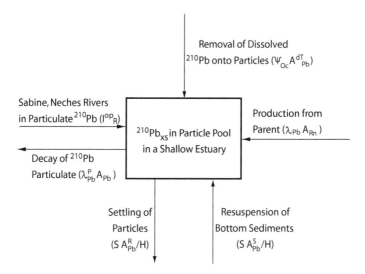

Figure 7.2 Box model for the inputs and removal of excess particulate ($^{210}\text{Pb}_{xs}$) in sediments and particles of a shallow estuary. (From Baskaran, 1999, with permission.)

(assumed to be negligible). Thus, the mass balance for particulate ^{210}Pb is as follows:

$$\lambda_{Pb}A_{Rn} + I_R^{op} + \psi_c^0 A_{Pb}^{Td} + S\,A_{Pb}^s/H = \lambda_{Pb}A_{Pb}^P + S\,A_{Pb}^R/H \qquad (7.19)$$

If we rearrange equation 7.19, we can write the equation for resuspension rate as:

$$S = H\left[\lambda_{Pb}A_{Pb}^P - \psi_c^0 A_{Pb}^{Td} - I_R^{op}\right]\Big/\left[A_{Pb}^s - A_{Pb}^R\right] \qquad (7.20)$$

Sediment Deposition and Accumulation/Accretion Rates

Sediment *deposition* is defined as the temporary emplacement of particles on the seabed. On the other hand, sediment *accumulation* is defined as the net sum of particle deposition and removal processes over a long period, which is distinctly different from the aforementioned sediment deposition that represents shorter time intervals (McKee et al., 1983). Similarly, *accretion*, commonly measured in marshes, is the net positive accumulation (when deposition is greater than removal). Both accumulation and accretion result in preservation of strata. Understanding the formation of strata in coastal margins requires an important distinction between the deposition and accumulation of particles (McKee et al., 1983). Finally, the term *sedimentation* refers to integrated particle transport to and emplacement on the seabed, as well as removal and preservation. Some of the early work in chemical and geological oceanography demonstrated that temporal changes in marine sedimentation could be measured using radionuclides (Goldberg and Bruland, 1974; Turekian and Cochran, 1978). As particles continue to accumulate on the seabed, there is accretion upward and continued burial and preservation of progressively older sediments at depth.

Due to preferential scavenging and lateral transport of a daughter radionuclide, the activity of daughter A_D can be greater than that of the parent A_P in sediments. The inputs of daughter radionuclides that are not directly from the in situ decay of the parent (*supported*) are termed *unsupported* or *excess* activity. The unsupported A_D is equal to the supported A_D minus the A_p, as shown in the theoretical radionuclide profiles in figure 7.3. Moreover, the curve for the unsupported A_D decreases with depth more than the supported A_D because it is not being produced in situ from the parent. Consequently, the excess activity of a radionuclide can be used to calculate the time elapsed since the particles with unsupported A_D were last at the surface, relative to a particular depth (x). However, to calculate this it must be assumed that the sedimentation rate and supply of unsupported A_D has remained constant over time.

Assuming that particle-reactive radionuclides and many other trace elements are scavenged by particles settling through the water column, and that these fluxes of sediment and the associated radionuclide have remained constant over time, an *accumulation rate* (A) can be calculated using the following equation:

$$A = \lambda x/\ln(C_0/C_x) \qquad (7.21)$$

where: C_0 = activity of radionuclide at the sediment surface; C_x = activity of the radionuclide at a depth (x) below the C_0; x = sediment depth (cm); and λ = decay constant.

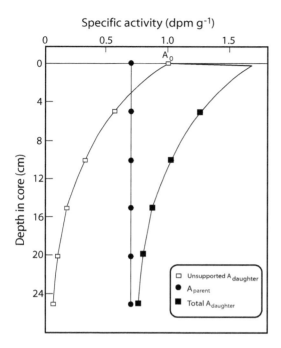

Specific activity (dpm g^{-1})

Figure 7.3 Theoretical radionuclide profiles of unsupported or excess, A_D, which is equal to the supported A_D minus the A_p. The curve for the unsupported A_D decreases with depth more than the supported A_D because it is not being produced in situ from the parent. (Modified from Libes, 1992.)

As mentioned above, the accumulation and burial of particles is usually not a simple process in shallow coastal systems like estuaries and continental shelves. In many cases, particles are reworked and mixed by physical and biological processes causing particles to move vertically and laterally (Nittrouer et al., 1984). Particle mixing is usually modeled as a one-dimensional vertical diffusion process (Goldberg and Koide, 1962; Guinasso and Schink, 1975; Nozaki et al., 1977; DeMaster and Cochran, 1982). Although diffusion may not be representative of mixing events on real timescales, when integrated over long periods it has more of a diffusive character. Goldberg and Koide (1962) first devised a method for quantitatively determining sediment mixing by benthic organisms in an attempt to observe the homogenization of radionuclide profiles in the upper bioturbated zone—more details are provided on animal–sediment interactions and diagenetic reactions in chapter 8. Since the flux of substance across a boundary is proportional to its concentration gradient, diffusive mixing is described by a proportionality factor, known as the diffusion or mixing coefficient (D). This coefficient is used to quantify sediment mixing (Goldberg and Koide, 1962) despite the fact that particles themselves are not diffusing (Matisoff, 1982). Many early studies used this parameter to quantify bioturbation and mixing (Aller and Cochran, 1976; Robbins et al., 1977; Benninger et al., 1979; Krishnaswami et al., 1980, 1984; Olsen et al., 1981; Aller and DeMaster, 1984). If particle mixing is assumed to be analogous to diffusion with sediment accumulation and radionuclide decay, the steady-state profile for excess activity of a nonexchangeable radionuclide is defined by the following *advective–diffusion* equation (Nittrouer et al., 1984):

$$D(\partial^2 C/\partial x^2) - A(\partial C/\partial x) - \lambda C = 0 \qquad (7.22)$$

The solution and rearrangement of equation 7.19 to calculate accumulation rates is as follows:

$$A = \lambda x / \ln(C_0/C_x) - D/x(\ln C_0/C_x) \qquad (7.23)$$

When mixing is not important (i.e., $D = 0$), equation 7.23 simplifies to equation 7.21. When mixing is present, using equation 7.23 will result in an overestimate of the true accumulation rate (Nittrouer et al., 1984). An empirical approach is used to calculate D, whereby in situ sediment profiles of radionuclides are used to integrate mixing processes over time; such profiles can be used when mixing is significant and accumulation is slow (i.e., $A^2 < \lambda D$). Under these conditions, equation 7.23 can be converted as follows:

$$D = \lambda(x/[\ln C_0/C_x])^2 \qquad (7.24)$$

In general, short-lived isotopes such as ^{234}Th and ^{7}Be are used to calculate mixing rates for the upper 10 cm. Using ^{234}Th and ^{210}Pb short-term mixing rates and accumulation rates, respectively, were calculated for Fourleague Bay (USA) (figure 7.4; Day et al., 1995). The inflection of ^{210}Pb at the base of the surface mixed layer (dashed line) reflects the logarithmic decrease in excess activities associated with decay, from which an accumulation rate can be calculated. Many studies have employed the use of both short- and long-term radionuclides to understand fully the dynamics of time-offset of deposition and accumulation rates (McKee et al., 1983; Smoak et al., 1996).

Uranium

In the U decay series we begin with the most abundant U isotope (^{238}U) and end with stable ^{206}Pb (figure 7.5). Many of the intermediate daughter products in this decay series have very short half-lives which are generally not useful in estuarine studies; however, there are a number of key daughter products that have been used in understanding geochemical rates in estuaries that will be discussed here. To date, most of the U work by geochemists over five decades has been conducted in oceanic environments (Starik and Kolyadin, 1957; Sackett et al., 1973; Klinkhammer and Palmer, 1991; Cochran, 1992). Uranium is released from continental weathering and delivered to the coastal ocean by rivers. The three isotopes of uranium (^{238}U, ^{235}U, ^{234}U) found in nature have longer half-lives ($>10^5$ y) than the oceanic mixing time (ca. 10^3 y). Since U is commonly found in a stable uranyl carbonate complex ($UO_2[CO_3]_3^{-4}$), the concentrations and distribution of U in seawater are relatively constant in the ocean (Nozaki, 1991; Moore, 1992). Conversely, the distribution and concentration of U in rivers, estuaries, and coastal regions is extremely variable and not well understood (Moore, 1967; Bhat and Krishnaswami, 1969; Bertine et al., 1970; Boyle et al., 1977; Martin et al., 1978a, b; McKee et al., 1987). The central questions involving U cycling in rivers and estuaries is whether transport through these systems is conservative or nonconservative and if not, what are the key physical/biological processes controlling such transformations. Earlier studies also tried using the ^{234}U/^{238}U ratio as a tracer in estuaries but found that the variability in the mixing zone was indiscernible (Martin et al., 1978a). Unfortunately, further advance using this isotope ratio have not progressed in estuarine work to date.

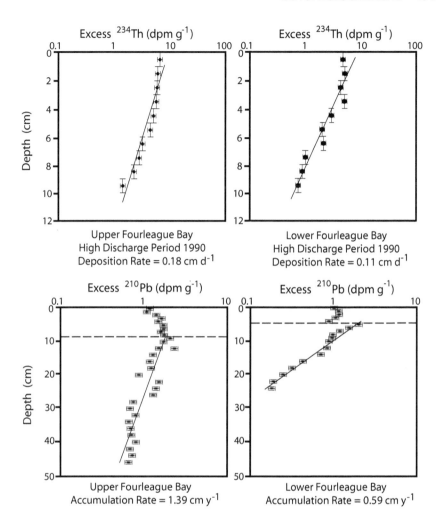

Figure 7.4 Down-core profiles of ^{234}Th and ^{210}Pb for Fourleague Bay (USA) Short-term deposition rates and accumulation rates are calculated using these profiles. The inflection of ^{210}Pb at the base of the surface mixed layer (dashed line) reflects the logarithmic decrease in excess activities associated with decay, from which an accumulation rate can be calculated. (Modified from Day et al., 1995.)

Much of the early work on U in estuaries concluded that U was mixed conservatively (Borole et al., 1977, 1982; Martin et al., 1978a,b). This was followed by work that suggested nonconservative mixing was the dominant pattern for U mixing in estuaries (Maeda and Windom, 1982; McKee et al., 1987; Cochran, 1992; Swarzenski et al., 1999). Uranium–salinity plots illustrate some of the examples of conservative and nonconservative mixing in estuaries (figure 7.6). Studies in large river estuarine systems, such as the

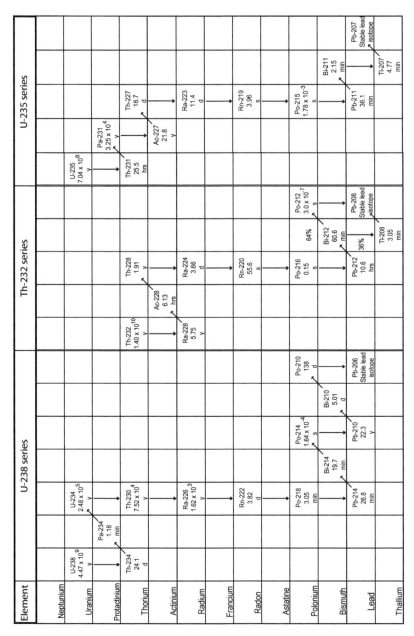

Figure 7.5 Uranium and thorium decay series and the half-lives of each isotope. The chart begins with the most abundant U isotope (^{238}U) and ends with stable lead isotopes (^{206,207,208}Pb). Alpha decay is denoted by the vertical arrows and beta decay by the diagonal arrows. (Modified from Griffin et al., 1963.)

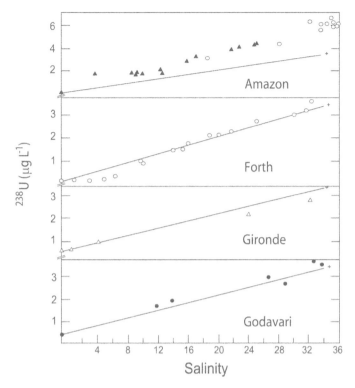

Figure 7.6 Uranium–salinity plots showing some examples of conservative and non-conservative mixing in the Amazon (Brazil), Forth (UK), Gironde (France), and Godavari (India) estuaries. (Modified from Cochran, 1992.)

Amazon and Mississippi, show the occurrence of both conservative and nonconservative behavior of U. For example, surface waters in the Amazon show a nonconservative behavior with decreasing salinity (figure 7.7a); the U removal at salinities less than 15 implies that hydrous metal oxides are likely responsible for adsorptive removal during flocculation and coagulation processes (McKee et al., 1987; Swarzenski and McKee, 1998). Conversely, U removal in the Mississippi River is not controlled by conventional sorption/desorption processes with carrier phase oxides. It is generally conservative in behavior except for exceptionally high discharge periods, such as the flood of 1993 (figure 7.7b).

McKee et al. (1987) suggested that there was a source of U at mid-salinities in the Amazon River estuary. These higher U concentrations were attributed to suboxic diagenesis, whereby U is released from Fe oxyhydroxide coating on sediment particles to pore waters. Moreover, McKee et al. (1987) suggested that the uranyl–carbonate complex inhibited reduction of soluble U(VI) to particle reactive U(IV), keeping U in solution and allowing for exchange between pore waters and the overlying water column. This has been further supported in other suboxic basins such as Framvaren (Norway) (Todd et al., 1988; McKee and Todd, 1993) and Sannich Inlet (Canada) (Todd et al., 1988). More recently, Swarzenski et al. (1999) also showed that the reduced U(IV) occurred at very low levels of detection and that chemical/biological reduction of U(VI) was largely inhibited across the redox transition in the stratified Framvaren Fjord (Norway) water column. The K_d values clearly show that there is not a significant amount of oxidation state transformation occurring from U(VI) to U(IV) in the anoxic water column below the O_2/H_2S interface (figure 7.8). However, other work has argued that suboxic waters may allow for the conversion of U(VI) to U(IV) in the Baltic Sea and other anoxic basins in the world (Anderson, 1987; Anderson et al., 1989). Thus, more work is clearly needed to further understand the complex interactions of active and carrier phase (Fe and Mn oxides) redox transformations, direct and indirect microbial transformations, and colloidal complexation that may be involved in the nonconservative behavior of U in estuaries.

Thorium

The four commonly used isotopes of thorium ($^{234}Th, ^{228}Th, ^{230}Th$, and ^{232}Th) are produced from the decay of uranium and radium parents (figure 7.5). Thorium is present in highly insoluble forms and can be rapidly removed by scavenging of particulate matter.

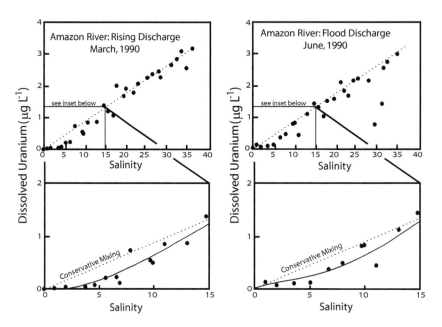

Figure 7.7a Surface waters in the Amazon River (in March and June 1990) showing nonconservative behavior with decreasing salinity; the U removal at salinities less than 15 implies that hydrous metal oxides are likely responsible for adsorptive removal during flocculation and coagulation processes. (Modified from Swarzenski and McKee, 1998.)

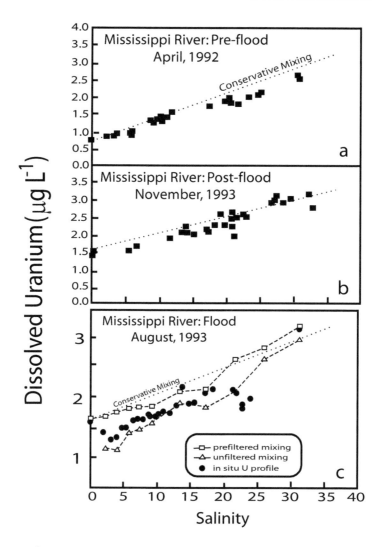

Figure 7.7b Surface waters in the Mississippi River showing U behavior in pre-flood, flood, and post-flood conditions (April, 1992, August, 1993, November, 1993). The U removal in the Mississippi River is not controlled by conventional sorption/desorption processes with carrier-phase oxides. It is generally conservative in behavior except for exceptionally high-discharge periods, such as the flood of 1993. (Modified from Swarzenski and McKee, 1998.)

Thorium radionuclides, particularly ^{234}Th ($t_{1/2}$ = 24.1 d) and ^{228}Th ($t_{1/2}$ = 1.91 d), have primarily been used in estuaries as indicators of mixing in recently deposited sediments (Aller and Cochran, 1976; Cochran and Aller, 1979; Santschi et al., 1979; Aller et al., 1980; McKee et al., 1983, 1986, 1995; DeMaster et al., 1985; Corbett et al., 2003) and

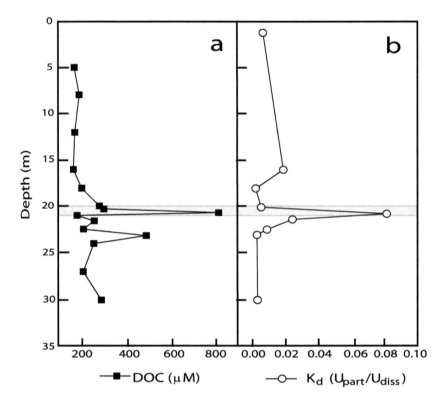

Figure 7.8 Water column profiles of (a) dissolved organic carbon (DOC) and (b) K_d values (particulate uranium/dissolved uranium) across the redox transition in the stratified Framvaren fjord (Norway). (Modified from Swarzenski et al., 1999.)

to constrain the cycling of dissolved organic carbon and other particle reactive elements (Santschi et al., 1979, 1980; Minagawa and Tsunogai, 1980; Kaufman et al., 1981; Guo et al., 1997; Quigley et al., 2002). Thorium-234 is produced continuously in seawater by alpha decay of its parent ^{238}U. Any deviation from secular equilibrium (*disequilibrium*) between ^{238}U and ^{234}Th is attributed to removal processes by biotic and abiotic particles. The residence times of ^{234}Th and ^{228}Th can be quite variable, with very short periods on the order of hours to days in particle-rich estuarine/river environments (Aller and Cochran, 1976; Santschi et al., 1979; McKee et al., 1984) to longer periods of days to months in coastal waters with lower suspended sediment particle concentrations (Santschi et al., 1979, 1980; Li et al., 1981).

The role of biotic versus abiotic particles in scavenging Th(IV) in the water column has received considerable attention (McKee et al., 1986; Baskaran et al., 1992, 1996, 1999). Unlike estuarine environments, lithogenic particles in oceanic surface waters are in low abundance, consequently, ^{234}Th scavenging is considered to be primarily controlled by phytoplankton/phytodetritus production (Bruland and Coale, 1986; Coale and Bruland,

1987; Fisher et al., 1987; Buesseler, 1998). Other work has suggested that there may be differential scavenging rates of Th(IV) among different classes of phytoplankton (Baskaran et al., 1992). This work also suggested that microparticles (e.g., colloids, 1 nm–0.2 μm) are a significant fraction of particles that Th radionuclides are associated with in coastal (Baskaran et al., 1992) and estuarine waters (Guo et al., 1997; Santschi et al., 1999). The binding ligand of ^{234}Th to colloidal organic matter (COM) has been linked to ligands that have an acidic polysaccharide character (Guo et al., 2002; Quigley et al., 2002). Phytoplankton are likely to be the source of such polysaccharides since it is well known that they are capable of producing exudates that are high in polysaccharides (Alldredge et al., 1993; Santschi et al., 1998).

The high abundance of colloidal organic matter and lithogenic materials in estuaries makes these particles key in controlling the scavenging and residence time of Th radionuclides in the water column (Baskaran and Santschi, 1993; Baskaran, 1999). The abundance of metal oxides, associated with changes in redox, has also been shown to affect Th scavenging rate. For example, some of the first information on Th scavenging in the Black Sea indicated that Fe–Mn carrier phases were important in controlling the distribution of Th across the oxic–anoxic interface in the water column (Wei and Murray, 1994). Quigley et al. (1996) has also demonstrated the importance of microparticles (e.g., hematite) in scavenging Th(IV).

Thorium–particle interaction box models have included both reversible (Clegg and Whitfield, 1993) and irreversible sorption kinetics (Honeyman and Santschi, 1989, 1991). For example, the colloidal pumping model (CPM) suggested that Th scavenging begins with the smallest size fraction of particles and then is converted to the larger fraction through coagulation and flocculation processes (Honeyman and Santschi, 1989). Recent work has corroborated these model predictions showing experimentally that ^{234}Th complexed with COM (1–10 kDa) is transferred to larger (>0.1 μm) particles, confirming that coagulation is a dominant step in the transfer of ^{234}Th to the particulate phase (Quigley et al., 2001). This further emphasizes the importance of COM in binding metals in estuarine systems with high concentrations of COM. More details are provided on the abundance, sources, and characterization of colloidal or high molecular weight dissolved organic matter (DOM) and the CPM in chapters 10–14.

The distribution of ^{234}Th in sediments has proven to be a useful tracer of short-term interactions with the overlying water column as well as transport processes in sedimentary deposits in estuaries/rivers (Aller and Cochran, 1976; Aller et al., 1980; Cochran and Aller, 1979; Santschi et al., 1979; McKee et al., 1983, 1986; DeMaster et al., 1985; Corbett et al., 2004). In particle-rich estuarine systems such as the Mississippi River plume region, the spatial distribution of ^{234}Th in sediments shows that the highest inventory ratios are in close proximity to the river mouth and that sediment focusing in this region is closely associated with the plume (figure 7.9). Other work in Long Island Sound (USA) showed that the distribution of ^{234}Th could be used to examine the effects of lateral inputs of newly deposited organic matter on chlorophyll a inventories (Cochran and Hirschberg, 1991; Sun et al., 1994). The correlation between ratios of *measured* ^{234}Th versus *expected* (based on overlying water source from ^{238}U) inventories indicated that the measured inventory was greater than expected for shallow stations relative to deep stations (figure 7.10; Sun et al., 1994). Since the concentrations of dissolved ^{234}Th are low in Long Island Sound, inputs from enhanced scavenging during resuspension events in these shallow regions are

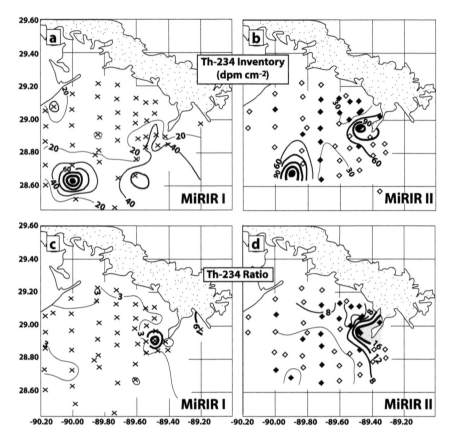

Figure 7.9 Spatial distribution of ^{234}Th in sediments in Mississippi River plume region showing that the highest inventory ratios are in close proximity to the river mouth. (Modified from Corbett et al., 2004.)

not likely to be as important as lateral inputs. Conversely, in the Hudson River estuary (USA), similarities in the ^{234}Th/^{7}Be activity ratios in bottom sediments and suspended particles indicate that resuspension controls the activity ratio of suspended particles during high-flow events (Feng et al., 1999).

Radium

Radium has four natural radionuclides in aqueous systems, ^{226}Ra ($t_{1/2} = 1620$ y), ^{228}Ra ($t_{1/2} = 5.76$ y), ^{224}Ra ($t_{1/2} = 3.66$ d), and ^{223}Ra ($t_{1/2} = 11.4$ d). Radium makes an excellent tracer in coastal systems because it has a highly particle-reactive parent (Th) (see figure 7.5 for decay pathways) and very different adsorption coefficients in fresh and salt waters (Cochran, 1980; Moore, 1992). In rivers and groundwaters Ra adsorbs strongly

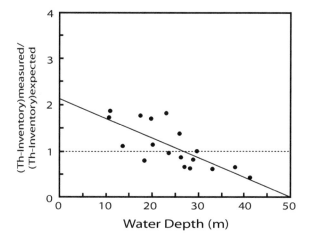

Figure 7.10 Correlation between ratios of measured ^{234}Th versus expected (based on overlying water source from ^{238}U) sediment inventories in Long Island Sound (USA). This shows that the measured inventory was greater than expected for shallow stations relative to deep stations. (Modified from Sun et al., 1994.)

to particles; however, with increasing salinity in mixing zones of estuaries desorption reactions predominate where it becomes entirely dissolved in full-strength seawater. Thus, estuarine sediments represent a constant source of Ra where it has been delivered on particles from rivers (Elsinger and Moore, 1983; Moore and Todd, 1993; Webster et al., 1995; Moore and Krest, 2004). Some of the earliest work on Ra in estuaries (Narragansett Bay, USA) indicated that ^{226}Ra and ^{228}Ra were found to be in linear concentrations in the mixing zone (Santschi et al., 1979). Other work in the Hudson River estuary and Long Island Sound showed that desorption processes in estuaries may provide a significant source of ^{226}Ra to the coastal ocean (Li et al., 1977; Li and Chan, 1979; Cochran et al., 1986). Concentrations of Ra in the mixing zone of Winyah Bay (USA) clearly demonstrate Ra activities that are greater than the conservative mixing line (figure 7.11). Waters draining salt marshes within this estuary were also shown to be significant source inputs of Ra (Bollinger and Moore, 1993; Rama and Moore, 1996). Similarly, desorption processes in sediments of large river systems, such as the Amazon River (Moore and Edmond, 1984; Key et al., 1985) and Mississippi River (Moore and Scott, 1986), have also been shown to be significant sources of Ra.

Recent studies have demonstrated that Ra radionuclides can be used to measure ground water inputs to coastal waters (Moore, 1996; Krest et al., 2000; Krest and Harvey, 2003). Other studies have used Ra radionuclides to model residence times in estuaries (Turekian et al., 1996; Charette et al., 2001; Kelly and Moran, 2002). The basic approach here is based on the balancing of ^{226}Ra inputs from groundwater with the removal of excess ^{226}Ra that results from tidal flushing. For example, in the Pettaquamscutt estuary (USA), the short-lived radionuclide ^{224}Ra was used to determine the residence time (8 ± 4 d) of

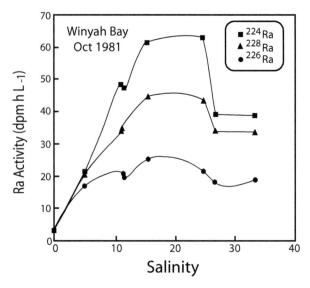

Figure 7.11 Concentrations of dissolved Ra in the mixing zone of Winyah Bay (USA), which clearly demonstrate higher Ra activities greater than the conservative mixing line. (Modified from Moore, 1992.)

this system (Kelly and Moran, 2002). Radium-derived groundwater fluxes were calculated using a simple box model; the highest fluxes occurred in summer (6.4–20 L m^{-2} d^{-1}) and the lowest in winter (2.1–6.9 L m^{-2} d^{-1}) (figure 7.12). These seasonal changes in Ra-derived groundwater fluxes were consistent with annual estimates of aquifer recharge—based on a tidal prism model and residence time of the estuary.

Radon

Radon-222 is an inert noble gas produced by alpha decay of ^{226}Ra with a short half-life of 3.85 d (see figure 7.5). It commonly occurs in disequilibrium with its parent at the sediment–water and water–atmosphere interfaces because of inert characteristics and loss by diffusion and gas bubble ebullitive stripping from sediments to overlying waters and then to the atmosphere (figure 7.13; Martens and Chanton, 1989). Thus, the primary application of ^{222}Rn in estuaries has been in estimating flux rates of pore waters across the sediment–water interface (Hammond et al., 1977, 1985; Martens et al., 1980; Smethie et al., 1981; Gruebel and Martens, 1984; Martens and Chanton, 1989) and the rate of gas exchange between the estuary and the atmosphere (Hammond and Fuller, 1979; Kipphut and Martens, 1982; Elsinger and Moore, 1983; Martens and Chanton, 1989) and more recently as a groundwater tracer (Bugna et al., 1996; Cable et al., 1996 a,b; Corbett et al., 1999, 2000).

Some of the earliest work on Rn in estuaries indicated that an enrichment in Rn in pore waters occurred from Rn recoil from solids and its overall inert natural character (Hammond et al., 1977). This work in the Hudson River estuary (USA) also concluded

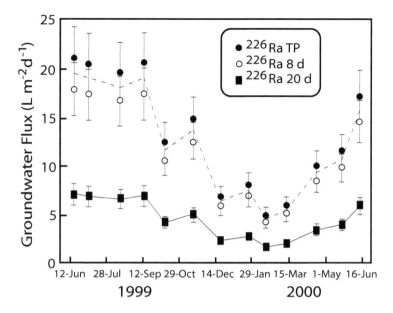

Figure 7.12 Radium-derived groundwater fluxes from Pettaquamscutt estuary (USA), calculated using a simple box model aquifer recharge, based on a tidal prism (TP) model and residence time (8 and 20 d) of the estuary. (Modified from Kelly and Moran, 2002.)

that a large percentage of the Rn in the water column was derived from molecular diffusion from sediments; advective inputs from sediment resuspension events and from tidal currents also affected these inputs. Once in the water column it was estimated that 40 to 65% of the Rn was lost to decay and evasion to the atmosphere. Similar work by Hartman and Hammond (1984) suggested that a significant amount of Rn in the water column of San Francisco Bay was lost to the atmosphere; flux to the atmosphere was largely dependent on wind speed, which was found to be the primary control mechanism in determining transfer coefficients.

Martens and Chanton (1989) showed that the disequilibrium between ^{226}Ra and ^{222}Rn in surface sediments is caused by simple diffusive loss, but can be enhanced by the stripping effects of biogenic gas bubbles produced at depth (e.g., CH_4) as they escape from sediments (ebullition). The relative percentage of diffusion versus ebullition on Rn fluxes in Cape Lookout Bight (USA) indicates that ebullition can account for as much as 48% d^{-1} of the transport of Rn from surface sediments to the atmosphere (figure 7.14). Corbett et al. (1999) reported that activities/concentrations of ^{222}Rn and CH_4 in selected regions of Florida Bay (USA) were high. The high activities/concentrations were attributed to a broad spectrum of processes such as turbulent mixing via bioturbation, bubble ebullition, plant-mediated transport, and groundwater discharge. Activities/concentration of ^{222}Rn and CH_4 in wells, solution holes, canals, and Florida Bay (USA) were also found to be strongly correlated (figure 7.15). Based on ^{222}Rn and CH_4 tracers produced from entirely

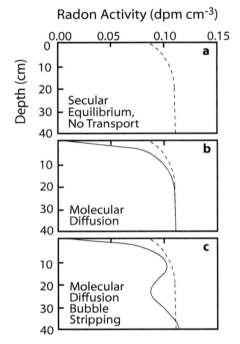

Figure 7.13 Theoretical down-core profiles of radon activity illustrating that ^{222}Rn is typically not in (a) secular equilibrium with its parent ^{226}Ra at the sediment–water and water–atmosphere interfaces because of inert characteristics and loss by (b) diffusion and (c) gas bubble ebullitive stripping from sediments to overlying waters and then to the atmosphere. (Modified from Martens and Chanton, 1989.)

different processes, it was concluded that groundwater circulation can provide significant nutrient inputs in the eastern region of Florida Bay, and that these inputs may rival nutrient inputs from surface flow across the Everglades (Corbett et al., 2000). An understanding of the mechanisms of groundwater transport is key in assessing many of the eutrophication problems currently plaguing estuarine systems—more details are provided on nutrient cycling in chapters 10–13.

Lead

Lead-210 is produced by radioactive decay of ^{222}Rn (figure 7.5) and can enter estuaries as a dissolved/complexed ion or in association with particles. Decay of ^{222}Rn to ^{210}Pb in the atmosphere is a source of atmospheric dissolved ^{210}Pb inputs to estuaries. Some of the ^{222}Rn formed by ^{226}Ra decay in soils escapes into the atmosphere where further decay through a series of short-lived radionuclides leads to ^{210}Pb ($t_{1/2} = 22.3$ y) (Appleby and Oldfield, 1992). The concentration of ^{210}Pb in the atmosphere over continents decreases in elevation due to a decrease in the concentration of ^{222}Rn from soils (Moore et al., 1973). Thus, the concentration of ^{210}Pb in the atmosphere strongly depends on longitude, depending on its location above the ocean or a continent. Lead-210 is primarily removed from the atmosphere from washout of wet and dry fallout. Seasonal variability in ^{210}Pb atmospheric fluxes has been well documented around the globe (Benninger, 1978; Krishnaswami and Lal, 1978; Nevissi, 1982; Turekian et al., 1983; Olsen et al., 1985;

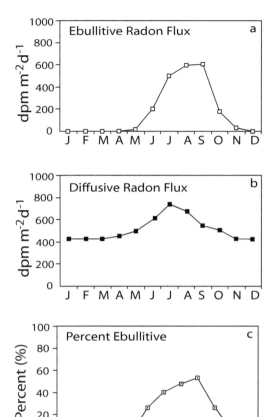

Figure 7.14 Radon-222 fluxes from (a) ebullitive versus (b) diffusive processes in Cape Lookout Bight (USA) showing that (c) ebullition can account for as much as 48% d^{-1} of the transport of Rn from surface sediments to the atmosphere. (Modified from Martens and Chanton, 1989.)

Baskaran, 1995). Lead-210 can also enter estuaries on particles from rivers as well as dissolved and particulate forms from offshore waters (Moore, 1992).

The relative importance of these different source inputs of ^{210}Pb to estuaries is a difficult process and has been the topic of numerous studies. For example, early work by Heltz et al. (1985) in Chesapeake Bay (USA) indicated that desorption of ^{226}Ra from particles entering the estuary could result in an apparent excess of ^{210}Pb—on the basis of equilibrium with respect to ^{226}Ra. Other work in Narragansett Bay (USA) demonstrated that dissolved ^{210}Pb could be scavenged from particles and that removal times were seasonally variable and linked with remobilization of ^{210}Pb in pore waters, resuspension rates, and stabilization in colloidal complexes (Santschi et al., 1979). It was also shown that 80% of the ^{210}Pb inputs to subtidal regions in Narragansett Bay were from atmospheric sources (Santschi et al., 1984). Studies on source functions have shown that atmospheric inputs versus offshore waters can be the dominant sources of ^{210}Pb to the overall budget

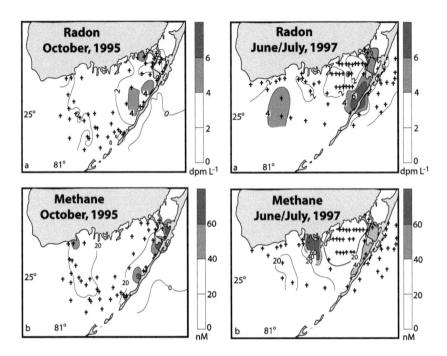

Figure 7.15 Activities/concentrations of ^{222}Rn and CH$_4$ in selected regions of Florida Bay (USA) are shown to have a strong positive correlation. (Modified from Corbett et al., 1999.)

in Long Island Sound (USA) (Benninger, 1978) and the Amazon shelf (DeMaster et al., 1986), respectively. Recent work in Framvaren Fjord (Norway) also indicated that inputs from atmospheric and terrestrial runoff were the dominant sources of ^{210}Pb (Swarzenski et al., 1999). Moreover, the vertical distribution of ^{210}Pb in the water column was largely controlled by the redox state of the fjord. Increases in dissolved ^{210}Pb at the zone of aerobic manganese reduction (AMR) indicated that Mn is more important than Fe as a carrier phase. This is primarily due to the reduction of MnO$_2$ occurring earlier than Fe oxides in the sequence of redox reactions.

Much of the work to date in estuaries and adjacent marsh/shelf environments using ^{210}Pb has been to determine sediment accumulation and accretion rates (Armentano and Woodwell, 1975; Krishnaswami et al., 1980; Church et al., 1981; Kuehl et al., 1982; Olsen et al., 1985; Paez-Osuna and Mandelli, 1985; McKee et al., 1986; Lynch et al., 1989; Bricker-Urso, 1989; Moore, 1992; Smoak et al., 1996; Dellapenna et al., 1998, 2001; Benoit, 2001; Corbett et al., 2003). Lead-210 is considered to be a reliable method for dating sediments deposited over the last 100 to 110 y (Krishnaswami et al., 1971). In the absence of bioturbation/mixing the activity gradient of excess ^{210}Pb in sediments, which is the net result of accumulation and radioactive decay, can provide information on the sedimentation rate of recent sediments. Unfortunately, in many estuaries (particularly shallow systems) physical mixing and bioturbation effects on the upper centimeters results in a degradation of the original depositional signal. To quantitatively deconvolute mixing effects and extreme changes in sedimentation rates a second particle radionuclide tracer has commonly been used in conjunction with ^{210}Pb.

One example of a dual radionuclide approach in the Sabine-Neches estuary (USA) [a shallow, turbid, high DOC (dissolved organic carbon) estuary] was the application of both ^{210}Pb and ^{239,240}Pu radionuclides (Ravichandran et al., 1995a). Plutonium has been effectively used earlier in conjunction with ^{210}Pb (Santschi et al., 1980, 1984; Jaakkola et al., 1983)—more details on Pu cycling are provided later in section 7.2.11. The lack of an exponential profile of excess ^{210}Pb (with depth) in the Sabine-Neches estuary was not from sediment mixing but was more controlled by its residence time and partitioning in the water column (Baskaran et al., 1997). More specifically, the relatively low K_d values, longer dissolved and particulate residence times for ^{210}Pb (due to binding with DOC), and shorter hydraulic residence times compared to other coastal systems, result in only partial removal of particle-reactive radionuclides from this estuary (table 7.3). In fact, the measured inventories of excess ^{210}Pb and ^{239,240}Pu in these sediments were only approximately 10 to 34% and 19 to 50%, respectively, of the total expected inventories (Ravichandran et al., 1995a). Unlike the ^{210}Pb profile, ^{239,240}Pu did correspond to the maximum fallout in 1963; this was used to estimate sedimentation rates of 4 to 5 mm y^{-1}.

McKee et al. (1983) demonstrated that ^{234}Th and ^{210}Pb can be used to determine 100-d deposition and 100-y accumulation rates, respectively, near the mouth of the Yangtze River estuary (China). The short-term deposition rate was approximately 4.4 cm per month (figure 7.16a) versus accumulation rates of approximately 5.4 cm y^{-1} (figure 7.16b), integrated over one century. Thus, the stratigraphic record over the last 100 y is incomplete

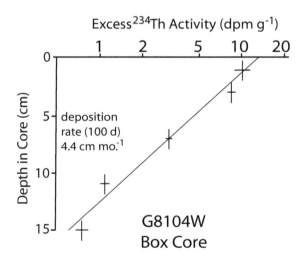

Figure 7.16a Excess ^{234}Th used to determine 100-d deposition rate near the mouth of the Yangtze River estuary (China). The short-term deposition rate was approximately 4.4 cmmo^{-1}. (Modified from McKee et al., 1983.)

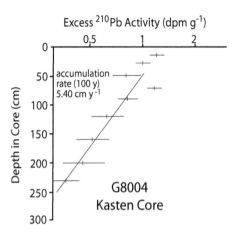

Figure 7.16b Excess ^{210}Pb used to determine 100 y accumulation rate near the mouth of the Yangtze River estuary (China). The accumulation rate was approximately 5.4 cm y^{-1} integrated over a century. (Modified from McKee et al., 1983.)

as it only represents a fraction of the sediment deposited there. As mentioned earlier, it is critical to distinguish between deposition and accumulation of particles in environments that are as dynamic as estuaries. In a similar study, Smoak et al. (1996) also demonstrated the effectiveness of using both ^{210}Pb and ^{234}Th radionuclides in understanding sedimentary processes on different timescales on the adjacent shelf of the Amazon River estuary. The general results showed that sediment inventories of the short-lived tracer (^{234}Th) were less than predicted by the inventories of the longer-lived tracer (^{210}Pb). They explained these trends by the higher abundance of ^{234}Th in mobile fluid muds and/or the water column relative to ^{210}Pb, and that supply of offshore water, scavenging efficiency, and deposition were lower in the past 2 y than the previous 100 y.

Excess ^{210}Pb activities have also been used to determine accretion rates in marsh sediments. For example, Bricker-Urso et al. (1989) tested the assumption that accretion rates of intertidal salt marshes on the coast of Rhode Island (USA) are approximately equal to rates of sea-level rise in this region. Despite having well-described exponential decay curves of excess ^{210}Pb in most marsh sediment cores, indicating a relatively constant rate of sediment accretion, some variations in the log excess ^{210}Pb indicated that accretion may not have been constant over time (figure 7.17). The sedimentation rate as described earlier is proportional to the slope of these plots. Accretion rates for these marshes indicate that both low and high marshes are maintaining pace with local sea level rise. Moreover, these accretion rates were within the range of accretion rates for other estuarine systems along the Atlantic and Gulf coasts of the United States (table 7.4). Despite having the highest accretion rates, Louisiana marshes are being rapidly lost because they experience the highest relative sea level of all coasts in the United States (Turner, 1991)—see chapter 2 for more details.

Polonium

Polonium-210 is a decay product of ^{210}Pb and is produced mainly in the water column, with some atmospheric inputs (Swarzenski et al., 1999). In the open ocean, it has been well documented that there is commonly a disequilibrium of ^{210}Pb/^{210}Po which has

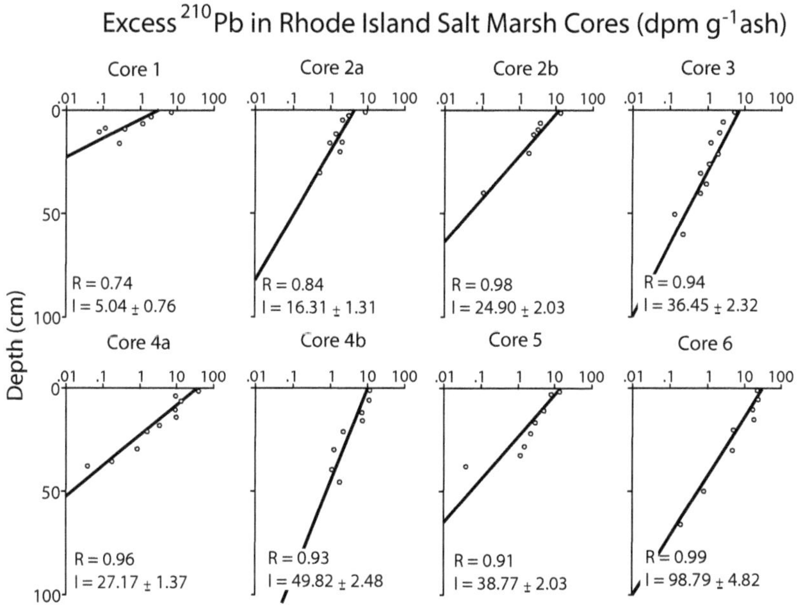

Excess ^{210}Pb in Rhode Island Salt Marsh Cores (dpm g^{-1}ash)

Figure 7.17 Estimated accretion rates of intertidal salt marshes on the coast of Rhode Island (USA). Despite having well-developed exponential decay curves of excess ^{210}Pb in most marsh sediment cores, indicating a relatively a constant rate of sediment accretion, some variations in the log excess ^{210}Pb indicate that accretion may not have been constant through time. R = correlation coefficient, I = inventories of excess ^{210}Pb (dpm cm^{-2}) (Benninger, 1979; Olsen et al., 1985). (Modified from Bricker-Urso et al., 1989.)

been attributed to the preferential uptake of Po by marine phytoplankton (Shannon et al., 1970; Fisher et al., 1983; Harada and Tsunogai, 1988). In estuaries, such as Narragansett Bay (USA), high seasonal variability in both ^{210}Pb and ^{210}Po was attributed to remobilization from sediments and formation of organic complexes (Santschi et al., 1979). Other work in central Florida has found that sulfide-rich groundwaters can have an excess of ^{210}Po (Harada et al., 1989), suggesting some linkage with redox reactions. More recent work has also demonstrated, in a permanently anoxic fjord (Framvaren, Norway), that ^{210}Po is largely controlled by the intense cycling of sulfur microorganisms. More specifically, the removal of ^{210}Po from the ^{210}Po-depleted upper water column is largely controlled by particle scavenging and settling. During particle settlement in the water column, ^{210}Po in the carrier-phase oxyhydroxides are dissolved in association with the activity of anoxygenic phototrophic microorganisms (*Chromatium/Chlorobium* spp.) at the O$_2$/H$_2$S interface (figure 7.18a; Swarzenski et al., 1999). Similar patterns are described for the vertical distribution of ^{210}Pb in this fjord; however, ^{210}Po was scavenged more rapidly than ^{210}Pb by organic and inorganic particles (figure 7.18b). Hence, there is clearly an important redox/microbial component in controlling the phase dynamics of both ^{210}Po and ^{210}Pb in this anoxic fjord (Swarzenski et al., 1999).

Table 7.4 Marsh accretion rates and sea-level rise rates.

Location (USA)	Marsh type[a]/ reference[b]	Accretion rate (cm y^{-1})	Method	SLR[c] (20)
Barnstable Harbor	S.a.[1]	0.15–0.27	[14]C	0.23
	S.a.[1]	0.34–0.79	Historical data	
	S.a.[1]	1.8	Stratigraphy	
Narragansett Bay	S.p.[2]	0.24	[210]Pb (constant flux)	0.26
	S.a.[2]	0.25–0.6	[210]Pb (constant flux)	
Delaware Bay	S.p.[3]	0.44–0.59	Concentration data (Pb)	0.30
	S.p.[4]	0.47	[210]Pb	
	S.a.[5]	0.42–0.78	[210]Pb	
	S.a.[6]	0.32–0.45	[210]Pb	
	S.a.[6]	0.26–0.43	[137]Cs	
	S.a.[6]	0.40	Pollen	
Long Island Sound	S.p.[7]	0.35	[210]Pb	0.22
	S.p.[8]	0.2–0.66	Glitter layers	
	S.p., S.a.[9]	0.54–0.81	Stratigraphy	
	S.a.[10]	0.47–0.63	[210]Pb	
	S.a.[11]	0.2–0.43	Glitter layers	
	S.a.[12]	0.25–0.47	Stratigraphy	
Chesapeake Bay	N.A.[13]	0.18–0.75	Pollen	0.35
Georgia	S.a.[14]	0.26–1.5	[210]Pb, [239,240]Pu	0.27
South Carolina	S.a.[15]	0.14–0.45	[210]Pb	0.34
	S.a.[15]	0.13–0.25	[137]Cs	
Louisiana	S.a.[16]	0.59–1.4	[137]Cs	0.92
	S.a.[17]	1.35	[137]Cs	
	S.a.[18]	0.75–1.35	[137]Cs	
	N.A.[19]	0.81–1.4	[210]Pb, [137]Cs	

[a]S.a. = *Spartina alterniflora* or low marsh; S.p. = *S. patens* or high marsh; N.A. = type of marsh not identified.
[b]References: [1]Redfield (1972); [2]Bricker-Urso et al. (1989); [3]Drier (1982); [4]Church et al. (1981); [5]Chrzastowski et al. (1987); [6]McCaffrey and Thomson (1980); [7]Harrison and Bloom (1977); [8]Siccama and Porter (1972); [9]Armentano and Woodwell (1975); [10]Richard (1978); [11]Flessa et al. (1977); [12]Ward et al. (1986); [13]Goldberg et al. (1979); [14]Sharma et al. (1987); [15]Hatton et al. (1983); [16]Delaune et al. (1981); [17]Delaune et al. (1978); [18]Delaune et al. (1987); [19]Hicks et al. (1983).
[c]SLR = relative sea-level rise rate in 1940–1980 (also in cm y^{-1}).
Modified from Bricker-Urso et al. (1989).

Beryllium

The cosmogenic radionuclide ^{7}Be ($t_{1/2}$ = 53.3 d) has effectively been used as a tracer of short-term particle cycling and depositional processes in estuarine systems (Aaboe et al., 1981; Martin et al., 1986; Olsen et al., 1986; Dibb and Rice, 1989a,b;

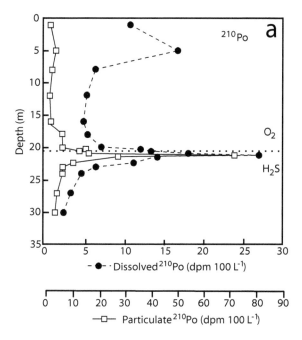

Figure 7.18a Dissolved and particulate ^{210}Po profiles in a permanently anoxic fjord (Framvaren, Norway), illustrating that ^{210}Po is largely controlled by the intense cycling of sulfur microorganisms at the O_2/H_2S interface. (Modified from Swarzenski et al., 1999.)

Baskaran and Santschi, 1993; Baskaran et al., 1993, 1997; Feng et al., 1999; Sommerfield et al., 1999; Allison et al., 2000; Corbett et al., 2003). Beryllium-7 is produced in the Earth's atmosphere by cosmic-ray spallation of nitrogen and oxygen nuclei and then rapidly scavenged by particles and aerosols in the atmosphere (Lal et al., 1958; Lal and Peters, 1967; Larsen and Cutshall, 1981; Olsen et al., 1981). It is flushed from the atmosphere and deposited on the Earth's surface by precipitation and dry deposition (Larsen and Cutshall, 1981). When reacting with acidic rainwater ^{7}Be is solubilized with Be^{2+} which is the predominant ion being delivered from the atmosphere; its small ionic radius (0.34 Å) and 2^+ cationic state allow it to become quickly scavenged by fine particles (Bloom and Crecelius, 1983; Olsen et al., 1985). Seasonal increases in the depositional flux of ^{7}Be are commonly attributed to injection of ^{7}Be-enriched stratospheric air to the tropospause (Olsen et al., 1985; Todd et al., 1989). Since ^{7}Be is produced cosmogenically, its depositional flux has a latitudinal dependence, and its concentration increases with increasing altitude (Lal and Peters, 1967). In general, seasonal variability in depositional fluxes of ^{7}Be were correlated with rainfall at collection stations near Galveston Bay (USA) (Baskaran et al., 1993; Baskaran, 1995), Cape Lookout Bight (USA) (Canuel et al., 1990), Long Island Sound (USA) (Turekian et al., 1983), and Chesapeake Bay (USA) (Olsen et al., 1985; Dibb, 1989).

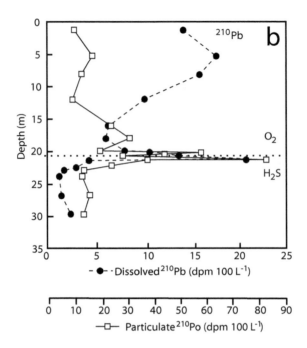

Figure 7.18b Vertical distribution of dissolved ^{210}Pb and particulate ^{210}Po in a permanently anoxic fjord (Framvaren, Norway) showing that there is clearly an important redox/microbial component in controlling the phase dynamics of both ^{210}Po and ^{210}Pb in this anoxic fjord. (Modified from Swarzenski et al., 1999.)

Past work has shown that ^{7}Be can be used as a tracer of river-borne sedimentary particles to estuaries (Dibb and Rice, 1989a,b; Baskaran et al., 1997) and continental shelves (Sommerfield et al., 1999; Allison et al., 2000; Corbett et al., 2003). For example, ^{7}Be sediment inventories at a series of stations in the main stem of Chesapeake Bay (figure 7.19a; Dibb and Rice, 1989a,b) indicate that lower inventories at the SUSQ (Susquehanna) station relative to the higher inventories at the BALT (Baltimore) and CALV (Calvert Cliffs) stations in April 1986 are likely due to inputs of ^{7}Be-rich sediments from the head of the Bay followed by redeposition further down in the estuary (figure 7.19b). Moreover, higher ^{7}Be inventories in upper bay stations are due to enhanced settlement of suspended particles at the head of the estuary during reduced river flow periods in the Susquehanna River. Riverine inputs of atmospherically derived ^{7}Be in estuaries of large rivers, such as the Mississippi, are even more pronounced. For example, ^{7}Be inputs from the drainage basin of the Mississippi River are significantly greater than any direct atmospheric inputs to the lower estuary; ^{7}Be in this region represents an indicator of terrestrially derived sediments (Corbett et al., 2003). Conversely, in other estuarine systems with considerably smaller drainage basins, such as the James River estuary (USA), the riverine input of atmospherically delivered ^{7}Be is less than 5% of the atmospheric inputs to the estuarine surface (Olsen et al., 1986).

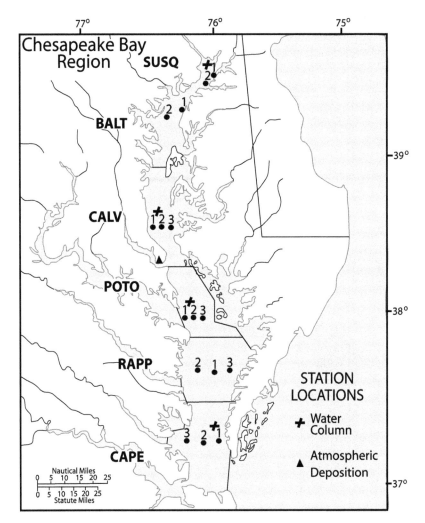

Figure 7.19a Series of stations in the main stem of Chesapeake Bay (USA) where ^{7}Be sediment inventories were measured. SUSQ = Susquehanna; BALT = Baltimore; CALV = Calvert; POTO = Potomac; RAPP = Rappahannock; CAPE = Cape Charles. (Modified from Dibb and Rice, 1989a,b.)

As discussed earlier, many of the particle-reactive nuclides like ^{7}Be can form complexes with dissolved organic ligands that can affect their residence time in the estuary as well as delivery from the water column to surface sediments. The Sabine-Neches estuary of southeast Texas has a short hydraulic residence time (ca. 10 d) and a low sediment inventory of particle-reactive nuclides such as ^{7}Be (Baskaran et al., 1997). The range of dissolved residence times of ^{7}Be (0.6–9.6 d) in this estuary was found to be relatively

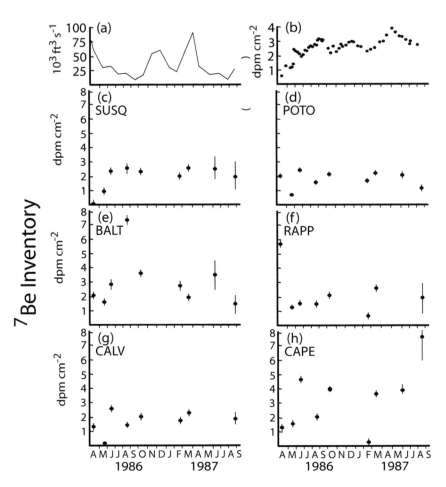

Figure 7.19b Susquehanna River discharge (a), ^{7}Be atmospheric deposition (b), and ^{7}Be inventories in Chesapeake Bay (USA) (c–h), showing lower inventories at the SUSQ station relative to the higher inventories at the BALT and CALV stations due to inputs of ^{7}Be-rich sediments from the head of the Bay followed by redeposition farther down in the estuary. (Modified from Dibb and Rice, 1989.)

long compared to those in other estuarine systems. Moreover, the range of distribution coefficients (K_d) of ^{7}Be was lower (0.15–8.7 10^4 cm^3 g^{-1}) than in most estuaries, indicating that most of the ^{7}Be was in the dissolved fraction. This estuary has also been shown to have high concentrations of DOC, capable of binding particle-reactive nuclides (Bianchi et al., 1997a). Thus, ^{7}Be is likely complexed in the DOC pool, which results in a longer residence time in the water column for dissolved ^{7}Be, allowing for the rapid removal of dissolved ^{7}Be in this estuary with a short hydraulic residence time. The rapid flushing of dissolved ^{7}Be (complexed in DOC) out of the estuary also explains for the low sediment inventory of ^{7}Be. The residence times and distribution coefficients of ^{7}Be in other estuarine systems shows a wide range of values (table 7.3), in part, due to the diversity of the concentration and sources of particulate and dissolved constituents across different estuarine systems.

Similar to ^{234} Th, downcore profiles of ^{7}Be can also be used to determine seasonal changes in sedimentation and sediment mixing rates in estuaries (Canuel et al., 1990). The basic assumption here, as described earlier, is that the nuclide (e.g., ^{7}Be) traces movements of particles during sediment accumulation and that the delivery and trapping of the nuclide to surface sediments is uniform across habitats within an estuary. The three basic processes controlling the depth distribution are: (1) supply rate from sedimentation; (2) radioactive decay; and (3) postdepositional particle mixing processes. Finally, it should be noted that using ^{7}Be for the aforementioned purposes also requires concurrent measurement of ^{7}Be in atmospheric fallout (Canuel et al., 1990).

Cesium

Nuclear weapons testing began in the mid-1940s with the first significant megaton yield in 1952 (Carter and Moghissi, 1977). This resulted in the first significant fallout of ^{137}Cs ($t_{1/2} = 30$ y) which occurred in the early 1950s followed by a maximum fallout in years 1962–1964 (peak 1963), then decreasing to essentially zero in the early 1980s. However, new inputs of ^{137}Cs occurred in association with the Chernobyl nuclear power plant accident in 1986. In estuaries that receive nuclear power effluent, such as the Hudson River estuary (USA), fallout ^{137}Cs can be distinguished from reactor-derived ^{137}Cs by its association with other reactor nuclides such as ^{134}Cs ($t_{1/2} \cong 2$ y) and ^{60}Co ($t_{1/2} \cong 5$ y) (Olsen et al., 1981). Cesium-137 has proven to be a useful tool for measuring sediment accumulation rates in estuaries (Simpson et al., 1976; Olsen et al., 1981; Ravichandran et al., 1995a,b) and accretion rates in wetlands (Delaune et al., 1978, 1983; Nixon, 1980; Hatton et al., 1983; Sharma et al., 1987; Lynch et al., 1989; Milan et al., 1995). Table 7.4 shows that ^{137}Cs compares well with other methods in determining marsh accretion rates.

Cesium-137 is an *impulse marker* where the subsurface peak of maximum activity is used to measure accumulation rates, and as a check against rates measured with other tracers. However, the resolution of this peak can be altered by physical and biological mixing processes. For example, feeding activities of benthic macrofauna can affect vertical movement of ^{137}Cs in sediments (Robbins et al., 1979). Thus, the best environments for using this tracer are high sedimentation environments with rapid burial and less disturbance of the peak depth. However, other work has shown that there are additional problems associated with the mobilization and downward diffusion of ^{137}Cs (Jaakkola et al., 1983), with more desorption occurring in higher-salinity regions of estuaries (Olsen et al., 1981). Despite further retrospective studies examining the associated remobilization problems with both ^{137}Cs and ^{210}Pb (Sugai et al., 1994), more work is needed to resolve these complex issues (Benoit et al., 2001). Using a dual-tracer approach (using both ^{137}Cs and ^{210}Pb) in sediment cores collected from the lower Chesapeake Bay, the maximum depth of ^{137}Cs was used to corroborate the physical mixing depths established with excess ^{210}Pb (figure 7.20; Dellapenna et al., 1998).

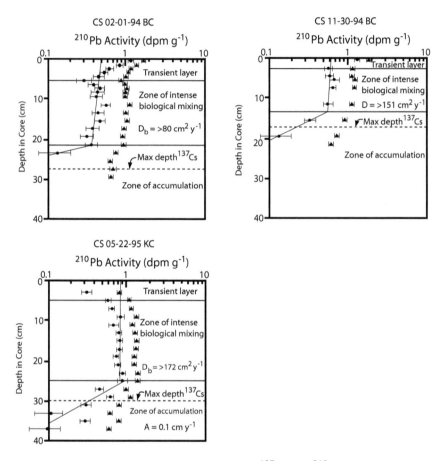

Figure 7.20 A dual-tracer approach (using both ^{137}Cs and ^{210}Pb) in sediment cores collected from lower Chesapeake Bay (USA). The maximum depth of ^{137}Cs was used to corroborate the physical mixing depths established with excess ^{210}Pb. ● = excess activity; ▲ = total activity. (Modified from Dellapenna et al., 1998.)

Plutonium

Isotopic ratios of plutonium have been used as effective geochemical tracers in estuaries over the past two decades (Benninger et al., 1979; Krishnaswami et al., 1980; Linsalata et al., 1980; Olsen et al., 1981a,b; Santschi et al., 1984; Baskaran et al., 1995; Ravichandran et al., 1995a,b). Of the seven long-lived plutonium isotopes, the four of primary concern in environmental studies have been ^{238}Pu ($t_{1/2} = 87.7$ y), ^{239}Pu ($t_{1/2} = 2.41 \times 10^4$ y), ^{240}Pu ($t_{1/2} = 6,571$ y), and ^{241}Pu ($t_{1/2} = 14.4$ y) (Baskaran et al., 1995). Major sources of Pu to both terrestrial and aquatic ecosystems are: (1) atomic weapons testing (Aarkrog, 1988); (2) burn-up of a ^{238}Pu-powered satellite (SNAP 9A) over the Indian Ocean in 1964 (Hardy et al., 1973); and (3) radioactive disposal and leakage from power plants (Sholkovitz, 1983). The activity ratio of $^{238/239,240}$Pu in the northern hemisphere fallout prior to the SNAP 9A burn-out was approximately 0.024 versus 0.05 afterward (Hardy et al., 1973). Similar to ^{137}Cs, the maximum subsurface concentration in ^{239,240}Pu represents the 1963 peak fallout layer (Krishnaswami et al., 1980; Santschi et al., 1980, 1984; Jaakkola et al., 1983). However, as mentioned earlier ^{239,240}Pu has been shown to be affected less than ^{137}Cs by mobility problems (Santschi et al., 1983). Moreover, ^{239,240}Pu has been used as a second tracer with ^{210}Pb to quantitatively separate mixing from sedimentation effects (Santschi et al., 1980, 1984). Using depth profiles of ^{239,240}Pu in the Sabine-Neches estuary, assuming the largest subsurface peak to be the 1963 fallout, the average sedimentation rates at three stations were estimated to range from 4 to 5 mm y^{-1} (figure 7.21; Ravichandran et al., 1995a). The time interval between the fallout peak in 1963 and the time these stations were sampled (1992) was 29 y. This sedimentation rate in the Sabine-Neches estuary is consistent with average sea-level rise in Gulf of Mexico estuaries (6 mm y^{-1}) at this time (Turner, 1991).

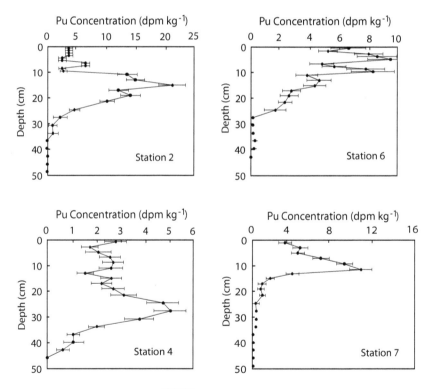

Figure 7.21 Depth profiles of ^{239,240}Pu from four stations in the Sabine–Neches estuary (USA), assuming the largest subsurface peak to be the 1963 fallout; the average sedimentation rate at three stations ranged from 0.3 to 10 mm y^{-1}. (Modified from Ravichandran et al., 1995a.)

Radiocarbon

The existence of radiocarbon (^{14}C) was not realized until 1934, when an unknown radionuclide was formed during exposure of nitrogen to neutrons in a cloud chamber (Kurie, 1934). In 1940, Martin Kamen confirmed the existence of ^{14}C when he prepared a measurable quantity of ^{14}C. Over the next few decades more details on the production rate of ^{14}C in the atmosphere and possible applications for dating archeological samples continued (Anderson et al., 1947; Arnold and Libby, 1949; Anderson and Libby, 1951; Kamen, 1963; Ralph, 1971; Libby, 1982).

Similar to the other cosmogenic radionuclides ^{14}C is produced by reaction of cosmic rays with atmospheric atoms such as N_2, O_2, and others to produce broken pieces of nuclei, which are called *spallation* products (Suess, 1958, 1968). Some of these spallation products are neutrons that can also interact with atmospheric atoms to produce additional products, which include ^{14}C and other radionuclides (^{3}H, ^{10}Be, ^{26}Al, ^{36}Cl, ^{39}Ar, and ^{81}Kr) (figure 7.22; Broecker and Peng, 1982). Thus, the reaction between neutrons and nitrogen ($^{14}N + n \rightarrow {}^{14}C + p$) is the dominant mechanism for the formation of ^{14}C in the atmosphere. Once formed, ^{14}C decay occurs by the following reaction, and has a half-life of 5730 ± 40 y:

$$^{14}C \rightarrow {}^{14}N + \beta^- + \text{neutrino} \tag{7.25}$$

The free ^{14}C atoms formed in the atmosphere become oxidized and form $^{14}CO_2$ which is rapidly mixed throughout the atmosphere (Libby, 1952). This ^{14}C then becomes incorporated into other reservoirs such as the biosphere, as plants fix carbon during the process of photosynthesis; the exchange between the atmosphere and the surface ocean is estimated to take approximately 5 y (Broecker and Peng, 1982).

Living plants and animals in the biosphere contain a constant level of ^{14}C; however, when they die there is no further exchange with the atmosphere and the activity

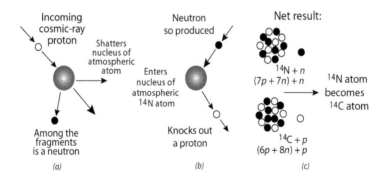

Figure 7.22 ^{14}C is produced by reaction of cosmic rays with atmospheric atoms such as N_2, O_2, and others to produce broken pieces of nuclei, which are called spallation products. The reaction between neutrons and nitrogen ($^{14}N+n \rightarrow {}^{14}C+p$) is the dominant mechanism for the formation of ^{14}C in the atmosphere. (Modified from Broecker and Peng, 1982.)

of ^{14}C decreases with a half-life of 5730 ± 40 y; this provides the basis for establishing the age of archeological objects and fossil remains. The assumptions associated with dating materials are: (1) that the initial activity of ^{14}C in plants and animals is a known constant and is independent of geographic location; and (2) that the sample has not been contaminated with modern ^{14}C (Faure, 1986). Unfortunately, measurements of ^{14}C in wood samples dated by dendrochronology indicated that there have been changes in the initial ^{14}C content over time (Anderson and Libby, 1951). Variations in the atmospheric ratio of ^{14}C/C in CO_2 (expressed in Δ notation, see below) over the last 1000 y are shown in figure 7.23 (Stuiver and Quay, 1981). This variation is believed to result from a combination of the following factors: (1) changes in cosmic-ray flux from activities in the Sun; (2) changes in the Earth's magnetic field; and (3) changes in the reservoir of carbon on Earth (Faure, 1986). Anthropogenic effects have both decreased and increased (by almost double) the atmospheric content of ^{14}C, due to combustion of fossil fuels over the past 100 y and nuclear explosions, respectively. Such variations in atmospheric ^{14}C content over the past 1000 y are quite evident. The dilution effect from inputs of fossil fuel combustion are clear after about 1850 in the early stages of the Industrial Revolution— this dilution is referred to as the *Suess effect* (Suess, E, 1906; Suess, H.E., 1958, 1968). The dilution stems from the fact that fossil carbon has been stored for such long periods underground, isolated from the atmosphere, that the ^{14}C signal has disappeared over time making it ^{14}C free. Two earlier times of anomalously high values of ^{14}C, at about 1710 and 1500, are referred to as the *de Vries effect*, and its causes are not known. The equation used for determining delta ^{14}C is defined as:

$$\Delta^{14}C = \left[\left(^{14}C/C\right)_{sample} - \left(^{14}C/C\right)_{standard} \right] / \left[\left(^{14}C/C\right)_{standard} \right] \times 1000 - IF$$

(7.26)

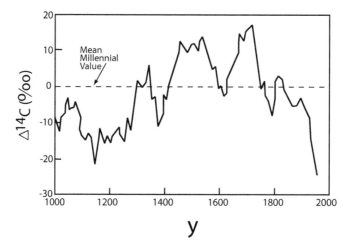

Figure 7.23 Variations in the atmospheric ratio of ^{14}C/C (per mil, ‰) in CO_2 over the last 1000 years (Modified from Stuiver and Quay, 1981.)

The ratio of a sample is measured in relation to a standard to improve the accuracy and precision of accelerator mass spectrometry measurements (Elmore and Phillips, 1987). Multiplying the ratio by 1000 results in the delta (del) values having units of parts per thousand, also know as per mil (‰). For standards, it is necessary to use wood from trees harvested before about 1850 pre-industrial, to avoid the Suess effects. The standard value for pre-industrialized atmospheric CO_2 is 13.56 dpm g^{-1} or $^{14}C/C$ equals 1.176×10^{-12} (Broecker and Peng, 1982). A correction term involving the effects of isotopic fractionation (*IF*) are also subtracted out of this equation. Isotopes are fractionated due to physical and chemical reactions (more details in the following section), thereby making the abundance of carbon isotopes (^{12}C, ^{13}C, and ^{14}C) different in plants (Faure, 1986). The National Bureau of Standards currently provides an oxalic acid ^{14}C standard that is used for this correction; however, there have been many problems associated with development of this standard (Craig, 1954, 1961; Stuiver and Polach, 1977).

The application of ^{14}C measurements in organic carbon cycling studies has been extensive in oceanic environments (Williams and Gordon, 1970; Williams and Druffel, 1987; Druffel et al., 1992; Wang et al., 1996; Bauer and Druffel, 1998; Bauer et al., 1998; Bauer, 2002). For example, Broecker and Peng (1982) used the ^{14}C in extracted CO_2 from oceanic water to determine the age and circulation patterns between surface and bottom waters. While only a few studies using such techniques in organic carbon cycling were applied early on in coastal and estuarine regions (Spiker and Rubin, 1975; Hedges et al., 1986), there has been a considerable increase in recent years (Santschi et al., 1995; Guo et al., 1996; Guo and Santschi, 1997; Cherrier et al., 1999; Mitra et al., 2000a; Raymond and Bauer, 2001a,b,c). The general paradigm emerging in river/estuarine systems is that DOC is always more ^{14}C-enriched (or younger) than particulate organic carbon (POC) (figure 7.24; Raymond and Bauer, 2001a). This is believed to be derived from fresh

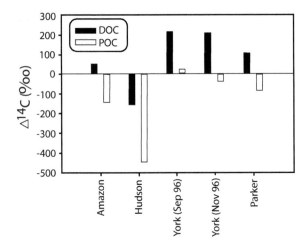

Figure 7.24 Radiocarbon data from the Amazon (Brazil), Hudson (USA), York (USA), and Parker (USA) river/estuarine systems showing that DOC is always more ^{14}C-enriched (or younger) than POC. (Modified from Raymond and Bauer, 2001a.)

leachate of surface soil litter (Hedges et al., 1986; Raymond and Bauer, 2001a). Under-standing the linkage between the carbon sources in drainage basin soils and river/estuarine systems is critical in determining the connection between terrestrial and aquatic systems. The residence time of organic carbon in soils has been examined using ^{14}C measurements (O'Brien, 1986; Trumbore et al., 1989; Schiff et al., 1990; Trumbore, 2000). These studies support the concept that ^{14}C-enriched DOC from surface soils is exported to streams and is younger than the soil organic carbon from which it is derived.

The heterogeneity of organic matrices in soils, sediments, and dissolved and suspended particulate materials complicates ^{14}C age-based determinations. There is considerable uncertainty in organic carbon measurements in the fractions that have ^{14}C less than contemporary. Bulk ^{14}C measurements mask the problems of heterogeneous samples that consist of a mixture of new and old carbon; different mixtures of these compo-nents can result in very different age determinations. Recently developed methods, such as automated preparative capillary gas chromatography (PCGC), now allow for sepa-ration of target compounds for ^{14}C analysis based on acceleration mass spectrometry (AMS) (Eglinton et al., 1996, 1997; McNichol et al., 2000). In very general terms, compound specific-isotope analysis (CSIA) allows for accurate determination of ages of compounds that are specific to a particular source (e.g., phytoplankton) within the heterogeneous matrix of other carbon compounds commonly found in sediments (e.g., terrigenous sources). Similarly, the application of CSIA to stable-isotope work has also proven to be useful in distinguishing between different types of organic carbon sources in estuarine systems (Goni et al., 1997, 1998). More details on how these compound and class-specific isotopic methods are used in characterizing organic carbon and nitrogen sources in estuaries are provided in chapter 9.

Stable Isotopes

The basic principle surrounding the application of stable isotopes in natural ecosystems is based on variations in the relative abundance of lighter isotopes from chemical rather than nuclear processes (Hoefs, 1980). Due to the faster reaction kinetics of the lighter isotope of an element, reaction products in nature can be enriched in the lighter isotope. These fractionation processes can be complex, as discussed below, but have proven to be useful in determining geothermometry and paleoclimatology, as well as sources of organic matter in ecological studies. The most common stable isotopes used in oceanic and estuarine studies are ^{18}O, ^{2}H, ^{13}C, ^{15}N, and ^{34}S. The preference for using such isotopes is related to their low atomic mass, significant mass differences in isotopes, covalent character in bonding, multiple oxidation states, and sufficient abundance of the rare isotope. In particular, studies focusing on the natural abundances of carbon and nitrogen isotopes have been successfully used in lake and coastal/estuarine systems to trace source inputs of terrestrial and aquatic organic matter (Sweeney et al., 1980; Peterson et al., 1985; Williams and Druffel, 1987; Cifuentes et al., 1988; Horrigan et al., 1990; Westerhausen et al., 1993) and sewage and nutrients (Voss and Struck, 1997; Caraco et al., 1998; Holmes et al., 2000; Hughes et al., 2000). The basis for using stable isotopes in food web analyses is that the organisms retain the isotope signals of the foods they assimilate; this phenomenon was demonstrated in the classic work by DeNiro and Epstein (1978, 1981). However, other studies have shown problems when attempting to distinguish stable isotopic signatures with seasonal changes in food resources (Cifuentes et al., 1988; Fry and Wainright, 1991; Fogel et al., 1992; Currin et al., 1995), effects of isotopic shifts between trophic levels (Fry and Sherr, 1984; Peterson et al., 1986), and decomposition effects in litter and during sediment burial on isotopic composition (DeNiro and Epstein, 1978, 1981; Benner et al., 1987; Jasper and Hayes, 1990; Meyers, 1994; Montoya, 1994; Sachs et al., 1999), making the application of stable isotopes in estuarine food webs very complex.

The average relative abundances of isotopes in the Earth's crust, oceans, and atmosphere, commonly expressed as stable isotope ratios, are shown in table 7.5. Small differences in the ratios of a particular element in natural samples can be detected using *mass spectrometry*; however, it cannot be achieved with high *precision* and *accuracy* (Nier, 1947). The solution to this problem, as explained earlier for ^{14}C measurements, is measuring isotope ratios in a sample concurrently with the standard; this does allow for adequate precision and accuracy. The equation used to describe this relative difference or del (δ) value is as follows:

$$\delta(\%o) = [R_{sample} - R_{standard}/R_{standard}] \times 1000 \qquad (7.27)$$

where: R = isotope ratio whereby the most abundant isotope is the denominator. As described earlier, since the equation is multiplied by 1000, δ has units of parts per thousand (‰). A sample is considered to be *enriched* when the heavy isotope is more abundant relative to the standard, resulting in positive del values. Conversely, del values will be negative, or *depleted*, when the sample has less of the heavy isotope than the standard.

Table 7.5 Relative abundances of some stable isotopes.[a]

Atomic number	Symbol	Mass number	Abundance (%)
1	H	1	99.99
		2	0.01
6	C	12	98.9
		13	1.1
		14[b]	10^{-10}
7	N	14	99.6
		15	0.4
8	O	16	99.8
		17	0.04
		18	0.2
16	S	32	95.0
		33	0.8
		34	4.2
		36	0.2

[a]These values are averages that are representative of the Earth's crust, ocean, and atmosphere. They have been rounded to the nearest 0.1%, except for the very rare isotopes.
[b]Radioactive.
Modified from Libes (1992).

Table 7.6 Internationally accepted stable isotope standards for hydrogen, carbon, oxygen, nitrogen, and sulfur.

Element	Standard	Abbreviation
H	Standard Mean Ocean Water	SMOW
C	*Belemnitella americana* from the Cretaceous Peedee formation, South Carolina (USA)	PDB
N	Atmospheric N_2	—
O	Standard Mean Ocean Water	SMOW
	B. americana from the Cretaceous Peedee formation, South Carolina	PDB
S	Triolite (FeS) from the Canyon Diablo iron meteorite	CD

Modified from Libes (1992).

Finally, the del value is zero when the sample and the standard have the same isotopic composition. The international standards for O, C, H, N, and S are Standard Mean Ocean Water (SMOW), Pee Dee Belemnite (PDB) carbonate fossil, SMOW, atmospheric nitrogen (N_2), and the Canyon Diablo Triolite (CDT) iron meteorite, respectively (table 7.6; Faure, 1986). More specifically, the $^{13}C/^{12}C$ of natural materials and standards is determined using isotope-ratio mass spectrometry (for details, see Hayes, 1983; Boutton, 1991).

Isotopic Fractionations

Isotopic fractionation can occur from both *kinetic* and *equilibrium* effects. In its simplest and most intuitive terms, physical processes affecting kinetic fractionation are largely due to the higher energy and more rapid diffusion and/or phase changes (e.g., evaporation) of the lighter isotope of a particular element. Thus, these fractionations are not in equilibrium. The energy of a molecule can be described by *electronic, nuclear spin, translational, rotational*, and *vibrational* properties (Faure, 1986; Fogel and Cifuentes, 1993; Chacko et al., 2001). Biologically mediated reactions involving enzymes kinetically fractionate isotopes, resulting in significant differences between substrates (or reactants) and products. Assuming that the substrate is not limiting to a reaction and the isotope ratio of the product is measured within a short period of time, the fractionation factor (α) can be defined as follows:

$$\alpha = R_p/R_s \qquad (7.28)$$

where: $R_p =$ isotope ratio for the product; and $R_s =$ isotope ratio for the substrate (or reactant).

Therefore, biologically mediated kinetic fractionation, where the products will have more of the lighter isotope in them due to differential partitioning from higher energy properties, will result in a net *depletion* of the heavy isotope in the product and a more negative del value. Moreover, kinetic fractionation occurs in reactions that are unidirectional, where the reaction rates actually depend on the isotopic composition of the substrates and products. Thus, the isotopic ratio of the substrate is related to the amount of substrate that has not been used (Mariotti et al., 1981)—this is described by the *Rayleigh* equation:

$$R_s/R_{s0} = f^{(\alpha - 1)} \tag{7.29}$$

where: R_{s0} = isotope ratio of the substrate at time zero; and f = fraction of unreacted substrate.

In equilibrium fractionation, where the isotopic effects are in equilibrium, there is an isotopic exchange involving the redistribution of isotopes of an element in different molecules (Faure, 1986). The equilibrium fraction can be defined as it relates to kinetic isotope effects:

$$\alpha_{eq} = k_2/k_1 \tag{7.30}$$

Carbon

Stable carbon isotopes have been commonly used to distinguish between *allochthonous* versus *autochthonous* organic carbon inputs to estuaries. One of the most important pieces of information that can be gathered from this information is the separation of C_3 and C_4 plant inputs (Peterson and Fry, 1989; Goni et al., 1998; Bianchi et al., 2002b). Other pioneering work on the application of stable carbon isotopes in estuaries showed that these isotopes can be used to understand *trophic interactions* in food chain dynamics (Fry et al., 1977; Fry and Parker, 1979, 1984; Fry and Sherr, 1984). In general, estuarine studies have shown that ^{13}C moves in a predictable way, where there is a 1 to 2‰ enrichment of ^{13}C for each step in a predator/prey interaction (Parsons and Lee Chen, 1995). Organic carbon is composed of a heterogeneous mixture of organic compounds having different ^{13}C values due to their biosynthetic pathways. Moreover, some classes of organic compounds (e.g., polysaccharides and proteins) are less decay resistant (labile) than others. For example, polysaccharides and proteins tend to be more ^{13}C-enriched than lipids (Deines, 1980; Hayes, 1993; Schouten et al., 1998). During microbial degradation of POC, removal of the more labile ^{13}C-enriched cellulose components results in the depletion of $\delta^{13}C$ in the remaining lignin-rich residue (Benner et al., 1987). Clearly, these general trends get considerably more complicated when factoring in different food sources (e.g., terrestrial versus aquatic) as well as size and physiology of the organisms involved (Incze et al., 1982; Hughes and Sherr, 1983; Goering et al., 1990; Megens et al., 2002). Other trends, such as seasonal depletion of $\delta^{13}C$ in estuarine food webs, have been attributed to variability in imports of DIC from freshwater sources (Simenstad and Wissmar, 1985). Isotopic fractionation distinguished on the basis of fractionation through photosynthetic pathways or trophic interactions provides the basis for discrimination of carbon sources. More specifically, Hayes (1993) concluded that the carbon isotopic composition of naturally synthesized compounds is controlled by: (1) carbon sources; (2) isotopic effects during assimilation processes in producer organisms; (3) isotope effects during metabolism and biosynthesis; and (4) cellular carbon budgets.

The classic work by Park and Epstein (1960) and others (see O'Leary, 1981, and Fogel and Cifuentes, 1993, for review) established that the enzyme ribulose 1,5-bisphosphate carboxylase (RuBP carboxylase) controls fractionation of carbon isotopes during photosynthesis in plants. The C_3 pathway is when atmospheric or dissolved CO_2 fixation results in production of two three-carbon compounds (3-phosphoglycerate) from the enzymatic conversion of RuBP. Through extensive laboratory studies, it was established that the isotopic fractionation between photosynthetic carbon ($\delta^{13}C = -27‰$) and atmospheric or dissolved carbon ($\delta^{13}C = -7‰$) in C_3 plants is approximately $-20‰$ (Stuiver, 1978; Guy et al., 1987; O'Leary, 1988). Based on this fraction information the following model for carbon isotopic fractionation in C_3 plants was established:

$$\Delta = a + (c_i/c_a)(b - a) \tag{7.31}$$

where: Δ = isotopic fractionation; a = isotope effects from diffusion ($-4.4‰$, O'Leary, 1988); b = combined isotope effect is from RuBP and phosphoenolpyruvate (PEP) carboxylases ($-27‰$); and c_i/c_a = ratio of CO_2 inside the plant to atmospheric.

The model assumes that the c_i/c_a ratio is important in determining the carbon isotopic composition of plant tissues (Farquhar et al., 1982, 1989; Guy et al., 1986). In general, when there is an unlimited supply of CO_2 that is available to the plant, enzymatic fractionation will dominate; however, when CO_2 is limiting, fractionation during diffusion will predominate (Fogel and Cifuentes, 1993). In conclusion, these models show that carbon fixation in C_3 plants is controlled by a dynamic process between atmospheric CO_2 availability—which is largely controlled by *stomatal conductance*—and enzymatic fractionation during photosynthesis. Opening stomata for CO_2 also allows water to be lost from plant tissues to the atmosphere; thus plants using the C_3 photosynthetic pathway have had to evolve a balance between carbon uptake and water loss in a diversity of environments.

Over time other photosynthetic systems evolved that allowed for more efficient carbon uptake in plants [C_4 and crassulacean acid metabolism (CAM)] without significant water loss in extreme environments (Ehleringer et al., 1991). In C_4 plants (e.g., corn, prairie grasses, and *Spartina* spp. marsh grasses), the first step in CO_2 fixation results in the formation of the four-carbon compound oxaloacetate; this phase of fixation is catalyzed by the PEP carboxylase enzyme. The fractionation during C_4 fixation is considerably less ($-2.2‰$) than that occurring with RuBP (O'Leary, 1988), making C_4 plants more enriched in the ^{13}C (-8 to $-18‰$) (Smith and Epstein, 1971; O'Leary, 1981). Other differences in the efficiency with which CO_2 is converted into new plant material in C_4 plants, via the *Calvin Cycle* in bundle-sheath cells of vascular plants, contribute to the minimization of carbon fractionation (Fogel and Cifuentes, 1993). A modified fractionation model for carbon isotopes in C_4 plants, as developed by Farquhar (1983) is as follows:

$$\Delta = a + (b_4 + b_3\phi - a)(c_i/c_a) \tag{7.32}$$

where: b_4 = isotopic effects from diffusion into bundle-sheath cells; b_3 = isotopic fractionation during carboxylation (-2 to $-4‰$) (O'Leary, 1988); and ϕ = leakiness of the plant to CO_2.

Unlike C_3 or C_4 land plants, diffusion of CO_2 limits photosynthesis in aquatic plants. Consequently, algae have adapted a mechanism that allows them to actively transport or "pump" either CO_2 or HCO_3^- across their membrane, thereby accumulating DIC in the cell (Lucas and Berry, 1985). Although carbon fixation is through RuBP carboxylase, there is considerably less fractionation in algae than in higher C_3 plants because much of the stored CO_2 pool does not leave the cell. Thus, the rationale for using carbon isotopes to distinguish between aquatic and land plant inputs to estuaries is based on the idea that phytoplankton generally use HCO_3^- as a carbon source for photosynthesis ($\delta^{13}C = 0\text{‰}$), as opposed to land plants which use CO_2 from the atmosphere ($\delta^{13}C = -7\text{‰}$) (Degens et al., 1968; O'Leary, 1981). This results in algae being more enriched than terrestrial plants. The model for carbon fractionation in aquatic plants is as follows:

$$\Delta = d + b_3(F_3/F_1) \tag{7.33}$$

where: $d =$ equilibrium isotope effect between CO_2 and HCO_3^-; $b_3 =$ isotope fractionation associated with carboxylation; and $F_3/F_1 =$ ratio of CO_2 leaking out of the cell to the amount inside the cell.

The carbon isotopic composition of phytoplankton has been shown to be strongly affected by the pCO_2 of surface waters in the ocean (Rau et al., 1989, 1992). Moreover, fractionation of carbon isotopes by phytoplankton is also correlated with cell growth rate, cell size, cell membrane permeability, and CO_2 (aq) (Laws et al., 1995; Rau et al., 1997).

In an extensive study on the biogeochemistry of the Delaware Bay estuary (USA), it was clearly established that the seasonal variability of $\delta^{13}C$ in DIC could not account for the $\delta^{13}C$ shifts in POC (Galimor, 1974; Cifuentes et al., 1988; Fogel et al., 1988; Pennock et al., 1988). The $\delta^{13}C$ of POC in the riverine end-member (~ -24 to -31) were more depleted than the marine end-member (~ -22 to -24); this in part, was due to the inputs of more depleted terrestrially derived organic carbon from the watershed of the Delaware River estuary (figure 7.25; Fogel et al., 1988). However, the more negative $\delta^{13}C$

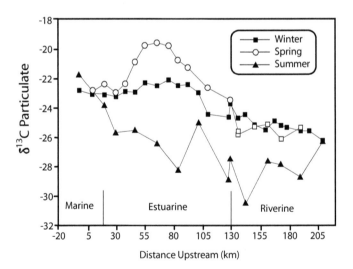

Figure 7.25 The $\delta^{13}C$ of POC in the riverine, estuarine, and marine end-members of the Delaware River estuary (USA) during winter, spring, and summer seasons. (Modified from Fogel et al., 1988.)

values were generally found in summer, during the highest rates of remineralization and productivity, and were attributed to preferential uptake of CO_2 versus HCO_3^- (Fogel et al., 1988). Other work has shown that remineralized CO_2 is isotopically light and similar to ambient phytoplankton in the water column (Jacobsen et al., 1970; Peterson and Fry, 1989). The uptake of CO_2 during photosynthesis also occurs faster than the isotopic equilibrium between CO_2, and HCO_3^-, further suggesting the importance of preferential CO_2 uptake in summer months.

The application of stable carbon isotopes in understanding the role of detritus in the diets of coastal fishes and macroinvertebrates has also proved to be quite important. The "outwelling hypothesis" of Odum (1968) suggested that salt marshes (and rivers) transport biologically available organic matter into near-shore waters, thereby enhancing secondary production on the shelf. While the importance of *Spartina alterniflora* marshes in supporting secondary productivity of a Georgia estuary was generally not validated using carbon isotopes alone (Haines, 1977), more recent studies using multiple stable isotope tracers have since proven that *Spartina* detritus is important to estuarine secondary production (Peterson and Howarth, 1987; Kwak and Zedler, 1997; Stribling and Cornwell, 1997). Recent work by Chanton and Lewis (2002) in Apalachicola Bay (USA), using multiple isotope tracers (^{34}S and ^{13}C), indicated that terrestrial and marine primary producers could be adequately separated before examining trophic linkages. When examining just the $\delta^{13}C$ values of POC, there is clearly a more depleted signal closer to land, indicative of terrestrial vascular plants, with a relatively more enriched signal further out in the bay, where plankton represent a more significant source (figure 7.26). Details on sulfur isotopes are provided later in the chapter.

Nitrogen

The largest reservoir of nitrogen exists in the atmosphere where it primarily occurs in diatomic molecular form (N_2). The two stable isotopes of nitrogen, ^{14}N and ^{15}N, occur in relative abundances of 99.64 and 0.36%, respectively (Bigeleisen, 1949; Sweeney et al., 1980). Similar to carbon, isotopic fractionation of N by biologically mediated reactions can be used to infer biogeochemical processes in ecosystems. For example, there is generally an increase in $\delta^{15}N$ with increasing trophic level (DeNiro and Epstein, 1981; Schoeninger and DeNiro, 1984). Fractionation by biological processes also leads to differences in $^{15}N:^{14}N$ ratios in DIN pools. The primary processes involved in such alterations include nitrification (Mariotti et al., 1981), denitrification (Miyake and Wada, 1971; Mariotti et al., 1981, 1982), reduction of NO_3^- to NH_4^+ (McCready et al., 1983), and assimilation by phytoplankton (Wada and Hattori, 1978).

Primary producers have N isotope signatures that reflect the $\delta^{15}N$ of their inorganic sources in addition to fractionation processes during uptake (Fogel and Cifuentes, 1993). The concentration of N affects the magnitude of fractionation as well as the enzymatic processes involved in assimilation (Mariotti et al., 1982; Pennock et al., 1996). In general, fractionation decreases or is absent when DIN uptake is the rate-limiting step (Wada and Hattori, 1978; Pennock et al., 1996; Granger et al., 2004). Food web studies have shown a +2 to +4‰ increase in the $\delta^{15}N$ in consumers relative to their food sources, caused by kinetic differences in light and heavy isotopes during metabolic processes (Minagawa and Wada, 1984). Despite these fractionation effects, dominant sources of PON can be

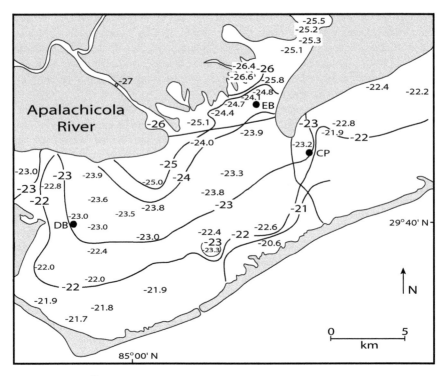

Figure 7.26 δ^{13}C values of POC in Apalachicola Bay (USA) showing that terrestrial and marine primary producers can be separated adequately before examining trophic linkages. (Modified from Chanton and Lewis, 2002.)

distinguished using N isotopes (Wada et al., 1990). For example, the δ^{15}N of marine phytoplankton is generally enriched by +8 to +10‰ relative to terrestrial plant sources in estuaries (O'Donnell et al., 2003). Finally, controlled ^{15}N-addition experiments have been used to understand fractionation processes but proved to be problematic by interfering with the natural DIN isotopic signals (Glibert et al., 1982). However, recent work has shown that ^{15}N additions can be effective in situations where natural DIN concentrations are high due to significant anthropogenic N loading (Hughes et al., 2000).

Fractionation of nitrogen isotopes in photosynthesis during primary assimilation of N_2, NO_3^- NO_2^-, and NH_4^+ involves both kinetic and equilibrium effects. For example, kinetic effects by assimilatory enzymes are also typically coupled with equilibrium effects between NH_4^+ and NH_3 (−19 to −21‰) (Hermes et al., 1985). Ammonium is assimilated into glutamate using the enzyme glutamate dehydrogenase; this is followed by the conversion of glutamate into glutamine using glutamine synthetase (Falkowski and Rivkin, 1976). Isotopic fractionation during the assimilation of NH_4^+ by certain algae ranges from 0 to −27‰ (Wada, 1980; Macko et al., 1987). Active transport occurs across the cell membrane when NH_4^+ concentrations in the water column drop below 100 μM (Kleiner, 1985). Changes in active versus passive transport can be accounted for in the following fractionation model of NH_4^+ assimilation:

$$\Delta = E_q + D + (C_i/C_o)[E_{enz} - D] \tag{7.34}$$

where: Δ = isotopic fractionation; E_q = equilibrium isotopic effects between NH_4^+ and NH_3; D = isotope effects of NH_3 diffusion in and out of the cell; C_i/C_o = ratio of the concentration of NH_3 inside and outside the cell; and E_{enz} = enzyme fractionation by either glutamine synthetase or glutamate dehydrogenase.

Oxidized forms of DIN (NO_3^- and NO_2^-) must also be converted into ammonia by either nitrate or nitrate reductases before being fixed into organic matter. Similarly, there is a nitrogenase reductase reaction involved in the fixation of N_2. The range of fractionation observed with NO_3^- assimilation is similar to that observed with NH_4^+, with more fractionation (by diatoms) occurring with higher ambient NO_3^- concentrations (Wada and Hattori, 1978). Active transport of NO_3^- has been observed by marine diatoms; however, details of the membrane-bound enzyme involved with this reaction remain unclear (Falkowski, 1975; Packard, 1979).

The typical isotopic compositions for particulate nitrogen in non-nitrogen fixing phytoplankton and macroalgae are -3 to $+18\%o$ (Schoeninger and DeNiro, 1984; Cifuentes et al., 1988; Fogel and Cifuentes, 1993). Nitrogen-fixing cyanobacteria and terrestrial plants are more enriched with ranges from -2 to $+4\%o$ and -6 to $+6\%o$, respectively (Fogel and Cifuentes, 1993). Non-nitrogen fixing terrestrial plants have $\delta^{15}N$ values that range from -5 to $+18\%o$ (Fogel and Cifuentes, 1993). Similarly, soils have $\delta^{15}N$ values that range from -6 to $+18\%o$ (Schoeninger and DeNiro, 1984). Drainage basins are also impacted by anthropogenic inputs of fertilizers in agricultural regions; fertilizer synthesized by the Haber–Bosch Process have isotopic signatures of approximately $0\%o$ (Black and Waring, 1977). Natural fertilizer, such as manure, is more enriched with values that range from $+18$ to $+35\%o$ (figure 7.27; Chang et al., 2002).

Stable nitrogen isotopes are good tracers of nitrification and denitrification processes in estuaries receiving anthropogenic N loading because both result in high $\delta^{15}N$ values in the external (watershed) nitrogen load (Mariotti et al., 1984). This is due to the more rapid processing of the light isotope (loss of ^{15}N-depleted N_2) and enrichment of the heavier isotope in soil residues (NH_4^+ and NO_3^-) in the watershed (Hoegberg and Johannisson, 1993). Consequently, incorporation of ^{15}N-enriched nutrients into estuarine food webs appears to result in a general ^{15}N enrichment in organic matter pools (Kwak and Zedler, 1997; Voss and Struck, 1997; McClelland and Valiela, 1998; Fry, 1999; Voss et al., 2000; Costanzo et al., 2001). Groundwater inputs to coastal marshes from agricultural sources have also been shown to enrich the $\delta^{15}N$ of emergent plant tissues (Page, 1995). Recent work has suggested that such changes in stable nitrogen isotopes (e.g., overall enrichment) should reflect the environmental condition of an estuary, as it relates to anthropogenic nutrient loading (McClelland et al., 1997; Fry et al., 2003).

However, these studies do warn of using stable nitrogen isotopes as indicators of external inputs because of ^{15}N-enrichment nitrification and denitrification processes that occur within the estuary (Cifuentes et al., 1988: Horrigan et al., 1990; Brandes and

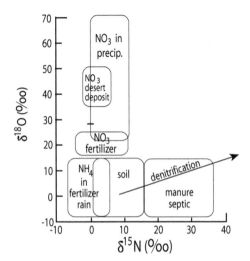

Figure 7.27 $\delta^{18}O$ and $\delta^{15}N$ values of different nitrogen sources to estuaries such as atmospheric (dry and wet), synthetic and natural fertilizer (e.g., manure), and septic tanks. (Modified from Chang et al., 2002.)

Devol, 1997). Recent work has also been able to effectively label seagrass tissues and their associated epiphytes and trace the ^{15}N label to consumer shrimp species (Mutchler et al., 2004); the ability to trace separate food resources to primary consumers shows great promise for labeling techniques in estuarine studies. Finally, the problems of effectively measuring $\delta^{15}N$ of NH_4^+ and NO_3^- sources in such studies have proven to be a serious problem in many past studies focused on nutrient loading to estuaries (Fry et al., 2000). Fortunately, recent technical advances in such isotopic measurements should prove useful in future studies (Sigman et al., 1997; Holmes et al., 1998).

Nitrogen isotopes have been used to distinguish between anthropogenic sources of fertilizer in freshwater/estuarine systems. For example, fertilizer nitrogen is isotopically depleted with a wide range of values depending on fertilizer type (Freyer and Aly, 1974; Heaton, 1986) compared to manure which is more enriched (Owens, 1985; Montoya, 1994). Recent work has shown that better separation of nitrogen sources can be made using both $\delta^{15}N$ and $\delta^{18}O$ (Kendall, 1998; Chang et al., 2002) (figure 7.27). Processes such as denitrification can also be identified in estuarine systems using this dual-tracer method because of the enrichment of both isotopes in remaining NO_3^- (Böttcher et al., 1990). Historical reconstruction of watershed N loading has also been useful in setting a baseline for current comparison and future management (Voss and Struck, 1997; Hodell and Schelske, 1998; Bianchi et al., 2000c; Ogawa et al., 2001). More details on these nitrogen cycling processes are provided in chapter 10.

Sulfur

Sulfur isotopes can effectively be used to examine important geochemical processes associated with redox changes in sedimentary environments. The speciation of sulfur is strongly affected by redox potential, pH, productivity, microbial sulfate reduction, and iron availability (Berner, 1984). More details are provided on the sulfur cycle in chapter 12. In general, during microbial dissimilatory sulfate reduction there is fractionation of sulfur isotopes which results in the enrichment of ^{32}S in H$_2$S relative to the residual SO$_4{}^{2-}$ (Kaplan and Rittenberg, 1964; Goldhaber and Kaplan, 1974; Jørgensen, 1979). In a system where there is unlimited SO$_4{}^{2-}$, the H$_2$S can become as depleted as $-40‰$. However, if the SO$_4{}^{2-}$ becomes limited the δ^{34}S of the H$_2$S and residual SO$_4{}^{2-}$ will become more enriched in ^{34}S as the SO$_4{}^{2-}$ is consumed (Nakai and Jensen, 1964). Sources of organic matter inputs can also affect sediment δ^{34}S of sulfide profiles in sediments. For example, in sediment cores collected from the Everglades, south Florida (USA), disulfide sulfur (DS) in the upper 10 cm of the sediments at the E1 site were found to be more enriched than at the U3 site, due to higher inputs of sawgrass and periphyton, which tend to incorporate isotopically light sulfide species (figure 7.28; Bates et al., 1998). Increases in the loading of organic matter generally enhance the rate of sulfate reduction which results in a concomitant decrease in the selectivity of stable sulfur isotopes (e.g., slower rate of reduction results in lower δ^{34}S values in the H$_2$S product) (Kaplan and Rittenberg, 1964; Ohmoto, 1992). Early work suggested that S isotope fractionation in sediments was conducted by only a few select species of bacteria (e.g., *Desulfotomaculum* spp. and *Delsufovibrio* spp.) (Chambers et al., 1975; Fry et al., 1988). However, more recent work has shown a greater diversity of sulfate-reducing bacteria in natural systems than previously thought (Habicht and Canfield, 1997; Llobet-Brossa et al., 1998; Böttcher et al., 1999; Sahm et al., 1999; Ravenschlag et al., 2000). Moreover, it has been suggested that an understanding of species-specific physiology of different sulfate reducers is needed to effectively understand S isotope fractionation during dissimilatory reduction (Detmers et al., 2001).

A significant fraction of H$_2$S produced by dissimilatory processes will also be oxidized by other reactions (Jørgensen, 1982). These intermediate sulfur species (e.g., elemental sulfur) from H$_2$S oxidation may be enriched in ^{34}S, also contributing to the overall ^{34}S signal of H$_2$S (Fry et al., 1988; Canfield and Thamdrup, 1994). Stable sulfur isotopes in

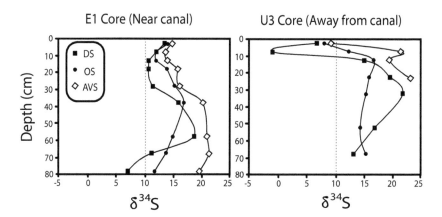

Figure 7.28 δ^{34}S values of different sulfur species [disulfide sulfur (DS), acid-volatile sulfur (AVS), and organic sulfur (OS)] in sediment cores collected from the Everglades, south Florida (USA). (Modified from Bates et al., 1998.)

authigenic minerals have also been used to interpret past redox conditions of estuarine systems, such as the Baltic Sea (Böttcher and Lepland, 2000). As dissolved sulfide reacts with Fe(II) to form iron sulfides there appears to be no fractionation, thereby allowing sedimentary sulfides to be used as indicators of the isotopic signature of H_2S produced during bacterial processes (Böttcher and Lepland, 2000).

The isotopic composition of S in estuarine plants can be affected by a broad spectrum of biotic and abiotic processes in estuaries. Sulfur uptake by photosynthetic organisms involves reduction of SO_4^{2-} to sulfide and then incorporation into cysteine (Fogel and Cifuentes, 1993). Other possibilities exist for direct uptake of sulfides in estuarine/coastal environments. For example, SO_4^{2-} reduction in salt marsh sediments yields isotopically depleted $\delta^{34}S$ porewater sulfide; the uptake of H_2S by the marsh plant *S. alterniflora* also results in isotopically depleted $\delta^{34}S$ plant tissues (Carlson and Forrest, 1982). Thus, the more H_2S that has been taken up by marsh plants from porewaters, the more depleted the signal of *S. alterniflora* plant tissues. Other studies have shown that seasonal shifts in the $\delta^{34}S$ of *S. alterniflora* tissues are correlated with the age, size, and activity of the associated microorganisms and fungi on marsh plants (Howarth and Giblin, 1983; Peterson et al., 1986; Currin et al., 1995; Chanton and Lewis, 2002). The more depleted $\delta^{34}S$ of estuarine marsh plants also provides a mechanism for separating terrestrial sources from upland plants that use SO_4^{2-} available from precipitation ($\delta^{34}S = +2$ to $+8‰$) and from local marsh plant sources in estuaries (Peterson and Howarth, 1987). Conversely, the $\delta^{34}S$ of phytoplankton in high-salinity regions of estuaries is expected to be nearly constant, because of the dominant role of SO_4^{2-} in buffering the sedimentary processes described earlier. When considering the concentrations of SO_4^{2-} in seawater (28 mM) and freshwater (<0.2 mM), the weighted effects of a seawater SO_4^{2-} isotopic signal ($+21‰$) tend to be dominant in the water column of estuarine waters when the salinity is < 1 (Fry and Smith, 2002). Plants are generally more depleted in $\delta^{34}S$ than SO_4^{2-} by approximately $-1.5‰$; thus the isotopic signature of most phytoplankton should be approximately $+19.5‰$ (Trust and Fry, 1992). Earlier work has shown a typical value of $+20.3$ for phytoplankton and macroalgae (Kaplan et al., 1963). However, benthic microalgae have been shown to be more depleted in $\delta^{34}S$ due to sedimentary sources of sulfides (Fry and Smith, 2002).

Hydrogen and Oxygen

There have been relatively few studies that have used hydrogen and oxygen isotopes in estuaries studies (Smith and Epstein, 1970; Estep and Dabrowski, 1980; Estep and Hoering, 1980; DeNiro and Epstein, 1981; Macko et al., 1983). Factors controlling the deuterium (D) to hydrogen ratios in organisms are largely governed by a diversity of metabolic and environmental processes (Smith and Epstein, 1970; Estep and Hoering, 1980; Stiller and Nissenbaum, 1980; Macko et al., 1983). The predominance of conservative hydrogen bonding in organismal tissues makes hydrogen isotopes useful tools in food web studies (Estep and Dabrowski, 1980). When water is taken up by plants there is no fractionation; however, once the water moves into leaves of higher plants it comes into contact with stomates, which open and close for gas exchange. It is at this time that the light isotope preferentially evaporates, resulting in an enrichment of D (40 to 50‰) in the remaining leaf water (Estep and Hoering, 1980). Although there may be potential for using hydrogen isotopes in distinguishing between C_3 and C_4 plants, due to differences in the stomatal conductance, no clear trends have been observed (Leaney et al., 1985). While there has not been an abundance of work with hydrogen isotopes, some work suggests that organically bound hydrogen may preferentially lose the light isotope during decomposition processes (Macko et al., 1983). Other work has suggested that, because of the complexity of hydrogen isotopic ratios observed in decomposing seagrass and macroalgal detritus, hydrogen isotopes cannot be used to discriminate between live and detrital organic matter in food web studies (Fenton and Ritz, 1988). Fractionation studies of oxygen isotopes in organic matter is limited to examining the $\delta^{18}O$ of cellulose and other carbohydrates (Fogel and Cifuentes, 1993). Unfortunately, no studies to date have been able to find significant differences in the $\delta^{18}O$ of cellulose that are linked to different photosynthetic pathways.

Summary

1. Radioactive isotopes (or radionuclides) provide useful tools for measuring rates of processes on Earth. The major sources of radionuclides are: (1) primordial (^{238}U, ^{235}U, and ^{234}Th-series radionuclides); (2) anthropogenic or transient (e.g., ^{137}Cs, ^{90}Sr, ^{239}Pu); and (3) cosmogenic (e.g., ^{7}Be, ^{14}C, ^{32}P). These isotopes can be further divided into two general groups, the particle-reactive and non-particle-reactive radionuclides.
2. Radioactivity is defined at the spontaneous adjustment of nuclei of unstable nuclides to a more stable state. Radiation (e.g., alpha, beta, and gamma rays) is released in different forms as a direct result of the changes in nuclei of these nuclides.
3. The experimentally measured rates of decay of radioactive atoms show that the decay is first order, where the number of atoms decomposing in a unit of time is proportional to the number present—this can be expressed in the following equation: $dN/dt = -\lambda N$. Another term used for characterizing rate of decay is half-life ($t_{1/2}$), the time required for half of the initial number of atoms to decay. Isotopes are considered to be in secular equilibrium, when the rate of decay of the parent is equal to that of its daughter.
4. The primary processes affecting the distribution of particle-reactive radionuclides in estuaries are as follows: (1) removal or scavenging of nuclides from the water column by precipitation, ion exchange, or hydrophobic interactions with particle surfaces; (2) complexation with organic materials that are adsorbed on particle surfaces; (3) flocculation of radionuclides in colloidal material; (4) sorption on Fe and Mn oxide coatings that undergo coprecipitation; (5) direct removal by biological processes; (6) desorption from suspended sediments (particularly from rivers); and (7) release from sediments via physical or biological mixing/uptake.
5. The removal rate of particle-reactive radionuclides on particles with a pore size of 0.4 μm or greater can be made using an empirically determined scavenging rate constant (λ_s), which equals 1/residence time—where the residence time (τ) of an element is defined as the ratio of the element's standing stock to the removal rate or supply.

6. Sediment deposition is defined as the temporary emplacement of particles on the seabed while sediment accumulation is defined as the net sum of particle deposition and removal processes over a long period. The term sedimentation refers to integrated particle transport to—and emplacement—on the seabed, as well as removal, and preservation.

7. If particle mixing is assumed to be analogous to diffusion with sediment accumulation and radionuclide decay, the steady-state profile for excess activity of a nonexchangeable radionuclide is defined by an advective–diffusion equation.

8. The three isotopes of uranium (^{238}U, ^{235}U, ^{234}U) found in nature have longer half-lives ($>10^5$ years) than the oceanic mixing time (ca. 10^3 years). The distribution and concentration of U in rivers, estuaries, and coastal regions are extremely variable and not well understood. More work is clearly needed to understand further the complex interactions of active and carrier phases (Fe and Mn oxides), redox transformations, direct and indirect microbial transformations, and colloidal complexation that may be involved in the nonconservative behavior of U in estuaries.

9. The four commonly used isotopes of thorium (^{234}Th, ^{228}Th, ^{230}Th, ^{232}Th) are produced from the decay of uranium and radium parents. Thorium is present in highly insoluble forms and can be rapidly removed via scavenging by particulate matter. The high abundance of colloidal organic matter and lithogenic materials in estuaries makes these particles key in controlling the scavenging and residence time of Th radionuclides in the water column.

10. Radium has four natural radionuclides in aqueous systems, ^{226}Ra ($t_{1/2} = 1,620$ y), ^{228}Ra ($t_{1/2} = 5.76$ y), ^{224}Ra ($t_{1/2} = 3.66$ d), and ^{223}Ra ($t_{1/2} = 11.4$ d). Radium makes an excellent tracer in coastal systems because it has a highly particle-reactive Th parent. Radium radionuclides can be used to measure groundwater inputs to coastal waters and to model residence times in estuaries.

11. Radon-222 is an inert noble gas produced by alpha decay of ^{226}Ra with a half-life of 3.85 d. The primary application of ^{222}Rn in estuaries has been in estimating flux rates of pore waters across the sediment–water interface and the rate of gas exchange between the estuary and the atmosphere.

12. Lead-210 is produced by radioactive decay of ^{222}Rn and can enter estuaries as a dissolved/complexed ion or in association with particles from the ocean, river, and atmosphere. Much of the work to date in estuaries and adjacent marsh/shelf environments using ^{210}Pb has been to determine sediment accumulation and accretion rates.

13. Polonium-210 is a decay product of ^{210}Pb and is produced mainly in the water column, with some atmospheric inputs. High variability in both ^{210}Pb and ^{210}Po in estuaries is generally attributed to remobilization from sediments and formation of organic complexes.

14. The cosmogenic radionuclide ^7Be ($t_{1/2} = 53.3$ d) has been effectively used as a tracer of short-term particle cycling and depositional processes in estuarine systems.

15. Cesium-137 is an impulse marker where the subsurface peak of maximum activity (or first appearance of activity) is used to measure accumulation rates. However, the resolution of this peak and depth of first appearance can be altered by physical and biological mixing processes. Feeding activities of benthic

macrofauna can affect vertical movement of ^{137}Cs in sediments. The best environments for using this tracer are areas with high sedimentation environments, rapid burial, and minimal disturbance of peak depth.

16. Of the seven long-lived plutonium isotopes, the four of primary concern in environmental studies have been ^{238}Pu ($t_{1/2} = 87.7$ y), ^{239}Pu ($t_{1/2} = 2.41 \times 10^4$ y), ^{240}Pu ($t_{1/2} = 6,571$ y), and ^{241}Pu ($t_{1/2} = 14.4$ y). The major sources of Pu to both terrestrial and aquatic ecosystems are: (1) atomic weapons testing; (2) burn up of a ^{238}Pu-powered satellite (SNAP 9A) over the Indian Ocean in 1964; and (3) radioactive disposal and leakage from power plants.

17. Similar to other cosmogenic radionuclides ^{14}C is produced by reaction of cosmic rays with atmospheric atoms such as N_2 and O_2. Living plants and animals in the biosphere contain a constant level of ^{14}C; however, when they die there is no further exchange with the atmosphere and the activity of ^{14}C decreases with a half-life of 5730 ± 40 y; this provides the basis for establishing the age of archeological objects and fossil remains.

18. Studies focusing on the natural abundances of carbon and nitrogen stable isotopes have been successfully used in lake and coastal/estuarine systems to trace source inputs of terrestrial and aquatic organic matter, sewage, and nutrients. The basis for using stable isotopes in food web analyses is that the organisms retain the isotopic signals of the foods they assimilate.

19. Using the equation $\delta(\%o) = [R_{sample} - R_{standard} / R_{standard}] \times 1000$, a sample is considered to be enriched when the heavy isotope is more abundant relative to the standard, resulting in positive del values. Conversely, del values will be negative, or depleted, when the sample has less of the heavy isotope than the standard. Isotopic fractionation can occur from both kinetic and equilibrium effects.

20. Stable carbon isotopes have been commonly used to distinguish between allochthonous versus autochthonous organic carbon inputs to estuaries. One of the most important pieces of information that can be gathered from this information is the delineation between C_3 and C_4 plant inputs.

21. Primary producers have N isotope signatures that reflect the δ^{15}N of their inorganic sources in addition to fractionation processes during uptake. Stable nitrogen isotopes are also good tracers of nitrification and denitrification processes in estuaries receiving anthropogenic N loading because both result in high δ^{15}N values in the external (watershed) nitrogen load.

22. Stable sulfur isotopes can be effectively used to examine important geochemical processes associated with redox changes in sedimentary environments. For example, SO_4^{2-} reduction in salt marsh sediments yields isotopically depleted δ^{34}S porewater sulfide; the uptake of H_2S by marsh plants also results in isotopically depleted δ^{34}S plant tissues.

23. There have been relatively few studies that have used hydrogen and oxygen isotopes in estuarine studies. Factors controlling the deuterium-to-hydrogen ratios in organisms are largely governed by a diversity of metabolic and environmental processes. Fractionation studies of oxygen isotopes in organic matter are limited to examining the δ^{18}O of cellulose and other carbohydrates.

Part IV

Organic Matter Sources and Transformation

Chapter 8

Organic Matter Cycling

Production of Organic Matter

In this chapter the general processes involved in controlling production and transformation of organic matter will be discussed as well as some of the associated *stoichiometric* changes of a few key biological elements (e.g., C, N, P, S). Stoichiometry is defined as the mass balance of chemical reactions as they relate to the *law of definite proportions* and conservation of mass (Sterner and Elser, 2002). For example, if we examine the average atomic ratios of C, N, and P in phytoplankton we see a relatively consistent ratio of 106:16:1 in most marine species. This is perhaps the best example of applied stoichiometric principles in natural ecosystems and is derived from the classic work of Alfred C. Redfield (1890–1983) (Redfield, 1958; Redfield et al., 1963). More specifically, Redfield compared the ratios of C, N, and P of dissolved nutrients in marine waters to that of suspended marine particulate matter (seston) (essentially phytoplankton) and found straight lines with equal slopes (figure 8.1; Redfield et al., 1963). This relationship suggested that marine biota were critical in determining the chemistry of the world ocean, clearly one of the most important historical findings linking chemical and biological oceanography (Falkowski, 2000). Moreover, the *Redfield ratio* has been further validated with recent data using improved analytical techniques (Karl et al., 1993; Hoppema and Goeyens, 1999). Other work has shown that there are predictable deviations from the Redfield ratio across a freshwater to open ocean marine gradient (figure 8.2; Downing, 1997). For example, N-to-P ratios in estuaries have commonly been shown to be lower and/or higher than the predicted Redfield ratio because of denitrification and anthropogenic nutrient enrichment processes, respectively.

Inputs of vascular plant organic matter (e.g., mangroves, salt marshes, seagrasses) to estuarine systems presents another problem in causing deviations of C:N:P from the Redfield ratio. Vascular plants have been shown to deviate from this ratio in part because of relatively high amounts of C and N compared to algae due to a higher abundance of structural support molecules (e.g., cellulose, lignin) and defense antiherbivory (secondary) compounds (e.g., tannins), respectively (Vitousek et al., 1988). Finally, the Redfield ratio

177

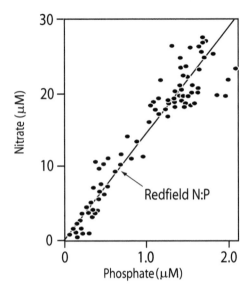

Figure 8.1 Regression of dissolved nitrate and phosphate from western Atlantic waters showing a ratio of approximately 16:1. (Modified from Redfield et al., 1963.)

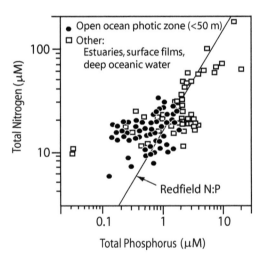

Figure 8.2 Regression of total (dissolved and particulate) nitrogen and phosphorus from estuaries, surface films, and oceanic surface (< 50 m) and deep waters. (Modified from Downing, 1997.)

is commonly used to infer resource limitation in phytoplankton (e.g., P versus N). Much of our thinking about resource limitation is derived from *Leibig's law of the minimum*, which simply states that organisms will become limited by the resource that is in lowest supply relative to their needs. More details on the cycling of P and N as it relates to the Redfield ratio and resource limitation are discussed in chapters 10 to 12. Based on the this ratio, Stumm and Morgan (1996) further modified the stoichiometry of the chemical reaction of *photosynthesis* (primary production) and oxidation (degradation) of organic matter by the following equation:

$$\leftarrow \text{oxidation (respiration)}$$

$$106\ CO_2 + 16\ HNO_3 + H_3PO_4 + 122\ H_2O \leftrightarrow (CH_2O)_{106}\ (NH_3)_{16}H_3PO_4 + 138\ O_2$$

$$\text{photosynthesis} \rightarrow \tag{8.1}$$

This equation offers a different perspective on how photosynthesis and degradation processes are linked to redox chemistry and the stoichiometric constraints on the availability of key elements in many biogeochemical cycles in aquatic ecosystems. As shown in equation 8.1, *primary production* is simply defined as the photosynthetic formation of organic matter. Other processes such as *chemosynthesis* involve the release of chemical energy from the oxidation of reduced substrates. There are different terms associated with describing components of this production. For example *gross primary production* (GPP) is the amount of carbon fixed (CO_2 converted into organic matter) by photosynthetic organisms (per unit time per unit volume). Similarly, *net primary production* (NPP) is the total carbon fixed minus the amount respired by the primary producer. *New production* and *old or recycled* production represent the NPP that is supported by nutrients from outside the estuary proper (e.g., river inputs) and production supported by nutrients regenerated from within the estuary (e.g., nutrient fluxes across the sediment/water boundary from remineralized organic matter in sediments), respectively. *Secondary production* is the turnover of organisms that range in size from bacteria to the largest mammals on Earth that rely on the consumption of organic matter produced by photosynthetic organisms as a food resource. Secondary production from both pelagic and benthic *consumers* is important in determining the fate of primary production in estuarine systems. There will be more discussion later in this chapter on the effects that animal–sediment interactions have on the decay dynamics of organic matter in sediments. However, this discussion will not provide any details on the many complex ecological processes, such as trophic/population dynamics, predator–prey models, or top-down versus bottom-up controls associated with secondary consumers, that affect the fate of primary production in estuaries (Valiela, 1995).

The general pools of organic matter in estuaries can be divided into *particulate* ($>0.45\ \mu m$) and *dissolved* ($<0.45\ \mu m$) fractions (*POM*, and *DOM*). Details on the cycling of dissolved/particulate and inorganic/organic C, N, P, and S will follow in chapters 10–13. As discussed earlier, the dominant living and nonliving (*detritus*) sources of organic matter to estuaries can be divided into *allochthonous* and *autochthonous* pools—sources produced outside and within the boundaries of the estuary proper, respectively.

These pools can further be divided into *heterotrophs* (e.g., animals and fungi) and *autotrophs* (e.g., vascular plants and algae) which are either *Prokaryotes* (unicellular organisms that do not possess a nuclear membrane or DNA in the form of chromosomes) or *Eukaryotes* (unicellular and multicellular organisms with nuclear membranes and DNA in the form of chromosomes). The taxonomic classification of all living organisms is shown in figure 8.3 (Engel and Macko, 1993). The old five-kingdom classification system has been replaced by a three-kingdom system, which is derived from base sequences of nucleic acids (Engel and Macko, 1993). The Prokaryotes, which constitute the *Eubacteria* (*Bacteria*) and *Archaebacteria*, represent important microbial groups essential in the decomposition

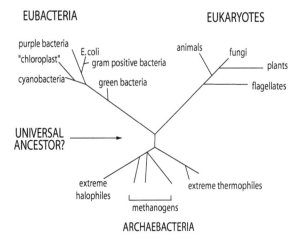

Figure 8.3 Taxonomic classification of all living organisms. (Modified from Engel and Macko, 1993.)

cycles of estuaries. The Archaebacteria, represent the third most recently recognized primary kingdom and are characterized as living in extreme environments such as anaerobic methanogens and halophilic bacteria, both of which are found in estuarine environments. Despite the fact that bacteria are a functionally diverse group of organisms, much of the literature that exists today in microbial ecology has ignored this taxonomic and functional diversity, treating bacterial processes as a "black box." Only recently have we begun to use molecular techniques to examine for such differences in the bacterial community in estuaries (Heidelberg et al., 2002). The phylogenetic composition of bacterial communities has been shown to vary across a temperature and salinity gradient in estuaries (Valencia et al., 2003). For example, alpha-proteobacteria (purple bacteria) (see figure 8.3) are generally found in the higher-salinity regions while beta-proteobacteria occur near the freshwater end-member (Yokokawa et al., 2004). Controls on the biomass of these different bacterial communities can also vary across different salinities, typically as a function of substrate supply and bacterial mortality (Jørgensen et al., 1998).

Bacterial production, typically measured from incorporation of ^{3}H-thymidine or ^{3}H-leucine (Fuhrman and Azam, 1982; Kirchman et al., 1985) has also been shown to vary considerably among different bacterial groups (Cottrell and Kirchman, 2003). The gamma-proteobacteria are typically the most abundant groups of bacteria in aquatic ecosystems (Eilers et al., 2000). *Viruses*, which are not classified as organisms, are important contributors to the DOM cycle because they can cause bacteria and phytoplankton cells to burst, via lysis, releasing dissolved cellular components into the DOM pool (Suttle, 1994; Wilhelm and Suttle, 1999; Wommack and Colwell, 2000; Gastrich et al., 2002). In fact, the abundance and production of virus-like particles (viriobenthos) in sediments, ranging from 10^8 to 10^9 mL sediment^{-1} and 0.13 to 1.6 $\times 10^8$ virus-like particles mL^{-1} sediment h^{-1}, respectively, are considerably higher than that found for water column virus particles (virioplankton) (Mei and Danovaro, 2004).

The dominant allochthonous inputs to estuaries are from riverine [e.g., terrestrial plant detritus and freshwater *plankton* (free floating and/or weak swimming organisms)], marine/estuarine plankton (e.g., *phytoplankton, zooplankton, bacterioplankton, virioplankton*), and bordering terrestrial wetland sources (mangroves and freshwater/salt marshes). Autochthonous sources typically include plankton, *benthic* and *epiphytic micro- and macroalgae, emergent* and *submergent* (e.g., *seagrasses*) aquatic vegetation (*EAV* and *SAV*) within the estuary proper, and *secondary production* (e.g., zooplankton, fishes, benthic animals).

The geomorphology of an estuary largely determines the extent to which shallow versus deep water primary production contributes to the overall biogeochemical cycling of an estuary. The horizontal gradient of estuarine habitats from open water, vegetated shoals, intertidal mud and sand flats, to fringing marshes are all connected by a dynamic exchange of dissolved and particulate constituents (Correll et al., 1992). For example, Chesapeake Bay (USA) has approximately 40% of the submerged bottom of the total estuarine area that is in water less than 2.0 m deep—characterized by a broad shoal and fringing wetland environments (Spinner, 1969). These types of shallow and intertidal environments have been recognized as important zones for the production of organic matter and biogeochemical cycling in estuaries (Roman et al., 1990; Childers et al., 1993, 2000; Buzzelli, 1998; Buzzelli et al., 1998).

Light attenuation is an important factor controlling primary production from intertidal to the deeper waters of estuaries (McPherson and Miller, 1987; Bledsoe and Phlips, 2000). Controls on light availability for growth of primary producers will vary between benthic and pelagic species but are generally controlled by *incident irradiance, photosynthetically active radiation* (PAR), and *mixing depth* in the water column. The *Lambert–Beer's law* or *light attenuation* (K_d) equation which describes the decrease of light of penetration with increasing water depth is as follows:

$$K_d = \ln (I_0/I_z) /Z \qquad (8.2)$$

where: I_0 = incident irradiance; I_z = light intensity at depth z; and Z = depth.

Shallow zones of DOM and POM production are inherently linked with the cycling and exchange of the adjacent deep channel environments of estuaries (Malone et al., 1986; Kuo and Park, 1995). Recent models have now included shallow environments in the wider-scale predictive models of organic matter cycling of estuaries (Pinckney and Zingmark, 1993; Madden and Kemp, 1996; Buzzelli et al., 1998).

Particulate and Dissolved Organic Matter in Estuaries

The photosynthetic fixation of inorganic carbon and nutrients into plant biomass is the primary source of organic matter within estuaries. Therefore, it is critical to have some basic understanding about the primary producer community in estuaries, in addition to the habitats and conditional constraints needed for growth. To this end, this section provides a general overview of the ecology of the principal taxa of primary producers (e.g., phytoplankton, benthic macroalgae, microphytobenthos, seagrasses, wetland plants) and their associated pools of POM and DOM in estuaries. Inputs of DOM, POM, and nutrients

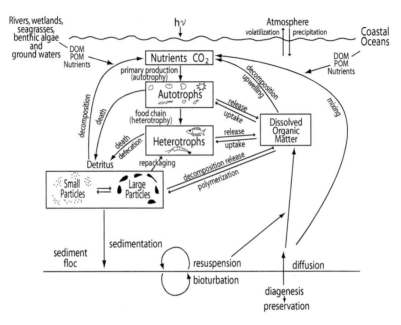

Figure 8.4 Biogeochemical cycling of nutrients, DOM, and POM, in the estuary proper bound by coastal ocean and riverine end-member exchange. (Modified from Wakeham and Lee, 1993.)

to estuaries can occur from both coastal ocean and riverine end-members which support both autotrophic and heterotrophic production in different estuarine regions (figure 8.4). Resuspension and diffusive flux from sediments can also provide significant inputs of dissolved materials to support production, via benthic–pelagic coupling, particularly in shallow estuaries. The conversion of dissolved materials by heterotrophs and autotrophs is a critical component of the "microbial loop" and is critically linked to sedimentation processes such as decomposition, aggregation, and flocculation which occur over different timescales in estuaries. Finally, the consumption and transformation of organic matter through metazoan and microbial (e.g., microbial loop) trophic levels is critical in the cycling of POM and DOM in estuaries (Wetzel, 1995); further details on these topics are provided later in the chapter.

Phytoplankton represent an important source of organic matter to most estuaries. The dominant classes of phytoplankton in estuaries are Bacillariophyta (diatoms), Cryptophyta (cryptomonads), Chlorophyta (green algae), Dinophyta (dinoflagellates), Chrysophyta (golden-brown flagellates, chrysophytes, raphidophytes), and Cyanophyta (cyanobacteria). Seasonal changes in phytoplankton abundance and composition in estuaries are primarily controlled by changes in riverine inputs, nutrients, tidal variability, algal respiration, light availability, horizontal exchanges, and consumption by grazers (Malcolm and Durum, 1976; Boynton et al., 1982; Mitchell et al., 1985; Malone et al., 1988; Boyer et al., 1993; Cloern, 1996; O'Donohue and Dennison, 1997; Thompson, 1998;

Lucas and Cloern, 2002). Temperature has also been shown to influence phytoplankton growth in laboratory conditions (Eppley, 1972) and more recently in the field (O'Donohue and Dennison, 1997). A temporal pattern of high production in summer and low production in winter is typical for temperate estuarine systems (Boynton et al., 1982). Nitrogen is generally regarded as the most limiting nutrient for phytoplankton in temperate estuaries (Fisher et al., 1988; Kemp et al., 1990; Boynton et al., 1982, 1996; Nixon, 1997; Nielsen et al., 2002); however, other studies have shown phosphorus limitation (Smith, 1984; Conley et al., 1995; Glibert et al., 1995, Conley, 2000) and possibly shifts between nitrogen and phosphorus along with co-limitation may be important as well (D'Elia et al., 1986; Malone et al., 1996; Conley, 1999; Fisher et al., 1999). Increased nutrient loading and alterations in nutrient ratios can alter the species composition and succession of phytoplankton (Schelske and Stoermer, 1971; Sanders et al., 1987; Oviatt et al., 1989), because of differences in the nutrient requirements of phytoplankton. For example, high inputs of nutrients typically result in a dominance of diatoms and small flagellate species (Tilman, 1977; Kilham and Kilham, 1984; Riedel et al., 2003). Finally, while it is well accepted that river-derived nutrients are important in controlling phytoplankton abundance and composition, other work has shown that in coastal regions where there is strong oceanic forcing on estuaries (e.g., U.S. Pacific coast) high phytoplankton productivity events can be associated with high-salinity events, due to an import of coastal phytoplankton (Hickey and Banas, 2003). More details on the dynamics of nutrient cycling are provided in chapters 10 and 11.

Benthic macroalgae and *microphytobenthos* are important sources of primary production in estuaries and have significant effects on the seagrass, tidal flat, and intertidal marsh habitats (Bianchi, 1988; Gould and Gallagher, 1990; Sullivan and Moncreiff, 1990; Rizzo et al., 1992; Pinckney and Zingmark, 1993; de Jonge and Colijn, 1994). Some benthic macroalgae commonly found in estuaries include: Chlorophyta (e.g., *Ulva latuca*, *Entermorpha intestinales*), Phaeophyta (e.g., *Fucus vesciculousus*), and Rhodophyta (e.g., *Gracilaria folifera*). The increase in anthropogenic loading of nutrients has resulted in numerous macroalgal blooms consisting primarily of the genera *Ulva*, *Entermorpha*, and *Gracilaria* spp. (Rosenberg and Ramus, 1984; Duarte, 1995; Kamer et al., 2001).

The microphytobenthos consist of an assemblage of benthic diatoms (principally pennate in shape) that typically migrate vertically in the sediments over a diurnal period (Serodiô et al., 1998). Enhanced turbidity in shallow regions from resuspension events can limit light penetration; thus, the most effective time for primary production occurs in intertidal sand and mud flats during daytime exposure periods (Guarini et al., 2000, 2002). Annual net primary production estimates for microphytobenthos cover a very broad range from 5 to 900 g C m^{-2} y^{-1} (Beardall and Light, 1994) to a somewhat narrower more recent estimate of 29 to 314 g C m^{-2} y^{-1} (Underwood and Kromkamp, 1999). Early studies showed that the patchy distribution of microphytobenthos was controlled by salinity (Admiraal, 1977), temperature/light (Admiraal and Pelletier, 1980; Hopner and Wonneberger, 1985; Bianchi, 1988; Bianchi and Rice, 1988), sediment topography (Colijn and Dijekma, 1981; Rasmussen et al., 1983), grain size (de Jonge, 1985), and current speed (Grant et al., 1986). There is increasing evidence that in shallow environments macroalgae and microphytobenthos are equally as important as phytoplankton in terms of total biomass and production (Cahoon, 1999; Webster et al., 2002; Dalsgaard, 2003). Even in the deeper coastal waters of the South Atlantic Bight (SAB) (USA) it has been estimated that benthic algal production supports a significant fraction of the total carbon production on the shelf (Marinelli et al., 1998; Nelson et al., 1999). In addition to being a carbon source for benthic and pelagic organisms, microphytobenthos are an important sink for nutrients that are fluxing across the sediment–water interface produced from remineralized organic matter. Inorganic nutrients (e.g., primarily NH_4^+, NO_3^-, and NO_2^-) in pore waters, generated from diagenetic transformation of organic matter, have been shown to be important sources of nitrogen for phytoplankton and benthic macroalgae/microphytobenthos in estuarine environments (Blackburn and Henriksen, 1983; Kemp and Boynton, 1984; Christensen et al., 1987; Cerco and Seitzinger, 1997; Risgaard-Petersen, 2003). For example, recent work in a temperate lagoon showed that microphytobenthos and macroalgae had significant effects on the sediment-water fluxes of dissolved organic and inorganic nitrogen (figure 8.5; Tyler et al., 2003).

The *benthic microlayer* contains actively photosynthesizing microphytobenthos which are also important in generating oxygen to the water column (Webster et al., 2002). Although high-energy coarse-grained permeable environments, such as sand flats, generally receive lower inputs of organic matter than lower-energy fine-grained systems, microphytobenthos are common in such environments and provide a "high quality" food resource for microbes and benthos, resulting in faster and more efficient remineralization of organic matter (Bianchi and Rice, 1988). Grazing by benthic meiofauna has also been shown to limit microphytobenthic biomass (Montagna et al., 1995; Carman et al., 1997; Goldfinch and Carman, 2000). In conclusion, there are many biogeochemical processes linked with redox chemistry, rates of organic matter decomposition, and elemental fluxes across the sediment–water interface that can be altered by the presence of

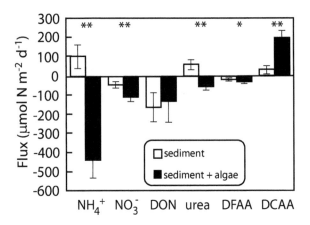

Figure 8.5 Experimental fluxes of NH_4^+, NO_3^-, dissolved organic nitrogen (DON), urea, dissolved free amino acids (DFAA), and dissolved combined amino acids (DCAA) across the sediment–water interface in cores collected in Hog Island Bay (USA). (Modified from Tyler et al., 2003.)

microphytobenthos in the microlayer of surface sediments. Periphyton represents another potentially important form of algal inputs to estuaries that can also affect sediment–water fluxes (Gleason and Spackman, 1974). These communities are particularly important in the Florida Everglades (USA) where calcareous periphyton consists of cynaobacterial filaments and/or diatoms (depending on the season) with inclusions of $CaCO_3$ crystals (Vymazal and Richardson, 1995; Rejmankova et al., 2004).

Seagrass meadows are prominent features of many shallow littoral habitats in estuarine systems (den Hartog, 1970; Phillips and McRoy, 1980; Green and Short, 2003) where they represent a major source of primary productivity (Hemminga and Duarte, 2000). The most well-studied seagrasses in temperate and subtropical/tropical environments are *Zostera* spp. (Nixon and Oviatt, 1973; Zieman and Wetzel, 1980; Dennison and Alberte, 1982; Wetzel and Penhale, 1983; Zieman and Zieman, 1989; Duarte, 1995; Nielsen et al., 2002) and *Thalassia* spp. (Day et al., 1989; Hillman et al., 1989; Czerny and Dunton, 1995). Other common marine species include *Halodule* spp. (Czerny and Dunton, 1995; Tomasko and Dunton, 1995; Lirman and Cropper, 2003), *Syringodium* spp. (Zieman et al., 1989), *Ruppia* spp. (Heck et al., 1995; Bortolus et al., 1998), *Cymodocia* spp. (Cebrian et al., 1997, 2000), *Posidonia* spp. (Pirc, 1985; Cebrian et al., 1997; Hadjichristophorou et al., 1997; Mateo et al., 2003), *Enhalus* spp. (McKenzie and Campbell, 2003), and *Amphibolis* spp. (Verduin et al., 1996). The rates of primary production for a select range of seagrasses are typically within 0.4 to 1.5 g C m^{-2} d^{-1} (table 8.1; Alongi, 1998). The submerged canopy of these systems reduces shallow-wave energy thereby allowing for more efficient trapping of suspended particulates derived from both autochthonous and allochthonous sources (Ward et al., 1984; Fonseca and Kenworthy, 1987). These systems have significant effects on the biogeochemical cycling of carbon, nitrogen, phosphorus, and oxygen in shallow water columns and sedimentary environments (Roman and Able, 1988; Caffrey and Kemp, 1990, 1991; Barko et al., 1991). Seagrasses are sensitive to changes in light attenuation that, typically, are usually related to water quality changes from alterations in the watershed (higher suspended sediment inputs) and/or within an estuary (e.g., dredging effects). As such, they are commonly considered to be indicators of environmental quality (Dennison et al., 1993). Shifts in composition from a seagrass dominated community to a macroalgal, microphytobenthos, and phytoplankton based food web are also typically associated with anthropogenic inputs of nutrients (Short and Burdick, 1996; Kaldy et al., 2002). These shifts to algal-dominated communities, caused by a competitive advantage of the epiphytic algae for light and space (Twilley et al., 1985; Tomasko and Lapointe, 1991; Hauxwell et al., 2001; Drake et al., 2003), also resulted in reduced support for secondary consumers at higher trophic levels (e.g., fishes) (Wyda et al., 2002). Self-shading within the seagrass canopy is another factor that has recently been incorporated into bio-optical models (Zimmerman, 2003). Much of the biomass produced in seagrass meadows is represented by leaves which typically become detached from the rhizomes at certain times of the year and are exported to other regions. For example, seagrass leaf litter export, which can represent as high as 50–60% of total leaf production, can accumulate in large piles of detritus (typically along sandy beaches)—commonly referred to as "banquettes" (Mateo et al., 2003).

Wetlands such as *freshwater/salt marshes* and *mangroves* have been shown to be major sources of primary production in estuarine systems (Kirby and Gooselink, 1976; Pomeroy and Wiegert, 1981). In fact, the *outwelling hypothesis* of Odum (1968) suggested that salt

Table 8.1 Rates of net primary production for a few selected seagrasses and seaweeds from various locations.

Genus/species	Location	Net primary production ($g\ C\ m^{-2}d^{-1}$)	References
Kelps			
Laminaria	North America	0.3–65.2	Mann (1982)
Macrocystis	South America, New Zealand, South Africa	1.0–4.1	Branch and Griffiths (1988) Schiel (1994)
Ecklonia	Australia, South Africa	1.6–6.2	
Rocky intertidal/subtidal Macroalgae			
Various seaweeds	Europe	0.5–9.0	Heip et al. (1995)
Enteromorpha	Hong Kong, USA	0.1–2.9	Thybo-Christensen et al. (1993)
Ascophyllum	USA, Europe	1.1	Niell (1977)
Distyopteris	Caribbean	0.5–2.5	
Fucus	North America	0.3–12.0	
Sargassum	Caribbean	1.4	
Ulva	Europe	0.6	
Gracilaria	Europe	0.3	
Cladophora	Europe	1.6	
Seagrasses			
Zostera marina	USA, Europe, Australia	0.2–8.0	
Thalassia	USA, Caribbean, Australia, South East Asia	0.1–6.0	Hillman et al. (1989) Pollard and Moriarty
Halodule	USA, Caribbean	0.5–2.0	Stevenson (1988)
Cymodocea	Mediterranean, Australia	3.0–18.5	Fortes (1992)
Posidonia	Mediterranean, Australia	2.0–6.0	
Enhalus	Southeast Asia	0.3–1.6	
Amphibolis	Australia	0.9–1.9	

Modified from Alongi (1998).

marshes transport biologically available organic matter into near-shore waters, thereby enhancing secondary production on the shelf. While this phenomenon was generally not supported by many of the earlier studies (Teal, 1962; Odum and de la Cruz, 1967; Haines, 1977), more recent work (e.g., Moran et al., 1991) has shown that 6 to 36% of near-shore

DOC off the coast of Georgia originates from coastal salt marshes, thereby enhancing secondary production on the shelf. A study by Bianchi et al. (1997b) along with other recent studies have demonstrated (with the use of more efficient biomarkers) that a significantly higher fraction of vascular plant materials is being transported as POC further out to the continental shelf and slope (Moran et al., 1991; Moran and Hodson, 1994; Trefry et al., 1994). Globally, export of organic carbon from mangroves and salt marshes varies significantly and ranges from approximately 2 to 420 g C m^{-2} y^{-1} and 27 to 1052 g Cm^{-2} y^{-1}, respectively (Alongi, 1998). In other cases, wetlands have been shown to import DOC at a much higher rate than POC release (Childers et al., 1999). One of the most important determinants of POM/DOM export and/or import in wetland systems is hydrology (Gosselink and Turner, 1978; Kadlec, 1990). Consequently, many of the coastal wetlands have been classified by different *typologies* primarily determined by geomorphology, landscape, and water flow (Brinson, 1993). The importance of the coastal geomorphology, geophysics, as well as large and small-scale hydrologic features of wetlands, have been shown to strongly influence the exchange between coastal and wetland systems (Twilley et al., 1985, 1997; Brinson, 1993; Twilley and Chen, 1998). More details on the exchange between estuarine and coastal systems are provided in chapter 16.

Salt marsh communities exhibit distinct vegetation patterns across a landscape of changing elevation and tidal influence (Bertness and Ellison, 1987; Wiegert and Freeman, 1990; Fischer et al., 2000). *Salt marshes* can be defined as stands of rooted vegetation which are flooded and drained by tidal forces. In general, the low marsh is largely dominated by a monoculture of *Spartina alterniflora* (Wiegert and Freeman, 1990). For more details on the geographic distribution of other dominant marsh plants such as *Juncus* spp. and *Distichlis* spp. see the work of Chapman (1960) and Webster and Benfield (1986). Over the past century, *Phragmites australis* has become the dominant wetland plant in many tidal wetlands, replacing *Spartina* spp. in the U.S.A (Chambers et al., 2003). This has altered many of the microbial reactions and accretion rates that are essential to the key biogeochemical transformations in marsh sediments (Harrison and Bloom, 1977; Ravit et al., 2003; Rooth et al., 2003). Consequently, there have been extensive and expensive efforts to restore *Spartina* spp. in marshes (Meyerson et al., 2000; Rice et al., 2000). Salt marshes also serve as sites for the introduction of "new" N via N$_2$ fixation (Valiela, 1983; Capone, 1988); however, most of this work has focused on rhizospheric fixation associated with the living *S. alterniflora* (Talbot et al., 1990; Newell et al., 1992). Standing dead shoots of *S. alterniflora* have also been shown to serve as substrates for N$_2$-fixing epiphytic bacteria (Day et al., 1973) and cyanobacteria (Green and Edmisten, 1974). For example, recent work has shown that the annual input of new N to a salt marsh in North Carolina (USA), from epiphytic cyanobacteria on dead *S. alterniflora* shoots, was estimated to be 2.6 g N m^{-2} y^{-1} (Currin and Paerl, 1998). This new N can have significant effects on the biogeochemical rates and pathways of organic matter cycling in estuaries. The release of reduced forms of N from intertidal marshes has also been well established in past studies (Nixon, 1980; Childers and Day, 1990; Childers et al., 1993). In conclusion, the breakdown of organic matter in marshes has been shown to have significant effects on the seasonal pulsing of nutrients in estuaries (Webster and Benfield, 1986).

Mangroves are defined as densely rooted forests bordering the lowlands of tropical and subtropical coastlines—generally occurring between 25°N and 25°S latitudes. In fact, 60 to 75% of tropical coastlines are bordered by these highly productive systems

(MacGill, 1958; Clough, 1998). There are eight different families of mangroves around the world, and some of the more well-studied genera are *Rhizophora, Avicennia, Laguncularia,* and *Bruguiera* spp. (Lugo and Snedaker, 1974; Robertson and Alongi, 1992). A "World Mangrove Atlas" has been constructed delineating the geographical distribution of existing mangroves around the world (Spalding et al., 1997). From a biogeochemical perspective, the major difference between organic matter in mangrove forests and marsh systems is the presence and high concentrations of woody detrital material. Wood production accounts for approximately 60% of the net primary production in mangroves (Alongi, 1998)—this has significant effects on rates of decomposition (as discussed in the following section). Otherwise, mangrove trees are similar to salt marsh plants in that they are: (1) adapted to grow in wet unstable soils generally low in oxygen; (2) occur in saline environments; and (3) experience frequent flooding and draining due to tidal fluctuations (Day et al., 1989).

The typical ranges of standing crop biomass for marshes and mangroves are 500 to 2000 g dry wt m^{-2} (Day et al., 1989) and 10,000 to 40,000 g dry wt m^{-2} (Twilley et al., 1992), respectively. When comparing estimates of above- and below-ground production of mangroves and salt marshes we see that mangroves are generally more productive than salt marshes above ground (table 8.2). However, in many cases the below-ground biomass of salt marshes exceeds that of above-ground production in salt marshes (Schubauer and Hopkinson, 1984). Conversely, the below-ground biomass of mangroves is usually about 50% of total forest biomass (Alongi, 1998). Other studies have shown that above-ground production of *S. alterniflora* is negatively correlated with latitude (Turner, 1976; Dame, 1989) and positively correlated with tidal range (Steever et al., 1976). The amount of

Table 8.2 Estimates of above- and below-ground net primary production (NPP) of some selected species/community types of salt marsh grasses and mangroves.

Community type	Location	Above-ground NPP (g dry wt m^{-2} y^{-1})	Below-ground NPP (g dry wt m^{-2} y^{-1})
Marsh grasses			
Distichlis spicata	Pacific coast	750–1500	—
	Atlantic coast	—	1070–3400
Juncus roemerianus	Georgia	2200	—
	Gulf coast	4250	1360–7600
Spartina alterniflora	Atlantic coast	500–2000	550–4200
	Gulf coast	3250	279–6000
S. patens	Gulf coast	7500	—
	Atlantic coast	—	310–3270
Mangroves			
Rhizophora apiculata	Southeast Asia	1900–390	—
Mixed *Rhizophora* spp.	New Guinea	1750–3790	—
	Indonesia	990–2990	—
Bruguiera sexangula	China	3500	—

Adapted from Kennish (1986) and Saenger (1994). Modified from Alongi (1998).

below-ground biomass in wetlands has also been shown to significantly alter the degree of anoxia in soils, thereby affecting pH, as well as nutrient and contaminant cycling (Howes et al., 1981; Boto and Wellington, 1988). For example, it is well known that mangrove and marsh plants *translocate* oxygen to their roots and rhizomes. Although there are conflicting reports concerning the temporal and spatial importance of such redox alterations in wetlands (McKee et al., 1988; Alongi, 1996, 1998), it is generally accepted that large amounts of underground biomass have long-term effects on sediment redox and the associated microbial/elemental cycling. In fact, mangrove forests may be similar to tropical rainforests in that a significant fraction of the organic matter pool is stored in living biomass as a mechanism for retaining nutrients (Archibold, 1995; Alongi et al., 2001). However, there remains considerable disagreement as to whether mangroves act as net sinks or sources of dissolved and suspended organic matter and nutrients to estuaries and adjacent coastal systems (Boto and Bunt, 1981; Twilley, 1985; Rivera-Monroy and Willey, 1996; Alongi, 1996; Alongi et al., 1998; Dittmar and Lara, 2001). In general, it appears that outwelling is more likely to occur in mangroves that have an excess of porewater nutrients, positive net sedimentation, and a macrotidal range (Dittmar and Lara, 2003).

Estuarine DOM consists of a diverse array of allochthonous and autochthonous sources (see reviews, Cauwet, 2002; Findlay and Sinasbaugh, 2003; Sinsabaugh and Findlay, 2003). These major sources primarily comprise riverine inputs, autochthonous production from algal and vascular plant sources, benthic fluxes, groundwater inputs, and exchange with adjacent coastal systems (figure 8.6). In coastal and open ocean waters, the DOM is generally positively correlated with phytoplankton biomass (Guo et al., 1994; Santschi et al., 1995; Hansell and Carlson, 2002). It has been estimated that phytoplankton release approximately 12% of total primary production as dissolved organics over a range of freshwater to marine environments (Baines and Pace, 1991). However, water column DOM in many estuarine systems does not show positive correlations with phytoplankton biomass because of allochthonous inputs from soils and vascular plants that are hydrologically linked to these wetland systems (Hedges et al., 1994; Argyrou et al., 1997; Guo and Santschi, 1997; Harvey and Mannino, 2001; Hernes et al., 2001; Jaffé et al., 2004).

Past work has shown that DOM from river/estuarine systems is primarily derived from terrestrial vegetation/soils (Malcolm, 1990; Opsahl and Benner, 1997). In fact, some of the highest DOM estuaries in the United States border the Gulf of Mexico (Guo et al., 1999; Engelhaupt and Bianchi, 2001), a region that also has some of the highest rates of fresh litter decomposition in soils (figure 8.7; Meentemeyer, 1978). Recent work has shown that DOC concentrations in groundwater near the forest recharge area of North Inlet (USA) range from 50 to 140 mg L^{-1}, but that much of this is lost within the freshwater portion of the aquifer through sorption and heterotrophic decay (Goni and Gardner, 2003). Despite these losses during transport to the coast, DOC discharge from saline groundwaters are high and can contribute as much as 600 mg C m^{-2} d^{-1} to the annual DOC budget of North Inlet (USA). Riverine/estuarine DOM is generally considered to be recalcitrant and transported conservatively to the ocean (Moore et al., 1979; Mantoura and Woodward, 1983; Prahl and Coble, 1994; van Heemst et al., 2000). However, spatial variability in the abundance of phytoplankton exudates (Aminot et al., 1990; Fukushima et al., 2001), uptake of DOM by bacteria (Findlay et al., 1991; Gardner et al., 1996; Zweifel, 1999; Pakulski et al., 2000), chemical removal processes (e.g., flocculation, deflocculation, adsorption, aggregation, precipitation) (Sholkovitz, 1976; Sholkovitz et al., 1978;

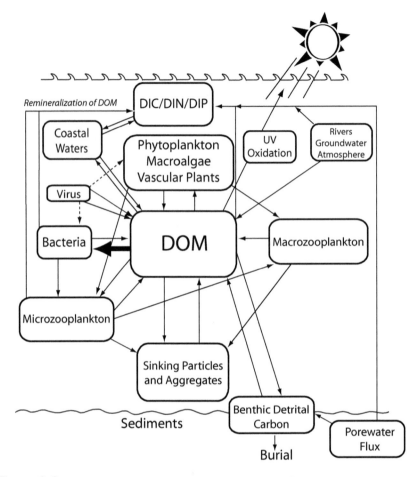

Figure 8.6 Major sources of dissolved organic matter (DOM) to estuaries, primarily composed of riverine inputs, autochthonous production from algal and vascular plant sources, benthic fluxes, groundwater inputs, and exchange with adjacent coastal systems. (Modified from Hansell and Carlson, 2002.)

Ertel et al., 1986; Lisitzin, 1995), inputs from porewaters during resuspension events (Burdige and Homstead, 1994; Middelburg et al., 1997), and atmospheric inputs (Velinsky et al., 1986) can all contribute to nonconservative behavior in estuaries.

Although mechanisms of DOM removal by physical/chemical processes in the mixing zone of estuaries are not well understood, they are believed to be important processes affecting the composition of riverine/estuarine DOM. Some of the earliest work on DOM removal processes noted that iron is important in the initial steps of flocculating humic substances across an estuarine salinity gradient and that much of this humic material was composed of humic acids (Swanson and Palacas, 1965; Eckert and Sholkovitz, 1976;

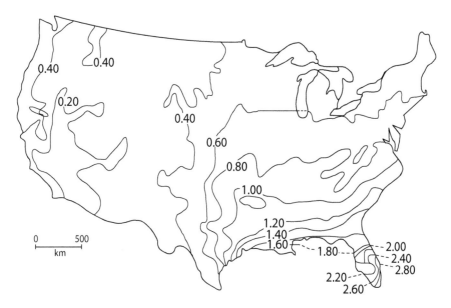

Figure 8.7 Rates of fresh litter decomposition (y^{-1}) in soils in the USA, using a simulation model based on evapotranspiration rates as a predictive variable. Contour lines represent loss rate (k) during an initial year of decay. (Modified from Meentemeyer, 1978.)

Sholkovitz, 1976; Sholkovitz et al., 1978; Fox, 1983, 1984). Other work showed that changes in tidal/mixing energy and total suspended load, within the turbidity maximum or from resuspension events, had significant effects on flocculation and deflocculation processes (Biggs et al., 1983; Eisma and Li, 1993). In the Dollard Estuary (The Netherlands), it was concluded that large flocs were formed during short durations of high suspended matter and that deflocculation of these large flocs occurred while settling or at the sediment–water interface (Eisma and Li, 1993). Recent laboratory experiments have also shown that sedimentary POC and DOC that is loosely bound to mineral surfaces can be released during resuspension events (Komada and Reimers, 2001). Thus, in many shallow estuarine systems where resuspension events are common occurrences, this may be another mechanism that allows for a relatively constant input of organic matter to the water column.

In estuaries where the input from riverine sources is highly refractory and terrestrial DOM dominates the bulk pool of DOM in the estuary (minor inputs from algal sources) conservative behavior is more likely to occur. For example, the Ob and Yenisei estuaries, located adjacent to the Kara Sea in the Arctic Ocean, show a conservative behavior of DOM versus salinity because of the highly refractory nature of the riverine DOM inputs combined with the rapid and selective removal of freshwater labile algal components during transport to the river mouth (figure 8.8; Köhler et al., 2003). Similarly, conservative mixing of DOM was recently observed in estuaries of southwest Florida (USA); this DOM was highly refractory and primarily derived from mangrove sources (Jaffé et al., 2004).

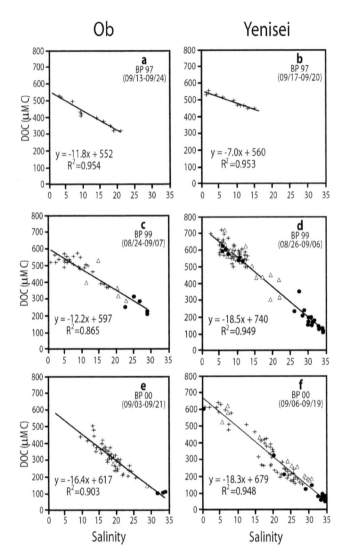

Figure 8.8 Conservative behavior of DOM versus salinity in the Ob and Yenisei estuaries, located adjacent to the Kara Sea in the Arctic Ocean. (Modified from Köhler et al., 2003.)

However, at salinities greater than 30 the mangrove signal was diluted by in situ inputs of DOM from marine phytoplankton.

It is well known that DOM exists in different size fractions in aquatic systems (Sharp, 1973) and that a large portion of estuarine DOM is composed of colloidal organic matter (Whitehouse et al., 1989; Filella and Buffle, 1993; Guo et al., 1995, 1999; Martin et al., 1995; Sempere and Cauwet, 1995; Guo and Santschi, 1997; Cauwet, 2002).

The characterization of different size classes of DOM is established by physical separation through filters/membranes of differing pore sizes; thus, colloids are an operationally defined fraction of the total DOM in the size range 0.001–1 μm (Vold and Vold, 1983). Unfortunately, using size as the primary criterion for characterizing colloids can be very misleading when considering the biochemical differences associated with these different size fractions and their reactivity with contaminants (Gustafsson and Gschwend, 1997; Benner, 2002).

Another term commonly used in association with DOM is *humic substances*; humic substances are typically defined as complex assemblages of molecules that have a yellow-to-brown color and are derived from plants and soils (Hatcher et al., 2001). Humic substances represent a large fraction of what is termed *chromophoric dissolved organic matter* (CDOM) in aquatic systems around the world (Blough and Green, 1995)—CDOM as a distinct component of DOM is discussed in more detail later in this chapter. Aquatic humic substances can be further categorized as *fulvic acids*, *humic acids*, and *humin* based on their solubility in acid and base solutions (Schnitzer and Khan, 1972; Aiken, 1988; Parsons, 1988; McKnight and Aiken, 1998). More specifically, humic acids typically have a molecular weight of greater than 100,000 daltons (Da) and are soluble above a pH of 2: fulvic acids which are smaller molecules ($\sim$500 Da) are soluble at any pH, and humin is not soluble across a full pH range (Sempere and Cauwet, 1995; McKnight and Aiken, 1998). For reasons of simplification, the term high molecular weight dissolved organic matter (HMW DOM) will be used to describe DOM greater than 1 kDa and less than 0.45 μm as being inclusive of colloidal and humic acid type substances, and low molecular weight compounds (LMW DOM) as being less than 1 kDa—unless further distinctions are otherwise required.

Numerous techniques have been developed over the past few decades to isolate DOM (see reviews, Benner, 2002; Hedges, 2002). Examples of some isolation procedures include solid-phase amberlite ion-exchange resins for humic substances (e.g., XAD-2, XAD-8, XAD-4) (Aiken et al., 1985; Thurman, 1985), gel filtration ultracentrifugation (Wells and Goldberg, 1991; Wells, 2002), flow field-flow fractionation (flow FFF) (Beckett et al., 1987; Hassellov et al., 1999; Gustafsson et al., 2001), silica-based C_{18} stationary phases (Louchouaran et al., 2000), size-exclusion chromatography (Chin and Gschwend, 1991), high-performance size-exclusion chromatography (Minor et al., 2002), gel filtration (Sakugawa and Handa, 1985) and ultrafiltration (Benner et al., 1992; Buffle et al., 1993; Buessler et al., 1996; Guo and Santschi, 1997). Some key problems associated with sold-phase extraction (SPE) procedures (e.g., XAD resins and C_{18} columns) are that the DOM needs to be pre-treated or acidified (pH $\sim$ <4) to enhance the extraction efficiency (adsorption onto the resin) and they only extract approximately 20% of the total DOM (Hedges et al., 1992). Moreover, there are concerns that during pretreatment the labile fractions of the DOM may be altered in composition (Benner, 2002). One of the more widely adopted methods for DOM isolation is tangential-flow (cross-flow) ultrafiltration (see review, Benner, 2002)—this is commonly referred to as *ultrafiltered* DOM or UDOM. This method allows for processing of large sample volumes but does not require pre-treatment of DOM (Benner et al., 1992). However, variability in the performance of cartridge membranes from different manufacture's on the same samples has created some concerns about the importance of interlaboratory comparisons (Buessler et al., 1996; Guo and Santschi, 1996). Furthermore, it is critical to perform system checks on the integrity

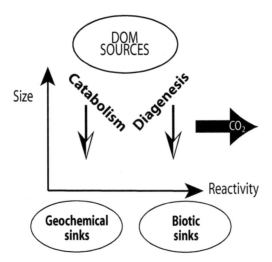

Figure 8.9 Relationship between size and reactivity of DOM largely depends on "age" of DOM inputs to a system as well as the relative importance of trophic versus diagenetic controls on DOM composition. (Modified from Sinsabaugh and Foreman, 2003.)

of the cartridge membrane with molecular probes to prevent possible "breakthrough," or leakage of HMW DOM during ultrafiltration (Guo et al., 2000b).

A considerable amount of work has shown that different size classes of DOM are utilized with different efficiencies by the microbial community (see reviews by Tranvik, 1998; Benner, 2002, 2003; Findlay, 2003). In fact, the size-reactivity continuum model states that larger size classes of organic matter are less degraded and more bioavialable than smaller size classes that are more refractory as a consequence of extensive processing over time (Amon and Benner, 1996). However, as pointed out by Sinsabaugh and Foreman (2003) this perspective of size and reactivity largely depends on the "age" of DOM inputs to a system as well as the relative importance of trophic versus diagenetic controls on DOM composition (figure 8.9). This model also contradicts the earlier *biopolymer degradation* (BD) and *abiotic condensation* (AC) models of humic substance formation (Hedges, 1988), whereby, the BD model states that newly released bioploymers from organic matter will eventually be broken down into more labile smaller molecules (figure 8.10; Hedges, 1988). Conversely, the AC model states that small molecules can repolymerize outside the cell through condensation reactions over time into larger more refractory macromolecules. The more recent work by Amon and Benner (1996) also added another component of reactivity to their model, which indicated that the HMW DOM fraction of DOM was more bioavailable to bacteria and more photoreactive than LMW DOM. While this model certainly applies quite well to open ocean systems, where the primary source of POM is phytoplankton, in many cases it may not be applicable to estuarine and freshwater systems where DOM is commonly derived from rivers (high humic-rich soil inputs), sediment porewaters (see below), and adjacent marsh systems, all of which have high lignocellulosic (particularly lignin) components that result in the HMW DOM fraction being more degraded (refractory) and older (Guo and Santschi, 2000; Mitra et al., 2000a,b;

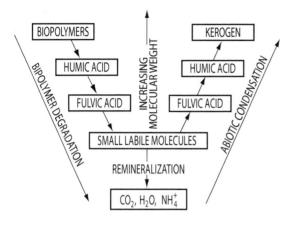

Figure 8.10 Humic substance formation depicted by the early biopolymer degradation and abiotic condensation models. (Modified from Hedges, 1988.)

McKnight et al., 2003; Wetzel, 2003). This issue of age and bioavailability in estuarine systems is complex and will be discussed in more detail specific to elemental components (e.g., N, C, P, S) of DOM cycling in chapters 10 to 13.

A considerable amount of work has focused on the importance of chemical composition, size, and age in controlling microbial metabolism of DOM (see reviews, Jansson, 1998; Tranvik, 1998; Benner, 2002, 2003; Findlay, 2003; Kirchman, 2003). The importance of microbes in DOM cycling was recognized and incorporated into the theory of the microbial loop, which first showed that bacteria are key in controlling the trophic linkages between DOM, POM, and inorganic nutrients in aquatic ecosystems (figure 8.11; Pomeroy, 1974; Azam et al., 1983). The initial step in remineralizing HMW DOM is through extracellular hydrolytic enzymes; once small enough, the LMW DOM can then be transferred across the outer membrane of bacteria (Arnosti, 2003). These DOM uptake and production processes clearly occur on shorter ecological timescales than those previously described during the formation of humic substances in soil horizons. However, they are similar to the timescales of flocculation/deflocculation processes in estuaries. While we know that bacterial utilization of monomers such as amino acids and glucose are quite high (see review, Kirchman, 2003), other small intermediate products (e.g., small organic acids) formed during macromolecular breakdown remain largely unknown. A large percentage of bacterial production can be supported by dissolved free amino acid (DFAA) uptake in estuaries (Keil and Kirchman, 1991a; Hoch and Kirchman, 1995). However, in estuarine and freshwater systems, the input of DOM with a relatively high abundance of aromatic compounds (tannins and lignins)—due to high vascular plant inputs from soils and wetlands—makes the DOM pool more refractory to microbes. The interrelationship between small monomeric and large macromolecular molecules in DOM and the roles they play in bacterial metabolism is an issue that requires further study in highly dynamic systems, such as estuaries, where allochthonous and autochthonous inputs are significant.

Another component of total DOM that has received considerable attention in recent years in estuaries, and particularly in shelf waters, is (CDOM) (see review, Blough and Del Vecchio, 2002). CDOM has been commonly referred to as Gelbstoff, gilvin, and yellow substance, because of its strong association with humic substances. In fact, CDOM is

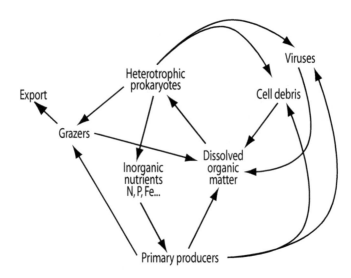

Figure 8.11 Microbial loop theory showing that bacteria are key in controlling the trophic linkages between DOM, POM, and inorganic nutrients in aquatic ecosystems. (Modified from Foreman and Covert, 2003.)

believed to be primarily composed of humic acids that absorb visible light, UV-A (wavelengths from 315 to 400 nm), and UV-B (wavelengths from 280 to 315 nm) (Blough and Del Vecchio, 2002). Absorption spectra for DOM typically decrease exponentially with increasing wavelength, as shown in figure 8.12, in CDOM samples collected from the Suromoni and Orinoco Rivers (Venezuela) in the Amazonian plains (Battin, 1998). The higher absorption values for the Suromoni, a small tributary that drains into the Orinoco, reflect the highly aromatic character of this terrestrially derived CDOM, which is different from the Orinoco which has some autochthonous contributions to the CDOM. The abundance of CDOM in estuarine and coastal waters has been shown to significantly affect biological and optical processes. For example, while CDOM can have beneficial effects on organisms by reducing their exposure to harmful UV-B, it can also reduce photosynthetically active radiation (PAR) to phytoplankton assemblages (Blough and Zepp, 1990; Bidigare et al., 1993; Vodacek et al., 1997; Neale and Kieber, 2000). CDOM also interferes with remote sensing of phytoplankton in estuarine and coastal waters; this has resulted in extensive efforts to improve algorithms that attempt to predict chlorophyll-a distribution and phytoplankton biomass from satellite imagery (Carder et al., 1999; Kahru and Mitchell, 2001).

In estuarine systems, CDOM abundance is typically controlled by the following: (1) riverine freshwater inputs of humic substances derived from soil organic matter in the watershed (Keith et al., 2002; Wang et al., 2004); (2) inputs of CDOM from porewaters during resuspension events (Coble, 1996; Burdige et al., 2004); (3) degradation and production of in situ algae and vascular plants (Del Castillo et al., 2000) and (4) anthropogenic inputs from the watershed (industrial and agricultural) (Bricaud et al., 1981).

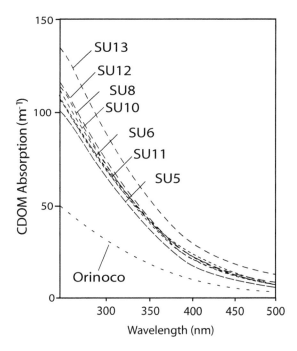

Figure 8.12 Absorption spectra for DOM with increasing wavelength in CDOM samples collected from the Suromoni (SU) and Orinoco Rivers (Venezuela) in the Amazonian plains. (Modified from Battin, 1998.)

Recent work has shown a significantly positive relationship between CDOM absorption and freshwater input in Narragansett Bay (USA) (figure 8.13; Keith et al., 2002). More specifically, there was a strong relationship between CDOM absorption and freshwater discharge during mid-spring and summer months ($r^2 = 0.88$ and 0.68). During late winter and early spring, the inputs of nutrients from the watershed result in formation of phytoplankton blooms which dominate over the absorption of CDOM. This dilution effect of phytoplankton on the relative absorption of CDOM has also been observed in the Gulf of Maine (USA) (Yentsch and Phinney, 1997).

Light absorption by CDOM (principally UV) can result in loss of its absorption and fluorescence properties (*photobleaching*) and the production of reactive intermediates/products. These products can have important implications for the following: (1) speciation of metals through the formation of reactive oxygen species (ROS) (e.g., H_2O_2 O_2^-, $\bullet OH$); (2) production of LMW DOM, bacterial substrates, DIC, and trace gases (e.g., CO, CO_2, COS); and (3) overall chemical composition of the total DOM pool in estuarine and coastal waters (Wetzel, 1990; Mopper et al., 1991; Wetzel et al., 1995; Bushaw et al., 1996; Moran and Zepp, 1997; Opsahl and Benner, 1998; Del Castillo et al., 1999, 2000; Moran et al., 2000; Engelhaupt et al., 2002; Keith et al., 2002; Moran and Covert, 2003; Yamashita and Tanoue, 2003). Light exposure of DOM collected from the

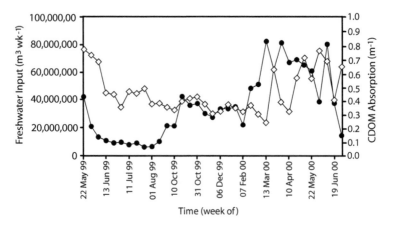

Figure 8.13 Positive relationship between CDOM absorption and freshwater input in Narragansett Bay (USA). (Modified from Keith et al., 2002.)

Satilla Estuary (USA) showed that > 50% of the original CDOM could be lost through photobleaching over a period of 51 days (Moran et al., 2000). Moreover, the degree of photobleaching was positively correlated with bacterial uptake of DOM—based on bacterial respiration (figure 8.14; Moran et al., 2000). These results provided evidence that biological degradation of DOM from the *blackwaters* of the Satilla Estuary, which is largely (~60%) composed of humic substances derived from soils (Beck et al., 1974), can be enhanced by *photodegradation*. Other work has shown that photochemical breakdown of refractory riverine DOM can result in rapid uptake by microbes (Amon and Benner, 1996). This work provides further support that there is significant transformation and utilization of terrestrially derived DOM across estuarine gradients that may be critical in understanding the deficiency of terrestrially derived DOM in the global ocean (Meyers-Schulte and Hedges, 1986; Hedges et al., 1997; Opsahl and Benner, 1997; Hedges, 2002).

Fluorescence is another means by which CDOM has been characterized in recent years due to its higher sensitivity and greater applicability to remote sensing than absorbance measurements (Vodacek et al., 1995). Over the past decade, new fluorescence techniques such as *three-dimensional excitation emission matrix* (3DEEM) spectroscopy have been applied for examining the fluorescence properties of DOM (Coble et al., 1990). Further work indicated that *fluorochromes* could be divided into the following three major classes: (1) *humic-like* (emission wavelength from 370 to 460 nm); (2) *protein-like* (emission wavelength from 305 to 340 nm); and (3) *chlorophyll-like* (emission wavelength of 660 nm) (Coble et al., 1998). Other work in freshwaters (Wu et al., 2001), estuaries (Mayer et al., 1999), and sediment pore waters (Coble, 1996), and on phytoplankton and bacteria indicated that the protein-like fluorochrome was commonly observed. More specifically, the protein-like fluorochrome can be divided into tyrosine-like (emission wavelength of approximately 300 nm) and tryptophan-like (emission wavelength of approximately 350 nm) categories (Coble, 1996). Recent work has also shown in the Ise Bay estuary (Japan) that, when comparing measured concentrations of these aromatic amino acids in relation

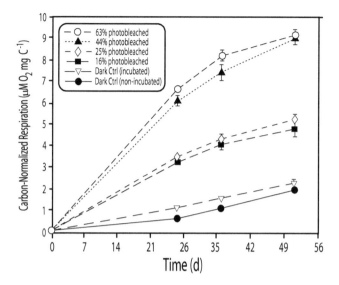

Figure 8.14 Bacterial respiration of Satilla Estuary (USA) DOM during dark incubations for 51 d as a function of prior photodegradation. Error bars represent 95% confidence limits (3 replicates). (Modified from Moran et al., 2000.)

to these specified fluorescence intensities, the dissolved amino acids were derived from small peptides and not protein molecules (Yamashita and Tanoue, 2003). Other work has recently shown that distinct fluorophores found in the humic-like fraction [in fulvic acids from the International Humic Substances Society (IHSS); Coble, 1996; Mobed et al., 1996] can be used to examine terrestrially/microbially derived aromaticity of DOM (McKnight et al., 2001). More details on the applications of chemical biomarkers in understanding the composition and sources of POM and DOM in estuaries will be discussed in chapter 9.

Sediments may also represent an important source of DOM to the water column of shallow estuarine systems (see review, Burdige, 2002). The accumulation of DOM in porewaters results from the diagenesis of sedimentary organic carbon into a complex mixture of macromolecules (e.g., humic substances) and smaller monomeric forms (e.g., amino acids) (Orem et al., 1986; Burdige, 2002). The flux of porewater DOM across the sediment–water interface has also been shown to represent an important source to the total DOM pool in estuaries (Alperin and Reeburgh, 1985; Alperin et al., 1992; Burdige et al., 1992; Martens et al., 1992; Burdige and Homstead, 1994; Argyrou et al., 1997; Middelburg et al., 1997; Burdige, 2001, 2002). It has been suggested that benthic fluxes are a significant source to coastal and estuarine systems (0.9×10^{14} g C y^{-1}) and may represent about the same magnitude as riverine inputs of DOC to the ocean (2–2.3×10^{14} C y^{-1}) (Burdige et al., 1992; Burdige and Homstead, 1994). The role of porewater DOM that gets entrained in the *benthic nepheloid layer* (BNL), from resuspension events of estuarine and shelf sediments, may also be important in affecting the composition and age of DOM in shelf and deeper slope waters (Guo and Santschi, 2000; Mitra et al., 2000a). Similarly, porewaters contribute to the protein or amino acid-like fluorescence signature of the water column; maximum concentrations of these signals have been observed at the sediment–water interface (Coble, 1996). More details on the early diagenesis of sedimentary organic matter (SOM) and DOM formation will be presented later in this chapter (p. 204).

Decomposition of Organic Detritus

Organic detritus has long been recognized for its importance as a food resource and its influence on overall biogeochemical cycling in coastal systems (Tenore, 1977; Rice, 1982; Tenore et al., 1982; Mann and Lazier, 1991). Major contributors of organic detritus in estuarine systems are decaying plant materials and animal fecal pellets. Vascular plant detritus is particularly important in many estuarine systems; this refractory material typically requires the activity of microbial communities to convert lignocellulosic polymers into more labile available food resources for higher consumers (Moran and Hodson, 1989a,b). The decay of aquatic organic detritus is generally divided into (1) *leaching*, (2) *decomposition*, and (3) *refractory* phases (Olah, 1972; Odum et al., 1973; Fell et al., 1975; Harrison and Mann, 1975; Rice and Tenore, 1981; Valiela et al., 1985; Webster and Benfield, 1986).

In the *leaching phase*, soluble compounds are rapidly lost from fresh detritus over a scale of minutes to weeks (figure 8.15; Wilson et al., 1986). In the case of the marsh plant *S. alterniflora*, as much as 20 to 60% of the original material can be lost during this phase (Wilson et al., 1986). These soluble DOM compounds released from detrital particles are rapidly used by a high abundance of free bacteria in the surrounding water column (Aneiso et al., 2003). Much of this leached material likely consists of short-chain carbohydrates, proteins, and fatty acids (Dunstan et al., 1994; Harvey et al., 1995).

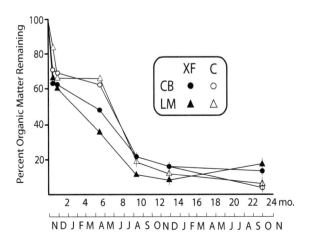

Figure 8.15 Percentage organic matter remaining during a 23-month decay experiment with *S. alterniflora* detritus. (Modified from Wilson et al., 1986.)

There has also been considerable debate about the rate of decay of these labile leached substrates in oxic versus anaerobic conditions, with some studies showing faster decay under oxic conditions (Bianchi et al., 1991; Lee, 1992; Sun et al., 1994; Harvey et al., 1995) and others showing no effects of redox (Henrichs and Reeburgh, 1987; Andersen, 1996)—more discussion on this later in the chapter as it relates to preservation in sediments. The decomposition phase involves the heterotrophic breakdown of detritus by microbes and metazoans (e.g., *detritivores* and *deposit feeders*). Nitrogenous compounds of micro- and macroalgae are more likely to contribute to the nutrition of detritivores than are vascular plants (Findlay and Tenore, 1982). As microorganisms colonize detritus during this phase, there is a relative increase in the nitrogen content of the "aging" detritus (Darnell, 1967; Tenore et al., 1982; Rice and Hanson, 1984). Earlier studies proposed that most of the nitrogen could be equated with "protein enrichment" from microbes (Newell, 1965; Odum et al., 1973). However, further work showed that the presence of nonprotein nitrogen in plant detritus (Suberkropp et al., 1976) and enrichment from complexation (physical and chemical) in *humic geopolymers* (Hobbie and Lee, 1980; Rice, 1982; Rice and Hanson, 1984) suggested that protein-N was not the only source. Adsorption of NH_4^+ by detritus may have also added to the total N (Mackin and Aller, 1984). The attached bacteria are considered to be more active than the surrounding free bacteria (Griffith et al., 1994) and are critical in the breakdown of refractory compounds. Moreover, there is a succession of microorganisms that occurs as attached bacteria solubilize POM using *exoenzymes* which then becomes available for free bacteria (Azam and Cho, 1987). Other work has shown that adsorption of NH_4^+ and attached bacteria were not important contributors to total nitrogen in decomposing *S. alterniflora* litter (Hicks et al., 1991). The next successional stage is typically occupied by protozoa (Biddanda and Pomeroy, 1988; Caron, 1987).

Vascular plants contain more ligneous and phenolic compounds, which can also reduce resource availability to consumers (Valiela et al., 1979; Rice and Tenore, 1981; Harrison, 1982). *Secondary plant compounds* such as phenols, alkaloids, tannins, organic acids, saponins, terpenes, steroids, essential oils, and glycosides can deter herbivory and detritivory (Swain, 1977; Rietsma et al., 1982). For example, tannins are phenolic compounds known to inhibit microbial activity through the precipitation of enzyme proteins (Janzen, 1974). These compounds are important in the genesis of humic and fulvic acids, which have also been shown to inhibit enzyme activity (Thurman, 1985; Wetzel, 2003). Similarly, C_4 plants (e.g., *S. alterniflora*, *S. patens*, *D. spicata*) produce cinnamic acids (e.g., ferulic and *p*-coumaric acids) which inhibit herbivory more effectively than comparable C_3 marsh plants (*Juncus* spp.) (Haines and Montague, 1979; Valiela et al., 1979; Valiela,1995). While the nitrogen enrichment by microbes certainly enhances the "quality" of vascular plant detritus for consumers, microbes also break down structural carbohydrates (e.g., cellulose, xylan) that many invertebrate detritivores/herbivores are incapable of digesting (Kofoed, 1975; Levinton et al., 1984). The presence of these secondary compounds generally results in slower decay rates for vascular versus nonvascular plants (Valiela, 1995). It should be noted that there are certain invertebrate species that have evolved the ability to digest this highly refractory plant materials. For example, sesarmid crabs (Grapsidae) have been shown to remove 20 to 40% of the mangrove litterfall in East African mangrove forests (Slim et al., 1997). The final refractory phase is characterized by detritus composed of lignin and cellulose which decay very slowly (Maccubbin and Hodson, 1980;

Wilson et al., 1986). The source of detritus will clearly affect the period of refractory decay. For example, the refractory phase of phytodetritus may last for only a few weeks compared to vascular plant detritus which may last for months to years (Valiela, 1995; Opsahl and Benner, 1999). Finally, temperature and size of the decomposing detrital particles also represent parameters affecting decay rate. Early work by Hodson et al. (1983) showed that lignocellulose remineralization rates were enhanced with decreasing particle size due to the higher relative abundance of microbes with an increasing surface-to-volume ratio. To no surprise, litter bag experiments indicate that increases in temperature also increase decay rates due to enhanced microbial activity (Wilson et al., 1986).

Berner (1980) first introduced the following concept of a "one-G" model for determining first-order decay constants (k) of organic matter decomposition:

$$G_t = G_0 e^{-kt} \qquad (8.3)$$

where: G_t = mass of detritus at time t; and G_0 = initial mass of detritus.

Although this model appeared to work well in calculating decay constants for predominantly labile sources of organic matter such as macro- and microalgae, there were problems with more refractory sources of detritus such as *Spartina* (Rice and Hanson, 1984). To better describe the decay dynamics of refractory detritus, which generally contain both labile and refractory biochemical components, the following "two-G" model was developed by Rice and Hanson (1984):

$$G_t = G^* + G_{10} e^{-k_1^t} \qquad (8.4)$$

where: G^* = constant mass of refractory material, defined by $G^* = G - G_1$, where G is the total mass of the detritus and G_1 is the labile material; G_{10} = initial mass of labile material; and k_1 = decay constant of labile material.

In a decomposition experiment comparing decay dynamics of macroalgal (e.g., *Gracilaria foliifera*) and vascular plant (e.g., *Spartina alterniflora*) detritus, it was found that the two-G model proved more precise in calculating decay constants over shorter time intervals (days) compared to the one-G model, which works relatively well from over weeks to months (Rice and Hanson, 1984). Other work has shown clear differences in the labile and refractory components of macroalgae and vascular plants, whereby the more abundant refractory lignocellulosic component in *S. alterniflora* persists over a period of 40 weeks as compared to similar losses of labile fatty acids in both over about 3 weeks (figures 8.16 and 8.17; Tenore et al., 1984). Similarly, Westrich and Berner (1984) performed laboratory experiments to show that decomposition of phytodetritus in sediments can be separated into two decomposable fractions—with significant differences in reactivity, and a highly refractory (nonmetabolizable) fraction. This work established, for the first time, experimental evidence that organic matter decomposition in sediments can be described by a multi-G model.

In estuarine wetlands, relatively small amounts of the above-ground production of higher plants are grazed by herbivores (Teal, 1962; Valiela, 1995; Alongi, 1998) or exported to coastal waters (Day et al., 1989; Alongi, 1998). Decomposition rates of organic detritus in wetlands are also relatively slow compared to those in regions with predominantly higher inputs of algal material. The generally slower rates of decay have been attributed to (1) anaerobic conditions, (2) acidity of the water, and (3) inhibition of

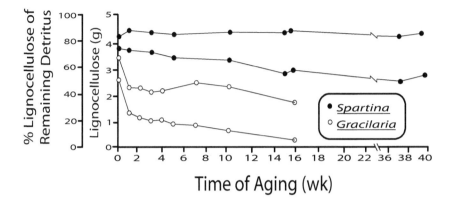

Figure 8.16 Percentage and abundance (g) of lignocellulose in remaining detritus of *S. alterniflora* and *G. foliifera* over a 40-week period. (Modified from Tenore et al., 1984.)

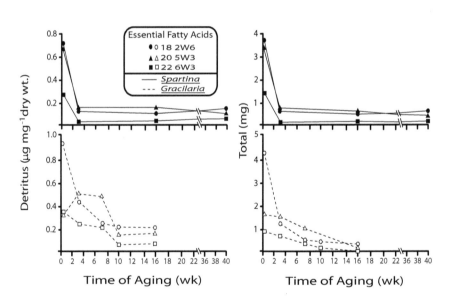

Figure 8.17 Relative and total abundance of essential fatty acids in remaining detritus of *S. alterniflora* and *G. foliifera* over a period of about 3 weeks. (Modified from Tenore et al., 1984.)

decomposition by dissolved humic substances and secondary compounds (as discussed above) (Qualls and Haines, 1990). While redox conditions and composition of detritus are more important in estuarine systems, freshwater wetlands and tidal freshwater regions of estuaries may be affected more by low pH values. These low pH values are primarily caused by high concentrations of dissolved humic substances containing carboxylic acid groups (Thurman, 1985; Wetzel, 2003).

The location of marsh detritus above ground or below ground (e.g., flooded or not) strongly affects decomposition rates of vascular plant detritus (Hackney and de la Cruz, 1980; Neckles and Neill, 1994). Recent work on marsh grass (*Sarcocornia fructicosa*) decomposition in the Ebre River estuary (Spain) showed that stems and roots (the most recalcitrant material in marsh grasses) represented 70 to 80%, respectively, of the remaining material in litter bags after one year (Curćo et al., 2002; figure 8.18). In a similar study on decomposition of salt marsh plants in the Po River estuary (Italy) (Scarton et al., 2000), it was shown that the percentage of root and stem material remaining after 65 weeks of decay in *S. fruticcosa* (66 and 64%, respectively) and *P. australis* (30 and 50%, respectively) was in general agreement with work by Curćo et al. (2002). However, the significantly high percentage of leaf material remaining in the *P. australis* (50%) versus the *S. fructicosa* (4%) further supported the idea that *P. australis* is more decay resistant than most marsh plants and may be more effective in sustaining marsh/sediment elevation with increasing sea-level rise. Similarly, *S. patens* has been shown to be important in vertical accretion in Louisiana marshes because of its slow decay rates (Foote and Reynolds, 1997). The range of marsh accretion rates for marshes along the Gulf of Mexico and Atlantic United States coast is 0.15 to 1.35 cm y^{-1}, as compared to a range of relative sea level rise estimates from 0.23 to 0.92 cm y^{-1} (table 7.4). Many mangrove swamps are also threatened by sea-level rise, particularly in areas where peats dominate and terrigenous soil inputs are low (Woodroffe, 1995; Kikuchi et al., 1999). Typical accretion rates in mangroves range from −0.82 to 1.6 cm y^{-1} (Bird, 1980; Spenceley, 1982). Recent work has shown that aerial root structures (e.g., prop roots and pneumatophores) enhance trapping and sedimentation of particles during tidal exchange due to greater surface area friction (Furukawa and Wolanski, 1996; Krauss et al., 2003).

Early Diagenesis

As discussed in chapter 2, most modern estuarine systems are have been filling with sediments since their formation about 5000 to 6000 y BP; this invokes long-term accumulation and storage of organic matter in these systems. The fate of SOM will be dependent on the amount of early diagenesis that occurs in the upper sediments, which is largely controlled by the "quality" of organic detrital inputs (discussed above) and redox conditions of the sedimentary environment. Leeder (1982, p. 32) defined diagenesis as "the many chemical and physical processes which act upon sediment grains in the subsurface," which is further distinguished from *halmyrolysis*, defined as "a more restricted aspect of chemical changes operative at the sediment–water interface." The full sequence transformation of SOM in the geosphere (down to several kilometers) is generally divided into the following stages of maturation: (1) *diagenesis*; (2) *catagenesis*; and (3) *metagenesis* (figure 8.19; Tissot and Welte, 1984). The stage of diagenesis (down to about 1000 m) involves the microbial and chemical depolymerization of the polymeric biological precursor molecules

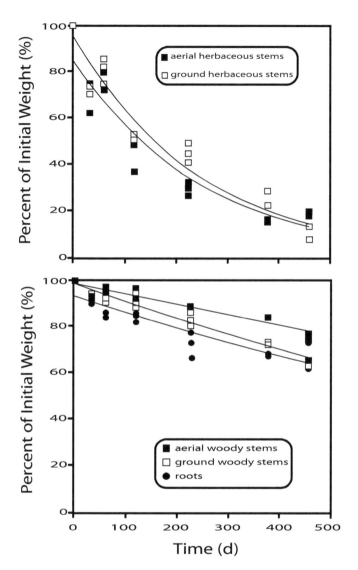

Figure 8.18 Percentage of initial weight of marsh grass (*Sarcocornia fructicosa*) remaining in litter bags in the Ebre River estuary (Spain) after one year. (Modified from Curćo et al., 2002.)

(e.g., polysaccharides and proteins) of SOM into monomeric molecules, which then recondense to form humic substances and eventually *kerogen*. During metagenesis and catagenesis, kerogen and select lipids are transformed into coal, gas, and oil at high temperatures and pressures. Since the biogeochemical processes covered in this book are not concerned with the metamorphic processes that occur at higher temperatures and pressure,

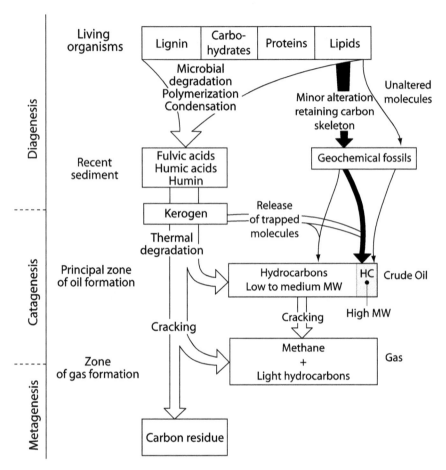

Figure 8.19 Full sequence transformation of soil organic matter in the geosphere (down to several kilometers) divided into the following stages of maturation: (1) diagenesis; (2) catagenesis; and (3) metagenesis. (Modified from Tissot and Welte, 1984.)

the focus here will be placed on diagenesis, or more specifically *early diagenesis*. Berner (1980 p. 38) defined early diagenesis "as the changes occurring during burial to a few hundred meters where elevated temperatures are not encountered and uplift above sea level does not occur." This is the stage where key microbial and chemical transformations occur at low temperatures in recently deposited sediments and have major effects on biogeochemical cycling in nearshore and estuarine environments.

Early diagenesis is typically described as a *steady-state phenomenon*; however, unless very long-term geological timescales are considered, steady-state conditions are generally not common in shallow turbid environments such as estuaries. There are many factors that contribute to these non–steady-state conditions, such as variations in sedimentation rate, inputs of organic matter, chemistry of bottom waters and sediments, bioturbation rates, and resuspension (Lasagna and Holland, 1976). Consequently, numerous attempts have been made to examine *non–steady-state diagenesis* over shorter periods in estuarine systems (Alongi et al., 1996; Luther et al., 1997; Mortimer et al., 1998; Anschutz et al., 2000; Deflandre et al., 2002).

As a result of particle settlement to the sediment–water interface, there is a mass accumulation of sediments which results in compaction of sediments and the physical upward transport or *advection* of solutes in porewaters to the overlying water. Similarly, solutes in porewaters can also move by *diffusion* as a result of concentration gradients. Thus, porewaters can be transported by advection from burials, molecular diffusion, and biological pumping or irrigation (Aller, 2001; Jørgensen and Boudreau, 2001). Diffusion in aqueous environments occurs according to *Fick's laws of diffusion* (Berner, 1980).

Fick's first law, used for steady-state conditions, is as follows:

$$J_i = -D_i \, (\delta C_i / \delta x) \qquad (8.5)$$

where: J_i = diffusional flux of component i in mass per unit area per time; D_i = diffusion coefficient of i in area per unit time; C_i = concentration of component i in mass per unit time; and x = direction of maximum concentration gradient (negative sign indicates flux is in opposite direction of gradient).

Fick's second law, used for non–steady-state conditions, is as follows:

$$\delta C_i / \delta t = Di \left(\delta^2 C_i \right) / \delta x \qquad (8.6)$$

To apply Fick's laws to solute fluxes in sediments, adjustments have to be made to these equations to account for the negative interference effects that sediment particles have on the diffusion of solutes in pore waters (Lerman, 1979; Berner, 1980). For example, *tortuosity*, defined as the length of the tortuous path that a solute travels around particles across a distance across a certain depth interval can be described by the following equation (Berner, 1980; Krom and Berner, 1980a):

$$\theta = \delta l / \delta x \qquad (8.7)$$

where: θ = tortuosity; δl = length of sinuous path that solute travels over; and δx = depth interval.

The effects of tortuosity on whole sediment diffusion can be further described as (Berner, 1980):

$$D_s = D/\theta^2 \qquad (8.8)$$

where: D_s = whole sediment diffusion coefficient (area of sediment per unit time).

Tortuosity is determined indirectly by measuring the electrical resistivity of natural sediments from which the porewaters have been extracted (Berner, 1980) using the following relationship:

$$\theta^2 = \varphi F \qquad (8.9)$$

where: F = formation factor, whereby $F = R/R_0$; typical values for F in fine-grained sediments are 1.5 to 3.0 (Ullman and Aller, 1982; Iverson and Jørgensen, 1993); R = electrical resistivity of the sediments; and R_0 = electrical resistivity of the pore waters.

After making these adjustments for diffusion in sediments, the mass balance and vertical concentration patterns of nonconservative solutes in saturated sediments can be described by the following one-dimensional advective–diffusive general diagenetic equation (GDE) (Berner, 1980; Aller, 2001; Jørgensen and Boudreau, 2001):

$$\delta\varphi C/\delta t = \varphi D_s \left(\delta^2 C/\delta x^2\right) - \delta\omega_p C/\delta x - \sum R_i \qquad (8.10)$$

where: C = concentration of solute; t = time; D_s = whole sediment diffusion coefficient of solute C; x = sediment depth relative to surface ($x = 0$); ω_p = pore water advection velocity relative to the sediment–water interface; and ΣR_i = sum of all reactions affecting solute C.

These types of diagenetic models have been commonly used to describe the distribution of redox-sensitive metals (e.g., Mn and Fe) because of their close association with the mineralization of organic matter in sediments (Burdige, 1993; Jørgensen and Boudreau, 2001). In particular, a number of steady-state models have been used to describe the diagenesis of Mn and Fe (more details on this later in chapter) in sediments (Burdige and Gieskes, 1983; Aller, 1990; Boudreau, 1996; Dhaker and Burdige, 1996; van Cappellen and Wang, 1996; Slomp et al., 1997; Overnell, 2002). In a recent study on sediments from the Loch Etive estuary (Scotland), diagenetic processes based on reactive Mn and Fe oxides as electron acceptors were compared using the diagenetic models of van Cappellen and Wang (1996) and Slomp et al. (1997). It was concluded that there was both considerable agreement and disagreement in the modeled sediment profiles of Mn and Fe, and that the application of multiple diagenetic models is strongly encouraged for comparison (Overnell, 2002).

Assuming that the transport (e.g., physical mixing and bioturbation), depositional inputs (e.g., sources/quality and amounts of POM), mass sediment accumulation rate, temperature, and decomposition are in steady state, as well as lateral homogeneity of the deposit, the concentration of POM will not change over time. Although these conditions are almost never met in estuarine systems (Berner, 1980), steady-state conditions will be used here for the general purposes of this discussion. Thus, assuming steady-state conditions, the GDE for POM decay is as follows (Rice and Rhoads, 1989):

$$\delta G_s/\delta t = D \left(\delta^2 G_s/\delta x^2\right) - \omega \left(\delta G_s/\delta x\right) - k G_s = 0 \qquad (8.11)$$

where: t = time; x = sediment depth relative to surface layer ($x = 0$); G_s = steady-state concentration (mass/sediment volume) of POM in sediment; D = random mixing coefficient for particles; ω = burial rate resulting from sedimentation of particles at the sediment surface; and k = apparent first-order decay rate of POM, whereby $\delta G_s/\delta t = -k G_s$.

As discussed earlier, not only does deposition of POM to the sediment surface typically vary over time in estuaries, but there are also issues of POM actively growing at the sediment–water interface, such as benthic microalgae. This is an important issue because the rate of supply of POM is assumed to be equal to the depositional flux from the water column (Berner, 1980). More details on diagenetic models addressing the surface boundary condition constraints as well as POM lability can be found in Aller (1982) and Rice and Rhoads (1989). When examining biological mixing as a one-dimensional diffusive process, bioturbation and biodiffusion coefficients are used in the GDE similarly to standard Fickian diffusivity (Wheatcroft et al., 1991). The justification for using a biodiffusion coefficient (D_B) is because it is assumed that all biological mixing activities are integrated over time, thereby making it a diffusion-like process. Estimates of D_B can be made using a regression of down-core distribution of radionuclides—as described in chapter 7. More specifically, biodiffusivity can be described by modification of the advective–diffusion equation 7.22 (Nittrouer et al., 1984), where sediment accumulation rate (A) is described by the following equation:

$$A = \lambda x / \ln(C_0/C_x) - D_B/x \, [\ln(C_0/C_x)] \qquad (8.12)$$

where: λ = decay constant (y^{-1}); C_0 = activity of radionuclide at upper level of the profile ($_0$); C_x = activity of radionuclide at distance x below the $_0$ (dpm g^{-1}); and D_B = biodiffusivity (cm^2 y^{-1}).

Typical D_B values for Long Island Sound estuary (USA) and Chesapeake Bay have been estimated to range from 0.5 to 110 cm^2 y^{-1} (Aller and Cochran, 1976; Aller et al., 1980) and 6 to >172 cm^2 y^{-1} (Dellapenna et al., 1998), respectively. Since mixing is clearly not a one-dimensional process, estimated biodiffusivity values may seriously underestimate mass transport when mixing is affected by horizontal advection (Wheatcroft et al., 1991).

When examining the rates of organic matter decay during early diagenesis, redox becomes an important controlling variable. Dissolved O_2 in the water column is a critical factor controlling benthic communities and elemental cycles in sediments (Aller, 1980; Yingst and Rhoads, 1980; Rosenberg et al., 2001). Redox in sediments is typically defined by E_h (redox potential) where positive values indicate oxidizing conditions (Stumm and Morgan, 1996). In redox chemistry, the half-reaction concept is used to understand E_h, which is the electron activity in (volts) relative to a standard hydrogen electrode (Chester, 2003). Thus, the activity of electrons in solution is expressed by electron activity ($p\varepsilon$), a dimensionless term, or E_h (volts) by the following equation:

$$p\varepsilon = F/2.3RT \, (E_h) \qquad (8.13)$$

where: F = Faraday's constant; R = gas constant; and T = absolute temperature.

E_h conditions are primarily controlled by production and decomposition of organic matter; positive values are oxidizing and negative values are reflective of reducing conditions. The depth of the *redox potential discontinuity* (RPD) layer is associated with distinct coloration, indicative of differences between oxic and suboxic conditions (Fenchel and Riedl, 1970; Santschi et al., 1990). The oxidized region in surface sediments of estuarine sediments (one to a few centimeters of sediment) is typically orange–brown in color (Fe and Mn oxides) followed by a reduced region that is gray–black in color (monopolysulfides). The depth of the RPD is largely controlled by the amount of organic matter loading, physical mixing, and bioturbation. Recent work has shown that the location of the RPD depth is very similar when measured by electrodes in sediment cores or sediment profile images (SPIs) (Rosenberg et al., 2001).

The decomposition of organic matter in marine/estuarine sediments proceeds through a sequence of *terminal electron acceptors* (e.g., O_2, NO_3^-, MnO_2, $FeOOH$, SO_4^{2-},

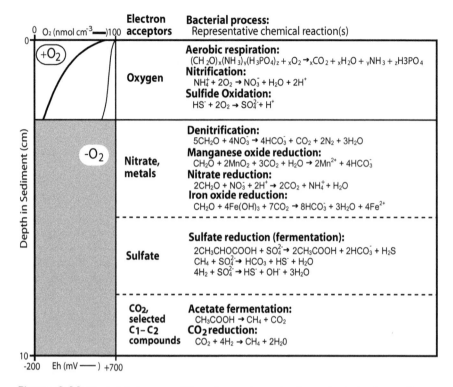

Figure 8.20 Bacterial decomposition of organic matter in marine/estuarine sediments through a sequence of terminal electron acceptors (e.g., O_2, NO_3^-, MnO_2, FeOOH, SO_4^{2-}, and CO_2) and changing redox. (Modified from Deming and Baross, 1993.)

and CO_2) (Richards, 1965; Froelich et al., 1979; Canfield, 1993) as shown in figure 8.20; Deming and Baross, 1993). The sequence of reactions are principally determined by the free energy yields (ΔG^0) (see chapter 4) per mole of organic C (table 8.3). In essence, the products of one electron acceptor become the electron donor for another electron acceptor such as the oxidation of Fe^{2+} and FeS by MnO_2 and NO_3^-. While this sequence does provide a basic framework from which to work, recent work has shown that the versatility of many bacteria may allow for several of these reactions to occur in the same zone (Brandes and Devol, 1995). This is likely because natural populations of bacteria are more adaptable compared to the pure cultures in which many of these specific redox reactions have been tested (Jørgensen and Boudreau, 2001). Moreover, it is believed that a consortia of mutualistic bacteria are responsible for anaerobic decomposition because any particular bacterium is not capable of complete remineralization (Fenchel et al., 1998), unlike aerobic microorganisms which do have the enzymatic capacity to perform total mineralization (Kristensen and Holmer, 2001). However, despite these differences in the ability of oxic and anaerobic bacterial consortia to break down organic complexes, other work has shown that the slower decay rates commonly observed with increasing sediment

Table 8.3 Pathways of organic matter oxidation in sediments and their standard free-energy yields, $\Delta G°$, per mole of organic carbon.

Pathway and Stoichiometry of Reaction	kJ mol^{-1}
Oxic respiration:	
$CH_2O + O_2 \rightarrow CO_2 + H_2O$	-479
Denitrification:	
$5CH_2O + 4NO_3^- \rightarrow 2N_2 + 4HCO_3^- + CO_2 + 3H_2O$	-453
Mn oxide reduction:	
$CH_2O + 3CO_2 + H_2O + 2MnO_2 \rightarrow 2\ Mn^{2+} + 4HCO_3^-$	-349
Fe oxide reduction:	
$CH_2O + 7CO_2 + 4Fe(OH)_3 \rightarrow 4Fe^{2+} + 8HCO_3^- + 3H_2O$	-114
Sulfate reduction:	
$2CH_2O + SO_4^{2-} \rightarrow H_2S + 2HCO_3^-$	-77
Methane production:	
$HCO_3^- + 4H_2 + H^+ \rightarrow CH_4 + 3H_2O$	-136
$CH_3COO^- + H^+ \rightarrow CH_4 + CO_2$	-28

depth are primarily due to stage of decomposition, origin of organic matter, and diminished exchange of metabolites and not to redox effects (Canfield, 1994; Kristensen et al., 1995; Aller and Aller, 1998; Hulthe et al., 1998). Recent work has corroborated this by showing that fresh organic materials decay at similar rates independent of redox conditions (Kristensen and Holmer, 2001). Thus, as we move across redox gradients with increasing sediment depth there exists a complex distribution of organic matter in different stages of decay ranging from large polymeric organic substrates, that are broken down by hydrolysis and fermentative processes, into smaller monomeric and water-soluble molecules (e.g., formate, acetate, propionate) (Kristensen and Hansen, 1995). These smaller moieties (e.g., acetate) are then consumed in the dominant pathways of anaerobic respiration (e.g., NO_3^- and SO_4^{2-} reduction).

A major difference in the cycling of DOM in porewaters, compared to water column DOM, is the relatively high amount of remineralization that occurs through anoxic pathways, such as fermentation. Much of the DOM in porewaters is in the < 3 kDa size fraction and is generally considered to be highly refractory (Burdige, 2002). This is in accordance with the size-reactivity continuum model proposed by Amon and Benner (1996) described earlier. Burdige and Gardner (1998) proposed a similar model called the *porewater size/reactivity model* (PWSR) to describe the cycling of DOM in sediments. Similar to remineralization of POM described earlier for the water column, sedimentary POM is initially hydrolyzed by bacteria in the benthos to HMW DOM; this is further hydrolyzed to monomeric LMW DOM (mLMW DOM)—these small molecules (e.g., amino acids, monosaccharides) are used in other respiratory processes by bacteria in sediments (figure 8.21; Burdige, 2002). Another fraction of the HMW DOM is converted into polymeric LMW DOM (pLMW DOM) which is believed to be more recalcitrant than the HMW DOM due to chemical alterations during hydrolysis (Guo and Santschi, 2000).

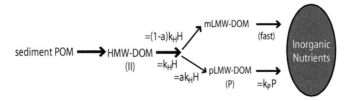

Figure 8.21 Porewater size/reactivity model (PWSR) model used to describe the cycling of DOM in sediments. Similar to remineralization of POM described earlier for the water column, sedimentary POM is initially hydrolyzed by bacteria in the benthos to HMW DOM; this is further hydrolyzed to monomeric LMW DOM (mLMW DOM)—these small molecules (e.g., amino acids, monosaccharides) are used in other respiratory processes by bacteria in sediments. H = HMW DOM, P = polymeric LMW DOM (pLMW DOM), k_H = rate constant for HMW DOM consumption, k_p = rate constant for pLMW DOM consumption, and a = fraction of HMW DOM consumption that passes through the pLMW DOM pool. (Modified from Burdige, 2002.)

Oxygen is the preferred terminal electron acceptor in the traditional view of unidirectional redox succession with increasing sediment depth (figure 8.20). During aerobic degradation other oxygen-containing intermediates (e.g., $^\bullet O_2{}^-$, H_2O_2, $^\bullet OH$) are formed that can also assist in the breakdown of refractory organic matter (Canfield, 1994). Similarly, $NO_3{}^-$ is relatively abundant in porewaters within the oxic zone from nitrification processes. As organic matter decomposition increases and sediment depth increases, organic matter mineralization changes to anaerobic processes (Henrichs and Reeburg, 1987; Gond and Hollander, 1997; Boudreau and Canfield, 1993; Roden et al., 1995). Although oxic decomposition is considered to be more efficient, a considerable amount of organic matter in estuarine systems is decomposed by suboxic and anaerobic processes. For example, in the suboxic zone $NO_3{}^-$ can be consumed by bacterial denitrification as well as oxidation of Mn^{2+} (Schultz et al., 1994; Luther et al., 1997). Similarly, $NH_4{}^+$ can be oxidized by nitrate to Mn oxides in anaerobic sediments (Hulth et al., 1999; Anschutz et al., 2000). The consumption of O_2 also changes from use by heterotrophic bacteria as they break down organic material to *chemolithotrophic bacteria* where inorganic compounds are oxidized. For example, as discussed earlier the consumption of O_2 from both chemical and bacterial Mn (III, IV) and Fe (III) oxides is another important sink for O_2 in the suboxic zone of sediments (Aller et al., 1986; Aller, 1990), with the majority of the reactions being mediated by chemolithotrophic bacteria (Aller and Rude, 1988; Nealson and Saffarini, 1994; Dollhopf et al., 2000). Metal oxides have also been shown to be important in controlling the distribution of DOM in porewater (Tipping, 1981; Gu et al., 1995; Chin et al., 1998; Jakobsen and Postma, 1999; Filius et al., 2000). Recent work in the Saguenay Fjord (Canada) showed that, there was strong evidence that, as Fe and Mn metal oxides are reduced, adsorbed DOC is released (Delflandre et al., 2000).

Down-core porewater profiles of Mn^{2+}, Fe^{2+}, and H_2S in sediments from Aarhus Bay estuary (Denmark) show sharp gradients in the subsurface peaks of Mn^{2+} and Fe^{2+}, indicating the *reduction of Mn and Fe oxides* in the upper 2 and 4 cm, respectively

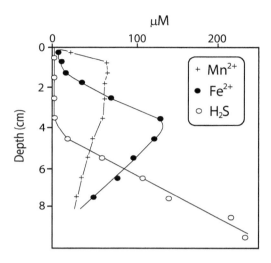

Figure 8.22 Down-core porewater profiles of Mn^{2+}, Fe^{2+}, and H_2S in sediments from Aarhus Bay estuary (Denmark). (Modified from Thamdrup et al., 1994.)

(figure 8.22; Thamdrup et al., 1994). The decrease in Mn^{2+} and Fe^{2+} near the sediment–water surface indicates that these reduced species are oxidized, as well as released to the water column. The sharp drop in Fe^{2+} as it diffuses downward between 4 and 8 cm results from the incorporation of Fe^{2+} and H_2S into FeS or FeS_2 (pyrite). The increase in H_2S with increasing depth results from SO_4^{2-} reduction. Bacterial SO_4^{2-} reduction represents the terminal pathway of organic C decay in anoxic sediments (Capone and Kiene, 1988). Sulfate is considered to be unlimiting in regions of maximal SO_4^{2-} reduction (Howarth, 1984). Conversely, the upward diffusion of H_2S results in the *reoxidation of sulfides*. It has been estimated that approximately half of the O_2 in estuarine systems is consumed by reoxidation of sulfides (Jørgensen, 1977; Boudreau and Canfield, 1993). In fact, low O_2 bottom waters in Chesapeake Bay have been shown to be significantly affected by the oxidation of reduced S compounds (e.g., H_2S) (Roden and Tuttle, 1993a,b; Roden et al., 1995). Many studies in nearshore and estuarine environments have shown that sulfate reduction accounts for the majority of organic matter in sediments (Howarth and Teal, 1979; Howarth and Giblin, 1983; Chanton and Martens, 1987a,b; Mackin and Swider, 1989; Roden et al., 1995). In Chesapeake Bay, SO_4^{2-} reduction accounted for 30 to 35% of the average net primary production and may have accounted for as much as 60 to 80% of the total sediment carbon metabolism (Roden et al., 1995). The relative importance of SO_4^{2-} reduction will largely depend on the amount of physical and biological mixing in the sediments as well as the amount of organic matter. Enhanced mixing of O_2 into sediments clearly results in a greater importance of O_2 in organic matter decomposition (Jørgensen and Revsbech, 1985).

The decomposition of organic matter in sediments has traditionally been described by diagenetic models of porewater and solid-phase profiles of measured rates of dominant decay processes such as sulfate reduction (Berner, 1980). Similar to the multi-G detrital decay models described earlier, models of sedimentary organic matter decay also use different pools of organic matter consisting of refractory (nonreactive) and labile (highly reactive) components (Burdige, 1991; Roden and Tuttle, 1996). The general concept of multiple-G pools of organic matter having different reactivities, first suggested by Jørgensen (1978), was first tested by Westrich and Berner (1984) in a set of laboratory experiments using estuarine sediments. This work showed that planktonic organic matter could be divided into G_1, G_2, and G_3 or G_{NR} pools, characterized by a freshest most labile metabolizable fraction, the second-most metabolizable material (10-fold slower in decay than G_1), and a nonreactive (NR) fraction that did not decay within the timeframe of their experiment (likely decays over a period of years), respectively (Westrich and Berner, 1984). Other models of organic matter decay, using sulfate reduction rates, have shown that decay in sediments occurs as a continuum from predepositional processes in the water column to change with depth in sediments (Middelburg, 1989). The reactivity of decomposing organic matter decreases with increasing depth in sediments (Burdige, 1991). For example, organic matter decay rates estimated from a multi-G or "mixture" model in sediment cores collected from Chesapeake Bay indicated that decay rates decreased in the top 2 m from 8.2 to 3.7 y^{-1} (G_1 material) and further decreased at depths from 12 to 14 cm to 2.1 to 0.2 y^{-1} (G_2 and G_3 materials) (Burdige, 1991). These rates were found to be within the range of decay rates observed using lab-slurries in earlier experiments (Martens and Klump, 1984; Westrich and Berner, 1984; Burdige and Martens, 1988; Middelburg, 1989). Pre-depositional decay processes have also been shown to be important in affecting rates of organic matter decay in sediments (Middelburg, 1989; Burdige, 1991). One additional feature of the Burdige (1991) mixture model is that it addresses the importance of sedimentary N and P mineralization, which was essentially ignored in the prior S and C based models (Westrich and Berner, 1984; Middelburg, 1989).

Animal–Sediment Relations and Organic Matter Cycling

The effects of macrofaunal (benthic invertebrates larger than 500 μm) *bioturbation* on the chemical and physical properties of sediments have been well documented in the literature (Fager, 1964; Rhoads, 1974; Aller, 1978; Yingst and Rhoads, 1978, 1980; McCall and Tevesz, 1982; Rhoads and Boyer, 1982). Early work established the effects of bioturbation on long- (years to decades) and short-term (weeks to months) sediment mixing, using particle-reactive radionuclides such as ^{210}Pb and [137]Cs (Nozaki et al., 1977; Robbins et al., 1977; Gardner et al., 1987) and [234]Th and [7]Be (Aller and Cochran, 1976; Krishnaswami et al., 1980), respectively. These changes in the properties of sediments have been shown to be linked to faster and more complete decay of organic matter in sediments than in sediments with no bioturbation (Andersen and Kristensen, 1992; Aller, 1994; Banta et al., 1999). Macrofaunal activities that produce such bioturbation effects are typically from tube irrigation, movement, fecal pellet egestion, and gallery digging (Francois et al., 2001). In fact, a new model of bioturbation using a functional approach was recently used to show physical and biological reworking mechanisms (figure 8.23) (Francois et al., 2001). Macrofauna can be divided into the following five functional groups: (1) biodiffusers—mix sediment randomly (e.g., amphipod, *Pontoporeia hoyi*) (Robbins et al., 1979); (2) upward conveyors—move sediments from depth to the surface through their guts (e.g., polychaete, *Leitoscoplos fragilis*) (Rice, 1986; Bianchi and Rice, 1988); (3) downward-conveyors—move sediments from the surface to depth through their guts (e.g., *Nereis diversicolor*); (4) regenerators—transfer sediment to the surface

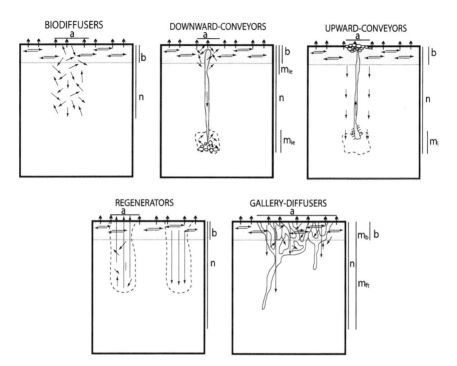

Figure 8.23 Model of bioturbation using a functional approach to show physical and biological reworking mechanisms. Other mixing parameters defined as follows: a = width of the organism mixing zone; n = depth of the organism mixing zone; m_{ie} = height of the ingestion–egestion zone of the downward-conveyor organism; m_b = height of the diffusion zone of the gallery-diffuser organism; and b = height of the first row of the matrix. Arrows represent movement of sediment particles. (Modified from Francois et al., 2001.)

through burrowing or digging activity (e.g., crab, *Uca pugilator*) (Gardner et al., 1987); and (5) gallery diffusers—build gallery systems in sediment leading to advective transport of particles due to egestion and animal movement (e.g., burrowing shrimp, *Callianassa* spp.) (Bianchi, 1991). On a larger scale, the depth at which particles and oxygen are mixed at depth through bioturbation is largely dependent on the successional stage of the benthic community (Rhoads et al., 1978; Rhoads and Boyer, 1982). For example, the early successional Stage I is typically represented by small opportunistic species with high population densities and growth rates resulting in a shallow RPD layer as opposed to the later Stage III which has more equilibrium species (climax community), smaller populations densities, slower growth rates, and a deeper RPD (figure 8.24; Zajac, 2001). Stage II is characterized by benthic species and RPD layers that are intermediate between Stage I and III communities. Recent work in Long Island Sound estuary showed that O_2 consumption rates in sediments were significantly altered after experimentally changing the composition of deposit-feeding polychaetes (Zajac, 2001). Disruption of the balance between the physical supply of O_2 from the water column and biotic demand demonstrated

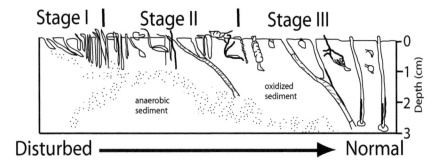

Figure 8.24 Successional stages of the benthic community. (Modified from Zajac, 2001.)

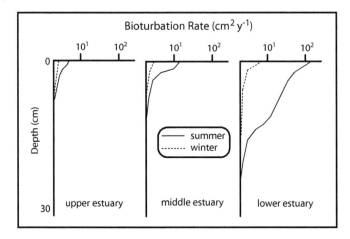

Figure 8.25 Depth of mixing and bioturbation rates as they relate to seasonal temperature changes in different regions of Chesapeake Bay (USA). (Modified from Schaffner et al., 2001.)

in this work further supports the importance of understanding the relationship between benthic functional diversity and biogeochemical cycles.

The composition of benthic communities and their potential roles in bioturbating sediments are largely controlled by salinity, grain size, and sediment depositional gradients in estuaries. For example, in Chesapeake Bay the macrofauna in the southern reaches (lower estuary) are characterized by large populations of head-down deposit feeders (late successional stage) in fine sands and high salinity (Schaffner, 1990; Dellapenna et al., 1998). This is contrasted with smaller populations of surface feeding tube-building species in the northern bay (upper estuary) where salinities are lower and sediments are finer grained (Dellapenna et al., 1998). These differences in benthic community composition largely are reflected quite well when examining the depth of mixing and bioturbation rates as they relate to seasonal temperature changes (figure 8.25; Schaffner et al., 2001).

Bioturbation has also been shown to influence fluxes of organic matter, nutrients, and contaminants in estuarine sediments and across the sediment–water interface (Kristensen and Blackburn, 1987; Aller, 1994; Kure and Forbes, 1997). Several models have been developed to determine the fate of particulate matter that settles on the sediment surface—most have used an advective–diffusion formulation (Goldberg and Koide, 1962; Guinasso and Schink, 1975; Robbins et al., 1979; Berner, 1980; Fisher et al., 1980; Aller, 1982; Rice, 1986; Boudreau, 1996; Aller et al., 2001; Boudreau and Jørgensen, 2001). Recent work by Aller (2001) described the following basic ways that solute transport can be enhanced by bioturbation—in the context of the parameters and assumptions made in the GDE: (1) biogenic turbulent diffusion; (2) biogenic advection; (3) generation of internal sources and sinks; and (4) geometric analogs (e.g., burrow structures) to the bio-irrigated zone. When modeling solute transport, it is important to recognize that such microscale features and structures occur within benthic community assemblages, because most GDE models are too simplistic and have unrealistic assumptions on the degree of sediment homogeneity (Wheatcroft et al., 1991; Aller, 2001).

Controls on the Preservation of Organic Matter in Estuarine Sediments

The preservation of organic matter in coastal and estuarine sediments is believed to be principally controlled by productivity, sedimentation accumulation rate, bottom water and sediment redox conditions, and sorption as a function of specific surface area of sediments (see review, Hedges and Keil, 1995). In this section the focus will be primarily on factors controlling preservation or organic matter in estuarine sediments; the use of chemical biomarkers for historical paleo-reconstruction of past estuarine environments is discussed in chapter 15.

In the case of productivity, the predicted positive relationship between burial rate and productivity was first discovered in estuaries and can be altered by the source and lability of the organic matter being produced. For example, it has been estimated that only 14 to 21% of the POC deposited (based on sediment trap data) in Chesapeake Bay is ultimately buried in sediments (Roden et al., 1995). Thus, despite high levels of net primary production (NPP) rates (e.g., 20–40 mol C m^{-2} y^{-1}) in plankton-dominated estuaries impacted by eutrophication, mineralization rates are also typically high (Jørgensen et al., 1990; Sampou and Oviatt, 1991; Roden et al., 1995). As mentioned earlier a significant fraction of mineralization occurs through SO_4^{2-} reduction, so separating out the effects of redox here also makes this complicated. Such high rates of mineralization are due to the lability of the organic matter sources (e.g., phytodetritus), as discussed in the preceding sections. However, in estuaries where macrophyte detrital inputs are large, such as Cape Lookout Bight (USA), significantly more of the organic matter ($\sim$ 70 %) is preserved in sediments (Martens and Klump, 1984). Therefore, in estuarine systems, where there are more diverse inputs of organic matter, compared to the coastal and deep ocean, it is critical to be able to identify changes in sources of organic matter; this requires the use of chemical biomarkers in concert with bulk elemental measurements (more details discussed in chapter 9).

The first evidence for enhanced preservation with high net sediment accumulation rates in oceanic environments was presented in the context of organic burial efficiency—to avoid dilution effects (Henrichs and Reeburgh, 1987; Cowie and Hedges, 1992). *Organic burial efficiency* is defined as the accumulation rate of organic matter below the active zone of diagenesis divided by the organic matter flux to surface sediments (Henrich and Reeburgh, 1987; Hedges and Keil, 1995). If we switch to carbon currency, we see that the positive relationship between percentage C preserved and net sedimentation rate across a broad spectrum of sedimentation regimes indicates that at the highest accumulation rates there appears to be a decrease in percentage C preserved (Aller, 1998; figure 8.26). In most cases high accumulation rates can result in such rapid burial of labile phytodetrital material that it escapes efficient oxic decay processes, as evidenced by low C/P ratios (Ingall and van Capellen, 1990). Estuaries that fall into the category of the highest net sediment accumulation rates (e.g., 0.1–1 m y^{-1}) are typically deltaic systems (Nittrouer et al., 1985; Harris et al., 1993; Aller, 1998; McKee et al., 2004).

   Once again the issue of covariables enters into the picture, since many of these high net sediment accumulation regions also typically have high productivity (Cadée, 1978; Redalje et al., 1994; DeMaster et al., 1996)—primarily due to riverine nutrient inputs and coastal upwelling, and coastal currents (Aller, 1998). Aller (1998) proposed that the lower C burial rates are due to the development of a *fluidized-bed reactor* in these high accumulation rate environments with a high mineralization capacity due to the following processes: (1) repetitive redox successions; (2) metabolite exchange; and (3) efficient decomposition linked to cometabolism and metal cycling. Frequent resuspension of sediments across these deltaic environments can produce a fluidized bed in the surface sediments which

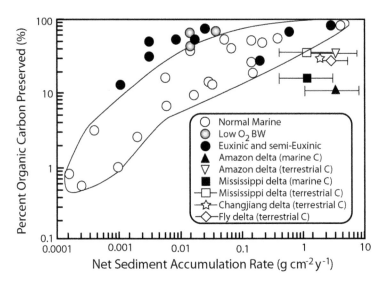

Figure 8.26 Positive relationship between % C preserved and net sedimentation rate across a broad spectrum of sedimentation regimes; BW = bottom waters. (Modified from Aller, 1998.)

has also been shown to be very efficient in the remineralization of organic matter due to repetitive redox successions (Aller, 1998). Cometabolism, a process whereby mixing of fresh labile organic carbon (algal sources) may enhance remineralization (via higher microbial turnover rates) of natural recalcitrant material (i.e., terrestrial carbon) (Lohnis 1926; Canfield et al., 1993; Canfield, 1994; Aller, 1998) may also be a major controlling variable on remineralization. Cometabolism processes in these deltaic environments may be high due to significant mixing of both labile phytodetritus and terrestrial carbon. High inputs of Fe and Mn oxides to these environments commonly result in frequent reduction-reoxidation cycling, thereby enhancing mineralization rates (Aller et al., 1996; Aller, 1998; McKee et al., 2004).

The role of O_2 in the preservation of organic matter in oceanic and coastal sediments has been intensely debated over the past 20 years—to date no consensus has been reached (Emerson, 1985; Henrichs and Reeburgh, 1987; Canfield, 1989; Lee, 1992; Calvert and Pedersen, 1992; Hedges and Keil, 1995; Dauwe et al., 2001). Due to the variability in redox conditions commonly found in nearshore systems, the term *diagenetic oxygen exposure time* was proposed as an indicator for organic matter preservation based on the amount of time sinking POM resides in the oxic porewaters of sediments (Hartnett et al., 1998; Hedges et al., 1999). Estuarine environments are prone to more frequent changes in redox due to their shallowness and dynamic mixing properties (see chapter 3). Moreover, many of these estuaries are in a state of eutrophication resulting in an expansion of hypoxic conditions (Nixon, 1995; Bricker et al., 1999; Rabalais and Turner, 2001). Estuaries such as the Long Island Sound (Parker and O'Reilly, 1991), Chesapeake Bay (Roden and Tuttle, 1996), Baltic Sea (Europe) (Elmgren and Larsson, 2001), and the Mississippi River plume (Rabalais and Turner, 2001) are a few examples of systems experiencing frequent hypoxia. The importance of seasonal changes in hypoxia on organic matter preservation in sediments has also been a topic of recent interest. In estuarine shallow systems where sometimes daily resuspension events occur depending on wind patterns (Baskaran et al., 1997), or in river-dominated deltaic systems where fluid muds also allow for frequent resuspension events (Aller, 1998), oxic conditions may prevail most of the year, promoting degradation. Similarly, bioturbation and/or bioirrigation can also result in relatively rapid and unpredictable changes in the anoxic or oxic conditions of sediments over a period of hours to minutes (Jumars et al., 1990; Aller, 1994). Thus, based on the aforementioned views of unidirectional changes in redox and biogeochemical rates with increasing sediment depth, much of this would not be applicable to sediments where bioturbation and/or physical mixing are the dominant factor. Only recently have we begun to design experiments that allow for such variability in pulsing of controlled redox conditions in laboratory experiments to examine such effects on organic matter decay (Sun et al., 2002)—more detail on such chemical biomarker approaches is described in chapter 9. Other cofactors affecting timescales of redox conditions involve positive feedback pathways whereby release of remineralized nutrients from sediments to the overlying water fuel more water column production, thereby maintaining high organic loading and low O_2 conditions (Ingall and Jahnke, 1994).

The strong association between organic matter and fine-grained sediments has long been speculated to be a result of sorption of organic matter to mineral surfaces (Weiler and Mills, 1965; Tanoue and Handa, 1979; Mayer, 1999). However, only more recently has this been further tested since a significant fraction of POC in riverine and estuarine suspensions is sorbed to mineral grains (Keil et al., 1994a,b, 1997; Mayer, 1994a,b; Bergamaschi et al., 1997; Ransom et al., 1998). More specifically, it was proposed that most SOM occurs as layers of organic matter within the surface features on mineral surfaces, and can be described as *monolayer-equivalent* amounts (Keil et al., 1994a; Mayer, 1994a,b). Proteins typically show organic C concentrations in a range that reflects adsorption to monolayer saturation (Arai and Norde, 1990), hence considered to be monolayer-equivalent within 95% confidence bands (Mayer, 1994a). More recent work contended that this organic matter existed as discontinuous "blebs" of bacterial and undifferentiated protoplasm (Ransom et al., 1997, 1998). The fact that much of this material is not easily removed from the mineral matrix has major ramifications for the role of sorptive preservation of organic matter in sedimentary environments. One reason for such stability is that organic matter may be lodged in small *mesopore-sized* spaces which are too small for access by microbial enzymes (Mayer, 1994a,b). However, more recent work has shown that while enhanced preservation via enzyme exclusion in mesopores may still be an important protective mechanism, other more complicated microfabric arrangements within the pores may need to be considered (Mayer et al., 2004). If we accept the use of the term monolayer-equivalents, the covariation of percentage organic C and surface area within a monolayer-equivalent has been shown to range from 0.5 to 1.0 mg OC m^{-2} (Keil et al., 1994a,b). *Submonolayer* (below the monolayer-equivalent range) sediments should reflect more efficient decomposition of organic matter on mineral surfaces while supermonolayer (above the monolayer-equivalent range) sediments should reflect lower efficiency in organic matter decomposition. Hedges and Keil (1995) depicted the association between percentage total organic matter burial and monolayer zones across a broad spectrum of marine environments (figure 8.27). The presence of *supermonolayer* zones in anoxic basins and coastal oxygen minimum zones (OMZ) clearly demonstrates the effect of redox on monolayers. Conversely, deltaic regions have submonolayers reflecting the efficient degradation of organic matter in such systems (Aller, 1998). Estuaries typically have monolayer-equivalents suggesting stabilization of organic in mineral matrix surfaces (Mayer, 1994a). However, more studies are needed to better constrain that broad spectrum of redox conditions found in estuaries.

When considering that the decay of POM in sediments is accompanied by a suite of DOM intermediates, it is important to consider also the role of DOM in organic matter preservation in sediments. As described earlier in the PWSR model, there is a mechanism to create highly refractory high molecular weight humic substances in sediments which would serve to enhance organic matter in sediments (Burdige, 2001, 2002). It has also been suggested that mesopores may act to enhance geopolymerization rates (Hedges and Keil, 1995). The overall size and refractory nature of sorbed DOM should also affect how tightly bound the material may be to the mineral matrix. For example, desorption and adsorption processes have been shown to be important in the exchange between POM and DOM (Thimsen and Keil, 1998). It was recently shown, using experimental laboratory sediments from the Hudson River and New York Bight estuaries, that only a small fraction of POM may be easily transferred from sediments to the water column during simulated resuspension events (Komada and Reimers, 2001).

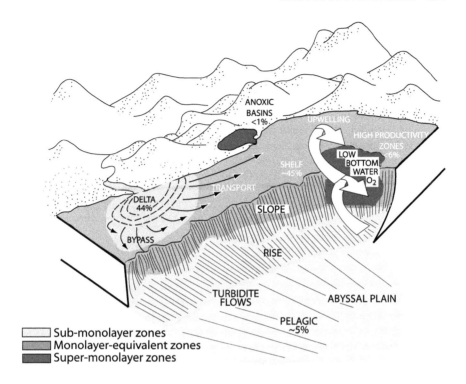

Figure 8.27 Association between percent total organic matter burial and monolayer zones across a broad spectrum of marine environments. (Modified from Hedges and Keil, 1995.)

Summary

1. Stoichiometry is defined as the mass balance of chemical reactions as they relate to the law of definite proportions and conservation of mass. The Redfield ratio provides the most well-known example of stoichiometric distinction where the average atomic ratios of C, N, and P in phytoplankton are relatively consistent (106:16:1) in most marine species.

2. Primary production is simply defined as the photosynthetic formation of organic matter. Other processes such as chemosynthesis involve the release of chemical energy from the oxidation of reduced substrates. Secondary production is the turnover of organisms that range in size from bacteria to the largest mammals on Earth that rely on the consumption of organic matter produced by photosynthetic organisms as food resource.

3. GPP is the amount of carbon fixed by photosynthetic organisms. Similarly, NPP is the total carbon fixed—minus the amount respired by the primary producer.

4. The general pools of organic matter in estuaries can be divided into particulate (>0.45 μm) and dissolved (<0.45 μm) fractions (POM and DOM). The dominant living and nonliving (detritus) sources of organic matter to estuaries can be divided into allochthonous and autochthonous pools—sources produced outside and within the boundaries of the estuary proper, respectively. These pools can further be divided into heterotrophs and autotrophs.

5. The dominant allochthonous inputs are from riverine, marine/estuarine plankton, and bordering terrestrial wetland sources. Autochthonous sources typically include plankton, benthic and epiphytic micro- and macroalgae, emergent and submergent (e.g., seagrasses) aquatic vegetation (EAV and SAV) within the estuary proper, and secondary production.

6. The dominant classes of phytoplankton in estuaries are Bacillariophyta (diatoms), Cryptophyta (cryptomonads), Chlorophyta (green algae), Dinophyta (dinoflagellates) Chrysophyta (golden-brown flagellates, chrysophytes, raphidophytes), and Cyanophyta (cyanobacteria).

7. Benthic macroalgae and microphytobenthos are important sources of primary production in estuaries and have significant effects on the seagrass, tidal flat, and intertidal marsh habitats.

8. Seagrass meadows are prominent features of many shallow littoral habitats in estuarine systems where they represent a major source of primary productivity. Wetlands such as freshwater/salt marshes and mangroves have been shown to be important major sources of primary production in estuarine systems.

9. These major sources estuarine DOM primarily consist of riverine inputs, autochthonous production from algal and vascular plant sources, benthic fluxes, groundwater inputs, and exchange with adjacent coastal systems.

10. Spatial variability in the abundance of phytoplankton exudates, uptake of DOM by bacteria, chemical removal processes (e.g., flocculation, deflocculation, adsorption, aggregation, precipitation), inputs from porewaters during resuspension events, and atmospheric inputs can all contribute to nonconservative behavior in estuaries.

11. Humic substances represent a large fraction of what is termed chromophoric dissolved organic matter (CDOM) in aquatic systems around the world. Aquatic humic substances can be further categorized as fulvic acids, humic acids, and humin based on theis solubility in acid and base solutions.

12. There has been a considerable amount of work that has focused on the importance of chemical composition, size, and age in controlling microbial metabolism of DOM. The importance of microbes in DOM cycling was recognized and incorporated into the theory of the microbial loop, which first showed that bacteria are key in controlling the trophic linkages between DOM, POM, and inorganic nutrients in aquatic ecosystems.

13. The decay of aquatic organic detritus is generally divided into (1) leaching, (2) decompositon, and (3) refractory phases.

14. Decomposition rates of organic detritus in wetlands is relatively slow compared with algal material. The generally slower rates of decay have been attributed to (1) anaerobic conditions, (2) acidity of the water, and (3) inhibition of decomposition by dissolved humic substances and secondary compounds.

15. Early diagenesis is defined as the changes occurring during burial to a few hundred meters where elevated temperatures are not encountered and uplift above sea level does not occur.

16. As a result of particle settlement to the sediment–water interface there is a mass accumulation of sediments which results in compaction of sediments and the physical upward transport or advection of solutes in porewaters to the overlying water. Similarly, solutes in porewaters can also move by diffusion as a result of concentration gradients.

17. Diffusion in aqueous environments occurs according to Fick's laws of diffusion.

18. The mass balance and vertical concentration patterns of nonconservative solutes in saturated sediments can be described by the one-dimensional advective–diffusive general diagenetic equation (GDE).

19. The decomposition of organic matter in marine/estuarine sediments proceeds through a sequence of terminal electron acceptors (e.g., O_2, NO_3^-, MnO_2, FeOOH, SO_4^{2-}, and CO_2).

20. The effects of macrofaunal bioturbation on the chemical and physical properties of estuarine sediments are well documented. The composition of benthic communities and their potential roles in bioturbating sediments are largely controlled by salinity, grain size, and sediment depositional gradients in estuaries.

21. The preservation of organic matter in coastal and estuarine sediments is believed to be principally controlled by productivity, sedimentation accumulation rate, bottom water and sediment redox conditions, and sorption as a function of specific surface area of sediments.

22. Due to the variability in redox conditions commonly found in nearshore systems, the term diagenetic oxygen exposure time was proposed as an indicator for organic matter preservation based on the amount of time sinking POM resides in the oxic porewaters of sediments.

Chapter 9

Characterization of Organic Matter

In chapter 8, a general overview was provided on the dominant sources of organic matter in estuarine systems. In general, estuarine organic matter is derived from a multitude of natural and anthropogenic allochthonous and autochthonous sources that originate across a freshwater to seawater continuum. Knowledge of sources, reactivity, and fate of organic matter are critical in understanding the role of estuarine and coastal systems in global biogeochemical cycles (Simoneit, 1978; Hedges and Keil, 1995; Bianchi and Canuel, 2001). Due to a wide diversity of organic matter sources and the dynamic mixing that occurs in estuarine systems, it remains a significant challenge in determining the relative importance of these source inputs to biogeochemical cycling in the water column of sediments. Temporal and spatial variability in organic matter inputs adds further to the complexity in understanding these environments. In recent years there have been significant improvements in our ability to distinguish between organic matter sources in estuaries using tools such as elemental, isotopic (bulk and compound/class specific), and chemical biomarker methods. This chapter will provide a general overview of the biochemistry of dominant organic compounds in organic matter and the techniques used to distinguish them in estuarine systems.

Bulk Organic Matter Techniques

The abundance and ratios of important elements in biological cycles (e.g., C, H, N, O, S, and P) provide the basic foundation of information on organic matter cycling. For example, concentrations of total organic carbon (TOC) provide the most important indicator of organic matter since approximately 50% of most organic matter consists of C. As discussed in chapter 8, TOC in estuaries is derived from a broad spectrum of sources with very different structural properties and decay rates. Consequently, while TOC provides essential information on spatial and temporal dynamics of organic matter it lacks any specificity to source or age of the material.

When bulk C information is combined with additional elemental information, as in the case of the C-to-N ratio, basic source information can be inferred about algal and terrestrial source materials (see review, Meyers, 1997). The broad range of C:N ratios across divergent sources of organic matter in the biosphere demonstrate how such a ratio can provide an initial proxy for determining source information (table 9.1). The basic reason for such differences in C:N ratios between vascular plants (>17) and microalgae (5–7) are simply due to the carbohydrate-rich (e.g., cellulose)/protein-poor and protein-rich/carbohydrate-poor nature of each source, respectively. The most abundant carbohydrates supporting this high C content in vascular plants are structural polysaccharides such as cellulose, hemicellulose, and pectin (Aspinall, 1970). Recent work estimated the relative importance of terrestrial versus marine sources in coastal sediments, using a simple mixing model based on C:N weight ratios of 6 for the marine (Müller, 1977) and 13 for the terrestrial (Parrish et al., 1992) end-members of organic matter (Colombo et al., 1996a,b). However, C:N ratios can be very misleading in determining organic matter sources in the absence of additional source proxies. In some cases, selective utilization of N, due to N limitation in a system, can result in artificially high C:N ratios resulting in misidentification of source materials. Similarly, the colonization of bacterial and fungal populations on "aging" vascular plant detritus can represent a significant fraction of the total N pool [due to the

Table 9.1 Approximate carbon-to-nitrogen ratios in some terrestrial and marine producers.[a]

	C/N
Terrestrial	
Leaves	100
Wood	1000
Marine vascular plants	
Zostera marina	17–70
Spartina alterniflora	24–45
S. patens	37–41
Marine macroalgae	
Browns (Fucus, Laminaria)	30 (16–68)
Greens	10–60
Reds	20
Microalgae and microbes	
Diatoms	6.5
Greens	6
Blue–greens	6.3
Peridineans	11
Bacteria	5.7
Fungi	10

[a]Data compiled in Fenchel and Jørgensen (1977), Alexander (1977), and Fenchel and Blackburn (1979), and data of Valiela and Teal (1976).
Modified from Valiela (1995).

typically low (e.g., 3 to 4) C:N ratios found in bacteria] (Tenore et al., 1982; Rice and Hanson, 1984), thereby decreasing the bulk C:N ratio of this material. Finally, artifacts from the standard procedure of removing carbonate carbon when measuring TOC can also alter C:N ratios (Meyers, 2003). Specifically, after removing carbonate a residual N made up of both inorganic and organic N remains, which is why the N used in C:N ratios is defined as total nitrogen (TN). In most cases the inorganic component of this residual N is relatively small in sedimentary and water column sources of organic matter. However, in sediments having very low concentrations of organic matter (e.g., < 0.3%), the relative importance of this residual inorganic N could be significant, resulting in underestimates of C:N ratios (Meyers, 2003). More specifically, this results from NH_4^+ adsorption on low organic content sediments. Fortunately, most estuarine sediments are greater than 1% organic matter, generally making this artifact relatively minor.

Hydrodynamic sorting of sediment particles has been shown to be an important mechanism for the distribution of C:N ratios in estuarine/coastal sediments (Prahl et al., 1994; Keil et al., 1998; Bianchi et al., 2002b; Galler et al., 2003). The basic premise is that large fragments of sometimes woody plant materials are entrained in coarse sediments because of settling rates higher than those of fine organic matter fractions, typically resulting in higher C:N ratios in coarse-grained sediments. However, there is usually more variability in the range of C:N ratios in coarse-grained sediments because of poor sorting that occurs in such sediments. Moreover, these sediments may have a significant fraction of the total carbon represented as benthic microalgae, depending on the light availability (Bianchi, 1988; Marinelli et al., 1998), that lowers C:N ratios, thereby adding to the variability commonly found in these permeable sediments. Fine-grained sediments typically have C:N ratios lower than those of coarse-grained sediments because of the higher surface area that allows for greater sequestration and sorption of terrestrially derived materials (in soils from the watershed) into mesopores (Mayer, 1994a,b; Keil et al., 1994b, 1998; Mayer et al., 2004). Differences in the sources (e.g., C_3 versus C_4 plants) of terrestrially derived material can also be associated with different particle size fractions (Goni et al., 1997, 1998)—more details on this are discussed in the section on Lignin (p. 282). Fine-grained sediments have a proportion of clay minerals greater than that of coarse sediments which, due to their negative charge, can adsorb NH_4^+ from porewaters resulting in a decrease in C:N ratio (Meyers, 1997). Despite the aforementioned limitations when using C:N ratios as indicators of organic source materials, recent work has shown that on a global scale, mean soil C:N accounts for 99.2% of the variability in riverine dissolved organic carbon (DOC) fluxes across different biomes (Aitkenhead and McDowell, 2000). Using this simple predictor, the total flux of carbon from land to the oceans was estimated to be 3.6×10^{14} g y^{-1}; additional work using chemical biomarkers is needed to better understand the relationship between soil C:N and DOC loading to estuaries/rivers.

Isotopic Mixing Models

End-member mixing models have been used to evaluate sources of dissolved inorganic nutrients (C, N, S) (Day et al., 1989; Fry, 2002) and particulate and dissolved organic matter (POM and DOM) (Raymond and Bauer, 2001a,b; Gordon and Goni, 2003; McCallister et al., 2004) in estuaries. However, due to significant overlap in stable isotopic signatures it has proven difficult to discern multiple sources of dissolved and particulate constituents, when using single and dual bulk isotopes in complex systems (Cloern et al., 2002). More recently, new approaches using end-member mixing models that utilize multiple-isotopic tracers coupled with chemical biomarker measurements have proven useful. We begin this section by first describing simple conservative mixing models that attempt to define sources of inorganic nutrients, followed then by examples of models that utilize a multiple-isotopic tracer and/or chemical biomarker approach to determine sources of organic matter.

In the case of nutrients, source components have been modeled as conservative mixtures (C_{mix}) of riverine and marine end-members using the following equation:

$$C_{mix} = fC_R + (1 - f)C_O \tag{9.1}$$

where: C = concentration; R = river end-member; O = ocean end-member; and f = fraction of freshwater (35 − measured salinity)/35.

End-member concentration mixing is typically represented by a straight line—based on concentration versus salinity. The concentration of dissolved inorganic carbon (DIC) versus salinity shows a linear conservative relationship for three different riverine:seawater mixing ratios (1:1, 5:8, 1:10) (figure 9.1). However, when plotting weighted end-member isotopic values of estuarine samples versus salinity, the conservative mixing curves are curvilinear rather than linear, which can be described by the following equation (Spiker, 1980; Fry, 2002):

$$\delta_{mix} = [fC_R \delta_R + (1 - f)C_O \delta_O]/C_{mix} \tag{9.2}$$

The curvilinear relationship between $\delta^{13}C$ versus salinity and the same riverine:seawater mixing ratios is shown in figure 9.1. Fry (2002) recently provided a summary of representative freshwater–marine mixing models for the stable isotopes of water (hydrogen and oxygen) as well as DIC, DIN, and SO_4^{2-} (figure 9.2). These models indicate that δD and $\delta^{18}O$ are linearly correlated with changes in salinity, and that the 1:1 C_R:C_O mixing ratio for these isotopes should be constant in estuaries. Thus, any relationships between salinity and food web dynamics should be linked to changes in the ratio. Conversely, the isotopes of SO_4^{2-} are relatively constant throughout an estuary with low riverine inputs (1:470) with virtually no change across a salinity gradient. As discussed earlier, this is due to the weighted effects of the seawater SO_4^{2-} isotopic signal (+21‰) which will dominate in the water column of estuarine waters, except in very low salinity waters (e.g., <1) (Fry and Smith, 2002). Therefore, we expect food webs linked with plankton production to have a sulfur isotopic signal that is uniform throughout the estuary. Exceptions to this may involve $\delta^{34}S$ depleted sulfide inputs from porewater inputs during resuspension events in shallow estuaries and direct uptake by benthic microalgae at the sediment–water interface (Deegan and Garritt, 1997; Chanton and Lewis, 1999). The DIC curve reflects the kinetic effects of selective uptake of the lighter ^{12}C in the DIC pool at higher salinities due to higher photosynthesis in these regions, which have higher irradiance. The high riverine inputs (15:1) in the DIN model show the effects of ^{15}N-enriched nitrogen loading from the riverine end-member. In conclusion, organisms in food webs in the lower-salinity regions of an estuary should have lower δD, $\delta^{34}S$, and $\delta^{13}C$ relative to the higher-salinity regions having higher δD, $\delta^{18}O$, $\delta^{13}C$, and $\delta^{34}S$ and lower $\delta^{15}N$ (Fry, 2002).

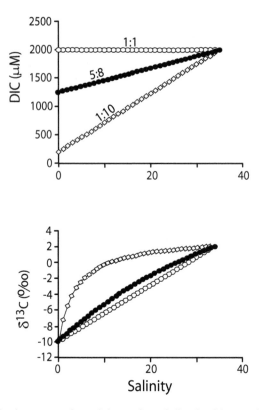

Figure 9.1 Freshwater–marine mixing ratios of dissolved inorganic carbon (DIC) and isotopic composition ($DI^{13}C$) across three different salinity gradients. Bottom: isotopic change between $-10‰$ at the freshwater end-member, and $+2‰$ at the marine end-member, both end-member values are based on concentration-weighted averages (data sources: Spiker and Schemel, 1979; Spiker, 1980). (Modified from Fry, 2002.)

As discussed in chapter 7, the application of stable isotopes as tracers of organic matter sources in aquatic systems has been quite extensive (Lajtha and Michener, 1994; Michener and Schell, 1994). Many studies have also used end-member values from a combination of stable and radiocarbon isotopes and biomarkers (e.g., lignin–phenols and lipids) to determine carbon sources in coastal systems (e.g., Smith and Epstein, 1971; Hedges and Parker, 1976; Prahl and Muehlhausen, 1989; Westerhausen et al., 1993; Goñi et al., 1997, 1998; Raymond and Bauer, 2001a,b; Gordon and Goni, 2003; McCallister et al., 2004). The range of isotopic values for organic matter sources in a number of estuaries are shown in table 9.2; many of these values have been used to establish end-member sources to coastal systems. Many of the early investigations were based on a two end-member mixing model (binary) in which the more depleted terrestrial $\delta^{13}C$ end-member could be used along with the more enriched marine phytoplankton end-member to establish their relative abundance throughout the estuary. For example, the following binary equation

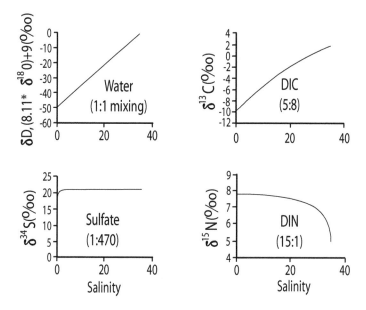

Figure 9.2 Freshwater–marine mixing models for hydrogen, oxygen, sulfate, and dissolved inorganic carbon (DIC) and nitrogen (DIN) across salinity gradients. Modeled gradients for sulfate are based on riverine–marine ratios of 60 μm versus 28 μm (data from Kendall and Coplen, 2001). (Modified from Fry, 2002.)

could be used to determine the percentage terrestrial organic matter in an estuary:

$$\%OC_{Terr} = (\delta^{13}C_{sample} - \delta^{13}C_{marine})/(\delta^{13}C_{riverine} - \delta^{13}C_{marine}) \qquad (9.3)$$

where: $\delta^{13}C_{sample}$ = isotopic composition of a sample; $\delta^{13}C_{marine}$ = published isotopic value of marine phytoplankton (see table 9.2); and $\delta^{13}C_{riverine}$ = published isotopic values of riverine POM (see table 9.2).

Recent work compared the effectiveness of binary and three-end-member models in determining the relative abundance of marine versus terrestrial sources off the Atchafalaya River estuary (Gordon and Goñi, 2003). The authors suggested that it was important to use a three-end-member model that combines both biomarkers and stable isotopes for three organic carbon (OC) source end-members (soils, riverine, and marine). This separates the terrestrial end-member into two different sources: vascular plants and soils. The three-end-member mixing was based on the following three equations:

$$\delta^{13}C_{sample} = \delta^{13}C_{marine} \times OC_{marine} + \delta^{13}C_{soil} \times OC_{soil}$$

$$+ \delta^{13}C_{vascular\ plant} \times OC_{vascular\ plant} \qquad (9.4)$$

Table 9.2 Published ranges of isotope values of potential organic matter sources to estuaries.

Source	δ13C (‰)	δ19N (‰)	Δ14C (‰)	References
Terrigenous (vascular plant)	−26 to −30	−2 to +2		Fry and Sherr (1984): Deegan and Garritt (1997)
Terrigenous soils (surface)/forest litter	−23 to −27	2.6 to 6.4	+152 to +310	Cloern et al. (2002): Richter et al. (1999)
Freshwater phytoplankton	−24 to −30	5 to 8		Anderson and Arthur (1983): Sigleo and Macko (1985)
Marine/estuarine phytoplankton	−18 to −24	6 to 9		Fry and Sherr (1984): Currin et al. (1995)
C–4 salt marsh plants	−12 to −14	3 to 7		Fry and Sherr (1984): Currin et al. (1995)
Benthic microalgae	−12 to −18	0 to 5		Currin et al. (1995)
C-3 Freshwater/brackish marsh plants	−23 to −26	3.5 to 5.5		Fry and Sherr (1984): Sullivan and Moncreiff (1990)
Specific to York River Estuary				
Freshwater grass leachate (*Peltandra virginica*)	−29.6			McCallister et al. (2004)
Marsh OM (0–6 cm)	−22.3 to −26.4		+45 to +58	Raymond and Bauer (2001a)
Marsh macrophytes	−23.3 to −28.9	5.3 to 11.0		Neubauer (2000)
Marsh microalgae (benthic)	−23.7 to −27.7	8.4 to 11.3		Neubauer (2000)
Phytoplankton (freshwater end member)[a]	−27.5 to −34.6		+110 to +164	Raymond and Bauer (2001a)
Phytoplankton (midsalinity)	−21.8 to −24.2		+56 to +72	Raymond and Bauer (2001a)
Phytoplankton (York River mouth)	−20.1 to −22.8		+47 to +62	Raymond and Bauer (2001a)
Chesapeake Bay DOM	−23.7		−77	Raymond and Bauer (2001a)
Terrigenous (leaf OM)			+100	Raymond and Bauer (2001a)
HMW DOM (0 salinity)	−27.8 to −28.1	4.0 to 4.7	+434	McCallister et al. (2004)
HMW DOM (10 salinity)	−24.0 to −24.5	5.5 to 7.5		McCallister et al. (2004)
HMW DOM (20 salinity)	−22.3 to −22.7	7.8 to 9.2		McCallister et al. (2004)
FW POM	−28.2 to −30.0	6.4 to 7.9	+24 to −190	Raymond and Bauer (2001a): McCallister et al. (2004)

	$\delta^{13}C$		$\Delta^{14}C$	Reference
Humics (resin extracted) Specific to Hudson River	−27.5		+111	McCallister et al. (2004)
POM (240 km)	−29.0	6.0	−101 to −156	McCallister et al. (2004): Raymond and Bauer (2001a)
POM (122 km)[b]	−27.1 to −27.4	2.8 to 3.2	96	McCallister et al. (2004): Raymond and Bauer (2001a)
DOC (240 km)	−27.0 to −27.2		−73 to −137	Bauer et al., unpubl. data
DOC (152 km)	−27.0		−110	Bauer et al., unpubl. data
Phytoplankton (240 km)	−30.0 to −31.1	8.0	−44 to −50	Bauer et al., unpubl. data: Caraco et al. (1998)
Phytoplankton (165 km)[c]	−24.2		−74	Caraco, unpubl. data
Phytoplankton (152 km)	−30.5	8.0	−52	Bauer et al., unpubl. data: Caraco et al. (1998)
Submerged macrophytes	−21.7 to −22.2	8.0	−37 to −38	Caraco et al. (1998); Caraco, unpubl. data
Emergent macrophytes	−26.0	8.0	+90	Caraco et al. (1998); Raymond and Bauer, (2001b)
Terrestrial (leaf OM)	−27.0	−2.0		Caraco et al. (1998)
Terrigenous (sedimentary rocks)[d]	−28.6 to −29.		−866 to −999	Petsch (2000)
Humics (resin extracted)	−27.2		+22	McCallister et al. (2004)

[a] Phytoplankton isotopic values for the York and Hudson (unless otherwise noted) were predicted from measured $\delta^{13}C$-DIC and $\delta^{14}C$-DIC values and assumed a kinetic fractionations of 20‰ for $\delta^{13}C$ values (Chanton and Lewis, 1999). Because $\Delta^{14}C$ values were normalized to $\delta^{13}C$ according to the principles of Stuiver and Polach (1977), no additional correction was applied.
[b] $\Delta^{14}C$ from 152 km.
[c] Phytoplankton signature from plankton net low.
[d] OM isotope values are from weathering profiles of Marcellus Shale (Hudson/Mohawk River valley at depths of 8, 57, and 170 cm.
Modified from McCallister et al. (2004).

$$\Lambda_{\text{sample}} = \Lambda_{\text{marine}} \times OC_{\text{marine}} + \Lambda_{\text{soil}} \times OC_{\text{soil}}$$

$$+ \Lambda_{\text{vascular plant}} \times OC_{\text{vascular plant}} \qquad (9.5)$$

$$N/C_{\text{sample}} = N/C_{\text{marine}} \times OC_{\text{marine}} + N/C_{\text{soil}} \times OC_{\text{soil}}$$

$$+ N/C_{\text{soil}} \times OC_{\text{soil}} \qquad (9.6)$$

where: OC_{marine}, OC_{soil}, and $OC_{\text{vascular plant}}$ were the fractions of organic carbon derived from marine, soil, and vascular plants; was the total lignin-phenols/100 mg org.C; the nitrogen:carbon (N:C) ratios were based on Redfield values for different sources. Soil N:C values are based on average values of suspended POM in respective rivers and their associated tributaries. Due to the heterogeneous composition of terrigenous end-members, it was concluded that the three-end-member mixing model was required to separate terrigenous sources in sediments of the Atchafalya River estuary (Gordon and Goñi, 2003). One of the fundamental problems with these mixing models is that it is assumed that the decay processes in the water column and sediments are not changing the isotopic signature of organic compounds between end-member source inputs. Issues concerning biomarker decay can be addressed because there are signature decay compounds that can be used many times as indices of decay—more provided on these biomarkers later in the chapter. Other work using compound specific isotope analysis (CSIA), as described in chapter 7, has shown that there are different fractionations observed for lipid biomarker compounds under different redox regimes (Sun et al., 2004). More work is clearly needed on this topic if we are to incorporate molecular-based isotopic studies in determining estuarine and coastal sources of organic matter.

The use of ^{13}C and ^{15}N measurements in bacterial biomass has proven useful for establishing organic matter sources of bacteria (Hopkinson et al., 1998; Coffin and Cifuentes, 1999). More recently, radiocarbon and ^{13}C and ^{15}N stable-isotope signatures of bacterial nucleic acids have been shown to effectively discern bacterial sources of organic matter (Cherrier et al., 1999; McCallister et al., 2004). Prior work has shown that the greater sensitivity and dynamic range of $^{14}\Delta C$ (~ -1000 to $+435$‰) versus $\delta^{13}C$ (~ -35 to 12‰) and ^{15}N (~ -2 to $+40$‰) in organic matter sources allows for better separation of allochthonous and autochthonous sources in estuarine systems (Raymond and Bauer 2001a,b; Bauer 2002). More specifically, recent work has shown that substantially greater amounts of POC and DOC are utilized by bacteria in the York River estuary (USA) compared to the Hudson River estuary (USA), possibly due to differences in the age of POC and DOC (e.g., both older in the Hudson system) in these systems (Raymond and Bauer, 2001b,c). Based on a three-end-member source mixing model it was determined that such differences resulted in greater consumption of "younger" allochthonous sources of DOC and POC by bacteria in the York River estuary, with the Hudson being "fueled" by older organic matter sources form soils (McCallister et al., 2004).

Nuclear Magnetic Resonance

Nuclear magnetic resonance (NMR) spectroscopy is a powerful and theoretically complex analytical tool that can be used to characterize organic matter. We will now briefly describe some of the basic principles of NMR before discussing its application in estuarine systems. In the most general terms, many nuclei (e.g., ^{1}H, ^{13}C, ^{31}P, ^{15}N) have a spin with a *spin state* that when placed in a strong *magnetic field*, simultaneously irradiated with electromagnetic (microwave) radiation, absorbs energy through a process called *magnetic resonance* (Solomons, 1980; Levitt, 2001). This resonance is detected by the instrument as an NMR signal. Most NMR instruments scan multiple magnetic field strengths that interact with the spinning nuclei that become aligned with the *external magnetic field* of the instrument. The microwave frequency of liquid- and solid-state NMR instruments generally range from 60 to 600 megahertz (MHz), with the application of higher frequency resulting in greater resolution. The actual magnetic field detected by the nucleus of a "naked" atom is not the same as one that has a different electronic environment; these differences provide the basis for determining the structure of a molecule. Differences in field strengths are caused by the *shielding* or *deshielding* of electrons associated with adjacent atoms—this electron interference is referred to as a *chemical shift*. The chemical shift of a nucleus will be proportional to the external magnetic field and since many spectrometers have different magnetic field strengths it is common to report the shift independently of the external field strength. Chemical shifts are typically small (in the range of Hz) relative to field strength (MHz range), so it has become convenient to express chemical shift in units of parts per million (ppm) relative to an appropriate standard. Charge polarized magic angle spinning (CPMAS) NMR is another commonly used method that takes molecules with high degrees of spin and imparts this spin to molecules that have none, making it extremely useful for obtaining signals from atoms that usually do not have a spin (Levitt, 2001).

Proton (^{1}H) and ^{13}C NMR have been the most common NMR tools for the nondestructive determination of functional groups in complex biopolymers in plants, soils/sediments, and DOM in aquatic ecosystems (Schnitzer and Preston, 1986; Hatcher, 1987; Orem and Hatcher, 1987; Benner et al., 1992; Hedges et al., 1992, 2002). Although ^{13}C NMR signals are generated in the same way as ^{1}H signals, the magnetic field strength is wider for the chemical shifts of ^{13}C nuclei. One of the key advantages in using ^{13}C NMR in characterizing organic matter in natural systems is to be able to determine the relative abundances of carbon associated with the major functional groups. Although ^{13}C NMR does not provide as much insight into the specific decay pathways of certain biopolymers through the use of chemical biomarkers, it does provide a nondestructive approach that allows a greater understanding of a larger fraction of the bulk carbon. The application of ^{31}P NMR (Ingall et al., 1990; Hupfer et al., 1995; Nanny and Minear, 1997; Clark et al., 1998) and ^{15}N NMR (Almendros et al., 1991; Knicker and Ludemann, 1995; Knicker, 2000) have also been useful in characterizing organic and inorganic pools of P and N in natural systems, respectively. More recently, two-dimensional (2D) ^{15}N –^{13}C NMR has been used to study the fate of protein in experimentally degraded algae (Zang et al., 2001).

One area where ^{13}C NMR first appeared as a useful tool for estuarine studies was its application in characterizing bulk chemical changes in decomposing wetland plant materials (Benner et al., 1990; Filip et al., 1991). For example, Benner et al. (1990) used four structural types of carbon (e.g., paraffinic, carbohydrate, aromatic, and carbonyl) to determine bulk C changes in mangrove leaves (*Rhizophora mangle*) during decomposition in tropical estuarine waters. The structural assignments for the major chemical shifts and ^{13}C NMR spectra for different stages of mangrove decay are shown in table 9.3

Table 9.3 Structural assignments for the major chemical shifts observed in the CPMAS ^{13}C NMR spectra of mangrove leaves.

33 ppm	Methylene carbon indicative of long-chain aliphatic and paraffinic structures
73 ppm	Oxygenated alkyl carbon indicative of polysaccharides (ring carbons C-2, C-3, C-5)
105 ppm	Anomeric C-1 (acetal carbon) of polysaccharides, ketal carbon of polysaccharides, nonprotonated aryl carbon in tannin, protonated aryl carbon in lignin (C-2, C-6 in syringyl units)
116 ppm	Protonated aryl carbon (e.g., C-3, C-5 in *p*-hydroxyl phenols and C-2, C-5, C-6 in guaiacyl units)
130 ppm	Alkyl substituted aryl carbon (e.g., C-1, C-2, C-6 in *p*-hydroxyphenols)
145 ppm	Oxygen-substituted aryl carbon (*ortho* to other aryl-O carbons) indicative of tannin (di- and trihydroxybenzenes) and lignin (C-4 in guaiacyl units)
154 ppm	Oxygen-substituted aryl carbon (*meta* to other aryl-O carbons) indicative of tannin (proanthocyanidins or procyanidins) and lignin (C-3, C-5 in syringyl units)
175 ppm	C=O in carboxyl groups, amides, aliphatic esters
200 ppm	C=O in aldehydes and ketones

Modified from Benner et al. (1990).

and figure 9.3. This work showed a preferential loss of carbohydrates and preservation of paraffinic polymers. Similarly, ^{13}C NMR analyses on DOM released from decomposing *Spartina alterniflora* showed preferential loss of carbohydrates from fresh tissues (Filip et al., 1991). Two-dimensional ^{15}N–^{13}C NMR of the remains of the decomposed microalgae, labeled with ^{13}C and ^{15}N isotopes, indicated that proteins and highly aliphatic compounds are preserved over time (Zang et al., 2001), supporting the concept of "encapsulation" where labile compounds are protected by macromolecular components such as *algaenan* (Knicker and Hatcher, 1997; Nguyen and Harvey, 2001; Nguyen et al., 2003).

Characterization of DOM using ^{13}C NMR indicates that river- and marsh-derived DOM is largely composed of aromatic carbon—reflective of lignin inputs from vascular plant materials (Lobartini et al., 1991; Hedges et al., 1992; Engelhaupt and Bianchi, 2001). Conversely, DOM in the Mississippi River was found to be primarily composed of aliphatic versus aromatic compounds, and is believed to be largely derived from freshwater diatoms growing under high-nutrient low-light conditions (Bianchi et al., 2004). Other sources of DOM to estuaries that can significantly affect bulk carbon composition are derived from sediment porewater fluxes and local runoff from soils. For example, ^{13}C NMR work indicated that under anoxic conditions porewaters were dominated by DOM primarily composed of carbohydrates and paraffinic structures derived from the decay of algal/bacterial cellulose and other unknown materials—with minimal degradation of lignin (Orem and Hatcher, 1987). Under aerobic conditions, porewater DOM had lower carbohydrate and higher abundance of aromatic structures. Recent work by Engelhaupt and

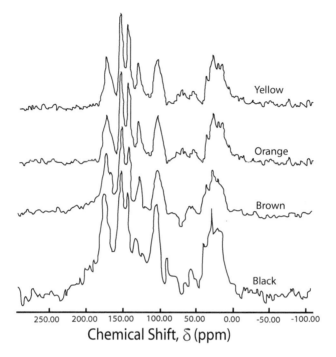

Chemical Shift, δ (ppm)

Figure 9.3 Conventional ^{13}C cross-polarization magic angle spinning (CP/MAS) solid-state NMR spectra of mangrove leaves at different stages of decomposition in estuarine waters. Differences in carbon functionality are defined by the following key characteristic resonances: carbohydrates (50–90 ppm); carboxyl (175 ppm); aromatic (90–170 ppm). (Modified from Benner et al., 1990.)

Bianchi (2001) further supported that lignin was not degraded efficiently under low-oxygen conditions in a tidal stream, adjacent to Lake Pontchartrain estuary (USA).

Molecular Biomarkers

Due to the complexity of organic matter sources in estuaries and the aforementioned problems associated with making only bulk measurements to constrain them, the application of chemical biomarkers has become widespread in estuarine research (see review, Bianchi and Canuel, 2001). The term *biomarker molecule* has recently been defined by Meyers (2003, p. 262) as "compounds that characterize certain biotic sources and that retain their source information after burial in sediments, even after some alteration." This molecular information is more specific and sensitive than bulk elemental and isotopic techniques in characterizing sources of organic matter, and further allows for identification of multiple sources (Meyers, 1997, 2003).

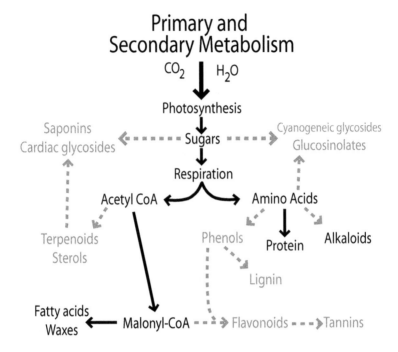

Figure 9.4 Pathways of the sum of all biochemical processes is called metabolism. It is distinguished between primary and secondary metabolism. Primary metabolism contains all pathways necessary to keep the cell alive while in secondary metabolism compounds are produced and broken down that are essential for the entire organism.

The *catabolic* and *anabolic* pathways that are responsible for the formation of many of the biomarker compounds discussed in this chapter occur through an "intermediary" metabolism via *glycolysis* and the *citric acid cycle* (Voet and Voet, 2004). The biosynthetic pathways of these compounds can be divided into primary and secondary metabolism (figure 9.4). Many of these compounds are not used as chemical biomarkers in estuarine research but are shown here simply to illustrate their relationship with the biomarkers discussed in this chapter. For more details on the biosynthetic pathways illustrated here, please refer to Voet and Voet (2004) and Engel and Macko (1993).

Lipids

Lipids are water-insoluble compounds that are soluble in nonpolar (chloroform, benzene) solvents. These energy-rich compounds represent a significant component of the carbon flux through trophic pathways in coastal/estuarine systems (Sargent et al., 1977). In fact, lipids typically account for 10 to 20% (dry weight) of the biomass in actively growing algal populations (Sargent and Falk-Petersen, 1988). Simple lipids (fats), composed of

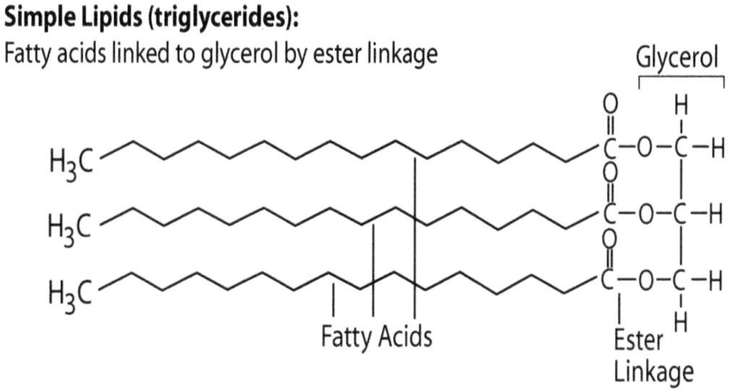

Simple Lipids (triglycerides):
Fatty acids linked to glycerol by ester linkage

Figure 9.5 Chemical structure of simple lipid, composed of three fatty acids bonded to a glycerol (C_3 alcohol), collectively known as a triglyceride.

three fatty acids (see details in section below) bonded to a *glycerol* (C_3 alcohol), are known as *triacylglycerols* (figure 9.5).

Complex lipids (e.g., *phospholipids and glycolipids*) differ in that they contain additional elements (e.g., P, S, and N) or small hydrophilic compounds (e.g., sugars and certain amino acids) (figure 9.6). Two fundamental lipid classes of interest in coastal/estuarine environments are the *triacylglycerols* and the *phospholipids* (Parrish et al., 2000). The triacylgylcerols store energy reserves in a highly reduced form of C as part of the hydrocarbon chain of fatty acids and when oxidized release significant amounts of energy. In fact, they are commonly used as a conditional or "health" index for organisms. Phospholipids represent an integral component of cell membranes and share similar functions with sterols. Wax esters have been shown to be important as energetic reserves as well as for buoyancy in grazer communities such as copepods (Wakeham et al., 1980; Wakeham and Lee, 1989). Lipids are an important component of organic matter with a diversity of compound classes (e.g., hydrocarbons, fatty acids, *n*-alkanols, and sterols) that have been used as effective biomarkers of organic matter in coastal and estuarine systems (Volkman and Maxwell, 1984; Prahl, 1985; Volkmann and Hallegraeff, 1988; Yunker et al., 1993, 1995; Canuel et al., 1995, 1997; Canuel, 2001; Sun et al., 2002). In fact, the term *geolipid* has been used to describe decay-resistant biomarkers in sediments because lipids are typically recalcitrant compared with other biochemical components of organic matter, making them more long-lived in the sedimentary record (Meyers, 1997). However, anthropogenic sources in estuaries pose another problem when interpreting geolipid data—as in the case of hydrocarbons. As discussed in chapter 15, the abundance of hydrocarbons derived from anthropogenic sources (petroleum hydrocarbons) has increased significantly in estuaries since the industrial revolution. Natural oil seeps and erosion of bitumen deposits can also contribute to hydrocarbon abundance and composition in river/estuarine systems (Yunker et al., 1993). These petroleum hydrocarbons can be distinguished from biological hydrocarbons by the absence of odd-carbon chain lengths commonly found in biological hydrocarbons and the greater structural diversity found in petroleum hydrocarbons (Meyers and Takeuchi, 1981). In fact, the complexity of structural

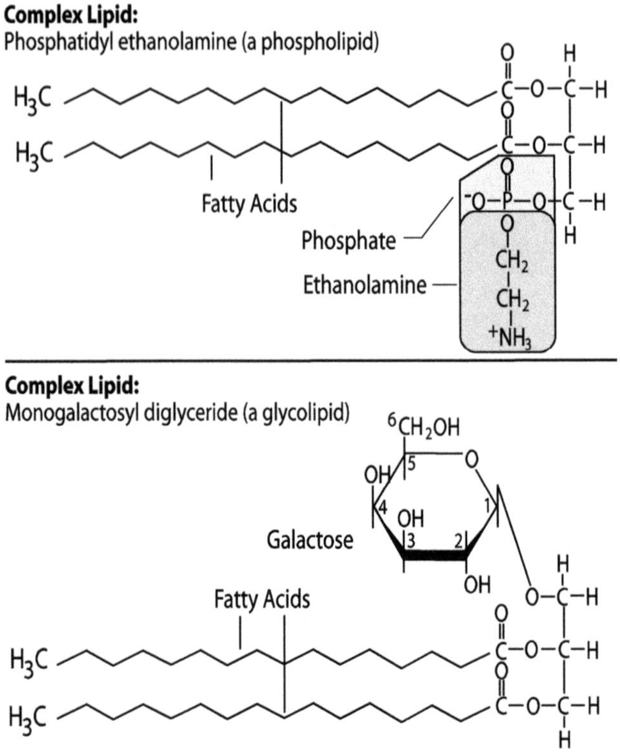

Figure 9.6 Chemical structure of complex lipids (e.g., phospholipids and glycolipids) which differ in that they contain additional elements (e.g., P, S, and N) or small hydrophilic compounds (e.g., sugars and certain amino acids).

compounds found in petroleum hydrocarbons has been referred to as an *unresolved complex mixture* (UCM) because of the difficulty in resolving extracts using chromatographic techniques (Meyers, 2003).

Hydrocarbons

Aliphatic hydrocarbons such as *n*-alkanes and *n*-alkenes have been successfully used to distinguish between algal, bacterial, and terrestrial sources of carbon in estuarine/coastal systems (Yunker et al., 1991, 1993, 1995; Canuel et al., 1997). Saturated aliphatic hydrocarbons are considered to be *alkanes* (or *paraffins*) and nonsaturated hydrocarbons which exhibit one or more double bonds are called *alkenes* (or *olefins*)—as indicated in the simple structures of hexadecane and 1,3-butadiene, respectively (figure 9.7). It should also be noted that, *n*-alkanes tend to be odd-numbered as they result from enzymatic decarboxylation of fatty acids. *Long-chain* n-*alkanes* (LCH) (e.g., C_{27}, C_{29}, and C_{31}) are generally considered to be terrestrially derived, originating from epicuticular waxes

$$CH_3-(CH_2)_{14}-CH_3$$
hexadecane

$$CH_2=CH-CH=CH_2$$
1,3 - butadiene

Figure 9.7 Chemical structure of a saturated aliphatic hydrocarbon or alkane (e.g., hexadecane) and a nonsaturated hydrocarbon or alkene (e.g., hexadecane and 1,3-butadiene.)

of vascular land and aquatic (e.g., seagrasses) plants. In contrast, *short-chain* n-*alkanes* (SCH) (e.g., C_{15}, C_{17}, and C_{19}) are derived from planktonic and benthic algal sources (Eglinton and Hamilton, 1963, 1967; Cranwell, 1973, 1984; Hostettler et al., 1989; Ficken et al., 2000; Jaffé et al., 2001) (table 9.4a). However, some compounds within the C_{20} to C_{28}range are likely produced by bacteria (Grimalt et al., 1985). Other interference problems arise from petrogenic sources of alkanes, such as the C_{14} to C_{36} range (Mazurek and Simoneit, 1984). The carbon preference index (CPI) has been used to distinguish between these alkane sources whereby a CPI > 1 is derived from biogenic sources compared to a CPI < 1 derived from anthropogenic sources (Simoneit et al., 1991). The presence of *long-chain* n-*alkenes* in estuarine sediments has been shown to be indicative of vascular plant sources. For example, high concentrations of the C_{28} homolog appear to be correlated with inputs of both mangrove (Jaffé et al., 2001) and salt marsh (Canuel et al., 1997) detritus. Similarly, $C_{21:6}$(*heneicosahexaene*) was found to be the major alkene in suspended particles of the Mackenzie River estuary (Canada) (Yunker et al., 1993, 1995), indicating the importance of phytoplankton as a major source of POC (Blumer et al., 1971; Lee and Loeblich, 1971; Schultz and Quinn, 1977) (table 9.4a). More specifically, freshwater phytoplankton are believed to be rich in the alkenes *heptadecadiene* and *heptadecacene* (Albaigés et al., 1984). Overall, long-chain hydrocarbons are considered to be more recalcitrant to decomposition than SCH (Prahl, 1985; Meyers and Ishiwatari, 1993; Canuel and Martens, 1996). Recent work has shown that aquatic macrophytes can be distinguished from phytoplankton inputs to sediments using the ratio of LCH (C_{27}, C_{29}, and C_{31}) to SCH (C_{15}, C_{17}, and C_{19}) n-alkanes (Silliman and Schelske, 2003) (table 9.4a). The proxy ratio (P_{aq}) of *midchain hydrocarbons* (MCH) (C_{23} and C_{25}) to LCH was used in the same study to distinguish between macrophytes and terrestrial land plants. Terrestrial inputs can be further separated using C_{31} as an indicator of grasses and $C_{27}$and C_{29} as sources of woody plant materials (Cranwell, 1973).

Branched and *cyclic hydrocarbons* have also been used as paleomarkers, principally in lake sediments, to interpret past changes in organic matter sources (Rowland and Robson, 1990; Meyers, 2003). In particular, *isoprenoid* hydrocarbons such as *pristane* (C_{19} isoprenoid hydrocarbon) and *phytane* (C_{20} isoprenoid hydrocarbon) (both derived from *phytol*) have been used as indicators of herbivorous grazing (Blumer et al., 1964) and methanogenesis (Risatti et al., 1984), respectively (table 9.4a; figure 9.8). *Terpenes* are a widespread group of natural plant compounds made up of *isoprene* (methylbuta-1,3-diene) units containing five carbons. The standard molecular formula for terpenes is $(C_5H_8)_n$, where n is the number of isoprene units. For example, *diterpenes* such as phytol contain four isoprene units and 20 carbon atoms (figure 9.9)—similarly, *triterpenes*, such as carotenoids (discussed later in this chapter, p. 248), have six isoprene units

Table 9.4a Hydrocarbon biomarkers used in aquatic systems.

Indicator	Carbon no./ structure	Terrigenous source[a]	Marine source[a]	Reference
n-C_{15}–C_{19} alkanes	15–19	Petrogenic, algae	Algae, bacteria	Comet and Eglinton (1987)
n-C_{23}–C_{33} alkanes	23–33	Higher plants	Bacteria	Eglinton and Hamilton (1967)
				Countway et al. (2003)
Isoprenoids/pristine	13–20	Petrogenic isoprenoids[b]	Zooplankton (pristine)	Blumer et al. (1964),
				Yunker et al. (1993)
Highly branched isoprenoids	(C_{20}, C_{25}, and C_{30})	Algae (diatoms)		Silliman et al. (1998)
n-$C_{21:6}$(heneicosahexaene)	21	Algae	Algae	Blumer et al. (1971)
Diploptene	30	Bacteria	Bacteria	Ourisson et al. (1987)
Diagenetic hopanes[c]	27–32	Petrogenic	—	Peters and Moldowan (1993)
Retene, cadalene, pimanthrene, simonellite	15–19	Higher plants	—	Simoneit and Mazurek (1982)
178 278 PAHs[d]	14–22	Petrogenic, atmosphere, peat	Atmosphere	Yunker et al. (1993)
Perylene		Diagenetic		Silliman et al. (1998)

[a]Petrogenic sources include eroded bitumens, oils seeps, etc.; algae refers to phytoplankton sources in ice algae, diatoms, and picoplankton.

[b]A well-defined series of seven isoprenoids that ranges from 2,6-dimethylundecane to phytane (Yunker et al., 1993).

[c]The 17 $\alpha(H)$, 21$\beta(H)$-hopanes with 27, 29, 30, 31, and 32 carbons and 17$\beta(H)$, 21$\alpha(H)$-normoretane.

[d]Parent PAH of molecular mass 178 to 278. Sterol classification is discussed in the text. These sterols follow 24-ethylcholest-5-en-3-β-ol as major constituents in the Mackenzie River at freshet; only 24-methylcholesta-5,22E-dien-3-β-ol is well known in higher plants (Volkman, 1986). Modified from Yunker et al. (1993).

Table 9.4b Allochthonous, plant triterpenoids used as biomarkers in aquatic systems.

Indicator	Carbon no./ structure	Terrigenous source[a]	Marine source[a]	Reference
Plant triterpenoids				
β-Amyrin (olean-12-en-3β-ol)	30	Higher plants	—	Volkman et al. (1987)
α-Amyrin (urs-12en-3β-ol)	30	Higher plants	—	Volkman et al. (1987)
24-Ethylcholesta-3,5-dien-7-one	29	Higher plants, peat	—	Robinson et al. (1987)
20:0–30:0 Even carbon n-alkanols	20–30	Higher plants	—	Cranwell (1981)
16:1–24:1 Even carbon n-alkanols	16–24	Higher plants, zooplankton(?)	Zooplankton	Sargent et al. (1977)

[a]Modified from Yunker et al. (1993).

241

Table 9.4c Sterols used as tracers in aquatic systems.

Indicator	Carbon no./structure	Terrigenous source[a]	Marine source[a]	Reference
Autochthonous, primarily marine sterols[b]				
24-Norcholesta=5,22E-dien-3β-ol	26Δ^{5,22}	—	Algae, zooplankton	Prahl et al. (1984) / Nichols et al. (1990)
Cholesta-5,22E-dien-3β-ol	27Δ^{5,22}	Algae, higher plants (?)[c]	Zooplankton, algae	Gagosian et al. (1968) / Nichols et al. (1986)
Cholest-5-en-3β-ol	27Δ^5	Algae, higher plants (?)[c]	Zooplankton, algae	Prahl et al. (1984) / Volkman (1986)
Cholesta-5,24(28)-dien-3β-ol	27Δ^{5,24}	—	Zooplankton, algae	Prahl et al. (1984)
24-Methylcholesta-5,22E-dien-3β-ol	28Δ^{5,22}	Diatoms, algae[c]	Algae	Volkman (1986) / Nichols et al. (1990)
24-Methylcholesta-5,24(28)-dien-3β-ol	28Δ$^{5,24(28)}$	Algae; diatoms	Algae	Volkman (1986) / Nichols et al. (1990)
24-Ethylcholesta-5,24(28)E-dien-3β-ol	29Δ$^{5,24(28)E}$	Algae, higher plants (?)[c]	Algae	Volkman et al. (1987)
24-Ethylcholesta-5,24(28)Z-dien-3β-ol	29Δ$^{5,24(28)Z}$	—	Algae	Volkman and Hallegraeff (1988)
Allochthonous, plant sterols[b]				
5α-Cholestan-3β-ol	27Δ^0	Stenol reduction	Stenol reduction	Cranwell and Volkman (1985)
24-Methylcholest-5-en-3β-ol	28Δ^5	Higher plants	Algae	Volkman (1986)
24-Ethylcholesta-5,22E-dien-3β-ol	29Δ^{5,22}	Higher plants	Algae	Volkman (1986)
24-Ethylcholest-5-en-3β-ol	29Δ^5	Higher plants	Algae	Volkman (1986) / Nichols et al. (1990)

[a]Petrogenic sources include eroded bitumens, oils seeps, etc.; algae refers to phytoplankton sources in ice algae, diatoms, and picoplankton.
[b]Sterol classification is discussed in the text.
[c]These sterols follow 24-ethylcholest-5-en-3β-ol as major constituents in the Mackenzie River at freshet; only 24-methylcholesta-5,22E-dien-3β-ol is well known in higher plants (Volkman, 1986).
Modified from Yunker et al. (1993).

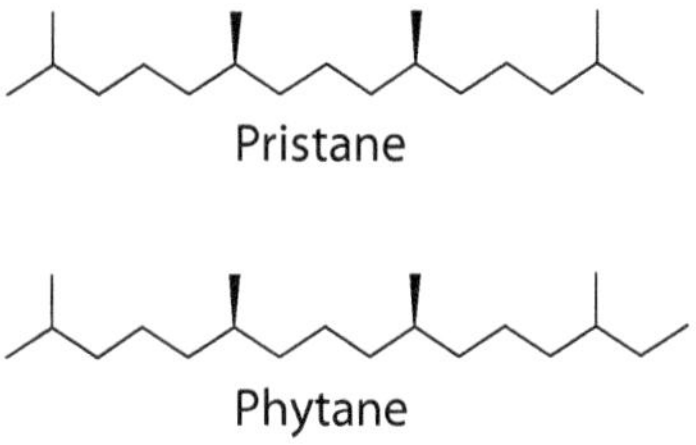

Pristane

Phytane

Figure 9.8 Chemical structures of the isoprenoid hydrocarbons pristane (C_{19} isoprenoid hydrocarbon) and phytane (C_{20} isoprenoid hydrocarbon).

and 40 carbon atoms. Although recent work reported the presence of pristane and phytane in sediments from the Florida Everglades (Jaffé et al., 2001), these biomarkers have not received much attention in estuarine systems. However, with the increasing application of biomarkers to paleoreconstruction in estuaries (see chapter 15) these biomarkers will likely prove useful in future work. Their potential as paleoredox biomarkers may be of particular interest since these compounds have been shown to be effective indicators of past (Didyk et al., 1978) and present (de Leeuw et al., 1977) redox conditions. However, recent work in Long Island Sound estuary has suggested this ratio can only be used as a redox indicator in the very early stages of diagenesis, due to complex differences in the pathways of phytol degradation (Sun et al., 1998).

Certain classes of *terpenoids* have been shown to be important terrestrial source tracers in rivers/estuaries. *Triterpenoids* are a large group of natural compounds which typically include steroids and sterols. Another subgroup of compounds within the triterpenoids, the *hopanoids,* are very abundant in bacteria where they have generally replaced cholesterol (Neunlist et al., 1988; Bravo et al., 2001). *Pentacyclic triterpenoids*, such as *diploptene* [17β(H), 21β(H)-hop-21(29) – ene] have been found to be useful diagenetic biomarkers for bacterial activity in peats and plant debris in the Mackenzie River estuary (Yunker et al., 1993) as well as for methane-oxidizing bacteria in the Columbia River estuary (USA) (Prahl et al., 1992; table 9.4a). The general *hopane* structure and the simplest C_{30} hopanoid, diploptene, are shown in figure 9.10. Although a few plants and fungal/plant associations (e.g., ferns, mosses, and lichens) can synthesize hopanes, these compounds are found to be ubiquitous in the cell walls of bacteria (Ourisson et al., 1987). Other hopanes, such as hop-17(21)-ene, are believed to be a decay product of diploptene and have been used as additional indicators of bacterial input (Brassell and Eglinton, 1983). In general, hopanoids primarily occur in aerobic bacteria such as methanotrophs, heterotrophs, and cyanobacteria, and in some cases anaerobic bacteria (Sinninghe-Damste et al., 2004).

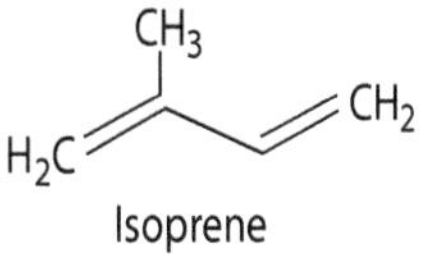

Isoprene

Figure 9.9 Chemical structure of an isoprene (methylbuta-1,3-diene) unit containing five carbons.

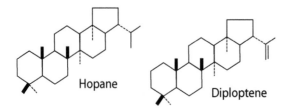

Figure 9.10 The general chemical structure of hopane and the simplest C_{30} hopanoid, diploptene.

Polycyclic aromatic hydrocarbons (PAHs) represent a specific source of hydrocarbons to estuarine systems (Simoneit, 1984) that are produced from incomplete combustion of organic matter and/or spillage from petroleum products (Stark et al., 2003). Their chemical structure consists of having two or more benzene rings fused together (figure 9.11). These compounds are highly lipophilic and hydrophobic, which will favor adsorption to suspended particulate matter and accumulation in sediments of aquatic systems—making them potential biomarkers for the delivery of organic matter to natural systems (Meyers and Quinn, 1973). Certain molecular criteria can be used to examine composition, distribution, and sources of PAHs, whereby natural versus anthropogenic sources can be distinguished (Yunker et al., 1993, 2002; Dickhut et al., 2000; Countway et al., 2003). Ratios of PAHs and molecular mass profiles may also be used to determine sources and diagenetic tracers of PAHs (Gschwend and Hites, 1981; Lipiatou and Saliot, 1991; Kennicutt and Comet, 1992; Yunker et al., 1993). Although PAHs are not produced naturally in organisms, certain PAH decay products derived from microbial breakdown, low-temperature diagenesis, and high-temperature alterations may be used as tracers of organic matter transport (Dahle et al., 2003; Meyers, 2003). For example, higher plant-derived materials have been traced in estuaries using *alkyl PAHs* such as *retene, pimanthrene,* and *cadalene* (Yunker et al., 1993; table 9.4a). Retene, produced from wood combustion and found in oils and coal (Barrick and Prahl, 1987), is also believed to be derived from pine resin (Simoneit, 1977; Wakeham et al., 1980). Studies in both lacustrine (Wakeham et al., 1980)

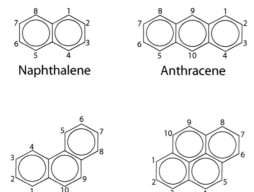

Figure 9.11 Chemical structure of polycyclic aromatic hydrocarbons (PAHs) consisting of two or more benzene rings fused together, as shown in the PAHs naphthalene, anthracene, phenanthrene, and pyrene.

and estuarine (Yunker et al., 1993) sediments have linked the presence of retene to dia-genetically altered terrestrial sources of higher plant materials. These alkyl PAHs have also been shown to have similar distribution patterns to hopanes in suspended particulates and sediments of the Mackenzie River estuary (Yunker et al., 1993). More details on the contaminant chemistry of PAHs in estuaries are provided in chapter 15.

Acyclic ketones, such as linear *alkane-2-ones,* are another class of lipids commonly found in aquatic systems (Volkman et al., 1983; Hernandez et al., 2001; Jaffé et al., 2001). Hernandez et al. (2001) found a shift in the alkane-2-one distribution from C_{27} to C_{31} in the upper estuary to C_{25} in the lower estuary and attributed these changes to tidal changes in the delivery of seagrass detritus. Another acyclic ketone recently found in estuarine sediments is 6,10,14-trimethylpentadecane-2-one, believed to be a decay product of phytol, which may be indicative of mangrove inputs (Jaffé et al., 2001).

Fatty acids

Fatty acids are building blocks of lipids and represent a significant fraction of the total lipid pool in aquatic organisms (Vance and Vance, 1996; Desvilettes et al., 1997; Abrajane et al., 1998; Feinss and Feulsen, 2002; Countway et al., 2003; Dalsgaard et al., 2003). Designation for the systematic names of fatty acid is A:BωC, where A is the number of carbon atoms, B is the number of double bonds, and C is the position of the double bond from the aliphatic end of the molecule. The most common fatty acids found in nature are saturated and unsaturated compounds with a chain length of C_{16} and C_{18} (Cranwell, 1982; Pulchan et al., 2003) as shown in figure 9.12. For example, palmitic acid is a 16-carbon sat-urated (no double bonds) fatty acid, oleic acid is an 18-carbon *monounsaturated* fatty acid (MUFA) (one double bond), and linoleic acid is a C_{18} *polyunsaturated* fatty acid (PUFA) (with two double bonds, 18:2ω6). Chain length and decomposability of fatty acids have been shown to be correlated, indicating a pre- and post-depositional selective loss of short-chain fatty acids (Kawamura et al., 1987; Canuel and Martens, 1993; Meyers and Eadie, 1993). Saturated fatty acids are also more stable and typically increase in relative propor-tion to total fatty acids with increasing sediment depth (Parker and Leo, 1965; Haddad et al., 1992; Sun et al., 1997). *Long-chain* ($>C_{22}$) saturated *fatty acids* are generally thought to be indicative of terrestrial (vascular plants and soils) organic matter sources (Tulloch, 1976; Brasell et al., 1980; Sargent et al., 1995; Shi et al., 2001; table 9.5).

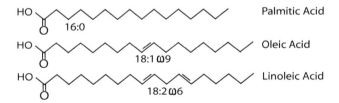

Figure 9.12 Chemical structures of some of the most common fatty acids found in nature which are typically saturated and unsaturated compounds with a chain length of C_{16} and C_{18} as shown in the case of palmitic, oleic, and linoleic acids.

Table 9.5 Fatty acid (FA) biomarkers for bacterial, algal, animals, and terrestrial sources in aquatic systems.

FATM	Reference
Bacterial markers	
Σ Odd carbon numbered + branched chain FA	Parkes and Taylor (1983); Gillan and Johns (1986); Kaneda (1991); Budge and Parrish (1998)
Σ Iso-and anteiso-C_{15} and C_{17}	Kaneda (1991); Viso and Marty (1993)
$18:1(\omega\text{-}7)/18:1\omega 9$	Volkman et al. (1980b)
Iso + anteiso 15:0/16:0	Mancuso et al. (1990); Kaneda (1991)
Iso + anteiso 15:0/15:0	Kaneda (1991)
Σ 15:0, *iso*- and *anteiso*-C_{15} and C_{17}, $18:1\omega\text{-}7$[2]	Najdek et al. (2002); Kaneda (1991)
$brC_{15}/15:0$[a]	Najdek et al. (2002)
Algal	
Fatty acids generally with chain lengths $<n\text{-}C_{20}$; polyunsaturated fatty acids (PUFAs); C_{27} and C_{28} sterols (e.g., diatoms – 16 : $1\omega 7$, 16 : 0, 14 : 0, 20: $5\omega 3$, S-5, S-6; dinoflagellates— 16:0, 18: $5\omega 3$, 22 : $6\omega 3$, 4-methyl sterols)	Canuel et al. (1995)
Animal	
16 : 0, 18 : 0, and 18 : $1w9$; cholest-5-en-3β-ol (S-3)	Canuel et al. (1995)
Terrestrial markers	
$18:2\omega\text{-}6$	Napolitano et al. (1997)
$18:2(\omega\text{-}6) + 18:3(\omega\text{-}3) >2.5$	Budge et al. (2001)
22:0 + 24:0	Budge and Parrish (1998),
Σ $C_{24:0}\text{–}$ $C_{32:0}$	Meziane et al. (1997); Canuel and Martens (1993)

[a]Used as a measure of bacterial growth in mucilaginous aggregates as bacteria experiencing favorable growth conditions yield higher proportions of branched-chain C_{15} over straight-chain $C_{15:0}$.
From Najdek et al. (2002).

However, long-chain saturated fatty acids are also found in marsh plants, seagrasses (Canuel et al., 1997), and even in some diatoms (Volkman et al., 1980a). *Short-chain fatty acids* can be derived from aquatic (algal and microbial) sources (C_{12}–C_{18}) (Simoneit, 1977) including (zooplankton, bacteria, benthic animals, and marsh plants) (Cranwell, 1982; table 9.5). Polyunsaturated fatty acids (PUFAs) are generally used as indicators of the presence of "fresh" algal sources (Canuel and Martens, 1993; Shaw and Johns, 1985), but a few PUFAs also occur in vascular plants. Higher plant sources of PUFAs are generally represented by $18:2(\omega 6)$ and $18:3(\omega 3)$ (Harwood and Russell, 1984; table 9.5). In general, C_{18} PUFAs are enriched in Prymnesiophytes, Dinophytes, Chlorophytes, and Cryptophytes while the C_{16} and C_{20} PUFAs are more enriched in diatoms (Volkman et al., 1989). Previous studies have shown that many estuarine organisms such as molluscs (Soudant et al., 1995) and polychaetes (Marsh and Tenore, 1990) cannot synthesize *de novo* many essential PUFAs (e.g., $\omega3$ and $\omega6$ PUFAs) required for their metabolism and need to obtain them through their diet (e.g., diatoms) (Canuel et al., 1995). Only plants and a few invertebrate protozoa are capable of synthesizing PUFAs (Dalsgaard et al., 2003). MUFAs are typically indicative of algal species (Volkman et al., 1989; Dunstan et al., 1994). Some monosaturated fatty acids, such as $18:1\omega7$, are found only in bacteria (Parkes and Taylor, 1983), while others (e.g., $18:1\omega9$) have been shown to be effective tracers of macrozooplankton and their fecal pellets (Wakeham and Canuel, 1988).

Branched-chain fatty acids (BrFAs) (*iso-* and *anteiso*) are believed to be primarily derived from sulfate-reducing bacteria (Perry et al., 1979; Cranwell, 1982; Canuel et al., 1995; table 9.5). However, it should be noted that BrFAs are not present in all sulfate-reducing bacteria or other heterotrophic bacteria (Kaneda, 1991; Kohring et al., 1994). The iso- and antesio- designation represents a branched fatty acid with the methyl group at the ω-1 position and the methyl group at the ω-2 position, respectively. The odd number (C_{15}, C_{17}, branched and normal) are believed to be derived from phospholipids, components of bacterial cell membranes (Kaneda, 1991). Even-numbered iso-branched fatty acids (e.g., C_{12}–C_{18}) are also found in algal sources (Schnitzer and Khan, 1972).

The distribution and abundance of n-*alkanols* (fatty alcohols) can be used to distinguish between terrestrial plant and algal/bacterial inputs to aquatic systems (Cranwell, 1982). The epicuticular waxes of vascular plants have been shown to have n-alkanols with an even number of carbon atoms (typically C_{22}–C_{30}) (Eglinton and Hamilton, 1967; Rieley et al., 1991). One of the most abundant n-alkanols in higher plants is *1-octacosanol* $[CH_3(CH_2)_{26}CH_2OH]$. Alkanols are synthesized by enzymatic reduction of fatty acids, as shown in the following example:

$$CH_3(CH_2)_nCOOH \rightarrow CH_3(CH_2)_nCHO \rightarrow CH_3(CH_2)_nCH_2OH \qquad (9.7)$$

$$\text{fatty acid} \quad \rightarrow \quad \text{aldehyde} \quad \rightarrow \quad \text{alcohol}$$

Saturated (C_{14}–C_{18}) alkanols and unsaturated (C_{16}–C_{24}) alkenols can also be used as indicators of zooplankton and as paleobiomarkers (table 9.4b). For example, isoprenoidal saturated alkanols, such as *pristanol* and *phytanol*, products of phytol, have been found in sediment cores (figure 9.13). Short-chain alkanols (even-numbered C_{14}–C_{18}) are believed to be derived from algae. Other n-alkanols (C_{26} and C_{28}) indicative of seagrass sources (Nichols and Johns, 1985), can be used to distinguish between inputs from other higher plant inputs. The relatively high concentrations of C_{30} n-alkanols in comparison with C_{26} and C_{28} in the Harney and Taylor River estuaries (USA) suggest a dominance of organic matter inputs from mangrove versus seagrass detritus (Jaffé et al., 2001). Bacteria and algae typically have n-alkanols with a C_{16} to C_{22} distribution (Robinson et al., 1984; Volkman et al., 1999). *Phytol* is one of the more prominent monounsaturated *isoprenoid alkenols* because of its source as an esterified side-chain within the chlorophyll molecule of plants (figure 9.14). Hence, chlorophyll decay is considered to be the dominant source of phytol and its degradation products in aquatic environments, especially since the *phytyl ester* is easily broken (Hansen, 1980). It is also common to find *dihydrophytol* at depth in

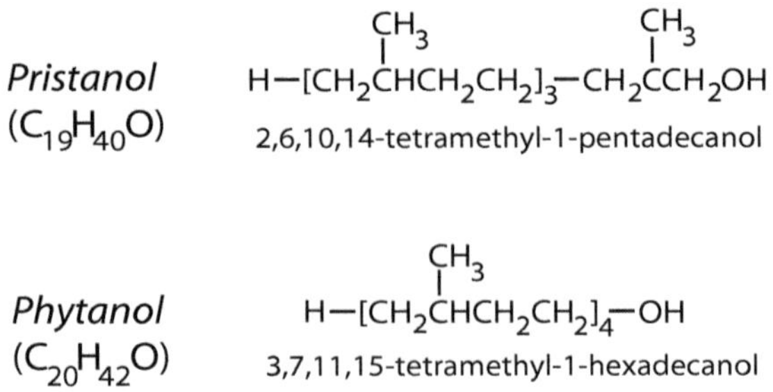

$$CH_3 \qquad\qquad CH_3$$

Pristanol $\quad H-[CH_2CHCH_2CH_2]_3-CH_2CCH_2OH$

$(C_{19}H_{40}O)$ 2,6,10,14-tetramethyl-1-pentadecanol

$$CH_3$$

Phytanol $\quad H-[CH_2CHCH_2CH_2]_4-OH$

$(C_{20}H_{42}O)$ 3,7,11,15-tetramethyl-1-hexadecanol

Figure 9.13 Chemical structures of the multibranched saturated alkanols, pristanol and phytanol.

sediments as a diagenetically reduced transformation product of phytol. Together, these two compounds have been used as effective biomarkers of biogeochemical processes (Volkman and Maxwell, 1984). For example, the ratio of dihydrophytol to phytol was shown to be significantly higher in zooplankton than in phytoplankton further supporting the use of dihydrophytol as an indicator of herbivorous grazing activity (Prahl et al., 1984; Sun et al., 1998). Recent work has shown that dominance of the n-C_{24} alkanol is indicative of cyanobacterial inputs in the sediment records of Mud Lake, Florida (Filley et al., 2001).

    Sterols and their respective derivatives have proven to be important biomarkers that can be used to estimate algal and vascular plant contributions as well as diagenetic proxies (Volkman, 1986; Sun and Wakeham, 1998; Canuel and Zimmerman, 1999). These compounds are a group of cyclic alcohols (typically between C_{26} and C_{30}) that are resistant to saponification. Sterols are biosynthesized from isoprene units using the mevalonate pathway and are classified as triterpenes (i.e., consisting of five isoprene units; figure 9.15; table 9.4). Terrestrial plants have been shown to have a high abundance of 24-ethylchloest-5-en-3β-ol (*sitosterol*) ($C_{29}\Delta^5$) and 24-methylcholest-5-en-3β-ol (*campesterol*) ($C_{28}\Delta^5$) sterols (Volkman, 1986; Jaffe et al., 1995; table 9.4c). Seagrasses also have characteristically high concentrations of $C_{29}\Delta^5$ (Nishimura and Koyama, 1977; Volkman et al., 1981; Nichols and Johns, 1985). Although C_{29} sterols are considered to be the dominant sterols found in vascular plants, certain epibenthic cyanobacteria and phytoplanktonic species have also been found in significant quantities (Volkman et al., 1981; Jaffé et al., 1995). The predominant sterols found inphytoplankton are C_{28}

Figure 9.14 Chemical structure of the monounsaturated isoprenoid alkenol, phytol. (Refer to figure 9.37 to see how the phytol side-chain is associated with the chlorophyll molecule.)

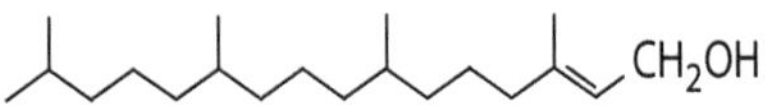

CH_2OH **Phytol**

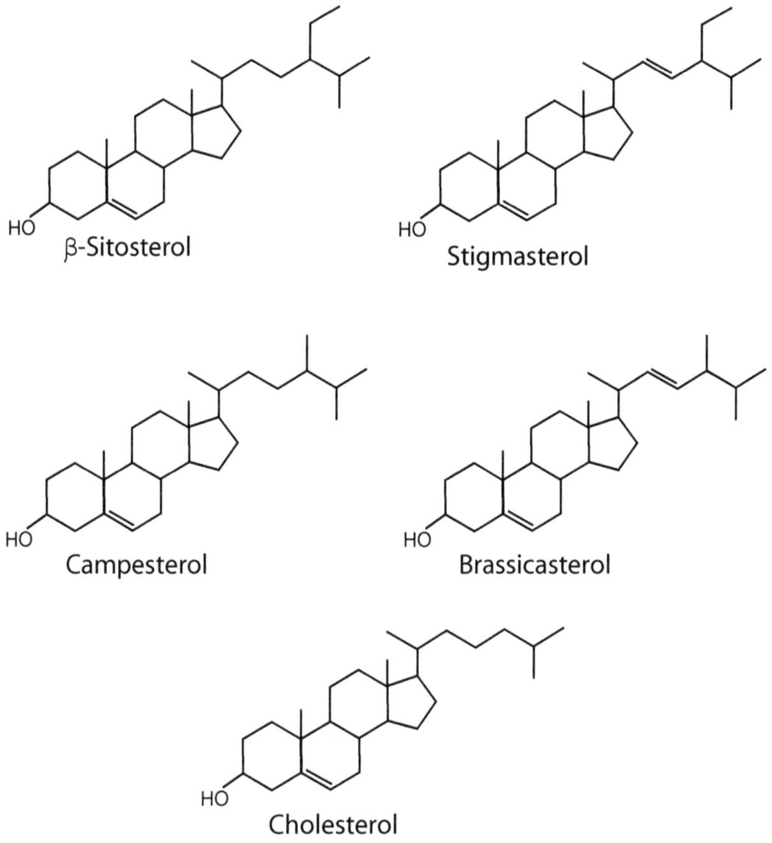

Figure 9.15 Chemical structures of some of the important sterols used as biomarkers in estuaries (e.g., β-sistosterol, stigmasterol, campesterol, brassicasterol, and cholesterol).

sterols (Volkman, 1986); 24-methylchloesta-5,22-dien-3β-ol (*brassicasterol*) ($C_{28}\Delta^{5,22}$) is particularly abundant in diatoms and Cryptomonads (Volkman et al., 1981; table 9.4C). Other phytoplankton markers include *dinosterol* (dinoflagellates) and 24-methylchloesta-5,24(28)-dien-3β-ol (*24-methylenecholesterol*) ($C_{28}\Delta^{5,24[28]}$) (diatoms, dinoflagellates, and Prasinophytes) (Volkman, 1986). Sterols can also be used to distinguish between wastewater and natural sources of organic matter using the ratios of 24-ethylcoprostanol to 24-ethylcholest-5-en-3β-ol (*sitosterol*) ($C_{29}\Delta^5$) and 5β-cholestan-3β-ol (*coprostanol*) ($C_{27}\Delta^0$) to cholest-5-en-3β-ol (*cholesterol*) ($C_{27}\Delta^5$) (Quemeneur and Marty, 1992). *Cholesterol* (cholest-5-en-3β-ol) ($C_{27}\Delta^5$) is generally considered to be derived from crustacean tissues (e.g., zooplankton and crabs) (Bergmann, 1949) as well as their fecal pellets. However, it is also found in trace amounts in some phytoplankton species (Gagosian et al., 1983), and marsh and seagrass plants, such as *S. alterniflora* and *Z. marina*, respectively

(Canuel et al., 1997). This recurring theme of overlapping sterol markers in different organic matter sources indicates that caution should be advised when using sterols solely to distinguish between land and aquatic sources (Volkman, 1986; Jaffé et al., 2001). Instead, biomarker source identifications should be corroborated across lipid compound classes and using bulk and compound-specific isotope analysis.

The *stanol:stenol* ratio has been commonly used as another molecular index of organic matter degradation in open ocean and coastal environments (Wakeham and Lee, 1989; Jaffé et al., 2001). In particular, microbial transformation of sterols has been effectively estimated using the ratio of $5\alpha(H)$-chloestan-3β-ol ($C_{27}\Delta^{0}$) to cholest-5en-3β-ol ($C_{27}\Delta^{5}$) (Wakeham and Lee, 1989; Canuel and Martens, 1993). Jaffé et al. (2001) also found significantly higher $C_{27:0}/C_{27:5}$, $C_{28:0}/C_{28:5}$, and $C_{29:0}/C_{29:5}$ ratios in the upper versus the lower regions of the Harney River estuary (USA) suggesting the importance of "fresher" organic matter inputs to the lower estuary, with inputs to the upper estuary being dominated by "reworked" soil-derived materials.

Cutins and suberins are lipid polymers in vascular plant tissues which serve as a protective layer (cuticle) and cell wall components of cork cells, respectively (Martin and Juniper, 1970; Holloway, 1973). The basic monomeric units of cutins and suberins are shown in figure 9.16. Cutins have been shown to be effective biomarkers for vascular plant inputs to coastal systems (Eglinton et al., 1968; Cardoso and Eglinton, 1983; Goñi and Hedges, 1990a,b). When cutin is oxidized, using the CuO method commonly used

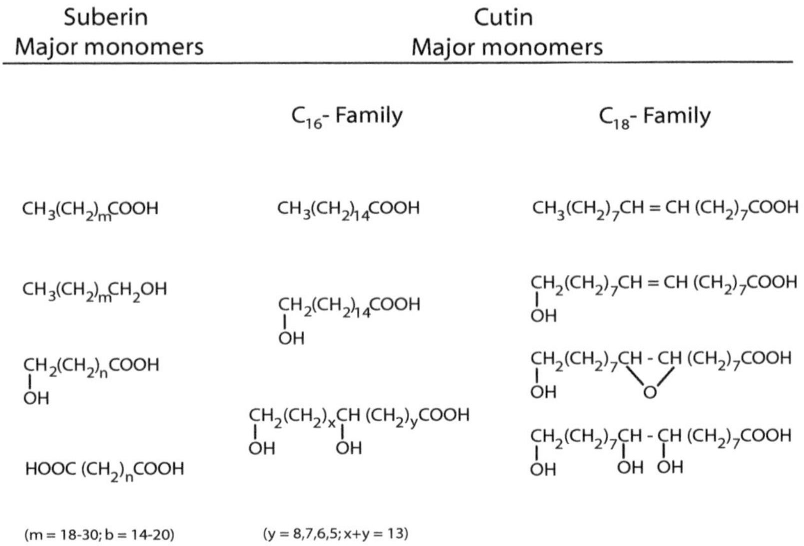

Figure 9.16 Chemical structure of cutin, a biopolyester mainly composed of interester-ified hydroxy and epoxy–hydroxy fatty acids with a chain length of 16 and/or 18 carbons (C_{16} and C_{18} class). Also, the chemical strcuture of the aliphatic monomers of suberin, derived from the general fatty acid biosynthetic pathway, namely from palmitic (16:0), stearic (18:0), and oleic acids.

for lignin analyses (Hedges and Ertel, 1982), a series of fatty acids are produced that can be divided into three groups: C_{16} hydroxy acids, C_{18} hydroxy acids, and C_n hydroxyl acids (Goñi and Hedges, 1990a). Since cutins are not found in wood, they are similar to the p-hydroxyl-lignin-derived cinnamyl phenols (e.g., $trans$-p-coumaric and ferulic acids) and serve as biomarkers of nonwoody vascular plant tissues. However, cutin acids were found to be more reactive than lignins in the degradation of conifer needles (Goñi and Hedges, 1990c) and are not considered to be as effective in tracing vascular plant inputs as are lignins (see below) (Opsahl and Benner, 1995). Although suberins have been found in peats and recent sediments (Cardoso and Eglinton, 1983), more work on their structure and function is needed for more effective applications as biomarkers (Hedges et al., 1997).

In the past decade there have been an increasing number of studies that have utilized lipid biomarkers to evaluate some of the central questions regarding controls on the relationships between organic fractions found in DOM and POM pools in estuaries, as well as on preservation in sediments. For example, recent work has shown that saturated fatty acids represented the dominant class of lipids in POM and DOM in the Delaware Estuary (USA) (figure 9.17; Mannino and Harvey, 1999). Concentrations of fatty acids were significantly higher in POM than in DOM and, within the DOM pool, comprised a higher percentage of OC in the very high molecular weight fraction (30 kDa–0.2 μm) (VHDOM) than the high molecular weight fraction (1–30 kDa) (HDOM). More recent work in a comparative study of Delaware Bay estuary (USA), San Diego Bay estuary (USA), Boston Harbor estuary (USA), and San Franciso Bay estuary (USA), also showed that the dominant lipids in HDOM (<0.2 μm to >1 kDa) were even-number saturated (C_{14}–C_{18}) fatty acids, followed by bacteria-specific odd-number branched (C_{15}–C_{17}) and normal fatty acids (Zou et al., 2004). Moreover, the absence of long-chain fatty acids (>C_{20}) in HDOM further supports the work by Mannno and Harvey (2000) that there is rapid removal of vascular-plant derived compounds along an estuarine gradient, presumably due to flocculation and photo-oxidation (Benner and Opsahl, 2001). From a cross-system comparison perspective, the high abundance of bacteria-specific fatty acids and amide-N compounds in HDOM supports the contention that a substantial fraction of oceanic DOM is also derived from bacterial membrane material (Liu et al., 1998; McCarthy et al., 1998), which can ultimately form in *liposomes* (Borch and Kirchman, 1999).

The absence of PUFAs and low concentrations of MUFAs (both sources of phytoplankton) in estuarine HDOM (Zou et al., 2004) supports other studies which show preferential loss of unsaturated over saturated fatty acids (Gomez-Belinchon et al., 1988; Wakeham and Lee, 1989; Sun and Wakeham, 1994; Harvey and Macko, 1997; Parrish et al., 2000). Thus, it appears that a significant fraction of estuarine HMW DOM is derived from degraded phytodetrital material that has been processed by bacteria in addition to bacterial membrane components. However, lipid composition of POM and the differential lability of lipids are also important in controlling what is processed by bacteria and ultimately partitioned into the DOM pool. Thus, other experimental work has shown rapid decay of unsaturated sterols under oxic conditions, suggesting that chemical structure alone (e.g., unsaturation) may not always be a good predictor of decomposability (Harvey and Macko, 1997). Nevertheless, the remarkable similarity in composition of HDOM from different estuarine systems emphasizes the importance of bacterial processing as an important control on the biochemical "signature" of lipids in HDOM (Zou et al., 2004). Finally, it

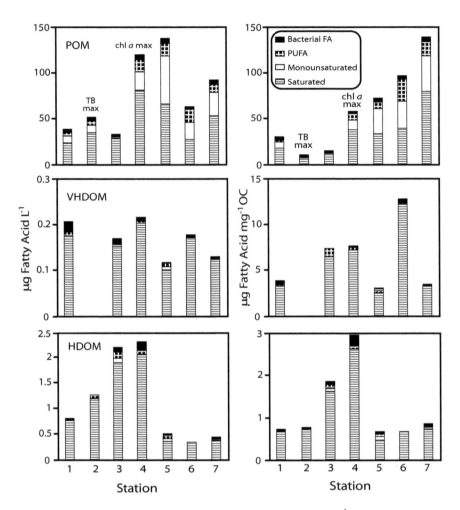

Figure 9.17 Distribution of major fatty acids (μg fatty acids L^{-1} and μg fatty acid mg^{-1} OC) in particulate organic matter (POM), very high molecular mass dissolved organic matter (VHDOM) (30 kDa–0.02 μm), and high molecular mass DOM (HDOM) (1-30 μm) in Delaware Bay estuary. Bacterial fatty acids (FA) included branched and normal saturated acids and 15:1Δ4 and PUFA = polyunsaturated fatty acids. (Modified from Mannino and Harvey, 1999.)

should be noted that physical processes such as solubility and adsorption/desorption are also important in controlling the partitioning of materials into HDOM.

As mentioned earlier, lipids tend to be more decay resistant than other biochemical constituents of organic matter, hence the name geolipid. In fact, Harvey et al. (1995) showed that under both oxic and anoxic conditions approximately 33% of the lipids in cyanobacterial (*Synechococcus* sp.) detritus remained after a 93-day incubation period—significantly higher than for protein or carbohydrates. It was further concluded that all cellular components of both diatom and cyanobacterial detrital material decomposed more rapidly under oxic conditions, which supports the importance of the redox-controlling mechanism in preserving organic matter discussed in chapter 8. Other laboratory experiments in estuarine sediment work have shown greater loss of sterols and fatty acids under oxic versus anoxic conditions (Sun et al., 1997; Sun and Wakeham, 1998). Harvey et al. (1995) proposed that the relatively longer turnover time of lipids can possibly be explained by the following mechanisms: (1) incorporation of free sulfide into lipids—particularly under anoxic conditions as shown in sediments (Kohnen et al., 1992; Russell and Hall, 1997), and (2) differences in structure noting fewer oxygen functionalized atoms compared to other components. More recent experimental sediments from Long Island Sound estuary proposed a mechanism involving the frequency of oscillating redox conditions (Sun et al., 2002). It was demonstrated that the rate and pathway of degradation of ^{13}C-labeled lipids were a function of the frequency of oxic to anoxic oscillations. The modes of oscillation used in this experiment are shown in figure 9.18. Lipid degradation rates increased significantly with increasing oscillation and overall oxygen exposure time; however, in some cases the rates are linear and in others they are exponential (figure 9.19). This supports other findings that oscillating redox conditions (Aller, 1998) and "oxygen exposure time" (Harnett et al., 1998) are important controlling variables for organic matter preservation in estuarine systems. The selective nature of degradative processes over changing redox conditions, and the fact that certain biochemical constituents can degrade over timescales of a few days, reflects the importance of using short-lived biomarker compounds in understanding diagenetic processes in shallow and dynamic estuarine systems (Canuel and Martens, 1996).

As indicated in the preceding sections, significant progress has been made using a multitude of chemical biomarkers and multivariate statistical methods in an attempt to identify the dominant factors controlling the sources, concentrations, and distribution of organic matter in estuaries. However, efforts to link such measurements across divergent estuarine systems remains a difficult task because of the unique regional signatures found in many estuaries. A paper by Canuel (2001) provides such a comparison whereby fatty acids are used to characterize POM in two divergent estuarine systems—the Chesapeake Bay (CB) and San Francisco Bay (SFB) estuaries. Canuel (2001) found a general enrichment of total fatty acids from the head (freshwater) to the mouth (high salinity) of both estuarine systems (figure 9.20). The fatty acid composition (14:0, 16:0, 161ω7, 18:1ω9, and 18:0) in both systems was indicative of phytoplankton, zooplankton, bacteria, and vascular plant sources. However, both systems were dominated by algal sources as indicated by the abundance of short-chain fatty acids. There was significantly more total fatty acids and PUFAs present in CB than in SFB possibly indicating more labile sources of POM even during nonbloom periods (figures 9.20 and 9.21). Northern regions of the CB were also different in that the fatty acids were enriched in 16:1ω7, n-branched odd-numbered acids and 18:1ω9—indicative of bacteria and freshwater/estuarine phytoplankton. While both systems have received significant increases in nutrient loading over the past century, phytoplankton production is limited in SFB due to light limitation and grazing by benthic bivalves (Alpine and Cloern, 1992). Thus, as indicated from the lipid composition the overall lability and abundance of POM in the CB is higher, particularly near the mouth and in the higher-salinity regions.

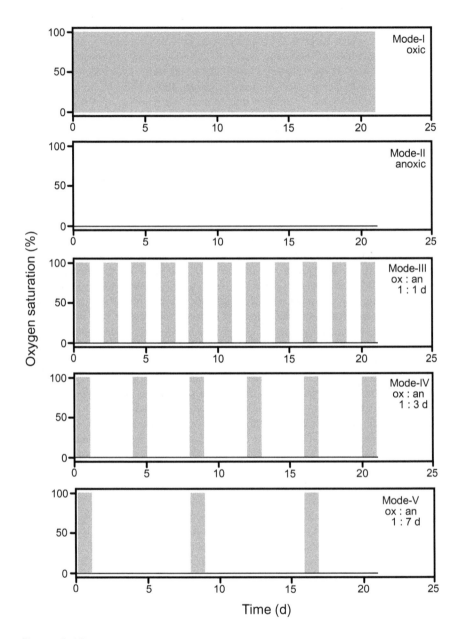

Figure 9.18 Illustration of the modes of oscillation of percent oxygen: (I) continuously oxic; (II) continuously anoxic; (III) 1 d oxic: I d anoxic; (IV) 1 d oxic: 3 d anoxic; and (V) 1 d oxic: 7 d anoxic. (Modified from Sun et al., 2002.)

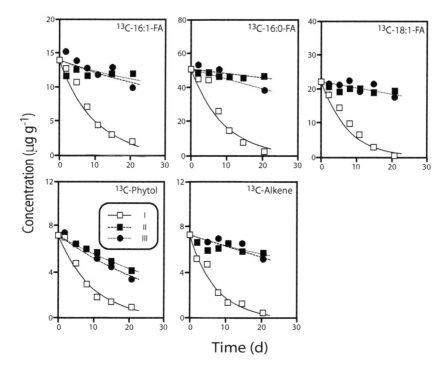

Figure 9.19 Variations in the concentration ($\mu g\ g^{-1}$) of five cell-associated ^{13}C-labeled compounds oxic (I: open) and anoxic (II: open, and VI: closed) incubations. Straight lines and curves were fit to a first-order degradation model of concentration data. (Modified from Sun et al., 2002.)

Carbohydrates

Carbohydrates are the most abundant class of biopolymers on Earth (Aspinall, 1983) and consequently represent significant components of water column POM and DOM in aquatic environments (Cowie and Hedges, 1984a,b; Arnosti et al., 1994; Aluwihare et al., 1997; Bergamaschi et al., 1999; Opsahl and Benner, 1999; Burdige et al., 2000; Hung et al., 2003a,b). Carbohydrates are important structural and storage molecules and are critical in the metabolism of terrestrial and aquatic organisms (Aspinall, 1970). The general formula for carbohydrate composition is $(CH_2O)_n$; these compounds can be defined more specifically as polyhydroxyl *aldehydes* and *ketones*—or compounds that can be hydrolyzed to them. Carbohydrates can be further divided into *monosaccharides* (simple sugars), *disaccharides*(two covalently linked monosaccharides), *oligosaccharides*(a few covalently linked monosaccharides), and *polysaccharides* (polymers made up of chains of mono- and disaccharides). Based on the number of carbons, monosaccharides (e.g., 3, 4, 5, and 6) or simple sugars are commonly termed trioses, tetroses, pentoses, and hexoses, respectively. Carbohydrates have *chiral* centers (usually more than one) making possible

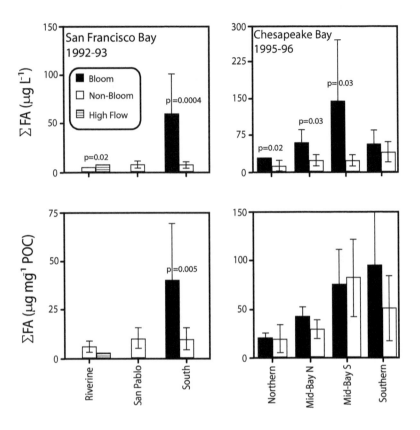

Figure 9.20 Total fatty acids (ΣFA) concentrations in water volume (μg L^{-1}) and POC (μg mg^{-1} OC) in San Francisco Bay (SFB) and Chesapeake Bays from 1992 to 1993 and 1995 to 1996, respectively. Bloom conditions are defined as Chl-a > 10 μg L^{-1}. High- and low-flow conditions were only tested for SFB. (Modified from Canuel, 2001.)

the occurrence of *enantiomers* (mirror images), whereby n chiral centers allows for the possibility of 2^n sterioisomers. For example, glucose has two sterioisomers—one that reflects plane-polarized light to the right [*dextrorotatory* (D)] and the other that reflects it to the left [*levorotatory* (L)]—as shown in the traditional *Fisher projection* (figure 9.22). Almost all naturally occurring monosaccharides occur in the D configuration; some of these *aldoses* and *ketoses* are shown in figures 9.22 and 9.24. Pentoses and hexoses can become cyclic when the aldehyde or keto group reacts with the distal hydroxyl on one of the distal carbons. For example, in glucose C-1 reacts with the C-5 forming a *pyranose ring*, as shown in a typical *Haworth projection* (figure 9.23). This results in the formation of two *diasteneomers* in equilibrium with each other around the asymmetric center at C-1 called *anomers;* when the OH group is above and below the plane of the ring at the anomeric C-1 it is referred to as β and α, respectively. Similarly, fructose can form a

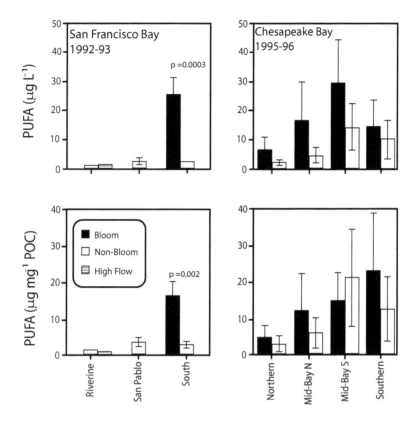

Figure 9.21 Polyunsaturated fatty acid (PUFA) concentrations in water volume ($\mu g\ L^{-1}$) and POC ($\mu g\ mg^{-1}$ OC) in San Francisco and Chesapeake Bay (SFB) from 1992 to 1993 and 1995 to 1996, respectively. Bloom conditions are defined as Chl-a > 10 $\mu g\ L^{-1}$. High and low flow conditions were only tested for SFB. (Modified from Canuel, 2001.)

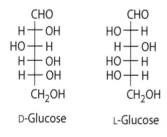

Figure 9.22 Traditional Fisher projection of the two glucose sterioisomers—one reflects plane-polarized light to the right [dextrorotatory (D)] and the other reflects it to the left [levorotatory (L)].

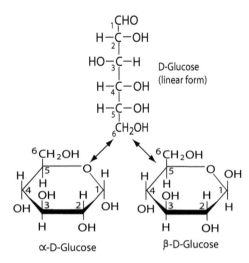

Figure 9.23 Typical Haworth projection of a pyranose ring of the aldose, glucose, formed between C-1 and C-5.

five-carbon ring called a *furanose* when the C-2 keto group reacts with the hydroxyl on C-5 (figure 9.24). *Cellulose*, a highly abundant polysaccharide found in plant cell walls, is composed of long linear chains of glucose units connected as *1,4'–β-glycoside linkages* (figure 9.25). These β-linkages enhance intrachain hydrogen bonding as well as van der Waal's interactions resulting in straight rigid chains of *microfibrils*, making cellulose very difficult to digest by many heterotrophs (Voet and Voet, 2004).

In phytoplankton, carbohydrates serve as important reservoirs of energy, structural support, and cellular signaling components (Lee, 1980; Bishop and Jennings, 1982).

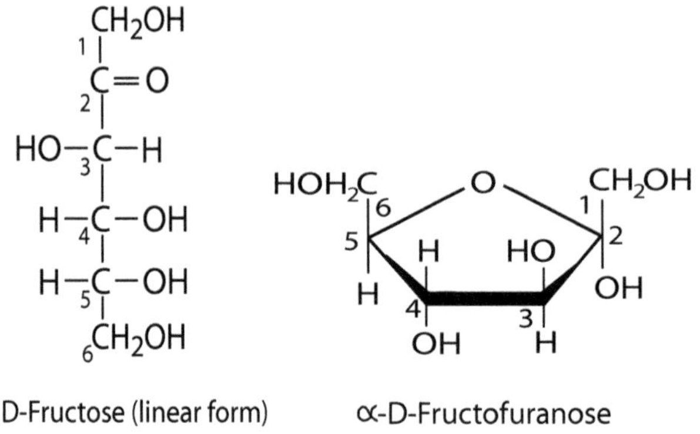

D-Fructose (linear form)   α-D-Fructofuranose

Figure 9.24 Traditional Fisher and Haworth projections of the ketose, fructose. Fructose can form a five-C ring called a furanose when the C-2 keto group reacts with the hydroxyl on C-5, as shown in α-D-fructofuranose.

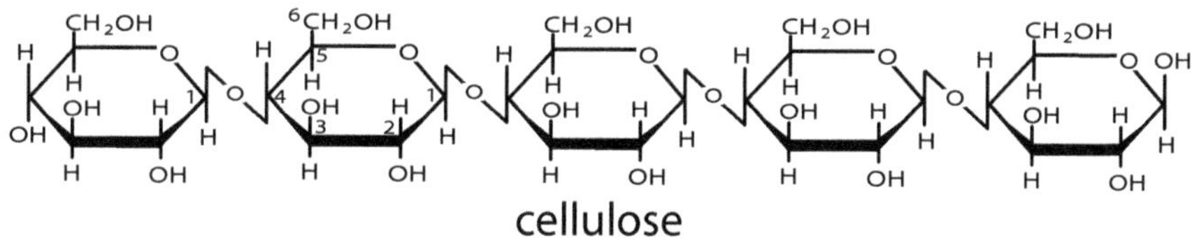

cellulose

Figure 9.25 Chemical structure of the polysaccharide cellulose, composed of long linear chains of glucose units connected as 1,4′–β-glycoside linkages.

Carbohydrates make up approximately 20 to 40% of the cellular biomass in phytoplankton (Parsons et al., 1984) and 75% of the weight of vascular plants (Aspinall, 1983). Structural polysaccharides in vascular plants, such as *α-cellulose, hemicellulose, and pectin,* are dominant constituents of plant biomass, with cellulose being the most abundant biopolymer (Aspinall, 1983; Boschker et al., 1995). Certain carbohydrates can also represent significant components of other structural (e.g., lignins) and secondary compounds (e.g., tannins) found in plants (Zucker, 1983). Carbohydrates serve as precursors for the formation of humic material (e.g., kerogen), a dominant form of organic matter in soils and sediments (Nissenbaum and Kaplan, 1972; Yamaoka, 1983). Extracellular polysaccharides are important in binding microbes to surfaces, construction of invertebrate feeding tubes and fecal pellets, and in microfilm surfaces (Fazio et al., 1982). For example, carbohydrates changed the mechanism of erodibility of sediments in a tidal flat in the Westerschelde estuary (The Netherlands) from individual rolling grains to clumps of resuspended materials (Lucas et al., 2003). Fibrillar material enriched in polysaccharides has been shown to be an important component of colloidal organic matter in aquatic systems (Buffle and Leppard, 1995; Santschi et al., 1998). Carbohydrates are also important in the formation of "marine snow" (Alldredge et al., 1993; Passow, 2002). *Polysaccharides* can be further divided into the dominant sugars (e.g., found in most organic matter) or *major sugars*, as well as other more source-specific forms, the *minor sugars*. While there have been problems with using some of these carbohydrate classes as effective biomarkers of organic matter in aquatic systems (due to a lack of specificity) there remains considerable interest in their application as biomarkers because they are dominant molecular constituents of organic matter.

There has been considerable attention to measuring the relative abundances of the *major sugars* (Mopper, 1977; Cowie and Hedges, 1984b; Hamilton and Hedges, 1988). The major sugars can be further divided as unsubstituted *aldoses* and *ketoses*. The aldoses or *neutral sugars (rhamnose, fucose, lyxose, ribose, arabinose, xylose, mannose, galactose, and glucose)* (figure 9.26) are the monomeric units of the dominant structural polysaccharides of vascular plants (e.g., cellulose and hemicelluloses). For example, cellulose is composed exclusively of glucose monomers, making it the most abundant neutral sugar in vascular plants. Since neutral sugars make up most of the majority of organic matter in the biosphere, molecular measurements of neutral sugars in natural organic matter are commonly thought of as total carbohydrates (Opsahl and Benner, 1999). Early work suggested that *fucose*, which is rarely found in vascular plants and is more abundant in

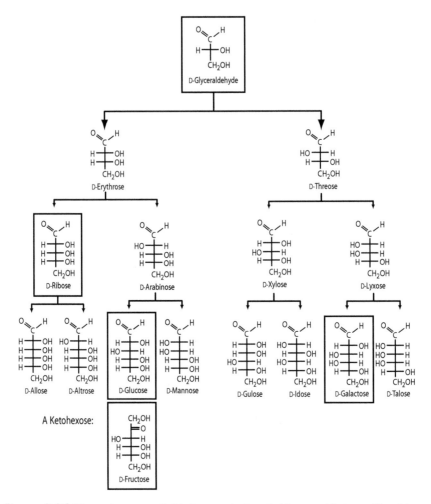

Figure 9.26 The major sugars divided as unsubstituted aldoses and ketoses. The aldoses or neutral sugars (rhamnose, fucose, lyxose, ribose, arabinose, xylose, mannose, galactose, and glucose) are the dominant monomeric units of the dominant structural polysaccharides of vascular plants (e.g., cellulose and hemicelluloses).

phytoplankton and bacteria, may be a good source indicator of aquatic organic matter (Aspinall, 1970; Percival, 1970, 1983). Other work in a hypersaline lagoon has suggested that *fucose, ribose, mannose*, and *galactose* are good source indicators of bacteria and cyanobacteria (Moers and Larter, 1993). Similarly, *xylose* or *mannose/xylose* were proposed as indicators for angiosperm versus gymnosperm plant tissues while *arabinose+ galactose* versus *lyxose + arabinose* provided a useful index for separating woody versus nonwoody vascular plant tissues (Cowie and Hedges, 1984a,b). In addition, *arabinose* has been found in association with mangrove detritus in lagoonal sediments (Moers and Larter, 1993), but has also been linked with the surface microlayers of brackish and marine waters (Compiano et al., 1993) and is believed to be derived from phytoplankton sources (Ittekkot et al., 1982). In general, all organisms contain the same complement of simple sugars, although the relative abundance may change as a function of source. Thus, the relative lack of source specificity in neutral sugar composition and differential stability (Hedges et al., 1988; Macko et al., 1989; Cowie et al., 1995) has in some cases compromised their effectiveness as biomarkers. For example, sediment trap and sediment samples in an anoxic fjord (Saanich Inlet, Canada) indicated that glucose, *lyxose,* and *mannose* in vascular plant detrital particles were lost in the upper sediments (Hamilton and Hedges, 1988). Conversely, *rhamnose* and *fucose* were produced in situ and were likely derived from bacteria known to be abundant in these deoxy sugars (Cowie and Hedges, 1984b). A long-term (4 y) incubation experiment that examined changes in the carbohydrate abundance and composition of five different vascular plant tissues [mangrove leaves and wood (*Avicennia germinans*), cypress needles and wood (*Taxodium distichum*), and smooth cordgrass (*Spartina alterniflora*)] showed that while glucose and xylose were degraded over time deoxy sugars increased in relative abundance, indicative of the importance of microbial biomass with increasing age of detritus (Opsahl and Benner, 1999; figure 9.27).

   Minor sugars, such as *acidic sugars, amino sugars,* and *O-methyl sugars* tend to be more source specific than major sugars and can potentially provide further information in understanding the biogeochemical cycling of carbohydrates (Mopper and Larsson, 1978; Klok et al., 1984a,b; Moers and Larter, 1993; Bergamaschi et al., 1999). In the case of amino sugars, such as glucosamine, an amino group substitutes for one of the hydroxyls and in some cases may be acetylated as in α-D-N-acetylglucosamine (figure 9.28). Minor sugars exist in a diversity of forms and are different from simple sugars by the substitution of one or more hydroxyl groups with an alternative functional group (Aspinall, 1983). The contribution of minor sugars as a percentage of total carbohydrates is shown in select sources of organic matter (figure 9.29; Bergamaschi et al., 1999). For example, acidic sugars such as *uronic acids* have been considered to be indicators for microbial biomass (Uhlinger and White, 1983) as well as indicators of bacterial sources of DOM and POM (Benner and Kaiser, 2003). Uronic acids are important extracellular polymers

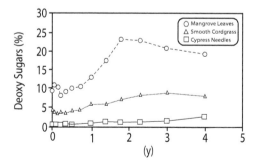

Figure 9.27 Relative abundance (%) of deoxy sugars as a function of total neutral sugar yields in decomposing mangrove leaves, smooth cordgrass, and cypress needles during a 4-year laboratory incubation study. (Modified from Opsahl and Benner, 1999.)

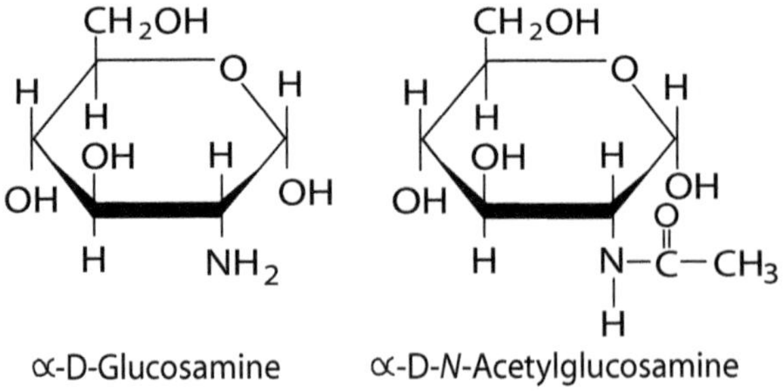

α-D-Glucosamine   α-D-N-Acetylglucosamine

Figure 9.28 Chemical structure of the amino sugar, glucosamine, and α-D-N-acetyl-glucosamine, where an acetylated amino group substitutes for one of the hydroxyls in glucosamine.

used by microbes to bind to surfaces, invertebrates to make feeding nets, and pelletization of fecal pellets (Fazio et al., 1982). Uronic acids are also important components of hemicellulose and pectin (Aspinall, 1970, 1983) and are therefore considered to be source indicators (biomarkers) for angiosperms and gymnosperms (Aspinall, 1970, 1983; Whistler and Richards, 1970). For example, *glucuronic* and *galacturonic acids* are the most abundant uronic acids in vascular plants (Danishefsky et al., 1970; Bergamaschi et al., 1999). Cyanobacteria, prymnesiophytes, and heterotrophic bacteria have been purported to be producers of acid polysaccharides in the upper water column of coastal waters (Hung et al., 2003a,b). In general, the dominant sources of acid polysaccharides are phytoplankton, prokaryotes, and macroalgae (Decho and Herndl, 1995; Biddanda and Benner, 1997; Hung et al., 2003a,b). Similarly, *O*-methyl sugars can be found in soils and sediments (Mopper and Larsson, 1978; Klok et al., 1984a,b; Bergamaschi et al., 1999), bacteria (Kenne and Lind berg, 1983), vascular plants (Stephen, 1983), and algae (Painter, 1983). In certain sources of organic matter, such as eelgrass (*Zostera marina*), minor sugars can represent a greater fraction of the total carbohydrate pool than neutral sugars (figure 9.29). Certain *O*-methyl sugars and amino sugars may have potential as source-specific indicators of filamentous cynaobacteria (e.g., *2/4/-OMe-xylose, 3- and 4-OMe-fucose, 3- and 4-OMe-arabinose, 2/4-OMe-ribose,* and *4-OMe-glucose*) and coccoid cyanobacteria and/or fermenting, SO_4^{2-} reducing, and methanogenic bacteria (e.g., *glucosamine, galactosamine, 2- and 4-OMe-rhamnose, 2/5/-OMe-galactose,* and *3/4-OMe-mannose*) (Moers and Larter, 1993).

Although carbohydrates are considered to be a significant fraction of the DOC pool in both oceanic (20–30%) (Pakulski and Benner, 1994) and estuarine waters (9–24%) (Senior and Chevolot, 1991; Murrell and Hollibaugh, 2000; Hung et al., 2001), very little is known about the uptake dynamics and general cycling of carbohydrates in estuaries. The turnover of polysaccharides (e.g., *xylan, laminaran, pullulan, fucoidan*) is also poorly understood, with recent work suggesting that molecular size may not be as important a variable in controlling the extracellular hydrolysis of these macromolecular compounds in coastal systems as previously thought (Arnosti and Repeta, 1994; Keith and Arnosti, 2001). While we do know that neutral sugars are a dominant component (40–50%) of the

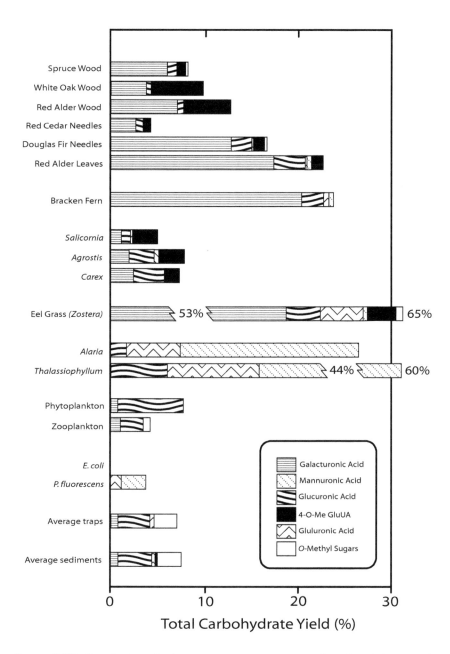

Figure 9.29 Contribution of minor sugars as a percentage of total carbohydrates in select sources of organic matter. (Modified from Bergamaschi et al., 1999.)

total carbohydrates in the water column (Borch and Kirchman, 1996; Mannino and Harvey, 2000), very little is known about differential uptake of these monosaccharides by bacteria (Rich et al., 1996; Kirchman, 2003, Kirchman and Borch, 2003). Many of the abundant sugars (e.g., glucose and rhamnose) typically occur as *dissolved combined neutral sugars* (DCNS) (Kirchman and Borch, 2003). Other studies have shown remarkable temporal and spatial stability in the composition of DCNS in the Delaware Bay estuary, despite significant variations in POC and DOC (Kirchman and Borch, 2003). The composition of DCNS was reflected as having glucose (23 mol%) and arabinose (6 mol%) as the most and least abundant, respectively. The relative abundance of neutral sugars, specifically glucose (Hernes et al., 1996), is typically viewed as an index of organic matter lability (Amon et al., 2001; Amon and Benner, 2003). Thus, it is remarkable that the composition of DCNS in Delaware Bay estuary did not show significant seasonal variation and was similar to the composition found in the open ocean (Borch and Kirchman, 1997; Skoog and Benner, 1997).

These similarities in DCNS composition have been attributed to the production of refractory heteropolysaccharides through bacterial processing of organic matter (Kirchman and Borch, 2003). Similarly, other work has shown high concentrations of *acyl-heteropolysaccharides* (APS) in HMW DOM from different aquatic systems believed to be derived from freshwater phytoplankton (Repeta et al., 2002). In fact, the relative abundance of seven major neutral sugars (rhamnose, fucose, arabinose, xylose, mannose, glucose, and galactose) in HMW DOM was found to be similar and the relative amounts in samples from different aquatic systems were nearly the same (figure 9.30).

Although marine and freshwater microalgae are both capable of producing APS (Aluwihare et al., 1997; Repeta et al., 2002), controls on the overall cycling of APS in aquatic systems remains largely unknown. Spectroscopic and molecular data indicate that APS in HMW DOM across divergent systems is very similar (Aluwihare et al., 1997). Multivariate analyses using direct temperature-resolved mass spectrometry (DT-MS) showed that HMW DOM (>1 kDa) from Chesapeake Bay and Oosterschelde estuary (The Netherlands) was largely composed of amino sugars, deoxysugars, and *O*-methyl sugars, suggestive of bacterial processing (Minor et al., 2001, 2002). These studies further support the hypothesis stated earlier that similar lipid composition or "signature" found in estuarine DOM (Zou et al., 2004) was ultimately determined by bacterial processing along an estuarine gradient. These molecular studies also support the notion that much of the DOC in estuaries is composed of both labile and refractory components (Raymond and Bauer, 2000, 2001a; Cauwet, 2002; Bianchi et al., 2004), with much of the labile fraction degraded before being exported to the ocean.

In sediments, carbohydrates have been estimated to be approximately 10 to 20% of the total organic matter (Hamilton and Hedges, 1988; Cowie et al., 1992; Martens et al., 1992). Like the water column, controls on the turnover and extracellular hydrolysis of polysaccharides to monosaccharides are not well understood for the dissolved and particulate carbohydrate fractions in sediments (Lyons, et al., 1979; Lions and Gaudette, 1979; Boschker et al., 1995; Arnosti and Holmer, 1999). Recent work has shown that neutral sugars represented 30 to 50% of the dissolved carbohydrates in Chesapeake Bay porewaters (Burdige et al., 2000). The percentage abundance of porewater neutral sugars ranged from glucose at the high end (28%) to rhamnose (6%) and arabinose (7%) at the low end (Burdige et al., 2000); these results agree with the relative percentages of DCNS

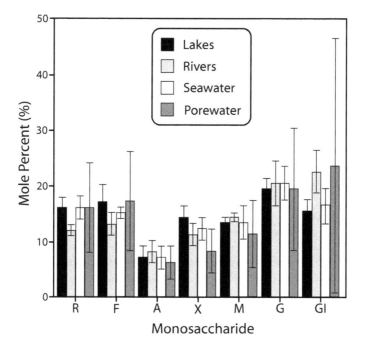

Figure 9.30 Relative abundance (mol%) of seven major neutral sugars (rhamnose, fucose, arabinose, xylose, mannose, glucose, and galactose) in the HMW DOM (> 1 kDa) in different aquatic systems. (Modified from Repeta et al., 2002.)

found in the water column of Delaware Bay (Kirchman and Borch, 2003). However, concentrations of particulate and dissolved carbohydrates in Chesapeake Bay sediments appeared to be decoupled with more of the dissolved carbohydrate fraction found in the HMW DOM of porewaters. Such differences between particulate and dissolved carbohydrate fractions have been reported in other studies (Arnosti and Holmer, 1999). A relatively higher abundance of dissolved carbohydrates in the HMW DOM fraction may reflect an accumulation of HMW intermediates generated during POC remineralization in sediments (Burdige et al., 2000).

Proteins

Proteins make up approximately 50% of organic matter (Romankevitch, 1984) and contain about 85% of the organic N (Billen, 1984) in marine organisms. Peptides and proteins comprise an important fraction of the particulate organic carbon (POC) (13–37%) and nitrogen (PON) (30–81%) (Cowie and Hedges, 1992; Nguyen and Harvey, 1994; van Mooy et al., 2002) as well as dissolved organic nitrogen (DON) (5–20%) and DOC (3–4%) in oceanic and coastal waters (Sharp, 1983). In sediments, proteins account for approximately 7 to 25% of organic carbon (Degens, 1977; Burdige and Martens, 1988; Keil et al., 1998, 2000) and an estimated 30 to 90% of total N (Henrichs et al., 1984; Burdige and Martens, 1988; Haugen and Lichtentaler, 1991; Cowie and Hedges, 1992). Other labile fractions of organic N include *amino sugars, polyamines, aliphatic amines, purines,* and *pyrimidines.*

Proteins are composed of approximately 20 α-amino acids which can be divided into different categories based on their functional groups (figure. 9.31). An α-amino acid consists of an amino group, a carboxyl group, a H atom, and a distinctive R group (side chain) that is bonded to a C atom—this C atom is the α-carbon because it is adjacent to the carboxylic group. Since amino acids contain both basic (-NH$_2$) and acidic (-COOH) groups they are considered to be *amphoteric.* In a dry state, amino acids are dipolar ions, where the carboxyl group exists as a carboxylate ion (-COO$^-$) and the amine group exists as an ammonium group (-NH$_3^+$)—these dipolar ions are called *zwitterions.* In aqueous solutions, there is an equilibrium between the dipolar ion and the anionic and cationic forms. Thus, the dominant form of an amino acid is strongly dependent on the pH of the solution; the acidic strength of these different functional groups is defined by its pK_a. Under highly acidic and basic conditions the amino acids occur as cations and anions, respectively. At an intermediate pH the zwitterion is at its highest concentration with the cationic and anioic forms in equilibrium—this is called the *isoelectric point.* Peptide bonds are formed from condensation reactions between the carboxyl group of one amino acid and the amine group of another (figure 9.32). As a result of the asymmetric α-carbon, each amino acid exists as two *enantiomers*; amino acids in living systems are almost exclusively L-isomers. However, the *peptidoglycan* layer in bacterial call walls has a high content of D-amino acids. Recent work has shown that the four major D-amino acids in bacterioplankton are alanine, serine, aspartic acid, and glutamic acid, and that under low carbon conditions bacteria may rely on the uptake of D-amino acids (Perez et al., 2003). Consequently, a high D/L ratio in DOM may indicate contributions from bacterial biomass and/or higher bacterial cycling of this material. Recently, the D/L ratio has been successfully used as an index of the diagenetic state of DON (Dittmar et al., 2001; Dittmar, 2004). Another process that can account for the occurrence of D-amino acids in organisms is *racemization,* which involves the conversion of L-amino acids into their mirror-image D-form. Estuarine invertebrates have been found to have D-amino acids in their tissues as a result of this process (Preston, 1987; Preston et al., 1997). Rates of racemization of amino acids in shell materials, calibrated against radiocarbon measurements, have also been used as an index for historical reconstruction of coastal erosion (Goodfriend and Rollins, 1998).

In oceanic and coastal waters, proteins are believed to be degraded by hydrolysis whereby other smaller components such as peptides and free amino acids are produced (Billen, 1984; Hoppe, 1991). Although many larger consumers can hydrolyze proteins in their food resources internally, bacteria must rely on external hydrolysis (via *ecto-* or *extracellular enzymes*) of proteins outside the cell (Payne, 1980). Extracellular hydrolysis allows for smaller molecules to be transported across the cell membrane—molecules typically larger than 600 Da cannot be transported across microbial cell membranes (Nikaido and Vaara, 1985). The digestibility of protein in different food resources has been examined using nonspecific *proteolytic enzymes* (protease-k), which can separate large polypeptides from smaller oligomers and monomers (Mayer et al., 1986, 1995). This results in another

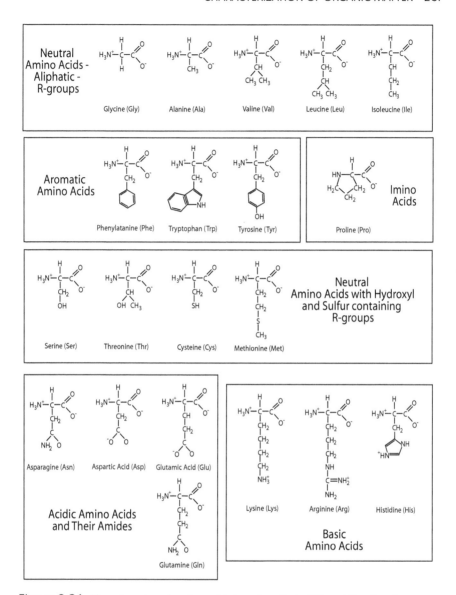

Figure 9.31 Six categories of amino acids based on their different functional groups.

category of groups of amino acids commonly referred to as *enzymatically hydrolyzable amino acids* (EHAAs) which can be used to determine what essential amino acids are limited to organisms (Dauwe et al., 1999). Only recently have we begun to understand the role of peptides in the cycling of amino acids in coastal systems. It appears that not all peptides are capable of being hydrolyzed and that peptides with greater than two amino acids are

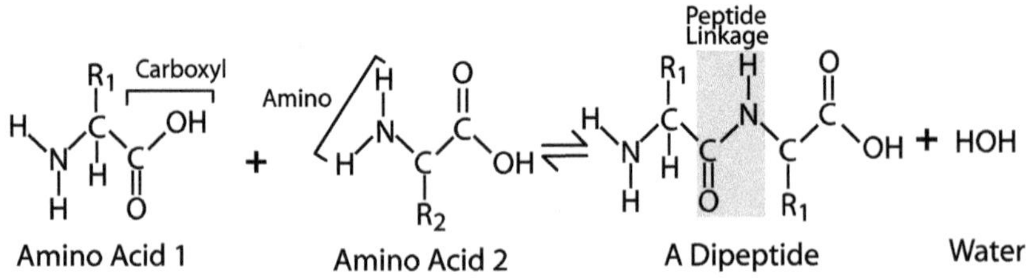

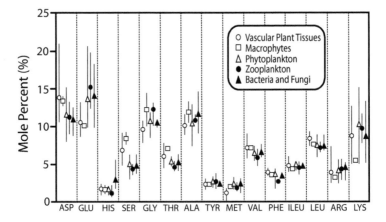

Figure 9.32 Illustration of a peptide linkage formed from a condensation reaction between the carboxyl of one amino acid and the amine of another.

hydrolyzed more rapidly than *dipeptides* (Pantoja and Lee, 1999). Extracellular hydrolysis of peptides and amino acids has also been known to occur by certain phytoplankton species (Palenik and Morel, 1991). During the extracellular oxidation of amino acids, NH_4^+, an important source of DIN for primary producers, is liberated and then becomes available for uptake. Recent work has shown that approximately 33% of the NH_4^+ uptake by the pelagophyte, *Aureococcus anophagefferens*, was provided by amino acid oxidation in Quantuck Bay estuary (USA) (Mulholland et al., 2002). This agrees with earlier work which documented elevated amino acid oxidation rates in Long Island estuaries during phytoplankton bloom events (Pantoja and Lee, 1994; Mulholland et al., 1998).

Amino acids are essential to all organisms and consequently represent one of the most important components in the organic nitrogen cycle. The typical mole percentage of protein amino acids in different organisms shows considerable uniformity in compositional abundance (figure 9.33; Cowie and Hedges, 1992). Hence, it has been argued that any observed differences in amino acid composition in the water column and/or sediments

Figure 9.33 The average and range (mol%) of protein amino acid compositions from different source organisms. (Modified from Cowie and Hedges, 1992.)

are likely to arise from degradation (Dauwe and Middelburg, 1998; Dauwe et al., 1999). In fact, the following "degradative index" (DI) was derived, from principal components analysis (PCA), using a diversity of samples ranging from fresh phytoplankton to highly degraded turbidite sediment samples:

$$DI = \sum_i [(var_i - avg\,var_i/std\,var_i)\,loading_i] \tag{9.8}$$

where: var_i = mole per cent (mol%) of amino acid; avg var_i = mean amino acid mol%; std var_i = standard deviation of amino acid mol%; and $loading_i$ = PCA derived loading of amino acid$_i$.

The more negative the DI values the more degraded the sample, with positive DI values indicative of fresh materials. Elevated mole percentages (e.g., > 1%) of *nonprotein amino acids* (NPAAs) have also been used as an index of organic matter degradation (Lee and Bada, 1977; Whelan and Emeis, 1992; Cowie and Hedges, 1994; Keil et al., 1998). For example, amino acids such as β-alanine, γ-aminobutyric acid, and ornithine are produced enzymatically during decay processes from precursors such as aspartic acid, glutamic acid, and arginine, respectively. β-Aminoglutaric is another NPAA found in marine porewaters (Henrichs and Farrington, 1979, 1987) with glutamic acid as its possible precursor (Burdige, 1989).

The decay of proteins and amino acids in phytoplankton cultures, under oxic and anoxic conditions, indicated very little selectivity (Nguyen and Harvey, 1997). However, 15 to 95% of the total particulate amino acid pool consisted of polypeptides/proteins in anoxic treatments versus 8 to 65% in oxic conditions. The similarity of particulate amino acid composition during phytoplankton decay agrees with other work in the Delaware Bay estuary (USA), which showed similar composition in the *dissolved combined* amino acid (DCAA) pool (Keil and Kirchman, 1993).

In aquatic systems, amino acids have typically been analyzed in characteristic dissolved and particulate fractions. For example, the typical pools of amino acids commonly measured in coastal systems are divided into the *total hydrolyzable amino acids* (THAAs) in the water column POM (Cowie and Hedges, 1992; van Mooy et al., 2002) and sedimentary organic matter (SOM) (Cowie and Hedges, 1992), and *dissolved free* amino acids (DFAA) and DCAA in DOM (Mopper and Lindroth, 1982; Coffin, 1989; Keil and Kirchman, 1991a). In general, DCAA concentrations are higher than DFAA concentrations in estuaries and may represent as much as 13% of the total DON (Keil and Kirchman, 1991a). Furthermore, DCAA can support approximately 50% (Keil and Kirchman, 1991a, 1993) and 25% (Middelboe et al., 1995) of the bacterial N and C demand, respectively, in estuaries. While most DFAA are derived from the heterotrophic breakdown of peptides and proteins in POM and DOM, they can be formed when NH_4^+ reacts with some carboxylic acids, in the presence of nicotinamide adenine dinucleotide phosphate (NADPH) (reduced form) (De Stefano et al., 2000). For example, glutamic acid can be formed from the reaction of NH_4^+ with α-ketoglutaric acid (De Stefano et al., 2000). Independent of their source, DFAAs have clearly been shown to be important sources of C and N to microbial communities in estuarine and coastal systems (Crawford et al., 1974; Dawson and Gocke, 1978; Keil and Kirchman, 1991; Middelboe et al., 1995). In fact, the extremely rapid cycling of amino acids observed in coastal waters (a few minutes) in the Hudson River plume (Ducklow and Kizchman, 1983) and Chesapeake Bay (Fuhrman, 1990) estuaries suggests strong coupling between DFAA sources (e.g., phytoplankton) and bacteria.

Amino acids are generally considered to be more labile than bulk C or N in the water column and sediments and have been shown to be useful indicators for the delivery of "fresh" fluxes of organic matter to surface sediments (Henrichs and Farrington, 1987; Burdige and Martens, 1988). In fact, 90% of the primary amino acids and N production was found to be remineralized during predepositional processes, particularly herbivorous grazing by zooplankton (Cowie and Hedges, 1992). The major fraction of amino acids in sediments occurs in the solid phase with very small fractions in the free dissolved state (1–10%) (Henrichs et al., 1984). Low concentrations of DFAA in pore waters may also reflect high turnover rates (Christensen and Blackburn, 1980; Jørgensen, 1987) since microorganisms preferentially utilize DFAA (Coffin, 1989). A considerable amount of the THAA fraction exists in resistant forms bound to humic compounds and mineral matrices (Hedges and Hare, 1987; Alberts et al., 1992). Thus, differential degradation among amino acids appears to be controlled by differences in functionality and intracellular compartmentalization. For example, aromatic amino acids (e.g., tyrosine and phenylalanine), along with glutamic acid and arginine, decay more rapidly than some of the more refractory acidic species such as glycine, serine, alanine, and lysine (Burdige and Martens, 1988; Cowie and Hedges, 1992). Serine, glycine, and threonine have been shown to be concentrated into the silicious exoskeletons of diatom cell walls which may "shield" them from bacterial decay processes (Kröger et al., 1999; Ingalls et al., 2003). Enhanced preservation of amino acids may also occur in carbonate sediments due to adsorption and incorporation into calcitic exoskeletons (King and Hare, 1972; Constanz and Weiner, 1988). Conversely, preferential adsorption of basic amino acids to fine particulate matter in river/estuarine systems occurs because of the attraction between the net positive charge of the amine group and the negative charge of the aluminosilicate clay minerals (Gibbs, 1967; Rosenfeld 1979; Hedges et al., 2000; Aufdenkampe et al., 2001). Molecular shape/structure (Huheey, 1983) and weight (higher van der Waals, attractive forces with increasing weight) also affect the adsorption of amines (Wang and Lee, 1993). Similarly, preferential adsorption of amino acids to clay particles can result in mixtures with other surface amino acids (Henrichs and Sugai, 1993; Wang and Lee, 1993, 1995) or reactions with other organic molecules, where *melanoidin polymers* are produced from condensation reactions of glucose with acidic, neutral, and basic amino acids (Hedges, 1978).

Redox conditions have also proven to be an important controlling factor in determining the decay rates and selective utilization of amino acids. Recent work found preferential utilization of nitrogen-rich amino acids over the non–amino-acid fraction of POM in suboxic waters (van Mooy et al., 2002). This selective utilization of the amino acid fraction was believed to be responsible for supporting water column denitrification processes. Other work in shallow coastal waters has demonstrated an oxygen effect on the degradation of amino acids (Haugen and Lichtentaler, 1991; Cowie et al., 1992). Using an experimental approach, more phytoplankton protein was found to be preserved under anoxic (15–95%) compared to oxic (8–65%) conditions (Nguyen and Harvey, 1997). However, there appeared to be only minor selective loss of amino acids within the THAA pool based on differences in functionality (figure 9.34). Patterns of minimal change in amino acid composition with sediment depth have also been observed in both oxic and anoxic coastal sediments (Rosenfield, 1979; Henrichs and Farrington, 1987). This suggests that

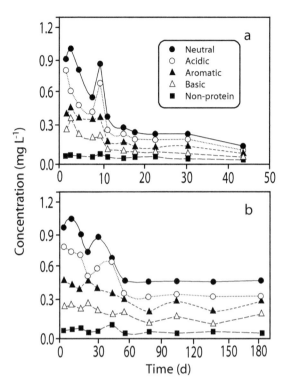

Figure 9.34 Total hydrolyzable amino acid (THAA) concentrations (mg L^{-1}) during (a) oxic and (b) anoxic decay of the dinoflagellate *P. minimum* over different time periods. Hydroxylamino acids not shown, and nonprotein amino acids = β-alanine and γ-aminobutyric acid. (Modified from Nguyen and Harvey, 1997.)

the two dominant groups of amino acids commonly found in phytoplankton (acidic and neutral amino acids) are utilized with similar efficiency during early diagenesis. Nguyen and Harvey (1997) did, however, observe selective preservation of glycine and serine during their experiment examining the decay of diatom material, further supporting the notion that these amino acids are more refractory because of their association with the silica matrix in diatom cell walls (see above).

The biogeochemical cycling of dissolved amino acids in organic rich anoxic pore waters has been shown to be controlled by the following "internal" transformations: (1) conversion of sedimentary amino acids to DFAA; (2) microbial remineralization (e.g., serve as electron donors or for direct uptake by sulfate-reducing bacteria) (Hanson and Gardner, 1978; Gardner and Hanson, 1979); and (3) uptake or reincorporation into larger molecules (e.g., geopolymerization) in sediments (Burdige and Martens, 1988, 1990; Burdige, 2002). The DFAAs in pore waters are considered to be intermediates of biotic and abiotic processing, as shown in figure 9.35 (Burdige and Martens, 1988). During predepositional settlement of POM to the sediment surface organic matter may be completely mineralized

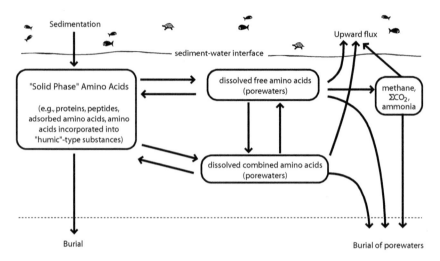

Figure 9.35 Cycling of particulate and dissolved [both free and combined amino acids (DFAA and DCAA)] in porewaters which are considered to be intermediates of biotic and abiotic processing. (Modified from Burdige and Martens, 1988.)

to inorganic nutrients or may be transformed to DOM. Postdepositional decay of organic matter results in selective loss of low molecular weight compounds such as DFAAs and simple sugars with increasing depth in the sediments (Henrichs et al., 1984)—as shown in DFAA profiles for Cape Lookout Bight (USA) sediments (figure 9.36; Burdige and Martens, 1990). The highest concentrations of DFAA in pore waters are typically in the top few centimeters of surface sediments (20–60 μM), compared to lower asymptotic values at depth (2–5 μM) and in overlying waters (<1 μM). The dominant DFAAs in these pore waters were NPAA (β-aminoglutaric acid, δ-aminovaleric, and β-alanine) which were primarily derived from fermentative processes. The high concentration gradient of DFAAs between the overlying waters and surface pore waters results in active fluxes across the sediment–water interface that are mediated by biological and diffusive processes. For Cape Lookout Bight, the highest estimated effluxes of total DFAAs were found in summer months with annual rates that ranged from 0.019 to 0.094 mol m^{-2} y^{-1} or 52 to 257 μmol m^{-2} d^{-1}(Burdige and Martens, 1990). These fluxes are considerably higher than are found in other nearshore environments, such as the Gullmar Fjord (Sweden) ($\sim$18 μmol m^{-2} d^{-1}) and the Skagerrak (northeastern North Sea) (-20 to 13 μmol m^{-2} d^{-1}) (Landen and Hall, 2000). Although bacterial uptake of DFAAs in surface sediments has been shown to decrease DFAA fluxes across the sediment–water interface (Jørgensen, 1984), the likely explanation for such high rates at the Cape Lookout site was significantly greater amounts of SOM. Finally, the interaction between the DFAA and DCAA pools in porewaters is likely to be quite important in understanding carbon preservation (Burdige, 2002). Unfortunately, there are relatively few studies have that examined DCAAs in porewaters (Caughey, 1982; Colombo et al., 1998; Lomstein et al.,

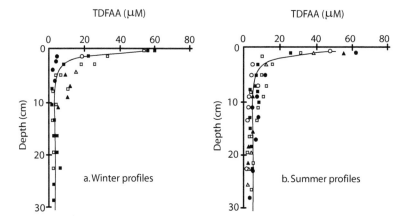

Figure 9.36 Concentrations (μM) of total dissolved free amino acids (TDAA) versus depth in (a) winter and (b) summer cores from Cape Lookout Bight (USA). (Modified from Burdige and Martens, 1990.)

1998; Pantoja and Lee, 1999)—with preliminary results indicating that the DCAA pool is one to four times the size of the DFAA pool.

Photosynthetic Pigments

The primary *photosynthetic pigments* used in absorbing photosynthetically active radiation (PAR) are *chlorophylls, carotenoids*, and *phycobilins,* with chlorophyll representing the dominant photosynthetic pigment. Although a greater amount of chlorophyll is found on land, 75% of the annual global turnover [($\sim 10^9$ Mg) occurs in oceans, lakes, and rivers/estuaries (Vesk and Jeffrey, 1987; Brown et al., 1991; Jeffrey and Mantoura, 1997)]. All of the light-harvesting pigments are bound to proteins, making up distinct carotenoid and chlorophyll–protein complexes. These pigment–protein complexes in algae and higher plants are located in the *thylakoid membrane* of chloroplasts. Chlorophylls are composed of cyclic *tetrapyrroles* coordinated around a Mg-chelated complex (figure 9.37; Rowan, 1989). Chlorophyll-*a* is the dominant pigment used to estimate algal biomass in aquatic ecosystems. Although chlorophylls are not as effective class-specific biomarkers as are carotenoids, they do provide some taxonomic distinctions (table 9.6). Chlorophylls-*a* and *b* contain a propionic acid esterified to a C_{20} phytol; chlorophylls c_1 and c_2, have an acrylic acid that replaces the propionic acid (figure 9.37). The chloropigments can also be used to resolve between contributions of eukaryotic and prokaryotic organisms. For example, eukaryotes and cyanophytes are found in oxygenated waters as opposed to green and purple sulfur bacteria, which are typically found in anoxic or hypoxic environments (Wilson et al., 2004). Bacteriochlorophyll *a* is the dominant light-harvesting pigment in purple sulfur bacteria (e.g., *Chromatiaceae*); its synthesis is inhibited by the presence of O_2 (Squier et al., 2004). Similarly, bacteriochlorophyll *e* has been shown to be indicative of green sulfur bacteria (e.g., *Chlorobium phaeovibroides* and *C. phaeobacteroides*)

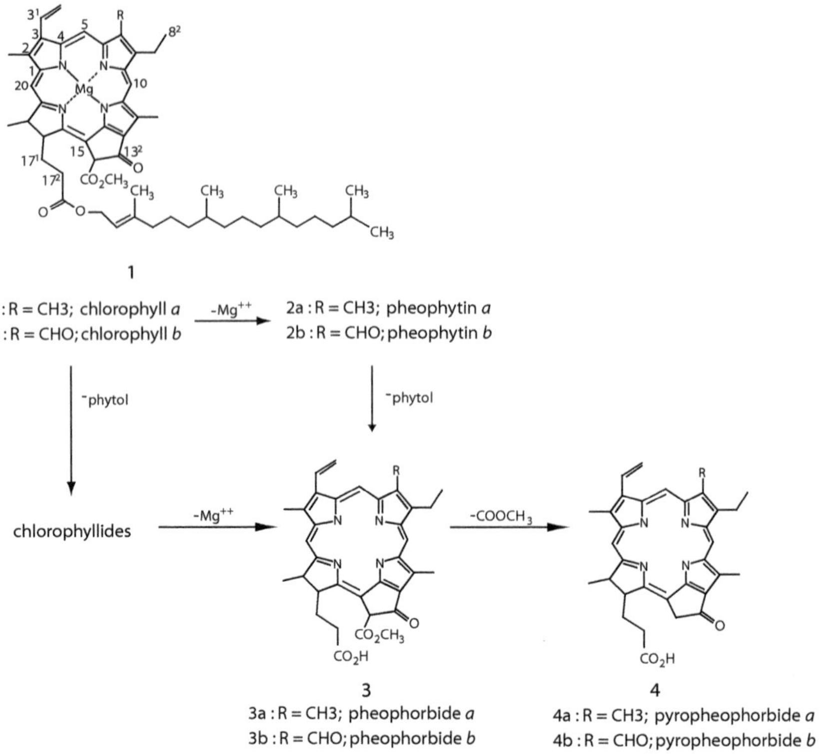

Figure 9.37 Chemical structures of chlorophylls-*a* and *b* which contain a propionic acid esterified to a C_{20} phytol; chlorophylls-c_1 and c_2 have an acrylic acid that replaces the propionic acid. Also included are the pheopigments, the four dominant tetrapyrrole derivatives of chloropigments (pheopigments) found in marine and freshwater/estuarine systems (chlorophyllide, pheophorbide, pheophytin, pyropheophorbide.) More specifically, chlorophyllase-mediated de-esterification reactions (loss of the phytol) of chlorophyll yield chlorophyllides. Pheophytins can be formed when the Mg is lost from the chlorophyll center. Pheophorbides are formed from removal of the Mg from chlorophyllide or removal of the phytol chain from pheophytin, and pyrolyzed pheopigments, such as pyropheophorbide and pyropheophytin, are formed by removal of the methylcarboxylate group (-COOCH$_3$) on the isocylic ring from the C-13 propionic acid group.

(Repeta et al., 1989; Chen et al., 2001), while bacteriochlorophylls-*c* and *d* can be found in other *Chlorobium* spp. As a result of their linkages to environmental redox conditions bacteriochlorophylls have been successfully used as an index for reconstructing paleoredox in aquatic systems (Chen et al., 2001; Squier et al., 2002, 2004)—more details on this in chapter 15.

Carotenoids are an important group of tetraterpenoids consisting of a C_{40} chain with conjugated bonds. These pigments are found in bacteria, algae, fungi, and higher plants and

Table 9.6 Summary of signature pigments useful as markers of algal groups and processes in the sea.

Pigment	Algal group or process	References
Chlorophylls		
Chl-*a*	All photosynthetic microalgae (except prochlorophytes)	Jeffrey et al. (1997)
Divinyl chl-*a*	Prochlorophytes	Goericke and Repeta (1992)
Chl-*b*	Green algae: chlorophytes, prasinophytes, euglenophytes	Jeffrey et al. (1997)
Divinyl chl-*b*	Prochlorophytes	Goericke and Repeta (1992)
Chl-*c* family	Chromophyte algae	Jeffrey (1989)
Chl-c_1	Diatoms, some prymnesiophytes, some freshwater chrysophytes, raphidophytes	Jeffrey (1976b); Stauber and Jeffrey (1988); Jeffrey (1989); Andersen and Mulkey (1983)
Chl-c_2	Most diatoms, dinoflagellates, prymnesiophytes, raphidophytes, cryptophytes	Jeffrey et al. (1975); Stauber and Jeffrey (1988); Andersen and Mulkey (1983)
Chl-c_3^c	Some prymnesiophytes, one chrysophyte, several diatoms and dinoflagellates	Jeffrey and Wright (1987); Vesk and Jeffrey(1987); Jeffrey (1989); Johnsen and Sakshaug (1993)
Chl-c_{CS-170}	One prasinophyte	Jeffrey (1989)
Phytylated chl-*c*-like^c	Some prymnesiophytes	Nelson and Wakeham(1989); Jeffrey and Wright (1994)
Mg3.8 DVP	Some prasinophytes	Ricketts (1966); Jeffrey (1989)
Bacteriochlorophylls	Anoxic sediments	Repeta et al. (1989); Repeta and Simpson (1991)
Carotenoids		
Alloxanthin	Cryptophytes	Chapman (1966); Pennington et al. (1985)
19-Butanoyloxyfucoxanthin^a	Some prymnesiophytes, one chrysophyte, several dinoflagellates	Bjørnland and Liaaen-Jensen (1989); Bjørnland et al. (1989); Jeffrey and Wright (1994)
β, ε-Carotene	Cryptophytes, prochlorophytes, rhodophytes, green algae	Bianchi et al. (1997c); Jeffrey et al. (1997)

Continued

Table 9.6 (*continued*)

Pigment	Algal group or process	References
β, β-Carotene	All algae except cryptophytes and rhodophytes	Bianchi et al. (1997c); Jeffrey et al. (1997)
Crocoxanthin	Cryptophytes (minor pigment)	Pennington et al. (1985)
Diadinoxanthin	Diatoms, dinoflagellates, prymnesiophytes, chrysophytes, raphidophytes, euglenophytes	Jeffrey et al. (1997)
Dinoxanthin	Dinoflagellates	Johansen et al. (1974); Jeffrey et al. (1975)
Echinenone	Cyanophytes	Foss et al. (1987)
Fucoxanthin[a]	Diatoms, prymnesiophytes, chrysophytes, raphidophytes, several dinoflagellates	Stauber and Jeffrey (1988); Bjørnland and Liaaen-Jensen (1989)
19-Hexanoyloxyfucoxanthin[a]	Prymnesiophytes, several dinoflagellates	Arpin et al. (1976); Bjørnland and Liaaen-Jensen (1989)
Lutein	Green algae: chlorophytes, prasinophytes, higher plants	Bianchi and Findlay (1990); Bianchi et al. (1997c); Jeffrey et al. (1997)
Micromonal	Some prasinophytes	Egeland and Liaaen-Jensen (1992, 1993)
Monadoxanthin	Cryptophytes (minor pigment)	Pennington et al. (1985)
9-*cis* Neoxanthin	Green algae: chlorophytes, prasinophytes, euglenophytes	Jeffrey et al. (1997)
Peridinin	Dinoflagellates	Johansen et al. (1974); Jeffrey et al. (1975)
Peridininol[a]	Dinoflagellates (minor pigments)	Bjørnland and Liaaen-Jensen (1989)
Prasinoxanthin	Some prasinophytes	Foss et al. (1984)
Pyrrhoxanthin[a]	Dinoflagellates (minor pigment)	Bjørnland and Liaaen-Jensen (1989)
Siphonaxanthin[b]	Several prasinophytes: one euglenophyte	Bjørnland and Liaaen-Jeasen (1989); Fawley and Lee (1990)
Vaucheriaxanthin ester[a]	Eustigmatophytes	Bjørnland and Liaaen-Jensen (1989)
Violaxanthin	Green algae; chlorophytes, prasinophytes: eustigmatophytes	Jeffrey et al. (1997)
Zeaxanthin	Cyanophytes, prochlorophytes, rhodophytes, chlorophytes, eustigmatophyes (minor pigment)	Guillard et al. (1985); Gieskes et al. (1988); Goerieke and Repeta (1992)

Biliproteins		
Allophycocyanins	Cyanophytes, rhodophytes	Rowan (1989)
Phycocyanin	Cyanophytes, cryptophytes, rhodophytes (minor pigment)	Rowan (1989)
Phycocrythrin	Cyanophytes, cryptophytes, rhodophytes	Rowan (1989)
Chlorophyll degradation products		
Pheophytin-a^d	Zooplankton fecal pellets, sediments	Vernet and Lorenzen (1987); Bianchi et al. (1988, 1991, 2000a)
Pheophytin-b^d	Protozoan fecal pellets	Bianchi et al. (1988); Strom (1991, 1993); Bianchi et al. (2000a)
Pheophytin-c^d	Protozoan fecal pellets	Strom (1991, 1993)
Pheophorbide-a	Protozoan fecal pellets	Strom (1991, 1993); Head et al. (1994); Welschmeyer and Lorenzen (1985)
Pheophorbide-b	Protozoan fecal pellets	Strom (1991, 1993)
Chlorophyllide-a	Senescent diatoms: extraction artefact	Jeffrey and Hallegraeff (1987)
Blue–green chl-a derivatives (lactones, 10-hydroxy chlorophylls)	Senescent microalgae	Hallegraeff and Jeffrey (1985)
Pyrochlorophyll-a	Sediments	Chen et al. (2003a,b)
Pyropheophytin-a	Sediments	Chen et al (2003a,b)
Pyropheophorbide-a	Copepod grazing: fecal pellets	Head et al. (1994)
Mesopheophorbide-a	Sediments	Chen et al. (2003a,b)

[a]Note that many carotenoids in trace quantities have wider distributions than those shown (Bjørnland and Liaaen-Jensen, 1989).
[b]Note that some "atypical" pigment contents reflect endosymbiotic events.
[c]Two spectrally distinct pigments have been found in these fractions (Garrido et al., 2000).
[d]Multiple zones found in each of these fractions (Strom, 1993).
Modified from Jeffrey et al. (1997).

277

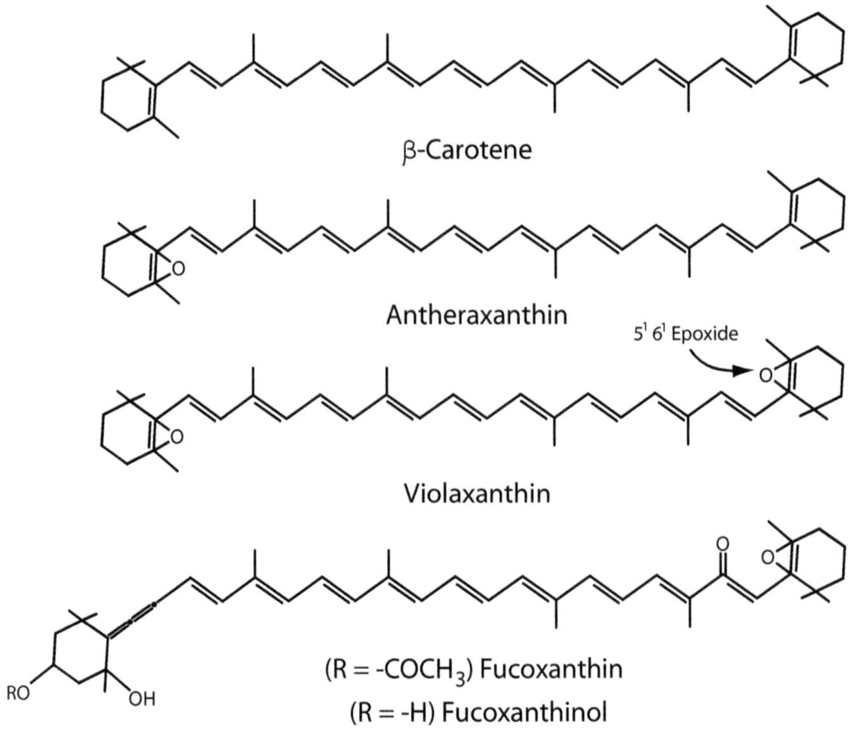

Figure 9.38 Chemical structures of select carotenoids can be divided into two groups: the carotenes (e.g., β-carotene) which are hydrocarbons and the xanthophylls (e.g., violaxanthin, fucoxanthin) which are molecules that contain at least one oxygen atom.

serve as the principal pigment biomarkers for class-specific taxonomic work in aquatic systems (table 9.6). Studies on carotenoid biosynthesis have shown lycopene to be an important precursor for the synthesis of most carotenoids (Liaaen-Jensen, 1978; Goodwin, 1980, Schutte, 1983). Carotenoids are found in pigment–protein complexes that are used for light harvesting as well as photoprotection (Porra et al., 1997). The carotenoids can be divided into two groups: the carotenes (e.g., β-carotene), which are hydrocarbons, and the xanthophylls (e.g., antheraxanthin, violaxanthin, fucoxanthin), which are molecules that contain at least one oxygen atom (Rowan, 1989; figure 9.38). Oxygenated functional groups, such as the $5'6'$-epoxide group, make certain xanthophylls like fucoxanthin more susceptible to bacterial decay than carotenes (Porra et al., 1997). Finally, there are two well-known "xanthophyll cycles" involved in photoprotection of photoautotrophs during excess illumination: (1) green algae and higher plants use the interconversion of zeaxanthin ↔ antherxanthin ↔ violaxathin where epoxide groups are gained or lost; (2) algae such as chrysophytes and pyrrophytes use the interconversion of zeaxanthin ↔ diatoxanthin ↔ diadinoxanthin (figure 9.39; Hager, 1980). Unlike the carotenoids and

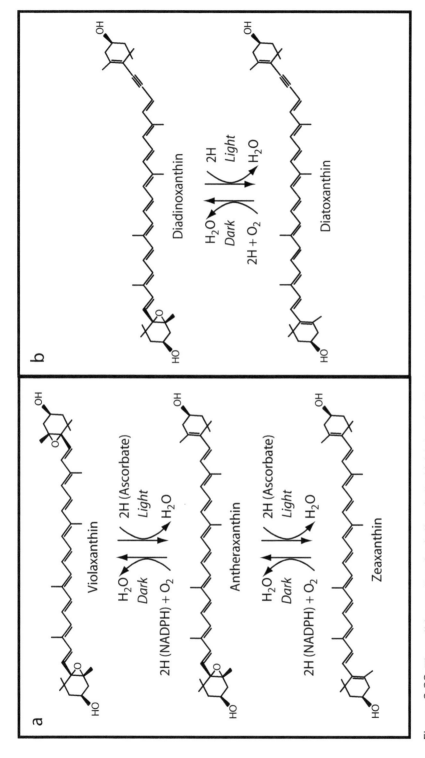

Figure 9.39 The well-known "xanthophyll cycles" which is involved in photoprotection of photoautotrophs during excess illumination: (a) green algae and higher plants use the interconversion of zeaxanthin ↔ antheraxanthin ↔ violaxathin where epoxide groups are gained or lost; (b) where algae such as chrysophytes and pyrrophytes use the interconversion of zeaxanthin ↔ diatoxanthin ↔ diadinoxanthin. (Modified from Hager, 1980.)

chlorophylls, phycobilins are water-soluble pigments that can be divided into four major types: phycocyanin, phycoerythrin, allophycocyanin, and phycoerythrocyanin (Rowan, 1989). These are light-harvesting pigments bound to apoproteins in the Cyanophyta, Rhodophyta, and Cryptophyta. They have not been used as extensively as carotenoids for taxonomic separation due to their more restrictive occurrence across different algal classes and the lack of well-developed HPLC methods for rapid analyses.

The degradation products of chlorophyllous pigments, formed primarily by heterotrophic processes, are the most abundant forms of pigments found in sediments (Daley, 1973; Repeta and Gagosian, 1987; Leavitt and Carpenter, 1990; Brown et al., 1991; Bianchi et al., 1993; Chen et al., 2001; Louda et al., 2002). These pigment degradation products may be used to infer the availability of different source materials to consumers. For example, the four dominant tetrapyrrole derivatives of chloropigments (pheopigments) found in marine and freshwater/estuarine systems (chlorophyllide, pheophorbide, pheophytin, pyropheophorbide) are formed during bacterial, autolytic cell lysis, and metazoan grazing activities (Sanger and Gorham, 1970, Jeffrey, 1974; Welschmeyer and Lorenzen, 1985; Bianchi et al., 1988, 1991; Head and Harris, 1996). More specifically, chlorophyllase-mediated de-esterification reactions (loss of the phytol side chain) of chlorophyll yield chlorophyllides when diatoms are physiologically stressed (Holden, 1976; Jeffrey and Hallegraeff, 1987; figure 9.37). Pheophytins can be formed from bacterial decay, metazoan grazing and cell lysis whereby the Mg is lost from the chlorophyll center (Daley and Brown, 1973). Pheophorbides, which are formed by removal of the Mg from chlorophyllide or removal of the phytol chain from pheophytin occurs primarily from herbivorous grazing activities (Shuman and Lorenzen, 1975; Welschmeyer and Lorenzen, 1985; Bianchi et al., 2000a). Pyrolized pheopigments, such as pyropheophorbide and pyropheophytin, are formed by removal of the methylcarboxylate group ($-COOCH_3$) on the isocylic ring from the C-13 proprionic acid group (Ziegler et al., 1988), primarily by grazing processes (Hawkins et al., 1986). These compounds can account for the dominant fraction of breakdown pigments in water column particulates (Head and Harris, 1994, 1996) as well as sediments (Keely and Maxwell, 1991; Chen et al., 2003a,b). Cyclic derivatives of chlorophyll a, such as cyclopheophorbide-a enol (Harris et al., 1995; Goericke et al., 2000; Louda et al., 2000), and its oxidized product, chlorophyllone a are also abundant in aquatic systems and are believed to be formed in the guts of metazoans (Harradine et al., 1996; Ma and Dolphin, 1996). Most chlorophyll produced in the euphotic zone is degraded to colorless compounds: the destruction of the chromophore indicates cleavage of the macrocycle (type II reactions according to Brown et al., 1991). The main mechanisms of type II reactions of chlorophylls are photo-oxidation, involving attack of excited-state singlet oxygen and enzymatic degradation (Gossauer and Engel, 1996). Another group of stable nonpolar chlorophyll-a degradation products are the steryl chlorin esters (SCEs) and carotenol chlorin esters (CCEs) (Furlong and Carpenter, 1988; King and Repeta, 1994; Talbot et al., 1999; Chen et al., 2003a,b). These compounds are formed by esterification reactions between pheophorbide a and/or pyropheophorbide a and sterols and carotenoids; these compounds are believed to be primarily formed by zooplanktonic and bacterial grazing processes (King and Repeta, 1994; Harradine et al., 1996).

Seasonal inventories of sedimentary chloropigments indicated that there are differences in timing between the maximum chlorophyll and pheopigment inventories in Long Island Sound estuary (Sun et al., 1994). These differences in inventories likely reflect

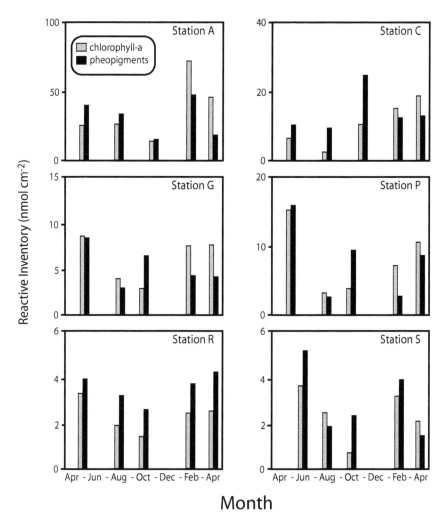

Figure 9.40 Seasonal inventories of sedimentary chloropigments (chlorophyll-a and pheopigments) in Long Island Sound estuary (USA). (Modified from Sun et al., 1994.)

differences in the growth patterns of phytoplankton (winter to early spring) and production of zooplankton feces (April and May) (figure 9.40). These results emphasize the importance of benthic–pelagic coupling in estuarine systems. Differential decay of plant pigment biomarkers can provide additional information on the importance of pre-and postdepositional decay processes and decoupling between benthic and pelagic processes. For example, chlorophyll-a decay rate constants in both anoxic and oxic sediments, with and without macrofauna have been shown to be approximately 0.07 d^{-1} (Leavitt and Carpenter, 1990; Bianchi and Findlay, 1991; Sun et al., 1994; Bianchi et al., 2000a).

Pigments "bound" to structural compounds such as lignins or surface waxes in higher plants have decay rate constants slower than those of similar pigments from nonvascular sources (Webster and Benfield, 1986; Bianchi and Findlay, 1990). Other work has shown the importance of pigment decay in the "free" versus "bound" state in estuarine sediments (Sun et al., 1994).

Plant pigments have been shown to be useful biomarkers of organic carbon in estuarine ecosystems (see review by Millie et al., 1993). For example, carotenoids and their degradative xanthophylls have been shown to be effective biomarkers for different classes of phytoplankton (Jeffrey et al., 1975; Jeffrey, 1976a, 1997), while chlorophyll has been used extensively as a means of estimating phytoplankton biomass (Jeffrey, 1997). Numerous studies have shown that photopigment concentration correlated well with microscopical counts of different species (Tester et al., 1995; Roy et al., 1996; Meyer-Harms and von Bodungen, 1997; Schmid et al., 1998), further corroborating that plant pigments are reliable chemosystematic biomarkers. For example, fucoxanthin, peridinin, and alloxanthin are the dominant accessory pigments in diatoms, certain dinoflagellates, and cryptomonads, respectively (table 9.6; Jeffrey, 1997). In general, pigment biomarkers have shown diatoms to be the dominant phytoplankton group in most estuarine systems (Bianchi et al., 1993, 1997a,b,c, 2002a; Lemaire et al., 2002). This is well illustrated when examining the concentrations of fucoxanthin in relation to other diagnostic carotenoids and chlorophyll-a in European estuaries over different seasons (figure 9.41; Lemaire et al., 2002). The matrix factorization program, Chemical Taxonomy (CHEMTAX), was introduced to calculate the relative abundance of major algal groups based on concentrations of diagnostic pigments (Mackey et al., 1996; Wright et al., 1996). For example, CHEMTAX was used to estimate the relative importance of different microalgal groups to total biomass and blooms in the Neuse River estuary (USA) (Pinckney et al., 1998). The estimated overall contribution of cryptomonads, dinoflagellates, diatoms, cyanobacteria, and chlorophytes in the Neuse River estuary from 1994 to 1996 was 23, 22, 20, 18, and 17%, respectively. This study revealed that there was significant variability in the relative abundance of phytoplankton and blooms over a three-year period and that loading of DIN was largely responsible for the blooms of cryptomonads, chlorophytes, and cyanobacteria (figure 9.42). The combined method of selective photopigment HPLC separation coupled with in-line flow scintillation counting using the chlorophyll-a radiolabeling (^{14}C) technique can also provide information on phytoplankton growth rates under different environmental conditions (Redalje, 1993; Pinckney et al., 1996, 2001). Fossil pigments have also been shown to be useful paleotracers of algal and bacterial communities (Swain, 1985; Leavitt, 1993; Chen et al., 2001; Bianchi et al., 2002). More details on the application of pigments and other chemical biomarkers in the historical reconstruction in estuaries are provided in chapter 15.

Lignin

Lignin has proven to be a useful chemical biomarker for vascular-plant inputs to estuarine/coastal margin sediments (Gardner and Menzel, 1974; Hedges and Parker, 1976; Goñi and Hedges, 1992; Hedges et al., 1997; Louchouaran et al., 1997; Bianchi et al., 1999b, 2002b). Cellulose, hemicellulose, and lignin generally make up > 75% of the biomass of woody plant materials (Sjöström, 1981). Lignins are a group of macromolecular heteropolymers (600–1000 kDa) found in the cell wall of vascular plants that are

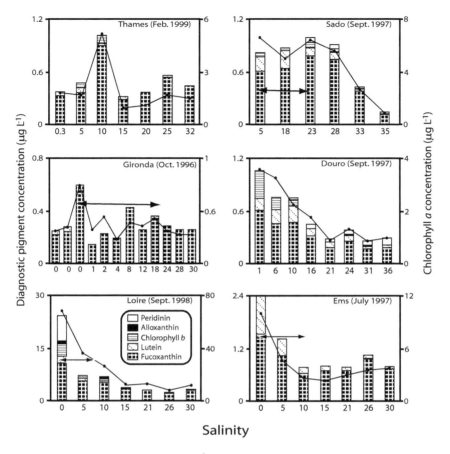

Figure 9.41 Concentrations (μg L^{-1}) of diagnostic carotenoids and chlorophylls-a and b in six [Thames (UK), Gironda (France), Loire (France), Sado (Portugal), Douro (Portugal), and Ems (The Netherlands)] European estuaries over different seasons. (Modified from Lemaire et al., 2002.)

made up of phenylpropanoid units (Sarkanen and Ludwig, 1971; de Leeuw and Largeau, 1993). The *shikimic acid pathway*, which is common in plants, bacteria, and fungi, is where aromatic amino acids (e.g., tryptophan, phenylalanine, tyrosine) are synthesized, thereby providing the parent compounds for the synthesis of the phenylpropanoid units in lignins. More specifically, the primary building blocks for lignins are the following monolignols: *p*-coumaryl alcohol, coniferyl alcohol, and sinapyl alcohol (Goodwin and Mercer, 1972; figure 9.43). These connecting units are cross-linked by carbon to carbon and the dominant β-*O*-4 aryl to aryl ether bonds, making lignins very stable compounds (figure 9.44). Oxidation of lignin using the CuO oxidation method (Hedges and Parker, 1976; Hedges and Ertel, 1982) yields 11 dominant phenolic monomers, which can be

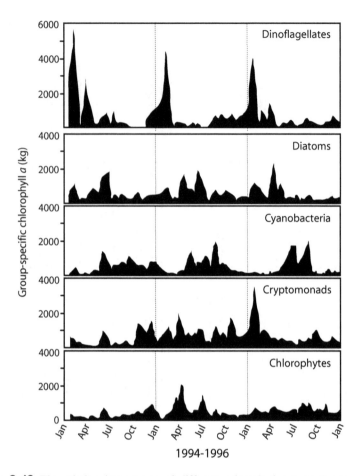

Figure 9.42 The relative importance of different microalgal groups (cryptomonads, dinoflagellates, diatoms, cyanobacteria, and chlorophytes) based on group-specific chlorophyll-*a* (kg) to total biomass, in the Neuse River estuary (USA) from 1994 to 1996. (Modified from Pinckney et al., 1998.)

separated into four families: *p*-hydroxyl, vanillyl (V), syringyl (S), and cinnamyl (C) phenols (Hedges and Ertel, 1982) (figure 9.45). Syringyl derivatives are unique to woody and nonwoody angiosperms, while cinnamyl groups are common to nonwoody angiosperms and gymnosperms (Hedges and Parker, 1976; Hedges and Mann, 1979). Hence, the S/V and C/V ratios can provide information about the relative importance of angiosperm versus gymnosperm sources and woody versus nonwoody sources, respectively (figure 9.46). However, all gymnosperm and angiosperm wood and nonwoody tissues yield vanillyl phenols as oxidation products. The *p*-hydroxyl phenols are generally not included when using lignins as indicators of vascular plant inputs to estuarine/coastal systems because these phenols can also be produced by nonlignin constituents (Wilson et al., 1985).

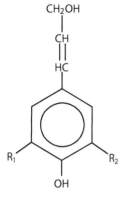

Figure 9.43 Structure of monolignol where $R_1 = R_2 = H$: p-coumaryl; $R_1 = H$, $R_2 = OCH_3$: confiferyl alcohol; $R_1 = R_2 = OCH_3$: sinapyl alcohol. (From de Leeuw and Largeau, 1993.)

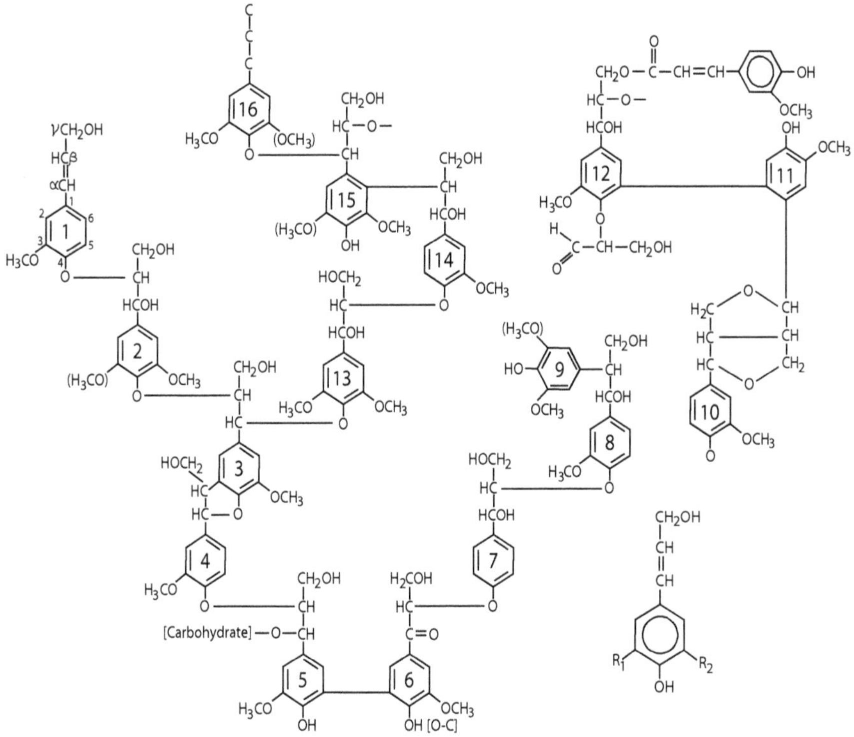

Figure 9.44 Lignin compounds, composed of connecting units of monolignols cross-linked by carbon to carbon and the dominant β-O-4 aryl to aryl ether bonds.

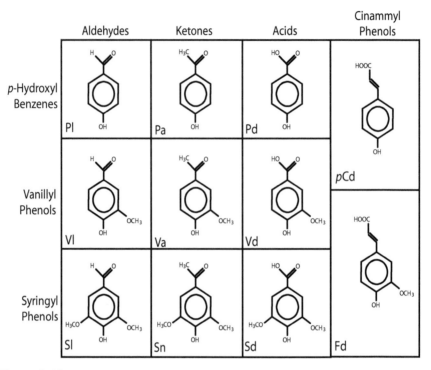

Figure 9.45 Eleven dominant phenolic monomers, yielded from the oxidation of lignin using the CuO oxidation method; these compounds can be separated into four families: *p*-hydroxyl, vanillyl (V), syringyl (S), and cinnamyl (C) phenols. (Modified from Hedges and Ertel, 1982.)

During the CuO oxidation procedure most, but not all, of the ether bonds in lignin are broken; however, many of the carbon-to-carbon bonds that link aromatic rings remain (Chang and Allen, 1971). Hence, these phenolic dimers have retained the characteristic linkages of ring to ring and side-chain to linkages, providing an additional 30 (or more) CuO products that can be used to identify taxonomic sources of vascular plants beyond the monomeric forms (Goñi and Hedges, 1990a). Moreover, the more numerous dimers may actually appear to be derived exclusively from polymeric lignin, unlike the monomeric which are found in soluble or ester-bound form. Thus, lignin dimers may better document inputs of polymeric lignin in natural systems (Goñi and Hedges, 1990a). However, using dimer-to-monomer ratios it was later found that monomers do adequately follow the decomposition of the lignin polymer in a long-term decay experiment of different vascular plant tissues (Opsahl and Benner, 1995).

A newly developed technique involving thermochemolysis in the presence of tetramethylammonium hydroxide (TMAH) has been shown to be successful in analyzing lignin in sediments (Clifford et al., 1995; Hatcher et al., 1995; Hatcher and Minard, 1996). The TMAH thermochemolysis procedure allows for effective hydrolysis and methylation of

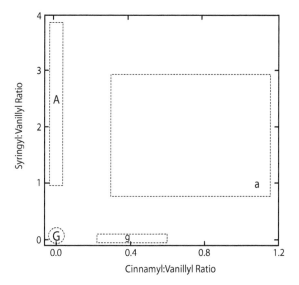

Figure 9.46 Source plot of syringyl-to-vanillyl (S/V) and cinnamyl-to-vanillyl (C/V) ratios, used to provide information about the relative importance of angiosperm versus gymnosperm sources and woody versus nonwoody plant sources.

esters and ester linkages, and results in the cleavage of the major β-O-4 ester bond in lignin (McKinney et al., 1995; Hatcher and Minard, 1996; Filley et al., 1999). Consequently, this technique has proven to be effective in examining abundances and sources of vascular plant inputs in both terrestrial (Martin et al., 1994, 1995; Fabbri et al., 1996; Chefetz et al., 2000; Filley, 2003) and aquatic (Pulchan et al., 1997, 2003; Mannino and Harvey, 2000; Galler et al., 2003) environments.

Lignins are considered to be some of the most stable compounds in vascular plant tissues where they are generally selectively preserved in terrestrial (Bates and Hatcher, 1989) and aquatic environments (Moran and Hodson, 1989a; Opsahl and Benner, 1995; Bianchi et al., 1999b, 2002b). The recalcitrance of lignin as a resource for microbes is believed to be due to its complex structure and diversity of chemical bonds (Martin and Haider, 1986). Early work suggested that lignin decomposition occurred primarily under oxidizing soil conditions and by fungi (e.g., white and brown rot fungi) and several bacteria (Crawford, 1981; Kogel-Knabner et al., 1991). The relative importance of side-chain oxidation versus aromatic ring cleavage has also been shown to be indicative of the mechanism of lignin decay by differences between fungi (Hedges et al., 1988). The cinnamyl phenols are considered to be more easily degraded than the syringyl and vanillyl phenols, because their linkages (e.g., ester-bond linkages) within the macromolecule are less stable than the linkages (carbon-to-carbon or β-aryl ether linkages) found in the other phenols (Haddad et al., 1992). Similar to other decay indices described in this chapter, Ertel and Hedges (1984) found a positive correlation between oxidation of lignin components and increasing acid-to-aldehyde ratios in vanillyl $(Ad/Al)_v$ and syringyl $(Ad/Al)_s$ phenols.

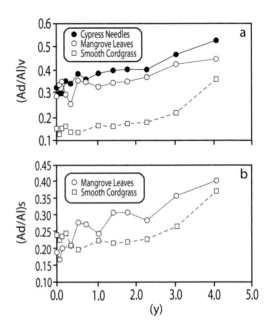

Figure 9.47 Concentrations of acid:aldehyde ratios in vanillyl $(Ad/Al)_v$ and syringyl $(Ad/Al)_s$ phenols during the subaqueous decay of lignin in vascular plants in a 4-year laboratory incubation study. (Modified from Opsahl and Benner, 1995.)

The subaqueous decay of lignin in vascular plants that are commonly sources of carbon in estuaries clearly shows significant increases in both ratios over time as with increasing diagenetic state (Opsahl and Benner, 1995) (figure 9.47). It has also been shown that different mineral phases in sediments (Fe, Mn, and Al oxides) can affect lignin decomposition. For example, Fe oxide reduced lignin in decomposition (Miltner and Zech, 1998)—this has major implications for the decay of terrestrially derived materials in estuaries and river plumes where Fe oxides are highly abundant.

In estuaries it is common to find a gradient of vascular plant inputs of POM with decreasing concentrations moving from the head to the mouth of the system (Bianchi and Argyrou, 1997; Goñi and Thomas, 2000). In the Baltic Sea, we can see such a gradient in lignin–phenol concentrations in water column POC (figure 9.48a; Bianchi et al., 1997d); the stations Luleälven River (Lu) to Baltic proper (Bp) are situated from the head (north) to the mouth (south) of the estuary. Hedges and Parker (1976) used lambda indices to quantify lignin in relation to a specified amount of carbon; for example, lambda-6 (Λ_6 or λ) is defined as the total weight in milligrams of the sum of vanillyl (vanillin, acetovanillone, vanillic acid) and syringyl (syringaldehyde, acetosyringone, syringic acid) phenols, normalized to 100 mg of organic carbon, while lambda-8 (Λ_8) includes the cinnamyl (p-coumaric and ferulic acid) phenols. These results support previous work that suggested greater inputs of allochthonous material in the northern Baltic where the greater freshwater inputs occur. The trend in S/V and C/V ratios indicate that the dominant sources of lignin were derived from woody gymnosperms (figure 9.48b). These results are consistent with the fact that most of the drainage basin bordering the Baltic Sea is

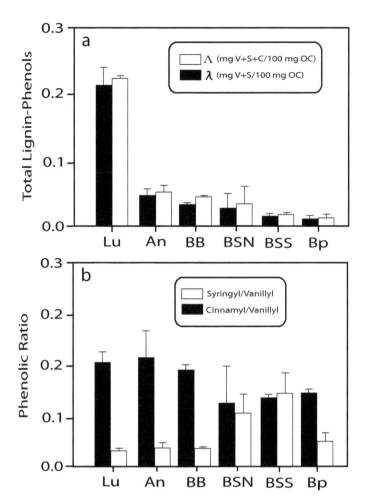

Figure 9.48 Lignin-phenol concentrations in (a) water column POC and (b) the syringyl-to-vanillyl (S/V) and cinnamyl-to-vanillyl (C/V) ratios of POC at stations in the Baltic Sea from Luleälven River (Lu) to Baltic proper (Bp)—situated from the head (north) to the mouth (south) of the estuary. Other stations are: Ån = Ångermanälven; BB = Bothnian Bay; BSN = Bothnian Sea north; BSS = Bothnian Sea south. (Modified from Bianchi et al., 1997.)

dominated by managed coniferous forests. Similarly, soil and sediment samples collected from a forest–brackish marsh–salt gradient in a southeastern United States estuary (North Inlet) also indicated that woody and nonwoody gymnosperm and angiosperm sources dominated SOM at the forest site while nonwoody marsh plants (*Spartina* and *Juncus*) dominated the marsh site (figure 9.49; Goñi and Thomas, 2000). The most extensive lignin

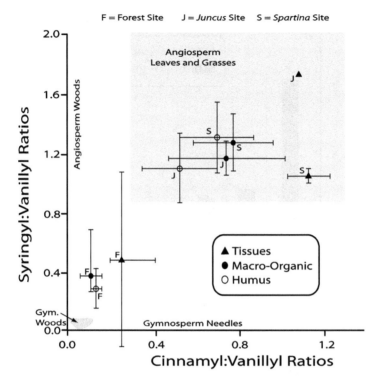

Figure 9.49 Syringyl-to-vanillyl (S/V) and cinnamyl-to-vanillyl (C/V) ratios in soils and sediment samples collected from a forest–brackish marsh–salt gradient in a southeastern U.S. estuary (North Inlet, USA). (Modified from Goñi and Thomas, 2000.)

degradation occurred at the forest site and the least amount at the salt marsh site where the occurrence of anoxic conditions dominated.

A corresponding gradient in freshwater and terrestrially derived inputs of lignin in DOM has also been documented in riverine/estuarine systems. For example, Benner and Opsahl (2001) found Λ_8 in HMW DOM in the lower Mississippi River and an inner plume region that ranged from 0.10 to 2.30, which is consistent with other HMW DOM measurements in river/coastal margin gradients (Opsahl and Benner, 1997; Opsahl et al., 1999). Controls on the abundance of lignin in DOM in estuarine/coastal systems may involve losses from in situ bacterial breakdown (discussed earlier), flocculation, sorption, and desorption processes with resuspended sediments (Guo and Santschi, 2000; Mitra et al., 2000a; Mannino and Harvey, 2000), and photochemical breakdown (Opsahl and Benner, 1998; Benner and Opsahl, 2001). In fact, HMW DOM in the bottom waters of the Middle Atlantic Bight (MAB), found to be "old" and rich in lignin, was believed to be derived from desorption of sedimentary particles largely derived from Chesapeake Bay and other estuarine systems along the northeastern margin of the United States (Guo and Santschi, 2000; Mitra et al., 2000a). Temporal and spatial variability in the relative importance of such processes can also commonly result in nonconservative behavior of total DOM and concentrations of lignin. The potential importance of these different mechanisms in removing terrestrially derived DOM in estuarine/shelf regions may be responsible for the large deficit in terrigenous DOM in the global ocean (Opsahl and Benner, 1997; Hedges et al., 1997).

In the past decade, *compound-specific isotopic analysis (CSIA)* has been used as an additional tool to bulk isotopic analyses to better identify sources and sinks of organic matter in sediments (Freeman et al., 1990; Hayes, 1993; Schoell et al., 1992). This technique has also been applied to water column as well as sedimentary sources of organic matter in coastal/estuarine ecosystems (Abrajano et al., 1994, 1998; Qian et al., 1996; Canuel et al., 1997; Ramos et al., 2003; Zou et al., 2004). This has provided more differentiation diversity of specific components of organic carbon (e.g., bacteria, zooplankton, algae) in isotopic analyses that have to date not been available using bulk isotopic analyses (Cifuentes and Salata, 2001). More specifically, CSIA provides information about the biogeochemical conditions whereby carbon fixation is occurring in the photic zone and the specific sources of organic matter (Bieger et al., 1997; Freeman, 2001). While much of the work using CSIA to date has been focused on carbon, more recent work has shown the application of nitrogen isotopes (Sachs et al., 1999).

Recent work has shown that the $\delta^{13}C$ of individual fatty acids allowed for differentiation of the $\delta^{13}C$ signature of water column food resources in two benthic mussels (*Mytilus edulis* and *Modiolus modiolus*) in Newfoundland (Canada) estuaries (Abrajano et al., 1994). During lipid synthesis there is generally a 3 to 5‰ fractionation relative to total biomass (Monson and Hayes, 1980, 1982; Freeman, 2001); in these Newfoundland estuaries, the isotopic range of fatty acids was found to be within that predicted from bulk $\delta^{13}C$ (Ostrom et al., 1997) in phytoplankton bloom materials (Abrajano et al., 1994; Ramos et al., 2003). The application of CSIA to better understand bacterial resources has also been effective in estuarine systems. For example, isotopic discrimination between bacterial phospholipid fatty acids (PLFA) from complex organic matter substrates in cultures is generally within the range of +5‰ or −5‰ (Abrajano et al., 1994; Canuel et al., 1997; Salata, 1999). Such differences allowed Boschker et al. (1999) to determine that *S. alterniflora* detritus was not the primary carbon source for bacteria in estuarine salt marsh sediments.

The $\delta^{13}C$ of fatty acids has also been used to determine sources of HMW DOM in estuaries from four different regions of the United States (Boston Harbor/Massachusetts Bay, Delaware/Chesapeake Bays, San Diego Bay, and San Francisco Bay) (Zou et al., 2004). This work demonstrated that a fraction of HMW DOM is derived from phytoplankton and bacteria, and that HMW DOM is formed from bacterial membranes and reworking of phytoplankton materials. Finally, isotopic ($\delta^{13}C$) discrimination between PLFA and smaller labile molecules (e.g., acetate, methane, methanol) is considerably greater (e.g., >20‰) than that found in comparison with complex substrates (Summons et al., 1994), making such distinctions even more apparent.

Bulk organic carbon and carbon isotope measurements, made on suspended river particulates and seabed samples within the Mississippi River dispersal system (Eadie et al., 1994; Trefry et al., 1994), indicated that river particulates were ^{13}C depleted (relative to marine organic carbon) and that 60 to 80% of the OC buried on the shelf adjacent to the river was marine in origin (as determined by the enriched ^{13}C values of the sediments).

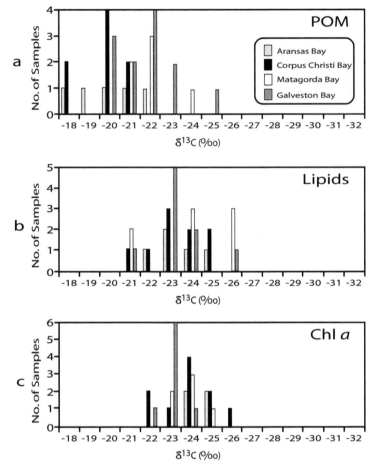

Figure 9.50a Compound-specific isotope analysis (CSIA) of suspended particulate organic matter (SPOM), lipids, and plant pigments in several estuaries on the Texas coast (USA). The range of $\delta^{13}C$ in these estuaries was -18 ‰ to -22 ‰. (Modified from Qian et al., 1996.)

Subsequently, it was suggested, using CSIA on lignin biomarkers, that much of the TOC delivered to the shelf is from C_3 and C_4 plants and materials from eroded soils in the northwestern grasslands of the Mississippi River drainage basin (Goñi et al., 1997, 1998). These investigators and others have also concluded that C_4 material (^{13}C-enriched values similar to marine organic carbon) was transported greater distances offshore because of its characteristically smaller grain size (Goñi et al., 1997, 1998; Onstad et al., 2000; Gordon and Goñi 2003). Based on this earlier work, Goñi et al. (1997) concluded that the C_3 (^{13}C depleted) terrestrial samples were being deposited on the shelf. However, recent analyses suggest that woody angiosperm material (^{13}C depleted) preferentially settles within the

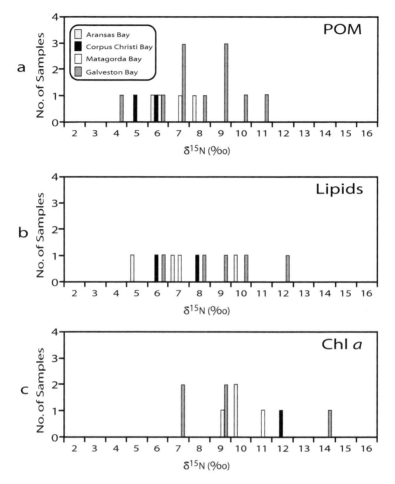

Figure 9.50b CSIA of SPOM, lipids, and plant pigments in several estuaries on the Texas coast (USA). The range of $\delta^{15}N$ in these estuaries was +4‰ to +14‰. (Modified from Quian et al., 1996.)

lower Mississippi and in the proximal portion of the dispersal system on the shelf (Bianchi et al., 2002). More recently, it has been demonstrated that erosion of relict marsh peats in transgressional facies of the lower Mississippi River can provide another source of "old" vascular plant detritus to the Louisiana coastal margin (Galler et al., 2003).

Another group of compounds well suited for CSIA because of their widespread occurrence are plant pigments (Bidigare et al., 1991; Kennicutt et al., 1992). Chorophylls and carotenoids are involved in photoautotrophic reactions that produce organic matter and can provide information on the nutritional and light conditions of algae (Welschmeyer and Lorenzen, 1985; Jeffrey et al., 1997). In fact, recent work has shown that a 5.1‰

isotopic depletion of chlorophyll *a* relative to total nitrogen in both field and culture experiments; this indicates that phytoplankton $\delta^{15}N$ can be determined from chlorophyll *a* in particulates and sediments with the simple addition of 5.1‰ (Sachs et al., 1999). The CSIA of suspended particulate organic matter (SPOM), lipids, and plant pigments in several estuaries on the Texas coast proved to be effective in determining the sources of organic carbon as well as their DIN sources (Qian et al., 1996). The range of $\delta^{13}C$ in these estuaries (-18 to -22‰) indicated that terrestrial inputs in these systems was minimal compared to phytoplankton sources of SPOM (figure 9.50a). The isotopic signature of the lipids and chlorophyll *a* was generally 2 to 4‰ more depleted than SPOM and more accurately reflected phytoplankton source inputs. The stable nitrogen isotope composition of SPOM, lipids, and chlorophylls ranged from $+4$ to $+14$‰ and did not show significant differences across estuaries (figure 9.50b); this suggested a common source of nitrogen throughout all these estuaries.

Summary

1. In recent years there have been significant improvements in our ability to distinguish between organic matter sources in estuaries using tools such as elemental, isotopic (bulk and compound/class specific), and chemical biomarker methods.

2. The broad range of C:N ratios across divergent sources of organic matter in the biosphere demonstrates how such a ratio can provide an initial proxy for determining source information. However, C:N ratios can be very misleading in determining organic matter sources in the absence of additional source proxies.

3. End-member mixing models have commonly been used to evaluate sources of dissolved inorganic nutrients (C, N, S) and organic matter (POM and DOM) in estuaries. Many investigations have been based on a two (binary) and more recently, three end-member mixing models in which the one or two more depleted terrestrial $\delta^{13}C$ end-members could be used along with the more enriched marine phytoplankton end-member to establish their relative abundance throughout the estuary.

4. Nuclear magnetic resonance (NMR) spectroscopy is a powerful and theoretically complex analytical tool that can be used to characterize organic matter. Proton (^1H) and ^{13}C-NMR have been the most common NMR tools for the nondestructive determination of functional groups in complex biopolymers in plants, soils/sediments, and DOM in aquatic ecosystems.

5. The term biomarker molecule has recently been defined as "compounds that characterize certain biotic sources and that retain their source information after burial in sediments, even after some alteration." The catabolic and anabolic pathways that are responsible for the formation of many of the biomarker compounds occur through an "intermediary" metabolism via glycolysis and the citric acid cycle.

6. Lipids are an important component of organic matter with a diversity of compound classes (e.g., hydrocarbons, fatty acids, *n*-alkanols, and sterols) that have been used as effective biomarkers of organic matter in coastal and estuarine systems. These compounds are water insoluble compounds that typically

account for 10 to 20% (dry weight) of the biomass in actively growing algal populations.

7. Aliphatic hydrocarbons such as n-alkanes and n-alkenes have been success-fully used to distinguish between algal, bacterial, and terrestrial sources of carbon in estuarine/coastal systems, and also anthropogenic fossil sources (e.g., petroleum hydrocarbons).

8. Branched and cyclic hydrocarbons have also been used as paleomarkers, prin-cipally in lake sediments, to interpret past changes in organic matter sources. In particular, isoprenoid hydrocarbons such as pristane (C_{19} isoprenoid hydro-carbon) and phytane (C_{20} isoprenoid hydrocarbon) (both derived from phytol) have been used as indicators of herbivorous grazing and methanogenesis, respectively.

9. n-Alkanols (fatty alcohols) can be used to distinguish between terrestrial plant and algal/bacterial inputs to aquatic systems. For example, the epicuticular waxes of land plants have been shown to have n-alkanols with an even number of carbon atoms (typically C_{22}–C_{30}).

10. Sterols and their respective derivatives have proven to be important biomarkers that can be used to estimate algal and terrestrial contributions as well as dia-genetic proxies. These compounds are a group of lipids (typically between C_{26} and C_{30}) that are resistant to saponification and can be classified as triterpenes.

11. The selective nature of degradative processes over changing redox conditions, and the fact that certain biochemical constituents can occur over timescales of a few days, reflects the importance of using short-lived biomarker compounds in understanding diagenetic processes in estuaries.

12. Carbohydrates are the most abundant class of biopolymers on Earth, and con-sequently represent significant components of water column POM and DOM in aquatic environments. Minor sugars, such as acidic sugars, amino sugars, and O-methyl sugars, tend to be more source specific than major sugars and can potentially provide further information in understanding the biogeochemical cycling of carbohydrates.

13. Proteins make up approximately 50% of organic matter, and contain about 85% of the organic N in marine organisms. More specifically, peptides and proteins comprise an important fraction of the POC (13–37%) and PON (30–81%), as well as DON (5–20%) and DOC (3–4%) in oceanic and coastal waters. Amino acids are essential to all organisms and consequently represent one of the most important components in the organic nitrogen cycle. Other labile fractions of organic N include amino sugars, polyamines, aliphatic amines, purines, and pyrimidines.

14. The primary photosynthetic pigments used in absorbing PAR are chlorophylls, carotenoids, and phycobilins, with chlorophyll representing the dominant pho-tosynthetic pigment. Although a greater amount of chlorophyll is found on land, 75% of the annual global turnover ($\sim 10^9$ Mg) occurs in oceans, lakes, and rivers/estuaries.

15. Bacteriochlorophyll a is the dominant light-harvesting pigment in purple sulfur bacteria (e.g., *Chromatiaceae*) and where its synthesis is inhibited by the presence of O_2. Similarly, bacteriochlorophyll e has been shown to be indicative of green sulfur bacteria (e.g., *Chlorobium phaeovibroides* and

C. phaeobacteroides), while bacteriochlorophylls *c* and *d* can be found in other *Chlorobium* spp.

16. Lignins are a group of macromolecular heteropolymers (600–1000 kDa) found in the cell wall of vascular plants that are made up of phenylpropanoid units. Lignin has proven to be a useful chemical biomarker for vascular-plant inputs to estuarine/coastal margin sediments. Cellulose, hemicellulose, and lignin generally make up > 75% of the biomass of woody plant materials.

Part V

Nutrient and Trace Metal Cycling

Chapter 10

Nitrogen Cycle

Sources of Nitrogen to Estuaries

Elemental nitrogen (N_2) makes up 80% of the atmosphere (by volume) and represents the dominant form of atmospheric nitrogen gas. Despite its high atmospheric abundance, N_2 is generally nonreactive, due to strong triple bonding between the N atoms, making much of this N_2 pool unavailable to organisms. In fact, only 2% of this N_2 pool is believed to be available to organisms at any given time (Galloway, 1998). Consequently, N_2 must be "fixed" into ionic forms such as NH_4^+ before it can be used by plants. Since N is essential for the synthesis of amino acids and proteins and because it is often in low concentrations, N is usually considered to be limiting to organisms in many ecosystems. Nitrogen has five valence electrons and can occur in a broad range of oxidation states that range from +V to -III, with NO_3^- and NH_4^+ being the most oxidized and reduced forms, respectively. Some of the most common N compounds that exist in nature, along with their boiling points, ΔH^0, and ΔG^0, are shown in table 10.1 (Jaffe, 2000); these thermodynamic data can be used to calculate equilibrium concentrations.

Fluxes in the global N cycle have been seriously altered by anthropogenic activities (Vitousek et al., 1997; Galloway et al., 2004). For example, fluxes of many nitrogen oxides, which are largely derived from burning fossil fuels, have increased significantly in the atmosphere resulting in photochemical smog and acid precipitation (table 10.2; Jaffe, 2000). Similarly, the advent of artificial N fertilizers (e.g., the *Haber–Bosch process,* where N_2 is fixed to NH_3 by industrial processes), which were developed to compensate for the general nonavailability of N_2 to most agricultural crops, has resulted in increased N loading from soils and sewage to rivers and estuaries around the world, and considerable *eutrophication* problems in these aquatic ecosystems. For example, biological N_2 fixation accounted for a major fraction of newly fixed N before the 1800s ($\sim$90–130 Tg N y^{-1}) (Galloway et al., 1995). However, since the beginning of the 1900s the amount of fixed N entering the atmosphere has doubled from processes such as synthetic fertilizer production, increased biological N_2 fixation from *legume* plants associated with agriculture, and NO_y deposition from fossil fuel combustion (Galloway et al., 1995). Model predictions

Table 10.1 Chemical data on important nitrogen compounds in the environment.

Oxidation state	Compound	b.p. (°C)	$\Delta H^0(f)$ (kJ mol^{-1}, 298 K)	$\Delta G^0(f)$
+5	N_2O_5(g)	11	115	
	HNO_3(g)	83	−135	−75
	$Ca(NO_3)_2$(s)		−900	−720
	HNO_3(aq)		−200	−108
+4	NO_2(g)	21	33	51
	N_2O_4		9	98
+3	HNO_2(g)		−80	−46
	HNO_2(aq)		−120	−55
+2	NO(g)	−152	90	87
+1	N_2O(g)	−89	82	104
0	N_2(g)	−196	0	0
−3	NH_3(g)	−33	−46	−16.5
	NH_4(aq)		−72	−79
	NH_4Cl(s)		−201	−203
	CH_3NH_2(g)		−28	28
	H_2O(g)	100	−242	−229

Modified from Jaffe (2000).

indicate that greater use of global fertilizer in *exoreic* watersheds, will increase from 74 to 182 Tg N y^{-1} between 1990 and 2050 (an increase of 145%) and that 90% of the dissolved inorganic nitrogen (DIN) in river export will be derived from anthropogenic sources (Kroeze and Seitzinger, 1998).

Nitrogen cycling in estuaries is in general affected by inputs of N from surface and groundwaters, atmospheric wet and dry fallout, as well as N recycling in both the water column and sediments (figure 10.1; Paerl et al., 2002). Dominant inputs of N to estuaries have been shown to be linked to freshwater inputs from rivers (Nixon et al., 1995, 1996; Boynton and Kemp, 2000; Seitzinger et al., 2002a; Bouwman et al., 2005). Many of these nitrogen inputs have increased in rivers and estuaries around the world as a direct result of human expansion (Smullen et al., 1982; Peierls et al., 1991; Cole et al., 1993; Howarth et al., 1996; de Jonge et al., 2002; Bouwman et al., 2005). Current inputs of N to Atlantic and Gulf coast U.S. estuaries are 2 to 20 times higher than in the pre-industrialized periods (Boynton et al., 1995; Howarth et al., 1996, Goolsby, 2000). In an intensive monitoring study of Skidaway River estuary from 1986 to 1996, increasing human population was shown to have a significant effect on the loading of dissolved nutrients to this system, which was ultimately incorporated into particulate organic matter (Verity, 2002a,b). Strong latitudinal gradients also exist in N inputs around the world. For example, the Northern Hemisphere accounts for 90% of the dissolved inorganic nitrogen (DIN) export to coastal systems, which is clearly linked to the high continental land mass draining in the Northern Hemisphere, synthetic fertilizer application (94%), NO_y deposition (80%), and human

Table 10.2 Fluxes in the global nitrogen cycle (units are Tg N y^{-1}).

	Stedman and Shetter (1983)	Jaffe (1992)	Galloway et al. (1995)	Range[a]
Terrestrial–Atmospheric				
1. Natural biological fixation (pre-agriculture)	110		90–130	90–170
2. Fixation due to planting of nitrogen-fixing plants			43	
3. Total biological fixation (1 + 2 above)		150		
4. Industrial fixation (fertilizer production)		40	78	30–78
5. Natural denitrification (pre-agriculture)	124.5	124	80–180	80–243
6. Additional denitrification due to agriculture		23	50–110	
7. Microbial NO$_x$ production	10	8	4	4–89
8. Microbial N$_2$O production (natural)	38	7	8	12–69
9. Ammonia volatilization	82[b]	122	68	16–244
10. Biomass burning N$_2$O production		12	2	5–15
11. Biomass burning NO$_x$ production	5	5	8	3–11
12. Anthropogenic N$_2$O production (all sources)	11		3.4	
Oceanic–Atmospheric				
13. Biological fixation	40	40	40–200	10–200
14. Denitrification	30.5	30	150–180	25–180
Atmospheric–atmospheric				
15. NO$_x$ production by lightning	3	5	3	0.5–10
16. NO$_x$ production by industrial combustion	20	20	21	15–40
Terrestrial–oceanic				
17. River run-off	34	34	34	14–40

[a] For additional flux estimates not reported here refer to Jaffe (1992).
[b] NH only.
Modified from Jaffe (2000).

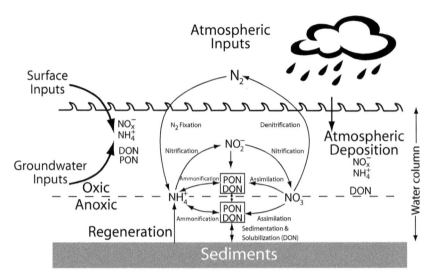

Figure 10.1 Schematic of nitrogen sources and cycling in estuaries. These sources range from a diverse group of both diffuse non-point agricultural, urban, and rural point sources (e.g., wastewater, industrial discharges, stormwater, and overflow discharges) across a broad spectrum of watersheds (e.g., urban, agricultural, upland and lowland forests). (Modified from Paerl et al., 2002.)

population (86%) in these watersheds (figure 10.2; Seitzinger et al., 2002a). Human population is perhaps the most critical factor here. For example, model predictions of DIN export for world regions in 2050 indicate that in regions such as South America and Africa, DIN export will potentially triple (compared to 1990 estimates) due to human population growth (figure 10.3; Seitzinger et al., 2002a). Similarly, areas in eastern and southern Asia will likely have dramatic increases in DIN export due to enhanced synthetic fertilizer use to support a growing population, and greater deposition of NO_y from a higher demand of fossil fuels in the region.

Nitrogen inputs to watersheds, which result in DIN export to coastal systems, often lead to enhanced primary production, since many estuaries are N limited (Nixon, 1986, 1995; D'Elia et al., 1992; Howarth et al., 2000). This can result in the formation of harmful algal blooms (HABs) as well as *hypoxia,* and in the worst case scenarios anoxic water columns (Valiela et al., 1990; Boynton et al., 1995; Paerl, 1997; Richardson, 1997). Other anthropogenic effects of N loading are discussed in chapter 15. The relationship between phytoplankton biomass (as chlorophyll-*a*) and total nitrogen, collected at 162 stations in 27 Danish fjords and coastal waters, illustrates the positive effects N inputs can have on phytoplankton biomass in estuaries (figure 10.4; Nielsen et al., 2002). High levels of N loading from the watershed and atmospheric inputs have also been shown to be positively correlated with eutrophication in estuaries, such as the Neuse River estuary (USA) (Paerl et al., 1990; Pinckney et al., 1998). While it has been argued that the chemical composition or form of N inputs can have significant effects on the composition of the phytoplankton

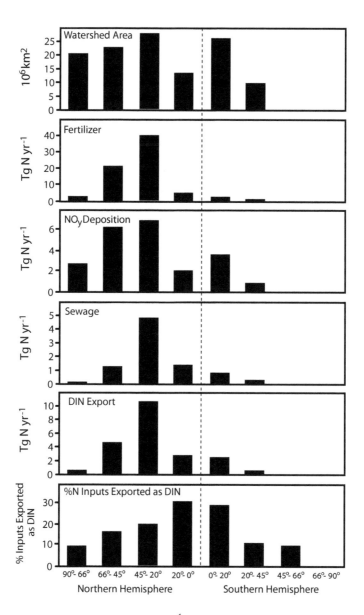

Figure 10.2 Latitudinal export (Tg N y^{-1}) of dissolved inorganic nitrogen (DIN) in Northern and Southern Hemispheres. (Modified from Seitzinger et al., 2002a.)

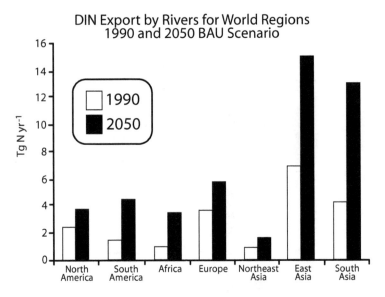

Figure 10.3 Model predictions of DIN export (Tg N y^{-1}) for world regions in 2050 compared with measured export in 1999. (Modified from Seitzinger et al., 2002a.)

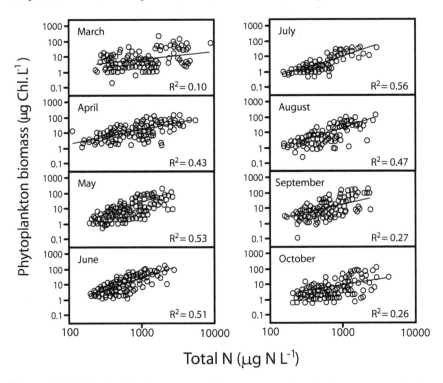

Figure 10.4 Relationship between monthly averages of total nitrogen concentration (μg N L^{-1} and phytoplankton biomass (μg Chl. a L^{-1}), during March to October, from 23 locations in Danish coastal waters. Lines represent least squares regression lines fitted on log-transformed data. (Modified from Nielsen et al., 2002.)

community (Collos, 1989), hydrodynamics and stratification of the system proved to be more important than the form of N in controlling phytoplankton community structure in the Neuse River estuary (Richardson et al., 2001). Toxic dinoflagellate blooms (e.g., *Pfisteria piscicida)* associated with fish kills and human health problems in this estuary are also believed to be linked with eutrophication (Burkholder et al., 1992; Burkholder and Glasgow, 1997; Glasgow and Burkholder, 2000). The occurrence of other toxic dinoflagellate blooms, such as *Alexandrium minutum,* has also been linked to eutrophication in the Penzé River estuary (France) (Maguer et al., 2004).

Nitrogen exported to Atlantic and Gulf coast urban estuaries shows a strong correlation with population density and N inputs, with sewage effluents accounting for 57% (averaged over all 11 watersheds) of the total (table 10.3; Castro et al., 2003). For the 20 agriculture estuaries, N sources (averaged over the 20 watersheds) were dominated by fertilizer (46%) and manure (32.3%) inputs. Damming in rivers can also change the dynamics of particulate N loading to coastal regions (Vörösmarty et al., 1997). In the case of the Mississippi River, where DIN has increased over several decades and the suspended load has decreased from damming, the particulate N load to the coast has decreased from 58 to 40% from 1950 to 1982 (Mayer et al., 1998). These findings in rivers and estuaries further support the concept of implementing specific watershed programs that make focused reductions in N loading (National Research Council, 2000; Paerl et al., 2002). It is also important to know where (e.g., reservoirs, streams, and rivers) and when nitrogen transformations are occurring in the watershed as part of the management plan (Seitzinger et al., 2002b). For example, models that detail the inputs and transformation of N in certain estuarine watersheds, such as the Waquoit Bay estuary (USA), have been shown to effectively improve management options of N loading that will fully restore water quality (Bowen and Valiela, 2004). Similarly, there are efforts to improve water quality in the Neuse River estuary through the modeling and monitoring programs (Reckhow and Gray, 2000); some of these modeling efforts now include ecosystem network analyses of N cycling (Christian and Thomas, 2003) in an approach first introduced by Ulanowicz (1987).

Submarine groundwater discharge (SGD) of N to estuaries has also been shown to be important—particularly, although not exclusively, in systems dominated by karst geomorphology, and where surface/river inputs are minimal. One such area is in northern Yucatan (Mexico), where concentrations of NO_3^- and NH_4^+ in SGD entering local lagoons range from 20 to 160 µM and 0.1 to 4 µM, respectively (Herrera-Silveira, 1994). Seasonal pulsing of SGD, driven by precipitation events, has not resulted in any extensive eutrophication in these lagoons (Pennock et al., 1999). However, there are differences in the trophic state (e.g., *oligotrophic, mesotrophic, and meso-eutrophic*) of these coastal lagoons which share similar geomorphological and climatological characteristics, but have different freshwater/seawater balances and hydrologic residence times—driven by SGD (Herrera-Silveira et al., 2002). Total system productivity and relative contribution of each primary producer (e.g., seagrasses, phytoplankton, and macroalgae) proved to be an effective approach in understanding the trophic dynamics of these coastal lagoons. In other carbonate systems such as Florida Bay (USA), N inputs from SGD may provide as much N (110 ± 60 mmol N $m^{-2}y^{-1}$) as surface waters from the Everglades (Corbett et al., 1999). Similarly, in siliciclastic systems, such as Great South Bay estuary (USA), SGD was found to represent greater than 50% of the NO_3^- inputs (Capone and Bautista, 1985; Capone and Slater, 1990).

Table 10.3 Nitrogen exported (kg N ha^{-1} y^{-1}) to estuaries from different watershed N sources and the percentage of N exported from different sources (in parentheses).

Watershed–estuary system	Watershed type	Agriculture runoff	Non-point source	Urban forest runoff	Upland human sewage	Atmospheric deposition	Total
Casco Bay, Maine	Urban	0.7 (12.6)	0.8 (15.0)	0.3 (5.5)	1.9 (36.3)	1.6 (30.6)	5.3
Great Bay, New Hampshire	Urban	1.3 (19.2)	1.4 (19.9)	0.3 (4.6)	2.5 (36.4)	1.4 (19.8)	6.8
Merrimack River, Massachusetts	Urban	0.8 (8.5)	0.5 (5.1)	0.6 (6.5)	5.7 (59.7)	1.9 (20.2)	9.5
Massachusetts Bay, Massachusetts	Urban	1.8 (3.7)	0.1 (0.3)	0.1 (0.2)	41.9 (85.5)	5.0 (10.3)	49.0
Buzzards Bay, Massachusetts	Urban	5.2 (23.8)	0.3 (1.2)	0.3 (1.5)	12.3 (56.4)	3.7 (17.1)	21.8
Narragansett Bay, Rhode Island	Urban	3.0 (11.1)	0.7 (2.7)	0.4 (1.5)	19.3 (70.7)	3.8 (13.9)	27.2
Long Island Sound, Connecticut	Urban	2.0 (15.4)	0.5 (4.1)	0.9 (6.8)	7.3 (56.7)	2.2 (17.0)	12.9
Hudson River–Raritan Bay, New York	Urban	2.8 (11.5)	0.8 (3.4)	0.8 (3.1)	15.7 (65.5)	3.9 (16.4)	24.0
Delaware Bay, Delaware	Urban	6.4 (31.9)	0.4 (2.1)	0.7 (3.6)	8.8 (43.8)	3.7 (18.5)	20.2
Charleston Harbor, South Carolina	Urban	3.9 (28.6)	0.2 (1.5)	0.2 (1.4)	8.4 (62.2)	0.9 (6.4)	13.5
Terrebonne–Timbalier Bays, Louisiana	Urban	1.4 (13.2)	0.2 (1.6)	0.002 (0.02)	5.1 (47.9)	3.9 (37.3)	10.6
Average		2.7 (16)	0.5 (5)	0.4 (3)	11.7 (57)	2.9 (19)	18.3
Great Bay, New Jersey	Agriculture	4.7 (47.4)	0.1 (1.2)	0.7 (6.8)	1.4 (14.1)	3.0 (30.5)	9.9
Chesapeake Bay	Agriculture	7.2 (53.4)	0.2 (1.5)	1.0 (7.5)	2.1 (15.3)	3.0 (22.3)	13.5
Pamlico–Pungo Sound, North Carolina	Agriculture	13.5 (74.1)	0.1 (0.3)	0.3 (1.7)	2.6 (14.4)	1.8 (9.6)	18.2
Wynah Bay, South Carolina	Agriculture	8.9 (70.2)	0.1 (0.6)	0.3 (2.1)	2.2 (17.5)	1.2 (9.5)	12.7
St. Helena Sound, South Carolina	Agriculture	4.7 (81.8)	0.0 (0.2)	0.1 (2.4)	0.1 (1.2)	0.8 (14.5)	5.7
Altamaha River, Georgia	Agriculture	6.6 (70.6)	0.1 (0.6)	0.3 (2.8)	1.6 (16.7)	0.9 (9.3)	9.4
Indian River, Florida	Agriculture	21.3 (73.1)	0.3 (1.1)	0.01 (0.05)	4.1 (13.9)	3.4 (11.8)	29.1
Charlotte Harbor, Florida	Agriculture	15.4 (85.2)	0.03 (0.2)	0.02 (0.12)	1.1 (6.2)	1.5 (8.4)	18.1
Tampa Bay, Florida	Agriculture	21.1 (78.3)	0.2 (0.7)	0.02 (0.07)	2.7 (9.9)	3.0 (11.0)	26.9
Apalachee Bay, Florida	Agriculture	4.5 (81.2)	0.1 (0.9)	0.2 (3.5)	0.2 (3.5)	0.6 (10.8)	5.6

Estuary	Watershed type						
Apalachicola Bay, Florida	Agriculture	7.2 (72.1)	0.1 (1.3)	0.2 (2.2)	1.6 (16.3)	0.8 (8.0)	10.0
Mobile Bay, Alabama	Agriculture	4.6 (54.3)	0.1 (1.3)	0.4 (4.7)	2.3 (26.9)	1.1 (12.8)	8.5
West Mississippi Sound, Mississippi	Agriculture	5.7 (63.1)	0.1 (1.3)	0.6 (6.8)	1.6 (18.1)	1.0 (10.7)	9.1
Calcasieu Lake, Louisiana	Agriculture	5.9 (50.7)	0.1 (0.55)	0.5 (4.5)	2.6 (22.3)	2.6 (22.0)	11.7
Sabine Lake, Texas	Agriculture	6.1 (66.3)	0.1 (0.79)	0.3 (3.8)	1.6 (17.3)	1.1 (11.9)	9.3
Galveston Bay, Texas	Agriculture	7.8 (47.2)	0.2 (1.36)	0.1 (0.8)	6.7 (40.4)	1.7 (10.2)	16.5
Matagorda Bay, Texas	Agriculture	3.0 (75.7)	0.04 (0.91)	0.0 (1.2)	0.5 (13.6)	0.3 (8.6)	4.0
Corpus Christi Bay, Texas	Agriculture	1.6 (67.5)	0.01 (0.35)	0.0 (1.7)	0.6 (25.0)	0.1 (5.4)	2.4
Upper Laguna Madre, Texas	Agriculture	0.7 (70.3)	0.01 (0.91)	0.0 (1.1)	0.1 (8.9)	0.2 (18.7)	1.0
Lower Laguna Madre, Texas	Agriculture	6.6 (76.7)	0.03 (0.34)	0.0 (0.2)	1.3 (15.0)	0.7 (7.8)	8.7
Average		7.9 (68)	0.1 (0.8)	0.3 (3)	1.8 (16)	1.4 (13)	11.5
Barnegat Bay, New Jersey	Atmospheric	2.6 (34.9)	0.1 (2.0)	0.6 (8.4)	0.3 (3.7)	3.7 (51.0)	7.3
St. Catherines–Sapelo, Georgia	Atmospheric	0.3 (14.7)	0.4 (15.5)	0.3 (11.3)	0.01 (0.64)	1.3 (57.9)	2.3
Barataria Bay, Louisiana	Atmospheric	2.3 (27.5)	0.2 (2.1)	0.003 (0.04)	2.6 (31.8)	3.2 (38.6)	8.3
Average		1.7 (26)	0.2 (7)	0.3 (7)	1.0 (12)	2.8 (49)	6.0

Watershed types are based on the primary sources of N to each estuary. Watersheds dominated by urban N sources (e.g., point, septic, and non-point source runoff) were classified as urban, watersheds dominated by agricultural N sources (e.g., fertilization, fixation, and manure) were classified as agriculture, and watersheds dominated by atmospheric N deposition were classified as atmospheric. Effluent from sewage treatment plants in the Barnegat Bay watershed is discharged offshore; sewage inputs to the Barnegat Bay estuary are from septic systems in the watershed. Modified from Castro et al. (2003).

Temperate estuaries located on the eastern oceanic coastlines of the Atlantic, such as the Galician rias on the northwest Iberian Peninsula (Spain), experience upwelling events that provide a significant source of ocean-derived N to the rias (Prego and Bao, 1997; Prego, 2002). During these upwelling events, which typically occur in winter, spring, and summer, residual velocities in the rias may be tripled (Prego and Fraga, 1992). In the case of Ria Vigo (Spain), enhanced circulation during upwelling events results mostly in export of N to the coast (Prego, 2002). For example, during upwelling there is a high net photosynthesis of organic N (10 mol N s^{-1}); however, much of it is exported (6.5 mol N s^{-1}) or sedimented (50 mg N d^{-1}). Thus, there is an effect of N fertilization during upwelling events from offshore that does result in eutrophication in the ria because of the concurrent high export of N during these events. This in contrast to western margin estuaries of the Atlantic, such as Chesapeake Bay and Narragansett Bay, where N inputs from fluvial and atmospheric sources have resulted in extensive eutrophication problems (Nixon et al., 1996).

Atmospheric N inputs have been shown to be a significant source of N to estuaries (Paerl, 1995; Fisher and Oppenheimer, 1991; Valigura et al., 1996; Paerl et al., 2002). In fact, atmospheric N represents as much as 40% of the new N to Pamlico Sound estuary (USA), via the Neuse River (Paerl and Fogel, 1994; Whitall and Paerl, 2001). Atmospheric N inputs can occur in the form of wet (e.g., N dissolved in rain or snow) or dry (e.g., settling dry particles of gaseous uptake) deposition (Paerl et al., 2002). The dominant forms of atmospheric N are inorganic and exist as oxidized gaseous forms (e.g., HNO_3, NO_2, and aerosol NO_3^-) and reduced forms (e.g., aerosol NH_4^+) that have largely been produced by anthropogenic activities (Likens et al., 1974; Galloway et al., 1994). Some of this atmospheric N is also in the form of organic N compounds that in some cases can represent a significant fraction of deposition in estuarine watersheds (Correll and Ford, 1982; Skudlark and Church, 1994; Whitall and Paerl, 2001). In fact, atmospheric deposition of NO_y was shown to be seven-fold better in predicting riverine N export to the North Atlantic Basin than agricultural N sources (Howarth et al., 1996, 2000; Howarth, 1998). This further supports the idea that controls on fossil-fuel sources of atmospheric N in the northeastern United States are critical in reducing N loading to watersheds and aquatic systems (Howarth et al., 2000). Unfortunately, the linkages between agricultural fertilizers and N inputs to aquatic systems remains poorly understood (Howarth et al., 2000). For example, on average about half of the N fertilizer added to agricultural soils is removed when it is harvested (Bock, 1984). A large fraction of this crop is then fed to animals which results in substantial volatilization of N to the atmosphere, as well as direct leaching to soils, making the entire food–agriculture system a more reliable base for comparison with N loading than just agricultural field plots (Howarth et al., 2000).

Dissolved organic nitrogen (DON) can represent a significant fraction of the total dissolved nitrogen (TDN) in rivers and estuaries, respectively (table 10.4; Berman and Bronk, 2003). Sources of DON to estuaries may be derived from actively growing phytoplankton communities, where small molecules such as dissolved free amino acids (DFAA) are released—these are highly available to bacteria, but represent only a small percentage of the total DON ($\sim$ <10%) (Anita et al., 1991; Bronk et al., 1994, 1998; Bronk and Ward, 1999; Diaz and Raimbault, 2000; Bronk, 2002). In fact, phytoplankton exudation rates of DON have been shown to be proportional to irradiance levels (Zlotnik and Dubinsky, 1989). Viral lysis and autolysis by phytoplankton during senescence should

Table 10.4 Total dissolved nitrogen (TDN) and organic nitrogen (DON) in estuaries and rivers.

Location	Depth	TDN (mL)	DON (μM-N)	References
Estuaries				
Shinnecock Bay, New York	Surface	2.0–4.9	0.6–4.3	Berg et al. (1997); Lomas et al. (1996)
Waquoit Bay, Massachusetts	Surface	140	40.0	Hopkinson et al. (1996)
Chesapeake Bay, mesohaline	Surface	34.1±12.3	21.3±16.0	Bronk et al. (1998)
Chesapeake Bay, mesohaline	Surface	42.5±3.7	2.3±9.2	Bronk and Glibert (1993)
Chesapeake Bay, mesohaline	Surface	23.1±1.7	22.2±1.6	Bronk and Glibert (1993)
Chesapeake Bay, mouth	Surface		16.3±5.5	Bates and Hansell (1999)
Apalachicola Bay	Surface	23	14.8±1.0	Mortazavi et al. (2000)
Delaware Estuary	Surface		40.8±29.3	Karl (1993)
Elba Estuary	Surface	72.2±17.6	65.0±12.2	Kerner and Spitzy (2001)
North Inlet, South Carolina	Surface	19.4–35.3	18.0–30.8	Lewitus et al. (2000)
Tomales Bay	Surface		5.8–12.6	Smith et al. (1991)
Rivers				
Russian rivers draining into Arctic	Surface		27.0±5.0	Gordeev et al. (1996); Wheeler et al. (1997)
Rivers entering the Baltic Sea	Surface	48.9±41.9	29.6±14.8	Stepanauskas et al. (2002)
Susquahanna River, Maryland	Surface	116	23.0	Hopkinson et al. (1998)
Satillia River, Georgia	Surface	62.6	59.0	Hopkinson et al. (1998)
Parker River, Maryland	Surface	37	26.0	Hopkinson et al. (1998)
Delaware River	Surface		29.7±23.7	Seitzinger and Sanders (1997)
Hudson River	Surface		33.5	Seitzinger and Sanders (1997)
Choptank River	Surface	41.3	26.9	Bronk (1993a)
Georgia and South Carolina Rivers	Surface		35.9±10.7	Alberts and Takacs (1999)
Streams in Sweden	Surface		24.3±7.4	Stepanauskas et al. (2000)
Wetland in Sweden (bulk DON)	Surface		98.0±68.0	Stepanauskas et al. (1999)
Laqunitas Creek (flows to Tomales Bay)	Surface		3.9–17.9	Smith et al. (1991)

Modified from Berman and Bronk (2003).

be also included in the pathways of phytoplankton-derived DON (Gardner et al., 1987; Agusti et al., 1998). In addition to direct exudation by phytoplankton, inputs of DON from "sloppy feeding" by zooplankton and/or protozoan grazers may be important (Strom and Strom, 1996). For example, correlations between regeneration rates of N (e.g., amino acids and DIN) were found to be highest in the midsalinity areas of highest phytoplankton and bacterioplankton production in the Mississippi River plume (Chin-Leo and Benner, 1992; Bronk and Glibert, 1993; Gardner et al., 1993, 1996, 1997)

Allochthonous DON sources from terrestrial runoff, plant detritus leaching, soil leaching, sediments, and atmospheric deposition may also represent important inputs to estuaries (Berman and Bronk, 2003). DON typically represents about 60 to 69% of the TDN in rivers and estuaries (Berman and Bronk, 2003). The major components of DON include urea, dissolved combined amino acids (DCAA), DFAA, proteins, nucleic acids, amino sugars, and humic substances (Berman and Bronk, 2003). However, less than 20% of DON is chemically characterized.

The major processes involved in the biogeochemical cycling of N in estuaries and the coastal ocean are: (1) biological N_2 fixation (BNF); (2) ammonia assimilation; (3) nitrification; (4) assimilatory NO_3^- reduction; (5) ammonification or N remineralization; (6) ammonium oxidation; (7) denitrification and dissimilatory NO_3^- reduction to NH_4^+; and (8) assimilation of DON (figure 10.5; Libes, 1992). These processes are essentially driven by bacteria, as discussed in the following section, and are in some cases energy producing, or may occur in *symbiosis* with another organism. These heterotrophic processes can result in both additions (e.g., N_2 fixation) and losses of N (e.g., denitrification) within estuaries.

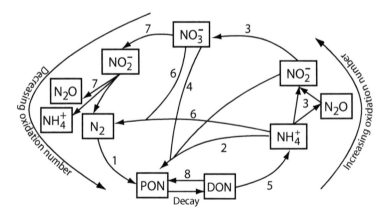

Figure 10.5 Major processes involved in the biogeochemical cycling of N in estuaries and the coastal ocean: (1) biological N_2 fixation; (2) ammonia assimilation; (3) nitrification; (4) assimilatory NO_3^- reduction; (5) ammonification or N remineralization; (6) ammonium oxidation (speculative at this time); (7) denitrification and dissimilatory NO_3^- reduction to NH_4^+; and (8) assimilation of dissolved organic nitrogen (DON). (Modified from Libes, 1992.)

When examining inputs and losses of N to estuaries, it has been shown that the net export of N from estuaries is essentially a function of the residence time of freshwater in an estuary (table 10.5; Nixon et al., 1996; Nowicki et al., 1997). In fact, a simple steady-state model, using only freshwater residence time, provided a good estimate of the fraction of upland N inputs that is exported or denitrified in estuaries (Dettmann, 2001). While external sources of N to estuaries from rivers, the atmosphere, and runoff can be quite high, a significant fraction of the N utilized by phytoplankton in estuaries is from internal recycling (Nixon, 1981; Wollast, 1993). Many of the processes involved in internal recycling, shown in figure 10.5, can occur in both the water column and sediments. The global budget of N in the coastal zone reveals that only 2.8% of the dissolved N from rivers is directly provided to phytoplankton in coastal waters—most is in the form of recycled N (figure 10.6; Wollast, 1993). This is because of the extensive internal recycling of N that occurs in estuaries. It has been estimated that 30 to 65% of the total N in estuaries is retained or lost through this cycling, thereby reducing inputs to the coastal ocean (Nixon et al., 1996). The contribution of protozoa and zooplankton to nutrient regeneration in the water column and sediments largely depends on the quality of bacterial substrates, as well as temperature (Gardner et al., 1997). It should also be noted that since rivers do deliver the bulk load of N to estuaries, rivers are in fact substantially more important as an "ultimate" source term than appears in figure 10.6. However, it is also important to remember that the amount of N delivered to estuaries from rivers is less than the N input to rivers because of removal processes in rivers (Seitzinger and Kroeze, 1998).

Transformations and Cycling of Inorganic and Organic Nitrogen

The relative importance of the biogeochemical processes that control estuarine N budgets (figure 10.5) varies considerably across estuaries because of other covariables such as redox conditions, organic carbon loading, and anthropogenic inputs—just to name a few. When comparing the general N pools in estuaries and coastal regions we see that the NH_4^+ represents a larger fraction of the total N budget because of the stronger linkage between sediments and the water column in estuaries compared to deeper shelf environments (figure 10.7; Anita et al., 1991; Berman and Bronk, 2003). Ammonium is typically the dominant form of DIN in sediments. We will now examine each of the biogeochemical processes listed in figure 10.5, as they relate to divergent estuarine systems, and will begin with biological N_2 fixation (BNF), since this is the only natural process that incorporates N_2 from atmospheric sources into estuaries.

Biological N_2 fixation is a process performed by prokaryotic (both heterotrophic and phototrophic) organisms (sometimes occurring symbiotically with other organisms) resulting in the enzyme-catalyzed reduction of N_2 to NH_3 or NH_4^+, or organic nitrogen compounds (Jaffe, 2000). Organisms that fix N_2 are also called *diazotrophs*. The high activation energy required to break the triple bond makes BNF an energetically expensive process for organisms, thereby restricting BNF to select groups of organisms. BNF is an important process in aquatic systems (Howarth et al., 1988a,b); some of the heterotrophic and phototrophic bacteria capable of BNF in coastal and estuarine systems are shown in table 10.6. Other nonbiological processes, such as lightning, can produce

Table 10.5 Nitrogen loading and denitrification in a variety of coastal marine systems (units = mmol N m^{-2}y^{-1}).

Coastal system	DIN	Total N	Denitrification	Percentage of DIN removed by denitrification	Percentage of total N removed by denitrification	Residence time[a] (months)	Note[b]
Delaware Bay	1368	1900	832	61	44	3.3	1
Boston Harbor	4320	9095	666	15	7	0.33	2
Galveston Bay	2350		321	14		2.3	3
Guadalupe Estuary	629	949	356	57	38	3	4
Norsminde Fjord	11,541		320	3		0.17	5
Ochlockonee Bay	1577	5991	700	44	12	0.15	6
Baltic Sea	95	191	88–161	93–169	46–84	240	7
Narragansett Bay	1442	1980	384–517	27–36	19–26	0.85	8
Potomac Estuary	1299	2095	330	25	16	5	9
Patuxent Estuary	695	902	282	41	31		10
Choptank Estuary	207	305	271	131	89		11
Chesapeake Bay (MD portion)	1027	1467	351	34	24		12
Chesapeake Bay (all)	657	938	243	37	26	7	13
Tejo Estuary	4468		2059	46			14
Scheld Estuary		13,400	5420		40	3	15

[a]Residence time from Nixon et al. (1996), unless noted

[b]Notes: 1, TN compiled in Nixon et al. (1996); DIN estimated assuming equals 72% of TN; denitrification from Seitzinger (1988b); 2, Nowicki (1994); 3, Zimmerman and Benner (1994); 4, Yoon and Benner (1992); residence time from Zimmerman and Benner (1994); 5, Nielsen et al. (1995); 6, Seitzinger (1987); TN, Seitzinger, unpublished measurements; 7, TN compiled in Nixon et al. (1996); DIN estimated assuming DIN = 50% TN (Granéli et al., 1990); denitrification range from Shaffer and Rönner (1984) (measurements of deep Baltic water) and Nixon et al. (1996) (from TN mass-balance calculations); 8, DIN and TN from Nixon et al. (1995) (DIN revised from earlier estimates of Nixon, 1981); denitrification from range of Nowicki (1994) (384) and Seitzinger et al. (1984) (517); 9, Boynton et al. (1995); DIN calculated assuming equal to 62% TN (USGS unpublished data); 10, Boynton et al. (1995); DIN calculated assuming equal to 77% TN (USGS unpublished data); 11, Boynton et al. (1995); DIN calculated assuming equal to 68% TN (USGS unpublished data); 12, Boynton et al. (1995); DIN calculated assuming equal to 70% TN (average Susquehanna, Potomac, Choptank, and Patuxent Rivers using USGS unpublished data); 13, Nixon et al. (1996); see 12 for DIN calculation; 14, Seitzinger (1988a); 15, Nixon et al. (1996). Modified from Nixon et al., 1996.

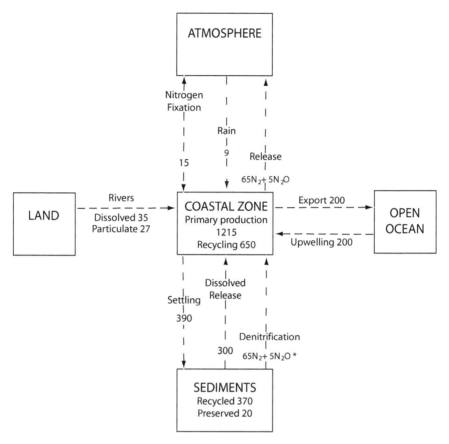

Figure 10.6 Global budget of N in the coastal zone (fluxes are in 10^{12} g N yr^{-1}).
* = More recent estimates of denitrification in estuaries and continental shelf sediments
range from $\sim$ 200 to 300 Tg N y^{-1} (Seitzinger and Giblin, 1996; Codispoti et al., 2001).
(From Wollast, 1993, with permission.)

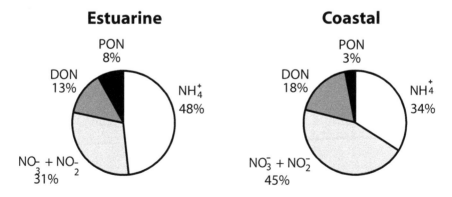

Figure 10.7 Average composition of nitrogen pools (excluding dissolved N_2) in estuarine waters DON = dissolved organic nitrogen; PON = particulate organic nitrogen. (Modified from Berman and Bronk, 2003.)

Table 10.6 Nitrogen-fixing bacteria isolated from marine environments.

Heterotrophs		Phototrophs	
Relation to O$_2$ and genus	Habitat	Group and genus	Habitat
1. Aerobes		1. Cyanobacteria	
Azotobacter spp.	Black Sea sediments	A. Chroococcacean	
.	Seagrass sediments	*Synechococcus* sp.	Snail shell, intertidal
	Macroalga *Codium*		Tropical sediments
	Intertidal sediments		Microbial mat
	Estuarine sediments	*Gloeocapsa* sp.	Marine aquarium
	Salt marsh sediments	B. Pleurocapsalean	
		Dermocarpa sp.	Marine aquarium
		Xenococcus sp.	Snail shell, intertidal
2. Microaerophiles		*Myxosarcina* sp.	Rock chip
Azospirillum spp.	*Spartina* roots	*Pleurocapsa* sp.	
	Zostera roots	C. Oscillatorian	
	Seawater	LPP[a] group	Snail shell, intertidal
Campylobacter spp.	*Spartina* roots	*Oscillatoria* sp.	Intertidal mat
Beggiatoa spp.	Sediment surface	*Microcoleus* sp.	Marsh mat
(?)	Shipworm	*Phormidium* sp.	Microbial mat
		D. Nostacalean	
3. Facultative anaerobes		*Anabaena* sp.	Algal mat
Enterobacter spp.	Beach sediment	*Calothrix* sp.	Microbial mat
	Intertidal sediments	*Nodularia* sp.	Microbial mat
Klebsiella spp.	Estuarine sediments	*Nostoc* sp.	Microbial mat
	Halodule roots		

Organism	Habitat
Vibrio spp.	Beach sediment
	Mangrove bark
	Sea urchin
	Seawater
	Mangrove bark
2. Prochlorales	
Prochloron[b]	Ascidians
3. Anoxyphotobacteria	
A. Rhodospirillaceae	
Rhodopseudomonas sp.	Harbor muds
B. Chromatiaceae	
Thiocapsa sp.	Anaerobic habitat
4. Strict anaerobes	
Desulfovibrio spp.	Seagrass sediments
	Intertidal sediments
	Salt marsh sediments
	Estuarine sediments
Clostridium spp.	Seagrass sediments
	Intertidal sediments
	Salt marshsediments
	Estuarine sediments

[a] *Lyngbya/Plectonema/Phormidium* group.
[b] Presumptive fixer, only active in association.
See Capone (1988) for data sources.
Modified from Howarth et al. (1988a,b).

NH_3-N N_2 fixation, but the most dominant form of nonbiological N_2 fixation occurring around the world is from industrial sources using the Haber–Bosch process (Jaffe, 2000). During BNF, prokaryotes use an enzyme complex called nitrogenase that is composed of the following *metalloproteins*: (1) an Fe protein, which serves as the source of electrons for reducing power; and (2) a Mo–Fe protein, which acts to bind N_2 to the enzyme (Jaffe, 2000). One limitation here is that the nitrogenase enzyme complex is highly sensitive to O_2, because of reactions with the Fe component of the complex (Burns and Hardy, 1975). Consequently, only organisms living in an anaerobic environment, or living in an aerobic environment capable of creating an anaerobic subenvironment, are able to fix N_2. For example, the cyanobacteria, *Nodularia* spp., commonly found in estuaries, forms specialized well-sealed cells called *heterocysts* capable of BNF (Moisander and Pearl, 2000). The effects of BNF on N limitation in freshwater and estuarine systems have also been an interesting topic of debate (Howarth et al., 1988a; Paerl, 1996; Vitousek et al., 2002). In freshwater systems, when the ratio of N:P is low, BNF by cyanobacteria will dominate the phytoplankton community, thereby adjusting N concentrations, making P limitation more common in high-productivity freshwater environments (Schindler, 1977; Howarth et al., 1988a,b). In contrast, when N:P ratios are low in temperate estuaries N-fixing cyanobacteria are not capable of reaching high enough numbers to offset the depletion in N (Howarth et al., 1988a,b). The primary causes for this are believed to be: (1) enhanced turbulence in estuaries which acts to break up cyanobacterial filaments exposing nitrogenase to O_2 (Paerl, 1985); (2) Mo limitation caused by stereochemical interference of molybdate uptake by SO_4^{2-} (Howarth and Cole, 1985; Marino et al., 1990); and (3) zooplankton grazing on cyanobacteria (Howarth et al., 1999). However, recent evidence from the Baltic Sea has shown that the highest water column TDN concentrations were correlated with the abundance of filamentous N-fixing cyanobacteria (figure 10.8; Larsson et al., 2001). The dominant diazotrophic filamentous bacteria in the Baltic Sea are *Aphanizomenon flos-aquae*, *Nodularia spumigena* (Wasmund, 1997), and the less common *Anabaena lemmermannii* (Laamanen, 1997). Increases in TDN each summer in the upper mixed layer of the northern Baltic proper are the result of N inputs from N-fixing cyanobacteria, which appear to control new production (*sensu* Dugdale and Goering, 1967). In fact, estimated N-fixation rates in the Baltic ranged from 2.3 to 5.9 mmol N m^{-2} d^{-1}, which sustained 30 to 90% of the total pelagic production in June through August (Larsson et al., 2001).

Biological N_2 fixation in benthic environments such as cyanobacterial mats (Carpenter et al., 1978; Savela, 1983; Bebout et al., 1993), the rhizosphere of salt marshes (Marsho et al., 1975; Teal et al., 1979; Hanson, 1983), seagrass beds (Capone, 1982), and mangrove root systems (Vitousek et al., 2002; Ravikumar et al., 2004) can be an important source of N to estuaries. Some of the highest N-fixation rates (100–150 mg N_2 m^{-2} d^{-1}) are found in these intertidal environments, most likely because of high availability of metals from reducing conditions in sediments, reduced grazing effects, and lower turbulence (Howarth et al., 1988a). Some of the common mat-forming N-fixing cyanobacteria in marine and estuarine systems are *Calothrix* sp., *Microcoleus* sp., *Lyngbya* sp., and *Oscillatoria* sp. (Vitousek et al., 2002). In the case of *Lyngbya* sp. cyanobacterial mats, it has been estimated that BNF supplies about 44% of the N required for production (Bebout et al., 1994). In some cases, denitrification may be a competing process deep within the cyanobacterial mats where 20% of the fixed N can be lost, making cyanobacterial

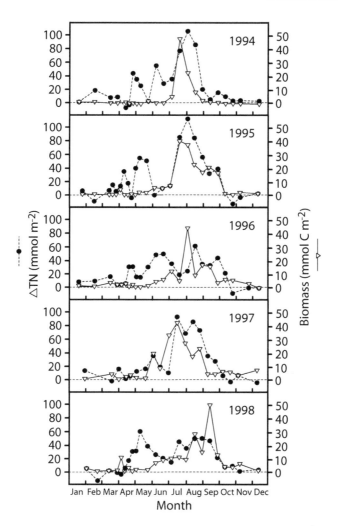

Figure 10.8 Increase in average total nitrogen concentration (mmol m^{-2}) in the upper 20 m of the water column relative to the average concentration at depths of 25 and 30 m, compared to biomass (mmol m^{-2}) of heterocystous cyanobacteria in the top 20 m (Station BY31) in the Lansort Deep region of the Baltic Sea—from 1994 to 1998. (Modified from Larsson et al., 2001.)

mats both sources and sinks of N (Joye and Paerl, 1993, 1994). Overall, denitrification rates in estuaries greatly exceed BNF rates (Nixon et al., 1996). Symbiotic associations between nitrogen-fixing bacteria and vascular plant root systems in marine environments (e.g., mangroves) are also fairly well documented (Ogan, 1990; Ravikumar et al., 2004). For example, recent work has shown the nitrogen-fixing *Azospirillum* bacteria associated with mangrove roots significantly enhanced growth and pigment production of mangrove seedlings. These bacteria are capable of synthesizing the *phytohormone*, indoleacetic acid (IAA), which is also believed to be an important component in promoting plant growth (Ravikumar et al., 2004).

Ammonia assimilation is where NH_3 or NH_4^+ is taken up and incorporated into organisms as organic nitrogen molecules (Jaffe, 2000), as shown below in the case of NH_4^+:

$$NH_4^+ \rightarrow N(NH_3)\text{-organic} \qquad (10.1)$$

The ability to take up these reduced forms of N is a distinct energetic advantage for organisms because it is a direct means of obtaining a reduced form of N—unlike the uptake of NO_3^- which requires the extra step of NO_3^- reduction. However, some phytoplankton prefer NO_3^- over NH_4^+ as a N source. In estuaries, NH_4^+ is generally the most dominant form of DIN in sediments; however, some pools of NH_4^+ are unavailable for uptake because they are bound within the internal matrix of minerals (Rosenfield, 1979; Krom and Berner, 1980b).

Nitrification is where NH_3 or NH_4^+ is oxidized to NO_2^- or NO_3^- through the following two energy-producing reactions (Delwiche, 1981):

$$NH_4^+ + 3/2(O_2) \rightarrow NO_2^- + H_2O + 2H^+, \Delta G^0 = -290\,kJ\,mol^{-1} \qquad (10.2)$$

$$NO_2^- + 1/2(O_2) \rightarrow NO_3^-, \Delta G^0 = -89\,kJ\,mol^{-1} \qquad (10.3)$$

Equation 10.2 involves the oxidation of NH_4^+ to NO_2^- and is primarily performed by bacteria of the genus *Nitrosomonas* (some species of the genus *Nitrocystis* spp. are also capable) (Day et al., 1989), and equation 10.3 is the continued oxidation of NO_2^- to NO_3^-, performed by bacteria of the genus *Nitrobacter* spp. During this process of NH_4^+ oxidation, these bacteria use CO_2 as their source of C to be fixed into organic compounds. Hence, these bacteria are considered to be *chemoautotrophs* because they gain their energy by chemical oxidation, and are autotrophs because they do not rely on external sources of organic matter. These bacteria also occur as mixed-species communities or *consortia* because of their codependency on the availability of reaction products (Capone, 2000). Finally, nitrification requires O_2; thus, the activity of these nitrifying bacteria in sediments is particularly sensitive to the absence of dissolved oxygen (DO) (Henriksen et al., 1981; Kemp et al., 1982).

Some intermediary products produced during nitrification, such as NH_2OH, NO, and N_2O (Jaffe, 2000), are important biogases in estuaries (as discussed in chapter 5). For example, N_2O was found to be primarily produced (not excluding denitrification) from nitrification in *R. mangle* mangrove forest sediments (Bauza et al., 2002). As shown in figure 10.9, there is a good correlation between NO_3^- and N_2O. Ammonium is negatively correlated with these oxidized forms of DIN because NH_4^+ is oxidized to NO_3^- under these oxic conditions; nitrification typically occurs when E_h values are greater than 200 mV (Smith et al., 1983). These results support other work which showed that N_2O is a significant source in the N budget of mangrove sediments, where there is substantial

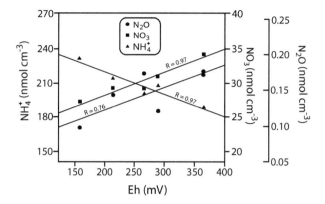

Figure 10.9 Correlation between concentrations (nmol cm^{-3}) of NH$_4$$^+$ (triangles), NO$_3$$^-$ (squares), and N$_2$O (circles) and redox potential (mV) in *Rhizophora mangle* sediments. (Modified from Bauza et al., 2002.)

root development by *R. mangle* (Koike and Terauchi, 1996). Much of the nitrification that occurs near the sediment–water interface in these systems is in the oxidized linings of animal burrows as well as the well-oxygenated zone of the rhizosphere (Boto, 1982). In general, both nitrification and denitrification control N$_2$O production such that benthic N$_2$O evolution in regional and global models has been estimated under the assumption that N$_2$O increases in proportion to both processes (Capone, 1996; Seitzinger and Kroeze, 1998).

Assimilatory NO$_3$$^-$ reduction to NH$_4$$^+$ (ANRA) involves the simultaneous reduction of NO$_3$$^-$ and uptake of N by the organism into biomass, as shown below:

$$NO_3^- + H^+ \rightarrow NH_3\text{-organic} \tag{10.4}$$

This pathway is particularly dominant when reduced forms of N are low, such as oxic water columns in estuaries. This represents an important pathway for the uptake of N by estuarine organisms capable of uptake of both reduced and oxidized forms of N (Collos, 1989).

Ammonification is the process where NH$_3$ or NH$_4$$^+$ is produced during the breakdown of organic nitrogen by organisms (also called N remineralization). In the case of the most dominant nitrogen-containing organic compounds, proteins, peptide linkages are broken down followed by *deamination* of amino acids to produce NH$_3$ or NH$_4$$^+$. Ammonification is the conversion of DON into ammonium (DON $\rightarrow$ NH$_4$$^+$). This process occurs through the decay of plants and animals mediated by heterotrophic bacteria (Blackburn and Henriksen, 1983; Jenkins and Kemp, 1984; Nowicki and Nixon, 1985; Henriksen and Kemp, 1988) or through excretion from animals (Rowe et al., 1975; Bianchi and Rice, 1988; Gardner et al., 1993). Ammonification occurs both in sediments and the water column; however, in shallow coastal environments a larger proportion of the mineralization takes place in sediments—as particles settle more rapidly to the sediment surface.

For example, NH_4^+ regeneration rates in the water column of coastal regions has been shown to range from 0 to 0.2 $\mu M\ h^{-1}$ with uptake values from 0.01 to 0.22 $\mu M\ h^{-1}$ (Selmer, 1988). Other work has also shown similar ranges for NH_4^+ regeneration rates in Delaware Bay (0.02–0.62 $\mu M\ h^{-1}$) (Lipschultz et al., 1986), Georgia coastal wetlands (0.02–0.35 $\mu M\ h^{-1}$), and the Mississippi River plume (0.08–0.75 $\mu M\ h^{-1}$) (Gardner et al., 1997).

Ammonium oxidation (anammox), which involves the anaerobic NH_4^+ oxidation with NO_3^-, may be another mechanism whereby N is lost from estuarine/coastal systems. This process has been shown to represent from 20 to 67% of N_2 production in temperate continental shelf sediments (Thamdrup and Dalsgaard, 2002); however, only limited knowledge currently exists on the pathways of anammox in natural systems. It does appear that the optimal conditions for anammox are different from those for denitrification (Rysgaard and Glud, 2004). Although there is currently no evidence for existence of anammox in estuaries, it has been found in a diversity of environments such as waste water treatment plants (Mulder et al., 1995), anoxic water columns (Dalsgaard et al., 2003; Kuypers et al., 2003), temperate shelf sediments (Thamdrup and Dalsgaard, 2002), Arctic Sea ice (Rysgaard and Glud, 2004), and Arctic shelf sediments (Rysgaard et al., 2004). Estuarine sediments are sites of intense organic matter remineralization which can yield high concentrations of NO_3^- and NH_4^+ (Blackburn and Henriksen, 1983; Nowicki and Nixon, 1985)—the required substrates for anammox. Other processes, such as Mn(II) oxidation by NO_3^- or Mn(IV) reduction by NH_4^+ (Luther et al., 1997), can also result in the evolution of N_2 in suboxic sediments—more discussion on this in chapter 14. Thus, anammox is likely to occur in estuarine systems and, therefore, it is included as a potential process of N loss in estuaries in this chapter.

Denitrification, also known as *dissimilatory* NO_3^- or NO_2^- reduction, is the reduction of NO_3^- or NO_2^- to any gaseous form of N such as NO, N_2O, and N_2 by microorganisms, as in the case of N_2 production shown below:

$$CH_2O + NO_3^- + 2H^+ \rightarrow CO_2 + 1/2N_2 + 2H_2O \qquad (10.5)$$

The denitrification pathway involves the production of intermediates such as NO_2^-, NO, and N_2O. Numerous methods on how to measure denitrification have been developed in recent years ranging from acetylene inhibition (Sørensen, 1978), labeled N tracers (Nishio et al., 1982; Rysgaard et al., 1993), direct N_2 flux measurements (Seitzinger et al., 1980; Nowicki, 1994), water column and sediment changes in N_2 (Devol, 1991), to core and chamber incubation with N isotopes (Nielsen, 1992)—more details on comparisons of these methods are provided in Lamontagne and Valiela (1995). Denitrification represents the major pathway from which N is lost from estuaries and has been shown to be proportional to the log mean water residence time (figure 10.10; Nixon et al., 1996). The positive linear relationship between denitrification and log mean water residence time ($r^2 = 0.75$), across a range of rivers and estuaries, suggests that with increasing residence time N will be recycled in the water column and sediments more extensively resulting in greater denitrification (Nixon et al., 1996). Other factors important in regulating denitrification rates in estuaries are sediment–water NH_4^+ exchange (Kemp et al., 1990), consumption of O_2 by sediments (Seitzinger, 1990), and external inputs of N (Seitzinger, 1988, 2000;

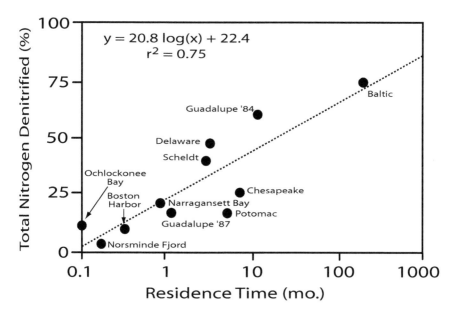

Figure 10.10 Fraction (%) of total N input from land and atmosphere that is denitrified in different estuaries as a function of residence time (months). (Modified from Nixon et al., 1996.)

Nixon et al., 1996). Inputs of N to estuaries from rivers are also reduced from denitrification in rivers; model estimates indicate that losses can range from 1 to 75% of external N inputs to rivers (Howarth et al., 1996; Alexander et al., 2000; Seitzinger et al., 2002a). Factors controlling denitrification in rivers include NO_3^- concentration, N loading, river length and depth, flow rate, water residence time, O_2 content, organic matter content of sediments, and seasonality. In estuaries, a survey of 14 systems revealed that denitrification removes 3 to 100% of the N loading with an average of 48% ± 39% (table 10.5; Seitzinger, 2000). Although there is no significant relationship between DIN loading and denitrification rate across these systems, the frequency distribution of these data indicated that N removal in about 50% of the estuaries ranged from 30 to 60%.

Seventeen genera of facultative anaerobic bacteria (e.g., *Pseudomonas* and *Bacillus*) can perform denitrification under anaerobic or low-oxygen conditions, where they use NO_3^- as an electron acceptor during anaerobic respiration (Jaffe, 2000). In fact, in many estuaries, denitrification is limited by the availability of NO_3^- (Koike and Sørensen, 1988; Cornwell et al., 1999). Sources of NO_3^- and NO_2^- for denitrification are from diffusive inputs from the overlying water column and nitrification in the sediments (Jenkins and Kemp, 1984). The activity of other bacterial processes under anoxic conditions has been shown to affect the activity of denitrifying bacteria. For example, SO_4^{2-} reduction occurs in anoxic sediments whereby SO_4^{2-} is reduced to sulfide (Morse et al., 1992)—more details are provided in chapter 12. Sulfide has been shown to be toxic to many bacteria including those involved in nitrification and denitrification (Joye and Hollibaugh, 1995).

There has been a long controversy in the literature concerning the lowest O_2 concentrations where denitrification can occur (Robertson and Kuenen, 1984). Some believe it can occur at aerobic–anaerobic interfaces (Christensen and Tiedje, 1988; Bonin et al., 1989), while others believe that low O_2 conditions inhibit synthesis of the essential enzymes for denitrification (Payne, 1976; Kapralek et al., 1982). It has also been speculated for some time that anoxic microsites may allow for the occurrence of denitrification in aerobic environments (Jannasch, 1960). Laboratory studies indicate that there are differential effects of O_2 on the different steps of denitrification, whereby the NO_3^- reduction step is less sensitive than NO_2^- or N_2O reduction (Bonin and Raymond, 1990).

In general, NO_3^- reduction leads to N_2, which reduces the total availability of N in a system (Howarth et al., 1988a,b). However, if we take another pathway of *dissimilatory NO_3^- reduction to NH_4^+* (DNRA), we find that N will be retained in a system, thereby increasing the total available N for organisms (Koike and Hattori, 1978; Jørgensen, 1989; Patrick et al., 1996). Although very little is known about the ecological consequences of DNRA (Sørensen, 1987; Cornwell et al., 1999), rates of DNRA can be as high as denitrification in shallow estuaries and tidal flats (Koike and Hattori, 1978; Jørgensen, 1989; Rysgaard et al., 1995; Bonin et al., 1998; Tobias et al., 2001). Recent work in Laguna Madre/Baffin Bay (USA) has shown that while sulfides can inhibit denitrification (figure 10.11), they may stimulate DNRA by providing an electron donor; this likely results in the retention of available N in estuaries (An and Gardner, 2002). Studies have reported the existence of chemolithotrophic bacteria that use sulfur compounds as an electron donor to convert NO_3^- into NH_4^+ (Schedel and Truper, 1980). Thus, the occurrence of a chemolithotrophic coupling between NO_3^- and sulfide may exist through a linkage of NO_3^--storing sulfur bacteria (e.g., *Thioplaca, Beggiatoa,* and *Thiomargarita*) in coastal sediments (Schulz et al., 1999; An and Gardner, 2002). The retention of N through DNRA plus N addition by BNF, in the shallow hypersaline conditions of Laguna Madre/Baffin Bay, has also been shown to contribute to the transition of N limitation to P limitation (Cotner et al., 2004).

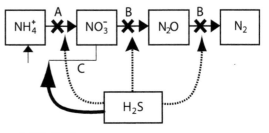

Figure 10.11 Relationship between sulfide and N cycling in sediments. Solid arrows indicate positive effects and dashed lines indicate negative effects. (Modified from An and Gardner, 2002.)

A: Nitrification
B: Denitrification
C: Dissimilatory nitrate reduction to ammonium

It is common for many temperate estuarine phytoplankton and macroalgae to be limited by N, despite high inputs from rivers and atmospheric deposition (Malone et al., 1988; Paerl et al., 1990, Peckol et al., 1994; Nixon, 1995; Tomasky and Valiela, 1995). One of the factors contributing to N limitation in estuaries is denitrification, which typically occurs via the consumption of NO_3^- from the water column (direct denitrification) as well as NO_3^- produced by nitrification in sediments (coupled nitrification–denitrification) (LaMontagne et al., 2002). Total nitrogen losses from denitrification in estuaries typically range from 25 to 40% (Benner et al., 1992; Nixon et al. 1996; Bronk, 2002). Increases in nutrient loading (Seitzinger, 1988) and sediment organic matter (Nowicki et al., 1999) have been shown to be important factors that enhance rates of denitrification. However, other factors such as sediment oxygen, carbon content, temperature, water depth, and bioturbation by the benthos (Kemp et al., 1990) have also been linked with controlling denitrification.

Assimilation of dissolved organic nitrogen (DON) involves the uptake and incorporation of organic forms of N, such as amino acids, into biomass by both heterotrophic and autotrophic organisms. The DON pool represents an important component of the N cycle in estuarine systems (Sharp, 1983; Jackson and Williams, 1985), with a diverse range of sources and sinks (Bronk and Ward, 2000; Bronk, 2002). However, much of the work to date has focused on DIN loading to estuaries. Increases in loading of DIN to estuaries have also been accompanied by increases in DON (Correll and Ford, 1982). In fact, rivers which are a critical source of N to estuaries can have over 80% of their total N represented as DON (Meybeck, 1982; Seitzinger and Sanders, 1997; Bronk, 2002). The major "sinks" or pathways of DON consumption are mediated by bacteria (Anita et al., 1991; Bronk, 2002), archeobacteria (Ouverney and Fuhrman, 2000), and to a lesser degree protists (Tranvik et al., 1993). Phytoplankton can also use some DON compounds (see below). The production of organic N *de novo* from DIN by bacteria as well as the fixation of atmospheric N_2 is evidence for the importance of particulate heterotrophic organic N (HON) in aquatic systems (Kirchman, 1994). Modeled controls on the HON, based on the ratio of respired allochthonous inputs (Alloc.:Resp.) to autochthonous production (AP), indicate that rivers and streams have the highest Alloc. Resp.:AP ratios (figure 10.12; Caraco and Cole, 2003). Consequently, the percentage of N derived from heterotrophic organic N (% HON) is relatively important in these systems compared to estuaries and oceanic systems. Estuaries have a broader range of per cent HON because of the diversity of source inputs. The HON of the Delaware River estuary is only 5% because of the dominance of AP, while Sapelo Island has values as high as 78%, because of the high Alloc.:Resp. supported by *Spartina* spp. detritus (Kirchman, 1994; Hoch and Kirchman, 1995).

After a significant fraction of DON is mineralized through the microbial food web, it then becomes available for phytoplankton uptake (Lewitus et al., 2000; Bronk, 2002; Seitzinger et al., 2002b). However, not all DON is bioavailable for microorganisms (Manny and Wetzel, 1973). For example, estuarine bacteria in the Delaware and Hudson River estuaries were able to assimilate 40 and 72% of the total DON, respectively (Seitzinger and Sanders, 1997). Bacteria have been shown to utilize both DIN and DON sources directly, with most of the DON uptake work focused on simple DCAA and DFAA (Gardner and Lee, 1975; Coffin, 1989; Fuhrman, 1990; Keil and Kirchman 1991a,b;

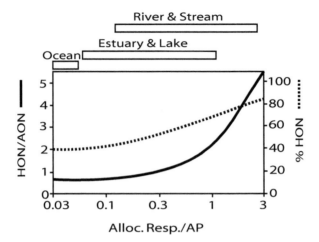

Figure 10.12 Modeled controls on the heterotrophic organic nitrogen (HON) and percentage of HON based on the ratio of respired allochthonous inputs (Alloc.:Resp.) in different aquatic systems. AON, allochthonous organic nitrogen; AP, allochthonous production. (Modified from Caraco and Cole, 2003.)

Jørgensen et al., 1993; Kirchman, 1994; Kroer et al., 1994; Middelboe et al., 1995; Gardner et al., 1998). It is well documented that urea is actively utilized by bacteria (Zobell and Feltham, 1935; Middelburg and Nieuwenhuize, 2000), and is even more extensively recycled by bacteria (Cho et al., 1996; Jørgensen et al., 1999). Other studies have also demonstrated that bacteria can utilize complex natural and anthropogenic DON sources (Carlsson et al., 1999; Stepanauskas et al., 2000; Wiegner and Seitzinger, 2001; Seitzinger et al., 2002c). Bioavailability of DON to bacteria has also been shown to be dependent on the sources (e.g., atmospheric, pasture soils, hardwood forest soils) (Timperley et al., 1985; Peierls and Paerl, 1997; Seitzinger and Sanders, 1999). For example, recent work has shown that urban/suburban storm water DON had the highest proportion of bioavailable DON compared to pasture and forest sources (figure 10.13; Seitzinger et al., 2002c). Similarly, atmospheric DON has been shown to selectively enhance growth rates of facultative heterotrophic and photoheterotrophic phytoplankton, such as dinoflagellates and cynaobacteria (Neilson and Lewin, 1974; Paerl, 1991). These results clearly demonstrate that if we are to effectively predict the impact of N inputs to estuaries and coastal systems, future N budgets will not only need to include DON but also to separate the bioavailable fractions as well.

Dissolved inorganic nitrogen sources, such as NO_3^- and NH_4^+, are typically the dominant forms of N utilized by phytoplankton and other algae. However, small DON sources (e.g., < 600 nominal molecular weight), and in some cases large molecules after further processing by exoenzymes, can also be utilized (Pantoja and Lee, 1994; Berg et al., 1997; Mulholland et al., 1998). In particular, as DIN becomes depleted during phytoplankton bloom succession, DON may actually be used during bloom formation (Butler et al., 1979; Berman and Chava, 1999; Anderson et al., 2002). For example, in

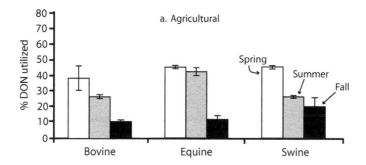

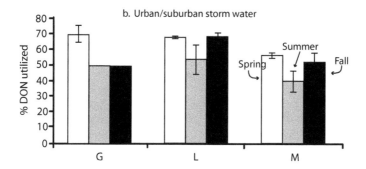

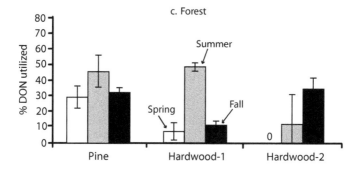

Figure 10.13 Proportion (%) of bioavailable DON from (a) agricultural; (b) urban/suburban storm water, (M and L were sites on streams with residential storm water flow, while G was located as a sewer grate receiving runoff from residential and business areas); and (c) pasture and forest sources, utilized by estuarine plankton over different seasons (e.g., spring, summer, and fall). (Modified from Seitzinger et al., 2002c.)

the Gulf of Riga (Baltic Sea), DON can represent as much as 85% of the total dissolved nitrogen (TDN) in spring and summer, and is a major source of N to phytoplankton and cyanobacteria (Berg et al., 2001). In particular, urea has been shown to be actively taken up by phytoplankton (Savidge and Hutley, 1977; Tamminen and Irmisch, 1996). Some of this enhanced phytoplankton production by bloom species was also shown to be linked with greater mineralization of N by heterotrophic bacteria, further supporting the contention that DOM stimulates heterotrophic bacteria resulting in greater production of bloom-forming phytoplankton (Berg et al., 1997; Maestrini et al., 1999). Inputs of DON from rainwater have also been shown to stimulate phytoplankton biomass and production (Timperley et al., 1985; Peierls and Paerl, 1997; Seitzinger and Sanders, 1999).

As discussed in chapter 8, photodegradation of DOM has been shown to release smaller organic and inorganic molecules (Kieber et al., 1990; Mopper et al., 1991; Miller and Moran, 1997). In particular, the production of DIN byproducts, amides, amines, and amino acids, through photodegradation of DON, was first demonstrated by Rao and Dhar (1934). Many other studies have followed in recent years showing the production of NH_4^+ (Bushaw et al., 1996; Gao and Zepp, 1998; Wang et al., 2000; Buffam and McGlathery, 2003), NO_3^- (Kieber et al., 1999; Buffam and McGlathery, 2003), and urea (Buffam and McGlathery, 2003). Photoproduction rates of NH_4^+ and DFAA are wide ranging, depending on the source of DON (table 10.7; Buffam and McGlathery, 2003). For example, humic-rich waters in the southeastern United States (e.g., Bushaw et al., 1996; Gao and Zepp, 1998) have been shown to release NH_4^+ consistently at rates as high as 0.3 μmol N L^{-1} h^{-1}. Concentrations of amino acids have also been shown to increase due to photoproduction within the range 0.03–9.5 nmol N L^{-1} h^{-1} (Tarr et al., 2001). This work also showed that numerous amino acids were photodegraded to NH_4^+ but did not provide a major source.

Sediment–Water Exchange of Dissolved Nitrogen

Estuarine and coastal sediments are important environments for the bacterial remineralization of nutrients (Rowe et al., 1975; Nixon, 1981; Blackburn and Henriksen, 1983; Boynton et al., 1982; Nowicki and Nixon, 1985; Sundbäck et al., 1991; Warnken et al., 2000) that also support neritic production, via fluxes across the sediment–water interface (Cowan and Boynton, 1996). Hence, the role of the depositional flux of organic matter to sediments has long been recognized as an important controlling factor on nutrient regeneration and fluxes (Nixon et al., 1976; Nixon, 1981; Kelly and Nixon, 1984; Billen et al., 1991; Blackburn, 1991; Twilley et al., 1999). The rates of remineralization and nutrient fluxes across the sediment–water interface have typically been shown to be highest with increasing water temperatures (Hargrave, 1969; Kemp and Boynton, 1984). Many other factors, such as redox status of the sediments and overlying water column, sorption/desorption processes, microbial respiration, and macromeiobenthic excretion, contribute to the variability of sediment–water exchange of nutrients (Henriksen et al., 1980; Kemp and Boynton, 1981; Kanneworff and Christensen, 1986; Cowan and Boynton, 1996).

When examining fluxes across the sediment–water interface across different estuarine systems, it becomes apparent that the role of DIN flux to the water column is important (table 10.8). Nixon (1981) showed that there was a good relationship between NH_4^+ flux

Table 10.7 Synthesis of results (photoproduction rates of NH_4^+ and DFAA) from this and other DOM photodegradation studies that emphasize nitrogen dynamics.

Sample water used	Light regime	NH_4^+ production rate (μmol N L^{-1} h^{-1})	Normalized NH_4^+ production rate[a] (μmol N L^{-1} m^{-1} h^{-1})	NH_4^+ production as fraction of initial DON (% h^{-1})	Rate of production of other measured amine groups (μmol N L^{-1} h^{-1})	Reference
Humic pond, swamp, river Waters, and isolated fulvics	Natural sunlight or simulator, 18 h Total = 860 W m^{-2}	0.05–0.34	0.0032–0.0042	0.25–0.92	Not measured	Bushaw et al. (1996)
Humic Satilla River water	Solar simulator, 4 h = mid-June sun at 34°N	0.1	0.011	0.25	Not measured	Gao and Zepp (1998)
Swedish clear-water lake, epilimnion and hypolimnion	Natural sunlight, 7 h Total = 345 W m^{-2}	No change	No change	No change	0.010–0.017[b] (DFAA) 0.057 + (DCSS, epilimnion only)	Jørgensen et al. (1998)
Concentrated humic acids from river and estuary	Natural sunlight, 7 h UV = 9.4 W m^{-2}	0.058–0.060	0.0010–0.0015	0.11–0.17	0.009–0.041 (DPA)	Bushaw-Newton and Moran (1999)
Humic lakes, streams, rivers in Sweden	UVA-340 lamps, 12 h 2.1 W m^{-2} UV-B 20 W m^{-2} UV-A 5 W m^{-2} PAR	No change	No change	No change	Not measured	Bertilsson et al. (1999)
Coastal waters from the Gulf of Riga	UVA-340 lamps, 12 h 0.3 W m^{-2} UV-B 20 W m^{-2} UV-A 6.1 W m^{-2} PAR	−0.03 – 0.0	Not reported	−0.16 – 0.0	−0.014 – 0.006 (DFAA) 0.0–0.17 (DCAA)	Jørgensen et al. (1999)

Continued

Table 10.7 (*continued*)

Sample water used	Light regime	NH$_4^+$ production rate (μmol N L^{-1} h^{-1})	Normalized NH$_4^+$ production rate[a] (μmol N L^{-1} m h^{-1})	NH$_4^+$ production as fraction of initial DON (% h^{-1})	Rate of production of other measured amine groups (μmol N L^{-1} h^{-1})	Reference
Humic bayou and river water and isolated riverine humics and fulvics	Solar simulator, 10 h Total = 600 or 765 W m^{-2}	0.11–1.9	0.002–0.056	0.8–2.6	Up to 0.011 (various individual amino acids)	Wang et al. (2000) Tarr et al. (2001)
Groundwater and DOM concentrated from estuarine water	Natural sunlight or solar simulator, 5–10 h	−0.31–0.13	−0.002–0.0015	−0.81–0.91	No change for most samples	Koopmans and Bronk (2002)
S. alterniflora leachate and samples from high-salinity lagoon system	UVA-340 lamps, 36 h 4 W m^{-2} UV-B 42 W m^{-2} UV-A 57 W m^{-2} PAR	0.006–0.032	0.0005–0.007	0.028–0.087	Up to 0.00094[c] (various individual amino acids)	Buffam and McGlathery (2003)
S. alterniflora leachate and samples from high-salinity lagoon system	UVB-314 lamps, 36 h 27 W m^{-2} UV-B 25 W m^{-2} UV-A 47 W m^{-2} PAR	0.001–0.046	0.0009–0.010	0.013–0.208	Up to 0.0011[c] (various individual amino acids)	Buffam and McGlathery (2003)

[a] Production rate normalized to absorptivity of water at 350 nm.
[b] Production rate estimated from graphical representation of data.
[c] DFAA measured in groundwater sample only.
Modified from Buffam and McGlathery (2003).

Table 10.8 System-wide average of exchange at the sediment–water interface in coastal marine environments.

Estuary	Note[a]	Nutrient Fluxes ($\mu mol\ m^{-2}\ h^{-1}$)						SOC ($g\ m^{-2}\ d^{-1}$)
		PO_4	NN	NO_3	NH_4	DIN	N_2	
Ochlockonee Bay	1		16.0		18.7	34.7	88.4	0.90
Apalachicola Bay	2		−30.3	−37.2	38.0	7.7		
Mobile Bay	3,4	3.9	16.1	14.2	62.8	76.9		0.55
Mississippi River Bight	5,6	17.5	−9.17	−15.8	126.3	117.3		0.84
Fourleague Bay	7,8	1.4	−54.7		141.7	86.9		1.03
Fourleague Bay	9	−8.0	−19		129	110		1.2
Trinity–San Jacinto estuary	10	0.6	−2.7		11.7	9.0	18.5	0.15
Gudalupe estuary	11	−3.1	176.2	183.1	−155.3	20.9		0.98
Gudalupe estuary	12		8–10		30–60		5–30	
Nueces estuary	11	−6.4	106.5	−16.8	54.5	29.1		0.73
Nueces estuary	12		0–10		8–50		4–59	
Laguna Madre–Upper	11	−0.2	0.0	0.1	1.6	1.5		1.68
San Francisco Bay	13	−4.2–54		0–33	17–208			0.35–0.70
Narragansett Bay	14	38–233	10		75–500		59	

Continued

329

Table 10.8 (continued)

Estuary	Note[a]	Nutrient Fluxes (μmol m^{-2} h^{-1})						SOC (g m^{-2} d^{-1})
		PO$_4$	NN	NO$_3$	NH$_4$	DIN	N$_2$	
Chesapeake Bay	15							
North Bay		−16.3		−117–9	−34–101			0.1–0.65
Mid Bay		0.0–148		−100–12	9–507			0.01–0.86
Lower Bay		−1.5–13		−8–19	15–181			0.3–0.75
Patuxent estuary	16	0–15		12.5–37.5	10–200		133	0.75–2.25
Neuse River estuary	17	−2.3–46		0–6.4	71–454			0.70–1.87
South River estuary	17	−8.3–23		0.0–5.8	0.0–267			0.71–2.72
Bay of Cadiz	18	21–379			258–1525			2.2–7.5

[a]Notes: 1, Seitzinger (1987); 2, Mortazavi, unpublished data; 3, Cowan et al. (1996); 4, Miller-Way, unpublished data; 5, Twilley and McKee (1996); 6, Bourgeois (1994); 7, Miller-Way (1994); 8, Twilley, unpublished data; 9, Teague et al. (1988); 10, Zimmerman and Benner (1994); 11, Montagna, unpublished data; 12, Yoon and Benner (1992); 13, Hammond et al. (1985); 14, Elderfield et al. (1981a,b); 15, Cowan and Boynton (1996); 16, Boynton et al. (1991); 17, Fisher et al. (1982); 18, Forja et al. (1994)
Modified from Cowan et al., 1996.

and organic matter production in estuaries. However, some of the more eutrophic temperate systems (e.g., Chesapeake and Patuxent) clearly show NH_4^+ fluxes to the water column higher than those of Gulf of Mexico estuaries, because of higher organic matter loading to sediments. In Mobile Bay (USA), moderate sediment oxygen consumption (SOC) and salinity stratification (not shown) allow for prolonged hypoxia/anoxia events (figure 10.14; Cowan et al., 1996). This results in almost exclusive release of NH_4^+ from the sediments, with only moderate fluxes of NO_2^- and NO_3^-, which alternate between release and uptake. Despite the continuous release of NH_4^+ from sediments throughout the year in Mobile Bay, the actual flux rates are relatively low to moderate compared to those of other estuaries (table 10.8), because the low loading of labile organic material to these sediments (Cowan et al., 1996). Efflux of NH_4^+ can be important to phytoplankton production in estuaries, particularly in shallow systems where benthic–pelagic coupling is enhanced. For example, 13 to 40% of the nutrient requirements for algal growth in the Chesapeake Bay (in August) were supported by NH_4^+ efflux to the water column (Boynton and Kemp, 1985). External loading of N may also result in higher fluxes of DON to the water column across the sediment–water interface. As mentioned earlier,

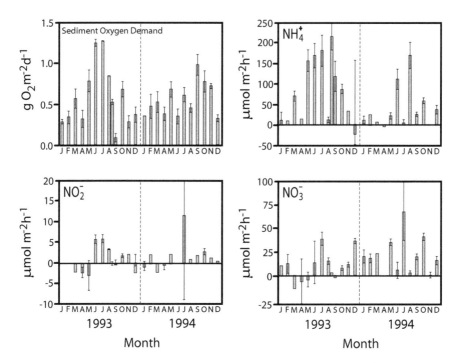

Figure 10.14 Monthly averaged concentrations of oxygen ($g\,O_2\,m^{-2}\,d^{-1}$) and nutrient (NH_4^+, NO_2^-, and NO_3^-) ($\mu mol\,m^{-2}\,h^{-1}$) fluxes at Station DR7 in Mobile Bay (USA) from January 1993 to December 1994. Positive and negative values represent fluxes out and into the sediments, respectively. (Modified from Cowan et al., 1996.)

urea is recognized as an important intermediate in the N cycle that is generally considered to be controlled by internal recycling of POM by bacteria (Jørgensen et al., 1999) and uptake by phytoplankton (Tamminen and Irmisch, 1996). A combination of both heterotrophic and autotrophic processes is believed to be responsible for an increase in urea in bottom water compared to surface waters of the Chesapeake Bay (Lomas et al., 2002). Fluxes of urea and primary amines to the water column across the sediment–water interface range from 140 to 1000 μmol m^{-2} d^{-1} (Lomstein et al., 1989; Giblin et al., 1997; Rysgaard et al., 1999) and can certainly result in significant concentration increases in bottom waters, especially in the absence of phytoplankton uptake due to light limitation (Lomas et al., 2002).

Shifts in the stoichiometry of regenerated nutrients in sediments from the expected Redfield ratio have been suggested as an important factor controlling the metabolism of shallow estuaries (Smith, 1991). For example, N:P ratios from Narragansett Bay were found to be <16 because of intense remineralization in sediments, which yields lower inorganic N relative to P (Nixon and Pilson, 1983). Lower N:P ratios primarily occur because of N losses (e.g., N_2 and N_2O) from denitrification (Seitzinger et al., 1980; Nixon, 1981; Kemp et al., 1990; Nixon et al., 1996). The O:N ratio of fluxes across the sediment–water interface should be 13 if NH_4^+ is the end-product of ammonification; however, if oxidation results in the formation of NO_3^- (ammonification coupled with nitrification) the ratio should be 17 (Twilley et al., 1999). Very high O:N ratios (e.g., >100) have been observed and may be from a coupling of nitrification and denitrification due to conversion of regenerated NH_4^+ to N_2, where N is lost from the system (Boynton and Kemp, 1985). Denitrification rates in four Gulf of Mexico estuaries along the Texas coast ranged from 4.5 to 9.0 g m^{-2} y^{-1}, which represented an average loss of 14 to 136% of the N input to these systems (Zimmerman and Benner, 1994). The upper range of these loss rates is considerably higher than the average N loss rates found for other estuaries (40–50%) (Seitzinger, 1988). The reason for such high loss rates along the Texas coast is due to low inputs of N from rivers, which causes more denitrification to be fueled by nitrification rather than $NO_2^- + NO_3^-$ uptake in sediments (Zimmerman and Benner, 1994; Twilley et al., 1999).

The major pathways of the N cycle in sediments are strongly influenced by the redox conditions in bottom waters and sediments (figure 10.15). Both diffusive and advective processes strongly control the distribution of O and N compounds which ultimately affect the coupling between nitrification and denitrification (Jørgensen and Boudreau, 2001). For example, anaerobic degradation of N-containing organic matter will contribute to the formation of DON and NH_4^+ (ammonification), which can then either efflux to the water column or be oxidized to NO_2^- and NO_3^- (nitrification). Depending on the diffusive gradient, NO_3^- may also move downward where it can support denitrification just below the oxic–anoxic interface in sediments (figure 10.15; Henrikson and Kemp, 1988; Kristensen, 1988; Rysgaard et al., 1994). Experimental loading of N in sediments showed that moderate loading increased N removal via denitrification but decreased rates of denitrification at high N loading (Sloth et al., 1995). The incorporation and/or loss of NO_3^- from sediments will also be dependent on the presence and type of bioturbation (Aller, 2001). Due to the fine-scale differences in the spacing of nitrification and denitrification in sediments, the relative importance of benthic structure (e.g., size and spacing of burrows)

a

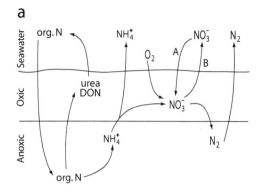

b

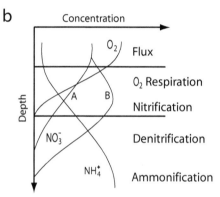

Figure 10.15 Major pathways of
the N cycle in sediments (a), as a
function of redox conditions in
bottom waters and sediments (b).
Both diffusive and advective
processes strongly control the
distribution of O and N compounds
which ultimately affect the
coupling between nitrification and
denitrification. (Modified from
Jørgensen and Boudreau, 2001.)

is critical. Ammonium also accounts for most of the DIN secreted by invertebrates, as
an end-product of protein catabolism (Le Borgne, 1986). Consequently, in addition to
production of remineralized N through the microbial cycle it is important to consider
the effects of metazoan-derived NH_4^+. For example, benthic macroinvertebrate excretion estimates in sediments near the Mississippi River plume ranged from 7 to 18 μmol
$NH_4^+ m^{-2} h^{-1}$, which accounted for as much as 50% of the net flux of NH_4^+ in these
sediments (Gardner et al., 1993). These NH_4^+ excretion rates are in agreement with those
found for other benthic invertebrates in Danish coastal waters (Blackburn and Henriksen,
1983) and in Buzzards Bay (USA) (Florek and Rowe, 1983).

Micro- and macrophytobenthos act to regulate nutrient fluxes across the sediment–
water interface and also serve to retain N in estuaries, unlike denitrification (Nixon et al.,
1996; An and Joye, 2000; Bronk, 2002) and anammox (Thamdrup and Dalsgaard, 2002),
both of which contribute to N losses in estuaries. Consequently, benthic micro- and
macroalgae are considered to be important regulators of benthic–pelagic coupling of DIN
(Sundbäck and Graneli, 1988; Caffrey and Kemp, 1990; McGlathery et al., 1997; Tyler
et al., 2003) and DON (Tyler et al., 2003) in estuaries. For example, in the shallow coastal

lagoon of Hog Island Bay (USA), where dense macroalgal mats commonly occur, sediments tend to be autotrophic with a net uptake of N (e.g., DIN, urea, DFAA, and DCAA) (Tyler et al., 2003). Conversely, algal-depleted sediments are net heterotrophic resulting in the release of DIN, particularly in the form of NH_4^+. Macroalgae in the Childs River estuary (USA) retain N by reducing denitrification, which can then create a positive feedback by coupling enhanced N availability and microalgal growth (LaMontagne et al., 2002). The coupling of nitrification–denitrification has also been shown to be, in part, regulated by microphytobenthos (Risgaard-Petersen, 2003; Sundback et al., 2004). Nitrification–denitrification coupling in sediments occurs from the transport of nitrification products, such as NO_3^- and NO_2^-, across the redox boundary (Jenkins and Kemp, 1984). Production of O_2 by microphytobenthos can enhance coupled nitrification–denitrification by stimulating nitrification (An and Joye, 2001). However, if N levels are too low, coupled denitrification is reduced because of competition for N between microphytobenthos and nitrifying bacteria (Rysgaard et al., 1995; Risgaard-Petersen, 2003). The problem of N limitation commonly occurs in sandy sediments, where N uptake by microphytobenthos is considerably higher than denitrification (Sundbäck and Miles, 2000). In contrast, N limitation is less of a problem in muddy sediments because microphytobenthic uptake of N is equal to or less than denitrification (Dong et al., 2000). While much of this work has been conducted on littoral sediments, recent work has also demonstrated a similar effect of microphytobenthos on N cycling in sublittoral sediments (Sundbäck et al., 2004).

Ammonium produced in sediments under anaerobic conditions generally equilibrates between porewaters and sediments, and diffuses to surface sediments (Rosenfield, 1979; Krom and Berner, 1980b). In fact, it is common to estimate NH_4^+ production using kinetic diffusive-reaction models (e.g., Berner, 1980; Aller, 2001), but only in the absence of O_2 (Ullman and Aller, 1989). This is because in the presence of O_2, NH_4^+ can be oxidized to NO_3^-, which may then be completely or partially denitrified to N_2 (Seitzinger et al., 1984). Although NH_4^+ has been documented to be one of the more important components of DIN flux across the sediment–water interface in marine systems (Blackburn and Henriksen, 1983; Nowicki and Nixon, 1985; Hopkinson et al., 2001; Laursen and Seitzinger, 2002), it remains a small component in freshwater systems (Gardner et al., 1987, 1991). It has been postulated that this difference in flux rate is based on chemical differences in the controls of NH_4^+ diffusion in these two types of sediments (figure 10.16; Gardner et al., 1991). This conceptual model by Gardner et al (1991) suggests that while adsorption or cation-exchange sites will reduce the mobility of NH_4^+ in both sediment types, seas salts in estuaries should in effect neutralize the cation-exchange sites, thereby enhancing NH_4^+ diffusion and ultimately flux rates. This "salt effect" was tested experimentally and showed that changing ionic composition in waters above intact freshwater and estuarine cores did enhance NH_4^+ flux by 30% (Gardner et al., 1991). In a similar study, it was further demonstrated that the higher exchangeable NH_4^+ concentrations were 3 to 6.5 times greater in sediments incubated in freshwater versus sea water (Seitzinger et al., 1991). This is because the concentrations of cations, which compete with NH_4^+ for exchange sites in porewaters (Berner, 1980; Simon and Kennedy, 1987), are considerably less in freshwater than in sea water. However, in high organic matter sediments, NH_4^+ adsorption constants may be more controlled by "clay–humic complexes" (Boatman and Murray, 1982). The higher exchangeable NH_4^+ concentrations in freshwater is also accompanied by a greater fraction of NH_4^+ that is nitrified, and subsequently denitrified,

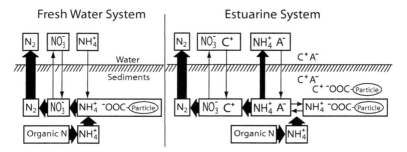

Figure 10.16 Conceptual model by Gardner et al. (1991) which suggests that while adsorption or cation-exchange sites will reduce the mobility of NH_4^+ in both sediment types, sea salts in estuaries should in effect neutralize the cation-exchange sites, thereby enhancing NH_4^+ diffusion and ultimately flux rates. (Modified from Gardner et al., 1991.)

contributing to the lower flux of NH_4^+ in freshwaters compared to marine and estuarine systems (Seitzinger et al., 1991).

Isotopic fractionation of O_2 and NO_3^- diffusing into sediments may prove useful in better understanding the role of diagenetic processes in sediments versus water column N cycling. For example, bacterially mediated denitrification displays a strong discrimination against ^{15}N and ^{18}O containing NO_3^- molecules in favor of ^{14}N (Cline and Kaplan, 1975; Mariotti et al., 1981) and ^{16}O (Lehmann et al., 2003). Thus, in suboxic waters isotopic discrimination during denitrification will result in ^{15}N- and ^{18}O-enriched NO_3^- compared to "normal" oxic waters (Cline and Kaplan, 1975). However, while fractionation of downward-diffusing NO_3^- during denitrification in coastal sediments appeared to be minimal, the isotopic signature of outward-diffusing NH_4^+ was enriched in ^{15}N by 4.5‰, compared to ambient organic matter and the NO_3^- in overlying waters (Brandes and Devol, 1997; Sebilo et al., 2003; Lehmann et al., 2004). This discrepancy in the isotopic signature of NH_4^+ and organic matter was attributed to fractionation of NH_4^+ during nitrification in the oxygenated zone. Recent isotopic work in the Sheldt estuary (The Netherlands) indicated that the $\delta^{15}N$ in NO_3^- in the upper regions of the estuary were enriched (18–20‰) and were relatively more depleted (2–5‰) in the lower regions (figure 10.17; Middelburg and Nieuwenhuize, 2001). The balance between nitrification and denitrification processes changes across the Sheldt estuary (Soetaert and Herman, 1995), making interpretation of isotopic data difficult. However, the more depleted $\delta^{15}NO_3^-$ values in the lower estuary likely reflect the effects of enhanced nitrification. Preferential utilization of $^{14}NH_4^+$ by nitrifying bacteria will result in outward flux of depleted $^{15}NO_3^-$ (Mariotti et al., 1984). Finally, other studies have shown that NH_4^+ produced from DNRA in sediments is strongly depleted in ^{15}N (McCready et al., 1983), allowing for further discrimination of N processes in sediments. Results from these studies underscore how understanding the isotopic signature of N in different forms of DIN can help understand the role of nitrification and denitrification processes, pre- and postdepositional, on N cycling in estuarine and coastal systems.

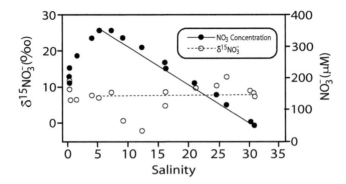

Figure 10.17 Relationship between NO_3^- (μM) and $\delta^{15}NO_3^-$ (‰) across a salinity gradient in the Sheldt estuary (The Netherlands). (Modified from Middelburg and Nieuwenhuize, 2001.)

Nitrogen Budgets for Selected Estuaries

Nutrient budgets provide a quantitative approach for comparing estuaries with both similar and divergent characteristics. They also provide a means of constraining many of the sources and sinks discussed earlier in this chapter, which creates a capacity to understand and predict the fate and transport of nutrient inputs to a system. The first attempt in constructing a mass balance for N in coastal systems was about a century ago in the North Sea (Johnstone, 1908). Since that time there have been many published estimates of nutrient budgets for temperate (e.g., Billen et al., 1985, 2001; Wulff and Stigebrandt, 1989; Boynton et al., 1995; Nixon et al., 1995, 1996; Kemp et al., 1997; Smith and Hollibaugh, 1997; Savchuk, 2000, 2002; Prego, 2002) and tropical and subtropical (Smith, 1984; Eyre, 1995; Eyre et al., 1999; McKee et al., 2000; Mortazavi et al., 2000; Brock, 2001; Eyre and McKee, 2002; Ferguson et al., 2004) estuaries. Empirical or first-order budgets that attempt to describe seasonal changes in nutrient inputs, based on direct measurements over different spatial and temporal scales, have been useful in describing the relative importance of external nutrient loading (e.g., Boynton et al., 1995; Nixon et al., 1996). However, this approach typically has problems when estimating the exchange of nutrients between estuaries and the coastal ocean (Nixon, 1987). These problems result from the simple approach of using the difference between inputs and retention of a particular constituent within an estuary to estimate exchange rates. High spatial and temporal variability of direct measurements along estuarine–ocean boundaries, along with the relatively small net exchange of constituents relative to maximum ebb and flow currents make first-order budgets problematic (Jay et al., 1997; Ferguson et al., 2004). Net inputs or losses of nutrients within estuaries have commonly been assessed with the use of salinity mixing diagrams to determine nonconservative behavior of nutrients (e.g., Officer, 1979; Smith, 1987).

Much of the research involving such modeling efforts has been in North America and Europe, with the general paradigm being high nutrient inputs from rivers in the spring accompanied by high sorption to particles in the oligohaline region and high phytoplankton production in the mesohaline region (Kemp and Boynton, 1984; Malone et al., 1996; Nixon et al., 1996). This is then followed by recycling of particle-bound nutrients in summer months showing a distinct seasonal gradient linked with physical mixing and light regimes (Kemp and Boynton, 1984; Malone et al., 1996). In comparison, subtropical and tropical estuaries, which are relatively understudied systems, have very different controlling parameters. For example, many of the Australian estuaries receive large episodic inputs of freshwater, which can then be followed by minimal input and variability over long periods (Eyre and Twigg, 1997). Furthermore, these systems are typically very shallow and are not limited by light throughout the year (Ferguson et al., 2004).

In this section, We begin with a comparison of N budgets compiled from both temperate and tropical estuaries from a paper published by Nixon et al. (1996). A more recent study with updated budgets for these same estuaries (table 10.9) was published by Dettmann (2001). We will then introduce some comparisons with additional papers on N budgets from both temperate and tropical/subtropical regions in an attempt to identify similarities and differences in the biogeochemical processes controlling N cycling in these estuaries.

In general, when examining the nitrogen budgets of temperate estuaries such as Chesapeake Bay, Narragansett Bay, and the Baltic Sea, N inputs are largely from upstream sewage and fertilizer sources (Nixon et al., 1996). There appears to be a net export of total nitrogen (TN) from the bay (figure 10.18; Boynton et al., 1995). These results agree with earlier estimates of TN export (~30%) to the coastal ocean (Nixon, 1987). These results clearly show that inorganic nutrient inputs from terrestrial and atmospheric sources are processed very rapidly in the Chesapeake Bay reflective of the dynamic cycling of elements in estuaries. Other temperate systems with long-term water quality records have proven to be useful in developing models that examine the response to human activities in the watershed. For example, N inputs to the Seine River estuary (France) over the last 50 years indicates that urban sources of N have essentially remained the same but that inputs from agricultural sources have increased 5-fold due to more extensive agricultural practices (figure 10.19; Billen et al., 2001). Although N inputs from wastewater to the Seine did increase during the rapid expansion of Paris, from 1950 to 1980, wastewater purification kept pace with such increases, essentially neutralizing these inputs. However, increases in the gross flux of N from the watershed to the Seine is currently much greater compared to the 1950s due to hydrologic changes in the watershed. This has resulted in less denitrification and riparian retention in the watershed. For example, as much as 70% of the N was removed in riparian wetlands in the watershed during the 1950s. Changes in natural and human-induced hydrological variation in the watershed have resulted in greater export of N to the estuary from diffuse sources compared to the 1950s.

Nixon (1982, 1992) found linkages between total fisheries yield and primary production in marine and estuarine waters. Although the response of primary production to biogeochemical parameters (e.g., nutrients, light) in estuaries has been well established, the response of higher trophic levels to primary production is not as well linked (Kremer et al., 2000). Fisheries yields may in some cases respond better to top-down controls. Network analysis showed very detailed interactions between benthic and pelagic trophic pathways in Chesapeake Bay, showing benthic food resources as important linkages in

Table 10.9 Nitrogen budget (mmol m^{-2} y^{-1}) for seven North American and four European estuaries.

Estuary export	Upland N inputs					N Storage	N Losses					Net
	Atmosphere	N fixation	Rivers and wastewater	Other sources	Total inputs	Water column	Denitrification	Sedimentation	Fish landings	Other	Total losses	
Boston Harbor	203[a]	0[i]	7554[a,b]	134[a,c]	7891	0[i]	683[d]	187[e]	small[e]	150[j]	1020	6871
Narragansett Bay	91[g]	nd	1775[g]	122[g]	1988	0[g]	384[h]	215[i]	138[g,j]	na	612	1376
Delaware Estuary	see total	see total	see total	see total	2100[k]	0[j]	825[l]	255[j]	small[j]	na	1080	1020
Chesapeake Bay	113[m]	nd	826[m]	nd	939	0[j]	245[m]	326[m]	83[m]	na	654	285
Potomac Estuary	113[m]	nd	1981[m]	nd	2094	0[j]	331[n]	836[m]	96[m]	na	1263	831
Ochlockonee Bay	82[o]	<3.5[p]	5910[i,p]	nd	5995	0	639[p]	115[q]	small[j]	na	754	5241
Guadalupe Estuary												
1984	71[r]	42[r]	407[r]	nd	520	24[r]	281[r]	1[r]	30[r]	na	312	184
1987	69[r]	42[r]	1743[r]	nd	1854	29[r]	281[r]	59[r]	74[r]	na	414	1411
Baltic Sea 20[t]	60[s]	26[s]	126[s]	na	212	14[t]	162[t]	11[s]	5[s]	na	178	
Norsminde Fjord 11,427[w]	77[u]	nd	11,795[u]	na	11,872	115[u]	207[u]	see other[v]	see other[v]	123[v]	330	
Westerschelde												
1970s 7,219[x]	nd	nd	8289[x]	5615[x]	13,904	nd	5615[x]	1070[x]	nd	na	6,685	
1980s 13637[y]	nd	nd	9,26[y]	8022[y]	17,648	nd	2941[y]	1070[y]	nd	na	4,011	

North Adriatic Sea											
537^z	46^z	8^z	861^z	na	915	nd	287^z	77^z	14^z	na	378

[a] Alber and Chan (1994).
[b] Includes rivers, sewage effluent, CSOs, and sewage sludge.
[c] Runoff and groundwater.
[d] Nowicki et al. (1997).
[e] Kelly and Nowicki (1992).
[f] Losses to dredging.
[g] Nixon et al. (1995).
[h] Nowicki (1994).
[i] Value 215 is the midpoint of the range 134–296 given in the summary by Nixon (1995).
[j] Nixon et al. (1996).
[k] Jaworski, pers. comm.
[l] The value 825 is at the lower end of the range 825–1,025 given by Nixon et al. (1996). This value was derived from data on sediment type (Biggs and Beasley, 1988), assuming that denitrification occurs only in sediment that is not gravel or gravelly sand.
[m] Boynton et al. (1995).
[n] Hendry and Brezonik (1980).
[p] Seitzinger (1987).
[q] Value 115 is middle of range given by Nixon et al. (1996).
[r] Brock et al. unpublished manuscript; Brock, pers. comm.
[s] Larsson et al. (1985).
[t] Wulff and Stigebrandt (1989). Denitrification calculated by difference.
[u] Nielsen et al. (1995).
[v] Total other losses (sedimentation and fisheries), calculated by difference (Nielsen et al. 1995).
[w] Directly determined (Nielsen et al. 1995).
[x] Billen (1985). Denitrification calculated by difference.
[y] Soetaert and Herman (1995).
[z] Degobbis et al. (1986). Denitrification calculated by difference.
For Delaware Estuary, inputs from all upland sources are aggregated as total inputs. Entries indicated as not determined (nd) or small are assumed to be zero in calculations; na = not applicable. Net export calculated by difference (net export = total inputs − N storage in water column − total losses). Modified from Dettmann (2001).

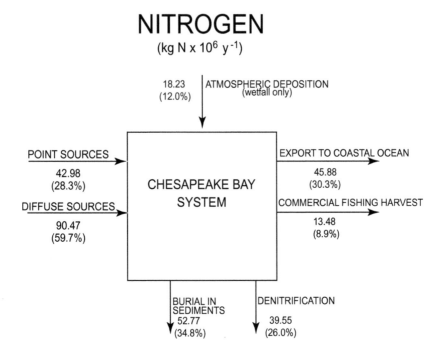

Figure 10.18 Total nitrogen budget (kg N × 10^6 y^{-1}) in the Chesapeake Bay system. (From Boynton et al., 1995, with permission.)

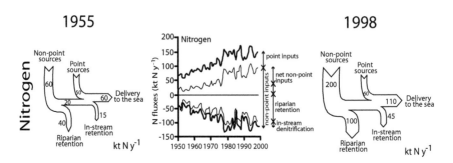

Figure 10.19 Nitrogen inputs (kt N y^{-1}) to the Seine River estuary (France) over the last 50 years. (Modified from Billen et al., 2001.)

fisheries production (Baird and Ulanowicz, 1989). Other work has shown that the trophic transfer efficiency from primary production to fisheries is more efficient in the Baltic Sea than in Chesapeake Bay (Ulanowicz and Wulff, 1991). Approximately 9% of the total N inputs to the Chesapeake Bay are taken up by fisheries (primarily menhaden) (figure 10.18; Boynton et al., 1995). This is likely an underestimate of N uptake by fisheries since migration of fish species is not considered here (Kremer et al., 2000). For example, 5 to 10% of the C from primary production is removed from inshore waters by offshore migration of menhaden in the Gulf of Mexico (Deegan, 1993).

In tropical/subtropical estuaries, such as the Brunswick Estuary (Australia), a simple model which used freshwater residence times and nutrient availability was found to effectively predict phytoplankton biomass (Ferguson et al., 2004). Based on budget modeling scenarios for different flow periods throughout the year, a conceptual model of N cycling in the Brunswick estuary was developed (figure 10.20). The four phases of nutrient cycling depicted here are: high flow, enhanced recycling, autumn wet season (medium flow), and low flow. During high flow (figure 10.20a), the residence time can be less than 1 day, resulting in the loss of particulate and dissolved materials from the estuary to local coastal waters. The only retention of N in the systems occurs in the form of total particulate nitrogen (TPN), which occurs in regions of low salinity with relatively large

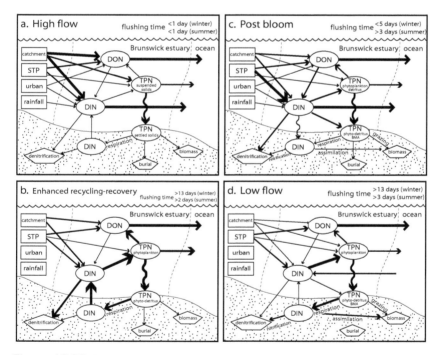

Figure 10.20 Conceptual model N cycling in the Brunswick estuary (Australia) based on budget modeling scenarios for different flow periods throughout the year. (Modified from Ferguson et al., 2004.)

particulate materials. Sorption of N on these particles is also likely important here, similar to that found in northern temperate systems (Kemp and Boynton, 1984). During the recycling phase (figure 10.20b), phytoplankton become more important in the uptake of DIN, whereby PN becomes an important pool with some of this recycled as DON. The high inputs of phytodetritus provide a labile pulse of organic matter which results in high NH_4^+ production in sediments, which then effluxes back to the water column providing nutrients for further primary production. During this phase a significant fraction of DON can be exported to coastal waters. During the wet season (median flow) (figure 10.20c), there is no significant bloom development as most of the N is exported to the coastal ocean. Also, denitrification and assimilation of DIN by bacteria becomes a significant mechanism for removal of N. Finally, in the low-flow period (figure 10.20d), DIN uptake by algal blooms and sediments accounts for most of the uptake of DIN loading, which is predominantly from sewage treatment plants (STP). Low flow also reduces suspended load which results in greater light availability and high benthic microalgal (BMA) uptake of N; this competition between BMA and nitrifying bacteria causes NO_3^- limitation, reducing denitrification. It should be noted that while these scenarios can occur at virtually any time of the year in tropical systems, nutrient supply and freshwater residence times appear to be the most important controlling factors of these scenarios (Eyre, 2000; Ferguson et al., 2004).

In another subtropical estuary in Australia (Moreton Bay), nitrogen inputs were dominated by BNF (9177 t y^{-1}) which further emphasizes differences from temperate systems, where N fixation is considered to be relatively insignificant (e.g., Boynton et al., 1995; Nixon et al., 1995; table 10.10). The second most important source of N in this system was atmospheric (1692 t y^{-1}); however, it represented only 11% of the total N inputs which is comparatively low in the more polluted regions of many temperate systems (Paerl, 1995). However, this area of southeast Queensland is one of the top five fastest regions of urban growth in the world so N loading will continue to increase in accordance with the strong positive relationship between population growth and N loading to estuaries (Caraco, 1995). About 41% of the N entering Moreton Bay is exported to the ocean with about 56% being lost through denitrification (Eyre and McKee, 2002; table 10.10). Thus, the most dramatic difference between Moreton Bay and many temperate systems is the dominance of biological processes (e.g., BNF and denitrification) compared to physical controls (e.g., river discharge, residence time) on inputs and losses in temperate systems (Nixon et al., 1996). The high denitrification rates in Moreton Bay may also be responsible for the considerably higher dependence of recycled N, compared to Chesapeake Bay, which has comparatively lower denitrification rates and utilization of recycled N (Eyre and McKee, 2002). Other subtropical estuaries along the Gulf of Mexico, Apalachicola Bay (USA) (Mortazavi et al., 2000) and Nueces Bay (USA) (Brock, 2001) estuaries, have denitrification as a major loss term for N. For example, 25 to 40% of the N losses in the Nueces estuary are from denitrification processes (Brock, 2001). However, these generalizations must be made with caution. For example, denitrification can be a major factor controlling N losses in temperate systems as a function of residence time (Nixon et al., 1996). Nitrogen fixation in temperate systems like the Baltic Sea may also represent a more important component of the total N budget than previously thought (Larsson et al., 2001).

Table 10.10 Nutrient budget for Moreton Bay (Australia) (t y^{-1}).

Budget components	Nitrogen
Standing stocks	
Sediment (solid phase)	68,081
Sediment (pore water)	92
Water column (total)	777
Biomass	33,695
Inputs	
Point sources	3383
Non-point sources	571
Atmosphere	1692
Groundwater	120
Primary production	—
N fixation	9177
Total	14,883
Outputs	
Denitrification	−8152
Pelagic respiration	—
Benthic respiration	—
Dredging	−187
Burial	−31
Fisheries harvest	−147
Pumice stone passage	−160
Ocean exchange	−6206
Recycling	
Benthic fluxes	5841
Biological uptake	73,000
Phytoplankton	21,469
Sedimentation	

Modified from Eyre and McKee (2002).

Summary

1. Elemental nitrogen (N$_2$) makes up 80% of the atmosphere (by volume) and represents the dominant form of atmospheric nitrogen gas. Despite its high atmospheric abundance, N$_2$ is generally nonreactive, due to strong triple bonding between the N atoms, making much of this N$_2$ pool unavailable to organisms.
2. Nitrogen sources to estuaries range from a diverse group of both diffuse non-point agricultural, urban, and rural point sources (e.g., wastewater, industrial discharges, stormwater, and overflow discharges) across a broad spectrum of watersheds (e.g., urban, agricultural, upland and lowland forests).

3. Nitrogen inputs to watersheds, which results in DIN export to coastal systems, often leads to enhanced primary production, since many estuaries are N limited. This can result in the formation of HABs as well as hypoxia, and in worst case scenarios anoxic water columns; see other effects in chapter 15.

4. Dissolved organic nitrogen can represent a significant fraction of the TDN in rivers and estuaries.

5. Sources of autochthonous DON to estuaries may be derived from actively growing phytoplankton communities, where small molecules such as DFAA are released, which are highly available to bacteria. Allochthonous DON sources from terrestrial runoff, plant detritus leaching, soil leaching, sediments, and atmospheric deposition may also represent important inputs to estuaries.

6. The major processes involved in the biogeochemical cycling of N in estuaries and the coastal ocean are: (1) BNF; (2) ammonia assimilation; (3) nitrification; (4) assimilatory NO_3^- reduction; (5) ammonification or N remineralization; (6) anammox; (7) denitrification and dissimilatory NO_3^- reduction to NH_4^+; and (8) assimilation of DON.

7. Biological N_2 fixation is a process performed by prokaryotic (both heterotrophic and phototrophic) organisms (sometimes occurring symbiotically with other organisms) resulting in the enzyme-catalyzed reduction of N_2 to NH_3 or NH_4^+, or organic nitrogen compounds.

8. Ammonia assimilation is where NH_3 or NH_4^+ is taken up and incorporated into organisms as organic nitrogen molecules.

9. Nitrification is where NH_3 or NH_4^+ is oxidized to NO_2^- or NO_3^-.

10. Assimilatory NO_3^- reduction to NH_4^+ involves the simultaneous reduction of NO_3^- and uptake of N by the organism into biomass.

11. Ammonification or N remineralization is the process where NH_3 or NH_4^+ are produced during the breakdown of organic nitrogen by organisms. In the case of the most dominant nitrogen-containing organic compounds, proteins, peptide linkages are broken down followed by deamination of amino acids to produce NH_3 or NH_4^+.

12. Anammox, which involves the anaerobic NH_4^+ oxidation with NO_3^-, may be another mechanism whereby N is lost form estuarine/coastal systems as N_2.

13. Denitrification, also known as dissimilatory NO_3^- reduction, is the reduction of NO_3^- to any gaseous form of N such as NO, N_2O, and N_2 by microorganisms.

14. Assimilation of DON involves the uptake and incorporation of organic forms of N, such as amino acids, into biomass by both heterotrophic and autotrophic organisms.

15. A significant fraction of DON is mineralized through the microbial food web, and becomes available for phytoplankton uptake.

16. Estuarine and coastal sediments are important environments for the bacterial remineralization of nutrients that also support neritic production, via fluxes across the sediment–water interface.

17. Shifts in the stoichiometry of regenerated nutrients in sediments from the expected Redfield ratio has been suggested as an important factor controlling the metabolism of shallow estuaries.

18. Micro- and macrophytobenthos act to regulate fluxes of nutrients across the sediment–water interface and also serve to retain N in estuaries, unlike denitrification, which regulates flux and does not retain N.

19. Isotopic fractionation of O_2 and NO_3^- diffusing into sediments may prove useful in better understanding the role of diagenetic processes in sediments versus water column N cycling. For example, bacterially mediated denitrification displays a strong discrimination against ^{15}N- and ^{18}O-containing NO_3^- molecules in favor of ^{14}N and ^{16}O.

20. Nutrient budgets provide a quantitative approach for comparing estuaries with both similar and divergent characteristics. They also provide a means of constraining many of the sources and sinks, which creates a capacity to understand and predict the fate and transport of nutrient inputs to a system.

21. Much of the research involving modeling efforts have been in North America and Europe with the general paradigm being high nutrient inputs from rivers in the spring accompanied by high sorption to particles in the oligohaline region and high phytoplankton production in the mesohaline region. This is then followed by recycling of particle-bound nutrients in summer months showing a distinct seasonal gradient linked with physical mixing and light regimes. In comparison, subtropical and tropical estuaries, which are relatively understudied systems, have very different controlling parameters.

Chapter 11

Phosphorus and Silica Cycles

Sources of Phosphorus to Estuaries

Phosphorus (P) is one of the most well-studied nutrients in aquatic ecosystems because of its role in limiting primary production on ecological and geological timescales (van Capellen and Berner, 1989; Holland, 1994; Tyrell, 1999; van Cappellen and Ingall, 1996). Other key linkages to biological systems include the role of P as an essential constituent of genetic material (RNA and DNA) and cellular membranes (phospholipids), as well as in energy-transforming molecules (e.g., ATP, etc.). Consequently, marine P has received considerable attention in recent decades, with particular emphasis on source and sink terms in budgets (Froelich et al., 1982; Meybeck, 1982; Ruttenberg, 1993; Sutula et al., 2004). Excessive loading of N to estuarine waters can result in P limitation in systems that are generally considered to be N limited. In such cases where primary production is limited by P, N:P ratios are expected to exceed the Redfield value of 16:1 but can be replenished by sediment efflux of P due to redox changes. For example, after the initial N loading of a system there will be an increase in primary production, which can cause the system to become P limited. Then, the phytodetritus from these early stages of N loading can be remineralized in sediments resulting in anoxic conditions in surface sediments, which can then enhance P release from sediments to the overlying waters where primary production is once again enhanced. Evidence for the role of sediment-derived P on primary production in estuaries with high N loading has been shown to occur particularly in shallow water systems (Timmons and Price, 1996; Cerco and Seitzinger, 1997). On the other hand, many coastal areas have also been subjected to high P loading from anthropogenic sources, where in some cases inputs of P are 10 to 100 times greater than in preindustrial times (Caraco et al., 1993). In many cases, P and N loading to estuarine systems will occur simultaneously and decoupling or isolating their individual effects can be difficult (e.g., HELCOM, 2001).

The cycling and availability of P in estuaries is largely dependent upon P speciation. Consequently, total P (TP) has traditionally been divided into *total dissolved P* (TDP) and *total particulate P* (TPP) fractions (Juday et al., 1927), which can further be divided

346

Table 11.1 Dissociation constants of phosphoric acid at 25°C.

	Distilled water[a] (pK)	Seawater[b] (pK)
$H_3PO_4 \leftrightarrow H^- + H_2PO_4^-$	2.2	1.6
$H_2PO_4^- \leftrightarrow H^+ + HPO_4^{2-}$	7.2	6.1
$HPO_4^{2-} \leftrightarrow H^+ + PO_4^{3-}$	12.3	8.6

[a] Stumm and Morgan (1981).
[b] Atlas (1975).

into *dissolved* and *particulate organic P* (DOP and POP) and *dissolved* and *particulate inorganic P* (DIP and PIP) pools. Another defined fraction within the TP pool is *reactive phosphorus (RP)*, which has been used to describe the *potentially bioavailable P (BAP)* (Duce et al., 1991; Delaney, 1998). Much of the work to date has focused on the *soluble reactive P* (SRP), which is characterized as the P fraction that forms a phospho-molybdate complex under acidic conditions (Strickland and Parsons, 1972). A significant fraction of the SRP is composed of *orthophosphate* ($H_2PO_4^-$) and acid-labile organic compounds such as simple phosphate sugars (McKelvie et al., 1995). The DIP fraction is composed of phosphate (PO_4^{3-}), phosphoric acid (HPO_4^{2-}), orthophosphate ($H_2PO_4^-$), and triprotic phosphoric acid (H_3PO_4); dissociation constants in freshwater and seawater are shown in table 11.1 (Atlas, 1975; Stumm and Morgan, 1981). The relative abundance of these species will vary with pH in aquatic systems, making $H_2PO_4^-$ and HPO_4^{2-} the more common species in freshwater and seawater, respectively (figure 11.1; Morel, 1983). The difference between TDP and SRP provides an estimate of the DOP pool, which has more recently been referred to as *soluble nonreactive P* (SNP) (Benitez-Nelson and Karl, 2002). This can represent a much larger pool than SRP and can also be an important source of P to oceanic organisms (Benitez-Nelson and Karl, 2002). Using ^{31}P nuclear magnetic resonance (NMR) spectroscopy, the dominant groups of P found in DOP in oceanic systems are phosphonates, phosphate monoesters, orthophosphate,

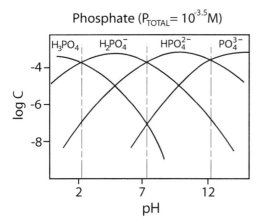

Phosphate ($P_{TOTAL} = 10^{-3.5}$M)

Figure 11.1 Relative abundance of different species of dissolved inorganic P as a function of pH in aquatic systems; $H_2PO_4^-$ and HPO_4^{2-} are the more common species in freshwater and seawater, respectively. (Modified from Morel, 1983.)

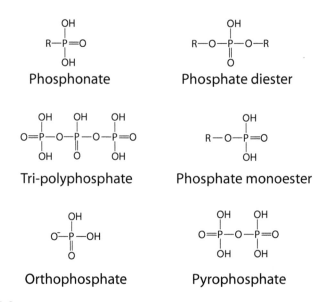

Figure 11.2 Dominant groups of P found in DOP (phosphonates, phosphate monoesters, orthophosphate, phosphate diesters, pyrophosphates, and tri- and tetrapolyphosphates) in oceanic waters. (Modified from Kolowith et al. 2001.)

phosphate diesters, pyrophosphates, and tri- and tetrapolyphosphates (figure 11.2; Clark et al., 1998; Kolowith et al. 2001), and very little work on this topic has been performed in river/estuarine systems. It should be noted that phosphonates are a group of compounds that have a C–P bond, often associated with phosphoproteins (Quin, 1967) and phospholipids (Hori et al., 1984).

Rivers are the major source of P to the ocean, via estuaries, where major chemical and biological transformations of P occur before it is delivered to the ocean (Froelich et al., 1982; Conley et al., 1995). The major source of P to rivers is from weathering of rock materials, and this is the major pathway from which P is lost from terrestrial systems (table 11.2; Jahnke, 2000). Phosphorus is the tenth most abundant element on Earth with an average crustal abundance of 0.1% (Jahnke, 2000). Apatite is the most abundant phosphate mineral in the Earth's crust, representing more than 95% of all the crustal P. Thus, the yield of P from weathering processes will vary depending on the rock type. For example, P is generally low in granites (e.g., 0.13–0.27%), higher in shales (0.15–0.40%), and highest in basalts (0.40–0.80%) (Kornitnig, 1978). The fact that the uptake of P into organic matter per year is greater than the amount lost by land or supplied by rivers emphasizes the importance of P recycling in natural systems (Berner and Berner, 1996). The total prehuman weathered P [DIP, DOP, POP (soil-derived), and Fe-bound PIP] flux from rivers to the ocean is estimated to be 9.8 to 16.8 × 10^{12} g y^{-1} (table 11.3a; Compton et al., 2000). The total prehuman DIP (orthophosphate, $H_2PO_4^-$, HPO_4^{2-}, and PO_4^{3-}) flux was 0.3 to 0.5 × 10^{12} g y^{-1}; this is based on an average phosphorus concentration of 7 to 10 μg L^{-1} in unpolluted rivers (Meybeck, 1982, 1993; Savenko

Table 11.2 Summary of phosphorus reservoir amounts, total fluxes, and residence time.

Reservoir	Reservoir amount, A (mol $\times 10^{-12}$)	$\sum$ Fluxes (mol area$^{-1} \times 10^{-12}$)	Residence time (years)
Atmosphere	0.00009	0.15	0.0006 (5.3 h)
Land biota	96.9	6.0	16.2
Land	6460	9.81	949
Surface ocean	87.5	34.2	2.56
Ocean biota	1.6-4.0	33.6	0.048–0.19 (18–69 d)
Deep ocean	2812	1.98	1420
Sediments	1.29×10^8	0.71	1.82×10^8
Total ocean system	2902	0.12	24,180

Modified from Jahnke (2000).

and Zakharova, 1995). The DOP flux of 0.2×10^{12} g y^{-1} was estimated from a DOC C/P weight ratio of 1000 (Meybeck, 1982) and the POP from a POM C/P ratio of 193 (Meybeck, 1983; Ramirez and Rose, 1992). As expected, the present-day total river P flux is higher and ranges from 0.8 to 1.4×10^{12} g y^{-1} (table 11.3b). This higher flux is primarily due to higher inputs of PIP and DIP; aeolian inputs, which are mostly particulate, have also increased due to human activities (Compton et al., 2000). The wide range in riverine fluxes of RP to the ocean has been attributed to the following: (1) dominance of particulate P compared to dissolved P, and (2) degree to which particulate P is solubilized

Table 11.3a Summary of prehuman phosphorus flux to the ocean.

	$\times 10^{12} \text{g}^{-1}\text{y}$
Riverine phosphorus	
DIP	0.3–0.5
DOP	0.2 (maximum)
POP (0.5 soil derived; 0.4 shale derived)	0.9 (maximum)
PIP, Fe–bound (P adsorbed to	
iron–manganese oxide/oxyhydroxides)	1.5–3.0
PIP, detrital	6.9–12.2
Total river phosphorus	9.8–16.8
Eolian phosphorus	1.0 (20% reactive)
Total river + Eolian phosphorus flux	10.8–17.8
Total prehuman potentially reactive	
phosphorus flux	3.1–4.8
(DIP+DOP+POP+iron-bound PIP+reactive Eolian)	

Modified from Compton et al. (2000).

Table 11.3b Summary of present-day phosphorus flux to the ocean.

	$(\times\ 10^{12}g^{-1}y)$
Riverine Phosphorus	
DIP	0.8–$1.4 \times 10^{12}\ g^{-1}y$
DOP	0.2 (average)
POP (0.5 soil derived; 0.4 shale derived)	0.9 (average)
PIP, Fe-bound (P adsorbed to	
iron-manganese oxide/oxyhydroxides)	1.3–7.4
PIP, detrital	14.5–20.5
Total river phosphorus	17.7–30.4
Eolian phosphorus	1.05 (20% reactive)
Total River + Eolian phosphorus flux	18.7–31.4
Total prehuman potentially reactive	
phosphorus flux	3.4–10.1
(DIP+DOP+POP+iron-bound PIP+20% Eolian)	

Modified from Compton et al. (2000).

as it passes through estuaries (Froelich, 1988). Thus, seasonal storage of RP in streams, rivers, and estuaries can have a significant effect on the composition and abundance of RP delivered from rivers to the coastal ocean. Approximately 90% of the total sediment eroded from land remains stored in the river systems over a period of decades (Meade and Parker, 1985). This has major implications for the storage and remobilization of P, as it relates to carbon loading, remineralization and deposition rates, and sediment redox—discussed later in this chapter.

Inputs of atmospheric sources of P are generally considered to be insignificant to coastal systems. In fact they only represent <10% of the riverine flux of reactive P (Duce et al., 1991; Delaney, 1998). Only in highly oligotrophic systems, such as oceanic gyres and the eastern Mediterranean (Krom et al., 1991, 1992), can such inputs have a significant impact on primary production. In fact, atmospheric deposition of DIP may account as much as 38% of the new production during summer and spring in the Levantine Basin, eastern Mediterranean (Markaki et al., 2003). While gaseous forms of N and S are important components of natural systems, no significant quantities of any stable gaseous forms of P have been found in aquatic systems. However, phosphine (PH_3), a volatile gas of the P cycle, has been measured in anoxic freshwater sediments, flooded wetlands, and sewage treatment facilities (Gassmann, 1994; Glindemann et al., 1996). The nanomolar concentrations PH_3 which can be found in aquatic sediments will not contribute significantly to the fluvial loading of P in rivers, but may serve to transport P from wetlands via evasion to the atmosphere to other regions (Wetzel, 2001). Dry fallout of aeolian P from dust, industry, and sea salts, on land represents 1.0 and $1.05 \times 10^{12}\ g^{-1}\ y^{-1}$ in prehuman and present-day P fluxes to the ocean, respectively (tables 11.3a and 11.3b; Compton et al., 2000).

Phosphorus Fluxes Across the Sediment–Water Interface

The release of P from estuarine sediments is a common and important process that varies spatially and temporally. Early studies showed, using in situ benthic flux chambers, that P fluxes ranged from 30 to 230 mg P m^{-2} d^{-1} in estuaries such as Narragansett Bay (USA) (Elderfield et al., 1981a, b), Potomac River estuary (USA) (Callender and Hammond, 1982), San Franciso Bay (USA) (Hammond et al., 1985), and Guadalupe Bay (USA) (Montagna, 1989). Large differences in P fluxes have also been found when comparing estuaries on the east coast of the United States and Europe to systems in the Gulf of Mexico (table 11.4; Twilley et al., 1999). Fluxes in the Gulf of Mexico, excluding the Mississippi River Bight, were significantly lower due to lower loading of P to these systems and a shorter residence time. Another key factor controlling the release of P from sediments is temperature; for example, in many temperate systems much of the regeneration of P in sediments occurs via microbial processes which are typically highest in summer months. Similarly, another fundamental pattern involving P release from sediments is that it decreases with increasing salinity. These spatiotemporal patterns are effectively illustrated across an estuarine transect, from upper to lower Galveston Bay, showing higher water column concentrations of orthophosphate at higher temperature and lower salinities (figure 11.3; Santschi, 1995). Spatial variability in the extent of subtidal and intertidal areas may also have an impact on the spatiotemporal variability in P concentrations. For example, intertidal mudflats are considered to be important reservoirs of P, which adds to the spatial variability of P sources in estuaries (Flindt et al., 1997; Coelho et al., 2004; Lillebo et al., 2004). Air-exposed sediments have been shown to have a greater capacity of phosphate sorption than subtidal sediments (Lillebo et al., 2004). In fact, 79% of the total DIP that can be potentially exported form Mondego estuary (Portugal) was stored in mudflats (Lillebo et al., 2004). Phosphorus release from sediments with salt marsh and seagrass plant coverage was also found to be substantially less due to adsorption of P in sediments with rhizomes and roots (Lillebo et al., 2004). Despite the overall predictability in seasonal and temporal patterns of P release from sediments to the overlying water column in estuaries, the specific mechanisms are still not well understood. In the next section we will compare and contrast some of the key mechanisms involved in the release of P from sediments that may explain these general patterns.

Numerous studies in freshwater (e.g., Roden and Edmonds, 1997; Wetzel, 1999; Hupfer et al., 2004) and marine (e.g., Krom and Berner, 1981; Sundby et al., 1992; Gunnars and Blomqvist, 1997; Anschutz et al., 1998; Rozan et al., 2002; Sutula et al., 2004) systems have examined mechanisms controlling the release and efflux of P from sediments. Since total sediment P does not reflect exchange capacity or bioavailability of P, sequential chemical extraction techniques have been useful in separating out different pools of P (Ruttenberg, 1992; Jensen and Thamdrup, 1993). Sedimentary pools of P have generally been divided into the following fractions: (1) organic P; (2) Fe-bound P; (3) *authigenic P minerals* [e.g., carbonate fluoroapatite (CFA), struvite, and vivianite]; and (4) *detrital P minerals* (e.g., feldspar) (see Ruttenberg, 1993; Ruttenberg and Berner, 1993). More specifically, it is the organic P (Ingall and Jahnke, 1997) and Fe-bound P (Krom and Berner, 1981) fractions that are considered to be the most reactive involving the release of P from sediments, via pore waters, during P regeneration.

Table 11.4 System-wide average of exchange at the sediment-water interface in coastal marine environments.

Estuary	Reference[a]	Nutrient fluxes (μmol m^{-2} h^{-1}) PO$_4$	Sediment oxygen consumption (SOC) (g m^{-2} d^{-1})
Gulf of Mexico estuaries			
Ochlockonee Bay	1		0.90
Apalachicola Bay	2		
Mobile Bay	3,4	3.9	0.55
Mississippi River Bight	5,6	17.5	0.84
Fourleague Bay	7,8	1.4	1.03
Fourleague Bay	9	−8.0	1.2
Trinity–San Jacinto estuary	10	0.6	0.15
Gudalupe estuary	11	−3.1	0.98
Gudalupe estuary	12		
Nueces estuary	8	−6.4	0.73
Nueces estuary	12		
Laguna Madre-upper	13	−0.2	1.68
Other U.S. Estuaries			
San Francisco Bay	13	−4.2–54	0.35–0.70
Narragansett Bay	14	38-233	
Chesapeake Bay	15		
North Bay		−16.3	0.1–0.65
Mid Bay		0.0–148	0.01–0.86
Lower Bay		−1.5–13	0.3–0.75
Patuxent estuary	16	0.0–15.0	0.75–2.25
Neuse River estuary	17	−2.3–46.0	0.70–1.87
South River estuary	17	−8.3–23	0.71–2.72

[a]References: 1, Seitzinger (1987); 2, Mortazavi, unpublished data; 3, Cowan et al. (1996); 4, Miller-Way, unpublished data; 5, Twilley and McKee (1996); 6, Bourgeois (1994); 7, Miller-Way (1994); 8, Twilley, unpublished data; 9, Teague et al. (1988); 10, Zimmerman and Benner (1994); 11, Montagna, unpublished data; 12, Yoon and Benner (1992); 13, Hammond et al. (1985); 14, Elderfield et al. (1981a,b); 15, Cowan and Boynton (1996); 16, Boynton et al. (1991); 17, Fisher et al. (1982). Modified from Twilley et al. (1999).

While it is generally accepted that much of the inorganic P is bound with Fe and Ca in sediments, the organic fraction is less clear. Some of the organic P components occur as phytic acid, nucleic acids, and humic substances (Ogram et al., 1978; de Groot, 1990). In the case of organic P release, PO$_4^{3-}$ is remineralized during diagenesis of organic matter. Significant decreases of P in settled organic matter particles further supports the importance of heterotrophic bacteria in converting organic P into PO$_4^{3-}$ (Wetzel, 1999; Kleeberg, 2002). In the Fe bound fraction of P, release of P can occur from Fe oxides as they become

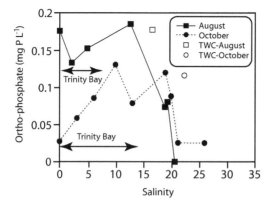

Figure 11.3 Spatiotemporal patterns of orthophosphate concentrations (mg P L^{-1}) across an estuarine transect, from upper to lower Galveston Bay (USA) in August and October 1989. Texas Water Commission (TWC) data were collected a week prior to sampling by Santschi (1995). (Modified from Santschi, 1995.)

reduced under anaerobic conditions (McManus et al., 1997). Much of the Fe-bound P is complexed with *amorphous Fe* and not *crystalline Fe* (Anschutz et al., 1998); this amorphous Fe is defined as being ascorbate leachable (ASC-Fe). Other experimental work has shown that P released from estuarine and freshwater sediments is largely controlled by the reduction of FeOOH (Gunnars and Blomqvist, 1997). Thus, the degree to which P is released through both of these pathways is further controlled by the stability of redox conditions as well as the extent of authigenic P mineral formation.

Authigenic P mineral formation has been shown to be important in controlling P removal in sediments (Ruttenberg and Berner, 1993; Slomp et al., 1996). Similarly, oxygenated bottom waters can reduce P efflux whereby much of the P in pore water becomes adsorbed on to Fe(III) oxides (Ingall and Jahnke, 1997). Increases in pH can enhance the release of adsorbed P that is hydrated to Fe and Al oxides, through ligand exchange mechanisms involving competition between OH^- and PO_4^{3-} (Lijklema, 1977). In carbonate systems, a decrease in pH resulted in the liberation of P that was adsorbed on calcite (Stumm and Leckie, 1971; Staudinger et al., 1990). In the case of apatite, there are two general pathways for its chemical precipitation (Krajewski et al., 1994). The first involves supersaturation of porewaters (as related to apatite) and nucleation of precursors such as octacalcium phosphate (OCP) or amorphous calcium phosphate (ACP) (Jahnke et al., 1983; van Cappellen and Berner, 1988), and the other involves direct nucleation of apatite in undersaturated porewaters with respect to precursors (van Capellen and Berner, 1989). More recent work has further supported the role of a precursor pathway involving OCP as opposed to direct nucleation (Gunnars et al., 2004). More work is clearly needed on the role of abiotic and biotic processes on the release of P from sediments estuarine systems.

When compared to freshwater systems, interactions between of Fe and S (Canfield, 1989; Kostka and Luther, 1995) in marine and estuarine systems can further complicate the mechanisms controlling P release. For example, experimental work by Gunnars and Blomqvist (1997) indicated that the reduction of FeOOH was most important in controlling the release and efflux of PO_4^{3-} across the sediment–water interface. Moreover, it was found that the dissolved Fe:P ratio of the efflux was equal to one in freshwater coming from freshwater sediments but less than one in being released from marine sediments. It was further shown that carbon-normalized P remobilization from sediments was approximately 5-fold higher in marine systems with higher SO_4^{2-} concentrations than in freshwater systems with low SO_4^{2-} concentrations (Caraco et al., 1990). These differences in ratios are likely explained by quick scavenging of Fe^{2+} by Fe sulfides in marine and estuarine systems, which can result in the precipitation of FeS and FeS_2 (pyrite) (Taillefert et al., 2000). The formation of FeS and FeS_2 can readily occur in the presence of H_2S (Luther, 1991; Rickard and Luther, 1997; Theberge and Luther, 1997). The concentration of sulfide in anoxic bottom waters shows a strong negative relationship with Fe^{2+} concentrations in marine and freshwater systems (figure 11.4; Gunnars et al., 2004); much of the sulfide required for the precipitation of Fe sulfides is derived from SO_4^{2-} reduction under anaerobic conditions (Postgate, 1984). More details on reactions between Fe and S are described in chapter 12. More recently, it was shown that that shifts in the dissolved and sold phases of Fe, S, and P were linked with seasonal changes in organic matter loading and redox in Rehoboth Bay (USA) (figure 11.5; Rozan et al., 2002). Controls on the release of PO_4^{3-} to the overlying water column is further complicated by macroalgal uptake of PO_4^{3-} at the sediment surface in summer months. In general, sediments are more reducing in summer and oxidizing in late fall and winter. During the oxidizing periods much of the P is bound to Fe(III) oxides; under more reducing conditions Fe-bound P is released as PO_4^{3-}, where it is taken up by benthic macroalgae. Bottom-up control of benthic detritus-based

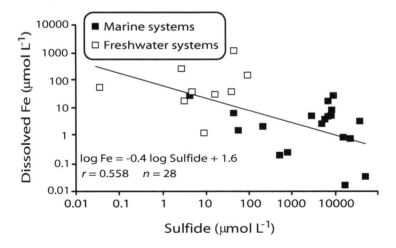

Figure 11.4 Negative log correlation between concentrations of sulfide ($\mu mol\ L^{-1}$) and of Fe^{2+} ($\mu mol\ L^{-1}$) in anoxic bottom waters in marine and freshwater systems. (Modified from Gunnars et al., 2004.)

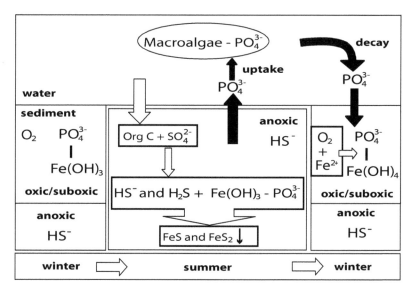

Figure 11.5 Dissolved and solid-phase shifts in concentrations of Fe, S, and P over different seasons and its relationship to changes in organic matter loading and redox in Rehoboth Bay (USA). (Modified from Rozan et al., 2002.)

tropical streams was also shown to be linked with landscape-scale changes in P concentration (Rosemond et al., 2002). Other microbially mediated processes can control the release of P from sediments. For example, in fresh to brackish waters Fe-reducing aerobic bacteria convert amorphous Fe(III) into Fe(II) releasing Fe-associated PO_4^{3-} (figure 11.6; Roden and Edmonds, 1997). Similarly, uptake of excessive amounts of P by aerobic bacteria can be stored as polyphosphates; under reducing conditions, these reserves can then be released as bacteria rapidly become degraded under anaerobic conditions (G'achter and Meyer, 1993; Hupfer et al., 1995). Phosphorus sorbed to manganese oxides can also be released under anaerobic conditions when reductive dissolution of these oxides occurs.

Differences in the mechanism controlling P release from sediments in freshwater and estuarine/marine systems has major implications for differences in N and P limitation observed in these systems (Caraco et al., *1989*, 1990). Nitrogen cycling is also different in these systems and contributes to this general paradigm (e.g., Paerl et al., 1987; Howarth et al., 1988a,b; Seitzinger et al., 2000, 2002b). It has been generally accepted that primary production in freshwater systems is P limited (Hecky and Kilham, 1988; Schindler, 1977) and more N limited in marine systems (Vince and Valiela, 1973; Caraco et al., 1987; Graneli, 1987). Early work showed that the N:P ($NH_4^+ + NO_3^-$)/RP release ratio from coastal sediments was about 8, which is only half of what is needed by phytoplankton in these systems. This N limitation is explained by the loss of N through high rates of denitrification in coastal systems (Seitzinger et al., 2000), with generally high return of P from sediments. Conversely, with lower rates of denitrification and high P immobilization

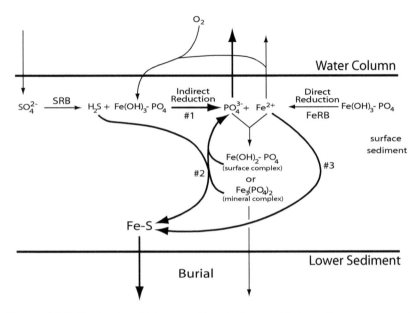

Figure 11.6 Fresh to brackish waters Fe-reducing aerobic bacteria convert amorphous Fe(III) into Fe(II), releasing Fe-associated PO_4^{3-}. (Modified from Roden and Edmonds, 1997.)

in freshwater sediments, it is reasonable to expect more P limitation in freshwater systems (Caraco et al., 1990).

Tropical carbonate estuarine systems tend to be more P limited than N limited compared to temperate systems, in part because of the widespread occurrence of N fixation in tropical systems (Carpenter and Capone, 1981; Smith, 1984; Short et al., 1985; Powell et al., 1989). For example, the availability of P over N, to tropical seagrasses such as *Syringodium filiforme*, was shown to be the primary determinant of seagrass production (Smith and Atkinson, 1984; Short et al., 1990). Another important difference in P dynamics between tropical carbonate and temperate siliciclastic systems is the high rates of phosphate ion sorption to the carbonate matrix (Berner, 1974; DeKanel and Morse, 1978; Gaudette and Lyons, 1980; Krom and Berner, 1980b). This is why carbonate sediments typically have low concentrations of DIP in porewaters (Berner, 1974; Morse and Cook, 1978). An excellent example of this was demonstrated on the Atlantic U.S. coast where intertidal macroalgae switched from N limitation to P limitation moving from temperate siliciclastic systems to tropical carbonate systems (Lapointe et al., 1992). Similarly, the breakdown of organic P in carbonate sediments has been shown to be tightly coupled with the release of DIP and rapid uptake by roots in seagrass communities (Short, 1987). Mangrove wetlands also have a general trend of net import of DIP and DOP (Nixon et al., 1984; Boto and Wellington, 1988), since these systems are generally P limited (Alongi et al., 1992; Rivera-Monroy et al., 1995; Davis et al., 2001). Finally, changes in N and P limitation can also occur on temporal scales in estuaries. For example, many estuaries are P limited in the spring, and then switch to N limitation in the summer (Conley, 1999). Seasonal storage of Fe-bound P in winter and spring, along with the temperature-controlled release of P in summer, can account for most of the temporal variability observed in estuarine systems (Jensen et al., 1995). Such temporal differences in N and P limitation have led to serious debates over what nutrient management strategies are needed in estuaries (Conley, 1999)—more on the topic of nutrient management is provided in chapter 15.

Cycling of Inorganic and Organic Phosphorus

In both freshwater and estuarine systems, concentrations of DIP have also been strongly linked with the suspended sediment load. In fact, a stable or "equilibrium" concentration range of DIP, between 0.5 and 2 μM, has been reported for a number of estuarine systems (Pomeroy et al., 1965; Liss, 1976; Froelich, 1988; Ormaza-Gonzalez and Statham, 1991). These stable DIP concentrations are believed to be controlled by a "buffering" of DIP through the adsorption and desorption onto metal oxide surfaces (Mortimer, 1941; Carritt and Goodgal, 1954; Stirling and Wormald, 1977). This "P buffering" is believed to balance the low availability of SRP in higher-salinity waters, which occurs from phytoplankton uptake and anionic competition for surface adsorption sites (Froelich, 1988; Fox, 1989). For example, TPP concentrations decrease with increasing salinity in the Delaware Bay (Lebo, 1991) and Scheldet (The Netherlands) estuaries, suggesting the importance of desorption of DIP from aluminum and iron oxides. Similarly, a decrease in the TPP content in waters of the lower Mississippi River and inner Louisiana shelf, with increasing salinity, are also indicative of P buffering (figure 11.7; Sutula et al., 2004). Conversely, Hobbie et al. (1975) reported that 60% of the DIP entering the Pamlico River estuary (USA) was scavenged by particulates and stored in sediments. Other work also suggested that in such large river systems such as the Amazon, DIP concentrations will largely be controlled by amorphous ferric hydroxide (Fox, 1989). More recent work in the Amazon River and estuary, indicate that while the release P from iron oxides/hydroxides, represents a significant source P to coastal ocean, bacterial decomposition of riverine organic matter is another important component (Berner and Rao, 1994). Changes in pH with increasing salinity may preclude phosphate from binding to FeOOH through a shift in the speciation of phosphate from $H_2PO_4^-$ to HPO_4^{2-} in addition to a change in surface charge on the FeOOH (Zwolsman, 1994). Finally, in the high-salinity reaches of estuaries, calcite can serve as a carrier phase for adsorbed P (de Jonge and Villerius, 1989). However, other work has shown that inorganic exchange processes were not able to "buffer" DIP concentrations across different regions of Chesapeake Bay (Taft and Taylor, 1976; Fisher et al., 1988; Conley et al., 1995). When examining the Fe:P ratio in citrate–dithionate–bicarbonate (CDB) extracts in surface and bottom samples of suspended particles, there is a clear decrease in the ratio with increasing salinity (figure 11.8; Conley et al., 1995). This decrease in CBD–Fe:P has also been observed in the St. Lawrence estuary (Canada) (Lucotte and d'Anglejan, 1983). However, the CBD–Fe:P fraction was a factor of 10 higher than in the Chesapeake Bay. Thus, it appears that particulate matter in the Chesapeake has a lower capacity to adsorb P through Fe interactions (Conley et al., 1995). Consequently, transformation of particle-bound P in the Chesapeake appears to be more controlled by biological productivity (e.g., phytoplankton). This agrees with the lower

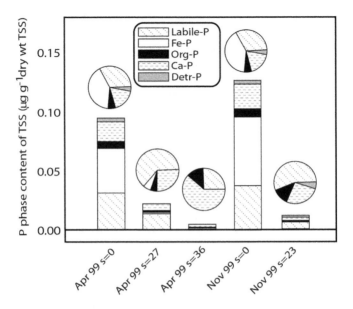

Figure 11.7 Total particulate phase content of phosphorus [e.g., labile, Fe-P, organic P (Org-P), calcium phosphate (Ca-P), and detrital phosphorus (Detr-p)] in total suspended solids (TSS) (μg g^{-1} dry wt. TSS) in waters of the lower Mississippi River and inner Louisiana shelf, versus seasonal variability and increasing salinity. (Modified from Sutula et al., 2004.)

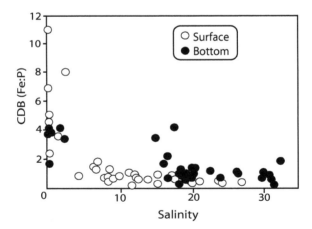

Figure 11.8 Fe:P molar ratio in citrate–dithionate–bicarbonate (CDB) extracts in surface and bottom samples of suspended particles in surface (open circles) and bottom (solid circles) samples across a salinity gradient in Chesapeake Bay (USA). (Modified from Conley et al., 1995.)

CBD–Fe:P ratios at the higher salinities in the mid and lower bay (figure 11.8), where DIP recycling and phytoplankton abundance are well correlated, particularly in summer months (Malone et al., 1988).

Another geochemical/physical mechanism controlling the P concentrations in estuarine waters may involve particle sorting and/or particle–colloid interactions. For example, negative correlations were found between particulate P and suspended particulate matter (SPM) and the partition coefficient (K_d) for orthophosphate $K_d =$ [TPP (mg kg^{-1})]/[orthophosphate (mg L^{-1})] and SPM in the Galveston estuary (USA) (figure 11.9; Santschi, 1995). Much of the total P in this system is composed of orthophosphate as indicated by the strong positive relationship between orthophosphate and TP. The negative correlation between K_d and SPM is commonly referred to as the "particle concentration effect," an effect well supported by radionuclide and trace metal work (e.g., Honeyman and Santschi, 1988; Baskaran et al., 1996; Benoit et al., 1994). This effect occurs when a fraction of the P, or trace elements and radionuclides, is associated with the colloidal fraction, which is less than 0.45 μm and greater than 1 kDa (Benoit et al., 1994). In fact, the colloidal P represented 30 to 80% of the filter-passing organic P concentration, an amount significant enough to account for such an effect. In both freshwater and marine systems, colloidal P has been shown to be associated with Fe, humic acids, and other organic molecules (e.g., Carpenter and Smith, 1984; Ridal and Moore, 1990; Hollibaugh et al., 1991). Thus, this represents another phosphate-buffering mechanism that regulates P concentrations, unlike that previously described in the literature.

As discussed earlier oxygen concentrations in sediments and bottom waters are important covariables in determining if sediments are a source or sink of P from sediments. Much of P buried in Baltic sediments is believed to be apatite bound (Carman and Jonsson, 1991). In the Baltic Sea, DIP in bottom waters was found to be correlated with an area of the bottom covered by hypoxic water (< 2.0 mL L^{-1}), but not with changes in TP loading over the past two decades (Conley et al., 2002). The expanse of hypoxic bottom waters are believed to be primarily related to reduced exchange of saltwater inflows from the North Sea over this period. Annual variations in P are primarily related to reduced water exchange, but eutrophication on the long term is probably still a real effect. Dissolved inorganic phosphorus in bottom water was also shown to be negatively correlated with oxygen concentration, indicating the release of DIP from sediments (figure 11.10; Conley et al., 2002). Organic loading of phytodetritus in experimental cores from the Baltic showed a significant increase in the efflux of DIP from sediments into the overlying water (Conley and Johnstone, 1995). Although there has been considerable debate about the need for reductions in P and/or N to reduce eutrophication in the Baltic (HEL-COM, 1993, 1998, 2001), this work indicates that changes in the residence time of bottom waters and the concomitant changes in oxygen concentration—from climatically driven phenomena–can control the delivery of P to the water column, an important find for future management decisions on nutrient loading in the Baltic Sea.

The spatial and temporal importance of DIP availability in trophic food webs of coastal and estuarine systems has been hinderd by methodology. Simply measuring DIP concentrations alone provides only limited information on how P affects primary production. To borrow once again from the coastal/open ocean literature, new P radioisotope techniques have proven to be a more effective means of understanding P dynamics (Lal and Lee, 1988; Waser et al., 1996; Benitez-Nelson and Buessler, 1999). The two P radioisotopes

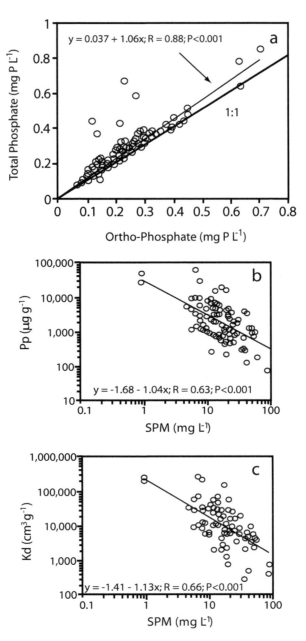

Figure 11.9 Negative correlations between particulate P and suspended particulate matter (SPM) and the partition coefficient (K_d) for orthophosphate (K_d = [TPP [mg kg^{-1}]]/[orthophosphate [mg L^{-1}]]) and (SPM) in the Galveston estuary (USA). (Modified from Santschi, 1995.)

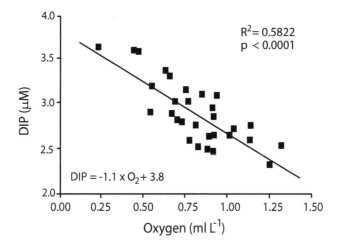

Figure 11.10 Negative correlations between dissolved inorganic phosphorus (μM) and oxygen (mL L^{-1}) concentrations in bottom waters of the Baltic Sea. (Modified from Conley et al., 2002.)

used here are ^{32}P ($t_{1/2} = 14.3$ d) and ^{33}P ($t_{1/2} = 25.3$ d); these decay rates are clearly on timescales certainly applicable to many biologically controlled processes in estuarine systems. Residence times of TDP in coastal waters of the East China Sea were found to be very short (3–4 d) indicating that even low P concentrations could significantly support high primary production (Zhang et al., 2004). This work further supports the fact that measuring concentrations of DIP is not an adequate approach in understanding the role of P in phytoplankton studies. Similar work is needed in estuarine systems where the spatial and temporal dynamic of DIP and DOP are even more dynamic.

Although the characterization of DOP in rivers and estuaries has been largely ignored, some work in the Mississippi River indicated that the composition of SNP primarily consisted of diester and monoester phosphates, phosphonates, orthophosphates, and/or tri and tetrapolyphosphates (Nanny and Minear, 1997)—essentially the same as that found in the ocean (Kolowith et al., 2001). In the open ocean, much of the DOP found in HMW DOM has been shown to be composed of P esters (75%) and phosphonates (25%) (Kolowith et al., 2001). However, the less reactive phosphonates represent a greater fraction of the total DOM, compared to the more bioavailable P esters, as DOM is degraded over time. Phosphonates and refractory P esters are also abundant in marine sediments and may represent a significant sink for organic P (Ingall et al., 1990). Conversely, recent work has shown that phosphonates may actually be preferentially removed relative to other bioavailable P esters in anoxic waters (Benitez-Nelson et al., 2004). From the perspective of soil inputs at the river end-member of estuaries, 90% of the total organic P in some soils was represented in the form of the monoester phosphate fraction (Condron et al., 1985), with phosphonates also being present (Hawkes et al., 1984). These comparisons of possible oceanic and terrigeneous end-member inputs to estuaries provide at least some insight into the potential composition of DOP in estuaries.

Phosphorus Budgets from Selected Estuaries

Due to the general paradigm of N limitation in estuaries versus P limitation in freshwater systems, the early classic work on P cycling began in limnology with an emphasis on the dynamics of P loading and anoxia (Einsele, 1936; Hutchinson, 1938; Mortimer, 1941). This was followed by the development of P budgets which centered on the concerns of eutrophication in lakes and the application of such budgets as management tools (e.g., Vollenweider, 1968, 1975; Janus and Vollenweider, 1984). Much of this early modeling work by Vollenweider was focused on export and retention of TP and used mass balance and empirical models to obtain predictions. More specifically, these studies were the first to establish a relationship between external loading of P and algal biomass in lakes. Much later, researchers began to focus on the role of P in estuaries (e.g., Smith, 1984; Smith et al., 1991). However, an important fundamental difference between the freshwater and estuarine models was the emphasis on TP retention in lakes compared to soluble P export in estuaries (see review by Harris, 1999).

Many of the P budgets for temperate estuaries in Europe and the United States indicate that there have been considerable increases in anthropogenic loading of P over the last century (Boynton et al., 1995; Nixon et al., 1996; Billen et al., 2001). For example, total P inputs to Chesapeake Bay are 13 to 24 times higher than in precolonial times; however, recent reductions have reduced these P inputs considerably in certain subestuaries of the Chesapeake Bay system. Just in the last 30 years, P inputs from domestic and industrial sources increased three-fold in the Seine River estuary, which has resulted in significant increases in algal biomass (figure 11.11; Billen et al., 2001). Unlike the majority of P budgets generated for temperate systems, most of the P inputs in the Seine River estuary are from point sources. More dramatic increases in N (five-fold) have resulted in this system becoming more P limited over time, a condition that likely existed before the high anthropogenic loading began. When examining the P budget for the Chesapeake Bay system, there is a net landward exchange of TP, similar to that found for TN (figure 11.12;

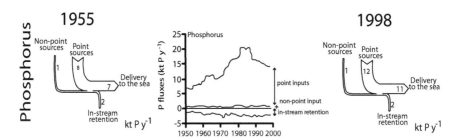

Figure 11.11 Phosphorus inputs (kt P y^{-1}) from domestic and industrial sources to the Seine River estuary over the past 30 years. (Modified from Billen et al., 2001.)

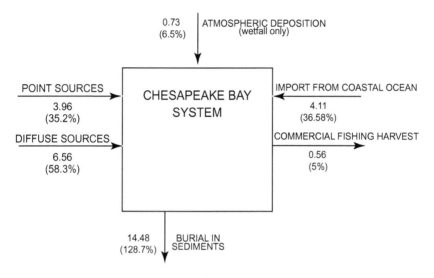

PHOSPHORUS
$(kg\,P \times 10^{6}\,y^{-1})$

0.73
(6.5%)

ATMOSPHERIC DEPOSITION
(wetfall only)

POINT SOURCES

CHESAPEAKE BAY
SYSTEM

IMPORT FROM COASTAL OCEAN

3.96
(35.2%)

4.11
(36.58%)

DIFFUSE SOURCES

COMMERCIAL FISHING HARVEST

6.56
(58.3%)

0.56
(5%)

14.48
(128.7%)

BURIAL IN
SEDIMENTS

Figure 11.12 Total phosphorus budget $(kg\,P \times 10^{6}\,y^{-1})$ in the Chesapeake Bay system. (Modified from Boynton et al., 1995.)

Boynton et al., 1995). Losses of TP from fisheries were generally low, but burial of PP was large and exceeded terrestrial and atmospheric inputs. While there was a net export of TN to the coastal ocean, TP was imported. Moreover, the magnitude of the TP import was positively correlated with terrestrial and atmospheric inputs. The retention of TP in the Chesapeake Bay system is believed to be related to estuarine morphology and circulation patterns (Boynton et al., 1995). As discussed in chapter 10, retention time is critical in controlling the net import or export of both N and P in estuaries (Nixon et al., 1996). The longer residence time of P relative to N in the Baltic Sea also complicates modeling efforts that have attempted to project the effects of P versus N reductions on phytoplankton productivity (Savchuk and Wulff, 2001). While much of the TP is retained in the Chesapeake, there remains the uncertainty of how available this P is to phytoplankton (Keefe, 1994); once again, more specific studies that examine different P species are needed to assess these questions. In Narragansett Bay inputs of P from rivers did not vary much over wet and dry years (Nixon et al., 1995). This suggests that much of the DIP and DOP is provided by upstream points sources to rivers; however, non-point sources appear to control much of the TPP (Nixon et al., 1995, 1996). Flux of P into Narragansett Bay from the coastal ocean is approximately equal to P fluxes into the bay from land drainage, upstream sewage, and fertilizer (Nixon et al., 1995). Sewage inputs represent about 20% of the TP inputs. Most of the N and P that enters Narragansett Bay is exported in the dissolved form to the coastal ocean along with 20% of the organic matter produced in the bay.

One of the major differences observed between tropical and temperate P budgets is the dominance of point sources in tropical estuaries, this is particularly evident in Moreton Bay (Australia) (table 11.5; Eyre and McKee, 2002). However, using the linear relationship between net transport of N and/or P from land to the ocean as a function of mean log residence time established by Nixon et al. (1996), there is good agreement in the estimated amount of P export ($\sim$ 70%) from Moreton Bay. The short residence time in Moreton Bay (46 d) is similar to that in other shallow coastal systems and is directly related to major losses of P from this system. Unlike N, it appears that there are similar biogeochemical processes controlling the fate of P in certain subtropical and temperate systems. Low DIN:DIP ratios in this system are explained by the rapid uptake of DIN, supplied mostly through N fixation, which is not enough to satisfy the DIP for primary production (Smith, 1991).

Table 11.5 Phosphorus budget for Moreton Bay, Australia (t y^{-1}).

Budget components	Phosphorus
Standing stocks	
Sediment (solid phase)	38,870
Sediment (pore water)	25
Water column (total)	172
Biomass	2282
Inputs	
Point sources	1182
Non-point sources	131
Atmosphere	95
Groundwater	2
Primary production	—
N fixation	—
Total	1429
Outputs	
Denitrification	—
Pelagic respiration	—
Benthic respiration	—
Dredging	−309
Burial	36
Fisheries harvest	−6
Pumice stone passage	−71
Ocean exchange	−1007
Recycling	
Benthic fluxes	2885
Biological uptake	9960
Phytoplankton	2974
Sedimentation	

Modified from Eyre and McKee (2002).

Thus, despite high rates of N fixation and denitrification, primary production in Moreton Bay is P limited (Eyre and McKee, 2002).

Sources of Silica to Estuaries

Although silicon (Si) is the second most abundant element in the Earth's crust it has relatively limited importance in biogeochemical cycles (Conley, 2002; Ragueneau et al., 2005a,b). Much of the work to date has focused on the weathering of Si (Wollast and Mackenzie, 1983) and the oceanic Si cycle (DeMaster, 1981; Tréguer et al., 1995); only recently has the cycling of Si in terrestrial ecosystems been shown to be important on a global scale (Conley, 2002). The majority of inputs to the oceans occur via rivers (80%), with much of the losses controlled by sedimentation biogenic silica or opal (figure 11.13; Tréguer et al., 1995). The average global concentration of dissolved SiO_2 (DSi) in rivers is 150 µmol L^{-1} (Conley, 1997); the majority of this DSi is in the form of silicic acid (H_4SiO_4) in rivers which typically have a pH in the range 7.3–8.0 (figure 11.14). However, it should be noted that recent studies have also suggested that uncertainties in the global Si budget clearly exist, and may need to be seriously re-evaluated. For example, the role of diatoms and phytoliths (see description below) in the particulate loading, as amorphous Si to the ocean, may be much larger than previously thought (Conley, 1997, Conley et al., 2000). Similarly, recent work has shown that biogenic Si burial in the Antarctic may be overestimated by 35% (DeMaster, 2002). Finally, other work has shown that reverse weathering of Si, where marine diatom frustules are rapidly converted into various forms of authigenic aluminosilicate phases during burial, does occur in restricted (deltaic) environments (e.g., Amazon and Mississippi) (Michalopoulos and Aller, 1995, 2004; Michalopoulos et al., 2000); however, it remains uncertain whether this process is of global importance to the Si budget. In the following sections of this chapter, the focus will be on the key biogeochemical processes controlling Si cycling in estuaries and rivers.

   The oceanic Si system was assumed to be in a steady state with inputs from rivers and other minor sources being balanced by outputs during burial (Tréguer et al., 1995). However, between these inputs and outputs there is a significant amount of recycling that occurs through biological processes. For example, riverine inputs of Si to the oceans are

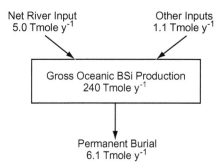

Figure 11.13 The silica cycle showing flux inputs (Tmol y^{-1}) and burial in the global ocean. (Modified from Tréguer et al., 1995.)

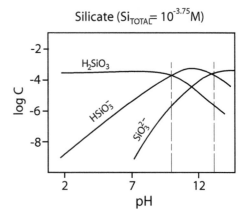

Figure 11.14 Based on an average global concentration of dissolved SiO_2 (DSi) in rivers of 150 μm L^{-1}; the relative abundance of dissolved S species (DSi) in rivers is in the form of silicic acid (H_2SiO_3), which typically has a pH in the range 7.3–8.0. (Modified from Conley, 1997.)

transformed into a large reservoir of biological SiO_2 (BSi) precipitates (via diatom production), of which only a small fraction (3%) gets buried. Diatoms (Bacillariophyceae) can represent as much as 50% of the global primary production and form exoskeletons composed of biogenic silica or opal (van Cappellen et al., 2002). While much of the BSi in oceanic sediments is considered to be derived from diatoms, radiolarians, and sponge spicules (DeMaster, 1981), other components such as *phytoliths* (opal accumulations in plant tissues) may be important in coastal and even open ocean systems (Conley, 2002). The low inputs of BSi to marine sediments occurs because most of the sinking diatomaceous opal dissolves and is recycled back into diatom production (Billet et al., 1983; Nelson et al., 1995). The sinking of BSi is an important process in the euphotic zone of the open ocean, because grazing remineralizes organic matter more quickly than Si, which results in sinking particles that are enriched in Si relative to N. This process is referred to as the "silicate pump" (Dugdale et al., 1995) or more recently the "silica pump," since it is BSi that is relevant here (Brzezinski et al., 2003). Amorphous Si dissolves more readily than refractory mineral silicates in suspended particulates (DeMaster, 1981; Hurd, 1983; M'uller and Schneider, 1993). The BSi that does make it to the sediments will also dissolve in porewaters (Jahnke et al., 1982; McManus et al., 1995). Thus, there is an exponential build-up of DSi in porewater profiles, with the lowest concentrations at the sediment–water interface. Differences in Al content and surface area of diatom exoskeletons, along with temperature and degree of saturation, and bioturbation, all affect dissolution rates of Si in sediments (e.g., van Cappellen et al., 2002). Recent laboratory (Bidle and Azam, 1999, 2001) and field studies (Bidle et al., 2003) have shown that removal of the organic layer on the diatom exoskeleton enhances rates of BSi dissolution. It should be noted that BSi has been used extensively as a paleoindicator of diatom biomass and productivity (Conley, 1988; Ragueneau et al., 1996)—more details on this in chapter 15. Finally, while there are is a wide variety of techniques that have been used to measure BSi in sediments (most wet alkaline methods), an interlaboratory comparison found comparable results between the two most widely used techniques (Conley, 1988).

The primary source of DSi (80%) to the global ocean, via estuaries, is riverine; however, anthropogenic alterations have begun to change the abundance and sources of riverine Si (Tréguer et al., 1995; Conley, 2002). For example, decreases in the delivery of DSi loading of Mississippi River to coastal waters were attributed to enhanced N loading to the river, which is believed to have increased diatom production and sedimentation in the watershed (Turner and Rabalais, 1991; Turner et al., 2003). However, increases in the number of dams can also account for such decreases (Conley et al., 1993). Decreases in DSi in the Danube River (Humborg et al., 1997) and Swedish rivers (Humborg et al., 2000, 2002) have also been attributed to enhanced uptake of DSi from enhanced diatom production. Very dramatic reductions in DSi (a 200 μM decrease) have been observed in the River Nile (Egypt) after construction of the Aswan High Dam (Wahby and Bishara, 1979). These areas of high sedimentation and primary production created by these dams in rivers have been referred to as an "artificial lake effect" (van Bennekom and Solomons, 1981). A second mechanism for the removal of DSi from dam building is loss of contact of water with vegetation and terrestrial processes, as recently shown in northern Sweden (Humborg et al., 2002, 2004). BSi carried in suspension by rivers as diatoms and phytoliths are also an important source to the global Si budget that needs further evaluation (Conley, 1997, 2002). Thus, losses of particulate matter, especially phytoliths carried in suspension, with the building of dams may represent a third mechanism for the loss of DSi (Conley, 2002). Unlike other polluted systems, changes in nutrient loading in the Seine over the past 50 years actually resulted in retention of Si with enhanced in situ diatom production (figure 11.15; Billen et al., 2001). More specifically, increases in diatom production in the Seine River, followed by sequestration of DSi in floodplain and sediments, did not change the amount of silica delivery to the coastal ocean. Other sources of riverine BSi in rivers are derived from benthic diatoms that grow in the "quiet zones" of small rivers, which can get dislodged and transported to larger river systems during high-discharge periods (Stevenson, 1990; Reynolds, 1995). Another factor that needs to be considered here, in the context of Si cycling, is that marine diatoms have one order of magnitude less in Si per biovolume than freshwater diatoms (Conley et al., 1989). It was hypothesized that these differences may stem from salinity differences or evolutionary adaptations by marine species to a lower DSi environment.

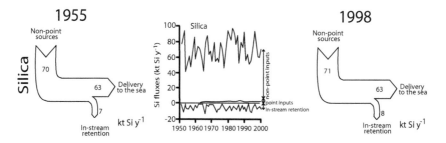

Figure 11.15 Changes in Si loading (kt Si y^{-1}) in the Seine River over the past 50 years. (Modified from Billen et al., 2001.)

Other less important sources of DSi to the estuaries and the coastal ocean include submarine groundwater discharge (SGD), atmospheric inputs, and in the case of estuaries, import of DSi from coastal upwelling (Ragueneau et al., 2005a,b). In the case of SGD, current estimates of the global flux range from 0.01 to 10% of surface runoff (Taniguchi et al., 2002). More work is needed on DSi in groundwaters before any conclusive statements can be made about its effects on estuarine biogeochemistry. As for atmospheric inputs of Si, fluxes to the global ocean have been estimated to be approximately 0.5 Tmol Si y^{-1} (Duce et al., 1991), although these inputs only account for 0.2% of the annual oceanic gross BSi production and 10% of the river inputs. Estuaries and coastal waters may also receive inputs of dust material containing phytoliths. Phytoliths have been shown to represent a significant fraction of dust particles (Romero et al., 2003), particularly if extensive burning of forest/marsh vegetation is occurring near coastal margins. Finally, inputs of DSi from coastal upwelling may provide a significant pulse of DSi to estuaries. For example, temperate estuaries located on the eastern oceanic coastlines of the Atlantic, such as the Galician rias on the northwest Iberian Peninsula (Spain), experience upwelling events that provide significant sources of ocean-derived N to the rias (Prego and Bao, 1997; Prego, 2002). Similar effects may be occurring in this region with DSi in the Pontevedra Ria (Dale and Prego, 2002).

Silica Cycling

Temporal and spatial variability in the annual cycles of DSi in estuaries primarily change as a function of river inputs and biological uptake. In Chesapeake Bay, sources of DSi in winter and spring were primarily from river inputs (figure 11.16; Conley and Malone, 1992). The variability in the maxima and minima of DSi decreased from the head to the mouth of the estuary, due to reduced linkages with freshwater inputs. In summer months, a significant fraction of DSi was supplied from benthic regeneration. The highest concentrations of DSi in bottom waters of the bay occurred in the summer because BSi dissolution is strongly temperature dependent (Kamatani, 1982). Much of the diatom demand for DSi can be supported by flux of this regenerated Si in sediments to the water column in Chesapeake Bay (D'Elia et al., 1983). From a spatial perspective, the highest DSi concentrations occurred in bottom waters in the mesohaline and lower bay regions, where regeneration rates were highest. However, maximum uptake rates of DSi by diatoms also occurred in the mesohaline region in spring, resulting in significant DSi limitation (figure 11.16; Conley and Malone, 1992).

The availability of DSi can regulate the composition of phytoplankton species (Kilham, 1971; Officer and Ryther, 1980; Egge and Aksnes, 1992). For example, diatom growth is dependent on the availability of DSi to nondiatom phytoplankton species which are Si independent. DSi uptake by diatoms can even occur in bottom waters or turbid surface water where light may be limiting (Nelson et al., 1981; Brzezinski and Nelson, 1989). However, once the supply of DSi has diminished diatom production will decrease and diatoms will be replaced by other phytoplankton species. Limitation of DSi in Chesapeake Bay resulted in a rapid decline in diatoms and an increase in cyanobacteria (Malone et al., 1991). Similarly, an estimated 50% decline in DSi from the 1950s to the 1980s in the Mississippi River (Turner and Rabalais, 1991) resulted in DSi limitation for diatoms on the

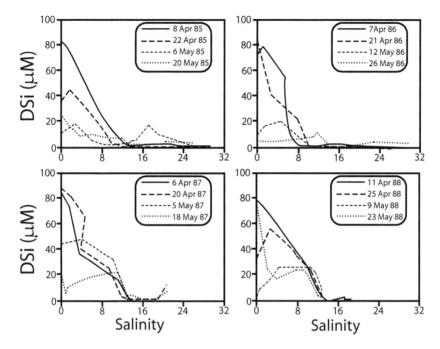

Figure 11.16 Temporal variability of concentrations of DSi (μM) in Chesapeake Bay across a salinity gradient. (Modified from Conley and Malone, 1992.)

Louisiana coast (Dortch and Whitledge, 1992). Harmful algal blooms (HABs) have been linked with decreases in the Si:N and Si:P ratios (Smayda, 1990; Anderson et al., 2002; Ragueneau et al., 2005a,b), many of the more common bloom species consist of toxic dinoflagellates (Steidinger and Baden, 1984), prymnesiophytes (Lancelot et al., 1987), and certain diatom species (e.g., *Pseudo-nitzchia australis*) (Scholln et al., 2000).

   Fluxes of DSi from sediments to the overlying waters in estuaries can represent a significant source of DSi for diatoms (D'Elia et al., 1983), particularly in shallow systems where benthic–pelagic coupling is more likely to occur. Possible mechanisms that affect the flux of DSi across the sediment–water interface in estuarine and coastal sediments include sediment permeability, biodeposition, and microphytobenthos (Ragueneau et al., 2005a,b). The typical range of DSi benthic fluxes in estuaries and coastal shelf systems is shown in table 11.6 (Ragueneau et al., 2005a). High sediment permeability, commonly found in coarse-grained sandy sediments, can result in high flushing rates and an abundance of oxidants (Middelburg and Soetaert, 2003; Precht and Huettel, 2003), which can enhance organic matter remineralization rates (Marinelli et al., 1998; Jahnke et al., 2000). High dissolution rates of BSi in sandy sediments are also believed to be linked to the high rates of organic matter turnover (Shum and Sundby, 1996; Ehrenhauss and

Table 11.6 Dissolved silica SiO$_2$ (DSi) benthic fluxes in various coastal ecosystems (all data in mmol m^{-2} d^{-1}).

Site	DSi flux	Period	Reference
Lake Michigan	2.2–10.1	April – August (1983-1985)	Conley et al. (1988)
Bay of Brest	0.8–2.6	April – June (1992)	Ragueneau et al. (1994)
Bay of Brest	0.11–6.25	May – November (2000)	Ragueneau et al. (2005b)
Baltic Sea	0.2–1.8	June (1995)	Ragueneau et al. (2005b)
San Nicolas Basin	0.9–1.3	8/83–4/85	Berelson et al. (1987)
San Pedro Basin	0.48–0.90	8/83–4/85	Berelson et al. (1987)
Potomac estuary	1–25	August 1979	Callender and Hammond (1982)
Chesapeake Bay	3.6–43.2	July 1980 and May 1981	D'Elia et al. (1983)
Skagerrak	0.55–3.97	March – September (1991–1994)	Hall et al. (1996)
N.W. Black Sea	0.2–6.7	May 1997 and August 1995	Friedrich et al. (2002)
Long Island Sound	0.8–1.1	Winter	Ullman and Aller (1982)
Bering Sea, outer shelf	0.3–0.8	1979–1982	Banahan and Goering (1986)
Amazon shelf	0.13–1.25	August 1989 – November 1991	DeMaster and Pope (1996)
Oosterschelde	7.2–112.8		Prins and Smaal (1994)
Pontevedra Ria	0.5–5.0	February – October 1998	Dale and Prego (2002)

Modified from Ragueneau et al. (2005a,b).

Huttel, 2004). Flushing of DSi from porewaters and possible linkages with bioturbation may be responsible for the higher dissolution rates (Ehrenhauss et al., 2004). Biodeposition from suspension or deposit-feeding organisms can lead to accumulation of flocculent material that is different in density and composition from surrounding sediments. These textural changes in the surface biodeposits of sediments have significant effects of the physical and chemical properties of sediments which may alter the cycling of Si (Asmus, 1986; Dame et al., 1991). In fact, it has been suggested that there is a seasonal retention of BSi in sediments caused by high densities of a benthic suspension feeder (Chauvaud et al., 2000). Whether the fluxing of DSi is driven by Fickian diffusion across concentration gradients or through the aforementioned advective processes, benthic diatoms have direct access to this available pool of DSi in overlying bottom waters or from porewaters (Facca et al., 2002). Uptake of DSi by benthic diatoms will also result in changes in DSi fluxes (Graneli and Sundbäck, 1986; Sundbäck, 1991).

Summary

1. Phosphorus (P) is one of the most well-studied nutrients in aquatic ecosystems because of its role in limiting primary production on ecological and geological timescales. Other key linkages to biological systems include the role of P as an essential constituent of genetic material (RNA and DNA) and cellular membranes (phospholipids), as well as in energy transforming molecules (e.g., ATP, etc.).

2. The cycling and availability of P in estuaries is largely dependent on P speciation. Consequently, total P has traditionally been divided into total dissolved P and total particulate P fractions, which can be further divided into dissolved and particulate organic P and dissolved and particulate inorganic P pools. Another defined fraction within the TP pool is reactive phosphorus, which has been used to describe the potentially bioavailable P. Much of the work to date has focused on the soluble reactive P, which is characterized as the P fraction that forms a phosphomolybdate complex under acidic conditions.

3. Rivers are the major source of P to the ocean, via estuaries, where major chemical and biological transformations of P occur before it is delivered to the ocean.

4. Inputs of atmospheric sources of P are generally considered to be insignificant to coastal systems and represent $<10\%$ of the riverine flux of reactive P.

5. The release of P from estuarine sediments is a common and important process that varies spatially and temporally.

6. Sedimentary pools of P have generally been divided into the following fractions: (1) organic P, (2) Fe bound P, (3) authigenic P minerals (e.g., CFA, struvite, and vivianite), and (4) detrital P minerals (e.g., feldspar).

7. When compared to freshwater systems, interactions between of Fe and S in marine and estuarine systems can further complicate the mechanisms controlling P release. Reduction of FeOOH is most important in controlling the release and efflux of PO_4^{3-} across the sediment–water interface.

8. Other microbially mediated processes can control the release of P from sediments. For example, in fresh to brackish waters Fe-reducing aerobic bacteria convert amorphous Fe(III) into Fe(II)-releasing Fe-associated PO_4^{3-}.

9. These stable DIP concentrations are believed to be controlled by a "buffering" of DIP through the adsorption and desorption onto metal oxide surfaces. This "P buffering" is believed to balance the low availability of SRP in higher-salinity waters, which occurs from phytoplankton uptake and anionic competition for surface adsorption sites.

10. Although the characterization of DOP in rivers and estuaries has been largely ignored, some work in the Mississippi River indicated that the composition of SNP primarily consisted of diester and monoester phosphates, phosphonates, orthophosphates, and/or tri- and tetrapolyphosphates.

11. One of the major differences observed between tropical and temperature P budgets is the dominance of point sources in tropical estuaries.

12. The majority of inputs to the oceans occur via rivers as both DSi and particulate BSi in suspended matter, and the losses are from sedimentation of BSi.

13. Riverine inputs of DSi to the oceans and estuaries are transformed by the growth of diatoms into a large pool of BSi.

14. Decreases in the delivery of DSi loading of Mississippi River to coastal waters have been attributed to enhanced N loading in the river, which is believed to have increased diatom production and sedimentation in the watershed.

15. Temporal and spatial variability in the annual cycles of DSi in estuaries primarily change as a function of river inputs and biological uptake.

16. Fluxes of DSi from sediments to the overlying waters in estuaries can represent a significant source of DSi for diatoms.

Chapter 12

Sulfur Cycle

Sources of Sulfur to Estuaries

Sulfur (S) is an important redox element in estuaries because of its linkage with biogeochemical processes such as SO_4^{2-} reduction (Howarth and Teal, 1979; Jørgensen, 1982; Luther et al., 1986; Roden and Tuttle, 1992, 1993a,b; Miley and Kiene, 2004), pyrite (FeS_2) formation (Giblin, 1988; Hsieh and Yang, 1997; Morse and Wang, 1997), metal cycling (Krezel and Bal, 1999; Leal et al., 1999; Tang et al., 2000), ecosystem energetics (King et al., 1982; Howarth and Giblin, 1983; Howes et al., 1984), and atmospheric S emissions (Dacey et al., 1987; Turner et al., 1996; Simo et al., 1997). The range of oxidations for S intermediates formed in each of these processes is between +VI and −II. Many of the important naturally occurring molecular species of S are shown in table 12.1. On a global scale, most of the S is located in the lithosphere; however, there are important interactions between the hydrosphere, biosphere, and atmosphere where important transfers of S occur (Charlson, 2000). For example, coal and biomass burning, along with volcano emissions inject SO_2 into the atmosphere, which can then be further oxidized in the atmosphere and removed as SO_4^{2-} in rainwater (Galloway, 1985). An example of biogenic sulfur formation is the reduction of seawater SO_4^{2-} to sulfide by phytoplankton and eventual incorporation of the S into dimethylsulfoniopropionate (DMSP). DMSP, in turn, is converted to volatile dimethyl sulfide (DMS; CH_3SCH_3)*m* which is emitted to the atmosphere. In the seawater, SO_4^{2-} represents one of the major ions, with concentrations that range from 24 to 28 mM, which is considerably higher than the concentrations found in freshwaters ($\sim$0.1 mM). This marked difference makes seawater the major input to estuaries and sets up an important gradient in estuarine biogeochemical cycling. In this chapter, the focus will be on the nonanthropogenic biogenic transformations of S that are relevant to biogeochemical cycling in estuarine and coastal waters.

Approximately 50% of the global flux of S to the atmosphere is derived from marine emissions of DMS. Oxidation of DMS in the atmosphere leads to production of SO_4^{2-} aerosols, which can influence global climate patterns (Charlson et al., 1987; Andreae and Crutzen, 1997). The key processes controlling DMS emissions from the euphotic zone

373

Table 12.1 Examples of some important naturally occurring sulfur compounds.

Oxidation state	Gas	Aerosol	Aqueous	Soil	Mineral	Biological
−II	H_2S, RSH RSR OCS CS_2		HS_2, HS^-, S^{2-} RS^-	S^{2-}, HS^- MS	S^2 HgS	Methionine $CH_3S(CH_2)_2CHNH_2COOH$ Cysteine $HSCH_2CHNH_2COOH$ Dicysteine
−I	RSSR		RSSR	SS^{2-} S_8	FeS_2	
0	$CH_3SOCH_3^+$					
II						
IV	SO_2	$SO_2\ H_2O$ HSO_3	$S_2O_3^{2-}$ SO_2, H_2O HSO_3 SO_3^{2-} $HCHOSO_2$	SO_3^{2-}		
VI	SO_3	H_2SO_4, HSO_4^- SO_4^{2-} $(NH_4)_2\ SO_4$, etc. Na_2SO_4 CH_3SO_3H	SO_4^{2-} HSO_4^-, SO_4^{2-} $CH_3SO_3^-$	$CaSO_4$ $ROSO_3$	$CaSO_4 \cdot H_2O$ $MgSO_4$	

Modified from Charlson (2000).

in the ocean are bacterial metabolism, water column mixing, and photochemistry (Kieber et al., 1996; Kiene and Linn, 2000). The major precursor of DMS in the ocean is the algal osmolyte DMSP [$(CH_3)_2S^+CH_2CH_2COO^-$] (Charlson et al., 1987; Dacey et al., 1987). DMS is formed by enzymatic cleavage of DMSP (Kiene, 1990). The release of dissolved DMSP (DMSPd) occurs from direct excretion, viral lysis, and grazing processes (Dacey and Wakeham, 1986; Malin et al., 1998). Once released, bacterioplankton can metabolize DMSPd where it can be converted into sulfate, DMS, and methanethiol (MeSH; CH_3SH) (Kiene and Linn, 2000). In fact, single species of bacterioplankton (e.g., *Rosebacter* spp.) may be largely responsible for most of the DMSPd metabolism and DMS generation at any given time in the ocean (Zubkov et al., 2001). Photochemical breakdown of DMS produces non–sea-salt sulfate, methanesulfonic acid (CH_3SO_3H), and sulfur dioxide (SO_2) (Andreae, 1986). Recent work also indicates that DMSP and its breakdown products are capable of scavenging hydroxyl radicals and may serve as an antioxidant system in marine algae (Sunda et al., 2002). While DMSPd is believed to be a major source of DMS (Turner et al., 1988), the budgetary transfer of DMSPd to DMS appears to occur with low efficiency (Kiene and Linn, 2000; Kiene et al., 2000).

Although there have been only a few studies to date, it has been suggested that coastal river plumes (Turner et al., 1996; Simo et al., 1997) and estuaries (Iverson et al., 1989; Cerqueira and Pio, 1999) may be important atmospheric sources of DMS. Other sulfur compounds such as carbonyl sulfide (COS) and carbon disulfide (CS_2) have also been shown to be possible sources of volatile S in estuaries. For example, significant concentrations of COS and CS_2 were found in four European estuaries, reaching as high as 220 ± 150 pM and 25 ± 6 pM, respectively (Sciare et al., 2002). Carbonyl sulfide is the most abundant sulfur compound in the atmosphere, and both COS and CS_2 may play an important role in the global radiation budget (Zepp et al., 1995). Recent estimates of CS_2 water-to-air fluxes from estuaries indicate that they may be comparable to open ocean fluxes, with both representing approximately 30% of the global budget (Watts, 2000).

Cycling of Inorganic and Organic Sulfur in Estuarine Sediments

Anaerobic sediment metabolism represents a significant pathway for carbon cycling in estuarine sediments (Jørgensen, 1977, 1982; Crill and Martens, 1987; Roden and Tuttle, 1992). In particular, SO_4^{2-} *reduction* (SR) is the terminal microbial respiration process in anaerobic sediments, when SO_4^{2-} is not limiting, leading to the formation of hydrogen sulfide (H_2S) (Capone and Kiene, 1988). Sulfate reduction has been shown to be particularly important in the S and C chemistry of highly productive shallow-water subtidal and salt marsh environments (Gardner, 1973; Howarth and Teal, 1979; King et al., 1985; Kostka and Luther, 1994; Ravenschl et al., 2000). In fact, some SO_4^{2-}-reducing bacteria (SRB) are closely associated with the *rhizosphere* of *S. alterniflora*, an interaction that is critical in controlling biogeochemical cycling in marsh sediments (Hines et al., 1989). Some of the dominant SRB genera within the families of *Desufovibrionaceae* and *Desulfobacteriaceae* are *Desulfovibrio desulfuricans*, *Desulfobulbus propionicus*, *Desulfobacter* spp., *Desulfococcus multivorans*, *Desulfosarcina variabilis*, and *Desulfobacterium* spp.

(Rooney-Varga et al., 1997; King et al., 2000). Some of these SRB are also important in CH_3Hg formation in contaminated sediments (more details provided in chapter 14). Sources of S to sediments are SO_4^{2-} and detrital sulfur inputs to the sediment–water interface. Organic matter decomposition via SR can be represented by the following equation (Richards, 1965; Lord and Church, 1983):

$$2[(CH_2O)_c(NH_3)_n(H_3PO_4)_p] + c(SO_4^{2-})$$

$$\rightarrow 2c(HCO_3^-) + 2n(NH_3) + 2p(H_3PO_4) + c(H_2S) \qquad (12.1)$$

where: c, n, and p represent the C:N:P ratio of the decomposing organic matter.

A significant fraction of the sulfides formed by SR are reoxidized to SO_4^{2-} at the oxic-anoxic interface in sediments. As shown in Chesapeake Bay sediments, SR leads to a decrease in SO_4^{2-} and an increase in H_2S in porewaters with increasing sediment depth over different seasons (figure 12.1; Marvin-DiPasquale et al., 2003). Higher depletion rates of SO_4^{2-} with depth at the mid-bay compared to upper and lower bay regions are primarily driven by differences in the amount and quality of organic matter, as well as temperature effects on microbial activity. This supports early work on SR that established organic matter as the primary controlling mechanism of SR rate (SRR) (Berner, 1964; Goldhaber and Kaplan, 1974; Lyons and Gaudette, 1979). A range of SRRs for estuarine and shallow coastal systems is shown in table 12.2 (Roden and Tuttle, 1993a); the wide range of SRRs observed here is due to the many factors that control SR in estuarine systems. For example, when examining the spatial and temporal variability of SR in Chesapeake Bay it was found that temperature accounted for 33 to 68% of the annual temporal variability in SRRs rates (Marvin-DiPasquale and Capone, 1998; figure 12.2). Although the kinetics of SR in laboratory experiments have been described as a first-order Monod-type saturation effect that involves concentrations of both SO_4^{2-} and organic matter, in many cases temperature may have the greatest influence (Nedwell and Abram, 1979). The effect of SO_4^{2-} limitation on SRR typically occurs at only very low concentrations (e.g., <3 mM) (Boudreau and Westrich, 1984). Spatial variability across the three regions of Chesapeake Bay was found to be controlled by the quality and quantity of organic matter, SO_4^{2-} availability, bioturbation, abundance of dissolved O_2 in overlying waters, rates of sulfide reoxidation, and availability of iron sulfide minerals (Marvin-DiPasquale and Capone, 1998; figure 12.2). Although SRR were lower in the upper bay, it should be noted that SRB in freshwater and oligohaline regions of estuaries have a greater affinity for SO_4^{2-} than do SRB in regions with typically higher concentrations of SO_4^{2-} (Lovley and Klug, 1983; Roden and Tuttle, 1993b). Using intact core incubations, kinetic experiments yielded a half-saturating SO_4^{2-} concentration (K_s) of 34 μM for the oligohaline regions of Chesapeake Bay; this was approximately 20-fold less than that found for cores from the mesohaline region (Roden and Tuttle, 1993b).

Dissolved sulfides (DS) ($= S^{2-} + HS^- + H_2S$) formed in porewaters during SR can diffuse into overlying bottom waters and contribute to O_2 depletion in estuaries (Tuttle et al., 1987). Sulfides can also be removed from porewaters via reactions with Fe oxyhydroxides to form pyrite (FeS_2) (Berner, 1970, 1984). In addition to vertical molecular diffusion as a controlling factor in the DS transport, gas bubble ebullition, derived from CH_4 production beneath the SR zone can be important in stripping DS from porewaters

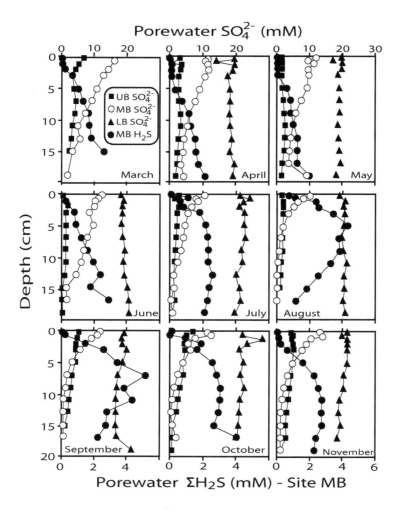

Figure 12.1 Depth profiles of SO_4^{2-} (mM) and H_2S (mM) in porewaters in the upper (UB), middle (MB), and lower (LB) Chesapeake Bay sediments. (Modified from Marvin-DiPasquale et al., 2003.)

(Roden and Tuttle, 1992). Based on studies in Chesapeake Bay (Roden and Tuttle, 1993a), Long Island Sound (Berner and Westrich, 1985), and Danish coastal sediments (Jørgensen, 1977), most of the sulfides formed in sediments over an annual cycle are not retained in sediments. In the middle and lower Chesapeake Bay, a comparison of annual SR with sulfide burial showed that <30% of the total sulfide production was permanently buried in mid-bay sediments (Roden and Tuttle, 1993a), with more recent estimates indicating it can be as low as 4 to 8% (Marvin-DiPasquale and Capone, 1998). However, >50% or of the reduced S was buried annually in the upper bay, due to higher amounts of reactive

Table 12.2 Comparison of depth-integrated sulfate reduction rates (upper 10–15 cm) in shallow-water subtidal and intertidal coastal sediments.

Location	Description	Water depth (m)	Temperature (°C)	Sulfate reduction (mmol m^{-2}d^{-1}) Measured	Corrected[a]	Source
Mid-Chesapeake Bay	Silt-clay ± bioturbated	10–40	25	10–90	10–90	Roden and Tuttle (1993a,b)
Lower Chesapeake Bay	Sand-silt bioturbated	12	25	25–125	25–125	Berner and Westrich (1985)
Long Island Sachem	Silt-clay lagoon	2	23	10	12	
Sound, NY, BH	Silt-clay lagoon	?	22	50	70	Goldhaber et al. (1977)
USA NWC	Silt-clay bioturbated	15	21	2	3	
FOAM	Silt-clay bioturbated	8	20	8	14	
DEEP	Silt-clay bioturbated	34	22	2	3	Aller (1980)
Buzzards Bay, MA, USA	Fine grained bioturbated	15	19	12	23	Novelli et al. (1988)
Town Cove, MA, USA	Fine grained lagoon	?	21	118	183	Sampou and Oviatt (1991)
MERL mesocosms, Narragansett Bay, RI, USA	Fine grained bioturbated	5	22	35–120	50–170	
Mud Bay, SC, USA	Fine grained bioturbated	2	22	9	13	Aller (1980)
Cape Lookout, Bight, NC, USA	Fine grained lagoon	10	25–27	90–180	90–180	Crill and Martens (1987)
Limfjorden, Denmark	Silt-clay bioturbated	4–12	15–20 (17.5)	7–15	16–34	Jørgensen (1977)
Danish coastal lagoons	Silt-sand	1	10	21.7	113	Thode-Andersen and Jørgensen (1989)

Location	Habitat					Reference
Baltic Sea	Eelgrass detritus	1	18	62.1	134	Bagander (1977)
	Silt with macroalgae	2	19	45.8	89	
Baltic Sea—north	Silt-clay bioturbated	10	10–12 (11)	4–8	19–37	Jørgensen (1989)
	Muddy sand/sandy mud	12–16	5–10	3–30	21–205	
Sea transition	Fine or medium sand	7–16	5–10	0.2–1.4	1–10	
	Silty mud	15–17	5–10 (7.5)	7–13	48–89	
NW Swedish coast	Fine-grained bioturbated	8–13	15	12	36	Gunnarsson and Ronnow (1982)
Easter Scheldt,	Under mussel culture	8–13	15	30	90	Oenema (1990a)
	Mussel beds	?	17–20 (18.5)	37–140	76–285	
SW-Netherlands	Abandoned channels (fine grained)	?	15–16 (15.5)	10–60	28–170	
Delaware Inlet, Nelson, New Zealand	Muddy intertidal		12	20	83	Mountfort et al. (1980)
Tay estuary, Scotland	Muddy intertidal		12.5	22.3	88	Parkes and Buckingham (1986)

[a] Corrected to a temperature of 25°C assuming a Q_{10} of 3; values in parentheses were used as basis for temperature correction when a range of temperatures was reported.
Modified from Roden and Tuttle (1993a).

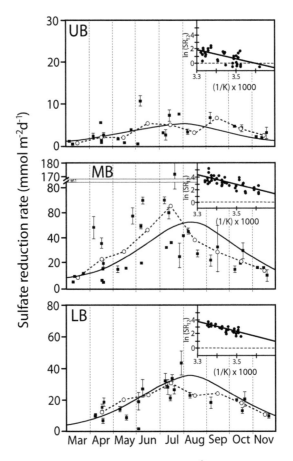

Figure 12.2 Spatial and temporal variability in SO_4^{2-} reduction rates (SRR) (mmol $m^{-2} d^{-1}$) across the three regions [upper, middle, and lower bay (UB, MB, and LB)] of Chesapeake Bay. (Modified from Marvin-DiPasquale and Capone, 1998.)

Fe and lower rates of bioturbation. This work supports earlier studies which suggested that the availability of easily reducible Fe (Berner, 1970; Pyzik and Sommer, 1981), bioturbation, and sedimentation rate (Chanton et al., 1987a) are key in controlling sulfide retention in sediments. The relatively rapid cycling of S accounts for a significant fraction of C mineralization as well as dissolved O_2 consumption in estuarine bottom waters.

Although SR has been viewed as the dominant pathway of organic matter oxidation in anaerobic salt marsh sediments (see reviews by Howarth, 1993, and Alongi, 1998), other work has suggested that microbial Fe(III) reduction (FeR) is also key in controlling the oxidation of organic C in these sediments and is inherently linked with dissolved sulfide formed from SR (Kostka and Luther, 1995). The role of Fe(III), Mn(IV), and U(VI) oxides in the nonenzymatic oxidation of organic matter has been well documented in marine and estuarine sediments (Aller et al., 1986; Sørensen and Jørgensen, 1987; Canfield et al., 1993). However, more recent work has demonstrated that enzymatic, via FeR bacteria (FeRB), in freshwater sediments and marine sediments is quantitatively more important than nonenzymatic processes (Lovley, 1991; Lovley et al., 1991, 1993). Using 16S rRNA phylogenetic analyses, previous work in freshwater sediments and waters has identified the dominant FeRB as *Geobacter metallireducens* (Lovley et al., 1987), *Shewanella putrefaciens* (Lovley et al., 1989), and *Pseudomonas* sp. (Balashova and Zavarzin, 1980); more recent work demonstrated that *Delsulfuromonas acetoxidans* is capable of reducing Fe(III) and Mn(IV) in marine and estuarine sediments (Roden and Lovley, 1993). In the breakdown of organic matter, large molecules generally undergo hydrolysis and subsequent fermentation to low molecular weight fatty acids, hydroxyl acids, alcohols, and molecular hydrogen (H_2) (Novelli et al., 1988). In particular, acetate (CH_3COO^-) and H_2 are considered to be the most abundant fermentation products utilized by SRB and FeRB (Jørgensen, 2000; Thamdrup, 2000; Kostka et al., 2002a). The stoichiometric equations of these two microbial respiratory pathways are as follows:

$$SO_4{}^{2-} + CH_3COO^- + 2H^+ \rightarrow 2CO_2 + 2H_2O + HS^- \tag{12.2}$$

$$8FeOOH + CH_3COO^- + 17H^+ \rightarrow 2CO_2 + 14H_2O + 8Fe^{2+} \tag{12.3}$$

It should be noted that there is a 1:2 ratio of $SO_4{}^{2-}$ reduced to C oxidized, which has been observed in field sediments (Jørgensen, 2000; Kostka 2002b). Sulfate-reducing bacteria that utilize CH_3COO^- appear to dominate over SRB that do not (Rooney-Varge et al., 1997; King et al., 2000). Using labeled ^{14}C-acetate it was estimated that 2 to 100% of the measured acetate pool was used in estuarine sediments and that the bioavailable pool decreased with sediment depth (Novelli et al., 1988). In the case of FeRB, while *amorphous Fe*(III) (e.g., oxyhydroxides) is considered to be the most predominant form reduced by FeRB, *crystalline Fe*(III) (e.g., goethite) has also been shown to be available for FeRB (Roden and Zachara, 1996)—more details provided on Fe (III) oxides later in this chapter.

Higher activities of FeRB have been linked with vegetated and bioturbated sediments (Kostka et al., 2002b). Bioturbation and macrophytes have also been shown to be important factors controlling SRR, whereby oxidants [e.g., O_2, Fe(III), and $SO_4{}^{2-}$] can be mixed to significant depths in sediments (Hines et al., 1989, Hines, 1991; Kostka et al., 2002a,b). For example, fiddler crabs have burrows that are 5 to 25 cm deep and can occur at densities ranging from 224 to 480 burrows m^{-2} (Bertness, 1985). Similarly, marsh plants (e.g., *Spartina*) have root zones that can enhance the exchange of O_2 between sediments and the overlying water and/or atmosphere via *evapotranspiration* as well as *passive diffusion* (Dacey and Howes, 1984). Macrophyte roots may also provide DOM and POM that can be used directly as a substrate for SR (Schubauer and Hopkinson, 1984). When examining the distribution of solid-phase Fe and S in Georgia (USA) marsh sediments with (BVL, bioturbated vegetated levee) and without (NUC, nonbioturbated unvegetated creek-bank) bioturbation (*Uca* burrows) and vegetation (*S. alterniflora*), there is clearly less total Fe(II) in the nonbioturbated unvegetated NUC site compared to the vegetated bioturbated BVL site (figure 12.3; Kostka et al., 2002b). Total reduced S (TRS) (= H_2S + S^0 + FeS + FeS$_2$) showed the opposite trends to Fe with increasing sediment depth,

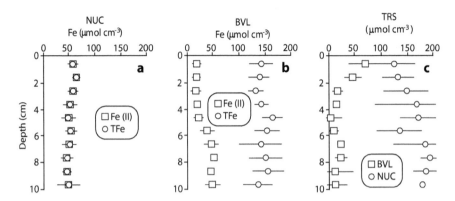

Figure 12.3 The distribution of solid-phase Fe(II), S, and total reduced S (TRS = H_2S + S^0 + FeS + FeS_2) in Georgia (USA) marsh sediments with (BVL) and without (NUC) bioturbation (*Uca* burrows) and vegetation (*S. alterniflora*). (Modified from Kostka et al., 2002b.)

with TRS increasing and decreasing in the NUC and BVL sites, respectively (figure 12.3). The FeRB were also two orders of magnitude higher in the BVL (10^7 cell g^{-1}) compared to NUC sediments. These results demonstrate that SR was the dominant respiration pathway in NUC sediments with FeR dominating in the BVL sediments. Thus, activities from bioturbation and vegetation can stimulate FeRB enough to outcompete SRB in salt marsh sediments (Kostka et al., 2002a,b). In the presence of Fe(III) oxyhydroxides, dissimilatory FeR is energetically favored over SR; however, since SRB produce sulfide which is a reductant of Fe(III) (Yao and Millero, 1996; von Gunten and Furrer, 2000), SRB can limit FeRB via abiotic reduction of reactive Fe(III) oxyhydroxides (Burdige, 1993; Wang and van Cappellen, 1996). The mechanism resulting in the inhibition of FeRB was observed in Sapelo Island marsh (USA) sediments during summer months, when high rates of sulfide production by SRB lead to the reduction of Fe(III) oxyhydroxides (Koretsky et al., 2003). While most prior work on SR has been in *Spartina*-dominated marshes, recent work has shown that some of the highest annual integrated SR rates (22.0 mol SO_4^{2-} $m^{-2}y^{-1}$) in wetland sediments were observed in a *Juncus roemerianus* (Needlerush) marsh on the Gulf coast (USA) (Miley and Kiene, 2004). Despite such rates of SR, porewater concentrations of DS were low (<73 μM), suggesting the importance of rapid oxidation or precipitation processes—more work is clearly needed in these subtropical systems.

The flux of DS across the sediment–water interface can in some cases be strongly influenced by the presence of chemoautotrophic bacterial mats in estuaries. These sulfur oxidizing bacteria occur at the oxic–anoxic interface. The dominant colorless bacteria that live within the microzone of the O_2 and H_2S interface (~1–2 mm) are *Beggiatoa* and *Thiovulum* (Jørgensen and Revsbech, 1983; Jørgensen and Des Marais, 1986). This early work demonstrated through the use of microelectrodes that these bacteria live within the microgradient of O_2 and H_2S (figure 12.4; Jørgensen and Revsbech, 1983). The steep H_2S and O_2 gradient typically occurs between 0 and 0.5 mm. In the presence of O_2, these bacteria can oxidize H_2S to S^0, which can be further oxidized to SO_4^{2-}

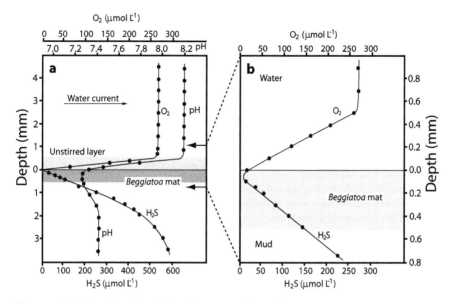

Figure 12.4 Depth profile of the microzone of O_2 and H_2S at the sediment–water interface ($\sim$ 1–2 mm) where *Beggiatoa* and *Thiovulum* thrive. (Modified from Jørgensen and Revsbech, 1983.)

(Nelson and Castenholz, 1981). These bacteria enzymatically oxidize H_2S in the presence of O_2, despite rapid abiotic oxidation of H_2S. The common pigmented sulfur bacteria are the green sulfur bacteria (GSB) (e.g., *Prosthecochloris aestuarii*), which are obligate anaerobes that utilize H_2S as the dominant electron acceptor for *photolithoautotrophy* (Massé et al., 2002). These GSB are commonly found in occurrence with brown-colored GSB (e.g., *Chlorobium vibriforme*) in brackish to hypersaline environments (Pfennig, 1989). Morphological and ultrastructure changes in GSB, particularly in *P. aestuarii*, have been observed in response to changing light levels in surface sediments (Guyoneaud et al., 2001). The brown-colored pelagic GSB are distinct in their carotenoid and chlorophyll pigments when compared to the other GSB (Repeta et al., 1989; Overmann et al., 1992). These pelagic species are commonly found in deep anoxic basins or shallow water environments, at a zone in the water column where there is an adequate light penetration and sulfide concentration (Repeta and Simpson, 1991; Chen et al., 2001). Another important group of pigmented S bacteria are the purple sulfur bacteria (PSB) (e.g., *Thiocapsa roseopersicina*) which are chemolithotrophic and can actually grow in the presence of O_2, which can be used as an electron acceptor. In addition to these S bacteria, there are non–S-oxidizing cyanobacterial mats (e.g., *Microcoleus chthonoplastes*) that can also have an indirect impact on the oxidation of sulfides in surface sediments through the production of O_2 during photosynthesis (Canfield and Des Marais, 1993).

The uptake of sulfides by rooted wetland plants living in anoxic sediments presents the problem of coping with toxicity since H_2S is highly toxic to many organisms (Howarth and Teal, 1979). Dissolved sulfides have also been shown to inhibit coupled nitrification–denitrification (Joye and Hollibaugh, 1995). Stable S isotopes can be used to determine trophic pathways and possible sources of S in estuaries. For example, stable isotope work indicates that there are four isotopically different pools of δ^{34}S in estuaries: (1) seawater SO_4^{2-} (+20‰), (2) sulfides formed by SR (−23 to −24‰), (3) porewater SO_4^{2-} (+15 to +17‰), and (4) rainfall inputs (+6.3‰) (Fry et al., 1982). For example, it has been well established that SRB preferentially utilize the lighter S isotope (Goldhaber and Kaplan, 1974; Chambers and Trudinger, 1978). Based on these available pools of S and the δ^{34}S values of different marsh plant tissues in estuaries, it appears that plants are likely converting sulfides into nontoxic forms during photosynthesis (Knobloch, 1966; Winner et al., 1981). However, the specific mechanisms of uptake and detoxification in marine algae and vascular marsh plants are not well understood and require further investigation. Another important factor that determines the availability of these different S isotopes to organisms concerns their differential advective and diffusive transport in sediments. For example, ^{32}S enrichment is commonly observed in many marine sediments (Goldhaber and Kaplan, 1980). In an attempt to obtain closure of the S cycle, a budget using S masses and isotopic fluxes was developed in coastal marine sediments (figure 12.5;

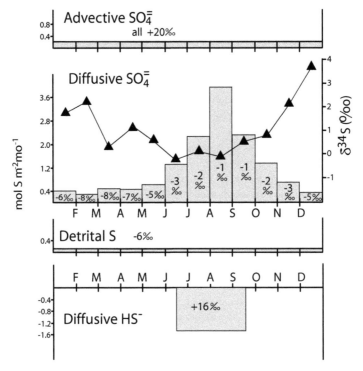

Figure 12.5 Sulfur isotopic budget based on mass isotopic fluxes (bar graphs) and their isotopic composition on a monthly basis in coastal marine sediments of cape Lookout Bight, NC (USA). Units are in mol S m^{-2} m^{-1}. (Modified from Chanton and Martens, 1987b.)

Chanton and Martens, 1987b). This budget successfully predicted the isotopic composition of buried S within 0.5‰, thereby confirming that isotopes diffuse in accordance to their concentration gradients across the sediment–water interface. Moreover, this work demonstrated that compounds with different isotopic constituents diffuse differently in response to concentration gradients and that isotopic composition of compounds diffusing across a particular horizon should not be expected to be the same for compounds located at the horizon (Jørgensen, 1979; Chanton and Martens, 1987b).

In sediments, Fe sulfides are typically divided into the following two groups: *acid volatile sulfides* (AVS), which are evolved via acid distillation and generally include amorphous forms [e.g., mackinawite (FeS), greigite (Fe_3S_4), and pyrrhotite (FeS)] (Morse and Cornwell, 1987), and pyrite (FeS_2). In some cases AVS may represent the dominant pool of sulfides in estuarine sediments (Oenema, 1990a). Pyrite (FeS_2) is the dominant Fe–sulfur mineral in most estuarine systems, particularly in salt marsh sediments (Hsieh and Yang, 1997). In sediments with high concentrations of sulfide, reductive dissolution of Fe(III) oxyhydroxide phases results in the formation of FeS_{aq} and FeS_2 as shown in the equation below:

$$Fe^{2+} + HS^- \rightarrow H^+ + FeS_{aq} \tag{12.4}$$

Pyritization can then occur under anoxic conditions relatively quickly by reaction of H_2S and S(0) [as S_8 or polysulfides (S_x^{2-})] with FeS, both aqueous and solid forms (Rickard, 1975,1997; Luther, 1991; Rickard and Luther, 1997; Rozan et al., 2002), in the following equations:

$$FeS_{aq} + H_2S_{aq} \rightarrow FeS_2 + H_2 \tag{12.5}$$

$$FeS_{aq} + S_x{}^{2-} \text{ or } [S_8] \rightarrow FeS_2 + S_{(x-1)}{}^2 \tag{12.6}$$

Pyrite is considered to be very stable under reducing conditions where it can be preserved over geological time thereby retaining high amounts of energy (Howarth, 1984); however, under oxidizing conditions FeS_2 decomposes rapidly. Although there can be considerable differences in the rates of pyritization of different precursor iron hydroxide minerals (Canfield and Berner, 1987; Canfield, 1989; Canfield et al., 1992; Raiswell and Canfield, 1996), the actual effects of mineralogy appear to only be important in the initial iron sulfidization and not in pyritization rates (Canfield et al., 1992; Morse and Wang, 1997). Variables such as pH, sulfide concentration, and organic matter abundance and composition were shown to be more important than different Fe hydroxide crystalline minerals [e.g., goethite (FeOOH), hematite (Fe_2O_3), and magnetite (Fe_3O_4)] in controlling rates of pyritization (Morse and Wang, 1997). *Euhedral* single crystals of FeS_2 are formed directly (Rickard, 1975; Luther et al., 1982), as opposed to widely abundant *framboidal* FeS_2 which are formed through Fe monosulfide and intermediates (Sweeney and Kaplan, 1973; Raiswell, 1982). Finally, in cases where there is a weak $\sum H_2S$ gradient, diffusion to the redox boundary layer is slow, and there is limited formation of reduced S species, the conversion of AVS into FeS_2 will be inhibited resulting in the build-up of AVS in sediments (Gagnon et al., 1995).

The degree of pyritization (DOP) is a parameter first used by Berner (1970) to distinguish environments where FeS_2 is Fe or carbon limited (Raiswell and Berner, 1985). It is a measure of the extent to which the original reducible or reactive Fe has been converted into FeS_2. The original equation defining DOP has been modified in recent years because of the operational definition of reactive nonsulfidic Fe; the following equation that incorporates some of these modifications was recently used by Rozan et al. (2002):

$$DOP = [FeS_2]/([FeS_2] + [AVS - Fe] + [Dithionite - Fe]) \qquad (12.7)$$

where: FeS_2 = concentration of reduced S as pyrite; AVS = acid volatile sulfur, composed of both aqueous and solid FeS; and Dithionite − Fe = a measure of nonsulfidic Fe.

It should be noted that "reactive Fe" is operationally defined as the dithionate-extractable Fe minus the Fe associated with FeS—this definition incorporates crystalline Fe(III) minerals and authigenic silicates that may become solubilized during the extraction procedures (Kostka and Luther, 1994). Much of the reactive Fe(III) in surface sediments appears to be represented by amorphous Fe(III) minerals as opposed to crystalline Fe(II) at depth (Kostka and Luther, 1995). For sediments that have high concentrations of AVS a different parameter called the degree of sulfidization (DOS) was first introduced by Boesen and Postma (1988) and recently modified by Rozan et al. (2002) as follows:

$$DOS = ([FeS_2] + [FeS])/[FeS_2] + [AVS - Fe] + [Dithionite - Fe]) \qquad (12.8)$$

Pyrite has been shown to recycle more rapidly in salt marsh sediments than in other coastal systems (Howarth, 1979; King, 1983; Lord and Church, 1983). The tidal rhythm, root metabolism, and atmospheric exposure time to oxygen creates dynamic redox conditions believed to be conducive for the formation of FeS_2 in marsh sediments (Oenema, 1990b). During certain seasons the redox conditions of the upper sediments in shallow subtidal and intertidal marsh systems becomes oxidizing resulting in the oxidation of sulfide and sulfide minerals to thiols and SO_4^{2-} (Lord and Church, 1983; Luther and Church, 1988). In fact, *Spartina* marshes have been shown to have a high oxidizing capability and ability to bind dissolved sulfides, which is coupled with evapotranspiration and/or water uptake in the root zone (Fry et al., 1982; Dacey and Howes, 1984). Thus, hydrology and marsh vegetation is important in controlling FeS_2 accumulation. For example, rates of FeS_2 formation and DOP in *Spartina* salt marshes in the eastern Schelde (The Netherlands) ranged from 2.6 to 3.8 mol FeS_2 m^{-2} y^{-1} over a depth range of 15 to 20 cm, which varied over marsh height (Oenema, 1990b). Pyrite oxidation was most dramatic in the upper 5 to 10 cm of the medium and high marshes (figure 12.6), where O_2 was likely introduced into the upper root zone through diffusion and air pores in marsh plant tissues (Mendelsohn et al., 1981). It was concluded that FeS_2 distribution in these sediments was primarily controlled by: (1) sedimentation of detrital FeS_2, (2) FeS_2 oxidation in the upper root zone, and (3) FeS_2 formation at the interface between suboxic and anoxic zones (Oenema, 1990b). Increases in the oxidation of FeS_2 in marsh sediments during spring and summer result from a positive oxidation potential created by the root zone of *S. alterniflora* (Giblin and Howarth, 1984; Gardner et al., 1988; Kostka and Luther, 1994). Other work has suggested that the unusually high concentrations of

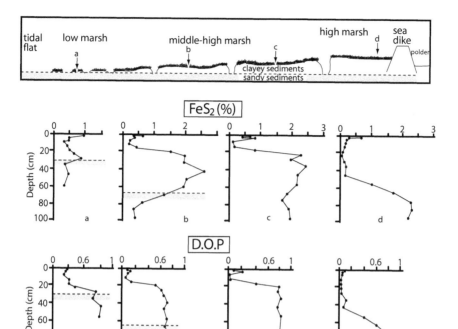

Figure 12.6 Depth distribution of FeS₂ formation and degree of pyritization (DOP) in relation to height in the Rattekaai *Spartina* salt marsh in the eastern Scheldt (The Netherlands). (Modified from Oenema, 1990b.)

thiols (e.g., glutathione) commonly found in salt marshes may be important in controlling the turnover of FeS_2 in these environments (Luther et al., 1986). In fact, FeS_2 may serve as a starting material in thiol production, where $S(0)$ and $S(-II)$ (both found in FeS_2) are used by chemosynthetic bacteria to form thiol (RSH) (e.g., glutathione) and SO_4^{2-} (figure 12.7). Pyrite can also be oxidized by the bacteria *Thiobacillus* spp. in the presence of O_2 or NO_3^- (King, 1983). This represents a critical step connecting inorganic and organic S in the dynamic recycling of FeS_2 in marsh sediments.

Early work identified salt marshes as localized sites of high S emission (Hitchcock, 1975; Goldberg et al., 1981; Steudler and Peterson, 1985). The two major S gases that make up the bulk of the flux from salt marshes, in highly vegetated areas, are H_2S and DMS (Steudler and Peterson, 1985; De Mello et al., 1987). Based on a comparison between different marsh plants, DMSP was found to be in highest concentration in the leaves of *Spartina alterniflora* (80–300 μmol g dry wt.$^{-1}$), with lower concentrations in the roots and rhizomes (20–60 μmol g dry wt.$^{-1}$) (Dacey et al., 1987). Other marsh grasses (e.g., *S. anglica*) (van Diggelen et al., 1986), macroalgae (e.g., *Ulva* sp.) (Jørgensen and Okholm-Hansen, 1985; Sørensen, 1988), and sea grasses (e.g., *Zostera marina*) (White, 1982) have also been shown to contain relatively high concentrations of DMSP. It should be noted that production of DMSP by *Z. marina* was likely affected

Figure 12.7 Pathway of $SO_4{}^{2-}$ reduction where FeS_2 may serve as a starting material in thiol production, where $S(0)$ and $S(-II)$ (both found in FeS_2) are used by chemosynthetic bacteria to form thiol (RSH) (e.g., glutathione) and $SO_4{}^{2-}$. (Modified from Luther et al., 1986.)

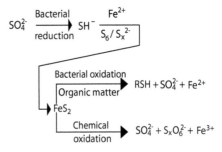

by algal epiphytes on the seagrass. The differences in turnover time of H_2S (0.1% d^{-1}) (King, 1988) and DMS ($100-30,000\%$ d^{-1}) (Howes et al., 1985) indicate that the sources and transformation processes controlling the cycling of these gases are clearly different. It was shown in a later study that these differences can be explained by the biological pathways in which these gases are generated and cycled. For example, the release of H_2S from marsh sediments predominantly occurs across nonvegetated sediments and is controlled by anaerobic decomposition processes. Conversely, DMS is controlled more by the distribution of DMSP in plants and their associated physiology (Dacey et al., 1987).

Cycling of Inorganic and Organic Sulfur in Estuarine Waters

The emergence of DS from sediments into stratified bottom waters has been shown to contribute significantly to bottom-water oxygen depletion in estuaries (Tuttle et al., 1987; Roden and Tuttle, 1992). For example, as discussed earlier DS release from mesohaline sediments in the Chesapeake Bay is important in controlling anoxia in bottom waters. Concentrations of H_2S can reach as high as 60 μM in bottom waters of the Chesapeake Bay, where it poses a threat to benthos (Seliger et al., 1985) and in some cases the overall fish/shellfish industry (Officer et al., 1984). More recent work in some of the dead-end canals and creeks of Rehoboth Bay (USA) reported major fish kills (2.5 million juvenile menhaden, *Brevoortia tyrannus*) and concentrations of H_2S that reached as high as 400 μM in surface waters (Luther et al., 2004). Hydrogen sulfide's toxicity stems from its ability to combine with the Fe–heme of blood cells, thereby replacing O_2 and inhibiting respiration (Smith et al., 1977). The persistence and concentrations of H_2S in the bottom waters of estuaries are controlled by many factors such as SRR, which is affected by organic matter/nutrient loading, temperature, and parameters controlling the flux of H_2S across the sediment–water interface and the transfer to surface waters (e.g., diffusive/advective processes and S-oxidizing bacteria).

Reduced sulfur compounds (e.g., sulfide and thiols) are some of the most important metal ligand groups in complexing B-type (transition) metals in aquatic systems (Krezel and Bal, 1999). In particular, glutathione is one of the most abundant thiols found in organisms and has been shown to be important in protecting cells against radiation damage as well as high levels of heavy metals (Giovanelli, 1987). The sulfhydryl (-SH) group in glutathione is responsible for most of the complexation of metals such

as Cu, Pb, Hg, Cd, and Zn (Krezel and Bal, 1999). Production of metal-binding peptides, or phytochelatins, has been shown to be important for phytoplankton under metal
stress (Ahner et al., 1994, 1995, 1997, 2002). The typical range of glutathione in estuar-
ine/coastal systems is 20 to 600 pM (Matrai and Vetter, 1988). Glutathione is the major low
molecular weight thiol in phytoplankton (Rijstenbil and Wijnholds, 1996), representing a
significant in situ source in aquatic systems. The bimodal distribution of glutathione in
dissolved, colloidal, and particulate materials in Galveston Bay likely suggests the importance of phytoplankton sources and not riverine sources (figure 12.8; Tang et al., 2000).

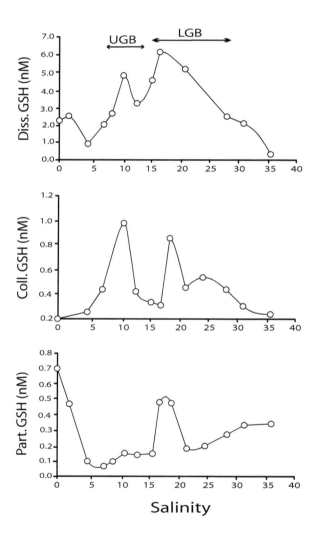

Figure 12.8 Bimodal distribution of glutathione (nM) in dissolved, colloidal, and particulate materials in Galveston Bay (GB) (USA) across a salinity gradient in upper and
lower GB (UGB and LGB). (Modified from Tang et al., 2000.)

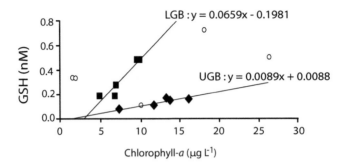

Figure 12.9 Relationship between chlorophyll-*a* (μg L^{-1}) and particulate thiol (nM) in upper and lower regions of Galveston Bay (UGB and LGB). Stations with open circles (river and seawater end-members) were not used in the regression. (Modified from Tang et al., 2000).

This is further supported by the positive correlation between glutathione and chlorophyll-*a* concentrations (figure 12.9). Release of glutathione from phytoplankton cells can occur through enzymatic control (Meister and Anderson, 1983) as well as grazing activities by zooplankton.

The lack of a strong correlation between chlorophyll-*a*, phytoplankton productivity/ biomass, and DMS (Barnard et al., 1984; Turner et al., 1988) has been attributed to seasonal variability in the taxonomic composition of phytoplankton, since only certain species of phytoplankton produce DMS (Andreae, 1986; Turner et al., 1988). Positive correlations between salinity and DMSPd and DMS concentrations in both Delaware and Chesapeake Bays reflected changes from low DMSP-producing phytoplankton (e.g., diatoms) in estuaries to more coastal and oceanic populations of phytoplankton [e.g., prymnesiophytes (*Phaeocystis* spp.)] that have higher DMSP production (Iverson et al., 1989). Concentrations of DMS in surface waters of the Baltic Sea revealed clear seasonal variation that ranged from 2 to 200 ng L^{-1} (Leck et al., 1990). While no correlation was found between DMS and chlorophyll-*a* concentration and/or phytoplankton productivity there were correlations on very select timescales with certain types of phytoplankton (figure 12.10). For example, diatoms were not found to be associated with DMS production in the Baltic Sea, while certain dinoflagellates and cyanobacteria were in some cases associated with DMS production (Leck et al., 1990). However, there was significant correlation between DMS and both copepods and total zooplankton (figure 12.11). Prior work has suggested that DMSP can be converted into DMS in the digestive tracts of zooplankton, as well as during the decay of fecal pellets (Dacey and Wakeham, 1986). Thus, zooplankton grazing, in combination with the accompanied lytic and bacterial effects, was most strongly correlated with the release of DMSP from phytoplankton cells and subsequent formation of DMS in the Baltic Sea (Leck et al., 1990). Five species of calanoid copepods from Long Island Sound were also found to have DMSP in gut contents; the carbon-specific

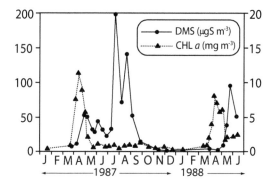

Figure 12.10 Annual variation in concentrations of dimethyl sulfide (DMS) (μg S m^{-3}) and chlorophyll-a (mg m^{-3}) in surface waters of the Baltic Sea from 1987 to 1988. (Modified from Leck et al., 1990.)

content was comparable to that found in phytoplankton-containing DMSP (Tang et al., 1999). These studies indicate that future work on the biogeochemical cycling of DMSP needs to continue to incorporate zooplankton as an important source of DMSP. Thus, removal processes of DMS in estuaries are believed to be largely controlled by biological consumption by bacteria and water column degassing (Froelich et al., 1985).

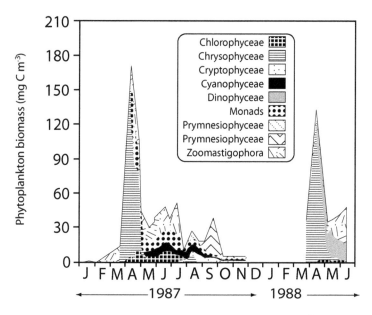

Figure 12.11 Annual variation in phytoplankton biomass (mg C m^{-3}) in surface waters of the Baltic Sea from 1987 to 1988. (Modified from Leck et al., 1990.)

Table 12.3 Comparison of seawater concentration of DMS, COS, and CS_2 in open ocean, coastal, shelf, and estuaries.

Area	DMS (nM)	COS (pM)	CS_2 (pM)	References
Open ocean (surrounding Western Europe)				
N. Atlantic Ocean (summer)	2.5			Berresheim et al. (1991)
Global (winter)	3.3			Andreae (1990)
(summer)	9.0			
N. Atlantic Ocean		5–19		Ulshofer et al. (1995)
N. Atlantic Ocean			8±4	Kim and Andreae (1987)
Shelf/coastal				
Southern North Sea (December/May)	1.8–5.9			Turner et al. (1996)
North Atlantic shelf			17±4	Kim and Andreae (1987)
Mediterranean Sea		18–23		Mihalopoulos et al. (1992)
Ionian Sea				Ulshofer et al. (1996)
N. Sea, English channel		142±90		Watts (2000)
N. American coastal waters	2.43–3.58			Iverson et al. (1989)
Estuarine				
6 Tidal European estuaries	(<0.02–10)	(60–1010)	(2–117)	Sciare et al. (2002)
	0.6	220	25	
Yarmouth, UK		130	263	Watts (2000)
Eastern seaboard USA			120	Bandy et al. (1982)
Canal de Mira (Portugal) (Winter/Summer)	2.9–5.3			Cerqueira and Pio (1999)
N. American estuaries	0.31–1.67	61–1466		Iverson et al. (1989) Zhang et al. (1998)

Modified from Sciare et al. (2002).

When comparing DMS sources in estuaries to the shelf and open ocean sources it is clear that estuaries are not a significant source to the atmosphere (table 12.3; Sciare et al., 2002). However, as mentioned earlier, COS and CS_2 may represent potentially important sources of S in estuaries (Watts, 2000; Sciare et al., 2002). Concentrations of both these compounds have been found to be 5 to 50 times higher than in coastal and open ocean environments (table 12.3). More work is clearly needed on these S compounds to better constrain spatial and temporal variability, and to evaluate their role in global S budgets.

Summary

1. On a global scale, most of the S is located in the lithosphere; however, there are important interactions between the hydrosphere, biosphere, and atmosphere where important transfers of S occur.

2. Approximately fifty percent of global biogenic flux of S to the atmosphere is derived from natural emissions of DMS which has been shown to have a significant effect on global climate patterns.

3. Although there have been only a few studies to date, it has been suggested that coastal plumes and estuaries may be important atmospheric sources of DMS. Other sulfur compounds such as carbonyl sulfide (OCS) and carbon disulfide (CS_2) have also been shown to possible sources of S in estuaries.

4. Sulfate reduction is the terminal microbial process in anaerobic sediments, when SO_4^{2-} is not limiting, leading to the formation hydrogen sulfide (H_2S). This process has been shown to be particularly important in the cycling of S and C chemistry of highly productive shallow-water subtidal and salt marsh environments.

5. Sulfate reduction occurs mainly in sediments, with rates controlled by the quality and quantity of organic matter, SO_4^{2-} availability, bioturbation, abundance of dissolved O_2 in overlying waters, rates of sulfide reoxidation, and availability of iron-sulfide minerals.

6. Dissolved sulfides DS $(DS = S^{2-} + HS^- + H_2S)$ formed in porewaters during SR can flux into overlying bottom waters and contribute to O_2 depletion in estuaries.

7. Although SR has been viewed as the dominant pathway of organic matter oxidation in anaerobic salt marsh sediments, other work has suggested that microbial Fe(III) reduction (FeR) is also key in controlling the oxidation of organic C in these sediments and is inherently linked with dissolved sulfide formed from SR.

8. The flux of DS across the sediment–water interface can in some cases be strongly influenced by the presence of chemoautotrophic bacterial mats in estuaries. These sulfur oxidizing bacteria occur at the oxic–anoxic interface and can occur as colorless or as pigmented forms (GSB and PSB).

9. The uptake of sulfides by rooted wetland plants living in anoxic sediments presents the problem of coping with toxicity since H_2S is highly toxic to many organisms.

10. Stable isotope work indicates that there are four isotopically different pools of $\delta^{34}S$ in estuaries: (1) seawater SO_4^{2-} (+20‰), (2) sulfides formed by SR (−23 to −24‰), (3) porewater SO_4^{2-} (+15 to +17‰), and (4) rainfall inputs (+6.3‰); it has been well established that SRB preferentially utilize the lighter S isotope, when reducing SO_4^{2-}.

11. In sediments, Fe sulfides are typically divided into the following two groups: acid volatile sulfides (AVS), which are evolved via acid distillation and generally include amorphous forms [e.g., mackinawite (FeS), greigite (Fe_3S_4), and pyrrhotite (FeS)], and pyrite (FeS_2).

12. FeS_2 distribution in sediments is primarily controlled by (1) sedimentation of detrital FeS_2, (2) FeS_2 oxidation in the upper root zone, and (3) FeS_2 formation at the interface between suboxic and anoxic zones.

13. The release of H_2S from marsh sediments predominantly occurs across nonvegetated sediments and is controlled by anaerobic decomposition processes. Conversely, DMS is controlled more by the distribution of DMSP in plants and their associated physiology.

14. The emergence of DS from sediments into stratified bottom waters has been shown to contribute significantly to bottom-water oxygen depletion in estuaries.

Chapter 13

Carbon Cycle

The Global Carbon Cycle

Carbon is the key element of life on Earth and exists in more than a million compounds (Holmén, 2000; Berner, 2004). The unique covalent long-chained and aromatic carbon compounds form the basis of organic chemistry and the "roadmap" for understanding life from the cellular to the ecosystem level. The oxidation states of C atoms range from +IV to −IV; methane (CH_4) is the most reduced form of C (−IV), with CO_2 and other carbonate forms existing in the most oxidized state (+IV). The major reservoirs of C are stored in the Earth's crust, with much of it as inorganic carbonate and the remaining as organic C (e.g., kerogen) (figure 13.1; Sundquist, 1993). The global C cycle can be divided into short- and long-term cycles based on the vast differences in the turnover times of different C pools (Berner, 2004). The carbonate reservoir can be divided into two primary subreservoirs: (1) dissolved inorganic carbon (DIC) in the ocean (H_2CO_3, HCO_3^-, and CO_3^{2-}), and (2) solid carbonate minerals [$CaCO_3$, $CaMg(CO_3)_2$, and $FeCO_3$] (Holmén, 2000). While the global C cycle is quite complex, it is perhaps the best understood of all the bioactive element cycles. In fact, there have been numerous review papers on this cycle (e.g., Keeling, 1973; Degens et al., 1984; Siegenthaler and Sarmiento, 1993; Sundquist, 1993; Schimel et al., 1995; Holmén, 2000). Much of the interest in the global C cycle in recent years stems from linkages with environmental issues concerning carbon-based greenhouse gases (e.g., CO_2 and CH_4) and their role in global climate change (Dickinson and Cicerone, 1986). As described in chapter 8, short-term controls on the C cycle are largely a function of the uptake of inorganic C by autotrophs to fuel fixation in photosynthesis, and the utilization of organic carbon as a food resource by heterotrophs recycling inorganic C back into the system. This short-term cycle, which allows for the transfer of C between the lithosphere, hydrosphere, biosphere, and atmosphere over periods of days to thousands of years, is relatively short in comparison to the more than 4 billion year age of the Earth (figure 13.1). In contrast, the long-term C cycle involves the transfer of C to and from rocks resulting in atmospheric changes in CO_2 that cannot be attributed to the short-term cycling effects (Berner, 2004).

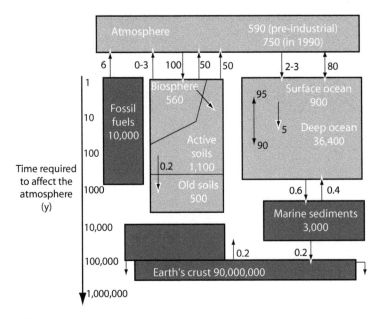

Figure 13.1 The major reservoirs and fluxes in the global carbon cycle, as they relate to chronology. The units for reservoirs and fluxes are Pg C (1 Pg = 10^{15} g) and Pg C^{-1}y^{-1}, respectively. (Modified from Sundquist, 1993.)

In this chapter the focus will be more on the overall modeling of C cycling in estuaries, since chapter 8 covered considerable details on the dominant autotrophs, and to a lesser degree heterotrophs, as well as the decomposition pathways in estuaries (figure 13.2). Similarly, chapter 9 contains a background on chemical biomarkers that also covers many of the important organic forms of C and their sources in estuaries. However, a brief section will be provided on certain components of organic C cycling in the water column and sediments of estuaries, not covered in any detail in prior sections. We will begin our discussion on the cycling of DIC, which has also not previously been covered.

Transformations and Cycling of Dissolved Inorganic Carbon

Our understanding of the DIC cycle in aquatic systems began with the development of analytical techniques that allowed for adequate detection and precision of DIC (Dyrssen and Sillén, 1967). When CO_2 dissolves in water it can then be hydrated to form H_2CO_3, which can then dissociate to HCO_3^- and CO_3^{2-} as shown in the following equations:

$$CO_2 + H_2O \leftrightarrow H_2CO_3 \tag{13.1}$$

$$H_2CO_3 \leftrightarrow H^+ + HCO_3^- \tag{13.2}$$

$$HCO_3^- \leftrightarrow H^+ + CO_3^{2-} \tag{13.3}$$

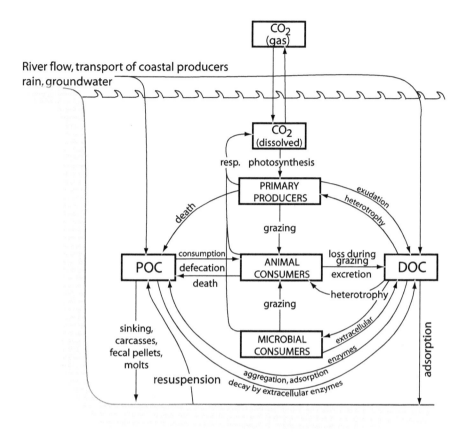

Figure 13.2 Transfer of carbon in aerobic estuarine environments. The boxes are reservoirs and the arrows are fluxes; the inorganic C component of this cycle is simplified in this figure. (Modified from Valiela, 1995.)

The HCO_3^- and CO_3^{2-} ions also dissociate into the following equilibria:

$$HCO_3^- + H_2O \leftrightarrow H_2CO_3 + OH^- \qquad (13.4)$$

$$CO_3^{2-} + H_2O \leftrightarrow HCO_3^- + OH^- \qquad (13.5)$$

$$H_2CO_3 \leftrightarrow H_2O + CO_2 \qquad (13.6)$$

Temperature-dependent equilibrium constants for carbonate are shown in table 13.1 (Larson and Buswell, 1942). Equations 13.4 and 13.5 can result in more alkaline waters due to the generation of OH^-; this is typical of what may be found in lakes and streams due to high carbonates in the drainage basin. The percolation of H_2O through soils results in the enrichment of CO_2 from plant and microbial decay processes forming H_2CO_3 which can

Table 13.1 Temperature dependence of some important carbonate equilibrium constants.

Reaction	Temperature, °C						
	5	10	15	20	25	40	60
1. $CO_{2(g)} + H_2O \leftrightarrow CO_{2(aq)}$; pK_h	1.20	1.27	1.34	1.41	1.47	1.64	1.8
2. $H_2CO_3 \leftrightarrow HCO_3^- + H^+$; $pK_{a,1}$	6.52	6.46	6.42	6.38	6.35	6.30	6.30
3. $HCO_3^- \leftrightarrow CO_3^{2-} + H^+$; $pK_{a,2}$	10.56	10.49	10.43	10.38	10.33	10.22	10.14
4. $CaCO_{3(s)} \leftrightarrow Ca^{2+} + CO_3^{2-}$; pK_{so}	8.09	8.15	8.22	8.28	8.34	8.51	8.74
5. $CaCO_{3(s)} + H^+ \leftrightarrow Ca^{2+} + HCO_3^-$; $p(K_{so}/K_{a,2})$	−2.47	−2.34	−2.21	−2.10	−1.99	−1.71	−1.40

Modified from Larson and Buswell (1942).

then dissolve calcium carbonate-enriched rock materials producing $[Ca(HCO_3)_2]$, which then ionizes to increase the amount of HCO_3^- and CO_3^{2-}. The solubility of CO_2 in water also increases with greater amounts of CO_3^{2-} (Wetzel, 2001). Much of the DIC in rivers is derived from the dissolution of carbonate rock (80%) with the remainder (20%) coming from weathering of aluminosilicates as shown in equation 6.1. Similarly, over 97% of the runoff has been classified as the $Ca(HCO_3)_2$ type, making HCO_3^-, Ca^{2+}, SO_4^{2-}, and SiO_2 the dominant dissolved constituents in global surface river waters (table 4.4). As mentioned earlier, another important factor controlling CO_2 in natural waters is respiration and photosynthesis. Although rivers have generally been considered net heterotrophic, increased damming, reduced suspended loads, and less light limitation has resulted in greater consumption of HCO_3^- by river phytoplankton (Humborg et al., 2000; Bianchi et al., 2004). Conversely, high decomposition rates in estuarine marsh systems may result in the export of DIC that rivals that of riverine export to coastal waters (Wang and Cai, 2004). More details on the effects of respiration and photosynthesis are provided in other sections of this chapter.

Oceanic, estuarine, and freshwaters are close to equilibrium with atmospheric CO_2; however, these equilibria are influenced by temperature and salinity. Carbon dioxide is 200 times more soluble than O_2; typical dissolved amounts of CO_2 based on exchange with the atmosphere at different temperatures are: 1.1 mg L^{-1} at 0°C, 0.6 mg L^{-1} at 15°C, and 0.4 mg L^{-1} at 30°C (Wetzel, 2001). The total DIC pool (ΣCO_2: CO_2, H_2CO_3, HCO_3^-, and CO_3^{2-}) in marine waters contains about 2 mmol C L^{-1} (mostly HCO_3^-), while in estuarine and freshwaters the ΣCO_2 is more variable and pH dependent (Wetzel, 2001). The dominance of one DIC species over another is largely dependent on the pH of natural waters (figure 13.3). For example, an average oceanic pH of 8.2 results in the predominance of HCO_3^-. More specifically, the *conjugate pairs* that largely control the pH-buffering capacity of seawater are HCO_3^-/CO_3^{2-} and $B(OH)_3/B(OH)_4^-$, in addition to minor contributions from silicate and phosphate, and S^{2-} and HS^- in anoxic waters (Holmén, 2000). For example, in high-salinity regions of estuaries, some of the most important buffering reactions are:

$$H^+ + B(OH)_4 \leftrightarrow B(OH)_3 + H_2O \tag{13.7}$$

$$H^+ + HPO_4^{2-} \leftrightarrow H_2PO_4^- \tag{13.8}$$

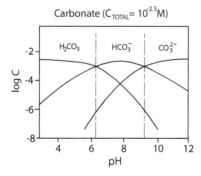

Figure 13.3 Dominant species of DIC over a range of pH values in natural waters. (Modified from Morel, 1983.)

$$H^+ + H_3SiO_4^- \leftrightarrow H_2SiO_4 \tag{13.9}$$

$$H^+ + NH_3 \leftrightarrow NH_4^+ \tag{13.10}$$

$$H^+ + OH^- \leftrightarrow H_2O \tag{13.11}$$

Since pH values less than 4.5 and greater than 9.5 are typically considered to be lethal to many organisms, the importance of buffering capacity of natural waters cannot be stressed enough. One term now used interchangeably with *acid neutralizing capacity* (ANC) of natural waters is *alkalinity* (Alk), defined as the concentration of negative charge that will react with H^+ and described by the following equation (Libes, 1992):

$$Alk = 2[CO_3^-] + [HCO_3^-] + [OH^-] - [H^+] \tag{13.12}$$

$$+ [B(OH_3)] + [H_3SiO_4^-] + [HPO_4^{2-}] + NH_3]$$

$$+ [other\ conjugative\ bases\ of\ weak\ acids]$$

Alkalinity is usually expressed as the total amount of base (in equilibrium with HCO_3^-, and CO_3^{2-}) that is determined by titration with a strong acid (Hutchinson, 1957). The units of total alkalinity (TAlk) are expressed as milliequivalents of acid needed to neutralize the negative charge in one liter of water (meq L^{-1}), see Stumm and Morgan (1996) for more details. In high-salinity waters where HCO_3^- and CO_3^{2-} represent the dominant forms of DIC, total alkalinity can be measured as *carbonate alkalinity* (CA) in the following equation:

$$Carbonate\ alkalinity = 2[CO_3^{2-}] + [HCO_3^-] \tag{13.13}$$

In fact, recent work has shown that the export of carbonate alkalinity in the Mississippi River has increased in the past half-century due to significant increases in rainfall and chemical weathering in the Mississippi Basin (Raymond and Cole, 2003). Moreover, it appears that watersheds with forested areas have less alkalinity transport than cropland systems. While the mechanisms of alkalinity export are not well understood, these results have major implications for management of land-use patterns in watersheds as it relates to carbon sequestration issues.

Alkalinity is generally assumed to be a conservative property in aquatic systems with high buffering capacity (Kempe, 1990). However, in systems that typically receive high inputs of organic matter and nutrients, the assumption of conservative behavior is no longer applicable because of the linkage between proton and electron transfers associated with redox reactions (Stumm and Morgan, 1996). For example, nitrification processes appear to have significant effects on alkalinity in the Schelde estuary (The Netherlands) (Frankignoulle et al., 1996). More recent work in the Shelde estuary showed a strong parallel between TAlk and NH_4^+ as well as NO_3^- and O_2 (figure 13.4; Abril and Frankignoulle, 2001). It was concluded from this work that temporal variability in NO_3^- and NH_4^+ concentrations, associated with ammonification, nitrification, and denitrification, could account for 28 to 62% of the alkalinity at all but one of the sampled stations.

SCHELDT RIVER

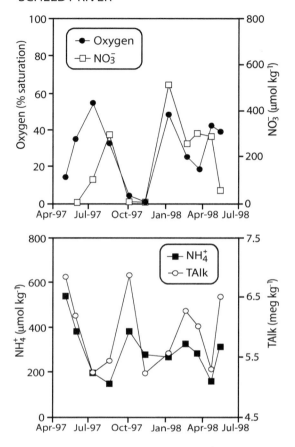

Figure 13.4 Seasonal variation of oxygen, NO_3^-, NH_4^+, and total alkalinity (TAlk) in the Scheldt basin from April 1997 to May 1998. (Modified from Abril and Frankignoulle, 2001.)

Other work has shown alkalinity production associated with Mn(IV), Fe(III), and SO_4^{2-} in an anoxic fjord (Yao and Millero, 1995).

In addition to alkalinity inputs from rivers, estuaries can also receive inputs from bordering wetlands. For example, Cai et al. (1999) have shown that respiration rates in rivers/estuaries of Georgia (USA) cannot account for O_2 consumption and CO_2 degassing. It has since been shown that the "missing" DIC source is from marshes in the Satilla estuary (USA) (Cai et al., 2000). Other studies have also shown that intertidal marshes are important sources of DIC in estuaries (Raymond et al., 1997, 2000; Neubauer and Anderson, 2003). It has been further suggested that CO_2 fixation of marsh grasses and the subsequent export of DIC and organic C to the coastal ocean can be described as a "salt marsh pump" that may have important implications for global carbon sequestration (Cai et al., 2004; Wang and Cai, 2004). Groundwater discharge may be an important pathway for DIC transport from marshes to rivers/estuaries (Cai et al., 2003). However, DIC and TAlk can be modified during their transport, primarily by diagenetic reactions, which can provide important geochemical signatures of groundwaters as indices of past interactions. The three dominant diagenetic reactions controlling DIC are aerobic respiration (see equation 8.1), SO_4^{2-} reduction (SR) (see equation 12.1), and *methanogenesis* (more details in the following section). Other studies have shown that certain stoichiometric ratios for these processes can provide useful signatures of diagenetic processes (Canfield et al., 1993; van Cappellen and Wang, 1996; Cai et al., 2002). For example, during aerobic respiration if 1 mole of organic carbon is respired, the $\Delta TAlk/\Delta DIC$ remains essentially unchanged. However, during anaerobic respiration through SR, the $\Delta DIC/\Delta SO_4^{2-}$ is approximately 2 and the $\Delta TAlk/\Delta DIC$ is approximately 1. Finally, when SO_4^{2-} is depleted and methanogenesis is the dominant form of anaerobic respiration, it can proceed through either of the following two pathways: *fermentation*, as shown below:

$$CH_3COOH \rightarrow CO_2 + CH_4 \qquad (13.14)$$

and *CO$_2$ reduction*:

$$CO_2 + 4H_2 \rightarrow CH_4 + 2H_2O \qquad (13.15)$$

The $\Delta TAlk/\Delta DIC$ for fermentation is essentially 0. Fermentation generally occurs more in freshwater systems, while CO_2 reduction is more commonly found in marine systems (Whiticar, 1999). Depending on the availability of $CaCO_3$ in the system, CO_2 produced during fermentation will be neutralized by $CaCO_3$ dissolution. Using these basic ratios as indicators in diagenetic models, recent work showed that the groundwaters in North Inlet (USA) are significantly altered by CO_2 inputs from SR across the entire salinity range of the estuary, while fermentation and $CaCO_3$ dissolution dominated in the low-salinity regions (Cai et al., 2003). It was also suggested that if groundwater DIC flux from North Inlet was extrapolated to other local marshes in the region, groundwater DIC (via salt marshes) represented a significant source of DIC to the South Atlantic Bight. More details are provided on estuarine export to the coastal ocean in chapter 16.

Carbon Dioxide and Methane Emissions in Estuaries

Recent work has shown high spatial and temporal variability of CO_2 and CH_4 emissions in estuaries (Abril and Borges, 2004). As mentioned in chapter 5, one of the more recent large-scale collaborative efforts examining controls on gas emissions in European estuaries is BIOGEST (Biogas Transfer in Estuaries, 1996–1999). A key objective of this work was to estimate gas transfer velocities (k) in estuaries, which are high relative to those in lakes and the ocean, and are critical in determining gas fluxes across the water–air interface. Due to the steep physicochemical gradients found across relatively short distances in estuaries, and the differences in techniques used to measure k, there has been considerable debate about the appropriate approach that should be used in such dynamic systems—more details are provided on this in chapter 5.

Recent work of BIOGEST further demonstrated that inner waters, tidal flats, and marshes are perhaps the most active sites of CO_2 emissions. For example, land- and marsh-derived organic carbon resulted in CO_2 atmospheric emissions of 10s to 1000s $mmol^{-2} d^{-1}$ from these estuarine habitats (Abril and Borges, 2004). A typical range of CO_2 fluxes for a select group of estuarine systems is shown in table 13.2. This supports the notion that organic matter derived from emergent/submergent macrophytes, microphytobenthos growing on intertidal mudflats, and land-derived vascular sources from these regions play a significant role in the overall net metabolism of estuaries. Estuaries are in many cases considered to be net heterotrophic (more details provided in the following section), as a result of this "excess" loading of allochthonous materials to the system (Smith and Hollibaugh, 1993, 1997; Frankignoulle et al., 1998; Cai et al., 1999; Raymond et al., 2000). Consequently, it is common to find O_2 deficits accompanied by supersaturation of CO_2 in estuaries, as shown for the Loire (France), Shelde (The Netherlands), and Thames (UK) estuaries (figure 13.5). Moreover, river waters entering at the head of an estuary typically show higher concentrations of CO_2 and lower concentrations of O_2. This is due to the loading of enriched pCO_2 waters from river inputs, which are high because they contain the mineralization "signatures" from soils, bordering freshwater wetlands, and resuspension from upper estuarine sediments (Raymond et al., 2000; Cole and Caraco, 2001; Richey et al., 2002). This upper oligohaline region of estuaries is also where the maximum turbidity zone (MTZ) or estuarine turbidity maximum (ETM) typically occurs, which further contributes to intense heterotrophic activity and recycling of organic matter resulting in consumption of O_2 and CO_2 production. For example, the MTZ in the Gironde Estuary (France) has been shown to be a site where labile DOC accumulates in mobile muds (Abril et al., 1999), where is it believed to aid in the cometabolism of refractory POC (Abril et al., 1999), making this zone analogous to the fluidized-bed reactor in deltaic muds (Aller, 1998). High remineralization activity at the oxic/anoxic interface, just above the fluid muds in the MTZ, results in higher suspended particulate matter (SPM), pCO_2, DIC, NH_4^+, and TAlk, with concomitant decreases in O_2, pH, and NO_3^- (figure 13.6; Abril et al., 1999). This enhanced heterotrophic activity in the MTZ of the Gironde clearly has an impact on the overall budget of DIC (table 13.3; Abril et al., 1999); the HCO_3^- is primarily derived from anaerobic decomposition and $CaCO_3$ dissolution. Although $CaCO_3$ dissolution may not be important in many estuaries, it has been shown to be responsible for large shifts in TAlk in the Loire (France) (Abril et al., 2003) and Godavari (India) (Bouillon et al., 2003) estuaries. The MTZ in the Changjiang (Yangtze) River estuary (China) (Zhang et al., 1999), the Mandovi-Zuari estuary on the Ganges delta (Bangladesh) (Mukhopadhyay et al., 2002), and the Pearl River estuary (China) (Zhai et al., 2005) also have high pCO_2 values. The outer estuary and river plume regions, which typically have greater light availability also have higher concentrations of phytoplankton biomass which, depending on the overall heterotrophic activity, can be a sink for CO_2. For example, the Amazon plume region takes up 0.014×10^{15} g C y^{-1} and represents almost as much of a sink as the rivers and wetlands systems in the Amazonian Basin (Richey et al., 2002). Similarly, area-integrated biological uptake rates in the Mississippi River plume, derived

Table 13.2 pCO_2 ranges and fluxes reported in inner estuaries.

Estuary	Number of cruises	Average pCO_2 range (μatm)	Average pCO_2 flux range (mmol·m^{-2}·d^{-1})	Method used for k[a]	References[b]
Altamaha (Georgia, USA)	1	380–7800		Floating chamber	1
Scheldt (Belgium/The Netherlands)	10	495–6650	260–660		2
Sado (Portugal)	1	575–5700	760	Floating chamber	2
Satilla (Georgia, USA)	2	420–5475	50	$k = 12$ cm h^{-1}	1
Seine (France)	2	826–5345	—		3
Thames (UK)	1	560–3755	210–290	Floating chamber	2
Gironde (France)	5	500–3535	50–110	Floating chamber	2
Loire (France)	3	770–2780	100–280	Floating chamber	4
Mandovi-Zuavi (India)	2	400–2500	11–67	W92	5
Douro (Portugal)	1	1330–2200	240	Floating chamber	2
York (Virginia, USA)	12	350–1895	12–17	C95 and C96	6
Tamar (UK)	2	390–1825	90–120	$k = 8$ cm h^{-1}	2
Rhine (The Netherlands)	4	570–1870	70–160	Floating chamber	2
Hudson (New York, USA)	6	515–1795	16–36	M and H93 and C95	7
Rappahannock (Virginia, USA)	9	474–1613	—		6
James (Virginia, USA)	10	284–1361	—		6
Elbe (Germany)	1	580–1100	180	Floating chamber	2
Columbia (Oregon, USA)	1	560–950	—		8
Potomac (Maryland, USA)	12	646–878	—		6

[a]Fluxes were either measured directly with a floating chamber or calculated using either a constant piston velocity (value indicated) or using various wind speed relationships: W92 is Wanninkhof (1992); C95 and C96 is a combined relationship of Clark et al. (1994) and Carini et al. (1996): M and H93 and C95 is a combined relationship of Marino and Howarth (1993) and Clark et al. (1994).

[b]References: 1, Cai and Wang (1998) and Cai et al. (1999); 2, Frankignoulle et al. (1998) and additional unpublished data from the BIOGEST project; 3, Abril et al., unpublished data; 4, Abril et al. (2003) and unpublished data; 5, Sarma et al. (2001); 6, Raymond et al. (2000); 7, Raymond et al. (1997); 8, Park et al., (1969).

The average pCO_2 range was obtained by averaging the lowest and highest values for each transect and gives information on the spatial variability from the upper part (highest pCO_2) to the estuarine mouth (lowest pCO_2). By contrast, the average CO_2 flux range is composed of the lowest and highest fluxes averaged over the estuarine surface for each cruise and gives information on the temporal variability from one cruise to another.
Modified from Abril and Borges (2004).

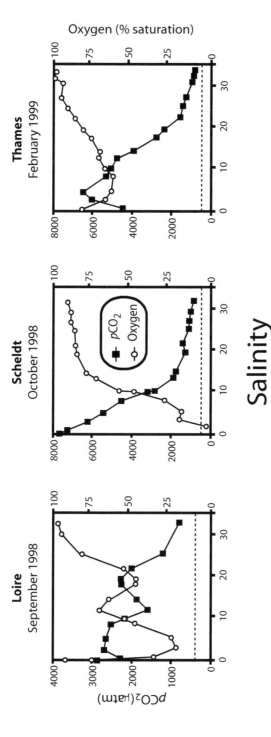

Figure 13.5 Concentrations of oxygen and pCO_2 across a salinity gradient in surface waters of the Loire, Scheldt, and Thames estuaries. (Modified from Abril and Borges, 2004.)

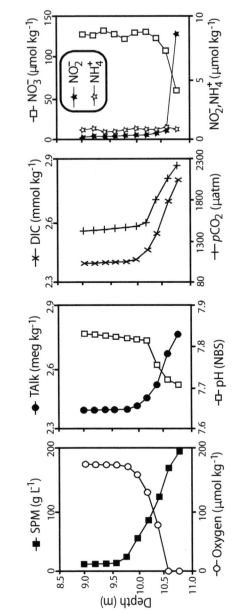

Figure 13.6 Vertical profiles of suspended particulate matter (SPM), oxygen, total alkalinity (TAlk), pH, calculated DIC, and pCO_2, NO_3^-, NO_2^-, and NH_4^+ in bottom water and fluid muds in the Gironde estuary. (Modified from Abril et al., 1999.)

Table 13.3 Inorganic carbon budget (t C d^{-1}) of the Gironde
Estuary (France) in September 1997.

HCO_3 production in the MTZ[a]	180
HCO_3^- output to the sea[b]	1234
CO_2 transfer to the atmosphere[c]	380

[a]Total alkanlinity (TAlk) increase of 0.32 meq kg^{-1} and river flow
of 500 m^3 s^{-1}. MTZ = Maximum turbidity zone.
[b]Linear extrapolation of total alkalinity to zero salinity 2.38 meq
kg^{-1} and river flow.
[c]Direct air–sea flux measurement (Frankignoulle et al., 1998).
Modified from Abril et al. (1999).

from DIC and TAlk (1.5–3 g C m^{-2} d^{-1}), were found to be some of the highest reported
values for rivers and estuaries (Cai, 2003). Conversely, the highly eutrophic Gironde estuary was found to be a net source of CO_2 to the atmosphere, with the waters remaining
in supersaturation (as high as 700 μatm) most of the year, due to high pollutant inputs
(Borges and Frankignoulle, 2002). So, the balance between autotrophic and heterotrophic
processes can be quite variable in these systems due to quality and quantity of organic
matter loading, as well as physical parameters such as residence time and turbulence.

Similar to CO_2, emission rates of CH_4 are highly variable in estuaries with the higher
rates being found in tidal flat and marsh environments (table 13.4). Recent estimates of
global contributions of CH_4 emissions to the atmosphere (1.8–3.0 × 10^{12} g CH_4 y^{-1})
represent less than 10% of the global oceanic emissions of CH_4, which is only 1 to 10%
of all global sources (Bange et al., 1994), making estuaries an insignificant source to the
global budget. As mentioned earlier, methanogenesis can occur in sediments through two
pathways (equations 13.14 and 13.15), when there are high inputs of labile organic carbon, and in the absence of O_2 and alternative electron acceptors such as SO_4^{2-} (Martens
and Berner, 1974; Magenheimer et al., 1996). Differences in the spatial and temporal
occurrences of SR versus methanogenesis, where bacteria (SRB) commonly outcompete
methanogenic bacteria (*methanogens*) for acetate (CH_3COOH) substrates (Capone and
Kiene, 1988), are principally determined by differences in the free energy yields (ΔG^0)
per mole of organic C (see table 8.3). This relationship between zones of SR and methanogenesis is nicely illustrated in the down-core profiles of CH_4 and SO_4^{2-} in the sediments
of the anoxic fjord, Sannich Inlet (Canada) (figure 13.7; Murray et al., 1978; Devol et al.,
1984). Methanogenesis is more common in freshwater than in marine environments; this
is readily apparent when comparing CH_4 emission rates in freshwater locations which
are two orders of magnitude higher than in salt water locations (table 13.4). Higher rates
of methanogenesis in marsh sediments are also due to inputs of labile organic matter
from plant roots at depth (van der Nat and Middelburg, 2000). For example, in addition
to CH_3COOH, other labile methyl substrates for methanogens might include methanol
(CH_3OH) and methylamine ($CH_3NH_3^+$).

Table 13.4 Methane fluxes from estuarine regions.

Sites	Comments	CH$_4$ emission (mmol m^{-2} d^{-1})	References[a]
Inner estuaries main channel			
Yaquina and Alsea (Oregon, USA)	Annual average	0.18	1
Hudson tidal river (New York, USA)	Annual average	0.35	2
Bodden (Germany)	Annual average, spatial range	0.03–0.21	3
Tomales Bay (California, USA)	Seasonal range, spatial average	0.007–0.01	4
European tidal estuaries	Median 9 estuaries, 18 cruises	0.13	5
Randers Fjord estuary (Denmark)	Annual average, spatial range	0.07–0.41	6
Estuarine plumes			
Rhine and Scheldt (Southern North Sea)	Spatial range in Marsh 1989	0.006–0.6	7
Amvrakikos bay (Aegean Sea)	Spatial average in July 1993	0.014	8
Danube (Northwestern Black Sea)	Spatial range in July 1995	0.26–0.47	9
Tidal flats			
White Oak (North Carolina, USA)	Annual average	1.2	10
(Freshwater sites)	Seasonal range	1–45	10
	Tidal variation	2.5–6.3	10
Choptank river estuary	Annual average (salinity 1–10)	2.4	11
Scheldt (Belgium/The Netherlands)	Annual averages		12
	Fresh water site	500	12
	Salt water site (salinity 25)	0.1	12

Tidal marshes			
White Oak (North Carolina, USA)	Annual averages		
(freshwater marsh)	Total	7.1	13
	Ebullitive	3.5	13
	Diffusive	3.5	13
Bay Tree Creek salt marsh (Virginia, USA)	Annual averages, three sites (salinity 5–23)	2.6–8.1	14
York River (Virginia, USA)	Annual averages		
	Salinity 2.6	3.0	15
	Salinity 5.5	3.8	15
	Salinity 8.8	0.9	15
Dipper Harbor	Annual averages		
(New Brunswick, Canada)	Salinity 20.6–23.5	0.13	16
	Salinity 31–35	0.03	16

[a]References: 1, De Angelis and Lilley (1987); 2, De Angelis and Scranton (1993); 3, Bange et al. (1998); 4, Sansone et al. (1998); 5, Middelburg et al. (2002); 6, Abril and Iversen (2002); 7, Scranton and McShane (1991); 8, Bange et al. (1996); 9, Amouroux et al. (2002); 10, Kelley et al. (1995). The seasonal average is from 4 bank stations and the seasonal range is the average of these stations in October–March and in August–September; the tidal variation is from a single bank station in August 1991. 11, Lipschultz (1981); 12, Middelburg et al. (1996) (the extremely high value is from an area polluted by sewage loads); 13, Chanton et al. (1989b); 14, Bartlett et al. (1985); 15, Bartlett et al. (1987); 16, Magenheimer et al. (1996).
Modified from Middelburg et al. (2002).

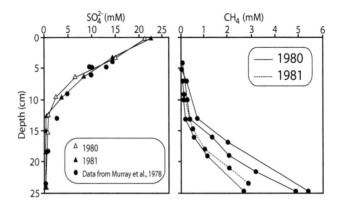

Figure 13.7 Vertical sediment profiles of $SO_4{}^{2-}$ and CH_4 in the anoxic fjord, Sannich Inlet. (Modified from Murray et al., 1978, and Devol et al., 1984.)

The presence of plant structures in sediments also mediates the pathways of CH_4 emission for sediments. In unvegetated sediments, 50 to 90% of the methane can be characterized by large volumes of gas bubbles (Chanton et al., 1989a,b). These bubbles are formed because the CH_4 raises the sum of partial pressures to above the hydrostatic pressure in the sediment. The equilibrium between dissolved and bubble CH_4 is controlled across different regions of estuaries by temperature, dissolved CH_4 concentration, and CH_4 bubble partial pressure (Chanton et al., 1989a). In unvegetated sediments mechanisms of gas transport are diffusion across the sediment–water interface and bubble *ebullition*; in the White Oak River estuary (USA) each mechanism was responsible for about 50% of the total flux from sediments (Kelley et al., 1990). In vegetated sediments, large volumes of CH_4 may also exist as bubbles but some of it may occur in the roots and rhizomes of plants (Chanton et al., 1989a; Sass et al., 1991; Schutz et al., 1991). The below-ground tissues of aquatic vascular plants are largely *aerenchymatous*, which characteristically have large air spaces. In fact, these air spaces have been shown to represent a significant volume (9–15 L m^{-3}) of gas space in the roots and rhizomes of *S. alterniflora* marshes (Dacey and Howes, 1984). In the Newport News swamp (USA), concentrations of CH_4 in vegetated sediments are reduced in both dissolved and bubble phases compared to unvegetated sediments, particularly in the root zone (figure 13.8; Chanton and Dacey, 1991). Thus, ebullition is a less important mechanism of transport in vegetated sediments than in unvegetated sediments. The mechanisms of CH_4 transport commonly found through plants are *simple molecular diffusion* (differential partial pressures between the sediment, plant, and atmosphere) (Lee et al., 1981), *convective throughflow ventilation* (via *lacunal* pressurization) (Dacey, 1981), and *effusion* (partial pressures differences in the pores of the plant) (Dacey, 1987). Diurnal patterns of CH_4 release from plants have also been shown to be primarily attributed to changes in temperature (Whiting and Chanton, 1992; Chanton et al., 1993) and light levels (King, 1990; Whiting and Chanton, 1996). Finally, there appears to be differential effects on the isotopic fractionation of CH_4

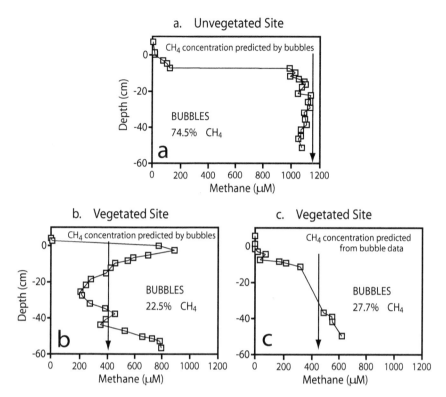

Figure 13.8 Concentrations of dissolved porewater CH_4 and sediment gas bubble CH_4 at an unvegetated (a) and vegetated (by *Peltandra* spp.) site (b,c) in Newport News Swamp, Virginia. The concentration of CH_4 at equilibrium is predicted from CH_4 solubility data from Yamamoto et al. (1976), and the CH_4 composition of the sediment gas bubbles is indicated by the arrow. (Modified from Chanton and Dacey, 1991.)

depending on the mechanism of transport. For example, during diffusion and/or effusion there is preferential loss of the light isotope (^{12}C), resulting in ^{13}C-enriched CH_4 remaining in plant tissues relative to sediments (Chanton et al., 1992); diurnal variations appear to affect these patterns of fractionation (Chanton and Whiting, 1996). In plants using convective throughflow ventilation, there were no differences in the ^{13}C isotopic signature of plants and sediments; however, in the absence of sunlight certain plants (e.g., *Typha*) may switch from convective throughflow ventilation to molecular diffusion (Chanton and Whiting, 1996). Thus, plant vegetation can play a central role in controlling CH_4 emissions from estuaries. Moreover, $\delta^{13}C$ isotopic signatures appear to provide a useful approach for determining plant-mediated transport mechanisms of CH_4 in estuaries.

Methane-oxidizing bacteria, or *methanotrophs*, can play a central role in reducing CH_4 emissions from estuaries, by converting CH_4 into bacterial biomass or CO_2 (Topp and Hanson, 1991). It has been estimated that methanotrophic bacteria in freshwater sediments can consume as much as 90% of the CH_4 produced (Reeburgh et al., 1993). Methanotrophs and NH_4-oxidizing bacteria have close evolutionary ties with linkages through their *monooxygenase* systems (Holmes et al., 1995). Both are obligate aerobic chemoautotrophs within the γ- and α-subdivisions of the Proteobacteria (Hanson and Hanson, 1996). Recent work in Galveston Bay (USA) shows high variability of both potential rates of CH_4 and NH_4^+ oxidation (figure 13.9; Carini et al., 2003). The low CH_4 concentrations in these sediments are typical for shallow, bioturbated, low organic matter, sandy systems. As expected, the highest activity of potential CH_4 oxidation was found in the top 10 cm. There are some interesting relationships between the methanotroph community and the NH_4-oxidizing bacteria that may indicate a sequestering of nitrogen from NH_4, which ultimately may restrict availability of NH_4 to NH_4-oxidizing bacteria (Carini et al., 2003)—further work is needed on these potential microbial interactions.

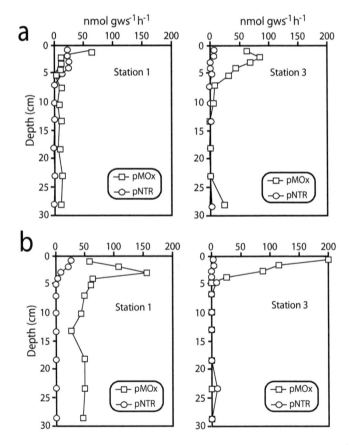

Figure 13.9 Seasonal and spatial variability in depth profiles of potential NH_4^+ (pNTR) and CH_4 (pMOx) rates in Galveston Bay sediments, for (a) August 1998 and (b) November 1998. (Modified from Carini et al., 2003.)

Methane concentrations in estuarine waters are usually higher than atmospheric equilibrium (2–3 nmol L^{-1}) and in most cases increase from the freshwater to the marine end-member (De Angelis and Lilley, 1987; Bange et al., 1998; Abril and Iverson, 2002; Middelburg et al., 2002; Vander borght et al., 2002; Abril and Borges, 2004). Data collected from BIOGEST at the Gironde (France), Thames (UK), Schelde (Belgium/The Netherlands), and Sado (Portugal) estuaries, indicate that spatial patterns of CH_4 can be quite variable (Middelburg et al., 2002; Abril and Borges, 2004). In the case of the Gironde and Thames estuaries, high inputs of CH_4 from riverine sources occur at the head of the estuary. Conversely, sources of CH_4 were found in the higher-salinity regions of the Schelde and Sado and were believed to be from bordering tidal flats (Middelburg et al., 2002), further emphasizing issues of spatial variability in estuaries.

Transformations and Cycling of Dissolved and Particulate Organic Carbon

Sources of particulate organic carbon (POC) and dissolved organic carbon (DOC) in estuaries consist of a diverse mixture of allochthonous and autochthonous sources (see chapter 8 for more details). Hydraulic residence time, river discharge, tidal exchange, and frequency of resuspension events are, to name a few, important physical controlling variables that determine the fate and reactivity of organic carbon in estuaries. Once again, in this section only a very brief overview of POC and DOC cycling dynamics in the waters and sediments of estuaries is provided, relying more on the background of organic carbon sources, chemical biomarkers, and bulk isotopic (single and dual) models provided in chapters 7–9.

Concentrations of POC in many estuaries are strongly coupled with suspended sediment particulates, which may depend on river discharge and/or resuspension events. For example, in the Sabine–Neches estuary (USA), there is a significant increase in POC when total suspended particulates are lower than 20 to 30 mg L^{-1} (figure 13.10; Bianchi et al., 1997a). This relationship has been found in large river systems; however, because of the higher suspended loads in many rivers, the increase in % POC of total organic carbon, typically occurs at less than 50 mg L^{-1} (Meybeck, 1982). This general pattern is attributed to a dilution effect of sediment load on % POC at high river discharge, and to an increase in phytoplankton production during high total suspended particulates (TSP) and low light availability. This relationship between TSP and phytoplankton biomass (as indicated by chlorophyll-*a* concentrations) across different salinity regions is well illustrated in the Delaware Bay estuary (figure 13.11; Harvey and Mannino, 2001). Seasonal ranges of POC concentrations in San Francisco Bay and Chesapeake Bay estuaries are typical of many estuaries and are not significantly different between systems (figure 13.12; Canuel, 2001). Seasonal and temporal differences in POC in both systems are generally controlled by river discharge and light availability. However, further analyses using fatty acid biomarkers reveals that phytoplankton represent a greater fraction of POC in Chesapeake Bay than in San Francisco Bay (Canuel, 2001) (see chapter 9 for more details). Thus, while bulk POC provides a general index of the overall loading of allochthonous and autochthonous C in the system, POC alone can be very misleading in terms of overall C cycling dynamics.

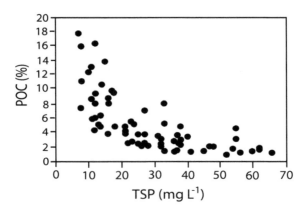

Figure 13.10 Percentage particulate organic carbon (POC) and total suspended particulates (TSP) in the water column from three regions of the Sabine–Neches estuary, sampled from March 1992 to October 1993. (Modified from Bianchi et al., 1997a.)

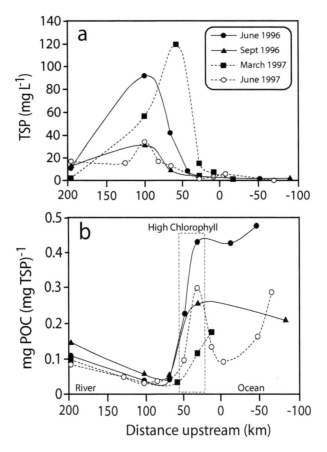

Figure 13.11 Seasonal distribution of (a) total suspended particulates (TSP) and (b) particulate organic carbon (POC) along a salinity gradient and distance from the mouth of the Delaware Bay estuary. (Modified from Harvey and Mannino, 2001.)

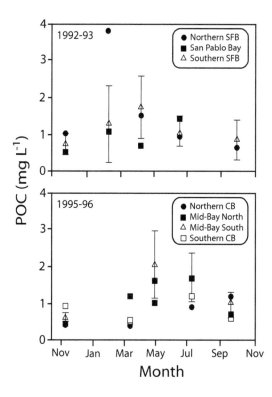

Figure 13.12 Seasonal range of particulate organic carbon (POC) in different regions of San Francisco Bay (1992–1993) and Chesapeake Bay (1995–1996) estuaries. Error bars represent standard deviations for mean concentrations when three sites were measured. (Modified from Canuel, 2001.)

Salinity gradients of DOC concentrations have commonly been used to examine conservative and nonconservative behavior of DOC in estuaries (see Guo et al., 1999). The typical DOC mixing gradients for six different estuaries in the Gulf of Mexico clearly indicate that decreases in DOC concentration with increasing salinity are occurring non-conservatively (figure 13.13). The concentrations of DOC in the lower-salinity regions are due to riverine inputs of DOC. Seasonal variability of phytoplankton blooms can at times change the distribution of DOC in estuaries, whereby higher concentrations of DOC co-occur with the blooms in the lower estuarine region due to lower light attenuation rates. The lower-salinity regions appear to be important sinks for DOC (figure 13.13); this is within the region of the ETM where fractions of DOC can be removed due to coagulation, flocculation, and other processes (see chapter 4 for more details). The range of DOC concentrations in a group of selected estuaries is shown in table 13.5. Other important subcomponents of the total DOC pool, described in chapter 8, are the high and low molecular weight dissolved organic carbon (HMW DOC and LMW DOC); the HMW DOC is sometimes referred to as colloidal organic carbon (COC) (table 13.5). If we exclude, for the moment, any effects that different techniques and filters may have on collecting the amounts and composition of HMW DOC, we see that there are significant differences in the relative importance of HMW DOC in estuaries. Sources of HMW DOC could be

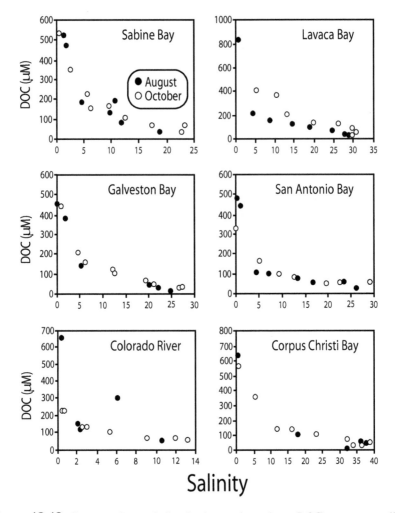

Figure 13.13 Concentrations of dissolved organic carbon (DOC) across a salinity gradient in six Texas estuaries in August and October. (Modified from Benoit et al., 1994.)

derived from old soil materials, fresh litter from terrestrial runoff, or perhaps more labile algal sources (both benthic and pelagic) in the estuary. Radiocarbon values of two size fractions of HMW DOC reveal that the smaller size fraction (>1 kDa) was younger than the larger (>10 kDa), with both being higher in abundance at the lower-salinity region of Galveston Bay estuary (figure 13.14; Guo and Santschi, 1997). These differences were attributed to inputs of older HMW DOC (more geopolymerized) from porewaters in resuspended sediments. The importance of older HMW DOC from porewaters to estuarine waters has also been shown to be important in coastal bottom waters along shelf

Table 13.5 Concentration of DOC and its mixing behavior in estuaries of the Gulf of Mexico.

Region	Estuary	[DOC] (μM)[a]	Mixing behavior of DOC	Reference
I (Texas)				
	Corpus Christi	560–630	Significant removal during mixing	Benoit et al. (1994)
	San Antonio Bay	330–480	Significant removal during mixing	Benoit et al. (1994)
	Lavaca Bay	N.A.	Significant removal during mixing	Benoit et al. (1994)
	Galveston Bay	460	Significant removal during mixing	Benoit et al. (1994)
	Sabine–Neches	530	Significant removal during mixing	Benoit et al. (1994)
		550	N.A.	Stordal et al. (1996)
		367–1350	N.A.	Bianchi et al. (1997a)
	Galveston Bay	420–480	Source input in low-salinity region in July	Guo and Santschi (1997)
II (Louisiana and Alabama)				
	Mississippi River plume	270–330	Conservative mixing during winter but with source input during summer	Benner et al. (1992)
	Lake Pontchartrain	425–483	Well mixed	Argyrou et al. (1997)
	Mobile Bay	424±105	Conservative	Pennock et al. (unpublished)
III (Florida)				
	Ochlockonee	~1050	Nonconservative mixing with inputs for LMW fraction and removal for HMW fraction	Powell et al. (1996)
	Rookery Bay	580–1250	N.A.	Twilley (1985)
IV (Mexico)				
	Laguna Madre	312[b]	N.A.	Amon and Benner (1996)
	Terminos Lagoon	60–330[c]	N.A.	Rivera-Monroy et al. (1995)
	Celestun Lagoon	[d]	N.A.	Herrera-Silveira and Remirez-Remirez (1996)

[a]Concentration of DOC from river end member or average in estuary.
[b]Data from samples with a salinity of ~22.
[c]Concentration of DOC was converted from that of DON and the Redfield ratio.
[d]Only tannic acid (a measure of natural phenolic material) concentrations were available (ranged from <1 to 18 mg L^{-1}).
N.A. = not available.
Modified from Guo et al. (1999).

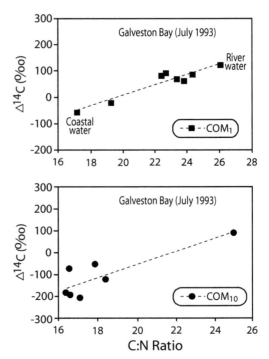

Figure 13.14 Relationship between $\Delta^{14}C$ and C:N ratios in macromolecular colloidal organic matter (COM) fractions (COM_1 = >1 kDa <0.02 μm, COM_{10} = >1 kDa <0.02 μm) in Galveston Bay, July 1993. (Modified from Guo et al., 1997.)

regions in the Mid-Atlantic Bight (Mitra et al., 2000a). This does not support the open ocean size–reactivity model proposed by Amon and Benner (1996), which simply stated, predicts that the "older" fraction of HMW DOC will reside in the lower molecular weight classes, typically at deeper water depths. However, the size-reactivity model is generally observed in oceanic waters because of the selective removal of labile "younger" C as POC produced in the euphotic zone sinks through the water column, accumulating the "older" smaller molecular weight DOC at depth (Amon and Benner, 1996; Guo et al., 1996).

Bulk radiocarbon and ^{13}C signatures of both total POC and DOC pools have also proven useful in constraining what their sources might be in estuaries (McCallister et al., 2004) (more details on this in chapters 7–9). For example, $\Delta^{14}C$-DOC is always more ^{14}C enriched compared to $\Delta^{14}C$-DOC along riverine margins of U.S. East Coast estuaries (see figure 7.24; Raymond and Bauer, 2001a). Since the ^{14}C-enriched DOC is also more depleted in ^{13}C, it appears that these differences are because of greater contribution of ^{14}C-enriched soil and litterfall organic matter that is leached from soils. Similarly, recent work has shown, using a dual-isotopic tracer approach of isotopic signatures ($\delta^{13}C$ and $\Delta^{14}C$) for bacterial nucleic acids collected from different regions and potential C source materials, that there was good delineation of C sources from aquatic and terrestrial systems in the York River estuary (figure 13.15; McCallister et al., 2004). In general, the results from these dual-isotopic tracer studies indicate that the broad classification of organic C, and the interchangeable use of the terms "old" and "refractory," are in many cases not valid.

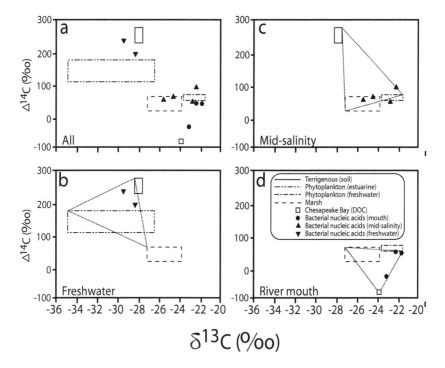

Figure 13.15 Comparison of Δ^{14}C versus δ^{13}C for bacterial nucleic acids and potential sources for the (a) entire York River estuary, (b) the freshwater, (c) mid-salinity, and (d) high-salinity (river mouth) regions in the estuary. Boxes are the 95% confidence intervals for the potential end-members in the York. Dotted lines represent the solution space from one run of a model. (Modified from McCallister et al., 2004.)

In fact, highly depleted ^{14}C (1000–5000 years old) in the Hudson River estuary appears to be an important labile source fueling heterotrophy (Cole and Caraco, 2001). Thus, the storage of organic matter for centuries and millennia in soils and rocks can actually become available to aquatic microbes over periods of weeks to months (Petsch et al., 2001), completing a unique linkage between river metabolism and the history of organic matter preservation in the drainage basin (Cole and Caraco, 2001).

In estuarine sediments, down-core concentrations of bulk C indices, such as total organic C (TOC), δ^{13}C, and C:N ratios can be used as a general index of the loading and sources of POC to sediments. As discussed in chapters 8 and 9, there are a number of problems with using the C:N ratio as the only diagnostic indicator of organic matter source; however, if coupled with stable isotope and chemical biomarker tools, sources are better constrained. If we simply use bulk C tools to examine down-core sediment profiles, we see that TOC and atomic C:N ratios in sediments from the York River estuary (USA) clearly reflect a dominance of phytoplankton inputs, with a typical %TOC in surface sediments of estuaries, decreasing with diagenetic "burn-off" at depth (figure 13.16;

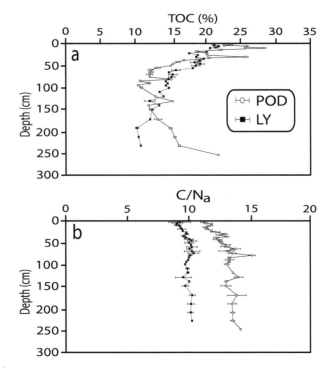

Figure 13.16 Down-core sediment profiles of (a) total organic carbon (TOC) and (b) atomic C:N ratios at two locations in the York River estuary. (Modified from Arzayus and Canuel, 2004.)

Arzayus and Canuel, 2004). The bulk δ^{13}C signal also reflects inputs of phytoplankton sources. For comparative purposes, when examining bulk down-core profiles of sediments collected off the Sinnamary River estuary (French Guiana), where there are large inputs of mangrove detritus, we see considerably higher C:N ratios and lower %TOC in surface sediments, reflecting slower remineralization of more refractory vascular plant inputs, in this case *Avicennia germinans* (figure 13.17; Marchand et al., 2004). The bulk δ^{13}C signal also shows a more depleted ^{13}C signature indicative of C_3 vascular plant sources. Down-core profiles of porewater bulk DOC and C:N ratios (of DOM) should also reflect changes in rates of POM remineralization. This is illustrated in down-core profiles of DOC, C:N ratios, and ΣCO_2 at three stations in Chesapeake Bay (figure 13.18; Burdige and Zheng, 1998). Differences in these DOC profiles are controlled primarily by differences in the physical and biogeochemical processes in sediments in the upper, middle, and lower regions of the bay. More specifically, the higher porewater DOC concentrations at the mid-Bay station are due to higher remineralization and a greater storage of the remineralization "signature" (e.g., DOC and ΣCO_2), due to lower O_2 conditions and less physical mixing than at stations in the upper and lower regions of the bay. Changes in the C:N ratios of porewater DOM at these stations are likely due to differences in the selective

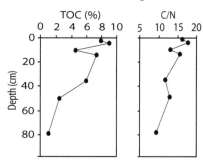

Figure 13.17 Down-core sediment profiles of total organic carbon (TOC) and atomic C:N ratios in the Sinnamary River estuary. (Modified from Marchland et al., 2004.)

utilization of N-rich DOM, and not due to differences in source inputs (e.g., terrestrial versus marine), as originally thought (Burdige, 2001). Thus, redox (more consistently anoxic versus "mixed" redox) and the presence/absence of macrofauna were the most important controlling factors in determining the remineralization of POM to DOM and overall storage of DOM in Chesapeake Bay sediments. Finally, the fraction of total DOC in sediments composed of HMW DOC and LMW DOC will largely depend on the sources of POC loading and the redox conditions of the sediments, this is best explained by the porewater size/reactivity (PWSR) model, described in chapter 8.

Signatures of bulk C profiles in sediments are also commonly used to make inferences about the efficiency of organic matter remineralization in sediments. For example, in sediments that are highly bioturbated (physically reworked) with high O_2 exposure, such as

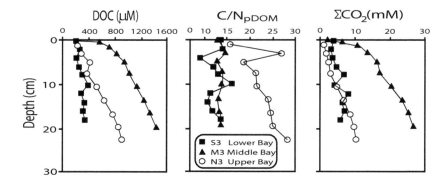

Figure 13.18 Down-core profiles of porewater dissolved organic carbon (DOC), C:N ratios, and total ΣCO_2 at three stations in Chesapeake Bay. (Modified from Burdige and Zheng, 1998.)

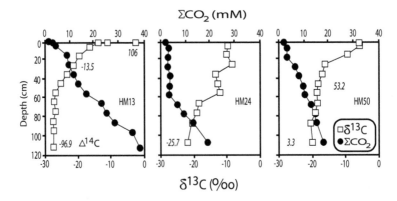

Figure 13.19 Depth profiles of ΣCO_2 and $\delta^{13}C$ in porewaters at three stations in sediments from the Gulf of Papua deltaic complex. ΣCO_2 from Aller et al. (2004). (Modified from Aller and Blair, 2004.)

the Gulf of Papua deltaic complex (Papua New Guinea), there is efficient remineralization of sedimentary organic matter (Aller and Blair, 2004). The asymptotic shape of the porewater ΣCO_2 and the terminal ^{13}C values of ΣCO_2 in the range of -27 to -19 suggest that both marine and terrestrial sources of organic matter are completely remineralized to CO_2 (figure 13.19). The oldest (most negative) ΣCO_2 $\Delta^{14}C$ was found at the nearshore station (HM13) (13 m water depth), reflecting more breakdown of older terrestrial organic matter than at the deeper stations. Estuarine deltaic environments represent some of the most diagenetically active sediments found in estuarine systems; anoxic sediments in fjords likely represent the opposite end-member having less diagenetic activity, with higher C preservation rates in sediments.

The Ecological Transfer of Carbon

The balance between production and consumption is critical in controlling the contribution of distinct biological components (e.g., individual to ecosystem level) to the overall carbon cycle. Before discussing such differences in estuaries we need to review a few old and new terms. In particular, systems are *net autotrophic* when production exceeds consumption and are *net heterotrophic* when consumption exceeds production. As described in chapter 8, *gross primary production* (GPP) is the amount of carbon fixed (CO_2 converted into organic matter) by photosynthetic organisms (per unit time per unit volume). Similarly, *net primary production* (NPP) is the total carbon fixed minus the amount respired only by the primary producers (R_a). *Net ecosystem production* (NEP) is derived from the difference between GPP and *ecosystem respiration* (R), which includes both heterotrophic and autotrophic processes. So, when NEP is negative, and GPP/R is <1, the system is said to be net heterotrophic. Consequently, NEP is of particular importance in determining the sources and sinks of carbon in coastal environments (Smith and Hollibaugh,

1993; Gattuso et al., 1998; Caffrey, 2004). Anthropogenic effects, such as increased nutrient loading, has in many cases increased in situ production over respiration of estuaries which results in a NEP, or also called net ecosystem metabolism (NEM), dominated by autotrophy (D'Avanzo et al., 1996). However, in estuaries where the organic carbon loading (allochthonous and autochthonous) is high the NEM is heterotrophic (Smith and Hollibaugh, 1993, 1997; Frankignoulle et al., 1998; Cai et al., 1999; Raymond et al., 2000). In fact, *extracellular release* (ER) of DOM from phytoplankton, which is constrained by total availability of *photosynthates*, was only able to supply less than half of the required C for bacterioplankton growth in a survey of marine/estuarine ecosystems (Baines and Pace, 1991). Thus, the magnitude of carbon and nutrient loading is very important in determining if a system is driven toward net autotrophy or heterotrophy (Kemp et al., 1997; Caffrey, 2004).

Spatial and temporal variability of NEM heterotrophy versus autotrophy has been commonly observed in estuaries. For example, increasing depth gradients can shift NEM from heterotrophic to autotrophic (Howarth et al., 1996; Raymond et al., 2000), with shallow shoal areas being more autotrophic than more turbid light-limited channels of estuarine marshes (Kemp et al., 1997; Caffrey et al., 1998). A recent study used dissolved O_2 data to determine primary production, respiration, and NEM from a diversity of habitats in 42 estuarine sites within the National Estuarine Research Reserves (NERR), from 1995 to 2000 (Caffrey, 2004). Results from this work indicated that temperature was the most important variable that explained within-site variability in metabolic rates. Moreover, all the sites except for three were found to be heterotrophic on an annual basis and the range of NEM varied from -7.6 to 0.9 g O_2 m^{-2} d^{-1} (table 13.6). In particular, freshwater sites were more heterotrophic than more saline sites, with more shallow regions (e.g., mangroves and marsh creeks) being the most heterotrophic, due to higher allochthonous inputs of carbon than in deeper regions of estuaries. Temporal variation in CO_2 fluxes in surface mixed layers at three stations in the Hudson River estuary revealed that heterotrophic processes dominate C fluxes in the Hudson (figure 13.20; Taylor et al., 2003). More specifically, these estimates on C flux were based on comparisons of microplanktonic respiration with NPP and bacterial net production (BNP) (Taylor et al., 2003). In all but five cases, respiration exceeded autotrophic production by factors of 1.1 to 340, emphasizing the importance of allochthonous inputs in supporting heterotrophic processes in this Hudson River estuary. This work supports other prior studies that have identified the tidal freshwater Hudson as being net heterotrophic (Kempe, 1984; Findlay et al., 1991; Howarth et al., 1996; Raymond et al., 1997; Cole and Caraco, 2001).

Carbon Budgets for Selected Estuaries

Estuaries and the mouths of large river systems are located at an important interface between land and ocean where terrestrially derived materials can be altered before entering continental shelves. Continental shelves provide an estimated net CO_2 sink of 0.1 Pg C y^{-1} (1Pg $= 10^{15}$g), with an export that may be as much as 20% of the oceanic biological pump (Liu et al., 2000). Unfortunately, ocean margins have only recently started to receive the appropriate attention they deserve in the context of their importance in the global C budget (Bauer and Druffel, 1998; Liu et al., 2000) (more details on this in chapter 16).

Table 13.6 Annual average rates of gross primary production, total respiration, and net ecosystem metabolism (NEM) (g O_2 m^{-2}d^{-1}) at National Estuarine Research Reserve (NERR) sites.

Region/reserve/site	Production		Respiration		NEM	
	Mean	SE	Mean	SE	Mean	SE
Caribbean and Gulf of Mexico						
Jobos Bay 10	4.2	0.3	6.8	0.5	−2.6	0.4
Jobos Bay 09	5.7	0.4	10.0	0.6	−4.3	0.4
Rookery Bay, Blackwater River	3.9	0.4	11.5	0.4	−7.6	0.3
Rookery Bay, Upper Henderson	5.6	0.3	11.6	0.4	−5.9	0.3
Apalachicola Bottom	3.1	0.2	5.6	0.4	−1.6	0.2
Apalachicola Surface	2.8	0.2	4.4	0.3	−2.5	0.3
Weeks Bay, Fish River	7.7	0.9	7.4	1.0	−2.2	0.3
Weeks Bay, Weeks Bay	6.9	1.3	7.0	1.2	−2.0	0.3
Southeast						
Sapelo Flume Dock	18.4	1.5	22.1	1.7	−3.7	0.3
Sapelo Marsh Landing	9.2	0.8	11.1	1.0	−1.9	0.3
ACE Big Bay Creek	12.4	0.7	17.9	0.9	−5.4	0.7
ACE St. Pierre	12.0	0.6	14.7	0.8	−2.6	0.3
North Inlet—Winyah Bay, Oyster Landing	7.0	0.3	7.9	0.4	−2.2	0.3
North Inlet—Winyah Bay, Thousand Acre Creek	4.7	0.3	5.6	0.3	−3.0	0.2
North Carolina, Masonboro Inlet	5.5	0.3	7.7	0.5	−0.9	0.2
North Carolina, Zeke's Island	3.5	0.3	6.4	0.4	−0.9	0.2
Mid-Atlantic						
Chesapeake Bay, Virginia Goodwin Island	5.2	0.4	4.7	0.5	0.5	0.2
Chesapeake Bay, Virginia Taskinas Creek	8.9	0.6	8.5	0.7	−2.1	0.2
Chesapeake Bay, Maryland Jug Bay	6.8	0.5	12.3	0.6	−5.6	0.4
Chesapeake Bay, Maryland Patuxent Park	8.2	1.6	10.2	1.4	−2.0	0.4

Site						
Delaware Bay, Blackwater Landing	11.2	1.0	13.9	1.2	−2.7	0.2
Delaware Bay, Scotton Landing	9.4	0.9	11.0	1.1	1.6	0.4
Mullica River Buoy 126	5.8	0.6	5.9	0.6	−0.03	0.2
Mullica River lower bank	2.7	0.3	4.8	0.5	−2.1	0.3
Northeast						
Old Woman Creek, State Route 2	2.3	0.2	6.4	0.3	−4.1	0.3
Old Woman Creek, State Route 6	2.7	0.2	6.3	0.4	−3.6	0.3
Hudson River, Tivoli South	3.0	0.3	4.6	0.1	−1.6	0.2
Narragansett Bay, Potters Cove	8.2	0.6	9.9	0.8	−1.7	0.3
Narragansett Bay, T-wharf	8.0	1.0	9.4	1.2	−1.3	0.4
Waquoit Bay, Central Basin	6.6	0.3	8.8	0.4	0.3	0.2
Waquoit Bay, Metoxit Point	5.6	0.4	7.2	0.5	−0.1	0.4
Great Bay, Great Bay Buoy	7.6	0.6	7.8	0.6	−0.2	0.2
Great Bay, Squamscott River	6.5	0.6	7.1	0.7	−0.6	0.3
Wells Head of Tide	3.3	0.4	6.9	0.8	−3.6	0.5
Wells Inlet	5.1	0.5	4.9	0.5	0.9	0.3
Pacific						
Padilla Bay, Bay View	11.4	1.1	11.7	1.0	−0.4	0.2
South Slough, Stengstacken Arm	14.4	1.4	16.5	1.4	−2.1	0.2
South Slough, Winchester Arm	10.0	0.9	11.3	1.0	−1.3	0.2
Elkhorn Slough, Azevedo Pond	11.0	0.5	13.3	0.5	−2.2	0.2
Elkhorn Slough, South Marsh	3.0	0.2	4.4	0.2	1.4	0.2
Tijuana River, Oneonta Slough	15.1	0.9	19.1	1.0	−4.0	0.3
Tijuana River, Tidal Linkage	28.1	2.1	32.3	2.3	−4.1	0.4

SE = standard error.

Modified from Caffrey (2004).

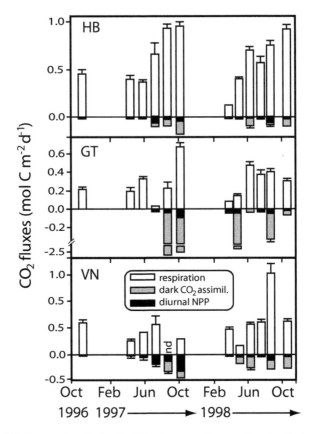

Figure 13.20 Temporal variation in CO_2 fluxes at three stations in the Hudson River estuary. Positive values represent production of CO_2 and negative values are consumption. CO_2 consumption represents the sum of net primary production (NPP) integrated over the photic zone and dark assimilation integrated over the surface mixed layer. Error bars represent ±1 standard deviation. (Modified from Taylor et al., 2003.)

Thus, an understanding of the biogeochemical cycling of C materials at the land–ocean boundary is critical if we are to better understand how these margins impact the global carbon cycle. A critical next phase in understanding exchange at land–ocean margins is to establish C budgets that attempt to quantify the fluxes and reservoirs of important biogeochemical and trophic processes. For example, carbon budgets in estuarine and river systems can help understand and constrain how much C is produced, buried, and/or exported to the adjacent shelf. Establishing reliable C budgets for estuarine systems is particularly difficult because of the high and continually changing anthropogenic inputs of nutrients (Howarth et al., 2000), which has resulted in the eutrophication of estuaries worldwide (Cloern, 2001).

As mentioned earlier, the tidal freshwater portion of the Hudson River estuary is highly net heterotrophic, with respiration governed largely by allochthonous inputs (non-point sources) of organic matter from the watershed (Cole and Caraco, 2001). A Generalized Watershed Loading Function (GWLF) Model was used to determine the role of natural versus anthropogenic processes in controlling fluxes of organic C to the estuary (Howarth et al., 2000). This model was originally designed to estimate fluxes of N and P from watersheds (Haith and Shoemaker, 1987), but was later adapted for estimating C fluxes (Howarth et al., 1991; Swaney et al., 1996). The structure of the GWLF model for the hydrologic controls on DOC inputs, as well as soil/sediment movement in forests, agricultural, urban, and suburban regions are shown in (figure 13.21; Howarth et al., 2000). More specifically, the GWLF is based on a simple mass balance approach for estimating the hydrologic fluxes of groundwater and surface flow (figure 13.21a). As for soil/sediment movement, the GWLF uses the universal soil loss equation to estimate erosion in forests and agricultural systems (figure 13.21b), and the STORM model (Hydraulic Engineering Center, 1977) for suburban and urban regions (figure 13.21c). The model results indicate that, compared to forests, agricultural systems export 10-fold more organic C per area (table 13.7). It is interesting to note that while the agricultural systems account for 74% of the organic C inputs from non-point sources to the Hudson River compared to only 18% from forested areas, the Hudson River watershed is mostly composed of forests (65% of total area) compared to only 28% as agricultural regions (table 13.7). This pattern of higher inputs from agricultural regions likely explains the relatively higher inputs of uniquely "older" POC in the Hudson River, compared to other rivers in the region (see figure 7.24); higher inputs from agricultural regions may have enhanced erosion and allowed for greater exposure of deep soil horizons, rich in ancient carbon (e.g., kerogen) (Raymond and Bauer, 2001a). Urban and suburban organic C inputs accounted for only 8% and made up approximately 7% of the area (table 13.7). Historically, the Hudson Valley was largely forested, followed by a period of deforestation and growth of agricultural land, which peaked in the early 1900s, and was then followed by a reforestation period to the present, due to abandonment of agricultural systems (Rod et al., 1989). Thus, it is fair to say that the earlier more forested pristine state of the Hudson River would have been less heterotrophic than it is today (Howarth et al., 1996).

In comparison with other estuarine systems, there has been a greater effort to construct food web and C models in the Baltic Sea (Elmgren, 1984; Wulff and Ulanowicz, 1989; Kuparinen et al., 1996; Donali et al., 1999; Sandberg et al., 2000, 2004), largely because of the availability of data, since the Baltic is one of the most extensively studied estuarine systems in the world. The first and most comprehensive modeling study compared the three major basins of the Baltic Sea, the Bothnian Bay (BB), Bothnian Sea (BS), and Baltic proper (Bp) (Elmgren, 1984). Although flows of detritus and losses from respiration were not included in this model, it was clearly determined that the highest overall production in food webs was in the Bp, followed by the BS, and then BB (Elmgren, 1984). An example of the mass balance model of carbon for the Bp is shown in figure 13.22. To be able to quantify a mass balance model in C units, it is necessary that the model be in a steady state; this early work by Elmgren (1984) was not. Recent improvements in modeling capabilities for food webs now allow for a more quantitative approach. For example, recent work has re-examined the carbon budget for the Baltic Sea, through network analysis (Ecopath II software), under the assumption of steady-state conditions, and it appears that the mass

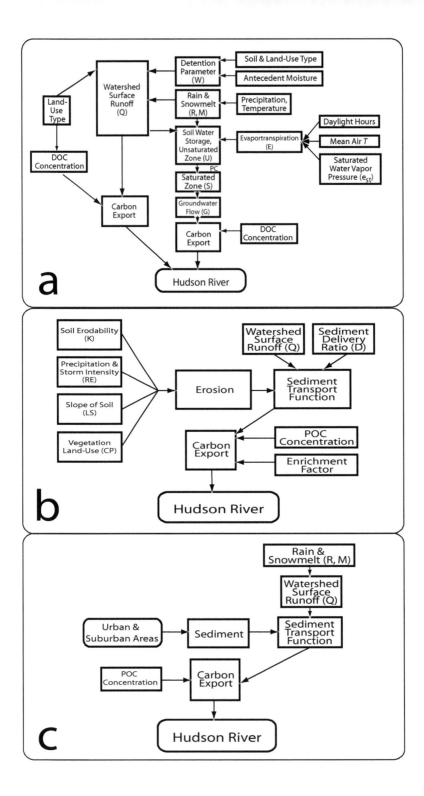

428

Table 13.7 Estimates of inputs of organic matter (g C m^{-2} y^{-1}) from non-point sources in the watershed to the tidal, freshwater Hudson River.

Regression analysis: average for all of Hudson River	3.1
Model result: average for all land uses	1.7
Model result: forest lands only	0.47
Model result: agricultural lands only	4.5
Model result: urban and suburban lands only	1.9

The regression analysis is based on Gladden et al. (1988) and uses average freshwater discharge for the Hudson River and a regression of log water discharge per area vs. log organic carbon export per area for various tributaries of the Hudson (Howarth et al. 1996). The model-derived estimates are from runs of GWLF reported by Swaney et al. (1996). All estimates are for years of average precipitation and discharge.
Modified from Howarth et al. (2000).

balances of carbon clearly deviate from steady state (Sandberg et al., 2000). For example, it was found that there was an excess of organic C of 45, 25, and 18 g C m^{-2} y^{-1} in the pelagic regions of the Bp, BS, and BB, respectively (Sandberg et al., 2000). Similar to the first model by Elmgren (1984), the overall C flow in the three basins in decreasing order was the Bp, BS, and BB (Sandberg et al., 2000). The surplus of organic C in each basin suggests that certain pools of C are not being consumed by organisms and that this excess C may in fact be coming from river inputs in the northern Baltic (Rolff, 2000).

A more recent study assessed the role of terrestrially derived DOC (TDOC) as a C source for secondary producers (e.g., bacteria) and its effect on the food web structure in the Gulf of Bothnia in the northern Baltic at three locations: the BB, Öre estuary (ÖE), and BS (Sandberg et al., 2004). The model showing the structure of food webs and organisms is shown in figure 13.23. Results from this work indicate that the bacterial-to-phytoplankton biomass (B/P) ratio was higher in the BB (44%) than in the ÖE (17%) or BS (24%), supporting the idea that C inputs from sources other than phytoplankton are controlling bacterial C demands. As stated earlier, a cross-system comparison of B/P ratios in marine/freshwater systems indicates that these ratios are around 30% (Cole et al., 1988; Baines and Pace, 1991). A comparison of all the sources of DOC available to bacteria

Figure 13.21 (a) Structure of the hydrological portion of the Generalized Watershed Loading Function (GWLF). Concentrations of dissolved organic carbon (DOC) are assigned to groundwater flows and to surface-water runoff as related to land use. (b) Structure of the GWLF for pathways of sediment in forests and agricultural systems; structure is based on the universal soil loss equation. Concentration of POC assigned to eroded soils as related to land use. (c) Structure of the GWLF model for pathways of sediment in urban and suburban areas. (Modified from Howarth et al., 1991.)

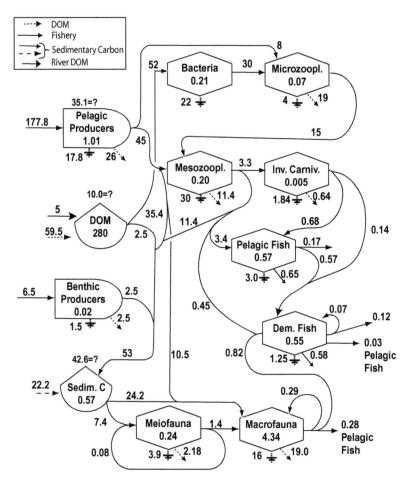

Figure 13.22 Mass balance model of carbon in the Baltic proper; carbon fluxes are expressed as $g\,C\,m^{-2}\,y^{-1}$ and reservoirs are in $g\,C\,m^{-2}$. Unbalanced fluxes are indicated by a question mark and the vertical arrow at the bottom of each organism compartment represents respiration (Elmgren, 1984). (Modified from and Sandberg et al., 2000, with permission.)

and the losses (e.g., advection) in the pelagic zones of each of these three systems are shown in figure 13.24. It should be noted that in comparison with previous Baltic Sea C budgets, this C budget includes three additional trophic links (flagellates) between the bacteria and microzooplankton as well as viral infections. It appears that there is a lack of 12 mmol C dm^{-2}, and a surplus of 16 and 17 mmol C dm^{-2} in the DOC budgets of BB, ÖE, and BS, respectively. Moreover, the contribution of TDOC from rivers was 37, 83, and 7% of the total C input in the BB, ÖE, and BS, respectively. Although the inputs to

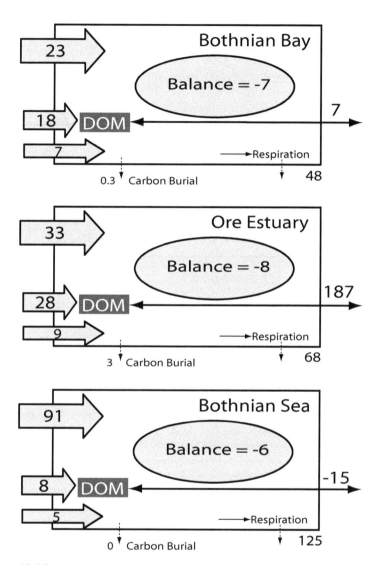

Figure 13.23 Model showing structure of food webs and organisms in the Gulf of Bothnia. Arrows indicate fluxes, shaded elliptical regions are microheterotrophic organisms, brackets inside the ellipse represent carbon flow to and from the microheterotrophs, and dotted flows represent flows to the detrital pool. All values are in mmol C dm^{-2} y^{-1}. Dissolved organic matter = DOM. (Modified from Sandberg et al., 2004.)

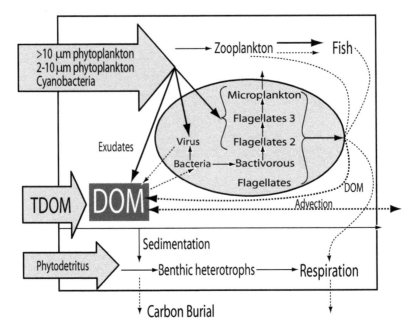

Figure 13.24 Structure of food webs and organisms in the Gulf of Bothnia at three locations. Model structure and arrows correspond to the outer component in figure 13.23. Total dissolved organic matter = TDOM. (Modified from Sandberg et al., 2004.)

the ÖE were highest, losses from advection and sedimentation resulted in TDOC having the greatest effect on the BB. The results from this work, which were well supported from earlier studies (Elmgren, 1984; Sandberg et al., 2000), showed that TDOC is important in controlling the abundance of plankton and food web structures in the Gulf of Bothnia.

Narragansett Bay (USA) is another estuarine system that has received considerable attention in past years, and was one of the first to have a constructed carbon budget (Nixon et al., 1995), albeit not as comprehensive as some of the more recent carbon budgets. The dominant forms of carbon inputs to this estuary are from phytoplankton, with an estimated 20% derived from land drainage and anthropogenic discharge (table 13.8). Sediments sequester 575 to 880 × 10⁶ mol of organic C each year, representing only 5 to 10% of the total organic C inputs going into storage or burial. Similarly, only 0.2% of the primary production is removed by the dominant fisheries in the system (table 13.8). Although carbon export was not directly measured in Narragansett Bay, preliminary estimates, based on stoichiometry of nutrient inputs (e.g., DIN and DIP from land, atmosphere, and offshore) and concentrations, indicate that export could be as low as 90 × 10⁶ mol C y⁻¹ to as high as 925 × 10⁶ mol C y⁻¹ to offshore waters (Nixon and Pilson, 1983). The upper extreme would account for only 10% of the primary production leaving the bay. Thus, unlike the Hudson (Findlay et al., 1991), Loire (Relexans et al., 1988), Baltic Sea (Sandberg et al., 2004), and Seine River estuary (France)

Table 13.8 The annual mass balance of C in Narragansett Bay (units are 10^6 mol y^{-1}).

	C
Inputs	
Atmospheric deposition	9600
Biological fixation	1815
Land drainage	235
Direct sewage	11650
Offshore	11650
Outputs	
Denitrification	7–14
Clam harvest	790–1565
Organic export offshore	
formed in the bay	1080
75% of river DOM input	1757–2634
Inorganic export[a]	8100–9200
Accumulation in sediments	575–880

[a]Calculated by difference; ranges are set by combinations of the low and high denitrification estimates (which lead to high or low estimates of autochthonous organic C) and upper and lower estimates of burial. Values have been rounded to two or three significant figures.
Modified from Nixon et al. (1995).

(Garnier et al., 2001), Narragansett Bay estuary appears to be net autotrophic (Nixon et al., 1995).

The C budget of the subtropical Moreton Bay estuary (Australia) indicates that most of the C loss was from atmospheric exchange of CO_2 associated with benthic algae and pelagic respiration (Eyre and McKee, 2002). As mentioned earlier, most of the nutrient budgets for estuarine systems have been constructed for temperate systems (Nixon et al., 1996). Similar to Narragansett Bay, the fisheries yield for Moreton Bay represented less than 0.1% of the total C (table 13.9). Based on the positive relationship between primary production and fisheries yield shown for other estuarine systems (Alongi, 1998), the expected yield (using total production) for Moreton Bay should be 71 kg ha^{-1} y^{-1}, significantly higher than the actual 26 kg ha^{-1} y^{-1}. Differences between the total C inputs and outputs yields a net export of 39,338 t of C (table 13.9) and represents 8% of the total primary production. Thus, more C is exported than imported in Moreton Bay

Table 13.9 Nutrient budget for Moreton Bay, Australia (t y^{-1}).

Budget components	Carbon
Standing stocks	
Sediment (solid phase)	116,872
Water column (total)	3667
Biomass	2,313,766
Inputs	
Point sources	4330
Non-point sources	6395
Atmosphere	5223
Groundwater	1
Primary production	501,000
Total	516,949
Outputs	
Pelagic respiration	−63,187
Benthic respiration	−465,632
Dredging	−2560
Burial	−1291
Fisheries harvest	−488
Pumice stone passage	−2776
Ocean exchange	−48,171
Recycling	
Phytoplankton	121,944
Sedimentation	

Modified from Eyre and McKee (2002).

making this a net autotrophic system. Based on the aforementioned estuaries, most of the net heterotrophy in estuaries is driven by high DOC inputs. Unfortunately, no direct measurements of DOC were available for the Moreton Bay C model.

Summary

1. The major reservoirs of C are stored in the Earth's crust, with much of it as inorganic carbonate and the remaining as organic C (e.g., kerogen).
2. When examining the aquatic chemistry of CO_2 the first important point is that when CO_2 dissolves in water it can then be hydrated to form H_2CO_3, which can then dissociate to HCO_3^-, and CO_3^{2-}.

3. The total DIC pool (ΣCO_2: CO_2, H_2CO_3, HCO_3^-, and CO_3^{2-}) in marine waters contains about 2 mmol C L^{-1} (mostly HCO_3^-), while in estuarine and freshwaters the ΣCO_2 is more variable and pH dependent.

4. One term now used interchangeably with acid neutralizing capacity (ANC) of natural waters is alkalinity (Alk)—defined as the concentration of negative charge that will react with H^+.

5. The three dominant diagenetic reactions controlling DIC are aerobic respiration, SO_4^{2-} reduction (SR), and methanogenesis.

6. When SO_4^{2-} is depleted and methanogenesis is the dominant form of anaerobic respiration, it can proceed through either of the following two pathways: fermentation or CO_2 reduction.

7. In addition to alkalinity inputs from rivers, estuaries can also receive inputs from bordering wetlands.

8. Estuaries are in many cases considered to be net heterotrophic as a result of an "excess" loading of allochthonous materials to the system.

9. River waters entering at the head of an estuary typically show higher concentrations of CO_2 and lower concentrations of O_2. This is due to the loading of enriched pCO_2 waters from river inputs, which are high because they contain the mineralization "signatures" from soils, bordering freshwater wetlands, and resuspension from upper estuarine sediments.

10. Similar to CO_2, emission rates of CH_4 are highly variable in estuaries with the higher rates being found in tidal flat and marsh environments.

11. The mechanisms of CH_4 transport commonly found in plants are simple molecular diffusion (differential partial pressures between the sediment, plant, and atmosphere), convective throughflow ventilation (via lacunal pressurization), and effusion (partial pressures differences in pore of plant).

12. Methane oxidizing bacteria, or methanotrophs, can play a central role in reducing CH_4 emissions from estuaries, by converting CH_4 into bacterial biomass or CO_2.

13. Concentrations of POC in many estuaries are strongly coupled with suspended sediment particulates, which may depend on river discharge and/or resuspension events.

14. The concentrations of DOC in the lower-salinity regions are due to riverine inputs of DOC. Seasonal variability of phytoplankton blooms can at times change the distribution of DOC in estuaries, whereby the higher concentrations of DOC co-occur with the blooms in the lower estuarine region due to lower light attenuation rates.

15. Net primary production (NPP) is the total carbon fixed minus the amount respired only by the primary producers (R_a). Net ecosystem production (NEP) is derived from the difference between GPP and ecosystem respiration (R), which includes both heterotrophic and autotrophic processes.

16. Estuaries and the mouths of large river systems are located at an important interface between land and ocean where terrestrially derived materials can be altered before entering continental shelves.

17. Establishing reliable C budgets for estuarine systems is particularly difficult because of the high and continually changing anthropogenic inputs of nutrients, which has resulted in the eutrophication of estuaries worldwide.

Chapter 14

Trace Metal Cycling

Sources and Abundance of Trace Metals

Like many other elements, natural background levels of trace elements exist in crustal rocks, such as shales, sandstones, and metamorphic and igneous rocks (Benjamin and Honeyman, 2000). In particular, the majority of trace metals are derived from igneous rocks, simply based on the relative fraction of igneous rocks in comparison with sedimentary and metamorphic rocks in the Earth's crust. The release of trace metals from crustal sources is largely controlled by the natural forces of physical and chemical weathering of rocks, notwithstanding large-scale anthropogenic disturbances such as mining, construction, and coal burning (release of *fly ash*). As discussed later in the chapter, adjustments can be made for anthropogenic loading to different ecosystems based on an *enrichment factor* which compares metal concentrations in the *ecosphere* to average crustal composition. Biological effects of weathering, such as plant root growth and organic acid release associated with respiration also contribute to these weathering processes. As some trace metals are more volatile than others, release due to volcanic activity represents another source of metals with such properties (e.g., Pb, Cd, As, and Hg). Just as Goldschmidt (1954) grouped elements (e.g., *siderophiles, chalcophiles, lithophiles,* and *atomophiles*) based on similarities in geochemical properties, trace metals also represent a group of elements with similar chemical properties. One particularly important distinguishing feature of these elements is their ability to bond reversibly to a broad spectrum of compounds (Benjamin and Honeyman, 2000). Thus, the major inputs of trace metals to estuaries are derived from riverine, atmospheric, and anthropogenic sources.

Although trace elements typically occur at concentrations of less than 1 ppb (part per billion) (or $\mu g\ L^{-1}$, also reported in molar units), these elements are important in estuaries because of their toxic effects, as well as their importance as micronutrients for many organisms. The fate and transport of trace elements in estuaries are controlled by a variety of factors ranging from redox, ionic strength, abundance of adsorbing surfaces, and pH, just to name a few (Wen et al., 1999). The highly dynamic nature of

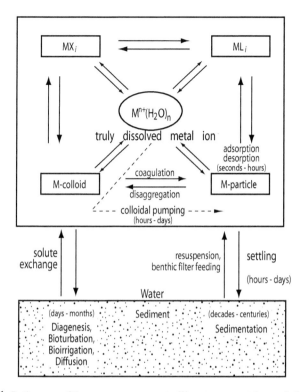

Figure 14.1 Pathways of key processes controlling trace metal speciation in aquatic systems, as they relate to the interchange of metals between water and sediments. (Modified from Santschi et al., 1997.)

estuarine systems, characterized by strong chemical and physical gradients, makes trace metal cycling considerably more complex in estuaries compared to other aquatic systems (Morel et al., 1991; Millward and Turner, 1995). For example, the partitioning of trace metals between the dissolved and particulate fractions in estuaries can be affected by variability of in situ processes such as coagulation and flocculation in the estuarine turbidity maximum (ETM), resuspension events (of sediments and porewaters), and sedimentation (figure 14.1; Santschi et al., 1997). All of these processes contribute to the complexity of trace metal speciation in estuaries (Boyle et al., 1977; Shiller and Boyle, 1987; Honeyman and Santschi, 1989; Buffle et al., 1990; Santschi et al., 1997, 1999; Wen et al., 1999). Larger scale internal and external processes such as storm events, tidal exchange, wind effects, and inputs from rivers and bordering wetlands also contribute to the overall partitioning of metals in estuaries. In this chapter, we will provide a general background of the key processes mentioned above that control metal ion chemistry and discuss more specific examples of trace metal cycling in the water column and sediments of selected estuaries.

Background on Metal Ion Chemistry

Metals commonly exist in multiple oxidation states and will generally have a positive *oxidation state* (e.g., $+1$ to $+6$) when bonded with other elements because of their *electrophilic* character (Benjamin and Honeyman, 2000). The common atoms that typically have free electron pairs (act as *Lewis bases*) to donate to metals are O, N, and S; the covalent bonds between these elements and metals are stronger than the *electrostatic* bonds that metals form with water molecules (see below). The thermodynamically stable form of a metal ion in the natural environment is primarily controlled by the ambient oxidation potential, which is essential in determining their fate and role in biological processes. For example, under anoxic conditions ferrous iron, Fe(II), can dissolve in water; however, if this water is exposed to oxygen, Fe(II) will oxidize to ferric iron, Fe(III), and will precipitate out of solution, dramatically lowering the concentration of dissolved Fe. Similarly, the toxicity of trace metals can change with redox, as is the case of the more toxic As(III) compared to As(V) (Santschi et al., 1997; Benjamin and Honeyman, 2000). The volatilization of trace metals can also significantly affect the thermodynamic equilibrium between gaseous and aqueous phases, which is controlled by Henry's law constant (see chapter 5). For example, while most trace metals typically have low vapor pressures and *fugacity*, there are some exceptions, such as Hg and *organometallic* compounds. In the case of methylmercury (CH_3Hg^+), there is a general tendency for this compound to be removed from the aqueous to the gaseous phase, due to the low (zero) partial pressure of this compound (Sunderland et al., 2004) (more on Hg cycling later in the chapter).

The complexation of metal ions with water molecules interferes with the hydrogen or electrostatic bonding between water molecules. As discussed in chapter 4, if different chemical species are to dissolve effectively in water, they need to disrupt the strong hydrogen bonding and clustering properties of water molecules. If the solute forms strong bonds with water, replacing water-to-water bonds, it will be thermodynamically stable and dissolution will occur (figure 14.2a; Benjamin and Honeyman, 2000). Conversely, if the solute-to-water bonds are weak, dissolution will be unstable, making the solute relatively insoluble (figure 14.2b). Thus, cationic metals form strong bonds with the oxygen atom of water, whereby most metals are surrounded by an "inner hydration sphere" of multiple water molecules (e.g., 4–8) as well as a weaker "outer sphere" of water molecules (Benjamin and Honeyman, 2000). The strength of these bonds will increase with increasing charge and decreasing size of the metal ion; this metal-to-water bond is typically illustrated as $Me(H_2O)_x^{n+}$. In this case, water is acting as an inorganic *ligand*, whereby oxygen atoms are donating electrons to the metal ion in this *single-ligand complex*. However, other dissolved species can replace water molecules as ligands in the hydration sphere, resulting in a *mixed-ligand complex* (figure 14.3; Benjamin and Honeyman, 2000). In cases where two or more different dissolved species have replaced water molecules, the complex is considered to be a *multidentate complex*. Ligands that form strong multidentate complexes are referred to as chelating agents e.g., ethylenediamimetetraacetic acid (EDTA), or the entire complex is known as a *chelate*. It is also important to remember that, while many ligands are anions, molecules such as NH_3 (which is neutral in charge) can serve as a strong complexing source for certain metals (see Benjamin and Honeyman '2000' for more details on this). While the complexation of metals with inorganic ligands have typically been evaluated theoretically on the basis of thermodynamic-association equilibrium

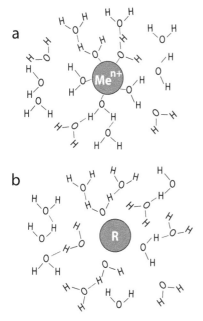

Figure 14.2 Schematic showing the effects of a charged (hydrophilic) solute on water structure and molecular orientation: (a) solute forms strong bonds with water where dissolution is favorable; (b) solute forms weak bonds where dissolution is not favorable (Modified from Benjamin and Honeyman, 2000.)

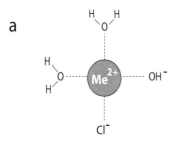

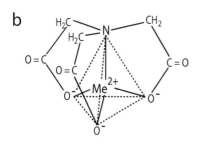

Figure 14.3 Schematic of (a) a hypothetical mixed ligand [Me(OH)Cl^0] complex, and (b) a multidentate nitrilotriacetate chelate of a divalent metal ion in a tetrahedral configuration. (Modified from Benjamin and Honeyman, 2000.)

models and the stability constants for the major complexes (e.g., Turner et al., 1981; Millero, 1985; Hering and Morel, 1989), this approach ignores the effects of organic ligands.

Over the past decade, there have been numerous studies that have shown the importance of organic ligands in complexing trace metals (Sunda and Ferguson, 1983; Coale and Bruland, 1988; Santschi et al., 1997, 1999), with particular emphasis on colloidal-sized particles (Benoit et al., 1994; Martin et al., 1995; Guentzel et al., 1996; Powell et al., 1996) (*colloids* are in the size range of 1 nm to 1 μm, see figure 4.7). These studies clearly show different trace metal behavior in colloidally complexed metals and truly dissolved metals (Buffle et al., 1990; Buffle and Leppard, 1995; Wen et al., 1999; Santschi et al., 1999). As discussed in chapter 8, there are three distinct pools of organic matter: POC (>0.2 to 0.45 μm), DOC (<0.2 to 0.45 μm), and HMW DOC/LMW DOC (<0.2 to 0.45 μm and >1 kDa); colloidal organic carbon (COC) is a term generally used by inorganic chemists in place of HMW DOC, which is commonly used by biochemists (see Wells, 2002, for further discussion on this nomenclature). Thus, a 1 nm spherical diameter is essentially equivalent to macromolecules with a nominal molecular weight of 1 kDa (Chin and Gschwend, 1991). Other important features about macromolecular organic matter, in the context of trace metal cycling, are that it is heterogeneous in composition (see chapters 8 and 9), size, and molecular weight, and is generally *polymeric, polyelectric, polyfunctional, amphiphilic*, and *polydisperse* (Santschi et al., 1997). Natural organic matter can also change its conformation according to pH and ionic strength. Santschi et al. (1999) described metal complexation in the context of organic ligand exchange, in its simplest form, using the following equation:

$$M_1L_1 + M_1L_2 = M_1L_2 + M_2L_1 \qquad (14.1)$$

where: M_1 = trace metal; M_2 = major metal or proton; L_1 = water or inorganic ligand; L_2 = organic ligand.

While dissolved and particulate inorganic compounds are clearly important in the complexation of free metals (Millward and Turner, 1995), organic complexation of metals has been shown to be a key process in estuarine waters (van den Berg, 1987; Kozelka and Bruland, 1998; Wells et al., 1998; Tang et al., 2001, 2002). Overall, the distribution and speciation of trace metals in estuaries will depend on their concentrations as well as the concentrations of dissolved complexing ligands and the associated coordination sites on colloids and particulates (Kozelka and Bruland, 1998). More specifically, in the dissolved phase, the metal can occur in three different phases as: (1) a free hydrated ion (M^{n+}), (2) an inorganic complex (M'), and (3) an organic complex (ML_i). Several analytical techniques have been developed to determine the free ion concentration and speciation of metals in natural waters. The two most common techniques, largely developed from work on Cu speciation, are *anodic stripping voltammetry* (ASV) and *competitive ligand equilibration/adsorptive cathodic stripping voltammetry* (CLE-ACSV) (see more details in Bruland et al., 2000). These electrochemical techniques, combined with metal titration analyses, to obtain ligand concentrations and conditional stability constants, can be used to determine how many ligands are involved in metal complexation as well as their relative differences in strength. The *conditional stability constant* is reported as follows:

$$K_{\text{MLi,M}'} = [\text{ML}]/[\text{M}'][\text{L}'] \qquad (14.2)$$

where: $[\text{L}']$ = total dissolved ligand $[\text{L}_T]$ – metal ligand $[\text{ML}]$, or the ligand not bound to metal.

Although many of the inorganic speciation syntheses of trace metals in natural waters have been well described (table 14.1; Stumm and Morgan, 1981; Turner et al., 1981; Millero and Hawke, 1992), sources and details on the functional groups of the organic ligands, detected by these techniques, remain largely unknown. Some of the dominant functional groups of organic matter that commonly form strong complexes, albeit at a slow rate (Hering and Morel, 1989), with trace metals are -COOH, -OH, -NR$_2$, and -SR$_2$ (R = -CH$_2$ or -H). The sulfhydryl (-SH) group in glutathione is responsible for most of the complexation of metals such as Cu, Pb, Hg, Cd, and Zn (Krezel and Bal, 1999; Tang et al.,

Table 14.1 Model results for metal speciation in natural waters.[a]

	Freshwater			Seawater	
	Inorganic		Inorganic and organic,	Inorganic,	Inorganic and organic,
	pH 6	pH 9	pH 7	pH 8.2	pH 8.2
Ag$^+$	72, Cl	65, Cl, CO$_3$	65, Cl	<1, Cl	<1, Cl
Al^{3+}	<1, OH, F	<,OH		<1, OH	
Cd^{2+}	96, Cl, SO$_4$	47, CO$_3$, OH	87, org, SO$_4$	3, Cl	1, Cl
Co^{2+}	98, SO$_4$	20, CO$_3$, OH		58, Cl, CO$_3$, SO$_4$	63, Cl, SO$_4$
Cr^{3+}	<1, OH	<1, OH		<1, OH	
Cu^{2+}	93, CO$_3$, SO$_4$	<1, CO$_3$, OH	<1, org	9, CO$_3$, OH, Cl	<1, org, CO$_3$
Fe^{2+}	99	27, CO$_3$, OH		69, Cl, CO$_3$, SO$_4$	
Fe^{3+}	<1, OH	<1, OH	<1, org, OH	<1, OH	<1, OH, org
Hg^{2+}	<1, Cl, OH	<1, OH		<1, Cl	
Mn^{2+}	98, SO$_4$	62, CO$_3$	91, SO$_4$	58, Cl, SO$_4$	25, Cl, SO$_4$
Ni^{2+}	98, SO$_4$	9, CO$_3$		47, Cl, CO$_3$, SO$_4$	50, org, Cl, SO$_4$
Pb^{2+}	86, CO$_3$, SO$_4$	<1, CO$_3$, OH	9, CO$_3$, org	3, Cl, CO$_3$, OH	2, CO$_3$, OH
Zn^{2+}	98, SO$_4$	6, OH, CO$_3$	95, SO$_4$, org	46, Cl, OH, SO$_4$	25, OH, Cl, org

[a]Data for inorganic freshwater and inorganic seawater from Turner et al. (1981). Data for systems with inorganics and organics from Stumm and Morgan (1981). Six organic ligands included corresponding to 2.3 mg/L total soluble organic carbon. Stability constants and inorganic composition of model water were not identical in the two studies, so comparisons are qualitatively valid but may have minor quantitative inconsistencies. Each entry has the percent of total metal present as the free hydrated ion, then the ligands forming complexed, in decreasing order of expected concentration. For instance, in inorganic freshwater at pH 9, Ag is present as the free aquo ion (65%), chloro-complexes (25%), and carbonato-complexes (9%).
Modified from Turner et al. (1981); Stumm and Morgan (1981); Millero and Hawke (1992).

2001, 2002). Some natural sources containing these functional groups are *phytochelatins* (*siderophores*), biopolymers (e.g., proteins), and humic substances (see chapters 8 and 9 for more details on these compounds). Compounds such as phytochelatins are used by algae (Donat and Bruland, 1995) and higher plants (Grill et al., 1985) to enhance metal uptake capability in the natural environment, particularly in systems where their concentrations are exceptionally low.

The conditional stability constants of different functional groups in organic matter can vary significantly with different trace metals and are critical in predicting trace metal speciation. The strength of binding constants is largely a function of ionic radius, atomic number, and valence (follows the *Irving–Williams series*) (Santschi et al., 1997). In general, the two methods for determining trace-metal binding constants to macromolecules involve using a *discrete model*, with only one of two types of functional groups, and a specific K for each is considered (e.g., carboxyl and phenol), or a *multi-site model*, which allows for a range of K values for each site (Perdue and Lytle, 1983; Dzomback et al., 1986). Other studies have suggested that complexation is controlled by low concentrations of organic ligands with high metal-binding specificity (Hering and Morel, 1989), as opposed to work which suggests that metal availability is controlled by *steric* constraints, whereby metals are considered to be "trapped" in colloidal aggregates (Honeyman and Santschi, 1989). In the second case, the primary limiting factor for metal availability is transport limitations such complexation models have involved the binding of metals in colloids (e.g., *colloidal pumping*) (Honeyman and Santschi, 1989), or the layering of metals between layers of macromolecules/colloids (e.g., *the onion model*) (Mackey and Zirino, 1994)—both clearly emphasize the importance of colloidal organic matter trace metal speciation.

The discrete model is still commonly used by electrochemists when determining trace metal complexation with organic matter (Buffle, 1990; Donat and Bruland, 1995). However, more recent approaches have led to the application of equilibrium speciation models, which in aquatic systems work best when the oxidation state remains relatively constant and when complexes formed from solution or absorption are reversible (Tipping et al., 1998). One example of such a model is the Windermere Humic–Aqueous Model (WHAM), which has been used to calculate equilibrium chemical speciation of alkaline Earth cations, trace metals, radionuclides in surface waters of rivers and estuaries, groundwaters, sediments, and soils (Tipping and Hurley, 1992; Tipping, 1993, 1994; Tipping et al., 1991, 1995a,b, 1998) (see chapter 4 for more details). Nevertheless, it remains very difficult to predict trace metal speciation in estuaries because of the high variability of organic matter composition and associated binding constants in estuaries.

Metal complexation with ligands is also important in controlling toxicity of metals. It has long been known that the toxicity of trace metals is more dependent upon their ionic activity than on their overall concentration (Sunda and Guillard, 1976; Anderson and Morel, 1978; Morel, 1983). As discussed earlier, factors such as pH, hardness, and DOM concentrations are key in controlling metal speciation and toxicity. A biotic ligand model (BLM) has been introduced to describe a characterized site where the concentrations of the metal reach a critical concentration at the metal–ligand site where, in the case of fish, would be located in the gills (Di Toro et al., 2001; Santore et al., 2001; Heijerick et al., 2002). Thus, the biotic ligand is defined as a specific receptor at some location on an organism where metal complexation leads to acute toxicity. In essence, the BLM is used to predict metal interactions at the biotic ligand in the context of other competing reactions in aqueous environments. Trace metals associated with colloids have been also shown to be different in their bioavailability and toxicity when compared to free aquo trace metal ions (Wright, 1977; Campbell, 1995; Doblin et al., 1999; Wang and Guo, 2000). While there remains considerable uncertainty whether colloidal bound metals have greater or less bioavailability, some studies have shown that colloidal Fe (e.g., Fe_2O_3 and FeOOH) is less available to marine phytoplankton (Wells et al., 1983; Rich and Morel, 1990), while other studies have shown enhanced rates of uptake by phytoplankton in the colloidal form (Wang and Guo, 2000). More details on metal toxicity as it relates to complexation processes in estuaries are provided in chapter 15.

Interactions between particles and trace metals are also important in controlling trace metal concentrations in estuaries. For example, processes such as adsorption, desorption, flocculation, coagulation, resuspension, and bioturbation are particularly important in controlling the interactions between dissolved (free aquo) trace metals and particulates in estuaries (Santschi et al., 1997, 1999; Benjamin and Honeyman, 2000). In particular, there are important binding sites on Fe and Mn oxyhydroxides, carbonates, clays, and POC/COC that are essential in controlling adsorption/desorption of trace metals. Both physical (i.e., *coulombic* interactions in the outer sphere) and chemical (covalent bonding with electrons in the inner sphere) forces are critical in controlling the binding of trace metals to particle surfaces (Santschi et al., 1997). In many cases, these solid-solution reactions have been described using metal oxides or hydroxides as the solid or absorbent phase. For example, when considering competition between metal ions and hydrogen ions in binding to an oxide surface, pH will be an important controlling variable. In general, the metal ion adsorption is enhanced at a higher pH and decreases at a lower pH; the adsorption edges indicate that binding by cations is "metal like" and for anions (in this case *metalloids*) is "ligand like" (figure 14.4; Santschi et al., 1997). In the case of metalloids, the curves are the inverse of the cation curves because these anionic metalloids are competing with OH^- for binding sites, making their binding more effective at a lower pH (Dzombak and Morel, 1990).

Particle–particle interactions, possibly involving metal oxides, clay minerals, and macromolecules (colloids) can have significant effects on trace metal behavior in estuaries. One such effect, commonly referred to as the *particle concentration effect* (PCE) has been defined by Santschi et al., 1997, p. 112) as "the physical effects which lead to a decreasing overall partition coefficient with increasing particle concentration; the effect is documented for both organic and inorganic species." For example, Benoit et al. (1994) demonstrated the PCE for Zn, Pb, Cu, and Ag in six Texas estuaries, where it was shown that particles were more enriched in these metals when particle concentrations were low (figure 14.5). While grain size (Duinker, 1983), colloids (Gschwend and Wu, 1985; Honeyman and Santschi, 1989) and competing particle surfaces may all contribute to the observed PCE, the colloidal effect has proven to be most important (Santschi et al., 1997). Thus, the major consequences of the PCE in estuaries are: (1) an increase in scavenging of trace metals at low particle concentrations, and (2) reduced desorption from particulates during resuspension events, as compared to the predicted partition coefficients of these chemical species at higher particle concentrations. Temporal and spatial variability of the PCE are also likely to be quite high in estuarine systems because of their dynamic character. For example, the kinetics of the PCE is likely to be higher in the upper regions of estuaries,

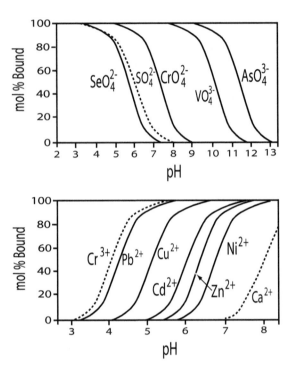

Figure 14.4 Metal binding on inorganic particles showing typical adsorption edges of selected anions and cations on Fe oxide particles reflective of "metal-like" and "ligand-like" complexes. (Modified from Santschi et al., 1997.)

near the ETM, where coagulation of colloidal-bound trace metals will occur over shorter periods (Stordal et al., 1996). As mentioned earlier, the heterogeneous nature of organic material found in estuaries may enhance how different organic constituents coalesce in colloidal material (Santschi et al., 1999). In fact, certain metals, such as Mg^{2+}, Ca^{2+}, and Mn^{2+}, may provide a mechanism for bridging these constituents in colloids (Chin et al., 1998). For example, studies in the inner regions of the Firth of Clyde (Scotland) have shown that colloid/solution partitioning of metal-selective ligands is central in controlling the behavior of trace metals in this region (Muller, 1998). More specifically, these weakly bound (divalent cation bridged) assemblages were found to exchange other metals (e.g., Cu, Pb, and Cd) with natural ligands due to charge effects (Muller, 1999).

Trace Metal Cycling in the Water Column

Trace metals that are particle reactive (e.g., Pb) or have a nutrient-like behavior (e.g., Cd) are typically removed from surface waters via adsorption in their vertical transport through

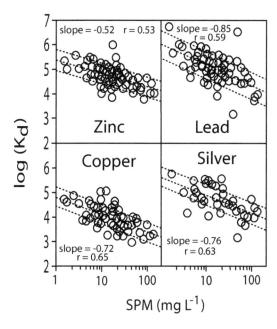

Figure 14.5 Particle concentration effect showing negative relationship between particle/solution partition coefficient (K_d) of selected trace metal ions and suspended particulate matter (SPM). Data are from six Texas estuaries (USA). (Modified from Benoit et al., 1994.)

the water column. These removal processes are more likely to occur in deeper estuaries, less affected by resuspension events, where particles can become trapped at the pycnocline or redox boundary, as found in the Baltic Sea (Pohl and Hennings, 1999). It is widely accepted that hydrous oxides of Fe and Mn are important in the sorptive removal of trace metals in estuaries (Perret et al., 2000; Turner et al., 2004). The lateral and vertical distribution of these carrier-phase metals in estuaries are largely controlled by particle dynamics, as opposed to other metals (e.g., Cu, Zn, and Co) which will be more affected by biotic uptake processes. Another metal that has been shown to behave in estuaries very much like the more well-known geochemically active metals (e.g., Fe, Mn, and Co) is Ti (Biggs et al., 1983; Church et al., 1986). In fact, as much as 65 to 80% of the dissolved Ti can be removed from the Delaware estuary (USA) above a salinity of 14 (figure 14.6; Skrabal et al., 1992). These patterns of Ti removal in high salinity regions have also been shown for Chesapeake Bay and Amazon River estuaries (Skrabal, 1995).

Differences in the pathways of trace metal cycling in the water column should be reflected in the overall vertical flux of particulate metals as they are transported through the water column. In fact, recent sediment trap work in the Baltic Sea has shown that the flux of particulate Mn and Fe was primarily controlled by the distance between the pycnocline and redoxcline (Pohl et al., 2004). More specifically, Fe and Mn concentrations in sinking particulates were found to be lowest in July and August, because the redox interface moved up to approximately 120 m water depth (where the sediment trap was located) (figure 14.7), allowing for reduction of Fe and Mn oxides. During the winter months, Al and Fe concentrations increased in trap materials, reflecting greater lithogenic

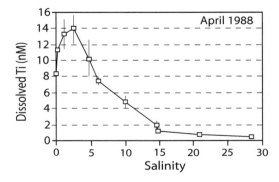

Figure 14.6 Concentrations of dissolved Ti as a function of salinity in the Delaware Bay estuary, in April 1988. (Modified from Skrabal et al., 1992.)

inputs, while in late summer particulates were enriched in Cd, Zn, and Ni, reflecting inputs from POM. In summer months, POM represented 63% of the total mass flux. Elevated concentration of Zn, As, Fe, Mn, and Pb in February 2001, was attributed to high lateral inputs from riverine sources. In fjord systems, poor circulation typically results in a well-defined persistent redoxcline, which occurs relatively shallow in the water column and is characterized by steep chemical gradients. These gradients have been well documented in suboxic basins such as Framvaren (Todd et al., 1988; McKee and Todd, 1993) and Sannich Inlet (Canada) (Todd et al., 1988). In Framvaren Fjord (Norway), the redoxcline occurs at a depth of 20 m in the water column and has a thickness of approximately 2 m (Hallberg and Larsson, 1999). Swarzenski et al. (1999) showed that the reduced U (IV) occurred at very low levels of detection, and that chemical/biological reduction of U (VI) was largely inhibited across the redox transition in the stratified Framvaren Fjord (Norway). Sharp concentration gradients of other metals (e.g., Cu, Cd, and Zn) at the redoxcline in Framvaren Fjord have been shown to be largely controlled by the chelation/complexation with organic ligands (Hallberg and Larsson, 1999).

Sorption-desorption from suspended particulates and sediment fluxes play a large part in controlling the nonconservative behavior of dissolved concentrations of Fe and Mn in estuaries (Klinkhammer and Bender, 1981; Yang and Sanudo-Wilhelmy, 1998). The mobilization of Mn from Mn-rich porewaters, followed by uptake on particles in the water column, has commonly been invoked as an important mechanism controlling seasonal changes of dissolved Mn in estuaries (Morris et al., 1987). The distribution of dissolved Mn across a salinity gradient in the Columbia estuary (USA) was very similar to what is found in many estuarine systems, indicating a source within the estuary that could possibly be from desorption of particulates, inputs from embayments, or in situ production (figure 14.8) (Klinkhammer and McManus, 2001). A vertical profile of the dissolved Mn, salinity, and suspended particulate matter (SPM) in the Columbia estuary showed that the Mn subsurface peak is more widespread than SPM (figure 14.9) (Klinkhammer and McManus, 2001). These lateral and vertical gradients of dissolved Mn have largely been attributed to in situ production via reduction of Mn-oxides in the Columbia River estuary, based on the following lines of evidence: (1) the Mn maximum occurs at the same depth

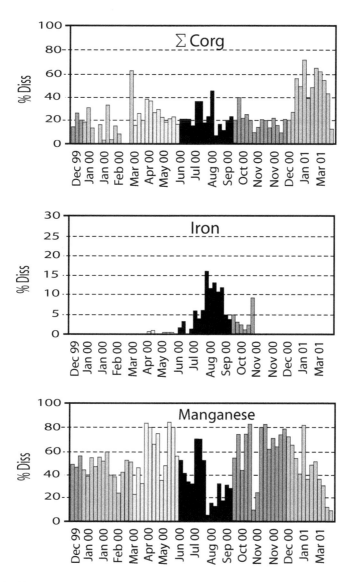

Figure 14.7 Particulate concentrations of Mn and Fe in sediment trap samples collected in the Baltic Sea from December 1999 to March 2001. The sediment trap was located at a water depth of approximately 120 m. (Modified from Pohl et al., 2004.)

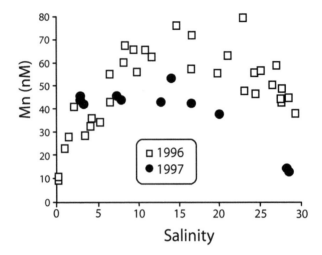

Figure 14.8 Distribution of dissolved Mn across a salinity gradient in the Columbia estuary (USA); filtered water samples collected in 1997 (closed circles) and in 1996 (open squares). (Modified from Klinkhammer and McManus, 2001.)

across different salinities, eliminating the likelihood of Mn inputs from embayments, (2) the maximum Mn concentration, which occurs at mid-depth, is consistent throughout the estuary, and (3) peak subsurface concentrations of Mn occur at the top of the ETM, where the highest microbial activity is found. Manganese reduction in the water column would require both suboxic conditions and Mn-oxides. Past work has suggested that particle aggregates provide an important mechanism for Mn transport (Sundby et al., 1981), and may also serve as a location for the Mn reduction, since aggregates have also been shown to have suboxic microsites within their inner matrices (Alldredge and Cohen, 1987). This work represents the first line of evidence for the importance of aggregates as sites for Mn reduction in the water column of an estuary.

While Fe and Mn are in many cases the ideal examples illustrating the importance of sorption/desorption processes in controlling dissolved metal concentrations, many other more bioactive metals follow similar trends, even in estuarine systems with very divergent properties. In one such study, comparisons on the cycling of Co were made between the Hudson River estuary (HRE) and San Francisco Bay (SFB) (Tovar-Sanchez et al., 2004). In both estuaries a clear nonlinear decrease in dissolved Co was observed with increasing salinity as a function of suspended particulates (figure 14.10). Despite differences in the concentrations of suspended particulates as well as large-scale differences in climate and hydrological regime, the relationship between particles and dissolved Co concentrations were similar. The enhanced desorption of Co in both systems at higher salinities is consistent with salinity effects on particle–solution exchange of Co (Turner et al., 2002). Since DOC was not measured in this study, the effects of organic ligands on Co partitioning could not be evaluated; however, both systems are not high in DOC. While biological uptake and release of bioactive metals, such as Se, can be significant in the presence of

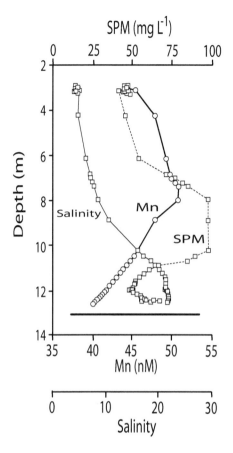

Figure 14.9 A vertical profile of the dissolved Mn, salinity, and suspended particulate matter (SPM) in the Columbia estuary, in the estuarine turbidity maximum. (Modified from Klinkhammer and McManus, 2001.)

in situ organic matter cycling (e.g., photosynthesis and respiration) (Baines et al., 2001), effects of these processes can often be masked by high loading of trace metals from anthropogenic inputs, such as occurs for Se in SFB (Cutter and Cutter, 2004).

The distribution and speciation of trace metals across an estuarine salinity/mixing gradient has been shown to be strongly affected by the abundance of inorganic and organic colloidal material (Dai et al., 1995; Millward and Turner, 1995; Rustenbil and Wijnholds, 1996; Santschi et al., 1997). As discussed earlier, the abundance and composition of inorganic (Sholkovitz et al., 1978) and more recently organic colloids (Wells et al., 2000) have long been considered to be important in controlling trace metal behavior. The distribution of colloidal trace metals across a salinity gradient in Galveston Bay (USA), showed a decrease with increasing salinity suggestive of a riverine source of colloids (Guo et al., 2000 a,b). In particular, trace metals such as Cu, Co, Ni, and Zn followed spatial trends with percentage organic matter in the water column, further supporting the possible role of colloidal complexation. Relative decreases and increases in concentrations of dissolved

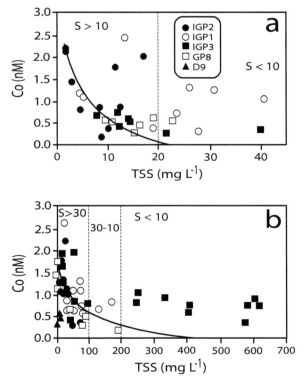

Figure 14.10 Dissolved Co concentrations versus total suspended solids (TSS) in different salinity regions of (a) Hudson River estuary and (b) San Francisco Bay estuary. Lines represent the best-fit regression for Co and TSS. (Modified from Tovar-Sanchez et al., 2004.)

and colloidal trace metals, respectively, with increasing salinity, also demonstrated colloidal partitioning in Narragansett Bay estuary (USA) (figure 14.11; Wells et al., 2000). It was further suggested that the transfer of Fe and Ni from the dissolved to the particulate was facilitated via colloidal pumping, where colloids transfer metals to the particulate phase (Farley and Morel, 1986; Honeyman and Santschi, 1989). Other studies have shown that Hg readily undergoes adsorption/desorption exchange between colloids and particulates (Stordal et al., 1996). Recent work in SFB estuary indicates that when riverine inputs were low, colloidal Hg was from internal sources (via resuspension); conversely, during high riverine flow greater than 50% of colloidal Hg was transported by river sources (Choe et al., 2003).

The importance of colloidal complexation clearly varies with the trace metal in question. For example, most of the Fe in rivers may be in colloidal form, which is critical in the coagulation/aggregation removal processes of Fe in estuaries (Millward and Turner, 1995). On the other hand, Cd and Ni have a low affinity for colloidal matter (Dai et al., 1995). In the Ochlocknee estuary (USA), Fe and Mn behaved nonconservatively and

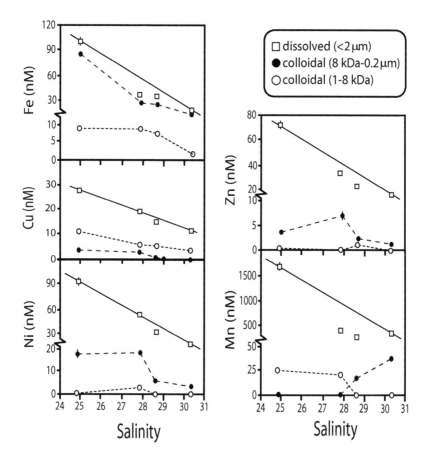

Figure 14.11 Concentrations of dissolved and colloidal metals across a salinity gradient in Narragansett Bay estuary (USA). (Modified from Wells et al., 2000.)

were largely associated with the colloidal fraction (Powell et al., 1996). In the Beaulieu estuary (UK), Ni was found to be largely unreactive during estuarine mixing (Turner et al., 1998). This was attributed to the limitation on particle–water interactions, due principally to the low particle affinity of Ni and its strong affinity for DOM. In particular, specific dissolved organic ligands and small suspended sediments were responsible for the unreactive character of Ni in the mixing zone. It is likely that much of the Ni was associated with low molecular weight DOM, based on work in the Ochlocknee estuary, which showed that Ni was initially found to be associated with colloidal materials in the river and then was converted into low molecular weight materials with increasing salinity (figure 14.12; Powell et al., 1996). The destabilization of colloidal metals has been shown to be linked with the release of biopolymers from greater phytoplankton abundance in the mid-to-lower estuary, which is very different from the colloidal destabilization that

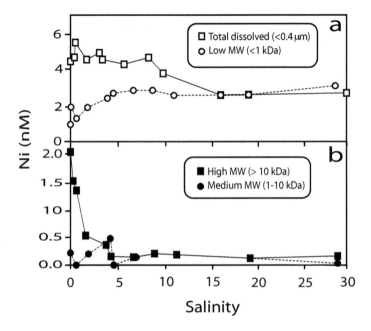

Figure 14.12 Concentrations of dissolved and colloidal Ni across a salinity gradient in the Ochlocknee estuary (USA). (Modified from Powell et al., 1996.)

occurs in the upper estuary caused by ionic changes (Wells et al., 2000). This pattern of reactivity with changing salinity was also observed for other metals such as Cu and Cd, and reflected a metal lability, relative to ligand exchange, that fit the predictions based on an Irving–Williams order. The kinetics of interactions between trace metals and different size fractions of dissolved and particulate inorganic and organic materials is essential to understand effectively the behavior in these highly dynamic environments.

The role and number of important ligand classes that control the complexation of different trace metals is highly variable across different estuarine systems. For example, Cd and Zn are controlled by three ligand classes in Narragansett Bay (Kozelka and Bruland, 1998), while Pb is controlled by only two ligand classes in SFB (Kozelka et al., 1997) and Narragensett Bay (Kozelka and Bruland, 1998) estuaries. In the case of Cu, concentrations of the stronger ligand class, commonly denoted [L_1], are equal or greater than total dissolved Cu [Cu_T], a trend commonly found in other estuarine systems (Kozelka and Bruland, 1998). Based on experimental laboratory and field measurements, production of [L_1] by phytoplankton occurs in response to [Cu^{2+}] concentrations. For example, when [L_1] > [Cu_T], CuL_1 will be the predominant species; however, in estuaries that have received anthropogenic inputs of metals, [Cu_T] could be greater than [L_1], in which case weaker ligands (e.g., [L_2] and [L_3]) may be produced by bacterioplankton/phytoplankton to "buffer" complex the remaining Cu (Kozelka and Bruland, 1998; Gordon et al., 2000).

Other studies have shown that the stronger-binding ligands are produced by microorganisms, while the weaker-binding ligands have a more humic character (Moffett et al., 1997; Vachet and Callaway, 2003). The typical range of difference between these conditional stability constants for Cu are $K'_{L1} = 10^{11-14}$ and $K'_{L2} = 10^{8-10}$ (Coale and Bruland, 1988; Donat and van den Berg, 1992). Dissolved metal and ligand concentrations and conditional stability constants for Cu, Pb, Zn, and Cd in Narragansett Bay are shown for comparison (table 14.2; Kozelka and Bruland, 1998).

A central question concerning the existence of chelators in the aquatic environments is whether these compounds are actively secreted by plankton to buffer metal concentrations outside the cell, or if they are the product of heterotrophic feeding processes whereby intracellular contents are released during "sloppy feeding" or via microbial breakdown of senescent cells (Wells et al., 1998). Further examination of the physicochemical kinetics of ligands of different size may shed some light on this question. Ligands that complex Cu, Pb, Cd, and Zn in Narragansett Bay were found to be associated with a diversity of dissolved and colloidal-sized complexes (Wells et al., 1998). For example, in the case of the strong ligand class [L_1], Pb and Cu speciation was controlled by both dissolved and colloidal-sized ligands in different regions of the bay, with a domination of colloidal for Pb–L_1 complexes. On the other hand, Cd and Zn speciation was largely associated with the dissolved complexes. However, in all cases the weak ligand classes were found in the colloidal with stronger more stable ligands in the dissolved. This agrees with the distribution of Pb in different size fractions in other estuarine waters (Benoit et al., 1994; Muller, 1999). Thus, while Pb, a nonbioactive metal, was bound in the colloidal-sized complexes, metals such as Zn, Cd, and Cu, which are essential bioactive metals, were found in the dissolved complexes. This may suggest that cross-membrane transport of bioactive metals are less restricted kinetically when associated with smaller ligand complexes (Wells et al., 1998).

To understand effectively the biogeochemistry of metal-binding ligands, it is important to be able to identify the active chemical functional groups and their possible sources in estuaries. While it is well established that Cu speciation in oceanic waters is strongly dominated by complexation with organic ligands (Coale and Bruland, 1988; Donat and Bruland, 1995), only recently have advances been made in understanding more about the functional group chemistry. In particular, -SH compounds are believed to represent the majority of organic ligands, in the form of thiols [e.g., glutathione (GSH)], that appear to be released by phytoplankton under Cu limitation (Leal et al., 1999). Other work showed that reduced S species could account for all the Cu-complexing ligands in Galveston Bay (Tang et al., 2001). More recent work has shown a linear relationship between dissolved Cd, Pb, and Cu concentrations and glutathione (figure 14.13; Tang et al., 2002), further demonstrating the importance of reduced S species in complexing metals in this estuary. Colloidal complexation is particularly evident in the oligohaline region of Galveston Bay (figure 14.13), where flocculation of riverine colloidal material is most likely to occur (e.g., Sholkovitz et al., 1978; Windom et al., 1989, 1991; Powell et al., 1996). Conversely, terrestrial metals, such as Al, Mn, and Ti show no visible trends with changing salinity or organic matter. Similarities in the metal:organic carbon ratios of colloids also suggested that these colloids were mostly organic and likely derived from humic substances and planktonic sources. The relative importance of colloidal size in complexing trace metals may also be altered by anthropogenic changes. For example, in tropical catchments (e.g., central

Table 14.2 Concentrations of Cu, Pb, Zn, and Cd and linearization results for dissolved ($<$0.2 μm) samples collected at three sites in the Narragansett Bay in June 1994.

Cu	Location	$[Cu_T]$ (nM)	L_1 (nM)	L_2 (nM)	$\log K_{cuL.2Cu'}$	L_3 (nM)	$\log K_{cuL.3Cu'}$
	Upper bay	27.9 ± 0.5	~38	40 ± 5	8.8 ± 0.4	100 ± 10	7.7 ± 0.4
	Mid-bay	16.1	~16	20 ± 2	8.8 ± 0.1	54 ± 4	7.7 ± 0.05
	Lower bay	12.7	~16	15 ± 3	9.2 ± 0.1	57 ± 17	7.5 ± 0.3

Pb	Location	$[Pb_T]$ (nM)	$[Pb']$ (nM)	L_1 (nM)	$\log K_{PbL.1Pb'}$	L_2 (nM)	$\log K_{PbL.2Pb'}$	Organic Pb	$[Pb^{2+}]$ (pM)
	Upper bay	0.32 ± 0.02	0.03	0.8 ± 0.2	10.0 ± 0.4	5.1 ± 0.8	8.8 ± 0.3	93.7%	1.2[a]
	Mid-bay	0.13	n.d.	0.6	10.2	6.0	8.6	81%	1[b]
	Lower bay	0.15	0.01	1.0	9.9	8.2	8.6	67%	0.4[a]

Zn	Location	$[Zn_T]$ (nM)	$[Zn']$ (nM)	$[L_T]$ (nM)	$\log K_{ZnL.Zn'}$	Organic Zn (%)	$[Zn^{2+}]$ (nM)[c]
	Upper bay	71.5	26.6	48.2	≥9	63	13.3
	Mid-bay	23.7 ± 0.8	0.6 ± 0.03	38.4 ± 0.9	9.4 ± 0.04	97	0.3
	Lower bay	16.3 ± 1.7	8.0 ± 0.8	10.6 ± 0.7	9.0 ± 0.04	51	4.0

Cd	Location	$[Cd_T]$ (nM)	$[Cd']$ (nM)	$[L_T]$ (nM)	$\log K_{CdL.Cd'}$	Organic Cd (%)	$[Cd^{2+}]$ (pM)[d]
	Upper bay	0.80 ± 0.03	0.22 ± 0.02	3.7 ± 0.4	8.9 ± 0.2	73	7
	Mid-bay	0.29	0.05	3.8	9.0	83	2
	Lower bay	0.30	0.07	3.6	9.1	77	2

[a] Calculated with $\alpha_{Pb'}$ = 25 and $[Pb']$ detected by DPASV.
[b] Calculated via $[L_1]\,K^{cond}_{PbL,Pb'}$; $\alpha_{Pb'}$ from Turner et al. (1981), Byrne et al. (1988) and sample pH.
[c] Calculated with $\alpha_{Zn'}$ = 2 (Turner et al., 1981; Byrne et al, 1988).
[d] Calculated with $\alpha_{Cd'}$ = 30 (Turner et al., 1981; Byrne et al., 1988).
Modified from Kozelka and Bruland (1998).

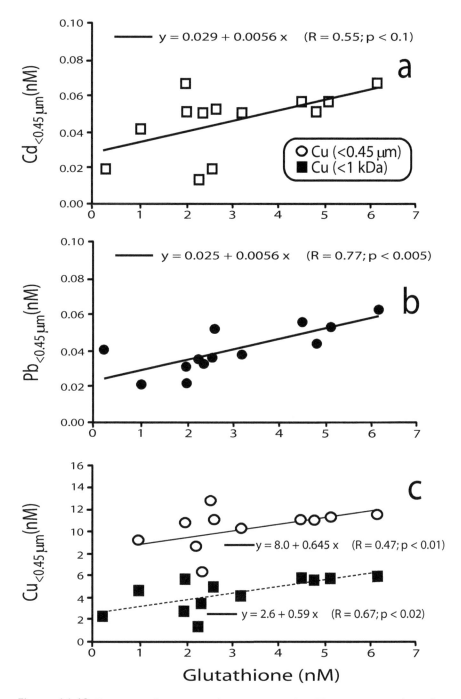

Figure 14.13 Trace metal concentrations versus glutathione concentrations in Galveston Bay (USA) for (a) Cd and (b) Pb in the filter-passing fraction (<0.45 μm), and (c) Cu in the filter-passing (<0.45 μm) and the ultrapermeate fraction (<1 kDa). (Modified from Tang et al., 2002.)

455

Amazonian forest) there has been uncontrolled deforestation which has led to enhanced *podzolization* in soils (Eyrolle et al., 1996). In most tropical environments, trace metals such as Cu and Al are complexed with low molecular weight DOC (<5 kDa); however, in regions of high podzolization the speciation of Ca, Mg, and Fe is largely controlled by higher molecular weight colloidal (>20 kDa) materials. These types of alterations clearly have important implications for the cycling of trace metals in the upper reaches of tropical estuaries.

As stated earlier, metal ligands may also be derived from sources other than phyto-plankton, in this case ligand abundance and production may not be as tightly coupled with phytoplankton in regions receiving high riverine inputs (Shank et al., 2004). For example, it was recently estimated that as much as 50% of the strong binding Cu ligands could be supplied to the South Atlantic Bight (USA) from DOM inputs from the Cape Fear River (Shank et al., 2004). Recent work on Fe complexation in the Mississippi River plume (MRP) indicates that only one ligand class is present (Powell and Wilson-Finelli, 2003). These results support some of the open ocean work on Fe complexing ligands, where evidence exists for both one (Powell and Donat, 2001) and two (Rue and Bruland, 1995) ligand classes. Once again, while sources of such ligands in the open ocean are likely derived directly from bacterioplankton/phytoplankton (Gonzalez-Davila et al., 1995; Moffett and Brand, 1996) or secondarily through bacterioplankton processing (Bruland et al., 1991), multiple sources of organic ligands from bacterioplankton, phytoplankton, and terrestrially derived organic matter (Bianchi et al., 2004) may be important in the MRP region (Powell and Wilson-Finelli, 2003). Finally, sources of ligands from sediments may represent another important input to estuarine systems, particularly in shallow estuaries (see more below).

Trace Metal Cycling and Fluxes in Sediments

Sediments can represent sources and sinks in estuaries. Factors determining the source versus sink pathways in estuaries will largely be determined by inputs from external and in situ processing in the water column as well as postdepositional processes in sediments. This section begins by examining some of the bulk sediment properties that affect accu-mulation of trace metals in sediments. Postdepositional processing of trace metals, which is fundamentally related to organic matter cycling and redox condition, is also examined. This section also examines the importance of inorganic and organic ligands as they relate to metal speciation and flux across the sediment–water interface. Finally, while the role of sulfides in binding trace metals (e.g., Fe) in sediments is briefly mentioned in this section, more details on this topic [as it relates to the degree of pyritization (DOP)], are provided in chapter 12.

Estuarine sediments provide a long-term record of the accumulation of trace metal inputs from riverine, atmospheric, and anthropogenic sources (Kennish, 1992; Windom, 1992). In many cases, anthropogenic inputs exceed natural background levels from weath-ering of rock materials described earlier, because of extensive human encroachment commonly found around these areas. Moreover, there are typically large variations in concentrations of trace metals in estuarine sediments due to the poorly sorted mixtures of sand, silt, and clay. Thus, there needs to be a way to separate background levels from anthropogenic inputs and to account for the natural variability in composition of sediments. One method has been to normalize trace elements to a carrier phase such as Al, Fe, Li, organic carbon, or grain size (Wen et al., 1999). In the case of elemental ratios, Al has often been chosen to normalize trace element concentrations because of high natural abundance in crustal rocks and generally low concentrations in anthropogenic sources. This metal:Al ratio has effectively been used as an indicator of pollution sources in rivers and coastal systems (Windom et al., 1988; Summers et al., 1996). Down-core profiles of normalized trace metal concentrations have also been effectively used to examine historical profiles of contaminant inputs to estuaries (Alexander et al., 1993). Grain-size normalization typically involves analyzing the <63 μm fraction since coarser grain-sized sediments (e.g., carbonates and sands) have a diluting effect on trace metal concentrations in sediments (Morse et al., 1993). Understanding the variability in horizontal and vertical distribution of trace metals can also be improved by coupling radiochemical tracers (e.g., ^{210}Pb and ^{137}Cs), commonly used for estimating sedimentation and particle-reworking rates (see chapter 7) in estuaries, with trace metals concentrations (Wen et al., 1999). These radionuclides can help constrain variability in the degree of reprocessing of trace metals across horizontal gradients as well as provide information on the historical accumulation of trace metals in estuaries (Ravichandran et al., 1995b).

Early investigations showed trace metal concentrations in pore waters to be generally higher than in overlying bottom waters in estuarine and shallow coastal systems (Presley et al., 1967; Elderfield et al., 1981a,b; Emerson et al., 1984). These differences result in a concentration gradient that allows trace metals in pore waters to diffuse from sediments to overlying waters (Elderfield and Hepworth, 1975). In addition to being released by diffusive mechanisms, porewater metals can become reincorporated into the sediments via adsorption, complex formation, and precipitation (Chester, 1990, 2003). Thus, the overall concentration of metals in sediments will reflect recent inputs from natural or anthropogenic sources as well as a recycling component that reflects longer-term diagenetic alterations (Chester, 2003). Thus, in order to determine effectively the behavior of metals in sediments, some assessment must be made in determining the role of sediment depositional sources and their diffusion across the sediment–water interface. While traditional sampling methods such as whole-core squeezers have commonly been used to extract down-core trace metal concentrations in porewaters (Presley et al., 1967, 1980; Presley and Trefrey, 1980), later work used a more direct chamber approach to monitor concentration changes over time (Rowe et al., 1992). The earlier gradient method only provides steady-state flux information, with diffusion of porewater based on Fick's first law (Berner, 1980) (see chapter 7), while the chamber method allows for more realistic non–steady-state information. When examining the role of sedimentary versus diffusive fluxes of metals in the Bang Pakong River estuary (Thailand), results showed a higher sedimentary (diagenetic) flux (Cu, Pb, Zn, Cd, Cr, and Ni) ($0.1 - 16.8$ μg cm^{-2} y^{-1}) compared to diffusive flux ($0.01 - 4.8$ μg cm^{-2} y^{-1}) (Cheevaporn et al., 1995). This work also showed that diagenetic flux represented 10 to 90% of the total flux contribution. These diffusive fluxes are within the range of values reported for the Conway Estuary (N. Wales), the Tees Estuary (England) (Elderfield and Hepworth, 1975), and the Trinity River estuary (USA) (Santschi et al., 1999). In many estuarine systems, the diagenetic remobilization of metals from sediments contributes significantly to the redeposition of metals into surface sediments.

Iron and Mn cycling in estuarine sediments have been shown to be strongly linked with redox and the diagenesis of organic matter (Overnell, 2002). As discussed in chapter 8, the decomposition of organic matter in marine/estuarine sediments proceeds through a sequence of *terminal electron acceptors* (e.g., O_2, NO_3^-, MnO_2, $FeOOH$, SO_4^{2-}, and CO_2) (Richards, 1965; Froelich et al., 1979; Canfield, 1993), as shown in figure 8.20 (Deming and Baross, 1993). The sequence of reactions are principally determined by the free energy yields (ΔG^0) (see chapter 4) per mole of organic C (table 8.3). Many of the color changes in sediments associated with *redox potential discontinuity* (RPD) are indicative of differences between oxic and reduced forms of metal complexes (Fenchel and Riedl, 1970; Santschi et al., 1990). For example, the oxidized region in surface sediments of estuarine sediments (one to a few centimeters of sediment) is typically orange–brown (*Fe and Mn oxides*) followed by a reduced region that is gray-black (*mono- and polysulfides*). The depth of the RPD is largely controlled by the amount of organic matter loading, physical mixing, and bioturbation. Many of these reactions involving Mn and Fe have been shown to be linked to microbially mediated (e.g., chemolithotrophic) processes (Srensen, 1982; Lovley and Phillips, 1988; Nealson and Myers, 1992; Nealson and Saffarini, 1994) in particular, bacterial reduction of Fe and Mn oxides (Burdige and Nealson, 1986; Lovely, 1991; Nealson and Myers, 1992). Iron and Mn oxides are also capable of oxidizing sulfides to SO_4^{2-} (Aller and Rude, 1988) and NH_3 to NO_3^- (Hulth et al., 1999) under anaerobic conditions. Other important reactions involving metal oxides include the oxidation of NH_3 to N_2 via Mn oxides in the presence of oxygen (Luther et al., 1997). Studies in sediments in the Amazon River estuary and rivers off the southern coast of the Gulf of Papua (New Guinea) have shown extensive Fe and Mn reduction at depth due to extensive suboxic diagenetic processes (Aller et al., 1986; Mackin and Aller, 1986; Alongi et al., 1992, 1993, 1996). An example of this is shown in the porewater profiles of dissolved Fe and Mn in suboxic sediments from the Gulf of Papua (figure 14.14; Alongi et al., 1996). These carrier-phase particles are clearly important components affecting both water column and sediment processes in estuaries.

Another oxidant near the sediment–water interface capable of oxidizing Mn(II) is iodate (IO_3^-), which occurs in the dissolved form as well as adsorbed onto particles (Ullman and Aller, 1985). Iodine is known to be released in porewaters during the remineralization of organic matter (Ullman and Aller, 1983, 1985; Kennedy and Elderfield, 1987). Once produced, iodine is then believed to be oxidized to IO_3^- through microbial processes, where it adsorbs onto metal oxides (Ullman and Aller, 1985). Recent work has shown that an iodide peak can be maintained in sediments through the reduction of IO_3^- by Mn(II), with reoxidation of iodide to IO_3^- above the iodide peak; thus, iodide production is adequate to account for the oxidation of all the upward diffusing Mn(II) via IO_3^- (Anschutz et al., 2000).

As described earlier, there are close linkages with the cycling of Cu and organic matter, which have largely been attributed to the recycling of metals through water column organisms (e.g., phytoplankton) (Morel and Hudson, 1985). Porewater concentrations of Cu are typically higher than those of bottom waters, provided that the potential benthic fluxes of Cu and other metals out of the sediment are low (Elderfield et al., 1981a,b; Skrabal et al., 1997). Many of the dominant inorganic mineral phases described above have an important impact on the cycling of Cu in sediments. For example, the partitioning of Cu between dissolved and solid phases in sediments can occur through the release of

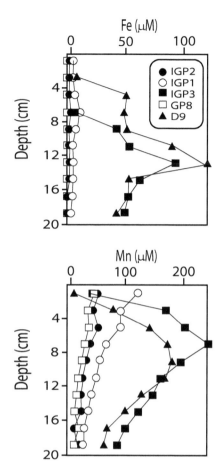

Fe (μM)

Mn (μM)

Depth (cm)

Depth (cm)

IGP2
IGP1
IGP3
GP8
D9

Figure 14.14 Porewater profiles of dissolved Fe and Mn in suboxic sediments from the Gulf of Papua. (Modified from Alongi et al., 1996.)

oxide-bound Cu during microbial reduction of Fe and Mn oxides (Morse and Arakaki, 1993), as well as the release of bound Cu from oxidized sulfide minerals (Mackin and Swider, 1989). However, the distribution of Cu in porewaters also appears to be strongly linked to organic matter cycling in sediments (Shaw et al., 1990; Widerlund, 1996). While there has been considerable research on the production of metal-complexing ligands in organisms and from inputs of riverine humic substances in the water column, the potential efflux of ligands from porewaters has been largely ignored. Based on the high diffusional fluxes of DOC in some estuaries (Burdige and Homstead, 1994; Alperin et al., 1999), it is likely that ligand inputs to the bottom water column may be significant.

Recent work in Chesapeake Bay showed concentrations of Cu-complexing ligands in porewaters that were orders of magnitude higher than dissolved Cu—with 87 to 99% of the Cu being complexed (Skrabal et al., 1997, 2000). Other work has shown that many of these porewater ligands are "strong" complexing ligands and only contribute significantly to the water column, via benthic fluxes, in estuaries with long residence times (Shank et al., 2004a,b). For example, short residence times in the Cape Fear estuary (USA) prevented sediment–water exchange from being an important source of Cu-complexing ligands to the water column. This contribution from sediments can be particularly important in estuaries such as the SFB (Donat et al., 1994) and Chesapeake Bay (Donat et al., 1994), where total dissolved Cu is commonly less than concentrations of L_1 ligands. These studies indicate that Cu-complexing ligands released from sediment porewaters can be an important factor influencing Cu speciation in estuaries (Skrabal at al., 2000).

Metal toxicity in estuaries has been shown to be correlated with certain sediment processes. For example, methylmercury(CH_3Hg) cycling in estuaries has been linked with numerous sedimentary phases such as clay minerals, sulfides, organic matter, and Fe and Mn oxyhydroxides (Huerta-Diaz and Morse, 1990; Bloom et al., 2003; Hammerschmidt et al., 2004). Coastal and estuarine sediments appear to represent a major source of CH_3Hg for food webs in the coastal environment (Gill et al., 1999; Mason et al., 1999; Hammerschmidt et al., 2004). In fact, nearshore sediments have provided an excellent record for the accumulation of Hg from pollution sources over the past 150 years ("legacy Hg") (Fitzgerald and Lamborg, 2003). The methylation of Hg to CH_3Hg has been shown to be primarily mediated by sulfate-reducing bacteria (SRB) (Gilmour et al., 1992). In fact the highest rates of Hg methylation in the Florida Everglades occur in the RPD layer at the transition where sulfidic porewaters become abundant (Gilmour et al., 1998). Sulfide concentration appears to control the speciation of Hg complexes (e.g., $HgHS_2^-$, $HgSH^+$, and HgS^0), with HgS^0 being the most available to bacteria (Benoit et al., 2001). While methylation is also controlled by the availability of Hg(II), other factors are important in controlling the partitioning of Hg. For example, in Long Island Sound (LIS) estuary (USA), sediment–water partitioning of both Hg (II) and CH_3Hg was found to be largely controlled by organic matter in the upper 10 cm (Hammerschmidt et al., 2004). Other work has shown that POM is more important than Fe and Mn oxides in the removal of Hg(II) in the Mersey estuary (UK) (Turner et al., 2004). Bioturbation has been shown to enhance SRB-mediated Hg methylation by increasing the availability of labile organic substrates and Hg for bacteria at depth and by flushing out metabolites at depth that can slow down bacterial metabolism (Maillacheruvu and Parkin, 1996; Benoit et al., 1999). After CH_3Hg is formed in the sediments, other factors such as location of the RPD layer (Gill et al., 1999) and abundance of Fe oxides (Bloom et al., 1999) and demethylating bacteria (Marvin-DiPasquale and Oremland, 1998) can affect the overall mobility and flux of CH_3Hg from sediments. Recent work has shown that the average diffusive sediment–water flux of CH_3Hg in LIS was 55 mol y^{-1}, which exceeded external sources (Hammerschmidt et al., 2004). In fact, most of the CH_3Hg in LIS phytoplankton is believed to be derived from sediment fluxes. The role of Hg methylation in coastal/estuarine sediments remains an important toxicological issue because the majority of fish consumed by humans of marine origin are from the coastal zone. More details on Hg toxicity are provided in chapter 15.

While the methylation of Hg essentially occurs in sediments, sources of Hg to estuarine systems can occur principally through atmospheric sources. For example, Hg deposition, was measured at different locations throughout the State of Florida (USA), in the Florida Atmospheric Mercury Study (FAMS) (Gill et al., 1995; Guentzel et al., 1998; Landing et al., 1998). This study occurred largely in response to the excessively high concentrations that were found in different trophic levels of the Florida Everglades (USA). Depositional fluxes of Hg in the summer months were found to be 5 to 8 times that of other seasons, primarily due to more rainfall. It was also shown that during summer seasons large convective thunderstorms reached high enough into the stratosphere to scavenge the reactive gaseous global pool of Hg. However, in winter months, inputs to the atmosphere from local power plants were largely responsible for Hg deposition to the region. The coupling of atmospheric inputs with rates of Hg methylation in the Florida Everglades (USA) (Gilmour et al., 1998), clearly makes this one of the most comprehensive Hg studies examining both regional and global inputs.

Summary

1. The majority of trace metals are derived from igneous rocks, simply based on the relative fraction of igneous rocks in comparison to sedimentary and metamorphic rocks in the Earth's crust.
2. The major inputs of trace metals to estuaries are derived from riverine, atmospheric, and anthropogenic sources.
3. The partitioning of trace metals between the dissolved and particulate fractions in estuaries can be affected by variability in river flow, tidal and wind energy, storms, coagulation, and flocculation in the estuarine turbidity maximum (ETM), resuspension events (of sediments and porewaters), and inputs from wetland and mudflat processes.
4. Metals commonly exist in multiple oxidation states and will generally have a positive oxidation state (e.g., $+1$ to $+6$) when bonded with other elements because of their electrophilic character.
5. The complexation of metal ions with water molecules interferes with the hydrogen or electrostatic bonding between water molecules.
6. The strength of these bonds will increase with increasing charge and decreasing size of the metal ion; this metal-to-water bond is typically illustrated as $Me(H_2O)_x{}^{n+}$. In this case, water is acting on an inorganic ligand, whereby oxygen atoms are donating electrons to the metal ion in this "single-ligand complex."
7. Over the past decade, there have been numerous studies that have shown the importance of organic ligands in complexing trace metals, with particular emphasis on colloidal-sized particles (colloids are in the size range of 1 nm to 1 µm).
8. The distribution and speciation of trace metals in estuaries will depend on their concentrations as well as the concentrations of dissolved complexing ligands and the associated coordination sites on colloids and particulates.
9. Some of the dominant functional groups of organic matter that commonly form strong complexes, albeit at a slow rate, with trace metals are -COOH, -OH, $-NR_2$, and $-SR_2$ (R = $-CH_2$ or -H). Some natural sources containing these functional groups are phytochelatins (siderophores), biopolymers (e.g., proteins), and humic substances.
10. Processes such as adsorption, desorption, flocculation, coagulation, resuspension and bioturbation are particularly important in controlling the interactions between dissolved (free aquo) trace metals and particulates in estuaries.

In particular, there are important binding sites on Fe and Mn oxyhy-droxides, carbonates, clays, and POC/COC that are key in controlling adsorption/desorption of trace metals.

11. The "particle concentration effect" has been defined as "the physical effects which lead to a decreasing overall partition coefficient with increasing particle concentration; the effect is documented for both organic and inorganic species."

12. The lateral and vertical distributions of these carrier-phase metals in estuaries are largely controlled by particle dynamics, as opposed to other metals (e.g., Cu, Zn, and Co) which will be more affected by biotic uptake processes.

13. Sharp concentration gradients of other metals (e.g., Cu, Cd, and Zn) at the redoxcline in certain fjords have been shown to be largely controlled by the chelation/complexation with organic ligands.

14. Sorption–desorption from suspended particulates and sediment fluxes play a large part in controlling the nonconservative behavior of dissolved concentrations of Fe and Mn in estuaries.

15. Ligands that complex trace metals are controlled by a diversity of dissolved and colloidal-sized complexes.

16. Factors determining the source versus sink pathway in estuaries will largely be determined by inputs from external and in situ processing in the water column as well as postdepositional processes in sediments.

17. Many of these reactions involving Mn and Fe have been shown to be linked with microbially mediated (e.g., chemolithotrophic) processes, in particular, bacterial reduction of Fe and Mn oxides are also capable of oxidizing sulfides to SO_4^{2-}, and NH_3 to NO_3^- under anaerobic conditions. Other important reactions involving metal oxides include the oxidation of NH_3 to N_2 via Mn oxides in the presence of oxygen.

18. Porewater ligands are "strong" complexing ligands, which only contribute significantly to the water column (via benthic fluxes) in estuaries with long residence times.

19. Metal toxicity in estuaries has been shown to be correlated with certain sediment processes. For example, CH_3Hg cycling in estuaries has been linked with numerous sedimentary phases such as clay minerals, sulfides, organic matter, and Fe and Mn oxyhydroxides.

20. The coupling of atmospheric inputs with rates of Hg methylation in the Florida Everglades clearly makes this one of the most comprehensive Hg studies examining both regional and global inputs.

Part VI

Anthropogenic Inputs to Estuaries

Chapter 15

Anthropogenic Stressors in Estuaries

Anthropogenic Change in Estuaries

Human demands on aquatic and terrestrial ecosystems are on the increase globally and have likely exceeded the regenerative capacity of the Earth since the 1980s. Demands on our aquatic resources will increase in coming decades as it is projected that 75% of the world's population (6.3 billion) will reside in coastal areas by 2025 (Tilman et al., 2001). The Earth's population is expected to reach 9 billion during this century, and the projected effects of contaminant loading and human encroachment on biodiversity still remain unclear. The disturbance on global coastal ecosystems and the threat it will have on the economically critical resources they provide, has been estimated to be valued at 12.6 trillion U.S. dollars (Costanza et al., 2001). It has become increasingly apparent that in many regions of the world, Earth systems, which have been viewed as being primarily controlled by natural drivers such as climate, vegetation, and lithology, are now controlled by social, societal, and economic drivers (e.g., population growth, urbanization, industrialization water engineering) (Meybeck, 2002, 2003). This replacement of natural drivers over the past 50 to 200 years has recently been referred to as the *Anthropocene era* (first postulated by Vernadski, 1926), as a next phase that follows the Holocene era (Crutzen and Stoermer, 2000). Other studies that have effectively made large-scale linkages between human effects on the Earth systems (Turner et al., 1990) and aquatic systems (Costanza et al., 1990, 1997; Meybeck, 2002, 2003; Meybeck and Vörösmarty, 2004) have all concluded that a more comprehensive and fine-scale interpretation of the Anthropocene is needed if we are to make future predictions and management decisions effectively.

The growth and movement of human populations have resulted in a significant stressor in the form of invasive species that has altered global biodiversity patterns. For example, the introduction of invasive species worldwide has changed the community composition and physical structure of many ecosystems (Elton, 1958; Vitousek et al., 1997). Estuarine systems, like the northern San Francisco Bay, have experienced serious declines in productivity at the base of the food web over recent decades after the introduction of the

465

Asian clam, *Potamocorbula amurensis*, in 1987 (Carlton et al., 1990). The invasive Asian Date mussel, *Musculista senhousia*, has invaded the west coast of the United States altering community dynamics, and has now spread to Western Australia, New Zealand, and the Mediterranean (Mistri, 2002). One of the more notable bivalve invasions, by the zebra mussel (*Dreissena polymorpha*), has proven to be as devastating and expansive across U.S. rivers, lakes, and estuaries as predicted in early studies (Strayer and Smith, 1993). For example, these mussels can filter a volume of water equivalent to the entire freshwater region of the Hudson River about every 2 d; this has dramatically altered the total suspended load and the phytoplankton community (Roditi et al., 1996). Similarly, invasion of wetland plant species, such as *Phragmites australis*, has resulted in displacement of dominant marsh *Spartina* spp. along the eastern U.S. coast (see more in chapter 8) (Chambers et al., 2003).

The effects of human encroachment on coastal systems has resulted in dramatic increases in the loadings of contaminants such as trace metals, hydrophobic organic contaminants (HOCs) (e.g., hydrocarbons and chlorinated hydrocarbons), and nutrients (Wollast, 1988; Schmidt and Ahring, 1994: Jonsson, 2000; Cloern, 2001; Elmgren, 2001). Inputs of these multiple stressors can interact to reduce, enhance, and/or mask the individual effects of each (Breitburg et al., 1999). In the case of HOCs, these compounds tend to be *lipophilic*, allowing for *bioaccumulation* and *biomagnification* and the transfer of these contaminants up through the food web, via trophic interactions. It is also well known that stressors such as nutrients and contaminants can alter phytoplankton community dynamics (Cloern, 1996; Riedel et al., 2003). While these stressors are commonly analyzed as separate entities in ecosystems, trace elements can act as micronutrients thereby affecting the uptake of nutrients (Riedel, 1984, 1985). For example, recent work has shown that in mesocosm experiments, phytoplankton did not respond to nutrient additions when trace elements were also added (Riedel et al., 2003). Thus, to manage estuarine systems effectively, it is necessary to develop models that incorporate the interactive effects of multiple stressors (Breitburg et al., 1999a,b). In 1993, the Swedish Environmental Protection Agency formed a new research program called EUCON which examined the interactions between eutrophication and contaminants (Skei et al., 2000). Results from this work have shown that eutrophication processes can increase the bioavailability of HOCs in the Baltic Sea.

Since considerable details on trace metal sources and transport are discussed in chapter 14, only a brief introduction on metals as contaminants is provided here. Many of the metals that have been studied in the context of contaminant cycling in estuaries are referred to as *heavy metals*, which have atomic weights that range from 63.546 to 200.590 and a similar electronic distribution in their outer electron shells (e.g., Cd and Zn) (Viarengo, 1989). These heavy metals can be further divided into *transitional metals* (e.g., Co, Fe, Mn), essential to metabolism at low concentrations but potentially toxic at high, and *metalloids* (e.g., As, Pb, Hg), not required for metabolism and toxic at low concentrations (Presley et al., 1980: Kennish, 1997). Some of the most common anthropogenic sources of heavy metals are mining, smelting, refining, electroplating, electric generating stations, automobile emissions, sewage sludge, dredged spoils, ash, and antifouling paints. While many of these sources result in more localized inputs to coastal systems, some metals like Hg are distributed atmospherically on a global scale. Despite reductions by as much as 50% in Hg emissions in North America emissions since the 1970s

(Sunderland and Chmura, 2000), declines in Hg concentrations in marine birds, fish, and shellfish from the same regions have not occurred. This decoupling between emissions and concentrations in organisms points to the need for further studies on Hg cycling, if effective management strategies are to be achieved (Sunderland et al., 2004).

The toxicity of metals varies significantly among organisms, but generally accumulate in tissues as a function of temperature, salinity, diet, spawning, and the ability of organisms to control metal concentrations. For example, certain low molecular weight metal-binding proteins called *metallothioneins*, found in prokaryotes and eukaryotes, have been shown to be able to sequester and detoxify metals (e.g., Roesijadi, 1994). Certain cellular organelles, *lysosomes*, also have the ability to sequester metals in organisms such as crustaceans, mollusks, annelids, and hydroids (e.g., Engel and Brouwer, 1993). The occurrence of the detoxifying mechanisms in aquatic invertebrates (Deeds and Klerks, 1999) can result in a developed resistance to these contaminants, which can then allow for these compounds to move trophically through food webs (Klerks and Lentz, 1998). An excellent example of developed metal resistance in estuaries involves Foundry Cove, in the Hudson River estuary, which was believed to be one of the most heavily Cd-polluted areas in the world; sediment Cd concentrations reached as high as 10,000 ppm (Klerks and Levinton, 1989). The most abundant invertebrate in Foundry Cove, the oligochaete *Limnodrilus hoffmeisteri*, was shown to have evolved resistance to the Cd, and consequently was thought to promote the transfer of the metal through the ecosystem. After a major clean-up effort (as part of an EPA Superfund site), it was also shown that *L. hoffmeisteri* subsequently lost its resistance following the clean-up in the mid-1990s (Levinton et al., 2003). This benchmark study illustrates how genetically adaptable some estuarine organisms can be to anthropogenic inputs over relatively short periods.

It has been estimated that 1.7 to 8.8×10^6 t of petroleum hydrocarbons are released into marine systems annually (National Research Council, 1985). The persistence of these contaminants, particularly in sediments, results in long-term impacts on recovery of benthic communities in estuaries (Elmgren et al., 1983). Of this, diesel fuel has been shown to be the most toxic because of its high polycyclic aromatic hydrocarbon (PAH) content (Kennish, 1992). This group includes highly carcinogenic compounds such as benzo[*a*]pyrene (Gelboin, 1980; Denissenko et al., 1996). The chemical structures of selected PAHs are shown in figure 15.1. PHAs are generated from anthropogenic or natural combustion processes, in addition to rapid transformation processes of biogenic precursors that occur in situ (Wakeham and Farrington, 1980; Wright and Welbourn, 2002). Many of the PAHs are associated with black carbon (BC), which can be formed from vapor-phase condensations of organic molecules (e.g., graphitic or soot material) or from residue of burned materials (e.g., char) (Goldberg, 1985). In fact, BC has not only been implicated as an important factor in controlling the distribution of HOCs, but may also represent a significant pool in the global carbon cycle (Kuhlbusch, 1998; Masiello and Druffel, 1998; Mitra et al., 2002). PHAs are also produced through biogenic precursors such as terpenes, pigments and steroids (Laflamme and Hites, 1979; Prahl and Carpenter, 1979, 1983; Wakeham et al., 1980a,b; Budzinski et al., 1997) (also see chapter 9). Many of the PAHs released into marine systems ($\sim 1.7 \times 10^5$ t y^{-1}) accumulate in estuaries; sedimentary records indicate that anthropogenic activities began introducing PAHs into the environment in the United States about 80 to 100 y ago (Gschwend and Hites, 1981). Another important class of HOCs that were once heavily used for industrial purposes,

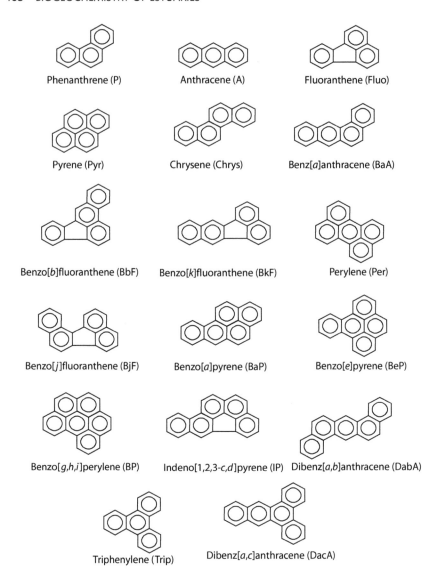

Figure 15.1 Chemical structures of selected polycyclic aromatic hydrocarbons (PAHs).

but have since been banned due to associated health problems, is the polychlorinated biphenyls (PCBs). These are very stable lipophilic compounds that biomagnify in aquatic food webs (Cairns et al., 1986). PCBs are a mixture of 209 possible *cogeners*; an example of the basic chemical structure of PCBs and the chemical formulas for selected PCBs are shown in figure 15.2.

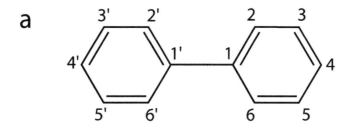

b

IUPAC no.	Structure
8	2,4'-dichlorobiphenyl
18	2,2',5-trichlorobiphenyl
16 ⎤	2,2',3-trichlorobiphenyl
32 ⎦	2,4',6-trichlorobiphenyl
28	2,4,4'-trichlorobiphenyl
31	2,4',5-trichlorobiphenyl
33 ⎤	2',3,4-trichlorobiphenyl
21 ⎦	2,3,4-trichlorobiphenyl
22	2,3,4'-trichlorobiphenyl
49	2,2',4,5'-tetrachlorobiphenyl
52	2,2',4,5'-tetrachlorobiphenyl
66 ⎤	2,3',4,4'-tetrachlorobiphenyl
95 ⎦	2,2',3,5',6-pentachlorobiphenyl
92 ⎤	2,2',3,5,5'-pentachlorobiphenyl
84 ⎦	2,2',3,3',6-pentachlorobiphenyl
101	2,2',4,5,5'-pentachlorobiphenyl
87	2,2',3,4,5'-pentachlorobiphenyl
110	2,3,3',4',6-pentachlorobiphenyl
149	2,2',3,4',5',6-hexachlorobiphenyl
118	2,3',4,4',5-pentachlorobiphenyl
138	2,2',3,4,4',5'-hexachlorobiphenyl
153	2,2',4,4',5,5'-hexachlorobiphenyl
170	2,2',3,3',4,4',5-heptachlorobiphenyl
174	2,2',3,3',4,5,6'-heptachlorobiphenyl
180	2,2',3,4,4',5,5'-heptachlorobiphenyl

Figure 15.2 The basic structural unit (a) of polychlorinated biphenyls (PCBs) and (b) selected chemical formulas of cogeners.

Despite the long history of contaminant loading in estuaries, there has been limited success in the clean-up of banned contaminants such as halogenated hydrocarbons, PCBs and dichloro-diphenyl-trichloroethane (DDT) in estuarine systems. For example, in the Baltic Sea, awareness of the environmental hazards of these substances in the 1960s (Jensen et al., 1969) resulted in strict bans in the 1970s, which led to significant reductions in the concentrations of DDT and PCBs in certain pelagic bird species in the Baltic Sea (Larsson et al., 2000; Olsson et al., 2000). Work in the Baltic illustrates that once an environmental problem has been identified, it may take decades for society to agree on a management plan and for that plan to become effective (Elmgren, 2001). It has taken, on average, about two decades for the complete recovery of DDT, and another two decades for an agreement on the action plans for eutrophication, which has yet to have been proven effective in the Baltic. In many cases, it is the scientific community that is unable to reach a consensus on, in many cases, the most basic approaches to an environmental problem; this frustration is effectively expressed by Elmgren (2001, p. 225), who wrote "fortunately, the politicians did not wait for scientific consensus before reducing emissions of mercury, PCB, and DDT and plant nutrients to the Baltic Sea."

In this chapter, we will discuss the basic controls on the transport of some of these contaminants mentioned above in the water column and sediments of estuaries. While there are five general classes of contaminants (petroleum hydrocarbons, halogenated hydrocarbons, heavy metals, radionuclides, and litter) considered to be critical in coastal environments (Waldichuk, 1989), the primary focus will be on the biogeochemical dynamics of a select group of contaminants as they relate to other stressors, such as nutrients, in this chapter. For a more comprehensive assessment of the actual statistics on toxicity and surveys of the distribution and loading rates of all the aforementioned contaminants in estuaries, please refer to Kennish (1997).

Partitioning and Toxicity of Trace Metals

The partitioning of metals between dissolved and particulate phases is strongly influenced by a diversity of factors (e.g., inorganic oxides, pH, and organic matter) in estuarine systems. One particularly important factor is the abundance of natural organic matter. Since many of the details involving the partitioning of metals were discussed in chapter 14, this section will focus more on the key parameters affecting toxicity and bioavailability of metals to aquatic organisms.

Past studies have shown a decrease in the toxicity of metals in the presence of organic matter, due to the binding of natural ligands that are believed to bind metals and reduce the concentration of free ionic species (Campbell, 1995; Carvalho et al., 1999; Doblin et al., 1999). The free ion activity model proposes that the free metal ion (as opposed to total metal ion concentration) is the dominant species available for organisms, and assumes that the colloidally complexed species should be less available (Campbell, 1995). Thus, the model does not account for chelated species (Sunda and Lewis, 1978). Moreover, this model also proposes that metal uptake is strongly controlled by the concentration of the metal–cell surface complex, which is dependent on the ambient free metal concentration. Toxic metals, such as Ag, can have as much as 60% of the dissolved fraction associated with colloidal organic matter (Wen et al., 2002). As described in chapter 14, a significant fraction of the binding is likely to be associated with sulfhydryl groups. There are considerable gaps in our knowledge on how colloidal organic matter complexation with metals affects *bioavailability* to aquatic organisms. For example, some studies have shown colloidally complexed metals to be more or less bioavailable to aquatic organisms (Guo et al., 2001).

Since colloidal material has both hydrophilic and *lipophilic* properties, making it permeable across membrane surfaces, lipid permeation may be an important mechanism for metal uptake. Lipophilic metals, such as organic Hg, have been shown to be taken up by

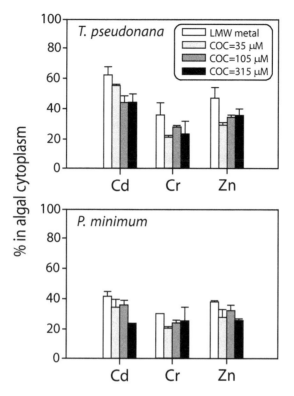

Figure 15.3 Percentages of Cd, Cr, and Zn in the cytoplasm of the diatom, *Thalassiosira pseudonana*, and the dinoflagellate, *Procentrum minimum*, after 1 day exposure to low molecular (LMW) bound metals and colloidal bound metals (COC). (Modified from Wang and Guo, 2000.)

aquatic organisms (Phinney and Bruland, 1994; Mason et al., 1996). Other colloidal bound metals, such as Cd, Cr, and Zn, have been shown to be able to penetrate into the cytoplasm of diatom and dinoflagellate cells (figure 15.3); (Wang and Guo, 2000). The bioavailability of metal complexed in inorganic versus organic colloidal materials also varies considerably. For example, colloidal-bound Fe in goethite (FeOOH) and hematite (Fe_2O_3) was not bioavailable to diatoms, but was in hydrous ferric oxide colloids (Wells et al., 1983). Further work is needed to understand better how metal uptake in phytoplankton can move up tropically in the food web. Past work on invertebrate uptake of metals has traditionally focused on "dissolved" versus "particulate" metals with a large emphasis on bivalve species (Hamelink et al., 1994; Wang and Fisher, 1997; Roditi and Fisher, 1999). This largely stems from the emphasis that has been placed on bivalves as pollution indicator organisms in monitoring programs (Goldberg et al., 1983; Rainbow and Phillips, 1993). Nevertheless, these earlier studies did not examine the role of colloidal material in their "dissolved" phase, which has now clearly been defined as a distinct fraction within which contaminants behave distinctly (Gustafsson and Gschwend, 1997).

While aquatic organisms can obtain metals from suspended particles or directly from solution (Luoma, 1989; Luoma et al., 1992), uptake of metals in solution by marine invertebrates is considered to be "passive," while uptake from particles is through "active" ingestion. Although the bioaccumulation of metals in organisms is largely dependent on solution speciation (Zamuda and Sunda, 1982), bioaccumulation is also largely controlled by pH (Hart and Scaife, 1977), salinity (Part et al., 1985), DOM abundance (Laegreid et al., 1983), and the abundance of other metals (Wright, 1977). Although free-ionic metal species can become complexed with organic ligands before being transported across cell membranes (Bruland et al., 1991), chelation can also result in a reduction of metal uptake. Charge and lipophilicity of organic–metal complexes are important in controlling the transfer of metals across cell membranes in the epithelium, gills, and guts of marine/estuarine organisms (Simkiss and Taylor, 1989; Carvalho et al., 1999). However, the amphiphilic character of organic matter makes it particularly difficult to predict what effect organic matter complexation will have on the bioavailability of metals. Recent work has shown that when comparing the uptake and bioaccumulation of colloidally complexed versus free-ionic radioactive metals (Ag, Cd, Ba, Fe, Sn, Zn, Co, Hg, and Mn) in juvenile brown shrimp (*Penaeus aztecus*), uptake kinetics were very similar (Carvalho et al., 1999). It was also noted that most of the colloidally complexed metals were bioaccumulated in the *hepatopancreas*, while the free-ionic metals were mostly found in the abdomen. Other more recent studies have also demonstrated bioavailability of colloidally complexed metals in invertebrates such as zebra mussels (*Dreissena polymorpha*) (Roditi et al., 2000) and oysters (*Crassostrea virginica*) (Guo et al., 2001, 2002).

A *biotic ligand model* (BLM) has been introduced to describe a characterized site where the metal reaches a critical concentration at the metal–ligand site, which in fish, would be located in the gills (Di Toro et al., 2001; Santore et al., 2001, Heijerick et al., 2002). In essence, the BLM is used to predict metal interactions at the biotic ligand in the context of other competing reactions in aqueous environments. This model is a generalization of the free-ion activity model, yet is different in that it has a competitive binding biotic ligand. Mortality will occur when the total concentration of metal bound to the ligand is above the threshold concentration. In this model, which is adapted from the gill–surface interaction model (GSIM) (Cleven and van Leeuwen, 1986), the fish gill is usurped by the ligand as the primary site of activity; this allows further application to other invertebrate species (figure 15.4; Di Toro et al., 2001). This model also allows free metal ions to compete with other cations (e.g., Na^+, Ca^{2+}), organic matter complexation, and inorganic ligands for the binding site of the ligand. The BLM was found to be effective in predicting the *lethal concentration* for 50% of the test organisms (LC50) for Cu, within a factor of 2 of measured values, in freshwater fish (Santore et al., 2001). Similarly, the BLM was found to be effective in predicting the *effective concentration* for 50% of the test organisms (EC50) for Zn in invertebrate species (Heijerick et al., 2002).

Partitioning and Toxicity of Hydrophobic Organic Contaminants

The fate and transport of HOCs in aquatic systems are largely controlled by sorptive interactions with particulates (McCarthy et al., 1989; Santschi et al., 1999). In particular,

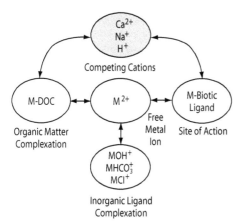

Figure 15.4 Schematic of the biotic ligand model. (Modified from Di Toro et al., 2001.)

POC was used for some time as a predictor of HOC distribution (Rutherford et al., 1992). When examining liquid–solid phase sorption interactions, the dissolved sorbing neutral HOC (NHOC) is called a *solute*, and is referred to as a *sorbate* when in contact with the solid (*sorbent*) surface. Sorption will occur when the attraction of forces between the sorbate and the sorbent, or the solute and solvent, are greater than the total repulsive forces (Adamson, 1976). The solubility of NHOCs is generally very low, making sorption a function of weak solute–solvent interactions rather than strong sorbate–sorbent interactions (Santschi et al., 1999). Since these molecules are characterized as having no hydrophilic properties, this sorption process involving weak solute–solvent and sorbate–sorbent interactions has been referred to as *hydrophobic sorption* (Karickhoff, 1984).

The fate and transport of both inorganic and organic compounds in estuarine systems, in a two-phase system, is dependent on their partitioning between particulate and dissolved phases. As described in chapters 4 and 7, this partitioning of the freely dissolved phase relative to that of the sorbed phase can be described using a partition coefficient (K_d). Since HOCs have a strong affinity for C, the fraction of organic C in the sorbent (for sorbents with a C content of greater than or equal to 0.1%) needs to be considered (Schwarzenbach et al., 1993). In the case of HOCs, equations are used to determine the partitioning coefficient between organic carbon and water (K_{oc}), based on aqueous solubility (S) or an *octanol/water partition coefficient* (K_{ow}) (Karickhoff et al., 1979; Means et al., 1980). The higher the K_{ow}, the greater affinity a particular HOC will have for TOC (see below). These equations are as follows:

$$\log K_{oc} = 1.00 \log K_{ow} - 0.317 \tag{15.1}$$

and

$$\log K_{oc} = -0.686 \log S + 4.273 \tag{15.2}$$

where: K_{ow} = octanol–water coefficient (concentration of chemical in octanol phase/concentration of chemical in aqueous phase); and S = aqueous solubility (μg mL^{-1}).

There are limitations in using these models primarily based on assumptions associated with homogeneity of reversible equilibration between liquid and solid phases, as well as problems associated with sorption capacity of the sorbent being exceeded—to name just a few (Santschi et al., 1999). Factors that limit the solubility of these compounds in solution are expected to increase partitioning onto particles; one such process is referred to as the *salting out* effect. This effect occurs with increases in dissolved solids in solution resulting in changes in the solubility of NHOCs (Means, 1995). Colloidal organic matter has also been shown to be important as a third sorbing phase in sorbing NHOCs (Means and Wijayaratne, 1982; Wijayaratne and Means, 1984a,b; Brownawall and Farrington, 1985; Periera and Rostad, 1990; Burgess et al., 1996; Mitra and Dickhut, 1999). Hydrogen bonding and hydrophobic interactions primarily control these associations between colloids and NHOCs (Means and Wijayaratne, 1982).

Many PAHs that are produced from combustion processes become associated with aerosols at ambient temperatures. Although the atmospheric transfer of HOCs, such as PAHs, has been shown to be important, this pathway remains poorly understood (Bouloubassi and Saliot, 1993). Fossil fuel combustion, coal gasification and liquification, petroleum cracking, waste incineration, and the production of coal tar pitch, coke, carbon black, and asphalt have all been shown to be possible sources of PAHs (McVeety and Hites, 1988). Sources of PAHs can be distinguished by using select ratios. For example, ratios of total methylphenanthrenes to phenanthrene are used to identify PAHs as *pyrolytic* or *petrogenic* sources (Laflamme and Hites, 1978; Prahl and Carpenter, 1983). Other source ratios, involving PAH isomers, are used to identify PAH transformations during transport; examples of ratios are phenanthrene/anthracene and fluoranthrene/pyrene (e.g., Mitra et al., 1999b; Dickhut et al., 2000). Other work using isomer ratios in Chesapeake Bay has shown that automobiles are the dominant source of carcinogenic PAHs (benzo[a]pyrene, benz[a]anthracene, and benzo[b]fluoranthrene) (Dickhut et al., 2000). Some examples of PAH isomer ratios and the associated emission sources are shown in table 15.1 (Dickhut et al., 2000). Moreover, it appears that there is a selective partitioning of PAHs that

Table 15.1 Polycyclic aromatic hydrocarbon (PAH) isomer ratios for major emisson sources.

Source[a]	BaA/chrysene	BbF/BkF	BaP/BeP	IP/BghiP
Automobiles	0.53±0.06	1.26±0.19	0.88±0.13	0.33±0.06
Coal/coke	1.11±0.06	3.70±0.17	1.48±0.03	1.09±0.03
Wood	0.79±0.13	0.92±0.16	1.52±0.19	0.28±0.05
Smelters	0.60±0.06	2.69±0.20	0.81±0.04	1.03±0.15

[a]At ambient temperatures, these PAHs remain >50% associated with aerosol particles in the atmosphere: benz[a]anthracene (BaA), benzo[b]fluoranthene (BbF), benzo[k]fluoranthene (BkF), benzo[a]pyrene (BaP), benzol[e]pyrene (BeP), indeno [123-cd]pyrene (IP), and benzo[ghi]perylene (BghiP).
Modified from Dickhut et al. (2000).

survive transport from the water column to the sediments. Coal-derived PAHs appear to be more abundant in surface sediments, with the carcinogenic PAHs remaining more in surface waters; this is believed to be related to greater decay of benzo[a]pyrene, benz[a]anthracene, and benzo[b]fluoranthrene in the water column and/or to dilution of these PAHs by previously deposited coal-combusted PAHs in the sediments.

While many of the high molecular weight (less volatile) PAHs generally remain irreversibly bound on particles (McGroddy and Farrington, 1995; McGroddy et al., 1996) and consequently unavailable for partitioning, lower molecular weight gaseous phase PAHs (with high hydrophobicity) appear to associate readily with plankton in surface waters (Countway et al., 2003). For example, most phenanthrene (>90%) is typically found in the gaseous phase at ambient temperature (Bidlemen, 1988). Other work in Chesapeake Bay has shown that the absorption of atmospheric sources of PAHs (e.g., phenanathrene and fluorene) represents a significant pathway to surface waters (Gustafson and Dickhut, 1997a,b,c). Unlike PAHs, PCBs have low vapor pressures and tend be associated with fine particulate organic matter, favoring partitioning into the sediments rather than the aqueous phase (Brownawall and Farrington, 1986; Pierard et al., 1996). However, PCB budgets for the Baltic Sea indicate that atmospheric inputs can be quite high (77%), as represented by wet deposition (7%), dry deposition (7%), and vaporphase adsorption (63%) (Axelman et al., 2000). More recent work has confirmed that the relatively volatile PAHs enter the Chesapeake Bay in the gaseous phase across the air–sea interface and that these compounds become associated with the autochthonous pool of phytoplankton carbon (Countway et al., 2003).

Due to complex partitioning dynamics of HOCs in natural waters, it is necessary to measure their phase distribution accurately to effectively understand their fate and transport. Mean concentrations of particulate and dissolved fractions of four different PAHs in the surface waters of southern Chesapeake Bay show how spatially variable PAHs can be (figure 15.5; Gustafson and Dickhut, 1997a). While there was a gradient for particulate PAHs from urban to remote regions, there were no seasonal trends in dissolved or particle-bound PAHs. It is commonly asserted that if the slope of log K_{oc} versus log K_{ow} is approximately 1, dissolved and particulate PAHs are considered to be in equilibrium, based on the equilibrium partitioning (EqP) theory, with PAHs in this region being at or near equilibrium (Schwarzenbach et al., 1993). When examining relationships between these distribution coefficients, it appears that PAHs are in equilibrium in surface waters of southern Chesapeake Bay (figure 15.5; Gustafson and Dickhut, 1997a). As mentioned above, the more volatile PAHs are associated with the more labile autochthonous pool of carbon in the surface waters of the lower Chesapeake Bay and York River; this work also showed that the soot-associated PAHs were more coupled with allochthonous carbon (Countway et al., 2003). Based on isomer ratios, it was further determined that the soot-derived PAH, perylene, was linked with inputs of terrestrially derived organic matter based on chemical biomarkers. This agrees with other studies that have shown interactions between perylene and terrestrially derived organic matter in coastal Washington sediments (Prahl and Carpenter, 1983) and the southern British Columbia coast (Yunker et al., 1999).

The fate of HOCs and overall dynamics of their potential equilibrium partitioning has also been shown to be affected by the biogeochemical characteristics of the sorbing matrix as well as inherent differences in HOCs. For example, higher concentrations of

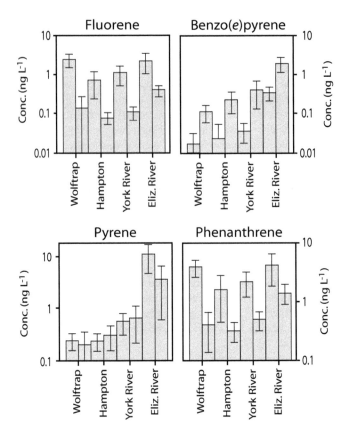

Figure 15.5 Mean concentrations of particulate and dissolved fractions of four different PAHs in the surface waters of southern Chesapeake Bay. (Modified from Gustafson and Dickhut, 1997c).

both PAHs and PCBs were observed in zooplankton feces compared to their microplankton food source (Baker et al., 1991). Particulates collected in traps were also found to have concentrations higher than those of surface water particulates; this was attributed to active sorption of HOCs during sedimentation of particulates. Some work has suggested that, as settling decomposing particulates age, their sorptive capacity increases (Koelmans et al., 1997). In fact, retention of PCBs in the water column of the Baltic Sea has been estimated to be less than one year (Jonsson, 2000). The sorption of HOCs to cultured phytoplankton also showed that sorptive equilibrium was not reached over one month, this was attributed to algal growth rates being more rapid than HOC sorption rates (Swackhamer and Skoglund, 1991, 1993). Controls on the cycling of HOCs (PAHs and PCBs) in Chesapeake Bay over short timescales (e.g., hours to days) were attributed to resuspension events and to a lesser degree biotic particles, while the incorporation of HOCs into

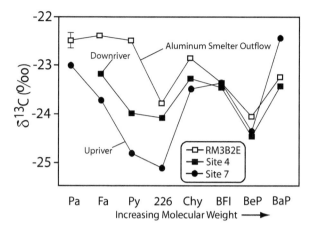

Figure 15.6 Relationship between the δ^{13}C signatures of different PAHs and their proximity to aluminum smelters in the St. Lawrence River (Canada). (Modified from Stark et al., 2003.)

organic-rich particles and ultimately burial in sediments was likely responsible for longer timescales (Ko and Baker, 1995).

Despite numerous attempts to determine the source apportionment (natural versus anthropogenic) of HOCs in sediments through the use of molecular criteria (Yunker et al., 2002; Hellou et al., 2002), overlaps in signatures continue to make source determination very difficult. Other work has resorted to the application of compound-specific isotope analysis (CSIA) (see chapter 9) to better constrain anthropogenic sources of PAHs (e.g., O'Malley et al., 1994; Mazeas and Budzinski, 2001). Recent work showed a strong relationship between the δ^{13}C signatures of different PAHs and their proximity to aluminum smelters in the St. Lawrence River (figure 15.6; Stark et al., 2003). Similarly, CSIA has been shown to be useful in determining the occurrence and type of PCB cogeners (Jarman et al., 1998; Drenzek et al., 2001; Yanik et al., 2003).

The accumulation of HOCs in estuarine sediments is largely determined by particle composition and reactivity, as well as short-term deposition and erosion dynamics (Bopp et al., 1982; Olsen et al., 1993). While their predepositional association with organic matter is critically important (Karickhoff, 1984), postdepositional changes in the HOC–organic matter complexes can also change due to diagenetic alterations (Berner, 1980; see chapter 8). Moreover, the extent to which these complexes are reversible with time is also a function of the amount and composition of organic matter (Brownawall and Farrington, 1986). Other estuarine work has also shown that carbon normalized sediment–porewater distribution of PAHs ($[K_{oc}]_{obs}$) was largely controlled by sedimentary carbon composition (McGroddy et al., 1996) and sediment deposition/resuspension (Mitra et al., 1999a,b). These types of PAH associations, along with many other factors, contribute to the commonly observed sorption disequilibria in natural systems (Socha and Carpenter, 1987). For example, an apparent disequilibrium between PAH concentrations in sediments and pore waters in Boston Harbor (USA) (figure 15.7; McGroddy and Farrington, 1995)

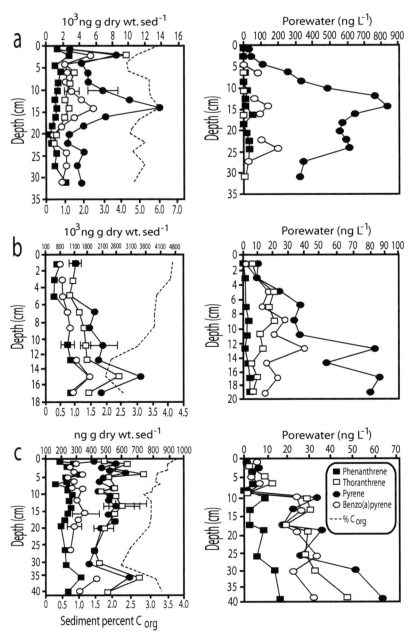

Figure 15.7 PAH concentrations in sediments and pore waters, in Boston Harbor (USA) at (a) Fort Point channel, (b) Spectacle Island, and (c) Peddocks Island. (Modified from McGroddy and Farrington, 1995.)

was attributed to (1) rapid biodegradation of PAHs in pore waters, (2) loss of PAHs to overlying bottom waters, and (3) limited availability of sedimentary PAHs required to reach equilibrium conditions.

Disequilibria between PCBs in porewaters and sediments have also been observed due to the presence of porewater colloids (Brownawall and Farrington, 1986). While many PCBs have been found in association with colloids in porewaters, significant changes in individual chlorobiphenyls with depth have been attributed to microbial decay. The two primary pathways of postdepositional decay are aerobic and anaerobic dechlorination (Yanik et al., 2003). In the aerobic decomposition pathway, cogeners are decomposed into chlorobenzoic acid (Williams and May, 1997), while in the anaerobic pathway the highly chlorinated cogeners are converted into lower chlorinated analogs (Pulliam-Holoman et al., 1998). Such postdepositional changes make the aforementioned newly developed source apportionment techniques (e.g., CSIA) particularly important when considering the complexity of signatures that are commonly observed in sediments.

Similar to metals, there is considerable debate on the effects of organic matter complexation with HOCs on bioavailability. Some sedimentary work has shown that uptake of HOCs by benthic invertebrates is enhanced (Kukkonen and Oikari, 1991; Haitzer et al., 1998), reduced (Landrum et al., 1985; Freidig et al., 1998), or not discernably effected (Lores et al., 1993) when HOCs are complexed with organic matter. More specifically, the carbon functionality of organic matter (e.g., aromaticity) was also shown to be important in determining bioavailability (McCarthy et al., 1989; Chin et al., 1997; Haitzer et al., 1999). Benthic organisms can take up HOCs from sediments and the water column depending on their feeding mode (e.g., deposit feeder versus suspension feeder); this factor has been shown to be quite important in controlling bioaccumulation of PCBs in benthos (Burgess and McKinney, 1999). In fact, EqP predicts that a benthic organism in equilibrium with its surrounding environment will have an HOC concentration (associated with its lipid pool) that is in steady state with the ambient concentrations of HOC in porewaters and sedimentary organic carbon (Schwarzenbach et al., 1993). Other work in the Seine River estuary (France) indicated that feeding preference, phytoplankton lipid fraction, and organic content of detritus were most important in determining bioaccumulation of PCBs in the food web (Loizeau et al., 2001). This is commonly expressed as the *biota–sediment accumulation factor* (BSAF) as shown below:

$$\text{BSAF} = (C_{\text{org}}/f_{\text{TLE}})/(C_{\text{sed}}/f_{\text{OC}}) \tag{15.3}$$

where: C_{org} = concentration of chemical in organism; f_{TLE} = lipid content in organism C_{sed} = concentration of chemical in sediments from which the organism is feeding; and f_{OC} = concentration of organic carbon in sediment. The BASFs for HOCs should be independent of particle and organism properties and vary as a consequence of HOC hydrophobicity—assuming that the assumptions of EqP are correct (DiToro et al., 1991). A range of BASFs from selected studies are shown in table 15.2). Recent work has shown that, after examining BASFs at two contaminated sites over time, bioaccumulation in estuarine systems should not be considered as an equilibrium process (Mitra et al., 2000b). These conclusions were based on the fact that the quality of POC and DOC, based on nutritional value and sorptive capacity, significantly affected bioaccumulation of PAHs in two estuarine benthic invertebrates.

Table 15.2 Biota–sediment accumulation factors (g OC g TLE^{-1}) from selected studies.

Compound	Organism	BSAFs	Reference
Dieldrin	*Lumbriculus variegates*	~0.25–2.5	Standley (1997)
PCBs (hexa- to deca-)	*Palaemontes pugio*	0.04–0.5	Maruya and Lee (1998)
^{14}C-tetrachlorobiphenyl	*Amphiura filiformis*	1.1–4.4	Gunnarsson et al. (1999)
PAH (total)	Crayfish[a]	0.01–0.1	Thomann and Komlos (1999)
PAH (total)	*Potamocorbula amurensis*	0.06–5.4	Maruya et al. (1997)
	Tapes japonica	0.01–2.1	
	Polychaetes[a]	0.4–2.0	
^{14}C-phen and ^{3}H-BaP	*Palaemonetes pugi*	0.18–1.5	Mitra et al. (2000b)
^{14}C-phen and ^{3}H-BaP	*Rangia cuneata*	0.13–1.7	Mitra et al. (2000b)

[a]Genus or species identity not provided in text.
Modified from Mitra et al. (2000b).

Past work has shown that PAHs in sediments have significant deleterious effects on benthic communities (Wakeham and Farrington, 1980; Bauer and Capone, 1985a,b; Bunch, 1987). In particular, low-energy environments, such as salt marshes are particularly susceptible to high accumulations of PAHs (Little, 1987). Microphytobenthos are particularly susceptible to PAHs due to the high sediment-binding and residence-time properties of PAHs (Plante-Cuny et al., 1993; Carman et al., 1997). However, in some cases PAHs have been shown to have a positive effect on microphytobenthos (Bennett et al., 2000). For example, enhanced abundance of microphytobenthos in Louisiana salt marsh sediments, at sites with higher levels of PAH contamination compared to other less-contaminated sites, was believed to be related to greater availability of NH_4^+ in porewaters in the contaminated sites, from enhanced remineralization of N-containing PAHs or natural geopolymers (Bennett et al., 1999, 2000). It may be that low molecular weight PAHs provide a food substrate for bacteria that promotes bacterial turnover— thereby enhancing remineralization. Estuarine coastal marshes in Louisiana (USA) have been exposed to high levels of petroleum-hydrocarbon contaminants due to the extensive offshore/nearshore drilling activities (e.g., Long, 1992). Other work has shown, using experimental treatments that hydrocarbon additions to salt marshes sediments, collected from Louisiana, resulted in an increase in the abundance of hydrocarbon-degrading bacteria (Nyman, 1999). More specifically, bacterial degradation of PAHs, such as phenenthrene, was readily degraded in Louisiana marsh sediments. Similarly, the carcinogenic PAH, benz[*a*]anthracene, was shown to be effectively metabolized by bacteria in sediments (Hinga et al., 1980). Collectively, this suggests that the bacterial community in this region is adapted to chronic exposure of petroleum hydrocarbons and that there is a direct link with enhanced remineralization of organic matter and contaminants in sediments.

In addition to the interactions between HOCs and microbial populations in sediments, there is evidence that contaminants affect the overall trophic structure of benthic food webs. As discussed in chapter 8, microphytobenthos are an important food source for many benthic invertebrates (Levinton and Bianchi, 1981; Levinton et al., 1984). In a follow-up to aforementioned work in Louisiana marsh sediments, other work has shown that diesel fuel contamination in sediments results in benthic diatom blooms due to reduced grazing pressure from *harpacticoid copepods* and from enhanced availability of DIN (Carman and Todaro, 1996; Carman et al., 1996, 2000). This is clearly illustrated in figure 15.8a, which shows the high accumulation of NH_4^+ in the porewaters in the dark treatments, when algal uptake is limited, but the highest in the dark treatment with diesel fuel additions (Carman et al., 2000). Similarly, the high positive change in chlorophyll-*a*, a measure of microphytobenthic biomass, was in the treatments with diesel additions (figure 15.8b). Many of the associations of PAHs with particulate and dissolved organic matter can be altered by diagenetic transformations in sediments (Prahl and Carpenter, 1983). While PAHs have been shown to be highly concentrated in sediments, resulting in significant toxicological effects on the life history (Bridges et al., 1994) and immune systems (Tahir et al., 1993) of benthic organisms, macrobenthos can have significant effects on the fate and transport of HOCs. For example, experimental manipulations showed that macrofauna were important in mediating the loss of HOCs from sediments (Schaffner et al., 1997). The importance of benthic macrofauna in benthic–pelagic coupling, via release of HOCs from sediments, has also been shown in the Baltic Sea (Gunnarsson et al., 1999).

Nutrient Loading and Eutrophication

This section will focus on *eutrophication* as a broad-scale process since the specific pathways and forms of nutrient inputs, associated with the N, P, S, and Si cycles, were discussed in prior chapters. Eutrophication is defined by Cloern (2001, p. 224) as "the myriad of biogeochemical and ecological responses, either direct or indirect, to anthropogenic fertilization of ecosystems at the land–sea interface." This more recent definition is not to be confused with the limnological description of eutrophication that is associated with aging lakes as they change from a nutrient-poor to a more nutrient-rich status, which results in the successional transition from lake, to a pond, and then a marsh (Wetzel, 2001). Currently, nutrient inputs represent the largest problem in rivers and estuaries (Nixon, 1995; Conley et al., 2000; Rabalais and Turner, 2001; Conley et al., 2002; Howarth et al., 2002). As discussed in previous chapters, N and P are the most important nutrients associated with eutrophication, with other elements such as Si, and in some cases micronutrients like Fe and Mo, also being potentially limiting. It has been estimated that approximately 33% of the 139 estuaries surveyed by the National Oceanic and Atmospheric Administration National Estuarine Assessment have symptomatic problems that stem from eutrophication (figure 1.3) and that if no changes are made, two out of every three estuaries will be affected by nutrient excessive loading by 2020 (Bricker et al., 1999). As the human population continues to grow, the use of fertilizers and fossil fuels, the major sources of nutrients, will also increase accordingly. These predictions are well supported by the historical linkages between the expansion of world fertilizer consumption, emission of

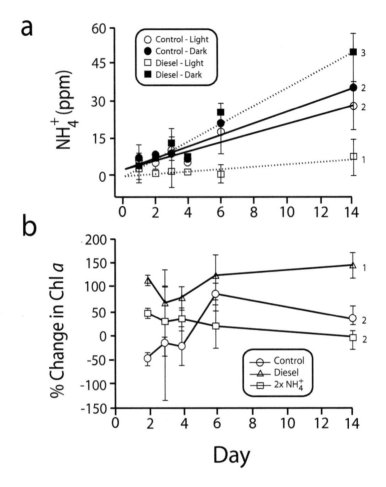

Figure 15.8 Concentration of (a) NH_4^+ and (b) percentage of chlorophyll-a over a 14-day period in a microcosm experiment. Control represents no diesel fuel added. Numbers at ends of lines indicate regression slopes that differ significantly. (Modified from Carman et al., 2000.)

nitrogen oxides (via fossil fuel combustion), and coastal eutrophication during, the period between 1960 and 1980 (figure 15.9; Boesch, 2002). Historical changes that date back well before this period also indicate that early settlement in the 17th century in the watershed of Chesapeake Bay estuary likely resulted in enhanced runoff into the estuary that probably peaked in the late 1800s (Cooper and Brush, 1991; Walker et al., 2000). Similarly, the first signs of eutrophication in Western Europe began in the 19th century with the enhanced delivery of N and P from rivers to coastal regions (Billen et al., 1999). This historical reconstruction on the chronology of eutrophication is based on down-core studies which

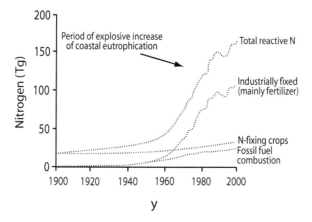

Figure 15.9 Historical linkages in the expansion of world fertilizer consumption, emission of nitrogen oxides (via fossil fuel combustion), and coastal eutrophication, between 1900 and 2000. (Modified from Boesch, 2002.)

use an assortment of paleotracers (e.g., Zimmerman and Canuel, 2000)—more details on this are described in the next section.

A paper by Cloern (2001) eloquently examines the evolving conceptual model of eutrophication over the past three decades, and offers some insightful alternative approaches for future management questions. In the late 1960s, the first conceptual eutrophication model (Phase I) was developed, largely from a limnological approach (e.g., Vollenweider, 1976; Schindler, 1987). This model essentially emphasized the fundamental signal of nutrient loading as it related to response in phytoplankton primary production and biomass accumulation, with the magnitude of the responses being proportional to nutrient loading. In a more updated model (Phase II), involving more contemporary views of coastal eutrophication, the following advances over the first model are discussed: (1) inclusion of a more comprehensive group of direct and indirect responses to nutrient inputs, (2) system-specific characteristics that can act as "filters" in buffering such responses, and (3) the application of management actions that can rehabilitate the estuarine system principally through reductions in nutrient inputs (Cloern, 2001). Finally, in Phase III, a conceptual model of coastal eutrophication is developed for future management decisions, that is focused on developing a better understanding of (1) the system attributes that help to buffer the response to nutrient inputs, (2) how nutrient inputs relate to other stressors in the system (assuming nutrient inputs are only one of the many possible stressors), (3) the global impact of coastal ecosystem change and how it relates to the sustenance of human populations, and (4) a larger management plan for the rehabilitation/restoration of coastal ecosystems that is based on a more universal scientific understanding of coastal eutrophication (figure 15.10; Cloern, 2001). The components of this model are to some extent based in a similar manner to the concept of global "syndromes," recently developed by the German Advisory Council on Global Change

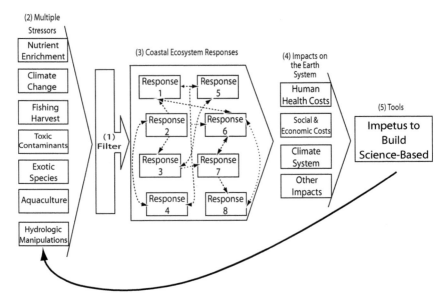

Figure 15.10 Conceptual model of eutrophication based on the concepts of (1) filter attributes, (2) nutrient enhancement as one of many stressors, (3) complex linkages between stressors, (4) impact of coastal change on global scale, and (5) application of broader concepts of eutrophication on management issues. (Modified from Cloern, 2001.)

(GACGC, 2000, p. 23), which are defined as "typical patterns of problematic people–environment interactions which can be found worldwide and can be identified as regional profiles of damage to human society." More specifically, this approach has been applied to rivers where the following syndromes were developed: (1) flow regulation; (2) fragmentation of river course; (3) sediment imbalance; (4) *neo-arheism* (drastic reduction in river flow); (5) chemical contamination; (6) acidification; (7) eutrophication; and (8) microbial contamination (Meybeck, 2003). Conceptual models of eutrophication and such typologies should prove useful in facilitating comparisons of aquatic systems at regional and global scales.

One area of intense interest associated with coastal eutrophication is the occurrence of hypoxia ($O_2 < 2$ mg L^{-1}) events (Rabalais and Turner, 2001). The adaptive response of nutrient enhancement by fast-growing phytoplankton and benthic micro- and macro-algal species leads to enhanced O_2 consumption in sediments. More specifically, benthic microbial metabolism increases in response to enhanced organic matter supply, which results in greater nutrient remineralization and release rates, O_2 consumption, and numerous other changes in sediments chemistry, such as the accumulation of metal sulfide (Jørgensen, 1996) (see chapter 12 for more details). The expansion of anoxic and hypoxic zones has dramatically increased in recent years and the ecological consequences on pelagic and benthic species can be severe (Stanley and Nixon, 1992; Diaz and Rosenberg, 1995; Cloern, 2001; Elmgren, 2001; Rabalais and Turner, 2001). Preliminary data on the northern Gulf of Mexico indicate that the production, recruitment, and population health of fisheries have not been affected by hypoxia in this region (Chesney and Baltz, 2001); however, more work is needed to corroborate these early findings. In some cases, the landing of planktivorous and demersal fishes in 14 European estuaries was found to be higher in eutrophic compared to oligotrophic systems (De Leiva Moreno et al., 2000). These results are likely explained by the occurrence of a mosaic of spatially variable oxygenated, hypoxic, and anoxic conditions in these "hypoxic regions" whereby nutrient enrichment in the oxygenated waters commonly results in enhanced prey abundance for fishes. Depending on the persistence and spatial changes in these multiple zones of O_2 availability, if fish can locate the prey in the more productive regions and stay away from the low O_2 regions, the overall effect of eutrophication can be positive (Breitburg, 2002). While the "search" for new regions of hypoxia in estuarine systems has reached almost "epidemic" proportions, it is important to remember that basic solubility changes in O_2 that occur in response to seasonal temperature changes can also reduce O_2 concentrations—something unfortunately overlooked in some recent surveys.

Although algal blooms are a naturally occurring phenomenon, increases in the global expanse and frequency of *harmful algal blooms* (HABs) over the last few decades has resulted in new questions about the possible relationship between HABs and eutrophication. There are two general types of HABs: the toxic forms that lead to the death of marine organisms and human seafood poisoning; and the nontoxic species that can have negative effects on the environment by modifying O_2 consumption, shading effects, and altering the stoichiometry of nutrients (Anderson et al., 2002). Many of the commonly observed noxious cyanobacterial blooms in freshwater lakes have been linked to P enrichment (Schindler, 1997). Similarly, in estuarine systems like the Baltic Sea, toxic cyanobacteria, such as *Nodularia spumigena*, have also been linked to P enrichment (Niemi, 1979; Kononen, 1992; Elmgren, 2001). Most of the coastal and estuarine HABs are microflagellates (e.g., *Alexandrium* spp., *Dinophysis* spp., *Gymnodinium* spp., and *Pfiesteria* spp.) (Anderson et al., 2002). One of the more infamous outbreaks of dinoflagellate HABs was from the highly toxic *Pfiesteria*, which most notably occurred in the Neuse, Pamlico, and New River estuaries of North Carolina (Burkholder et al., 1995, 1997; Glasgow et al., 2001). These organisms have one of the many competitive adaptations found in many dinoflagellates, which is the ability to use mixotrophic nutrition (or *mixotrophy*) or the simultaneous uptake of both inorganic and organic nutrients (Burkholder et al., 2001). However, many factors other than nutrients, such as weather conditions, algal species present, flushing residence time, presence and abundance of grazers, and differential success of different HABs species at a specific point in time, are important in determining if and when bloom conditions can occur. Many of the pristine waters found in the Gulf of Maine and northern U.S. west coast cannot use nutrient enrichment to explain for the expansion of the toxic *Alexandrium* spp., which causes paralytic shellfish poisoning (PSP) (Anderson et al., 2002). Thus, while in many cases it has been difficult to establish strong linkages between HABs and nutrient enrichment in coastal systems, linkages have been made in many other systems. For example, studies involving *Pseudo-nitzchia* spp., in both laboratory mesocosm and field studies (Dortch et al., 1997, 2000) in the Mississippi River plume, have proven to be some of the best examples documenting linkages between nutrient enrichment and HABs.

Historical Reconstruction of Environmental Change

Laminated sediment chronologies of organic and inorganic P, N, C, and biogenic Si have been used to study long-term trends in eutrophication in lacustrine systems (Conley et al., 1993). Moreover, chemical biomarkers, such as plant pigments (see chapter 9) have also been utilized as effective paleotracers of historical change in phytoplankton assemblages in lacustrine systems, where laminated sediments provide an excellent environment for preservation of chemical biomarkers (Watts and Maxwell, 1977). Fossil pigments have also been used in paleolimnology to understand better historical changes in bacterial community composition, trophic levels, redox change, lake acidification, and past UV radiation (Leavitt and Hodgson, 2001). In estuarine systems, highly variable physical conditions, extreme spatial gradients, and bioturbation have limited the application of these techniques (Nixon, 1988). In spite of these limitations, past studies have shown increased preservation of total organic carbon (TOC) in estuaries during the 20th century (Cooper and Brush, 1993; Eadie et al., 1994; Gong and Hollander, 1997; Louchouarn et al., 1997; Bianchi et al., 2000c, 2002a; Tunnicliffe, 2000; Zimmerman and Canuel, 2000). In fact, recent work compared long-term ecological data (e.g., phytoplankton biomass composition, nutrients, and benthos) and plant pigments preserved in laminated sediments of the Himmerfjarden Bay (Baltic proper). This bay has been monitored for the effects of municipal wastewater and runoff from agricultural land on coastal eutrophication since the 1980s (Elmgren and Larsson, 2001). This work essentially concluded that despite high-resolution sampling, using freeze-core sampling of sediment layers (annual varves), there was no correlation between pigments preserved in sediments and phytoplankton biomass (Bianchi et al., 2002a). However, when sedimentary pigments were averaged over longer time intervals (5 y), averages of annual diatom biomass were positively correlated to down-core concentrations of fucoxanthin (figure 15.11). This work demonstrated that, under favorable conditions for preservation in sediments (e.g., presence of laminations), plant pigments can be used to trace past changes in phytoplankton in estuaries. More recent work has also shown significant variability in the effectiveness of using plant pigments for ecological reconstruction in four European estuaries (Reuss et al., 2005).

The eutrophication of the Baltic Sea has been a topic of study for many years (Fonselius, 1972; Cederwall and Elmgren, 1980; Larsson et al., 1985; HELCOM, 1993, 1996, 1998, 2001). The expanse of laminated sediments and increases in TOC content in sediments since the early 1960s have been used as general tracers of recent eutrophication in the Baltic Sea (Jonsson et al., 1990). In fact, massive accumulations of nitrogen-fixing cyanobacteria (mostly *Nodularia* spp.) in surface waters have been observed in the Baltic Sea since 1982, particularly in summer (Niemi, 1979; Kononen, 1992), and have been attributed to warm weather (Kahru et al., 1994; Kahru, 1997). Such blooms have also been documented in the Peel-Harvey estuary (Australia) (Lenanton et al., 1985). The genus *Aphanizomenon* also commonly occurs in the water column, during the Baltic surface blooms (Wasmund, 1997). There is some anecdotal evidence that large cyanobacterial blooms did exist in the 19th century, prior to any anthropogenic nutrient inputs (Lindström, 1855). However, since the early studies of cyanobacterial blooms did not begin until they were already relatively common in the 1960s (see Finni et al., 2001) linkages with anthropogenic nutrient inputs prior to this period remain speculative. However, recent work has shown, using the down-core distribution of carotenoids (echinenone, myxoxanthophyll,

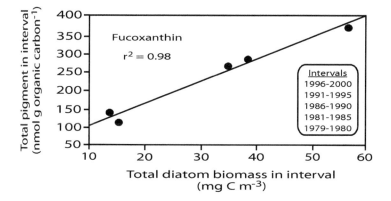

Figure 15.11 Regression between fucoxanthin concentrations in annual sediment layers versus average annual measured diatom biomass in Himmerfjärden (Sweden), both averaged over 5 y. (Modified from Bianchi et al., 2002a.)

and zeaxanthin) as chemical biomarkers for cyanobacteria, that *Nodularia spumigena* and *Aphanizomenon* spp. blooms started to become more common in the early 1960s (Poutanen and Nikkila, 2001). These pigment data also show that their occurrence prior to the mid-1940s was minimal. These results suggest that these blooms are in some way linked to enhanced anthropogenic nutrient loading, which also began during the mid-1960s in the Baltic Sea (Finni et al., 2001).

Other work has shown that large cyanobacterial blooms in the Baltic Sea may be a natural feature of this system that has existed for thousands of years. For example, recent work has demonstrated that, using down-core variations in cyanobacterial pigment biomarkers and $\delta^{15}N$ values, nitrogen-fixing blooms are nearly as old as the brackish water phase of the Baltic Sea, dating back to 7000 y BP—soon after the last freshwater phase of the Baltic, the Ancylus Lake, changed into the final brackish water phase, the Litorina Sea (Bianchi et al., 2000c). Intrusions of salt water into the Ancylus Lake began to occur because of eustatic sea level rise, which created the Danish straights in the southern Baltic and the current brackish water Baltic Sea system. These cyanobacterial blooms are believed to have been initiated by increased availability of P, from an inflow of P-rich seawater and increased release of P from sediments during periods of anoxia, caused by the formation of a halocline from the seawater intrusion (Bianchi et al., 2000c; Westman et al., 2003). The Ancylus–Litorina boundary has also been traced by changes in microfossils and authigenic minerals (e.g., FeS_2) in the extensive laminations found in sediments at this transition (Andren and Sohlenius, 1995; Sternbeck et al., 2000). Thus, efforts to restore the Baltic proper to its more natural state need to consider that these blooms have been a part of the Baltic for many years.

Sedimentary tracers such as diatom microfossils, TOC, degree of pyritization (DOP), and nutrient chemistry of sediments, have been used to determine past environmental change in the Chesapeake Bay watershed (Cooper and Brush, 1991, 1993). Early settlement in the 17th century in the watershed of the Chesapeake Bay estuary likely resulted in enhanced runoff into the estuary that probably peaked in the late 1800s (Cooper and Brush, 1991; Walker et al., 2000). Enhanced nutrient loading probably started in the late 19th century, with more dramatic increases with the use of fertilizers in the early 20th century (Wines, 1985; Cornwell et al., 1996). In fact, deoxygenated waters have been documented in Chesapeake Bay as early as the 1930s (Newcombe et al., 1938). Recent work that provided a more precise chronology of the occurrence of anoxia as related to eutrophication indicates that the major environmental change in the mesohaline region of Chesapeake Bay occurred between 1934 and 1948 (Zimmerman and Canuel, 2000). In particular, the first occurrence of enrichment from plankton and microbially derived organic matter is just after 1934 (figure 15.12). Based on inorganic proxies, this time also corresponds to increases in productivity (BSi [biological silica] and TOC) and hypoxia (AVS/NAVS [non-acid volatile sulfides]) (Zimmerman and Canuel, 2000). These changes correspond to increases in the use of inorganic fertilizers in the state of Maryland as well as the overall United States (figure 15.12); synthetic N fertilizers were first introduced in the 1940s (Vitousek et al., 1997). Since 1975, there have been 4- to 12-fold

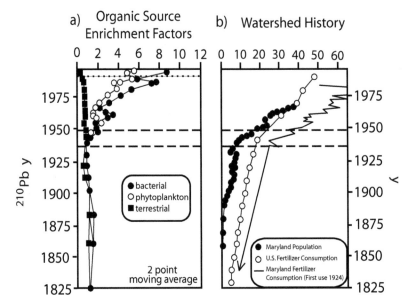

Figure 15.12 Down-core enrichment factors of organic matter derived from phytoplankton, bacterial, and terrestrial sources in Chesapeake Bay. Historical records of commercial fertilizer consumption in the U.S.A ($\times$ 10^6 t y^{-1}), and from the state of Maryland ($\times6000$ t y^{-1}), as well as human population growth ($\times10^5$). (Modified from Zimmerman and Canuel, 2000.)

enrichments in organic matter, which have been accompanied by 5- to 8-fold and 13- to 24-fold increases in N and P loading, respectively (Boynton et al., 1995).

One of the largest zones of oxygen-depleted coastal water in the entire western Atlantic is in the northern Gulf of Mexico (Rabalais and Turner, 2001). While enhanced nutrient input through the Mississippi–Atchafalaya system has been proposed as the primary cause of hypoxia events in this region (Turner and Rabalais, 1994), specific mechanisms associated with these events remain unclear. For example, the absence of long-term measurements of oxygen in this region has made it unclear if these hypoxia events occurred prior to anthropogenic inputs of nutrients from the Mississippi River. Some work has shown that down-core concentrations of acid volatile sulfides (AVS) began to increase about 50 years ago when nutrient loading increased from the Mississippi River to the Louisiana coast (Morse and Rowe, 1999). This work is also supported by the use of microfossils of benthic foraminifera as indicators of past redox in shelf sediments on the Louisiana coast (Sen Gupta et al., 1981, 1996; Sen Gupta and Machain-Castillo, 1993; Osterman et al., 2005). This work shows that during severe hypoxic events, since the 1950s, low-oxygen tolerant species such as *Ammonia parkinsonniana* over *Elphidium* become less abundant in comparison with *Fursenkoina*, known to be tolerant of low-oxygen conditions (Rabalais et al., 1996). More recent work has demonstrated, using the down-core distribution of pigment biomarkers, that occurrence of anoxygenic phototrophic brown-pigmented green sulfur bacteria (e.g., *Chlorobium phaeovibroides* and *C. phaeobacteroides*) on the Louisiana coast is correlated with increasing nutrient loading (Chen et al., 2001). Both *Chlorobium* species produce bacteriochlorophyll-*e* as their primary photosynthetic chlorophyll (Gloe et al., 1975). They also require reduced S (H_2S, SO) compounds for growth along with sunlight; thus, they commonly occur in the water column where there is a sufficient supply of both. Moreover, blooms of *C. phaeobacteroides* are commonly observed in the water column of shallow lakes (Fry, 1986) and deep suboxic basins such as the Black Sea (Repeta, 1993), where H_2S produced in anoxic sediments penetrates into bottom waters. The down-core distribution of bacteriochlorophyll-*e* homologs, along with the more stable bacteriopheophytin-*e*, indicated that highest concentrations occurred between 1960 and the present (figure 15.13; Chen et al., 2001). This period coincides with the increased nutrient loading from the Mississippi River. All of these studies support the idea that low-oxygen bottom waters have persisted for at least the past 40 years on the Louisiana shelf.

Another group of elements that have been used to reconstruct past changes in anthropogenic inputs in estuaries and other coastal systems are the *rare Earth elements* (REEs). These elements occur in the periodic table from lanthanum (La) to lutetium (Lu) and have similar chemical and physical properties due to their electronic configurations (Henderson, 1984). The two classes commonly used to separate REEs are the light REEs (LREEs) (from La to Sm) and heavy REEs (HREEs) (from Gd to Lu) (Henderson, 1984). The major ores of REEs used as *cracking catalysts* (catalysts used to breakdown a complex substance, especially the "breaking" of petroleum molecules into shorter molecules to extract low-boiling fractions such as gasoline) are monazite and bastnasite, which compared to average shales are highly enriched in LREEs relative to HREEs (Olmez et al., 1991). Despite their name, REEs are in many cases more abundant than trace metals in coastal sediments. The use of REEs increased in the 1950s, when they were used in *zeolite* to improve fluid cracking catalytic processes in the production of gasoline and

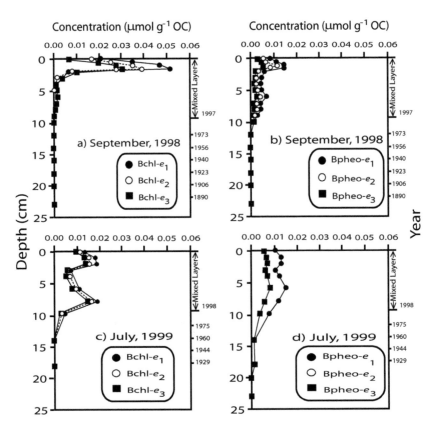

Figure 15.13 Down-core distribution of bacteriochlorophyll-*e* homologs, along with the more stable bacteriopheophytin-*e*, in sediments collected in the hypoxic region of the Louisiana shelf in 1998 and 1999. (Modified from Chen et al., 2001.)

fuel oil. This increase in demand for REEs continued through the 1970s, but dropped off in the 1980s when leaded gasoline was introduced (Olmez and Gordon, 1985). The enriched signal of LREEs from refinery emissions and/or waste products (e.g., *fly ash*), first observed by Olmez and Gordon (1985) and Kitto et al. (1992), was later observed as a recorded signal in sediments off the California coast (Olmez et al., 1991). Thus, similar to the way pulse radionuclide tracers (e.g., ^{137}Cs) are used (see chapter 7), historical changes in inputs can be used to determine the chronology of select anthropogenic inputs. More recent work has shown that LREEs, were unpredictably not enriched in sediments, deposited in the 1960s and 1970s, of the Sabine–Neches estuary (USA) (Ravichandran, 1996). This estuary is a highly industrialized system that is located in one of the most developed oil refinery regions in the United States; thus, it was expected that the abundance of REEs in these sediments would be high. The inefficiency of REE removal to the sediments in this estuary were likely due to the high complexation of REEs from effluents with DOC, followed by flushing of these DOC complexes out of the estuary. This estuary has been shown to have high DOC concentrations coupled with short hydraulic residence times (Ravichandran et al., 1995a,b; Bianchi et al., 1996). Other work has shown that REEs are normally removed from the water column during estuarine mixing (Sholkovitz, 1993, 1995); however, under such high DOC conditions the transport mechanisms are altered.

Summary

1. Demands on our aquatic resources will increase in coming decades as it is projected that 75% of the world's population (6.3 billion) people will reside in coastal areas by 2025.
2. The introduction of invasive species worldwide has changed the community composition and physical structure of many ecosystems.
3. The effects of human encroachment on coastal systems has resulted in dramatic increases in the loadings of contaminants such as trace metals, HOCs (e.g., hydrocarbons and chlorinated hydrocarbons), and nutrients.
4. Polycyclic aromatic hydrocarbons are generated from anthropogenic or natural combustion processes, in addition to rapid transformation processes of biogenic precursors that occur in situ.
5. Some of the most common anthropogenic sources of heavy metals are mining, smelting, refining, electroplating, electric generating stations, automobile emissions, sewage sludge, dredged spoils, ash, and antifouling paints.
6. Past studies have shown a decrease in the toxicity of metals in the presence of organic matter, due to the binding of natural ligands that are believed to bind metals and reduce the concentration of free ionic species. The free ion activity model proposes that the free metal ion (as opposed to total metal ion concentration) is the dominant species available for organisms, and assumes that the colloidally complexed species should be less available.
7. Since colloidal material has both hydrophilic and lipophilic properties, making it permeable across membranes surfaces, lipid permeation may be an important mechanism for metal uptake.
8. A biotic ligand model (BLM) has been introduced to describe a characterized site where the metal reaches a critical concentration at the metal–ligand site, which in fish, would be located in the gills. In essence, the BLM is used to predict metal interactions at the biotic ligand site in the context of other competing reactions in aqueous environments.
9. When examining liquid–solid phase sorption interactions, the dissolved sorbing neutral HOC (NHOC) is called a "solute," and is referred to as a "sorbate" when in contact with the solid (sorbent) surface.
10. In the case of HOCs, equations are used to determine the partitioning coefficient between organic carbon and water (K_{oc}), based on aqueous solubility (S) or an octanol/water partition coefficient (K_{ow}).
11. Factors that limit the solubility of these compounds in solution are expected to increase partitioning onto particles; one such process is referred to as the

"salting out" effect. This effect occurs with increases in dissolved solids in solution resulting in changes in the solubility of NHOCs.

12. Sources of PAHs can be distinguished by using select ratios. For example, ratios of total methylphenanthrenes to phenanthrene is used to identify PAHs as pyrolytic or petrogenic sources. Other work has resorted to the application of compound-specific isotope analysis (CSIA), to constrain better anthropogenic sources of PAHs.

13. Eutrophication is defined by Cloern (2001, p. 224) as "the myriad of biogeo-chemical and ecological responses, either direct or indirect, to anthropogenic fertilization of ecosystems at the land–sea interface." As it stands today, nutrient inputs represent the largest problem in rivers and estuaries.

14. The expansion of anoxic and hypoxic zones has dramatically increased in recent years and the ecological consequences on pelagic and benthic species can be severe.

15. Many factors other than nutrients, such as weather conditions, algal species present, flushing residence time, presence and abundance of grazers, and differential success of different HABs species at a specific point in time, are important in determining if and when bloom conditions can occur.

16. Laminated sediment chronologies of organic and inorganic P, N, C, and bio-genic Si have been used to study long-term trends in eutrophication systems. Fossil pigments have also been used in paleolimnology and more recently estuaries, to understand better historical changes in bacterial community composition, trophic levels, redox change, and past UV radiation.

17. Another group of elements that have been used to reconstruct past changes in anthropogenic inputs in estuaries and other coastal systems are the rare Earth elements.

Part VII

Global Impact of Estuaries

Chapter 16

Estuarine–Coastal Interactions

Rivers, Estuaries, and the Coastal Ocean

The *coastal ocean* is a dynamic region where the rivers, estuaries, ocean, land, and the atmosphere interact (Walsh, 1988; Mantoura et al., 1991; Alongi, 1998; Wollast, 1998). Coastlines extend over an estimated 350,000 km worldwide, and the coastal ocean, typically defined as a region that extends from the high water mark to the shelf break (figure 16.1; Alongi, 1998), covers approximately 7% (26×10^6 km^2) of the surface global ocean (Gattuso et al., 1998). Although relatively small in area, this highly productive region (30% of the total net oceanic productivity) supports as much as 90% of the global fish catch (Holligan, 1992). In recent years, the coastal ocean has been recognized for its global importance with both national and international programs such as the Land–Ocean Interactions in the Coastal Zone (LOICZ) program, a subprogram of the International Global Change Program (IGBP) started in 1993 (Pernetta and Milliman, 1995), the European Union coastal core project (European Land–Ocean Interaction Studies, ELOISE) (Cadée et al., 1994), and in the U.S. Shelf Edge Exchange Processes Program (SEEP I and SEEP II) (Walsh et al., 1988; Anderson et al., 1994), the Coastal Ocean Processes (CoOP) program, Ocean Margins Program (OMP), and Land–Margin Ecosystem Research (LMER), to name a few. SEEP I and SEEP II were designed to test the Walsh et al. (1985) hypothesis that increased anthropogenic nutrient supply to the coastal ocean would result in enhanced burial of organic matter in continental margins due to higher offshore export of new productivity in the nearshore waters. While the hypothesis of offshore transport and burial was shown to be valid along certain regions of the eastern U.S. coast, other regions showed a more along-shelf transport (Walsh, 1994). More recent work in the OMP, which revisited some of the objectives of SEEP I and SEEP II, found that the accumulation of organic matter in upper slope sediments was only <1% of the total primary production in the entire continental margin of North Carolina (DeMaster et al., 2002). There are many factors that will ultimately determine if and how much nearshore production is exported offshore from the coastal ocean. Some margins receive their terrestrial inputs from estuarine line-sources (numerous estuaries) with very little direct effects from rivers,

495

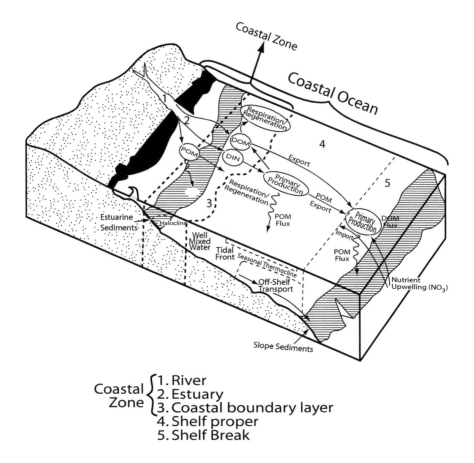

Coastal
Zone
{
1. River
2. Estuary
3. Coastal boundary layer
4. Shelf proper
5. Shelf Break

Figure 16.1 Generalized view of the idealized zones within the coastal ocean, in addition to some key biogeochemical processes that are essential in linking land and ocean. (Modified from Alongi, 1998.)

while others may receive large direct inputs from rivers, such as deltaic regions; these difference will have serious consequences on the amount of terrestrial material recycling that has occurred before entering the coastal zone, as well as how these particulate and dissolved materials will be transported offshore.

The world's 25 largest rivers transport approximately 40% of the fresh water and particulate materials entering the ocean (Milliman and Meade, 1983; Meade, 1996) (also see tables 4.2 and 4.3). Rivers transport an estimated 20 Pg y^{-1}(1 Pg = 10^{15} = 1 gigaton) of fluvial sediments to the coastal margin (Meybeck, 1982; Meade, 1996); associated with this sediment loading is an estimated 0.21 Pg of POC y^{-1} (Hedges and Keil,

1995; McKee et al., 2004). Global estimates of riverine flux of DOC to the oceans range from about 0.25 to 0.36 Pg y^{-1} (Meybeck, 1982; Degens et al., 1991; Aitkenhead and McDowell, 2000). In the case of P and N, about 91% of the 0.02 Pg y^{-1} of the P delivered to the ocean is in particulate form (Berner and Berner, 1996; Compton et al., 2000). Similarly about 50% of the total N is in particulate form (Meybeck, 1993); the percentages of N can vary considerably (40–86%) across different river systems (Mayer et al., 1998). Fluxes of dissolved inorganic nutrients (e.g., N and P) have increased worldwide by more than a factor of two (Meybeck, 1998). Conversely, other dissolved nutrients, such as dissolved Si (silicate), have decreased worldwide in the coastal ocean (Conley et al., 1993). These decreases have been attributed to higher removal of DSi, from enhanced diatom growth that has resulted from increases in N and P loading, coupled with greater flux of biogenic Si (via sedimentation) from the water column to the sediments (Billen et al., 1991; Ittekkot et al., 2000). Selective changes in the relative abundance of bioactive elements also results in changes in the Redfield ratio, which can in some cases cause significant shifts in the composition and abundance of coastal phytoplankton (Dortch and Whitledge, 1992; Turner and Rabalais, 1994; Nelson and Dortch, 1996). While the coastal ocean has been largely ignored in global carbon budgets, recent work has shown that shelf regions may actually be CO_2 sinks, recently referred to as the "continental shelf pump," that may account for as much as -0.95 Pg C y^{-1} (Tsunogai et al., 1999). Conversely, other work has shown that shelf-wide heterotrophy in marsh-dominated estuarine regions acts to enhance the source of CO_2 to the atmosphere (Cai et al., 2003; Wang and Cai, 2004). Despite these differences, the coastal ocean is nevertheless a dynamic region for the global transfer of C that is clearly in need of further study.

Estuaries serve as important interfaces between rivers and the coastal ocean where inputs of terrestrial materials can be significantly modified and recycled before entering the coastal zone. Unlike direct inputs from rivers in large deltaic regions, the hydrological forcing to the coastal zone in estuaries is significantly reduced and the residence time is higher, allowing for more recycling of materials (Wollast, 1983). For example, the estuarine turbidity maximum (ETM) (see chapter 6) can significantly enhance the residence time of riverine particles in estuaries before being delivered to the coast. In fact, large fractions of particulate materials from the continents can become trapped in estuarine regions never reaching the continental shelf. However, there is considerable variability in the export properties of estuaries across different regions. When examining the model-predicted latitudinal distribution of estuarine denitrification in the North Atlantic Basin, about 50% of the denitrification occurs in high-latitude estuaries (figure 16.2a; Seitzinger, 2000). Moreover, the actual export of N decreases in the high-latitude systems, because large rivers at tropical latitudes allow 70% of the TN to bypass the estuary (e.g., Amazon) (figure 16.2b). As discussed in chapters 10, 11, and 13, the fate of N, P, and C in high-latitude estuaries of the North Atlantic Basin is largely determined by the residence time in the system, which is considerably longer than that found in large river systems (Boynton et al., 1995; Nixon et al., 1995, 1996). This is particularly evident in large river systems, such as the Mississippi River, which has received high levels of anthropogenic nutrient loading, resulting in the occurrence of eutrophication and hypoxia events (e.g., Rabalais and Turner, 2001).

The role of estuaries in the exchange of CO_2 within the coastal zone has received considerable attention in recent years (Gattuso et al., 1998; Cai et al., 2004; Borges, 2005).

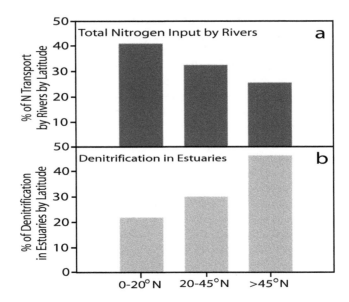

Figure 16.2 Model-predicted latitudinal distribution in the North Atlantic Basin: (a) estuarine denitrification; (b) total nitrogen transport by rivers. (Modified from Seitzinger, 2000.)

However, there remains considerable uncertainty in the role of estuarine systems, in part because of problems defining estuarine boundaries. As discussed in chapter 2, many deltaic estuaries have their boundaries extended beyond the coastline, where biogeochemical gradients in the water column become the actual boundaries of what is referred to as the *extended estuary*. Ketchum (1983, p. 3) first defined these outer reaches of estuaries as "plumes of freshened water which float on the more dense coastal sea water and they can be traced for many miles from the geographical mouth of the estuary." Inclusion of coastal ocean estimates of CO_2 uptake for high-latitude regions into global ocean estimates increases the total uptake by 57% (-1.93 compared to -1.56 Pg C y^{-1}) (Borges, 2005). While there are no available estimates of the outer estuarine contribution to the global CO_2 flux, inner estuarine regions contribute approximately 0.09 Pg C y^{-1} (Abril and Borges, 2004)—assuming a 50% contribution from remineralization of riverine POC (Abril et al., 2002). Lateral inputs of total organic carbon (TOC) and DIC from adjacent wetlands as well as anthropogenic sources have been shown to be more important in estuarine systems than in rivers (Cai and Wang, 1998; Neubauer and Anderson, 2003). For example, in tropical and subtropical estuaries export of C from mangroves contributes significantly to CO_2 emissions in the coastal ocean (Borges et al., 2005).

River-Dominated Ocean Margins

River-dominated ocean margins (RiOMars) are dynamic regions that receive inputs of organic carbon (OC) derived from both terrestrial and marine sources. The importance of

RiOMar to global OC burial (Hedges and Keil, 1995) is indicative of the tremendous magnitude of material fluxes in these river-dominated regions (Dagg et al., 2004; McKee et al., 2004) (see also the National Science Foundation RiOMar initiative: http://www.tulane.edu/~riomar). In these environments, the input of vascular plant-derived organic matter from land is significant (Hedges and Parker, 1976; Hedges and Ertel, 1982; Hedges, 1992), and marine primary productivity is high due to high nutrient inputs associated with riverine discharge (Lohrenz et al., 1990, 1994, 1997, 1999; Turner and Rabalais, 1991; Redalje et al., 1994; Hedges and Keil, 1995). While large rivers play an important role in the organic and terrestrial-marine linkage, 40 to 70% of global sediment input to the oceans is transported by smaller more abundant mountainous rivers (Milliman and Syvitski, 1992). When examining mass balance calculations of organic carbon it appears that much of this allochthonous input is not preserved in marine sediments (Hedges, 1992). Recent work has suggested that much of it may be remineralized at the coastal margin; however, the mechanisms remain largely unexplored (Aller, 1998). This lack of understanding results from the high degree of spatial and temporal variability in sources of OC at this coastal boundary (Sullivan et al., 2001), hydrodynamic sorting of OC sources (Bianchi et al., 2002b), and a wide spectrum of sedimentary regimes with differing rates of C remineralization (Aller, 1998; McKee et al., 2004).

As lithogenic and biogenic particles are delivered to the coastal zone (via rivers and estuaries), they are exposed to many dynamic processes (e.g., aggregation, flocculation, and desorption) that result in steep biogeochemical gradients on the continental shelf (Dagg et al., 2004). In these estuarine/river plume regions, temporal and spatial dynamics will vary considerably, depending on the freshwater forcing function—whether the freshwater source is directly from a river mouth (e.g., deltaic region) or more indirect via a line-source of smaller integrated inputs from a series of closely spaced estuaries. Seasonal changes in wind forcing, river discharge, temperature, and solar radiation are important physical controlling variables that affect plume dynamics. As particles settle out of the near-field plume, light availability will increase in the mid-salinity regions resulting in greater primary production in the far-field plume (Dagg et al., 2004). Enhanced light availability not only increases primary production, but also affects the usual trophic cascades through the microbial food, where bacterioplankton production is increased; resulting in increases in the release of dissolved inorganic nutrients and DOM. Seasonal changes in river discharge can shift where the peak regions of productivity occur in plumes. For example, in March 1991 during high discharge, the peak in primary production in the Mississippi River plume was pushed offshore and occurred at higher salinities than during lower discharge periods such as July 1990 (figure 16.3; Lohrenz et al., 1999; Dagg et al., 2004). Higher light availability also results in greater photochemical processing of DOM in the mid-salinity regions of plumes (Kieber et al., 1990; Miller and Zepp, 1995; Benner and Opsahl, 2001). As discussed in chapter 8, inputs of riverine CDOM (chromophoric dissolved organic matter) in plume regions are significantly altered by dilution with coastal waters as well as by photochemical bleaching.

The high vertical flux of particulates in river/estuarine plume regions commonly results in the accumulation of particles in the formation of a *benthic boundary layer* (BBL) and/or *mobile* and *fluid muds* (see chapter 6 for more details). The BBL is defined by Boudreau and Jørgensen (2001, p. 1) as "those portions of sediment and water columns that are affected directly in the distribution of their properties and processes by the presence of

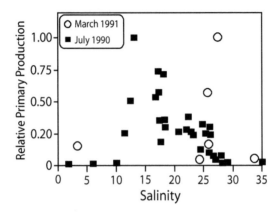

Figure 16.3 Relationship between primary production and salinity in the Mississippi River plume in July 1990 and March 1991. (Modified from Lohrenz et al., 1999.)

the sediment–water interface." Mobile muds, commonly occur in RiOMars, within the lower river estuary (salt-wedge region), and on the adjacent shelf, and are regions where diagenetic transformation processes are enhanced (Aller, 1998; Chen et al., 2003a,b; McKee et al., 2004). The occurrence of fluid muds may also allow for increased transport of materials across the continental margin. For example, fluid muds were shown to be important in transporting river-derived organic materials from Atchafalaya Bay estuary to the Louisiana shelf at certain times of the year (Allison et al., 2001; Gordon et al., 2001; Gordon and Goñi, 2003).

Deltaic regions contain about 50% of all the OC buried in coastal margins (Berner, 1989; Hedges and Keil, 1995). Depending on the amount of anthropogenic loading of nutrients as well as the geology of the watershed, there is a diversity of organic matter sources in these regions, usually containing a mixture of riverine-derived terrestrial materials, as well as more labile new primary production sources (Blair et al., 2003, 2004). It has been estimated that while 0.10 Pg of ancient OC (e.g., *kerogen*) may be oxidized each year in the coastal margin (Hedges, 1992; Berner, 2004), a comparable amount (0.08 Pg y^{-1}) may escape oxidation (Meybeck, 1993), allowing for further transport into the global ocean. Recent estimates suggest that 0.04 Pg y^{-1} of ancient OC may be transported to the global ocean (Blair et al., 2003). As discussed in chapters 8 and 9, organic matter preservation in coastal margins is largely determined by productivity, bottom-water O_2 levels, sediment accumulation rates, organic matter quality, and mixing rates (Hedges and Keil, 1995). Similarly, the role of organic matter protection by association with mineral matrices (Mayer, 1994a,b; Keil et al., 1997) has also been shown to be important in controlling the burial efficiency of OC in coastal margins. Conversely, as some of these terrestrially derived materials on sediments are removed through oxidation, sorption of more labile marine OC to clay fractions occurs (Blair et al., 2003, 2004).

The replacement of terrestrial OC by marine OC in coastal margins can vary considerably among regions. For example, comparisons between the Eel River (USA) and Amazon River nicely illustrate the relative importance of storage and transit time on the composition of POC delivered to the continental margin (Blair et al., 2004). In the Eel River system, there is mass wasting in the watershed that delivers bedrock and vascular vegetation (as indicated by the δ^{13}C depleted values) from soils, with minimal transformation during transport to the coast (figure 16.4a; Blair et al., 2003). These materials are rapidly buried with evidence of flood deposits; also, the age of the OC in offshore sediments increases with and reflects inputs of ancient OC in the form of kerogen. This rapid burial of ancient OC offshore with minimal transformation is typical of what is expected from short rivers on active margins, where inputs throughout the years are "flashy" or temporally extreme. On the other hand, the Amazon, which has a much larger watershed, with extensive storage and processing time of OC in lowland soils, allows for enough time

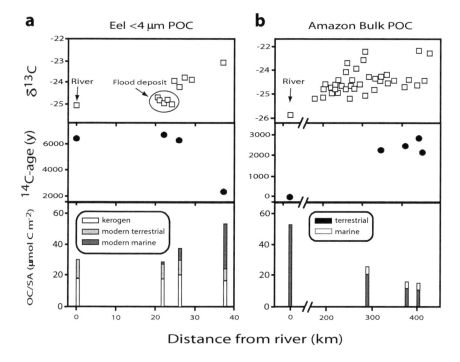

Figure 16.4 Comparison of the character [δ^{13}C, ^{14}C age, and organic carbon (OC)/surface area ratio] of POC collected in the (a) Eel (USA) and (b) Amazon rivers. In the Eel, sorption of marine OC on riverine particles, which have lost very little kerogen as they enter the coast, account for the decrease in age of particles with increasing distance from shore. Conversely, as particles selectively lose "young" labile terrestrial OC on the coast there is an increase in the age of these particles with increasing distance, since they are not completely reloaded with marine OC.

for the OC signature to be altered before being deposited and buried on the continental margin (figure 16.4b). The age of this material is also not as old, because it is derived from surface soils and marine OC. In fact, the age of the OC decreases moving offshore due to replacement of terrestrial OC by sorption of marine OC; this is also supported by the enrichment of $\delta^{13}C$ in the offshore direction. Thus, the Amazon estuarine system has a longer residence time, allowing for more alteration of the terrestrially derived OC than in the Eel River, even though both systems are considered to be RiOMars.

Groundwater Inputs to the Coastal Ocean

Submarine groundwater discharge (SGD) has received considerable attention in recent years because of its potential importance in the flux of materials to the coastal ocean (Bokuniewicz et al., 2003; Burnett et al., 2003; Moore, 2003; McKenna and Martin, 2004). As discussed in chapter 3, Burnett et al. (2003, p. 6) recently defined SGD as "any and all flow of water on continental margins from the seabed to the coastal ocean, regardless of fluid composition or driving force" (figure 16.5). As shown here, the key driving forces for SGD are convection, hydraulic head, tidal pumping, and wave set-up. The flow of freshwater across the sea floor can be described as SGD, which is the flow discharge across the sea floor, and a recharging flow across the sea floor, which is referred to as submarine groundwater recharge (SGR); the net discharge is the difference between

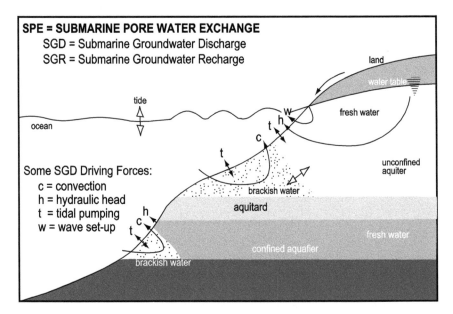

Figure 16.5 Illustration of fluid transport processes associated with submarine groundwater discharge and recharge. Arrows denote fluid movement (Modified from Burnett et al., 2003).

SGD and SGR (figure 16.5). The water displaced by these processes in the sediments is referred to as the submarine porewater exchange (SPE). A combination of nutrient-rich porewaters mixed with SGD from bioirrigation, wave and tidal pumping, and convection have been shown to enhance the effects of SGD in estuaries (Martin et al., 2004). In fact, porewater advective transport represented as much as 90% of the total SGD in the Indian River Lagoon (USA) (Cable et al., 2004). Assuming a mean river flow of 37,500 km^3 y^{-1}, SGD represents approximately 0.3 to 16% of the global river discharge (Burnett et al., 2003). Many of the estimates of terrestrially derived fresh SGD are in the range of 6 to 10% of the surface water inputs. Using benthic chambers, estimates of SGD into the Great South Bay (USA) were estimated to be 5 cm d^{-1} and 1.5 cm d^{-1} near the shoreline and 100 m from the shore, respectively (Bokuniewicz et al., 2004).

One of the early benchmark papers that recognized the potential importance of groundwater inputs to the coastal ocean was that of Johannes (1980), followed by the important work of Valiela and D'Elia (1990). Both of these papers reflected the early belief that the importance of groundwater inputs remained speculative and that documenting such inputs would be difficult. Since that time, documenting inputs of SGD has advanced considerably, with inputs being distinguished by water quality parameters such as color, temperature, salinity, and other geochemical signatures (Burnett et al., 2003). Anomalies in salinity have been widely used in many regions to document SGD inputs (e.g., Valiela et al., 1990; Matciak et al., 2001). Other chemical tracers used to identify SGD have included concentration gradients of CH_4, H_2S, dissolved Si, and CO_2 (e.g., Cable et al., 1996; Moore, 2003; Bokuniewicz et al., 2003). As described in chapter 3, radiochemical tracers have also been used; for example, using excess ^{222}Rn as a groundwater tracer in the Delaware Estuary, Schwartz (2003) calculated a groundwater flux of 14.5 to 29.3 m^3 s^{-1}, equivalent to the discharge of the Schuylkill and Brandywine Rivers, the second and third largest tributaries to the Delaware River and Bay. This discharge occurred along 12 km of the estuary and is equal to an upward flow velocity of 5 to 10 cm d^{-1}. Another region that has received considerable attention using Ra to trace SGD inputs is along the Florida shoreline of Apalachee Bay, where linkages between coastal inputs and the Floridian aquifer, located just a few meters below the surface, have been documented (Bugna et al., 1996; Cable et al., 1996, 1997; Rutkowski et al., 1999). Along the South Carolina and Georgia coasts, N and P inputs to offshore waters, from groundwater fluxes and salt marshes, were found to be higher than riverine inputs for that region (Simmons, 1992; Krest et al., 2000). Coastal aquifers where SGD can actually bypass the extensive recycling of terrestrially derived materials that commonly occur in estuaries have been referred to as "subterranean estuaries" (Moore, 1999). Examples of a select group of estuarine systems where fluxes of DIN and DIP from SGD have been measured are shown in table 16.1.

Some of the biological responses to SGD inputs of nutrients to the coastal margin include eutrophication, the occurrence of HABs, and macroalgal invasions. For example, macroalgae such as *Cladophora* spp., *Ulva* spp., *Entermorpha* spp., and *Gracilaria* spp., which have been generally used as indicators of anthropogenic inputs (e.g., sewage) (Lapointe and Matzie, 1996; Hadjichristophorou et al., 1997). In particular, the Florida Keys have experienced macroalgal expansion that has been linked to nutrients in SGD (Lapointe et al., 1990). The occurrence of "brown tides" and displacement of seagrass beds, due to advancement of macroalgal mats, have also been attributed to SGD inputs in

Table 16.1 Nutrient fluxes from submarine groundwater discharge.

River/estuary	DIN flux (mmol $m^{-2} y^{-1}$)	DIP flux (mmol $m^{-2} y^{-1}$)	Reference
Eastern Florida Bay	110 ± 60	0.21	Corbett et al. (1999)
Waquoit Bay	61.0^a		Valiela et al. (1992)
Pettaqumascutt River	61–80	4.4 ± 13	Kelly and Moran (2002)
Elizabeth River	$1.64 \times 10^3 \pm$ 1.68×10^3	58 ± 62	Charette and Buessler (2004)

[a]Nitrate only.

estuaries and along coastal margins (Valiela et al., 1990; LaRoche et al., 1997). Fin and shellfish kills in several New England estuaries have also been attributed to eutrophication caused by SGD (Valiela and D'Elia, 1990). To understand better the global impact of SGD on the coastal ocean, future studies should focus on key environmental signals for SGD inputs (e.g., temperature and precipitation), in target regions known to receive high inputs of SGD (e.g., deltas, karst, and coastal plains) (Burnett et al., 2003).

Summary

1. The coastal ocean is a dynamic region where the rivers, estuaries, ocean, land, and the atmosphere interact. Coastlines extend over an estimated 350,000 km worldwide, and the coastal ocean is typically defined as a region that extends from the high water mark to the shelf break.
2. Some margins receive their terrestrial inputs from estuarine line-sources (numerous estuaries) with very little direct effects from rivers, while others may receive large direct inputs from rivers, such as deltaic regions; these differences will have serious consequences on the amount of terrestrial material recycling that has occurred before entering the coastal zone, as well as how these materials (particulate and dissolved) will be transported offshore.
3. Rivers transport an estimated 20 Pg y^{-1} (1 Pg = 10^{15} = 1 gigaton) of fluvial sediments to the coastal margin; associated with this sediment loading is an estimated 0.21 Pg of POC y^{-1}. Global estimates of riverine flux of DOC to the oceans range from about 0.25 to 0.36 Pg y^{-1}.
4. While the coastal ocean has been largely ignored in global carbon budgets, recent work has shown that shelf regions may actually be CO_2 sinks recently referred to as the "continental shelf pump," that may account for as much as −0.95 Pg C y^{-1}.
5. Selective changes in the relative abundance of bioactive elements delivered by rivers/estuaries also results in changes in the Redfield ratio, which can in some cases cause significant shifts in the composition and abundance of coastal phytoplankton.

6. Export of N decreases in the high-latitude estuarine systems, with 70% of the TN in tropical latitudes bypassing estuaries occurring mostly in large river systems (e.g., Amazon).

7. Inclusion of coastal ocean estimates of CO_2 uptake for high-latitude regions into global ocean estimate increases the total uptake by 57% (-1.93 compared to -1.56 Pg C y^{-1}).

8. In RiOMar environments, the input of vascular plant-derived organic matter from land is significant, and marine primary productivity is high due to high nutrient inputs associated with riverine discharge.

9. While large rivers play an important role in this terrestrial–marine linkage, 40 to 70% of global sediment input to the oceans is transported by smaller more abundant mountainous rivers.

10. As lithogenic and biogenic particles are delivered to the coastal zone (via rivers and estuaries), they are exposed to many dynamic processes (e.g., aggregation, flocculation, desorption) that result in steep biogeochemical gradients on the continental shelf.

11. Recent estimates suggest that 0.04 Pg y^{-1} of ancient OC may be transported to the global ocean.

12. Assuming a mean river flow of 37,500 km^3 y^{-1}, SGD represents approximately 0.3 to 16% of the global river discharge.

13. Coastal aquifers where SGD can actually bypass the extensive recycling of terrestrially derived materials that commonly occur in estuaries have been referred to as "subterranean estuaries."

14. Some of the biological responses to SGD inputs of nutrients to the coastal margin include eutrophication, the occurrence of HABs, and macroalgal invasions.

Appendices

Appendix 1 Atomic weights of elements.

Atomic number	Name	Symbol	Atomic weight
1	Hydrogen	H	1.0079
2	Helium	He	4.0026
3	Lithium	Li	6.9410
4	Beryllium	Be	9.0122
5	Boron	B	10.811
6	Carbon	C	12.011
7	Nitrogen	N	14.007
8	Oxygen	O	15.999
9	Fluorine	F	18.998
10	Neon	Ne	20.180
11	Sodium	Na	22.990
12	Magnesium	Mg	24.305
13	Aluminum	Al	26.982
14	Silicon	Si	28.086
15	Phosphorus	P	30.974
16	Sulfur	S	32.066
17	Chlorine	Cl	35.453
18	Argon	Ar	39.948
19	Potassium	K	39.098
20	Calcium	Ca	40.078
21	Scandium	Sc	44.956
22	Titanium	Ti	47.956
23	Vanadium	V	50.942
24	Chromium	Cr	51.996
25	Manganese	Mn	54.938
26	Iron	Fe	55.845
27	Cobalt	Co	58.933
28	Nickel	Ni	58.693
29	Copper	Cu	63.546
30	Zinc	Zn	65.392
31	Gallium	Ga	69.723
32	Germanium	Ge	72.612
33	Arsenic	As	74.922
34	Selenium	Se	78.963

Continued

Appendix 1 (*continued*)

Atomic number	Name	Symbol	Atomic weight
35	Bromine	Br	79.904
36	Krypton	Kr	83.800
37	Rubidium	Rb	85.468
38	Strontium	Sr	87.520
39	Yttrium	Y	88.906
40	Zirconium	Zr	91.224
41	Niobium	Nb	92.906
42	Molybdenum	Mo	95.940
43	Technetium	Te	98.906
44	Ruthenium	Ru	101.07
45	Rhodium	Rh	102.91
46	Palladium	Pd	106.42
47	Silver	Ag	107.87
48	Cadmium	Cd	112.41
49	Indium	In	114.82
50	Tin	Sn	118.71
51	Antimony	Sb	121.76
52	Tellurium	Te	127.60
53	Iodine	I	126.90
54	Xenon	Xe	131.29
55	Cesium	Cs	132.91
56	Barium	Ba	137.33
57	Lanthanum	La	138.91
58	Cerium	Ce	140.12
59	Praseodymium	Pr	140.91
60	Neodymium	Nd	144.24
61	Promethium	Pm	146.92
62	Samarium	Sm	150.36
63	Europium	Eu	151.96
64	Gadolinium	Gd	157.25
65	Terbium	Tb	158.93
66	Dysprosium	Dy	162.50
67	Holmium	Ho	164.93
68	Erbium	Er	167.26
69	Thulium	Tm	168.93
70	Ytterbium	Yb	173.04
71	Lutetium	Lu	174.97
72	Hafnium	Hf	178.49
73	Tantalum	Ta	180.95
74	Tungsten	W	183.84
75	Rhenium	Re	186.21
76	Osmium	Os	190.23
77	Iridium	Ir	192.22

Continued

Appendix 1 (*continued*)

Atomic number	Name	Symbol	Atomic weight
78	Platinum	Pt	195.08
79	Gold	Au	196.97
80	Mercury	Hg	200.59
81	Thallium	Tl	204.38
82	Lead	Pb	207.20
83	Bismuth	Bi	208.98
84	Polonium	Po	209.98
85	Astatine	At	209.99
86	Radon	Rn	222.02
87	Francium	Fr	223.02
88	Radium	Ra	226.03
89	Actinium	Ac	227.03
90	Thorium	Th	232.04
91	Protactinium	Pa	231.04
92	Uranium	U	238.03
93	Neptunium	Np	237.05
94	Plutonium	Pu	239.05
95	Americium	Am	241.06
96	Curium	Cm	244.06
97	Berkelium	Bk	249.08
98	Californium	Cf	252.08
99	Einsteinium	Es	252.08
100	Fermium	Fm	257.10
101	Mendelevium	Md	258.10
102	Nobelium	No	259.10
103	Lawrencium	Lr	262.11

Appendix 2 SI units and conversion factors.

SI prefix units	
atto (a)	$= 10^{-18}$
femto (f)	$= 10^{-15}$
pico (p)	$= 10^{-12}$
nano (n)	$= 10^{-9}$
micro (μ)	$= 10^{-6}$
milli (m)	$= 10^{-3}$
centi (c)	$= 10^{-2}$
deci (d)	$= 10^{-1}$
deca (da)	$= 10^{1}$
hecto (h)	$= 10^{2}$
kilo (k)	$= 10^{3}$
mega (M)	$= 10^{6}$
giga (G)	$= 10^{9}$
tera (T)	$= 10^{12}$
peta (P)	$= 10^{15}$
exa (E)	$= 10^{18}$

Conversion factors

Force
1 newton (N) $= \text{kg m s}^{-2}$
1 dyne (dyn) $= 10^{-5} \text{N}$

Pressure
1 pascal (Pa) $= \text{kg m}^{-1} \text{s}^{-2}$
1 torr (torr) $= 133.32$ Pa
1 atmosphere (atm) $= 760$ torr
                 $= 12.5$ psi

Temperature
$^{\circ}\text{C} = 5/9(^{\circ}\text{F} - 32)$
$\text{K} = 273.15 + \,^{\circ}\text{C}$

Energy
1 kcal $= 1000$ cal
1 cal $= 4.184$ joules (J)

Speed
1 knot $= 1$ nautical mile h^{-1}
             $= 1.15$ statute miles h^{-1}
             $= 1.85$ kilometers h^{-1}
Velocity of sound in water (salinity $= 35$) $= 1507 \text{ m s}^{-1}$

Continued

Appendix 2 (*continued*)

Conversion factors

Volume

1 cubic kilometer (km^3)	$= 10^9$ m^3
	$= 10^{15}$ cm^3
1 cubic meter (m^3)	$= 1000$ liters
1 liter (L)	$= 1000$ cm^3
	$= 1.06$ liquid quarts
1 milliliter (ml)	$= 0.001$ liter
	$= 1$ cm^3

Length

1 kilometer (km)	$= 10^3$ m
1 centimeter (cm)	$= 10^{-2}$ m
1 millimeter (mm)	$= 10^{-3}$ m
1 micrometer (μm)	$= 10^{-6}$ m
1 nanometer (nm)	$= 10^{-9}$ m
Angstrom (Å)	$= 10^{-10}$ m
1 fathom	$= 6$ feet
	$= 1.83$ m
1 statute mile	$= 5280$ feet
	$= 1.6$ km
	$= 0.87$ nautical mile
1 nautical mile	$= 6076$ feet
	$= 1.85$ km
	$= 1.15$ statute miles

Mass

1 kilogram (kg)	$= 1000$ g
1 milligram (mg)	$= 0.001$ g
1 ton (tonne) (t)	$= 1$ metric ton
	$= 10^6$ g

Area

1 square centimeter (cm^2)	$= 0.155$ square inch
	$= 100$ square millimeters
1 square meter (m^2)	$= 10.8$ square feet
1 square kilometer (km^2)	$= 0.386$ square statute miles
	$= 0.292$ square nautical miles
	$= 10^6$ square meters
	$= 247.1$ acres
1 hectare	$= 10,000$ square miles

Appendix 3 Physical and chemical constants.

Avogadro's number (N)	$= 6.022137 \times 10^{23}$ mol^{-1}
Boltzmann's constant (k)	$= 1.380658 \times 10^{-23}$ J K^{-1}
Faraday's constant (F)	$= 9.6485309 \times 10^4$ C mol^{-1}
Gas constant (R)	$= 8.3145$ J mol^{-1} K^{-1}
Planck's constant (h)	$= 6.6260755 \times 10^{-34}$ J s^{-1}

Appendix 4 Geologic timetable.

	Era	Approximate age (millions of y ago (Ma))	Subdivisions	Approximate duration (millions of y)
PHANEROZOIC	QUATERNARY	10,000 years	HOLOCENE	10,000 years
		2	PLEISTOCENE	2
	TERTIARY	5	PLIOCENE	3
		24	MIOCENE	19
		37	OLIGOCENE	13
		57	EOCENE	20
		66	PALEOCENE	9
	MESOZOIC	144	CRETACEOUS	78
		208	JURASSIC	64
		245	TRIASSIC	37
	PALEOZOIC	286	PERMIAN	41
		360	CARBONIFEROUS	74
		408	DEVONIAN	48
		438	SILURIAN	30
		505	ORDOVICIAN	67
		545	CAMBRIAN	65
CRYPTOZOIC	PROTEROZOIC	2500		
	ARCHAEAN	3800	Oldest rock dated 3800 Ma	
	HADEAN	4600	Age of the Earth 4600 Ma	

Glossary

Abandonment   destructional phase of deltaic development caused by avulsion (channel switching within the delta plain) of the distributory channel.

Abiotic condensation model   states that small molecules can repolymerize outside the cell through condensation reactions over time into larger more refractory macro-molecules.

Absolute temperature   temperature measured on a scale with absolute zero as 0. This is conventionally measured in kelvin, where absolute zero corresponds to $0\,K$ ($-273.15°C$ or $-459.67°F$).

Accretion   slow addition to land by deposition of water-borne sediment and/or an increase of land along the shores of a body of water, as by alluvial deposit.

Accuracy   the quality of nearness to the truth or the true value.

Acid neutralizing capacity   the equivalent sum of all bases or base-producing materials, solutes plus particulates, in an aqueous system that can be titrated with acid to an equivalence point.

Acidolysis   the process of reacting an acid with an ester or an ester exchange.

Acid volatile sulfides   compounds evolved via acid distillation and generally include amorphous forms [e.g., mackinawite (FeS), greigite (Fe_3S_4), and pyrrhotite (FeS)] and pyrite (FeS_2).

Activation energy   the minimum amount of energy required to initiate a reaction.

Active margin   a margin consisting of a continental shelf, a continental slope, and an oceanic trench.

Activity coefficient   the ratio of the activity (of an electrolyte) as measured by some property, such as the depression of the freezing point of a solution, to the true concen-tration (molality). It is usually less than 1 and increases as the solution becomes more dilute, when the attractive forces between oppositely charged ions become negligible.

Adsorption   the attachment of molecules to a surface by attractions of the strength of reactants and products.

Advection   the transport of a property by organized flow whereby a gradient exists in the property with a component of flow across the gradient.

Aerenchymatous tissues common in aquatic vascular plants which characteristically have large air spaces.

Aerosols condensed phases of solid or liquid particles, suspended in state, that have stability to gravitational separation over periods of observation.

Agglomerates organic and inorganic mater bound by weak surface tension forces.

Aggregates inorganic particles bound by strong inter or intramolecular forces.

Albedo the fraction of light reflected by a body or surface.

Aldehydes any of a class of highly reactive organic chemical compounds obtained by oxidation of primary alcohols, characterized by the common group CHO.

Aldoses carbohydrates with an aldehyde in the open-chain form.

Algaenans highly aliphatic, insoluble, nonhydrolyzable compounds found in the cell walls of certain microalgae.

Aliphatic hydrocarbons compounds that include branched alkanes/alkenes, n-alkanes, and n-alkenes.

Alkalinity the concentration of negative charge that will react with H^+.

Alkyl polycyclic aromatic hydrocarbons PAHs such as retene, pimanthrene, and cadalene, used to trace higher plant-derived materials.

Allochthonous produced outside the system.

Alluvial feeder a valley within the drainage basin that supplies the water and sediment to the delta.

Alpha decay loss of an alpha particle (nucleus of a helium-4 atom) from the nucleus which results in a decrease in the atomic number by two (two protons) and the mass number by four units (two protons and two neutrons).

Alpha ray a stream of alpha particles or a single high-speed alpha particle.

α-Amino acid compounds consisting of an amino group, a carboxyl group, a H atom, and a distinctive R group (side chain) that is bonded to a C atom—this C atom is the α-carbon because it is adjacent to the carboxylic group.

Ammonia assimilation where NH_3 or NH_4^+ is taken up and incorporated into organisms as organic nitrogen molecules.

Ammonification process where NH_3 or NH_4^+ is produced during the breakdown of organic nitrogen by organisms.

Ammonium oxidation (anammox) anaerobic NH_4^+ oxidation with NO_3^-; this may be another mechanism whereby N is lost from natural systems.

Amorphous lacking structure and noncrystalline in nature.

Amorphous Fe noncrystalline Fe that is ascorbate leachable (ASC-Fe) (e.g., greigite).

Amphiphilic a molecule having a polar, water-soluble group attached to a nonpolar, water-insoluble hydrocarbon chain.

Amphoteric compounds that contain both basic (e.g., -NH_2) and acidic (e.g., -COOH) groups, such as amino acids.

Anabolic the phase of metabolism in which simple substances are synthesized into the complex materials of living tissue.

Anodic and cathodic stripping voltammetry electrochemical techniques, combined with metal titration analyses, for obtaining ligand concentrations and conditional stability constants, can be used to determine how many ligands are involved in metal complexation as well as their relative differences in strength.

Anthigenic minerals   minerals formed from within a sedimentary deposit, and not carried into it.

Anthropocene era   the replacement of natural by anthropogenic drivers of ecosystems over the past 50 or so years.

Anthropogenic   created by people or resulting from human activities.

Archaebacteria   the third most recently recognized primary kingdom and are characterized as living in extreme environments, such as anaerobic methanogens and halophilic bacteria.

Arrhenius equation   describes the relationship between the rate constant for a chemical reaction and the temperature at which the reaction occurs.

Assimilation   the uptake and incorporation of organic matter.

Assimilatory NO_3^- reduction to NH_4^+ (ANRA)   simultaneous reduction of NO_3^- and uptake of N by an organism into biomass.

Atom   the smallest component of an element having the chemical properties of the element.

Atomic number   number of protons (Z) in a nucleus.

Atomophiles   the material of the gaseous atmosphere of the Earth (e.g., O, H) and is the lightest phase in the separation of elements based on chemical affinities.

Authigenic minerals   minerals formed in situ such as carbonate fluoroapatite and pyrite.

Autochthonous   produced within a system.

Autotrophs   organisms (e.g., vascular plants and algae) that make their own food for energy.

Avulsion   the abandonment of a part or the whole of a channel belt in a deltaic system by a stream in favor of a new course.

Bacterioplankton   the bacterial fraction of the plankton that drifts in the water column.

Bar-built estuaries   inland shallow bodies of water that usually run parallel to the coast and are isolated from the sea by a barrier island where one or more small inlets allow for connection with the ocean—also called coastal lagoon estuaries.

Baroclinic   atmospheric condition when isobaric and constant-density surfaces are not parallel; in a baroclinically stratified fluid total potential energy can be converted into kinetic energy.

Barotropic   a condition in a water mass whereby the surfaces of constant pressure are parallel to the surfaces of constant density; in a state of barotropic stratification, no potential energy is available for conversion into kinetic energy.

Baseflow   the continual contribution of groundwater to rivers and is an important source of flow between rainstorms.

Bedload transport   body of coarse particles that moves along the bottom of a stream.

Benthic boundary layer   those portions of sediment and water columns that are affected directly in the distribution of their properties and processes by the presence of the sediment–water interface.

Benthic microlayer   actively photosynthesizing microphytobenthos which are also important in generating oxygen to the water column.

Beta (negatron) decay   occurs when a neutron changes into a proton and a negatron (negatively charged electron) is emitted, thereby increasing the atomic number by one unit.

Beta ray   a stream of beta particles.

Bioaccumulation   uptake and concentration of environmental chemicals by living systems.

Biogeochemical budget   essentially a "checks and balances" of all the sources and sinks as they relate to the material turnover in reservoirs.

Biofilms   defined as surface layers of microbes entrained in a matrix of extracellular polymeric substances (EPS)—same as microbial mats.

Bioflocculation   aggregation by microorganisms.

Biological N_2 fixation   a process performed by prokaryotic (both heterotrophic and phototrophic) organisms (sometimes occurring symbiotically with other organisms) resulting in the enzyme-catalyzed reduction of N_2 to NH_3 or NH_4^+, or organic nitrogen compounds.

Biomagnification   the sequence of processes in an ecosystem by which higher contaminant concentrations are attained in organisms of higher trophic level.

Biomarker molecule   molecules that are indicative of source materials as well as processes of material cycling.

Biopolymer degradation model   states that newly released bioploymers from organic matter will eventually be broken down into more labile smaller molecules.

Biotic ligand model (BLM)   model used to describe a characterized site where the metal reaches a critical concentration at the metal–ligand site,

Bioturbation   the effects of organisms on the chemical and physical properties of sediments.

Bitumen   brown or black viscous residue from the vacuum distillation of crude petroleum—includes asphalt and tar.

Blackwaters   aquatic waters that contain high concentrations of colored dissolved organic matter (e.g., tannins).

Branched fatty acids   iso- and anteiso fatty acids believed to be primarily derived from sulfate-reducing bacteria; the iso- and antesio designation represents a branched fatty acid with the methyl group at the ω-1 position and the methyl group at the ω-2 position, respectively.

Branched hydrocarbons   organic compounds that contain only carbon and hydrogen in a branched configuration.

Brownian motion   random motion of small particles from thermal effects.

Carbohydrate   the general formula for carbohydrate composition is $(CH_2O)_n$; these compounds can be defined more specifically as polyhydroxyl aldehydes and ketones.

Carbonate alkalinity   alkalinity based on the presence of the carbonate ion (CO_3^{2-}).

Carotenoids   are an important group of tetraterpenoids consisting of a C_{40} chain with conjugated bonds that are found in bacteria, algae, fungi, and higher plants, and serve as the principal pigment biomarkers for class-specific taxonomic work in aquatic systems.

Catabolic   a series of reactions in which complex molecules are broken down into simpler ones.

Catagenesis   the alteration of organic matter that begins at temperatures between 60° and 100°C.

Catchment basin   the geographical area draining into a river or reservoir.

Cellulose   a highly abundant polysaccharide found in plant cell walls that is composed of long linear chains of glucose units connected as $1,4'$-β-glycoside linkages.

Chalcophiles   elements that tend to combine with sulfur.

Chelate   the entire chelating complex.

Chelating agent   ligands that form strong mulitdentate complexes [e.g., ethylenediamimetetraacetic acid (EDTA)].

Chelation   the combination of a metallic ion with heterocyclic ring structures whereby the ion is held by bonds from each of the rings.

Chemical shift   differences in field strengths are caused by the shielding or deshielding of electrons associated with adjacent atoms.

Chemical weathering   chemical alteration of minerals that is enhanced as the surface area of exposed rocks increases.

Chemoautotrophs   organism that obtains its energy from the oxidation of chemical compounds and only uses organic compounds as a source of carbon.

Chemolithotrophs   organism capable of using CO_2 or carbonates as the sole source of carbon for cell biosynthesis, and deriving energy from the oxidation of reduced inorganic or organic compounds.

Chemosynthesis   the release of chemical energy from the oxidation of reduced substrates.

Chiral   having different left-handed and right-handed forms which are not mirror symmetric.

Chlorinity   the mass in grams of halides (expressed as chloride ions) that can be precipitated from 1000 g of seawater by Ag^+.

Chlorophylls   compounds consisting of cyclic tetrapyrroles coordinated around a Mg-chelated complex, and in the case of chlorophyll-a, represents a dominant pigment used to estimate algal biomass in aquatic ecosystems.

Choked   systems that have only one long narrow inlet.

Chromophoric dissolved organic matter   organic compounds such as humic and fluvic acids that form from decomposing organic matter, and absorb light at the blue end of the spectrum giving water a yellowish–brownish color at high concentrations.

Citric acid cycle   series of chemical reactions that are critical in all living cells that use oxygen as part of cellular respiration; also known as the tricarboxylic acid cycle or the Krebs cycle.

Coastal lagoon   inland body of water that runs parallel to the coast and is isolated from the sea by a barrier island where one or more small inlets allow for connection with the ocean.

Coastal ocean   a dynamic region where the rivers, estuaries, ocean, land, and the atmosphere interact.

Coastal plain estuary   formed by flooding of river valleys in the Pleistocene–Holocene during the last interglacial transgression, and occurs on low-relief coasts due to high sedimentation and in-filling of a river valley

Cogener   chemical substances composed of the same elements in the same proportions but having different properties because of different structures.

Cohesion   the result of interparticle surface attraction between clay minerals as well as the binding forces from organic mucous layers.

Colloids   particles in the size range of 1 nm to 1 μm.

Complexes   ions held by covalent bonds.

Conservative elements   changes in the concentrations of these elements reflect the addition or loss of water through physical processes.

Consortia   mixed-species communities.

Consumer   an organism that acquires its energy demands by eating other organisms.

Contact ion pairs   ions linked electrostatically with no covalent bonds.

Convective throughflow ventilation   biogas (e.g., CH_4) transport that occurs via lacunal pressurization.

Coordination number   number of bonds formed by the central atom in a metal–ligand complex.

Cosmogenic   formed in space, or in the atmosphere as it relates to certain nuclide production.

Coulombic   interaction between charged particles in close proximity.

Crevasse   breach in a levee on the bank of a river through which floodwater may flow.

Crystalline Fe   stable crystalline forms of Fe found in minerals such as goethite.

Cutins   lipid polyester polymers in vascular plant tissues which serve as a protective layer cuticle.

Darcy's law   the laminar flow of fluid through a porous medium.

Dalton's law of partial pressure   law which states that the total atmospheric pressure (P_t) is the sum of all partial pressures (P_i) exerted by each of the gases in the entire mixture of air.

Deamination   removal of an amine functional group from a molecule.

Decomposition decay phase   the heterotrophic breakdown of detritus by microbes and metazoans (e.g., detritivores and deposit feeders).

Delta   coastal accumulations, both subaqueous and subaerial, of river-derived sediments adjacent to, or in close proximity to, the source stream, including the deposits that have been secondarily molded by various marine agents, such as waves, currents, or tides.

Delta front   where the seaward edge of the delta plain merges with the subtidal or subdelta region of the delta.

Delta-front estuary   systems found in sections of deltas that are affected by tidal action and/or salt intrusion.

Delta plain   composed of an upper delta plain which is an older section of the delta that is not currently affected by tidal processes and a lower delta plain, composed of subaerial and intertidal zones dominated by channels and tidal inlets and their deposits.

Denitrification or dissimilatory NO_3^- reduction   the reduction of NO_3^- to any gaseous form of N such as NO, N_2O, and N_2 by microorganisms.

Densiometric froude   a dimensionless number that describes the ratio of nonlinear advection to the pressure gradient acceleration associated with the variation of fluid depth.

Deposit feeders   benthic organisms that ingest and extract the organic matter associated with sedimentary particles for their nutrition.

Desorption   the release of matter from the adsorption medium; opposite of adsorption.

Detrital minerals minerals introduced by weathering of parent rock materials (e.g., feldspar).

Detritivores organisms that feed on nonliving organic matter (detritus) for their nutrition.

Detritus nonliving organic matter.

De Vries effect anomalously high values of ^{14}C activity (2% greater than in 1900s), at around 1710 and 1500 AD; discovered in 1958 and causes not known.

Dextrorotatory stereoisomer that reflects plane-polarized light to the right.

Diagenesis processes that alter the structure, texture, and mineralogy of a sediment, turning it progressively into solid hard rock; early diagenesis occurs immediately after deposition or burial of the sediment.

Diagenetic oxygen exposure time an indicator for organic matter preservation based on the amount of time sinking POM resides in the oxic porewaters of sediments.

Diasteromers stereoisomeric chemical structures that are not enantiomers (mirror images) of one another.

Diazotrophs organisms that fix N_2.

Dielectric constant ratio of electric flux density to electric field.

Diffusion the spread of properties or material by random motions which occur on molecular scales and are accounted for in the equations of motion including the gradient in the property.

Diploptene pentacyclic triterpenoid $(17\beta(H),21\beta(H)$-hop-21(29)-ene) that has been found to be a useful diagenetic biomarker for bacterial activity.

Dipolar molecule opposing positive and negative charge distribution in a molecule.

Direct precipitation direct atmospheric input of precipitation to the area surface of an ecosystem.

Disaccharides two covalently linked monosaccharides.

Dissimilatory NO_3^- reduction to NH_4^+ pathway similar to denitrification only with more N being retained in a system.

Dissolution process of going into solution.

Dissolved organic matter dissolved (usually less than 0.45 or 0.2 μm) organic compounds that range from macromolecules to low molecular weight compounds (e.g., simple organic acids).

Dissolved organic phosphorus a subcomponent of all the phosphorus-containing dissolved organic compounds in dissolved organic matter (e.g., phosphonates).

Distributary a branch of a river that flows away from the mainstem and does not join again.

Dry deposition the settlement of particulate matter to the Earth's surface in the absence of precipitation.

Early diagenesis the changes occurring during burial to a few hundred meters where elevated temperatures are not present and uplift above sea level does not occur.

Ebullition process causing the release of a gas from a medium that is more erratic and unpredictable than simple diffusion.

Ecosphere areas on and within contact of the Earth that are habitable by living organisms.

Effusion biogas (e.g., CH_4) transport that occurs via the partial pressures differences in the pores of the plant.

Electron capture where a proton is changed to a neutron after combining with the captured extranuclear electron (from the K shell); the atomic number is decreased by one unit.

Electrophile an atom, molecule, or ion that is able to accept an electron pair.

Electrostatic bond a chemical bond whereby one atom loses an electron to form a positive ion, and the other atom gains an electron to form a negative ion.

Electrostriction process in which water molecules aggregate in higher densities than predicted near salt ions (e.g., Na^+); these "pockets" of water molecules have higher densities than the bulk water, resulting in a compression or reduction of the solvent.

Emergent aquatic vegetation (EAV) rooted vegetation that grows under and above the water surface in shallow zones where light penetrates.

Enantiomers two isomers that are related as mirror images to one another.

Enthalpy (H) is a parameter of the system used to describe energy as heat flow at a constant pressure.

Enthalpy of formation change in the enthalpy that occurs during a chemical reaction that leads to the formation of a compound from its elements in their most thermodynamically stable states.

Entropy the disorder in system.

Equilibrium when a chemical reaction and its reverse reaction proceed at equal rates.

Equilibrium fractionation when the isotopic effects are in equilibrium there is an isotopic exchange involving the redistribution of isotopes of an element in different molecules.

Erosion alteration of solids (soil, mud, rock, and so forth) by the forces of wind, water, ice, or movement in response to gravity.

Estuarine turbidity maximum a region where the suspended particulate matter (SPM) concentrations are considerably higher (10–100 times) than in adjacent river or coastal end-members in estuaries.

Estuary a semi-enclosed coastal body of water that extends to the effective limit of tidal influence, within which seawater entering from one or more free connections with the open sea, or any other saline coastal body of water, is significantly diluted with fresh water derived from land drainage, and can sustain euryhaline biological species from either part or the whole of their life cycle.

Eubacteria all prokaryotes excluding the Archaea.

Euhedral a crystal that is completely bounded by well-developed crystal faces whereby its growth is not restrained by adjacent crystals.

Eukaryotes unicellular and multicellular organisms with nuclear membranes and DNA in the form of chromosomes.

Eustacy a uniform worldwide change in sea level.

Eutrophication the myriad of biogeochemical and ecological responses, either direct or indirect, to anthropogenic fertilization of ecosystems at the land–sea interface.

Evapotranspiration evaporation of water from the land surface as well as that lost through transpiration in plants.

Exoenzyme an enzyme that is secreted by a cell and that works outside that cell.

Facies an accumulation of deposits that exhibits specific characteristics and grades laterally into other sedimentary accumulations that were formed at the same time but exhibit different characteristics.

Fatty acids   building blocks of lipids that represent a significant fraction of the total lipid pool in aquatic organisms.

Felsic   term used to describe the amount of light-colored feldspar and silica minerals in an igneous rock.

Fermentation   process in which an organism (e.g., microbe) causes an organic substance to break down into simpler substances, in particular the anaerobic breakdown of sugar into alcohol.

First-order decay   where the amount of material (e.g., number of atoms) decomposing in a unit of time is proportional to the number present.

Fischer projection   a simplified way to depict stereochemistry.

Fjard   low-relief system formed by glacial scouring.

Fjord   high-relief system formed by glacial scouring.

Flickering clusters   water molecules that are arranged into ice-like structures, which are believed to be transient features that change in statistical frequency of occurrence as a function of temperature and pressure.

Flocculation   a condition in which clays, polymers, or other small charged particles become attached to form a fragile structure called a floc; flocs form because of attractions between negative face charges and positive edge charges.

Floccules   nonliving biogenic material bound by electrochemical forces.

Fluidized-bed reactor effect   an effect in high accumulation rate environments that results in a high mineralization capacity due to repetitive redox successions, metabolite exchange, and efficient decomposition linked to cometabolism and metal cycling.

Fluid mud   high concentration of sediment suspended in mass concentrations of greater than 10 g L^{-1}.

Fluorochrome   a fluorescent molecule.

Flux   the amount of material that is transported from one reservoir to another over a particular time period (mass/time or mass/area/time).

Fly ash   the airborne combustion residue from burning coal or other fuels; consists mainly of various oxides and silicates.

Foreset   grade into silt and clay, slope seaward, will contain marine fossils, but also some material washed in from terrestrial environment.

Former fluvial valley   formed by flooding of river valleys in the Pleistocene–Holocene during the last interglacial transgression.

Former glacial valley   formed by flooding of river valleys in the Pleistocene–Holocene during the last interglacial transgression.

Free ions   ions that are so well hydrated that they do not interact (except for nonspecific interactions).

Fugacity   in thermodynamics, the tendency of a fluid to escape or expand isothermally; for gases at low pressures where the ideal gas law holds, fugacity is equal to pressure.

Fulvic acids   humic substances that are soluble at any pH.

Furanose   five-carbon ring sugar.

Gamma ray   highly penetrating, short-wavelength electromagnetic radiation emitted during the radioactive decay of many nuclides.

Geolipid   decay-resistant biomarkers in sediments because lipids are typically recalcitrant compared with other biochemical components of organic matter.

Gibbs free energy   thermodynamic function as defined by the equation: $G = H - TS$.

Glycolysis   the breakdown of glucose in a series of metabolic steps that yield energy in the form of ATP, even in the absence of oxygen.

Gravitational circulation   circulation of air and water masses driven primarily by differences in density.

Gross primary production   the amount of carbon fixed (CO_2 converted into organic matter) by photosynthetic organisms (per unit time per unit volume).

Haber process   also known as the Haber–Bosch process, this industrial process uses nitrogen (N_2) and hydrogen (H_2) to produce NH_3.

Half-life   the time required for half of the initial number of atoms of a radionuclide to decay.

Halmyrolysis   a more restricted aspect of chemical changes operative at the sediment-water interface.

Harmful algal blooms (HABs)   two general HAB forms exist—the toxic forms that lead to the death of marine organisms and human seafood poisoning, and the nontoxic species that can have negative effects on the environment by modifying O_2 consumption, shading effects, and altering the stoichiometry of nutrients

Haworth projection   common way of representing the cyclic structure of monosaccharides using a three-dimensional perspective.

Heavy metals   have atomic weights that range from 63.546 to 200.590 with a similar electronic distribution in their outer electron shells (e.g., Cd and Zn).

Hemicellulose   any of several heteropolymers (matrix polysaccharides) present in almost all cell walls along with cellulose.

Henry's law of equilibrium distribution   the equilibrium concentration of a gas in the aqueous phase is directly proportional to the pressure of that gas.

Hepatopancreas   digestive gland in the mid-gut.

Heterocysts   specialized well-sealed cells capable of biological N_2 fixation.

Heterotrophs   organisms (e.g., animals and fungi) that require external food resources for energy.

Highly mixed estuaries   estuaries with minimal stratification in the water column due to extensive tidal and wind mixing.

Hopanes   organic molecules found in the cell walls of bacteria.

Hopanoids   subgroup of compounds within the triterpenoids that are very abundant in bacteria where they have generally replaced cholesterol.

Horton overland flow   when precipitation rate exceeds infiltration capacity and overland flow of water occurs.

Humic acids   humic substances that are precipitated in water upon acification.

Humic substances   complex assemblages of molecules that have a yellow-to-brown color and are derived from plants and soils.

Humins   humic substances that are insoluble in water and dilute solutions at any pH.

Hydration   process by which water molecules become associated (or react) with other substances without destruction of the water molecule.

Hydraulic conductivity   a measure of the ability of groundwater to flow through the subsurface environment or a soil or rock formation.

Hydraulic gradient   change in hydraulic head (pressure) per unit distance in a given direction (dimensionless); it is the primary directional force of fluid flow in a porous medium.

Hydrogen bonding   dipole–dipole electrostatic interactions between the negatively and positively charged ends of different water molecules.

Hydrophobic sorption   sorption process involving weak solute–solvent and sorbate–sorbent interactions.

Hypoxia   low O_2 concentrations (or hypoxic = <2 mg L^{-1}), due to excessive nutrient loading in these systems.

Ideal solution   when an ion's activity is equal to its concentration.

Impact law   the settling velocity of particles typically greater than 1 mm in diameter varies according to the square root of the diameter.

Impulse marker   the subsurface peak of maximum activity of a radionuclide that was produced during a specific time period in the atmosphere (e.g., ^{137}Cs during bomb testing) and can used to measure accumulation rates.

Incident irradiance   density flux of radiance incident upon a given area.

Infiltration capacity (or rate)   a soil characteristic determining or describing the maximum rate at which water can enter the soil under specified conditions, including the presence of excess water; it has the dimensions of velocity (i.e., cm^3 cm^{-2} s^{-1} = cm s^{-1}).

Interflow   fraction of rainfall that infiltrates into soils and moves laterally through the upper soil horizons until intercepted by a stream channel, or until it returns to the surface at some point downslope from its point of infiltration.

Ionic strength   a measure of interionic interactions that are primarily derived from electrical attractions and repulsions.

Ion pairs   ions that are not well hydrated can get close enough to other ions to allow for electrostatic interactions.

Irving–Williams series   when the strength of metal-binding constants is largely a function of ionic radius, atomic number, and valence.

Isoelectric point   the intermediate pH where a zwitterion is at its highest concentration with the cationic and anionic forms in equilibrium.

Isoprenoids   class of naturally occurring chemicals similar to terpenes, derived from five-carbon isoprene units.

Isostacy   the state of being isostatic whereby there is a general equilibrium in the Earth's crust, supposed to be maintained by the yielding or flow of rock material beneath the surface under gravitative stress.

Isostatic rebound   the upward movement of the Earth's crust following isostatic depression.

Isotopes   different forms of an element that have the same Z value but a different N.

Isotopic depletion   when a sample has less of the heavy stable isotope than the standard.

Isotopic enrichment   when the heavy stable isotope is more abundant relative to the standard, resulting in positive del values.

Isotopic fractionation   selective removal of the lighter or heavier stable isotope in a sample due to both kinetic and equilibrium effects.

Isotopic mixing models models used to evaluate sources of dissolved inorganic nutrients (C, N, S) and organic matter (POM and DOM).

Kerogen organic matter in rocks in the form of a mineraloid which is of indefinite composition, insoluble in petroleum solvents.

Ketones organic compounds that contains a carbonyl group.

Ketoses sugars containing one ketone group per molecule.

Kinetic effect energy effect on a system or suite of chemical reactions.

Kinetic fractionation the higher energy and more rapid diffusion and/or phase changes (e.g., evaporation) of the lighter isotope of a particular element.

Kinetics examination of the rates of reactions.

Lacunal pressure pressure of gas buildup (e.g., CO_2, O_2) in air spaces of aquatic plants; this allows roots to be aerated.

Lambert–Beer's law describes the decrease in light of penetration with increasing water depth.

Laminar flow streamlined or viscous flow.

Latent heat amount of heat required to melt a unit mass of substance at the melting point.

Law of definite proportions states that, in a pure compound, the elements combine in definite proportions to each other.

Leaching decay phase soluble compounds are rapidly lost from fresh detritus over a scale of minutes to weeks.

Leaky lagoon systems with many inlets separated by small barrier islands.

Legumes plants such as the soybean that contain nitrogen-fixing bacteria on their roots; this results in a net increase in soil nitrogen content.

Leibig's law of the minimum states that organisms will become limited by the resource that is in lowest supply relative to their needs.

Lethal concentration-50 (LC50) concentration of a potentially toxic substance in an environmental medium that causes death of 50% of the experimental organisms over a certain period of exposure.

Levorotary stereoisomer that reflects plane-polarized light to the right.

Lewis base substance capable of donating an electron pair.

Ligands electron donors—usually negatively charged.

Light attenuation the decrease in light of penetration with increasing water depth.

Lignins a group of macromolecular heteropolymers (600–1000 kDa) found in the cell wall of vascular plants that are made up of phenylpropanoid units.

Line source the collective inputs of dissolved and particulate materials to a coastal zone from a series of estuaries in close geographic proximity.

Lipid water-insoluble compound that is soluble in nonpolar (chloroform, benzene) solvents. Lipids are an important component of organic matter with a diversity of compound classes (e.g., hydrocarbons, fatty acids, *n*-alkanols, and sterols) that have been used as effective biomarkers of organic matter.

Liposomes closed lipid vesicle surrounding an aqueous interior.

Lithophilic elements such as Si and Ca that tend to be associated with rock materials on Earth.

Littoral transport   the movement of littoral drift in the littoral zone by waves and currents including movement parallel (longshore transport) and perpendicular (on/offshore transport) to the shore.

Loess   deposits of silt (sediment with particles 2–64 μm in diameter) that have been laid down by wind action.

Long-chain fatty acids   saturated fatty acids ($>C_{22}$) that are generally thought to be indicative of terrestrial organic matter sources.

Long-chain *n*-alkanes   alkanes (e.g., C_{27}, C_{29}, and C_{31}) generally considered to be terrestrially derived sources of epicuticular waxes from vascular land and aquatic (e.g., seagrasses) plants.

Macroflocs   particles composed of microflocs.

Macrotidal   coastal and oceanic systems with a high mean tidal range greater than 4 m.

Mafic   term used to describe the amount of dark-colored iron and magnesium minerals in an igneous rock.

Magnetic field   a region in which magnetic forces can be observed.

Magnetic resonance   the absorption of energy that occurs when certain nuclei (e.g., ^1H, ^{13}C, ^{31}P, ^{15}N), which have a spin state, are placed in a strong magnetic field and simultaneously irradiated with electromagnetic (microwave) radiation.

Major sugars   sugars that are most abundant in natural organic matter and can be further divided as unsubstituted aldoses and ketoses. The aldoses or neutral sugars include rhamnose, fucose, lyxose, ribose, arabinose, xylose, mannose, galactose, and glucose.

Mangrove   densely rooted forest bordering the lowlands of tropical and subtropical coastlines—generally occurring between 25°N and 25°S latitudes.

Marcet's principle   constituents of seawater are found in relatively constant proportions in the oceans, indicating that the residence times of these elements are long (thousand to millions of years)—highly indicative of nonreactive behavior.

Mass spectrometry   an instrumental method used in conjunction with gas and liquid chromatography that provides accurate information about the molecular mass and structure of complex molecules.

Mean-life   the average life expectancy of a radioactive nuclide.

Melanoidin polymers   compounds produced from condensation reactions of glucose with acidic, neutral, and basic amino acids.

Mesopores   spaces which are too small for access by microbial enzymes.

Mesoscale   scales of motion of order of magnitude 100 km in the horizontal.

Mesotidal   coastal and oceanic systems with an average tidal range of 2 to 4 m.

Mesotrophic   intermediate between oligotrophic (low nutrient) and eutrophic (high nutrient) systems.

Metagenesis   organic matter alteration in the graphite stage during high-grade metamorphism.

Metalloids   subgroup of heavy metals (e.g., As, Pb, Hg) not required for metabolism and toxic at low concentrations.

Metalloproteins   metal–protein complexes such as Fe–protein and Mo–Fe protein, which serve as the source of electrons for reducing power, and can act to bind N_2 to an enzyme, respectively.

Metallothioneins metal-binding proteins found in prokaryotes and eukaryotes that have been shown to be able to sequester and detoxify metals.

Methanogen methane-producing prokaryote.

Methanogenesis biological production of methane.

Methanotroph methane-oxidizing bacteria.

Microbial mats defined as surface layers of microbes entrained in a matrix of extra-cellular polymeric substances (EPS).

Microfibrils small fibers visible only at the high magnification of the electron microscope.

Microflocs particles $<150 \, \mu m$ in diameter.

Microphytobenthos microscopic plants (e.g., benthic diatoms) that grow on solid Surfaces in aquatic systems.

Microtidal coastal and oceanic systems with a high mean tidal range less than 2 m.

Mid-chain hydrocarbons hydrocarbons with a carbon chain length that ranges from C_{23} to C_{25}.

Minor sugars sugars such as acidic sugars, amino sugars, and O-methyl sugars that tend to be more source specific than major sugars.

Mixing depth depth that is actively mixed in surface waters corresponding to the depth in which there is strong turbulence which is driven by surface forcing.

Mobile mud high-porosity BBL–upper seabed layer where diagenetic transformation processes are enhanced and commonly occur in river-dominated ocean margins.

Monolayer equivalents the amount of sedimentary organic matter occuring as layers of organic matter within the surface features on mineral surfaces.

Monosaccharides simple sugars.

Monounsaturated fatty acid fatty acid with one double bond (e.g., oleic acid).

Moraine till or rock fragments that range from clay to boulder size and randomly arranged without bedding.

n-Alkanes/paraffins saturated aliphatic hydrocarbons.

n-Alkanols fatty alcohols that can be used to distinguish between terrestrial plant and algal/bacterial inputs to aquatic systems.

n-Alkenes/olefins nonsaturated hydrocarbons which exhibit one or more double bonds.

Negatron elementary particle with negative charge.

Neo-Arheism drastic reduction in river flow.

Net autotrophic when production exceeds consumption.

Net ecosystem production the difference between GPP and *ecosystem respiration* (R), which includes both heterotrophic and autotrophic processes.

Net heterotrophic when consumption exceeds production.

Net primary production (NPP) the total carbon fixed minus the amount respired by the primary producer.

New production the NPP that is supported by nutrients from outside the estuary proper (e.g., river inputs) and production supported by nutrients regenerated from within the estuary (e.g., nutrient fluxes across the sediment/water boundary from remineralized organic matter in sediments), respectively.

Nitrification where NH_3 or NH_4^+ is oxidized to NO_2^- or NO_3^- through two energy-producing reactions.

Nonconservative elements elements that do not remain in constant proportion due to biological (e.g., uptake via photosynthesis) or chemical (e.g., hydrothermal vent inputs) processes. In estuaries, as well as other oceanic environments (e.g., anoxic basins, hydrothermal vents, and evaporated basins), the major components of seawater can be altered quite dramatically due to numerous processes (e.g., precipitation, evaporation, freezing, dissolution, and oxidation).

Nonideal solution discrepancy between ion activity and concentration.

Non-point sources contaminant sources which are diffuse and do not have a single point of origin (e.g., surface runoff).

Nonprotein amino acids proteins that are produced from the degradation of other primary amino acids; these amino acids are not used in the synthesis of proteins.

Normal atmospheric equilibrium concentrations (NAEC) concentrations based on expected equilibrium conditions, between water and atmosphere, at a particular pressure, temperature, salinity, and humidity.

Nuclear magnetic resonance spectroscopic technique used to elucidate chemical structure and molecular dynamics.

Nuclear spin intrinsic property of certain nuclei that results in an associated characteristic angular momentum and magnetic moment.

Nuclide nucleus of an atom having a specific atomic number and atomic mass number, which may be radioactive.

Octanol/water partition coefficient (K_{ow}) the partitioning coefficient between organic carbon and water (K_{oc}), based on aqueous solubility (S).

Oligosaccharides a few covalently linked monosaccharides.

Oligotrophic low-nutrient conditions.

Organic burial efficiency the accumulation rate of organic matter below the active zone of diagenesis divided by the organic matter flux to surface sediments.

Organometallic organic compound/complex containing metal atom.

Orthophosphate chemical structure is $H_2PO_4{}^-$ and it represents a significant fraction of the soluble reactive phosphorus (SRP) pool.

Osmolyte neutral solute in cells used in response to salinity changes.

Outwelling hypothesis suggestion by Odum (1968) that salt marshes transport biologically available organic matter into nearshore waters, thereby enhancing secondary production on the shelf.

Oxidants compounds that donate electrons to other compounds.

Oxidation chemical reaction with oxygen with a certain material that usually results in degradation of that material.

Oxidation state the sum of negative and positive charges in an atom.

Partially mixed estuaries vertical mixing is inhibited to some degree.

Partial pressure pressure exerted by a single component of a gas within a gas or liquid mixture.

Particle reactive the tendency for elements and molecules to sorb onto particle substrates.

Passive diffusion net movement of molecules from higher to lower concentration gradient in the absence of any facilitative process.

Passive margin   continental margin that is not along a plate boundary.

Pectin   plant polysaccharide capable of producing a gel-like material.

Peptidoglycan   cross-linked polysaccharide in a bacterial cell wall—also known as murein.

Petrogenic   produced from unburned petroleum products.

Phospholipids   the major structural lipids of most of our cellular membranes.

Photobleaching   photochemical breakdown of chemical compounds which results in a loss of their fluorescent properties.

Photodegradation   chemical breakdown of a substance caused by exposure to light.

Photosynthates   chemical products of photosynthesis used for growth and respiration.

Photosynthesis   process by which plants use light energy to make simple carbohydrates from carbon dioxide and water.

Photosynthetically active radiation   electromagnetic energy in the 400–700 nm wavelength range.

Photosynthetic pigments   compounds such chlorophylls, carotenoids, and phycobilins, used in absorbing photosynthetically active radiation (PAR).

Physical weathering   the fragmentation of parent rock materials and minerals through processes such as freezing, thawing, heating, cooling, and bioturbation (e.g., endolithic algae, fungi, plant roots, and earthworms).

Physiography   a description of the surface features of the Earth, with an emphasis on the origin of landforms.

Phytane C_{20} isoprenoid hydrocarbon derived from phytol—shown to be a useful indicator of herbivorous grazing.

Phytochelatins   organic compounds, produced by plants that have chelating properties.

Phytohormone   compounds such as indole acetic acid (IAA), which are capable of promoting plant growth.

Phytol   an alcohol ($C_{20}H_{40}O$) that is an important component of chlorophyll.

Phytoliths   opal accumulations in vascular plant tissues.

Phytoplankton   microscopic plants capable of carrying out photosynthesis that drift with the currents.

Plankton   free floating and/or weak swimming organisms.

Podzolization   soil-forming process that commonly results in the formation of humic acids and leaching of acids.

Point bar   accumulations of sand and gravel deposited in slack water on inside of a winding or meandering river; their characteristic half-ellipse shape is easily recognizable.

Point source   source of pollutants that can be traced to a discrete point of emission.

Polycyclic aromatic hydrocarbon   a hydrocarbon having two or more benzene rings fused together, produced from incomplete combustion of organic matter and/or spillage from petroleum products.

Polymeric   consisting of a long, repeating chain of atoms, formed through the linkage of many molecules called monomers.

Polysaccharides   polymers made up of chains of mono- and disaccharides.

Polyunsaturated fatty acid   fatty acids with more than one double bond (e.g., linoleic acid).

Porosity   ratio of void space to the bulk volume of rock or sediment containing that void space.

Positron   elementary particle with a positive electric charge.

Precision   degree of reproducibility among several independent measurements of the same true value.

Primary estuary   formed from terrestrial and/or tectonic processes with minimal changes from the sea—these systems have essentially preserved their original characteristics.

Primary minerals   minerals that are formed below the Earth's surface under high pressures and temperatures.

Primary producers   organisms that use light to synthesize organic matter from CO_2.

Primary production   the photosynthetic formation of organic matter.

Primordial   existing at or from the beginning, primeval.

Pristane C_{19} isoprenoid hydrocarbon derived from phytol—shown to be a useful indicator of herbivorous grazing.

Prodelta   an area seaward of the delta front where most of the fine-grained sediments are deposited along a steep gradient within the delta slope.

Progradation   lateral outbuilding of strata in a seaward direction. Progradation can occur as a result of a sea-level rise accompanied by a high sediment flux.

Prokaryotes   unicellular organisms (e.g., bacteria) that do not possess a nuclear membrane or DNA in the form of chromosomes.

Protein   compounds made of approximately 20 α-amino acids which can be divided into different categories based on their functional groups.

Proteolytic enzymes   enzymes that chemically break down proteins.

Purines   nitrogenous bases (adenine and guanine) found in DNA and RNA.

Pyrimidines   nitrogenous bases (cytosine and thymine) found in nucleotides.

Pyritization   the formation of pyrite (FeS_2) under anoxic conditions relatively quickly between H_2S and $S(0)$ [as S_8 or polysulfides (S_x^{2-})] and FeS—both aqueous and solid forms.

Pyrolytic   formed from pyrolysis.

Racemization   conversion of L-amino acids to their mirror-image D-forms.

Radiative forcing   increase or decrease in the amount of radiation flux.

Radioactivity   the spontaneous adjustment of nuclei of unstable nuclides to a more stable state.

Radionuclide   any radioactive isotope.

Radionuclide disequilibrium   any deviation from secular equilibrium.

Rare Earth elements (REEs)   elements that occur in the periodic table from lanthanum (La) to lutetium (Lu)—have similar chemical and physical properties due to their electronic configurations.

Rate constant   proportionality constant that appears in a rate law.

Rayleigh effect   states that the isotopic ratio of the substrate is related to the amount of substrate that has not been used.

Reactive phosphorus   the potentially bioavailable phosphorus (BAP).

Redfield ratio   a ratio originally used compared the ratios of C, N, and P of dissolved nutrients in marine waters to that of suspended marine particulate matter (seston) (essentially phytoplankton).

Redox reduction–oxidation term used to describe chemical reactions that involve loss of one or more electrons by one molecule (oxidation) and simultaneous gain by another (reduction).

Redox potential discontinuity layer a distinct layer in sediments associated with distinct coloration, indicative of difference between oxic and suboxic conditions.

Reductant electron donor in an oxidation–reduction reaction.

Refractory decay phase characterized by detritus composed of lignin and cellulose which decay very slowly.

Regression lateral migration of the shoreline away from the center of a land mass which is usually a continental interior.

Relative sea-level rise factoring in these regional differences in rates of sea level.

Remineralization the conversion of organic materials from the particulate to the dissolve state (e.g., PON → DON or DIN).

Reservoir the amount of material, as defined by its chemical, physical, and/or biological properties.

Residence time the ratio of the mass of a scalar (e.g., salinity) in a reservoir to the rate of renewal of the scalar, under steady-state conditions.

Respiration process by which oxygen is taken in and used by tissues in the body and carbon dioxide is released.

Restricted lagoon system with very few inlets or having a wide mouth.

Resuspension rate rate at which surface sediments are resuspended in the water column of aquatic systems.

Reynold's number dimensionless number used in scaling fluid systems and in determining the transition point from laminar to turbulent flow.

Rhizosphere area of soil in the vicinity of plant roots.

Ria formed by flooding of river valleys in the Pleistocene–Holocene during the last interglacial transgression, and occurs in regions of high relief (mountains and cliffs).

Richardson number dimensionless number that expresses the ratio of potential to kinetic energy, also known as the Froude number.

River-dominated estuary formed in high river discharge regions where the valley is presently not drowned by the sea.

River-dominated margin (RiOMar) dynamic regions that receives inputs of organic carbon (OC) derived from both terrestrial and marine sources.

Rule of constant proportions constituents of seawater are found in relatively constant proportions in the oceans, indicating that the residence times of these elements are long (thousand to millions of years)—highly indicative of nonreactive behavior. This is the same as Marcet's principle

Salinity a measure of the quantity of dissolved salts in seawater.

Salinometer instrument used to measure salinity, where the conductivity of water is measured; in essence the electrical current is controlled by the movement and abundance of ions, the more dissolved salts, the greater the conductivity.

Saltation the leaping movement of sand and/or soil particles as they move across an uneven bed or fluid medium.

Salting-in effect the solubility of other mineral salts may be enhanced with the formation of ion pairs,

Salting-out effect   factors that limit the solubility of compounds in solution (e.g., increasing salinity) are expected to increase partitioning of those compounds onto particles.

Salt marshes   stands of rooted vegetation which are flooded and drained by tidal forces.

Salt-wedge estuaries   river-dominated estuaries where the freshwater flow is high enough to create a sharp wedge-like stratification in the water column.

Scalar   quantity in mathematics consisting of a single real number used to measure magnitude.

Schmidt number   a dimensionless number, characteristic of each gas, which varies strongly with temperature and weakly with salinity, and is used to account for viscosity effects on the diffusion of gases.

Seagrasses   marine and estuarine flowering plants which generally attach to the substrate with roots.

Secondary estuary   an estuary that has been modified more by marine than river discharge processes, since the time that sea level reached its current position.

Secondary minerals   minerals resulting from the decomposition of a primary mineral and/or from the reprecipitation of the products of primary mineral alteration.

Secondary plant compounds   compounds such as phenols, alkaloids, tannins, organic acids, saponins, terpenes, steroids, essential oils, and glycosides which can deter herbivory and detritivory.

Secondary production   turnover of organisms that range in size from bacteria to the largest mammals on Earth that rely on the consumption of organic matter produced by photosynthetic organisms as a food resource.

Secular equilibrium   when the rate of decay of the parent radionuclide is equal to that of its daughter.

Sediment accumulation   the net sum of particle deposition and removal processes over a long period.

Sedimentation   integrated particle transport to and emplacement on the seabed, as well as removal and preservation.

Sediment deposition   the temporary emplacement of particles on the seabed.

Seston   discrete biological particles.

Setchenow salting-out and van't Hoff equations   equations used to describe the effects of temperature and salinity on the solubility of gases in seawater.

Shear stress   holding capacity of forces which tend to cause one portion of a body to move with respect to another in a direction parallel to their plane of contact.

Shikimic acid pathway   chemical pathway common in plants, bacteria, and fungi, where aromatic amino acids (e.g., tryptophan, phenylalanine, tyrosine) are synthesized, thereby providing the parent compounds for the synthesis of the phenylpropanoid units in lignins.

Short-chain fatty acids   fatty acids thought to be derived from aquatic sources (C_{12}–C_{18}) to others (C_{14}–C_{18}) from multi-sources (zooplankton, bacteria, and benthic animal and marsh plants).

Short-chain n-alkanes   alkanes (e.g., C_{15}, C_{17}, and C_{19}), which are derived from planktonic and benthic algal sources.

Siderophiles   elements that prefers to combine with iron rather than some other element.

Siderophores   iron-chelating compounds secreted by microorganisms.

Simple molecular diffusion where solutes are transported due to variations in the solute concentrations within fluid phases.

Sink flux of material out of the reservoir (many times proportional to the size of the reservoir).

Soil texture term commonly used to designate the proportionate distribution of the different sizes of mineral particles in a soil.

Soil–Vegetation–Atmosphere Transfer Schemes (SVATs) simulated models that use parameters such as vegetation cover, soil texture, water-holding capacity of soils, surface roughness, and albedo, to make predictions on soil moisture, runoff, evapotranspiration, and runoff.

Solubility product constant constant K_{sp} that indicates the degree to which a compound dissociates in water.

Soluble reactive phosphorus (SRP) the P fraction which forms a phosphomolybdate complex under acidic conditions.

Solvent substance that can dissolve other substances.

Solvent-separated ion pairs ions linked electrostatically, separated by more than one water molecule.

Solvent-shared ion pairs ions linked electrostatically, separated by a water molecule.

Sulfate reduction the terminal microbial process in anaerobic sediments, when SO_4^{2-} is not limiting, leading to the formation of hydrogen sulfide (H_2S).

Sorbate the sorbed neutral hydrophobic organic carbon when in contact with a solid (sorbent) surface.

Sorption general term for physical and chemical absorption and adsorption.

Source the flux of material into a reservoir.

Spallation the production of radionuclides such as ^{14}C, produced by reaction of cosmic rays with atmospheric atoms such as N_2, O_2, and others to produce broken pieces of nuclei.

Specific activity the observed counting rate in a sample.

Specific heat amount of heat required to raise the temperature of a unit mass of substance $1°C$.

Stagnant film model kinetic model used to calculate rates of gas exchange across the atmosphere–water boundary.

Standard free energy of formation free energy change that occurs when one mole of a substance is formed from its elements in their most stable state—at 1 bar and at a specified temperature.

Steady state state in which equilibrium has been achieved.

Steric relating to the spatial relationships of atoms in a molecular structure.

Sterols compounds that are a group of lipids (typically between C_{26} and C_{30}) that are resistant to saponification and can be classified as triterpenes.

Stoichiometry the mass balance of chemical reactions as they relate to the law of definite proportions and conservation of mass.

Stoke's law sinking speed for spherical, slow sinking, single particles.

Stomata microscopic pores found on the underside of leaves in the epidermal tissue.

Stomatal conductance numerical measure of the maximum rate of passage of either water vapor or carbon dioxide through the stomata (small pores on plant surfaces).

Strike-slip fault fault whose relative displacement is purely horizontal.

Structural estuary   formed by processes such as faulting, volcanism, postglacial rebound, and isostacy that have occurred since the Pleistocene.

Suberins   lipid polyester polymers in cell-wall components of cork cells.

Submarine groundwater discharge   any and all flow of water on continental margins from the seabed to the coastal ocean, regardless of fluid composition or driving force.

Submergent aquatic vegetation (SAV)   rooted vegetation that grows under water in shallow zones where light penetrates.

Submonolayer   below the monolayer-equivalent range in sediments, reflecting more efficient decomposition of organic matter on mineral surfaces.

Suess effects   the dilution effect on ^{14}C from inputs of fossil fuel combustion are clear after about 1850 in the early stages of the Industrial Revolution.

Supermonolayer   below the monolayer-equivalent range in sediments, reflecting less efficient decomposition of organic matter on mineral surfaces.

Surface microlayer   layer at the air–water interface that is composed of films 50 to 100 μm in thickness.

Surface tension   strength of a liquid surface.

Suspended load   fine solid particles, typically of sand, clay, and silt, that travels with stream water without coming into contact with the stream bed.

Symbiosis   mutually beneficial relationship between two organisms.

Tectonism   movement and deformation of the crust on a large scale, including epeirogeny, metamorphism, folding, faulting, and plate tectonics.

Terpenes   widespread group of natural plant compounds made up of isoprene (methylbuta-1,3-diene) units containing five carbons.

Tetrapyrrole   chemical substance consisting of four pyrrole rings, joined together into a single large ring known as a macrocycle, serves at the basic structure of chlorophyll.

Thalweg   path of maximum depth in a river or stream; this normally follows a meandering pattern back and forth across the stream.

Thermal expansion   increase of dimensions or volume of a specimen due to an increase in its temperature.

Thermodynamics   central theory in the physical sciences dealing with conversion of energy from one form into another.

Thylakoid membrane   internal membrane system of chloroplasts folded repeatedly into a stack of disks called grana.

Tidal prism   volume of water that flows in and out of an area between higher high tide and lower low tide.

Tidal river estuary   associated with large rivers systems that are influenced by tidal action with the salt front usually not well developed at the mouth.

Tidal trapping   when parcels of water are advected into lateral "traps" during one phase of the tidal cycle and then moved out in another results in an overall upstream transport.

Tortuosity   the length of the tortuous path that a solute travels around particles across a distance across a certain depth interval.

Total atmospheric pressure   sum of all partial pressures (P_i) exerted by each of the gases in the entire mixture of air.

Transfer velocity   velocity scale used in the transport of trace constituents across the air–sea interface.

Transgression   lateral migration of the shoreline toward the center of a land mass which is usually a continental interior.

Transitional metals   a subgroup of heavy metals (e.g., Co, Fe, Mn) that are essential to metabolism at low concentrations but potentially toxic at high.

Translocate   movement of a substance within a plant from one site to another.

Transpiration   transport of water by plants from soils to the atmosphere, whereby water is released through pore-like structures (stomata) in the leaves to the atmosphere.

Triglyceride   composed of three fatty acids bonded to a glycerol (C_3 alcohol).

Triterpenoids   a large group of natural compounds which typically include steroids and sterols.

Trophic interactions   ecological interactions between organisms relating to food or nutrition.

Turbulent mixing   mixing that results from the destabilizing forces of tides and/or winds.

Turnover time   time required to remove all the materials in a reservoir, or the average time spent by elements in a reservoir.

Type 1 estuary   well-mixed estuaries with mean flow in the seaward direction and the salt balance being maintained by diffusive processes, via tidal transport.

Type 2 estuary   partially mixed estuaries where the net flow reverses at depth and the salt flux is maintained by both diffusive and advective processes.

Type 3 estuary   these estuaries include fjords with two distinct layers and advection accounting for the majority of salt flux.

Type 4 estuary   salt-wedge estuaries where freshwater flow is out over stable denser bottom layer.

Typology   system classification based on types.

Universal gas constant   proportionality constant in the ideal gas law.

Unresolved complex mixture   complexity of structural compounds found in petroleum hydrocarbons that cannot be fully resolved using current chromatographic techniques.

Unsupported/excess activity   the inputs of daughter radionuclides that are not directly from the in situ decay of the parent (*supported*).

van der Waals force   weak physical force that holds two molecules or two different parts of the same molecule together.

Virioplankton   viruses that live in the water column and are vulnerable to the movement caused by wind and currents.

Viruses   group of particles that do not have a cellular structure and that consist of a molecule of DNA or RNA surrounded by a protein coat.

Viscosity   resistance to distortion or flow of a fluid.

Washover fan   a fan-like deposit of sand washed over a barrier island during a storm.

Weathering   physical breakdown of rocks from exposure to the elements of rain, wind, and ice.

Well-mixed estuaries   minimal vertical stratification in salinity.

Wet deposition   removal of atmospheric particles to the Earth's surface by rain or snow.
Wetlands   habitats that are regularly saturated by surface water or groundwater and subsequently characterized by distinct vegetation.

Xylan   polysaccharide of xylose and a component of hemicellulose.

Yield strength   the force required to break any cohesive bonds.

Zeolites   hydrous aluminum–sodium silicate in porous granules.
Zooplankton   microscopic to very small free-floating animals.
Zwitterions   dipolar ions.

References

Aaboe, E., Dion, E.P., and Turekian, K.K. (1981) [7]Be in Sargasso Sea and Long Island Sound waters. J. Geophys. Res. 86, 3255–3257.

Aarkog, A. (1988) Worldwide data on fluxes on [239,240]Pu and [238]Pu to the oceans. In Inventories of Selected Radionuclides in the Ocean. IAEA-TECDOC-481, pp. 103–138.

Abelson, P.H., and Hoering, T.C. (1960) The biogeochemistry of stable isotopes of carbon. Carnegie Inst. Wash. 59, 158–165.

Abelson, P.H., and Hoering, T.C. (1961) Carbon isotope fractionation in formation of amino acids by photosynthetic organisms. Proc. Natl. Acad. Sci. USA. 47, 623–632.

Abrajano, Jr., T.A., Bieger, T., and Hellou, J. (1998) Reply to Grossi and de Leeuw. Org. Geochem. 28, 137–142.

Abrajano, Jr., T.A., Murphy, D.E., Fang, J. Comet, P., Brooks, J.M. (1994) ^{13}C/^{12}C ratios in individual fatty acids of marine mytilids with and without bacterial symbionts. Org. Geochem. 21, 611–617.

Abramopoulos, F., Rosensweig, C., and Choudhury, B. (1988) Improved ground hydrology calculations for global climate models (GCMs): soil water movement and evapotranspiration. J. Climate 1, 921–941.

Abril, G., and Borges, A.V. (2004) Carbon dioxide and methane emissions from estuaries. In Greenhouse Gas Emissions: Fluxes and Processes, Hydroelectric Reservoirs, and Natural Environments (Tremblay, A., Varfalvy, L., Roehm, C., and Garneau, M., eds.), pp. 187–212, Springer, Berlin.

Abril, G., Etcheber, H., Delille, B., Frankignoulle, M., and Borges, A.V. (2003) Carbonate dissolution in the turbid and eutrophic Loire estuary. Mar. Ecol. Prog. Ser. 259, 129–138.

Abril, G., Etcheber, H., Le Hir, P., Bassoullet, P., Boutier, B., Frankignoulle, M. (1999) Oxic–anoxic oscillations and organic carbon mineralization in an estuarine maximum turbidity zone (The Gironde, France). Limnol. Oceanogr. 44, 1304–1315.

Abril, G., and Frankignoulle, M. (2001) Nitrogen–alkalinity interactions in the highly polluted Scheldt Basin (Belgium). Wat. Res. 35, 844–850.

Abril, G., and Iverson, N. (2002) Methane dynamics in a shallow, non-tidal, estuary (Randers Fjord, Denmark). Mar. Ecol. Prog. Ser. 230, 171–181.

Abril, G., Nogueira, E., Hetcheber, H., Cabecadas, G., Lemaire, E., and Brogueira, M.J. (2002) Behaviour of organic carbon in nine contrasting European estuaries. Estuar. Coastal Shelf Sci. 54, 241–262.

Abril, G., Riou, S.A., Etcheber, H., Frankigoulle, M., deWitt, R., and Middelburg, J.J. (2000) Transient, tidal time-scale, nitrogen transformations in an estuarine turbidity maximum-fluid mud system (The Gironde, south-west France). Estuar. Coastal Shelf Sci. 50, 703–715.

Adamson, A.W. (1976) Physical Chemistry of Surfaces. Wiley, New York.

Admiraal, W. (1977) Salinity tolerance of benthic estuarine diatoms as tested with a rapid polarographic measurement of photosynthesis. Mar. Biol. 39, 11–18.

Admiraal, W., and Pelletier, H. (1980) Distribution of diatoms on an estuarine mudflat and experimental analysis of the selective factor of stress. J. Exp. Mar. Biol. Ecol. 46, 157–175.

Agusti, S., Satta, M.P., Mura, M.P., and Benavent, E. (1998) Dissolved esterase activity as a tracer of phytoplankton lysis: evidence of high phytoplankton lysis rates in the northeastern Mediterranean. Limnol. Oceanogr. 43, 1836–1849.

Ahner, B.A., Kong, S., and Morel, F. (1995) Phytochelatin production in marine algae. 1. An interspecies comparison. Limnol. Oceanogr. 40, 649–657.

Ahner, B.A., Morel, F.M.M., and Moffet, J.W. (1997) Trace metal production of phytochelatin production in coastal waters. Limnol. Oceanogr. 42, 601–608.

Ahner, B.A., Price, N.M., Morel, F. M.M. (1994) Phytochelatin production by marine phytoplankton at low metal ion concentrations: laboratory studies and field data from Massachusetts Bay. Proc. Natl. Acad. Sci. USA. 91, 8433–8436.

Ahner, B.A., Wei, L., Oleson, J.R., and Ogura, N. (2002) Glutathione and other low molecular weight thiols in marine phytoplankton under metal stress. Mar. Ecol. Prog. Ser. 232, 93–103.

Aiken, G.R. (1988) A critical evaluation of the use of macroporous resins for the isolation of aquatic humic substances. In Humic Substances and their Role in the Environment (Frimmel, F.H., and Christman, R.F., eds.), pp. 4–15, John Wiley, New York.

Aiken, G.R., McKnight, D.M., Wershaw, R.L., and MacCarthy, P. (1985) Humic Substances in Soil, Sediment, and Water. John Wiley, New York.

Aitkenhead, J.A., and McDowell, W.H. (2000) Soil C:N ratio as a predictor of annual riverine DOC flux at local and global scales. Global Biogeochem. Cycles 14, 127–138.

Albaigés, J. Grimalt, J., Bayona, J.M., Risebrough, R., and de Lappe, B., and Walker, W. (1984) Dissolved, particulate and sedimentary hydrocarbons in deltaic environments. Org. Geochem. 6, 237–248.

Alber, T., and Chan, J. (1994) Sources of contaminants to Boston Harbor: Revised loading estimates. In Massachusetts Water Resources Authority Environmental Quality Department Technical Report Series No. 94-1. Massachusetts Water Resources Authority, Boston, MA.

Alberts, J.J., Filip, Z., Price, M.T., Hedges, J.I., and Jacobsen, T.R. (1992) CuO-oxidation products, acid hydrolysable monosaccharides and amino acids of humic substances occurring in a salt marsh estuary. Org. Geochem. 18, 171–180.

Alberts, J.J., and Takacs, M. (1999) Importance of humic substances for carbon and nitrogen transport into southeastern United States estuaries. Org. Geochem. 30, 385–395.

Alexander, C.R., Calder, F.D., and Windom, H.L. (1993) The historical record of metal enrichment in two Florida estuaries. Estuaries 16, 627–637.

Alexander, M. (1977) Introduction to Soil Microbiology. John Wiley, New York.

Alexander, R.B., Smith, R.A., and Schwarz, G.E. (2000) Effect of stream channel size on the delivery of nitrogen to the Gulf of Mexico. Nature 403, 758–761.

Alldredge, A., and Cohen, Y. (1987) Can microscale chemical patches persist in the sea? Microelectrode study of marine snow, fecal pellets. Science 235, 689–691.

Alldredge, A., Passow, U., and Logan, B. (1993) The abundance and significance of a class of large, transparent organic particles in the ocean. Deep Sea Res. 40, 1131–1140.

Allen, G.P., Salmon, J.C., Bassoulet, P., Du Penhoat, Y., and De Grandpre, C. (1980) Effects of tides on mixing and suspended-sediment transport in macrotidal estuaries. Sediment. Geol. 26, 69–90.

Allen, J.R.L. (1985) Principles of Physical Sedimentology. George Allen and Unwin, London.

Aller, R.C. (1978) Experimental studies of changes produced by deposit-feeders on pore water, sediment and overlying water chemistry. Am. J. Sci. 278, 1185–1234.

Aller, R.C. (1980) Diagenetic processes near the sediment–water interface of Long Island Sound. II. Fe and Mn. Adv. Geophys. 22, 351–415.

Aller, R.C. (1982) The effects of macrobenthos on chemical properties of marine sediment and overlying water. In Animal–Sediment Relations (McCall, P.L., and Tevesz, M.J.S., eds.), pp. 53–102, Plenum Press, New York.

Aller, R.C. (1990) Bioturbation and manganese cycling in hemipelagic sediments. Phil. Trans. Royal Soc. London 331, 51–68.

Aller, R.C. (1994) Bioturbation and remineralization of sedimentary organic matter: effects of redox oscillation. Chem. Geol. 114, 331–345.

Aller, R.C. (1998) Mobile deltaic and continental shelf muds as suboxic, fluidized bed reactors. Mar. Chem. 61, 143–155.

Aller, R.C. (2001) Transport and reactions in the bioirrigated zone. In The Benthic Boundary Layer (Boudreau, B.P., and Jørgensen, B.B., eds.), pp. 269–301, Oxford University Press, New York.

Aller, R.C., and Aller, J.Y. (1998) The effect of biogenic irrigation intensity and solute exchange on diagenetic reactions rates in marine sediments. J. Mar. Res. 56, 905–936.

Aller, R.C., Aller, J.Y., and Kemp, P.F. (2001) Effects of particle and solute transport on rates and extent of remineralization in bioturbated sediments. In Organism-Sediment Interactions (Aller, J.Y., Woodin, S.A., and Aller, R.C., eds.), pp. 315–334, University of South Carolina Press, Columbia, SC.

Aller, R.C., Benninger, L.K., and Cochran, J.K. (1980) Tracking particle-associated processes in nearshore environments by use of $^{234}Th/^{238}U$ disequilibrium. Earth Planet. Sci. Lett. 47, 161–175.

Aller, R.C., and Blair, N.E. (2004) Early diagenetic remineralization of sedimentary organic C in the Gulf of Papua deltaic complex (Papua New Guinea): Net loss of terrestrial C and diagenetic fractionation of carbon isotopes. Geochim. Cosmochim. Acta 68, 1815–1825.

Aller, R.C., Blair, N.E., Xia, Q., and Rude, P.D. (1996) Remineralization rates, recycling and storage of carbon in Amazon shelf sediments. Cont. Shelf Res. 16, 753–786.

Aller, R.C., and Cochran, J.K. (1976) ^{234}Th–^{238}U disequilibrium in nearshore sediment: particle reworking and diagenetic time scales. Earth Planet. Sci. Lett. 29, 37–50.

Aller, R.C., and DeMaster, D.J. (1984) Estimates of particle flux reworking at the deep-sea floor using $^{234}Th/^{238}U$ disequilibrium. Earth Planet. Sci. Lett. 67, 308–318.

Aller, R.C., Hannides, A., Heilbrun, C., and Panzeca, C. (2004) Coupling of early diagenetic processes and sedimentary dynamics in tropical shelf environments: the Gulf of Papua deltaic complex. Cont. Shelf Res. 24, 2455–2486.

Aller, R.C., Mackin, J.E., and Cox, R.T. (1986) Diagenesis of Fe and S in Amazon inner shelf muds: apparent dominance of Fe reduction and implications for the genesis of ironstones. Cont. Shelf Res. 6, 263–289.

Aller, R.C., and Rude, P.D. (1988) Complete oxidation of solid phase sulfides by manganese and bacteria in anoxic marine sediments. Geochim. Cosmochim. Acta 52, 751–765.

Allison, M.A. (1998) Historical changes in the Ganges–Brahmaputra delta front. J. Coast. Res. 14, 1269–1275.

Allison, M.A., Kineke, G.C., Gordon, E.S., and Goni, M.A. (2001) Development and reworking of a seasonal flood deposit on the inner continental shelf off the Atchafalaya River. Cont. Shelf Res. 20, 2267–2294.

Allison, M.A., Nittrouer, C.A., and Kineke, G.C. (1995) Seasonal sediment storage on mudflats adjacent the Amazon River. Mar. Geol. 125, 303–328.

Almendros, G., Frund, R., Gonzalez-Vila, F.J., Haider, K.M., Knicker, H., and Ludemann, H.D. (1991) Analysis of ^{13}C and ^{15}N CPMAS NMR-spectra of soil organic matter and composts. Fed. Europ. Biochem. Soc. 282, 119–121.

Alongi, D.M. (1996) The dynamics of benthic nutrient pools and fluxes in tropical mangrove forests. J. Mar. Res. 54, 123–148.

Alongi, D.M. (1998) Coastal Ecosystem Processes. CRC Press, New York.

Alongi, D.M., Ayukai, T., Brunskill, G.J., Clough, B.F., and Wolanski, E. (1998) Effect of exported mangrove litter on bacterial productivity and dissolved organic carbon fluxes in adjacent tropical nearshore sediments. Mar. Ecol. Prog. Ser. 56, 133–144.

Alongi, D.M., Boto, K.G., and Robertson, A.I. (1992) Nitrogen and phosphorus cycles. In Tropical Mangrove Systems (Robertson, A.I., and Alongi, D.M., eds.), pp. 251–292, American Geophysical Union, Washington, DC.

Alongi, D.M., Boyle, S.G., Tirendi, F., and Payn, C. (1996) Composition and behavior of trace metals in post-oxic sediments of the Gulf of Papua, Papua New Guinea. Estuar. Coast. Shelf Sci. 42, 197–211.

Alongi, D.M., Tirendi, F., and Christoffersen, P. (1993) Sedimentary profiles and sediment–water solute exchange of iron and manganese in reef- and river-dominated shelf regions of the Coral Sea. Cont. Shelf Res. 13, 287–305.

Alongi, D. M., Wattayakorn, G., Pfitzner, J., Tirendi, F., Zagorskis, I., Brunskill, G. J., Davidson, A., and Clough, B. F. (2001) Organic carbon accumulation and metabolic pathways in sediments of mangrove forests in southern Thailand. Mar. Geol. 179, 85–103.

Alperin, M.J., Blair, N.E., Albert, D.B., Hoehler, T.H., and Martens, C.S. (1992) Factors that control isotopic composition of methane produced in an anaerobic marine sediment. Global Biogeochem. Cycles 6, 271–329.

Alperin, M.J., Martens, C.S., Albert, D.B., Suayah, I.B., Benninger, L.K., Blair, N.E., and Jahnke, R.A. (1999) Benthic fluxes and porewater concentration profiles of dissolved organic carbon in sediments from the North Carolina continental slope. Geochim. Cosmochim. Acta 63, 427–448.

Alperin, M.J., and Reeburgh, W.S. (1985) Inhibition experiments on anaerobic methane oxidation. Appl. Environ. Microbiol. 50, 940–945.

Alperin, M.J., Reeburgh, W.S., and Devol, A.H. (1992) Organic carbon remineralization and preservation in sediments of Skan Bay, Alaska. In Productivity, Accumulation, and Preservation of Organic Matter in Recent and Ancient Sediments (Whelan, J.K., and Farrington, J.W., eds.), pp. 99–122, Columbia University Press, New York.

Alpine, A.E., and Cloern, J.E. (1992) Trophic interactions and direct effects control phytoplankton biomass and production in an estuary. Limnol. Oceanogr. 37, 946–955.

Aluwihare, L.I., Repeta, D.J., and Chen, R.F. (1997) A major biopolymeric component to dissolved organic carbon in surface sea water. Nature 387, 166–169.

Aminot, A., El-Sayed, M.A., and Kerouel, R. (1990) Fate of natural and anthropogenic dissolved organic carbon in the macro-tidal Elorn estuary. Mar. Chem. 29, 255–275.

Amon, R.M.W., and Benner, R. (1996) Photochemical and microbial consumption of dissolved organic carbon and dissolved oxygen in the Amazon River system. Geochim. Cosmochim. Acta 60, 1783–1792.

Amon, R.M.W., and Benner, R. (2003) Combined neutral sugars as indicators of the diagenetic state of dissolved organic matter in the Arctic Ocean. Deep-Sea Res. I 50, 151–169.

Amon, R.M.W., Fitznar, H.P., and Benner. R. (2001) Linkages among the bioreactivity, chemical composition, and diagenetic state of marine dissolved organic matter. Limnol. Oceanogr. 46, 287–297.

Amouroux, D., Roberts, G., Rapsomanikis, S., and Andreae, M. O. (2002) Biogenic gas (CH_4, N_2O, DMS) emission to the atmosphere from near-shore and shelf waters of the north-western Black Sea. Estuar. Coastal Shelf Sci. 54, 575–587.

An, S., and Gardner, W.S. (2002) Dissimilatory nitrate reduction to ammonium (DNRA) as a nitrogen link, versus denitrification as a sink in a shallow estuary (Laguna Madre/Baffin Bay, Texas). Mar. Ecol. Prog. Ser. 237, 41–50.

An, S.M., and Joye, S.B. (2001) Enhancement of coupled nitrification–denitrification by benthic photosynthesis in shallow estuarine sediments. Limnol. Oceanogr. 46, 62–74.

Andersen, F.O., and Kristensen, E. (1992) The importance of benthic macrofauna in decomposition of microalgae in a coastal marine sediment. Limnol. Oceanogr. 37, 1392–1403.

Andersen, R.A., and Mulkey, T.J. (1983) The occurrence of chlorophylls c_1 and c_2 in the Chrysophyceae. J. Phycol. 19, 289–294.

Anderson, D.M., Glibert, P.M., and Burkholder, J.M. (2002) Harmful algal blooms and eutrophication: nutrient sources, composition, and consequences. Estuaries 25, 704–726.

Anderson, D.M., Glibert, P.M., and Burkholder, J.M. (2002) Harmful algal blooms and eutrophication: nutrient sources, composition, and consequences. Estuaries 25, 704–726.

Anderson, D.M., and Morel, F.M. (1978) Copper sensitivity of *Gonyaulax tamarensis*. Limnol. Oceanogr. 23, 283–295.

Anderson, E.C., and Libby, W.F. (1951) Worldwide distribution of natural radiocarbon. Phys. Rev. 81, 64–69.

Anderson, E.C., Libby, W.F., Weinhouse, S., Reid, A.F., Kirshenbaum, A.D., and Grosse, A.V. (1947) Natural radiocarbon from cosmic radiation. Phys. Rev. 72, 931–936.

Anderson, R., Rowe, G., Kemp, P., Trumbore, S., and Biscaye, P. (1994) Carbon budget for the mid-slope depocenter of the Middle Atlantic Bight. Deep-Sea Res. II 41, 669–703.

Anderson, R.F. (1987) Redox behavior of uranium in an anoxic marine basin. Uranium 3, 145–164.

Anderson, R.F., Fleisher, M.Q., and LeHuray, P. (1989) Concentration, oxidation state, and particulate flux of uranium in the Black Sea. Geochim. Cosmochim. Acta 53, 2215–2224.

Anderson, T.F., and Arthur, M.A. (1983) Stable isotopes of oxygen and carbon and their application to sedimentologic and paleoenvironmental problems. *In* Stable Isotopes in Sedimentary Geology (Arthur, M.A., Anderson, T.F., Kaplan, I.R., Veizer, J., and Land, L.S., eds.), pp. 1–151, Soc. Econ. Paleontol. Mineral.

Andreae, M.O. (1986) The ocean as a source of atmospheric sulfur compounds. *In* The Role of Air–Sea Exchange in Geochemical Cycling (Buat-Menard, P., ed.), pp. 331–362, Riedel, Dordrecht.

Andreae, M.O., (1990) Ocean–atmosphere interactions in the global biogeochemical sulfur cycle. Mar. Chem., 30, 1–29.

Andreae, M.O., and Crutzen, P.J. (1997) Atmospheric aerosols: Biogeochemical sources and role in atmosphere chemistry. Science 276, 1052–1058.

Andren, T.A., and Sohlenius, G. (1995) Late Quaternary development of the north-western Baltic proper—results from the clay-varve investigation. Quart. Intl. 27, 5–10.

Aneiso, A.M., Abreu, B.A., and Biddanda, B.A. (2003) The role of free and attached microorganisms in the decomposition of estuarine macrophyte detritus. Estuar. Coastal Shelf Sci. 56, 197–201.

Angel, M.V., and Fasham, M.J.R. (1983) Eddies and biological processes. *In* Eddies in Marine Science. (Robinson, A.R., ed.), pp. 492–524, Springer Verlag, Berlin.

Anita, N.J., Harrison, P.J., and Oliveira, L. (1991) Phycological reviews: the role of dissolved organic nitrogen in phytoplankton nutrition, cell biology, and ecology. Phycologia 30, 1–89.

Anschutz, P., Sundby, B., Lefrancois, L., Luther, III, G.W., and Mucci, A. (2000) Interaction between metal oxides and species of nitrogen and iodine in bioturbated marine sediments. Geochim. Cosmochim. Acta 64, 2751–2763.

Anschutz, P., Zhong, S., and Sundby, B. (1998) Burial efficiency of phosphorus and the geochemistry of iron in continental margin sediments. Limnol. Oceanogr. 43, 53–64.

Appleby, P.G., and Oldfield, F. (1992) Application of lead-210 to sedimentation studies. *In* Uranium-series Disequilibrium: Applications to Earth, Marine, and Environmental Problems. (Ivanovich, M., and Harmon, R.S. eds.), pp. 731–778, Clarendon Press, Oxford, UK.

Archibold, O.W. (1995) Ecology of World Vegetation. Chapman and Hall, London, UK.

Argyrou, M.E., Bianchi, T.S., and Lambert, C.D. (1997) Transport and fate of dissolved organic carbon in the Lake Pontchartrain estuary, Louisiana, USA. Biogeochemistry 38, 207–226.

Ari, T., and Norde, W. (1990) The behaviour of some model proteins at solid–liquid interfaces: adsorption from single protein solutions. Colloids Surfaces 51, 1–15.

Armentano, T.V., and Woodwell, G.M. (1975) Sedimentation rates in a Long Island marsh determined by ^{210}Pb dating. Limnol. Oceanogr. 20, 452–455.

Arnold, J.R., and Libby, W.F. (1949) Age determinations by radiocarbon context: checks with samples of known age. Science 110, 678–680.

Arnosti, C. (2003) Microbial extracellular enzymes and their role in dissolved organic matter cycling. *In* Aquatic Ecosystems: Interactivity of Dissolved Organic Matter (Findlay, S.E.G., and Sinsabaugh, R.L., eds.), pp. 316–337, Academic Press, New York.

Arnosti, C., and Holmer, M. (1999) Carbohydrate dynamics and contributions to the carbon budget of an organic-rich coastal sediment. Geochim. Cosmochim. Acta 63, 393–403.

Arnosti, C., and Repeta, D.J. (1994) Oligosaccharide degradation by an aerobic marine bacteria: characterization of an experimental system to study polymer degradation in sediments. Limnol. Oceanogr. 39, 1865–1877.

Arnosti, C., Repeta, D.J., and Blough, N.V. (1994) Rapid bacterial degradation of polysaccharides in anoxic marine systems. Geochim. Cosmochim. Acta 58, 2639–2652.

Arpin, N., Svec, W.A., and Liaaen-Jensen, S. (1976) A new fucoxanthin-related carotenoid from *Coccolithus huxleyi*. Phytochemistry 15, 529–532.

Arzayus, K.M., and Canuel, E.A. (2004) Organic matter degradation of the York River estuary: effects of biological vs. physical mixing. Geochim. Cosmochim. Acta 69, 455–463.

Asmus, R. (1986) Nutrient flux in short-term enclosures of intertidal sand communities. Ophelia 26, 1–18.

Aspinall, G.O. (1970) Pectins, plant gums and other plant polysaccharides. *In* The Carbohydrates; Chemistry and Biochemistry, 2nd edn. (Pigman, W., and Horton, D., eds.), pp. 515–536, Academic Press, New York.

Aspinall, G.O. (ed.) (1983) The Polysaccharides. Academic Press, New York.

Aston, S.R. (1980) Nutrients, dissolved gases, and general biogeochemistry in estuaries. *In* Chemistry and Biogeochemistry of Estuaries (Olausson, E., and Cato, I., eds.), pp. 233–257, John Wiley, New York.

Atlas, E.L. (1975) Phosphate equilibria in seawater and interstitial waters. Ph.D Thesis, Oregon State University.

Atwater, B.F. (1979) Ancient processes at the site of Southern San Francisco Bay: movement of the crust and changes in sea level. *In* San Francisco Bay: The Urbanized Estuary (Conomos, T.J., ed.), pp. 31–45, Pacific Division, America Association Advancement Science, California Academy of Science, San Francisco.

Atwater, B.F., Hedel, C.W., and Helley, E.J. (1977) Late Quaternary depositional history, Holocene sea-level changes, and vertical crustal movement, southern San Francisco Bay, California. U.S. Geological Survey Professional paper 1014. Washington, DC.

Aufdenkampe, A., Hedges, J.I., and Richey, J.E. (2001) Sorptive fractionation of dissolved organic nitrogen and amino acids onto fine sediments within the Amazon Basin. Limnol. Oceanogr. 46, 1921–1935.

Axelman, J., Broman, D., and Naf, C. (2000) Vertical flux of particulate/water dynamics of polychlorinated biphenyls (PCBs) in the open Baltic Sea. Ambio 29, 210–216.

Azam, F., and Cho, B. (1987) Bacterial utilization of organic matter in the sea. *In* Ecology of Microbial Communities (Fletcher, M., Gray, T.R., and Jones, G, eds.), pp. 261–281, Society for General Microbiology, Cambridge University Press, New York.

Azam, F., Fenchel. T., Field, J.G, Gray, J.S., Meyer-Reil, L.A., and Thingstad, F. (1983) The ecological role of water-column microbes in the sea. Mar. Ecol. Prog. Ser. 10, 257–263.

Bach, S.D., Thayer, G.W., and Lacroix, M.W. (1986) Export of detritus from eelgrass (*Zostera marina*) beds near Beaufort, North Carolina, USA. Mar. Ecol. Prog. Ser. 28, 265–278.

Bagander, L. E. (1977) Sulphur fluxes of the sediment and water interface. In situ study of closed systems, Eh, pH. Ph.D. Thesis, University of Stockholm.

Baines, S.B., Fisher, N.S., Doblin, M.A., and Cutter, G.A. (2001) Uptake of dissolved organic selenides by marine phytoplankton. Limnol. Oceanogr. 46, 1936–1944.

Baines, S.B., and Pace, M.L. (1991) The production of dissolved organic matter by phytoplankton and its importance to bacteria: patterns across marine and freshwater systems. Limnol. Oceanogr. 36, 1078–1090.

Baird, D., and Ulanowicz, R. (1989) The seasonal dynamics of the Chesapeake Bay ecosystem. Ecol. Monogr. 59, 329–364.

Baker, J.E., Eisenreich, S.J., and Eadie, B.J. (1991) Sediment trap fluxes and benthic recycling of organic carbon, polycyclic aromatic hydrocarbons, and polychlorinated-biphenyl cogeners in Lake Superior. Environ. Sci. Technol. 25, 500–509.

Balashova, V.V., and Zavarzin, G.A. (1980) Anaerobic reduction of ferric iron by hydrogen bacteria. Microbiology 48, 635–639.

Baldock, J.A., and Skjemstad, J.O. (2000) Role of soil matrix and minerals in protecting natural organic material against biological attack. Org. Geochem. 31, 697–710.

Banahan, S., and Goering, J.J. (1986) The production of biogenic silica and its accumulation on the southeastern Bering Sea shelf. Cont. Shelf Res. 5, 199–213.

Bandy, A. R., Scott, D.L., Blomquist, B.W., Chen, S.M., and Thornton, D.C. (1982) Low yields from dimethyl sulfide oxidation in the marine boundary layer, Geophys. Res. Lett. 19, 1125–1127.

Bange, H.W., Bartell, U.H., Rapsomanikis, S., and Andreae, M.O. (1994) Methane in the Baltic and North Seas and a reassessment of the marine emissions of methane. Global Biogeochem. Cycles 8, 465–480.

Bange, H.W., Dahlke, S., Ramesh, R., Meyer-Reil, L.A., Rapsomanikis, S., and Andrea, M.O. (1998) Seasonal study of methane and nitrous oxide in the coastal waters of the southern Baltic Sea. Estuar. Coastal Shelf Sci. 47, 807–817.

Bange, H.W., Rapsomanikis, S., and Andreae, M.O. (1996) The Aegean Sea as a source of atmospheric nitrous oxide and methane. Mar. Chem. 53, 41–49.

Banta, G.T., Holmer, M., Jensen, M.H., and Kristensen, E. (1999) Anaerobic effect of two polychaete worms, *Nereis diversicolor* and *Arenicola marine*, on aerobic and anaerobic decomposition in organic-poor marine sediment. Aquat. Microb. Ecol. 19, 189–204.

Barko, J.W., Gunnison, D., and Carpenter, S.R. (1991) Sediment interactions with submerged macrophyte growth and community dynamics. Aquat. Bot. 41, 1–65.

Barnard, W.R., Andreae, M.O., and Iverson, R.L. (1984) Dimethylsulfide and *Phaeocystis poucheti* in the south eastern Bering Sea. Cont. Shelf Res. 3, 103–113.

Baross, J.A., Crump B., and Simenstad, C.A. (1994) Elevated "microbial loop" activities in the Columbia River estuarine turbidity maximum. *In* Changes in Fluxes in Estuaries: Implications from Science to Management (Dyer, K.R., and Orth, R.J., eds.), pp. 459–464, Olsen and Olsen, Fredensborg, Denmark.

Barrick, R.C., and Prahl, F.G. (1987) Hydrocarbon geochemistry of the Puget Sound Region III. Polycyclic aromatic hydrocarbons in sediments. Estuar. Coastal Shelf Sci. 25, 175–191.

Barrie, C.Q., and Piper, D.J.W. (1982) Late Quaternary Geology of Makkovik Bay, Labrador, Canada. Geol. Survey Canada, Paper 81–17, 1–37.

Barry, R.G., Crane, R.C., Locke, C.W., and Miller, G.H. (1977) The coastal environment of southern Baffin Island and northern Labrador–Ungava. Final Report, Project 136, to Imperial oil Limited, Aquitaine Co. of Canada, Ltd., and Canada-Cities Services Ltd. Inst. of Arctic and Alpine Res., University of Colorado, Boulder.

Bartlett, K.B., Bartlett, D.S., Harris, R.G., and Sebacher, D.I. (1987) Methane emissions along a salt marsh salinity gradient. Biogeochemistry 4, 183–202.

Bartlett, K.B., Harris, R.C., and Sebacher, D.I. (1985) Methaneflux from coastal salt marshes. J. Geophys. Res. 90, 5710–5720.

Baskaran, M. (1995) A search for the seasonal variability on the depositional fluxes of ^{7}Be and ^{210}Pb. J. Geophys. Res. 100, 2833–2840.

Baskaran, M. (1999) Particle-reactive radionuclides as tracers of biogeochemical processes in estuarine and coastal waters of the Gulf of Mexico. *In* Biogeochemistry of Gulf of Mexico Estuaries (Bianchi, T.S., Pennock, J.R., and Twilley, R.R., eds.), pp. 381–404, John Wiley, New York.

Baskaran, M., Asbill, S., Santschi, P.H., Davis, T., Brooks, J.M., Champ, M.A., Makeyev, V., and Khlebovich, V. (1995) Distribution of ^{239,240}Pu and ^{238}Pu concentrations in sediments from the Ob and Yenisey rivers and the Kara Sea. Appl. Radiat. Isot. 46, 1109–1119.

Baskaran, M., Coleman, C.H., and Santschi, P.H. (1993) Atmospheric depositional fluxes of ^{7}Be and ^{210}Pb at Galveston and College Station, Texas. J. Geophys. Res. 98, 20, 555–20, 571.

Baskaran, M., and Naidu, S. (1995) ^{210}Pb-derived chronology, and the fluxes of ^{210}Pb and ^{137}Cs isotopes into continental shelf sediments, East Chukchi Sea, Alaskan Arctic. Geochim. Cosmochim. Acta 59, 4435–4448.

Baskaran, M., Ravichandran, M., and Bianchi, T.S. (1997) Cycling of ^{7}Be and ^{210}Pb in a high DOC, shallow, turbid estuary of southeast Texas. Estuar. Coastal Shelf Sci. 45, 165–176.

Baskaran, M., and Santschi, P.H. (1993) The role of particles and colloids in the transport of radionuclides in coastal environments of Texas. Mar. Chem. 43, 95–114.

Baskaran, M., Santschi, P.H., Benoit, G., and Honeyman, B.D. (1992) Scavenging of Th isotopes by colloids in seawater of the Gulf of Mexico. Geochim. Cosmochim. Acta 56, 3375–3388.

Baskaran, M., Santschi, P.H., Guo, L., Bianchi, T.S., and Lambert, C. (1996) ^{234}Th:^{238}U disequilibria in the Gulf of Mexico: the importance of organic matter and particle concentration. Cont. Shelf Res. 16, 353–380.

Bates, N., and Hansell, D.A. (1999) Hydrographic and biogeochemical signals in the surface ocean between Chesapeake Bay and Bermuda. Mar. Chem. 67, 1–16.

Bates, A.L., and Hatcher, P.G. (1989) Solid-state ^{13}C NMR studies of a large fossil gymnosperm from the Yallourn Open Cut, Latrobe Valley, Australia. Org. Geochem. 14, 609–617

Bates, A.L., Spiker, E.C., and Holmes, C.W. (1998) Speciation and isotopic composition of sedimentary sulfur in the Everglades, Florida, USA. Chem. Geol. 146, 155–170.

Battin, T.J. (1998) Dissolved organic matter and its optical properties in blackwater tributary of the upper Orinoco River, Venezuela. Org. Geochem. 28, 561–569.

Bauer, J.E. (2002) Carbon isotopic composition of DOM. In Biogeochemistry of Marine Dissolved Organic Matter (Hansell, D.A., and Carlson, C.A., eds.), pp. 405–453, Academic Press, Elsevier Science, New York.

Bauer, J.E., and Capone, D.G. (1985a). Effects of four aromatic organic pollutants on microbial glucose metabolism and thymidine incorporation in marine sediments. Appl. Environ. Microbiol. 49, 828–835.

Bauer, J.E., and Capone, D.G. (1985b) Degradation and mineralization of the polycyclic aromatic hydrocarbons anthracene and naphthalene in intertidal marine sediments. Appl. Environ. Microbiol. 50, 81–90.

Bauer, J.E., and Druffel, E.R.M. (1998) Ocean margins as a significance source of organic matter to the deep ocean. Nature 392, 335–246.

Bauer, J.E., Wolgast, D.W., Druffel, E.R.M., Griffin, S., and Masiello, C.A. (1998) Distributions of dissolved organic and inorganic carbon and radiocarbon in the eastern North Pacific continental margin. Deep Sea Res. II. 45, 689–714.

Bauza, J.F., Morell, J.M., and Corredor, J.E. (2002) Biogeochemistry of nitrous oxide production in red mangrove (Rhizophora mangle) forest sediments. Estuar. Coastal Shelf Sci. 55, 697–704.

Beardall, J., and Light, B. (1994) Biomass, productivity and nutrient requirements of microphytobenthos. Tech. Rep. 16. Port Phillip Bay Environmental Study, Commonwealth Scientific and Industrial Research Organization, Canberra, Australia.

Bebout, B.M., Fitzpatrick, M.W., and Paerl, H.W. (1993) Identification of the sources of energy for nitrogen fixation and physiological characterization of nitrogen-fixing members of a marine microbial mat community. Appl. Environ. Microbiol. 59, 1495–1503.

Bebout, B.M., Paerl, H.W., Bauer, J.E., Canfield, D.E., and Des Marais, D.J. (1994) Nitrogen cycling in microbial mat communities: the quantitative importance of N-fixation and other sources of N for primary productivity. In Microbial Mats (Lucas, J., and Caumette, P., eds.), pp. 32–42, Springer-Verlag, Berlin.

Beckett, R., Jue, Z., and Giddings, J.C. (1987) Determination of molecular weight distributions of fulvic and humic acids using field flow fractionation. Environ. Sci. Technol. 21, 289–295.

Belknap, D.F., and Kraft, J.C. (1977) Holocene relative sea-level changes and coastal stratigraphic units on the northwest flank of the Baltimore Canyon trough geosyncline. J. Sed. Petrol. 47, 610–629.

Benedetti, M.F., Milne, C.J., Kinniburgh, D.G., van Riemsdijk, W.H., and Koopal, L.K. (1995) Metal-ion binding to humic substances—application of the non-ideal competitive adsorption model. Environ. Sci. Technol. 29, 446–457.

Benitez-Nelson, C.R., and Buessler, K.O. (1999) Temporal variability of inorganic and organic phosphorus turnover rates in the coastal ocean. Nature 398, 502–505.

Benitez-Nelson, C.R., and Karl, D.M. (2002) Phosphorus cycling in the North Pacific Subtropical Gyre using cosmogenic ^{32}P and ^{33}P. Limnol. Oceanogr. 47, 762–770.

Benitez-Nelson, C.R., O'Neill, L., Kolowith, L.C., Pellechia, P., and Thunell, R. (2004) Phosphonates and particulate organic phosphorus cycling in an anoxic marine basin. Limnol. Oceanogr. 49, 1593–1604.

Benjamin, M.M., and Honeyman, B.D. (2000) Trace Metals. In Earth System Science—from Biogeochemical Cycles to Global Change (Jacobson, M.C., Charlson, R.J., Rodhe, H., and Orians, G.H., eds.), pp. 377–411, Academic Press, New York.

Benner, R. (2002) Chemical composition and reactivity. In Biogeochemistry of Marine Dissolved Organic Matter (Hansell, D.A., and Carlson, C.A., eds.), pp. 59–85, Academic Press, New York.

Benner, R. (2003) Molecular indicators of bioavailability of dissolved organic matter. In Aquatic Ecosystems: Interactivity of Dissolved Organic Matter (Findlay, S.E.G., and Sinsabaugh, R.L., eds.), pp. 122–135, Academic Press, New York.

Benner, R., Fogel, M.L., Sprague, E.K., and Hodson, R.E (1987) Depletion of ^{13}C in lignin and its implications for stable carbon isotope studies. Nature 329, 708–710.

Benner, R., Hatcher, P.G., and Hedges, J.I. (1990) Early diagenesis of mangrove leaves in a tropical estuary: Bulk chemical characterization using solid-state ^{13}C NMR and elemental analyses. Geochem. Cosmochim. Acta 54, 2003–2013.

Benner, R., and Kaiser, K. (2003) Abundance of amino sugars and peptidoglycan in marine particulate and dissolved organic matter. Limnol. Oceanogr. 48, 118–128.

Benner, R., and Opsahl, S. (2001) Molecular indicators of the sources and transformations of dissolved organic matter in the Mississippi River plume. Org. Geochem. 32, 597–611.

Benner, R., Pakulski, J.D., McCarthy, M., Hedges, J.I., and Hatcher, P.G. (1992) Bulk chemical characteristics of dissolved organic matter in the ocean. Science 255, 1561–1564.

Bennett, A. Bianchi, T.S., and Means, J.C. (2000) The effects of PAH contamination and grazing on the abundance and composition of microphytobenthos in salt marsh sediments (Pass Fourchon, LA, USA): II: the use of plant pigments as biomarkers. Estuar. Coastal Shelf Sci. 50, 425–439.

Bennett, A. Bianchi, T.S., Means, J.C., and Carman, K.R. (1999) The effects of PAH contamination and grazing on the abundance and composition of microphytobenthos in salt marsh sediments (Pass Fourchon, LA, USA): I. A microcosm experiment. J. Exp. Mar. Biol. Ecol. 242, 1–20.

Benninger, L.K. (1978) ^{210}Pb balance in Long Island Sound. Geochim. Cosmochim. Acta 42, 1165–1174.

Benninger, L.K., Aller, R.C., Cochran, J.K., and Turekian, K.K. (1979) Effects of biological sediment mixing on the ^{210}Pb chronology and trace metal distribution in a Long Island Sound sediment core. Earth Planet. Sci. Lett. 43, 241–259.

Benoit, G. (2001) ^{210}Pb and ^{137}Cs dating methods in lakes: a retrospective study. J. Paleolimnol. 25, 455–465.

Benoit, G., Gilmour, C.C., and Mason, R.P. (1999) Sulfide controls on mercury speciation and bioavailability to methylating bacteria in sediment pore waters. Environ. Sci. Technol. 33, 951–957.

Benoit, G., Gilmour, C.C., and Mason, R.P. (2001) The influence of sulfide on solid-phase mercury bioavailability for methylation by pure cultures of Desulfobulbus propionicus. Environ. Sci. Technol. 35, 127–132.

Benoit, G., Oktay-Marshall, S., Cantu, A., hood, E.M., Coleman, C., Corapcioglu, O., and Santschi, P.H. (1994) Partitioning of Cu, Pb, Ag, Zn, Fe, Al, and Mn between filter-retained particles, colloids, and solution in six Texas estuaries. Mar. Chem. 45, 307–336.

Berelson, W. M., Hammond, D.E., and Johnson, K. (1987) Benthic fluxes and the cycling of biogenic silica and carbon in two southern California borderland basins. Geochim. Cosmochim. Acta 51, 1345–1363.

Berg, G.M., Glibert, P.M., Jørgensen, N.O.G., Balode, M., and Purina, I. (2001) Variability in inorganic and organic nitrogen uptake associated with riverine nutrient input in the Gulf of Riga, Baltic Sea. Estuaries 24, 204–214.

Berg, G.M., Glibert, P.M., Lomas, M.W., and Burford, M. (1997) Organic nitrogen uptake and growth by the Chrysophyte *Aurecoccus anophagefferens* during a brown tide event. Mar. Biol. 129, 377–387.

Bergamaschi, B.A., Tsamakis, E., Keil, R.G., Eglinton, T.I., Montlucon, D.B., and Hedges, J.I. (1997) The effect of grain size and surface area on organic matter, lignin, and carbohydrate concentration, and molecular compositions in Peru margin sediments. Geochim. Cosmochim. Acta 61, 1247–1260.

Bergamaschi, B.A., Walters, J.S., and Hedges, J.I. (1999) Distributions of uronic acids and *O*-methyl sugars in sinking and sedimentary particles in two coastal marine environments. Geochim. Cosmochim. Acta 63, 413–425.

Berger, U., and Heyer, J. (1989) Utersuchungen zum Methankreislauf in der Saale. J. Basic Microb. 29, 195–213.

Bergmann, W. (1949) Comparative biochemical studies on the lipids of marine invertebrates with special reference to the sterols. J. Mar. Res. 8, 137–176.

Berman, T., and Bronk, D.A. (2003) Dissolved organic nitrogen: a dynamic participant in aquatic ecosystems. Aquat. Microb. Ecol. 31, 279–305.

Berman, T., and Chava, S. (1999) Algal growth on organic compounds as nitrogen sources. J. Plank. Res. 21, 1423–1437.

Berner, R.A. (1964) Iron sulfide formed from aqueous solution at low temperatures and atmospheric pressure. J. Geol. 72, 293–306.

Berner, R.A. (1970) Sedimentary pyrite formation. Am. J. Sci. 268, 1–23.

Berner, R.A. (1974) Kinetic models for early diagenesis of nitrogen, sulfur, phosphorus, and silicon in anoxic marine sediments. *In* The Sea (Goldberg, E.D., ed.), pp. 427–450, John Wiley, New York.

Berner, R.A. (1980) Early Diagenesis: A Theoretical Approach. Princeton University Press, N.J.

Berner, R.A. (1984) Sedimentary pyrite formation. Am. J. Sci. 268, 1–23.

Berner, R.A. (1989) Biogeochemical cycles of carbon and sulphur and their effect of atmospheric oxygen over Phanerozoic time. Paleogeogr. Paleoclim. Paleoecol. 75, 97–122.

Berner, R.A. (2004) The Phanerozoic Carbon Cycle: CO_2 and O_2. Oxford University Press, New York.

Berner, E.K., and Berner, R.A. (1987) The Global Water Cycle: Geochemistry and Environment. Prentice Hall, New York.

Berner, R.A., and Berner, R.A. (1996) Global Environment: Water, Air, and Geochemical Cycles. Prentice Hall, New York.

Berner, R.A., and Rao, J. (1994) Phosphorus in sediments of the Amazon River and estuary: implications for the global flux of phosphorus to the sea. Geochim. Cosmochim. Acta 38, 2333–2339.

Berner, R.A., and Westrich, J.T. (1985) Bioturbation and the early diagenesis of carbon and sulfur. Am. Sci. 285, 193–206.

Berresheim, H., M.O. Andreae, R.L. Iverson, and Li, S.M. (1991) Seasonal variations of dimethylsulfide emissions and atmospheric sulfur and nitrogen species over the western north Atlantic Ocean. Tellus 43, 353–372.

Berthelot, M. (1862) Essai d'une Theorie sur la formation des ethers. Ann Chim. Phys. 66, 110–128.

Bertilsson, S., Stepanauskas, R., Cuadros-Hansson, R. Graneli, W., Wilkner, J., and Tranvik, L.J. (1999) Photochemically induced changes in bioavailable carbon and nitrogen pools in a boreal watershed. Aquat. Microb. Ecol. 19, 47–56.

Bertine, K.K., Chan, L.H., and Turekian, K.K. (1970) Uranium determination in deep-sea sediments and natural waters using fission tracks. Geochim. Cosmochim. Acta 34, 641.

Bertness, M.D. (1985) Fiddler crab regulation of *Spartina alterniflora* production on a New England salt-marsh. Ecology 66, 1042–1055.

Bertness, M.D. and Ellison, A.M. (1987) Determinants of pattern in a New England salt marsh plant community. Ecol. Monogr. 57, 129–147.

Bhat, S.G., and Krishnaswami, S. (1969) Isotopes of uranium and radium in Indian rivers. Proc. Indian Acad. Sci. 70, 1–17.

Bianchi, T.S. (1988) Feeding ecology of subsurface deposit-feeder *Leitoscoloplos fragilis* Verrill I. Mechanisms affecting particle availability on an intertidal sandflat. J. Exp. Mar. Biol. Ecol. 115, 79–97.

Bianchi, T.S. (1991) Density-dependent consumer effects on resource quality in carbonate sediments. Tex. J. Sci. 43, 283–295.

Bianchi, T.S., and Argyrou, M.E. (1997) Temporal and spatial dynamics of particulate organic carbon in the Lake Pontchartrain estuary, southeast Louisiana, USA. Estuar. Coastal Shelf Sci. 45, 557–569.

Bianchi, T.S., Argyrou, M., and Chipett, H.F. (1999b) Contribution of vascular-plant carbon to surface sediments across the coastal margin of Cyprus (eastern Mediterranean). Org. Geochem. 30, 287–297.

Bianchi, T.S., Baskaran, M., Delord, J., and Ravichandran, M. (1997a) Carbon cycling in a shallow turbid estuary of southeast Texas: the use of plant pigments as biomarkers. Estuaries 20, 404–415.

Bianchi, T.S., and Canuel, E.A. (2001) Organic geochemical tracers in estuaries. 32, 451–452.

Bianchi, T.S., Dawson, R., and Sawangwong, P. (1988) The effects of macrobenthic deposit-feeding on the degradation of chloropigments in sandy sediments. J. Exp. Mar. Biol. Ecol. 122, 243–255.

Bianchi, T.S., Engelhaupt, E., McKee, B.A., Miles, S., Elmgren, R., Hajdu, S., Savage, C., and Baskaran, M. (2002a) Do sediments from coastal sites accurately reflect time trends in water column phytoplankton? A test from Himmerfjarden Bay (Baltic Sea proper). Limnol. Oceanogr. 47, 1537–1544.

Bianchi, T.S., Engelhaupt, E., Westman, P., Andren, T., Rolff, C., and Elmgren, R. (2000c) Cyanobacterial blooms in the Baltic Sea: natural or human-induced? Limnol. Oceanogr. 45, 716–726.

Bianchi, T.S., Filley, T., Dria, K., and Hatcher, P.G. (2004) Temporal variability in sources of dissolved organic carbon in the lower Mississippi River. Geochim. Cosmochim. Acta 68, 959–967.

Bianchi, T.S., and Findlay, S. (1990) Plant pigments as tracers of emergent and submergent macrophytes from the Hudson River. Can. J. Fish. Aquat. Sci., 47, 492–494.

Bianchi, T.S., and Findlay, S. (1991) Decomposition of Hudson estuary macrophytes: photosynthetic pigment transformations and decay constants. Estuaries 14, 65–73.

Bianchi, T.S., Findlay, S., and Dawson, R. (1993) Organic matter sources in the water column and sediments of the Hudson River estuary: the use of plant pigments as tracers. Estuar. Coastal Shelf Sci. 36, 359–376.

Bianchi, T.S., Findlay, S., and Fontvieille, D. (1991) Experimental degradation of plant materials in Hudson River sediments. I. Heterotrophic transformations of plant pigments. Biogeochemistry 12, 171–187.

Bianchi, T.S., Freer, M.E., and Wetzel, R.G. (1996) Temporal and spatial variability and the role of dissolved organic carbon (DOC) in methane fluxes from the Sabine River floodplain (Southeast Texas, USA). Arch. Hydrobiol. 136, 261–287.

Bianchi, T.S., Johansson, B., and Elmgren, R. (2000a) Breakdown pf phytoplankton pigments in Baltic sediments: effects of anoxia and loss of deposit-feeding macrofauna. J. Exp. Mar. Biol. Ecol. 251, 161–183.

Bianchi, T.S., Kautsky, K., and Argyrou, M. (1997c) Dominant chlorophylls and carotenoids in macroalgae of the Baltic Sea (Baltic Proper): their use as potential biomarkers. Sarsia 82, 55–62.

Bianchi, T.S., Lambert, C.D., Santschi, P.H., and Guo, L. (1997b) Sources and transport of land-derived particulate and dissolved organic matter in the Gulf of Mexico (Texas shelf/slope): the use of lignin-phenols and loliolides as biomarkers. Org. Geochem. 27, 65–78.

Bianchi, T.S., Mitra, S., and McKee, M. (2002b) Sources of terrestrially-derived carbon in the Lower Mississippi River and Louisiana shelf: Implications for differential sedimentation and transport at the coastal margin. Mar. Chem. 77, 211–223.

Bianchi, T.S., Pennock, J.R., and Twilley, R.R. (eds.) (1999a) Biogeochemistry of Gulf of Mexico estuaries: implications for management. In Biogeochemistry of Gulf of Mexico Estuaries, pp. 407–421, John Wiley, New York.

Bianchi, T.S., and Rice, D.L. (1988) Feeding ecology of *Leitoscoloplos fragilis*. II. Effects of worm density on benthic diatom production. Mar. Biol. 99, 123–131.

Bianchi, T.S., Rolfe, C., and Lambert, C.D. (1997d) Sources and composition of particulate organic carbon in the Baltic Sea: the use of plant pigments and lignin-phenols as biomarkers. Mar. Ecol. Prog. Ser. 156, 25–31.

Bianchi, T.S., P. Westman, C. Rolff, E. Engelhaupt, T. Andren, and Elmgren, R. (2000b) Cyanobacterial blooms in the Baltic Sea: Natural or human-induced? Limnol. Oceanogr. 45, 716–726.

Biddanda, B.A., and Benner, R. (1997) Carbon, nitrogen, and carbohydrate fluxes during the production of particulate and dissolved organic matter by marine phytoplankton. Limnol. Oceanogr. 42, 506–518.

Biddanda, B.A., and Pomeroy, L.R. (1988) Microbial aggregation and degradation of phytoplankton-derived detritus in seawater. I. Microbial succession. Mar. Ecol. Prog. Ser. 42, 79–88.

Bidigare, R.R., Kennicutt II M.C., and Keeney-Kennicutt, W.L. (1991) Isolation and purification of chlorophylls *a* and *b* for the determination of stable carbon and nitrogen isotope compositions. Anal. Chem. 63, 130–133.

Bidigare, R.R., Ondrusek, M.E., and Brooks, J.M. (1993) Influence of the Orinoco River outflow on distributions of algal pigments in the Caribbean Sea. J. Geophys. Res. 98, 2259–2269.

Bidle, K., and Azam, F. (1999) Accelerated dissolution of diatom silica by marine bacterial assemblages. Nature 397, 508–512.

Bidle, K.D., and Azam, F. (2001) Bacterial control of silicon regeneration from diatom detritus: significance of bacterial ectohydrolases and species identity. Limnol. Oceanogr. 46, 1606–1623.

Bidle, K., Brzezinski, M.A., Long, R.A., Jones, J.L., and Azam, F. (2003) Diminished efficiency in the oceanic silica pump caused by bacteria-mediated silica dissolution. Limnol. Oceanogr. 48, 1855–1868.

Bidleman, T.F. (1988) Atmospheric processes: wet and dry deposition of organic compounds are controlled by their vapor-particle partitioning. Environ. Sci. Technol. 22, 361–367.

Bieger, T., Abrajano, Jr., T.A., and Hellou, J. (1997) Generation of biogeneic hydrocarbons during a spring bloom in Newfoundland coastal (NW Atlantic) waters. Org. Geochem. 26, 207–218.

Bigeleisen, J. (1949) The relative reaction velocities of isotopic molecules. J. Chem. Phys. 17, 675–678.

Biggs, R. B., and Beasley, E.L. (1988) Bottom and suspended sediments in the Delaware River and estuary. In Ecology and restoration of the Delaware River Basin. (S.K. Majumdar, S.K., Miller, E.W., and Sage, L.E., eds.), pp. 116–151, Pennsylvania Academy of Sciences, Philadelphia.

Biggs, R.B., and Howell, B.A. (1984) The estuary as a sediment trap: alternate approaches to estimating its filtering efficiency. In The Estuary as a Filter (Kennedy, V.S., ed.), pp. 107–129, Academic Press, New York.

Biggs, R.B., Sharp, J.H., Church, T.M., and Tramontano, J.M. (1983) Optical properties, suspended sediments, and chemistry associated with the turbidity maxima of the Delaware estuary. Can. J. Fish. Aquat. Sci. 40, 172–179.

Billen, G. (1984) Heterotrophic utilization and regeneration of nitrogen. In Heterotrophic utilization and regeneration of nitrogen (Hobbie, J.E., and Williams, P.J., eds.), pp. 313–355, Plenum Press, New York.

Billen, G., Garnier, J., Deligne, C., and Billen, C. (1999) Estimates of early industrial inputs of nutrients to river systems: Implications for coastal eutrophication. Sci. Total Environ. 243/244, 43–52.

Billen, G., Garnier, J., Ficht, A., and Cun, C. (2001) Modeling the response of water quality in the Seine River estuary to human activity in its watershed over the last 50 years. Estuaries 24, 977–993.

Billen, G., Lancelot, C., and Meybeck, M. (1991) N, P, and Si retention along the aquatic continuum from land to ocean. In Ocean margin Processes in Global Change (Mantoura, R.F.C., Martin, J.M., and Wollast, R., eds.), pp. 19–44, John Wiley, New York.

Billen, G., Somville, M., de Becker, E., and Servais, P. (1985) A nitrogen budget of the Scheldt hydrographical basin. Neth. J. Sea Res. 19, 223–230.

Billet, D.S.M., Lampitt, R.S., Rice, A.L., and Mantoura, R.F.C. (1983) Seasonal sedimentation of phytoplankton to the deep sea benthos. Nature 302, 520–522.

Bird, E.C.F. (1980) Mangroves and coastal morphology. Victorian Naturalist 97, 48–58.

Bird, E.C.F., and Schwartz, M.L. (1985) The World's Coastline. Van Nostrand Reinhold, New York.

Birkeland, P.W. (1999) Soils and Geomorphology, 2nd ed. Oxford University Press, New York.

Bishop, C.T., and Jennings, H.J. (1982) Immunology of polysaccharides. In The Polysaccharides (Aspinall, G.O., ed.), pp. 292–325, Academic Press, New York.

Bjørnland, T., and Liaaen-Jensen, S. (1989) Distribution patterns of carotenoids in relation to chromophyte phylogeny and systematics. In The Chromophyte Algae: Problems and Perspectives (Green, J.C., and Diver, B.S.C., eds.), pp. 37–60, Clarendon Press, Oxford, UK.

Black, A.S., and Waring, S.A. (1977) The natural abundance of ^{15}N in the soil–water system of a small catchment area. Aust. J. Soil Res. 15, 51–57.

Blackburn, T.H. (1991) Accumulation and regeneration: Processes at the benthic boundary layer. In Ocean margin Processes in Global Change (Mantoura, R.F.C., Martin, J.M., and Wollast, R., eds.), pp. 181–195, John Wiley, New York.

Blackburn, T.H., and Henriksen, K. (1983) Nitrogen cycling in different types of sediments from Danish waters. Limnol. Oceanogr. 28, 477–493.

Blair, N.E., Leithold, E.I., and Aller, R.C. (2004) From bedrock to burial: the evolution of particulate organic carbon across coupled watershed–continental margin systems. Mar. Chem. 92, 141–156.

Blair, N.E., Leithold, E.I., Ford, S.T., Peeler, K.A., Holmes, J.C., and Perkey, D.W. (2003) The persistence of memory: the fate of ancient sedimentary organic carbon in a modern sedimentary system. Geochim. Cosmochim. Acta 67, 63–73.

Blake, A.C., Kineke, G.C., Milligan, T.G., and Alexander, C. (2001) Sediment trapping and transport in the ACE basin, South Carolina. Estuaries 24, 721–733.

Bledsoe, E.L., and Phlips, E.J. (2000) Relationships between phytoplankton standing crop and physical, chemical, and biological gradients in the Suwannee river and plume region, USA. Estuaries 23, 458–473.

Bloom, N.S., and Crecelius, E.A. (1983) Solubility behavior of atmospheric ^{7}Be in the marine environment. Mar. Chem. 12, 323–331.

Bloom, N.S., Gill, G.A., Cappellino, S., Dobbs, C., McShea, L., Drsicoll, C., Mason, R., and Rudd, J. (1999) Speciation and cycling of mercury in Lavaca Bay, Texas, sediments. Environ. Sci. Technol. 33, 7–13.

Bloom N.S., Preuss, E., Katon, J., and Hiltner, M. (2003) Selective extractions to assess the biogeochemically relevant fractionation of inorganic mercury in sediments and soils. Anal. Chim. Acta 479, 233–248.

Blough, N.V., and Del Vecchio, R. (2002) Chromophoric DOM in the Coastal Environment. In Biogeochemistry of Marine Dissolved Organic Matter (Hansell, D.A., and Carlson, C.A., eds.), pp. 509–546, Academic Press, New York.

Blough, N.V., and Green, S.A. (1995) Spectroscopic characterization and remote sensing of non-living organic matter. In The Role of Non-living Organic Matter in the Earth's Carbon Cycle (Zepp, R.G., and Sonntag, C., eds.), pp. 23–45, john Wiley, Chichester, UK.

Blough, N.V., and Zepp, R.G. (eds.) (1990) Effects of Solar Ultraviolet Radiation on Biogeochemical Dynamics in Aquatic Environments. Woods Hole Oceanographic Institution Technical Report, WHOI-90-09.

Blumer, M., Guillard, R.R.L., and Chase, T. (1971) Hydrocarbons of marine phytoplankton. Mar. Biol. 8, 183–189.

Blumer, M., Mullin, M.M., and Thomas, D.S. (1964) Pristane in zooplankton. Science 140, 974.

Boatman, C.D., and Murray, J.W. (1982) Modeling exchangeable NH_4^+ adsorption in marine sediments: process and controls of adsorption. Limnol. Oceanogr. 27, 99–110.

Bock, B.R. (1984) Efficient use of nitrogen in cropping systems. In Nitrogen in Crop Production (Hauck, R.D., ed.), pp. 273–294, American Society of Agronomy, Madison, WI.

Boesch, D.F. (2002) Challenges and opportunities for science in reducing nutrient over-enrichment of coastal ecosystems. Estuaries 25, 886–900.

Boesch, D.F., Burger, J., D'Elia, C.F., Reed, D.J., and Scavia, D. (2000) Scientific synthesis in estuarine management. In Estuarine Science: A Synthetic Approach to Research and Practice (Hobbie, J.E., ed.), pp. 507–526, Island Press, Washington, DC.

Boesen, C., and Postma, D. (1988) Pyrite formation in anoxic sediments of the Baltic. Am. J. Sci. 288, 575–603.

Boicourt, W.C. (1990) The influences of circulation processes on dissolved oxygen in Chesapeake Bay. In Oxygen Dynamics in Chesapeake Bay, USA, A Synthesis of Recent Research (Smith, D.E., Leffler, M., and Mackiernan, eds.), pp. 1–59, Maryland Sea Grant, College Park, MD.

Bokuniewicz, H. (1995) Sedimentary systems of coastal-plain estuaries. In Geomorphology and Sedimentology of Estuaries. Developments in Sedimentology 53 (Perillo, G.M.E., ed.), pp. 49–67, Elsevier Science, New York.

Bokuniewicz, H., Pollack, M., Blum, J., and Wilson, R. (2004) Submarine ground water discharge and salt penetration across the sea floor. Ground Water 42, 983–989.

Bollinger, M.S., and Moore, W.S. (1993) Evaluation of salt marsh hydrology using radium as a tracer. Geochim. Cosmochim. Acta 57, 2203–2212.

Bonin, P., Gilewicz, M., and Bertrand, J.C. (1989) Effects of oxygen on each step of denitrification on *Pseudomonas nautica*. Can. J. Microbiol. 35, 1061–1064.

Bonin, P., Omnes, P., and Chalamet, A. (1998) Simultaneous occurrence of denitrification and nitrate ammonification in sediments of the French Mediterranean coast. Hydrobiologia 389, 169–182.

Bonin, P., and Raymond, N. (1990) Effects of oxygen on denitrification in marine sediments. Hydrobiologia 207, 115–122.

Bopp, R.F., Simpson, H.J., Olsen, C.R., Trier, R.M., and Kostyk, N. (1982) Chlorinated hydrocarbons and radionuclide chronologies in sediments of the Hudson River and estuary, New York. Environ. Sci. Technol. 15, 210–218.

Borch, N.H., and Kirchman, D.L. (1999) Protection of labile organic matter from bacterial degradation by submicron particles. Aquat. Microb. Ecol. 16, 265–272.

Borges, A.V. (2005) Do we have enough pieces of the jigsaw to integrate CO_2 fluxes in the coastal ocean? Estuaries 28, 3–27.

Borges, A.V., Delille, B., Schiettecatte, L., Gazeau, F., Abril, G., and Frankignoulle, M. (2004) Gas transfer velocities of CO_2 in three European estuaries (Randers Fjord, Scheldt, and Thames). Limnol. Oceanogr. 49, 1630–1641.

Borges, A.V., and Frankignoulle, M. (2002) Distribution and air–water exchange of carbon dioxide in the Scheldt plume off the Belgian coast. Biogeochemistry 59, 41–67.

Borole, D., Krishnaswami, S., and Somayajulu, B.L.K. (1977) Investigations on dissolved uranium, silicon, and particulate trace elements in estuaries. Estuar. Coastal Shelf Sci. 5, 743–754.

Borole, D., Krishnaswami, S., and Somayajulu, B.L.K. (1982) Uranium isotopes in rivers, estuaries and adjacent coastal sediments of western India: their weathering, transport and oceanic budget. Geochim. Cosmochim. Acta 46, 125–137.

Borsuk, M., Stowe, C., Luettich, R.A., Paerl, H., and Pinckney, J. (2001) Modeling oxygen dynamics in an intermittently stratified estuary: estimation of process rates using field data. Estuar. Coastal Shelf Sci. 52, 33–49.

Bortolus, A., Iribarne, O. O., and Martinez, M.M. (1998) Relationship between waterfowl and the seagrass *Ruppia maritima* in a southwestern Atlantic coastal lagoon. Estuaries 21, 710–717.

Boschker, H.T.S., Bertilsson, S.A., Dekkers, E.M.J., and Cappenberg, T.E. (1995) An inhibitor-based method to measure initial decomposition of naturally occurring polysaccharides in sediments. Appl. Environ. Microbiol. 61, 2186–2192.

Boschker, H.T.S., de Brower, J.F.C., and Cappenberg, T.E. (1999) The contribution of macrophyte-derived organic matter to microbial biomass in salt-marsh sediments: Stable carbon isotope analysis of microbial biomarkers. Limnol. Oceanogr. 44, 309–319.

Boto, K.G. (1982) Nutrient and organic fluxes in mangroves. *In* Mangrove Ecosystems in Australia (Clough, B.F., ed.), pp. 239–257, Australian National University Press, Canberra.

Boto, K.G., and Bunt, J.S. (1981) Tidal export of particulate organic matter from a northern Australian mangrove system. Estuar. Coastal Shelf Sci. 13, 247–255.

Boto, K.G., and Wellington, J.T. (1988) Seasonal variations in concentrations and fluxes of dissolved organic materials in a tropical, tidally-dominated, mangrove waterway. Mar. Ecol. Prog. Ser. 50, 151–160.

Böttcher, M.E., and Lepland, A. (2000) Biogeochemistry of sulfur in a sediment core from the west-central Baltic Sea: evidence from stable isotopes and pyrite textures. J. Mar. Syst. 25, 299–312.

Böttcher, M.E., Sievert, S.M., and Kuever, J. (1999) Fractionation of sulfur isotopes during assimilatory reduction of sulfate thermophilic gram-negative bacterium at 60 degrees C. Arch. Microbiol. 172, 125–128.

Böttcher, M.E., Strebel, O., Voerkelius, S., and Schmidt, H.L. (1990) Using isotope fractionation of nitrate-nitrogen and nitrate-oxygen for evaluation of microbial denitrification in a sandy aquifer. J. Hydrol. 114, 413–424.

Boudreau, B.P. (1996) The diffusive tortuosity of fine-grained unlithified sediments. Geochim. Cosmochim. Acta 60, 3139–3142.

Boudreau, B.P., and Canfield, D.E. (1993) A comparison of closed- and open-system models for porewater pH and calcite dissolution. Geochim. Cosmochim. Acta 57, 317–334.

Boudreau, B.P., and Jørgensen, B.B. (eds.) (2001) The Benthic Boundary Layer: Transport Processes and Biogeochemistry, Oxford University Press, Oxford, UK.

Boudreau, B.P., and Westrich, J.T. (1984) The dependence of bacterial sulfate reduction on sulfate concentration in marine sediments. Geochim. Cosmochim. Acta 48, 2503–2516.

Bouillon, S., Frankignoulle, M., Dehairs, F., Velimirov, B., Eiler, A., Abril, G., Etcheber, H., and Borges, A.V. (2003) Inorganic and organic carbon biogeochemistry in the Gautami Godavari estuary (Andhra Pradesh, India) during pre-monsoon: the local impact of extensive mangrove forests. Global. Biogeochem. Cycles 17, 1114.

Bouloubassi, I., and Saliot, A. (1993) Dissolved, particulate and sedimentary naturally derived polycyclic aromatic hydrocarbons in coastal environment: geochemical significance. Mar. Chem. 42, 127–143.

Bourgeois J. (1994) Patterns of benthic nutrient fluxes on the Louisiana continental shelf. M.S. thesis, University of Southwestern Louisiana.

Boutton, T.W. (1991) Stable carbon isotope ratios of natural materials: II. Atmospheric terrestrial, marine, and freshwater environments. In Carbon Isotope Techniques (Coleman, D.C., and Fry, B., eds.), pp. 173–185, Academic Press, NY.

Bouwman, A.F., van Drecht, G., Knoop, J.M., Beusen, A.H.W., and Meinardi, C.R. (2005) Exploring changes in river nitrogen export to the worlds oceans. Global Biogeochem. Cycles 19, doi:1029/2004GB002314.

Bowden, K.F. (1967) Circulation and diffusion. In Estuaries (Lauff, G.H., ed.), pp. 15–36, American Association for the Advancement of Science, Washington, DC.

Bowden, K.F. (1980) Physical factors: salinity, temperature, circulation, and mixing processes. In Chemistry and Biogeochemistry of Estuaries (Olausson, E., and Cato, I., eds.), pp. 38–68, John Wiley, New York.

Bowen, J.L., and Valiela, I. (2004) Nitrogen loads to estuaries: using loading models to assess the effectiveness of management options to restore estuarine water quality. Estuaries 27, 482–500.

Boyd, R., and Penland, S. (1988) A geomorphic model for Mississippi delta evolution. Trans. Gulf Coast Assoc. Geol. Soc. 38, 443–452.

Boyer, J. N., Christian, R. R., and Stanley, D. W. (1993) Patterns of phytoplankton primary productivity in the Neuse River estuary, North Carolina, USA. Mar. Ecol. Prog. Ser. 97, 287–297.

Boyle, E.A., Edmond, J.M., and Sholkovitz, E.R. (1977) The mechanism of iron removal in estuaries. Geochim. Cosmochim. Acta 41, 1313–1324.

Boynton, W.R., Boicourt, W., Brant, S., Hagy, J., Harding, E. Houde, E., Holliday, D.V., Jech, M., Kemp, W.M., Lascara, C., Leach, S.D., Madden, A.P., Roman, M.,

Sanford, L., and E.M. Smith. (1997) Interactions between physics and biology in estuarine turbidity maximum (ETM) of Chesapeake Bay, USA. CM 1997/S:11. International Council for the Exploration of the Sea, Copenhagen, Denmark.

Boynton, W.R., Garber, J.H., Summers, R., and Kemp, W.M. (1995) Inputs, transformations, and transport of nitrogen and phosphorus in Chesapeake Bay and selected tributaries. Estuaries 18, 285–314.

Boynton, W.R., and Kemp, W.M. (1985) Nutrient regeneration and oxygen consumption by sediments along an estuarine gradient. Mar. Ecol. Prog. Ser. 23, 45–55.

Boynton, W.R., and Kemp, W.M., (2000) Influence of river flow and nutrient loads on selected ecosystem processes—a synthesis of Chesapeake Bay data. In Estuarine Science: A Synthetic Approach to Research and Practice (Hobbie, J.E., ed.), pp. 269–298, Island Press, Washington, DC.

Boynton, W.R., Kemp, W.M., and Keefe, C.W. (1982) A comparative analysis of nutrients and other factors influencing estuarine phytoplankton production. In Estuarine Comparisons (Kennedy, V.S., ed.), pp. 69–90, Academic Press, New York.

Boynton, W.R., Matteson, L.L., Watts, J.L., Stammerjohn, S.E., Jasiniski, D.A., and Rohland, F.M. (1991) Maryland Chesapeake Bay water quality monitoring programs: ecosystem processes component level 1 interpretive report No. 8. UMCEES, CBL Ref. No. 91–110.

Boynton, W.R., Murray, L., Hagy, J.D., Stokes, C., and Kemp, W.M. (1996) A comparative analysis of eutrophication patterns in a temperate coastal lagoon. Estuaries 19, 408–421.

Branch, G.M., and Griffiths, C.L. (1988) The Benguela ecosystem. Part V. The coastal zone. Oceanogr. Mar. Biol. Ann. Rev. 26, 395.

Brandes, J.A., and Devol, A.H. (1995) Simultaneous nitrate and oxygen respiration in coastal sediments: evidence for discrete diagenesis. J. Mar. Res. 53(5), 771–797.

Brandes, J.A., and Devol, A.H. (1997) Isotopic fractionation of oxygen and nitrogen in coastal marine sediments. Geochim. Cosmochim. Acta 61, 1793–1801.

Brasse, S., Nellen, M., Seifert, R., and Michaelis, W. (2002) The carbon dioxide system in the Elbe estuary. Biogeochemistry 59, 25–40.

Brassell, S.C., Comet, P.A., Eglinton, G., Isaacson, P.J., McEvoy, J., Maxwell, J.R., Thomson, I.D., Tibbetts, P.J., and Volkman, J.K., (1980) The origin and fate of lipids in the Japan Trench. In Advances in Organic Geochemistry (Douglas, A.G., and Maxwell, J.R. eds.), pp. 375–392, Pergamon Press, Oxford, UK.

Brassell, S.C., and Eglinton, G. (1983) The potential of organic geochemical compounds as sedimentary indicators of upwelling. In Coastal Upwelling: Its Sediment Record (Suess, E., and Thiede, J., eds.), pp. 545–571, Plenum Press, New York.

Bravo, J.M., Perzi, M., Hartner, T., Kannenberg, E.L., and Rohmer, M. (2001) Novel methylated triterpenoids of the gammacerane series from the nitrogen-fixing bacterium Bradyrhizobium japonicum USDA 110. Eur. J. Biochem. 286, 1323–1331.

Breitburg, D.L. (2002) Effects of hypoxia, and balance between hypoxia and enrichment, on coastal fishes and fisheries. Estuaries 25, 767–781.

Breitburg, D.L., Sanders, J.G., Gilmour, C.C., Hatfield, C.A., Osman, R.W., Riedel, G.F., Seitzinger, S.P., and Sellner, K.P. (1999) Variability in responses to nutrients and trace elements, and transmission of stressor effects through an estuarine food web. Limnol. Oceanogr. 44, 837–863.

Brenon, I., and Le Hir, P. (1999) Modelling the turbidity maximum in the Seine estuary (France): Identification of formation processes. Estuar. Coastal Shelf Sci. 49, 525–544.

Brezonik, P.L. (1994) Chemical Kinetics and Process Dynamics in Aquatic Systems. Lewis Publishers, London.

Bricaud, A., Morel, A., and Prieur, L. (1981) Absorption by dissolved organic matter in the sea (yellow substance) in the UV and visible domains. Limnol. Oceanogr. 26, 43–53.

Bricker, S.B., Clement, C.G., Pirhalla, D.E., Orlando, S.P., and Farrow, D.G.G. (1999) National Estuarine Eutrophication Assessment: Effects of Nutrient Enrichment in the Nation's Estuaries. National Ocean Service, NOAA, Silver Springs, MD.

Bricker-Urso, S., Nixon, S.W., Cochran, J.K., Hirschberg, D.J., and Hunt, C. (1989) Accretion rates and sediment accumulation in Rhode Island salt marshes. Estuaries 12, 300–317.

Bridges, T.S., Levin, L.A., Cabrera, D., and Plaia, G. (1994) Effects of sediment amended with sewage, algae, or hydrocarbons on growth and reproduction in two opportunistic polychaetes. J. Exp. Mar. Biol. Ecol. 177, 99–119.

Brinson, M. M. (1993) A hydrogeomorphic classification for wetlands. Waterways Experiment Station, Vicksburg, MS. NTIS No. AD A270 053.

Brock, D.A. (2001) Nitrogen budget for low and high freshwater inflows, Nueces Estuary, Texas. Estuaries 24, 509–521.

Broecker, W.S., and Peng, T.H. (1974) Gas exchange rates between the air and the sea. Tellus 26, 21–35.

Broecker, W.S., and Peng, T.H. (1982) Tracers in the Sea, LDGEO Press, New York.

Bronk, D.A. (2002) Dynamics of organic nitrogen. In Biogeochemistry of Marine Dissolved Organic Matter (Hansell, D.A., and Carlson, C.A., eds.), pp. 153–231, Academic Press, San Diego, CA.

Bronk, D.A., and Glibert, P.M. (1993) Contrasting patterns of dissolved organic nitrogen release by two size fractions of estuarine plankton during a period of rapid NH_4^+ consumption and NO_2^- production. Mar. Ecol. Prog. Ser. 96, 291–299.

Bronk, D.A., Glibert, P.M., Malone, T.C., Banahan, S., and Sahlsten, E. (1998) Inorganic and organic nitrogen cycling in Chesapeake Bay: autotrophic versus heterotrophic processes and relationships to carbon flux. Aquat. Microb. Ecol. 15, 177–189.

Bronk, D.A., Glibert, P.M., and Ward, B.B. (1994) Nitrogen uptake, dissolved organic nitrogen release, and new production. Science 265, 1843–1846.

Bronk, D.A., and Ward, B.B. (1999) Gross and net nitrogen uptake and DON release in the euphotic zone of Monterey Bay, California. Limnol. Oceanogr. 44, 573–585.

Bronk, D.A., and Ward, B.B. (2000) Magnitude of DON release relative to gross nitrogen uptake in marine systems. Limnol. Oceanogr. 45, 1879–1883.

Brown, S.B., Houghton, J.D., and Hendry, G.A.F. (1991) Chlorophyll breakdown. In Chorophylls (Scheer, H., ed.), pp. 465–489, CRC Press, Boca Raton, FL.

Brownawall, B.J., and Farrington, J.W. (1985) Partitioning of PCBs in marine sediments. In Marine and Estuarine Geochemistry (Sigleo, A.C., and Hattori, A., eds.), pp. 97–120, Lewis Publishers, Boca Raton, FL.

Brownawall, B.J., and Farrington, J.W. (1986) Biogeochemistry of PCBs in interstitial waters of a coastal marine sediment. Geochim. Cosmochim. Acta 50, 157–169.

Bruland, K.W., and Coale, K.H. (1986) Surface water $^{234}Th/^{238}U$ disequilibria: spatial and temporal variations of scavenging rates within the Pacific Ocean. In Dynamic Processes in the Chemistry of the Upper Ocean (Burton, J.D., ed), pp. 159–172, Plenum Publications, New York.

Bruland, K.W., Donat, J.R., and Hutchins, D.A. (1991) Interactive influences of bioactive trace metals on biological production in oceanic waters. Limnol. Oceanogr. 36, 1555–1577.

Bruland, K.W., Rue, E.L., Donat, J.R., Skrabal, S.A., and Moffett, J.W. (2000) Intercomparison of voltammetric techniques to determine the chemical speciation of dissolved copper in a coastal seawater sample. Anal. Chim. Acta 405, 99–113.

Brzezinski, M.A., Jones, J.L., Bidle, K.D., and Azam, F. (2003) The balance between silica production and silica dissolution in the sea: insights from Monterey Bay, California, applied to the global data set. Limnol. Oceanogr. 48, 1846–1854.

Brzezinski, M.A., and Nelsen, D.M. (1989) Seasonal changes in the silicon cycle within a Gulf Stream warm-core ring. Deep-Sea Res. 36, 1009–1030.

Budge, S.M., and Parrish, C.C. (1998) Lipid biogeochemistry of plankton, settling matter and sediments in Trinity Bay, Newfoundland. II. Fatty acids. Org. Geochem. 29, 1547–1559.

Budzinski, H., Jones, I., Bellocq, C., and Garrigues, P. (1997) Evaluation of sediment contamination by polycyclic aromatic hydrocarbons in the Gironde estuary. Mar. Chem. 58, 85–97.

Buessler, K.O. (1998) The decoupling of production and particulate export in the surface ocean. Global Biogeochem. Cycles 12, 297–310.

Buessler, K., Bauer, J., Chen, R., Eglinton, T., Gustafsson, O., Landing, W., Mopper, K., Moran, S.B., Santschi, P., Vernon Clark, R., and Wells, M. (1996) An intercomparison of cross-flow filtration techniques used for sampling marine colloids: overview and organic carbon results. Mar. Chem. 55, 1–32.

Buffam, I., and McGlathery, K.J. (2003) Effect of ultraviolet light on dissolved nitrogen transformations in coastal lagoon water. Limnol. Oceanogr. 48, 723–734.

Buffle, J. (1990) Complexation Reactions in Aquatic Systems. An Analytical Approach. Ellis Horwood, New York.

Buffle, J., Altmann, R.S., Filella, M., and Tessier, A. (1990) Complexation by natural heterogeneous compounds: site occupation distribution functions, a normalized description of metal complexation. Geochim. Cosmochim. Acta 54, 1535–1554.

Buffle, J., and Leppard, D.L. (1995) Characterization of aquatic colloids and macromolecules: 2. key role of physical structures on analytical results. Environ. Sci. Technol. 29, 2169–2175.

Buffle, J., Perret, D., and Newman, M. (1993) The use if filtration and ultrafiltration for size fractionation of aquatic particles, colloids, and macromolecules. In Environmental Particles (Buffle, J., and van Leeuwen, P., eds.), pp. 171–230, Lewis Publishers, Boca Raton, FL.

Bugna, G.C., Chanton, J.P. Cable, J.E., Burnett, W.C., and Cable, P.H. (1996) The importance of groundwater discharge to the methane budgets of nearshore and continental shelf waters of the northwestern Gulf of Mexico. Geochim. Cosmochim. Acta 60, 4735–4746.

Bunch, J.N. (1987) Effects of petroleum releases on bacterial numbers and microheterotrophic activity in the water and sediment of an Arctic marine ecosystem. Arctic 40, 172–183.

Burdige, D.J. (1989) The effects of sediment slurrying on microbial processes, and the role of amino acids as substrates for sulfate reduction in anoxic marine sediments. Biogeochemistry 8, 1–23.

Burdige, D.J. (1991) The kinetics of organic matter mineralization in anoxic marine sediments. J. Mar. Res. 49, 727–761.

Burdige, D.J. (1993) The biogeochemistry of manganese and iron reduction in marine sediments. Earth Sci. Rev. 35, 249–284.

Burdige, D.J. (2001) Dissolved organic matter in Chesapeake Bay sediment pore waters. Org. Geochem. 32, 487–505.

Burdige, D.J. (2002) Sediment pore waters. In Biogeochemistry of Marine Dissolved Organic Matter (Hansell, D.A., and Carlson, C.A., eds.), pp. 612–653, Academic Press, New York.

Burdige, D.J., Alperin, M.J., Homstead, J., and Martens, C.S. (1992) The role of benthic fluxes of dissolved organic carbon in oceanic and sedimentary carbon cycling. Geophys. Res. Lett. 19, 1851–1854.

Burdige, D.J., and Gardner, K.G. (1998) Molecular weight distribution of dissolved organic carbon in marine sediment pore waters. Mar. Chem. 62, 45–64.

Burdige, D.J., and Gieskes, J.M. (1983) A pore water/solid phase diagenetic model for manganese in marine sediments. Am. J. Sci. 283, 29–47.

Burdige, D.J., and Homstead, J. (1994) Fluxes of dissolved organic carbon from Chesapeake Bay sediments. Geochim. Cosmochim. Acta 58, 3407–3424.

Burdige, D.J., Kline, S.W., and Chen, W. (2004) Fluorescent dissolved organic matter in marine sediment pore waters. Mar. Chem. 89, 289–311.

Burdige, D.J., and Martens, C.S. (1988) Biogeochemical cycling in an organic-rich marine basin: 10. The role of amino acids in sedimentary carbon and nitrogen cycling. Geochim. Cosmochim. Acta 52, 1571–1584.

Burdige, D.J., and Martens, C.S. (1990) Biogeochemical cycling in an organic-rich coastal marine basin: 11. The sedimentary cycling of dissolved, free amino acids. Geochim. Cosmochim. Acta 54, 3033–3052.

Burdige, D.J., and Nealson, K.H. (1986) Chemical and microbiological studies of sulfide-mediated manganese reduction. Geomicrobiol. J. 4, 361–387.

Burdige, D.J., Skoog, A., and Gardner, K. (2000) Dissolved and particulate carbohydrates in contrasting marine sediments. Geochim. Cosmochim. Acta 64, 1029–1041.

Burdige, D.J., and Zheng, S. (1998) The biogeochemical cycling of dissolved organic nitrogen in estuarine sediments. Limnol. Oceanogr. 43, 1796–1813.

Burgess, R.M., and McKinney, R.A. (1999) Importance of interstitial, overlying water and whole sediment exposures to bioaccumulation by marine bivalves. Environ. Pollut. 104, 373–382.

Burgess, R.M., McKinney, R.A., Brown, W.A., and Quinn, J.G. (1996) Isolation of marine sediment colloids and associated polychlorinated biphenyls: An evaluation of ultrafiltration and reverse-phase chromatography. Environ. Sci. Technol. 30, 1923–1932.

Burkholder, J.M., and Glasgow, H.B. (1997) *Pfiesteria piscicida* and other toxic *Pfiesteria*-like dinoflagellates: Behavior, impacts, and environmental controls. Limnol. Oceanogr. 42, 1052–1075.

Burkholder, J.M., Glasgow, H.B., and Deamer-Melia, N.J. (2001) Overview and present status of the toxic *Pfiesteria* complex. Phycologia 40, 186–214.

Burkholder, J.M., Glasgow, H.B., and Hobbs, C.W. (1995) Fish kills linked to a toxic ambush-predator dinoflagellate: distribution and environmental conditions. Mar. Ecol. Prog. Ser. 124, 42–61.

Burkholder, J.M., Mallin, M.A., Glasgow, H.B., Larsen, L.M., McIver, M.R., Shank, G.C., Deamer-Melia, N.J., Briley, D.S., Springer, J., Touchette, B.W., and Hannon, E.K. (1997) Impacts to a coastal river and estuary from rupture of a large swine waste holding lagoon. J. Environ. Qual. 26, 1451–1466.

Burkholder, J.M., Noga, E.J., Hobbs, C.W., Glasgow, H.B., and Smith, S.A. (1992) New "phantom" dinoflagellate is the causative agent of major estuarine fish kills. Nature 358, 407–410.

Burnett, W.C., Bokuniewicz, H., Huettel, M., Moore, W.S., and Taniguchi, M. (2003) Groundwater and pore water inputs to the coastal zone. Biogeochemistry 66, 3–33.

Burnett, W.C., Taniguchi, C.M., and Oberdorfer, J. (2001) Measurement and significance of the direct discharge of groundwater into the coastal zone. J. Sea Res. 46, 109–116.

Burns, R.C., and Hardy, R.W.F. (1975) Nitrogen Fixation in Bacteria and Higher Plants. Springer-Verlag, Heidelberg.

Burton, J.D., and Liss, P.S. (1976) Basic properties and processes in estuarine chemistry. *In* Estuarine Chemistry (Burton, J.D., and Liss, P.S. eds.), pp. 1–36. Academic Press, New York.

Busch, P.L., and Stumm. W. (1968) Chemical interactions in the aggregation of bacteria bioflocculation in waste treatment. Environ. Sci. Technol. 2, 49–53.

Bushaw, K.L., Zepp, R.G., Tarr, M.A., Schulz-Jander, D., Bourbonniere, R.A., Hodson, R.E., Miller, W.L., Bronk, D.A., and Moran, M.A. (1996) Photochemical release of biologically available nitrogen from aquatic dissolved organic matter. Nature 381, 404–407.

Bushaw-Newton, K.L., and Moran, M.A. (1999) Photochemical formation of biologically available nitrogen from dissolved humic substances in coastal marine systems. Aquat. Microb. Ecol. 18, 285–292.

Butcher, S.S., and Anthony, S.E. (2000) Equilibrium, rate, and natural systems. In Earth System Science, from Biogeochemical Cycles to Global Change (Jacobson, M.C., Charlson, R.J., Rodhe, H, and Orians, G.H., eds.), pp. 85–105, International Geophysics Series, Academic Press, New York.

Butler, E.I., Knox, S., and Liddicoat, M.I. (1979) The relationship between inorganic and organic nutrients in sea water. J. Mar. Biol. Assoc. UK 59, 239–250.

Butman, C.A., and Grassle, J.P. (1992) Active habitat selection by Capitella-sp. I. larvae. Two-choice experiments in still water and flume flows. J. Mar. Res. 50, 669–715.

Buzzelli, C.P. (1998) Simulation modeling of littoral zone habitats in lower Chesapeake Bay. I. An ecosystem characterization related to model development. Estuaries 21, 659–672.

Buzzelli, C.P., Wetzel, R.I., and Meyers, M.B. (1998) Dynamic simulation of littoral zone habitats in lower Chesapeake Bay. II. Seagrass habitat primary production and water quality relationships. Estuaries 21, 673–689.

Byrne, R. H., Kump, L. R., and Cantrell, K. J. (1988). The influence of temperature and pH on trace metal speciation in seawater. Mar. Chem. 25, 166–181.

Cable, J.E., Bugna, G., Burnett, W., and Chanton, J. (1996b) Application of CH_4 for the assessment of groundwater discharge to the coastal ocean. Limnol. Oceanogr. 41, 1347–1353.

Cable, J.E., Burnett, W.C., Chanton, J.P., Corbett, D.R., and Cable, P.H. (1997) Field evaluation of seepage meters in the coastal marine environment. Estuar. Coastal Mar. Sci. 45, 367–375.

Cable, J.E., Burnett, W.C., Chanton, J.P., and Weatherly, G.L. (1996a) Estimating ground-water discharge into the northeastern Gulf of Mexico using radon-222. Earth Planet. Sci. Lett. 144, 591–604.

Cable, J.E., Martin, J.B., Swarzenski, P.W., Lindenberg, M.K., and Steward, J. (2004) Advection within shallow pore waters of a coastal lagoon, Florida. Ground Water 42, 1011–1020.

Cacchione, D., Drake, D.E., Kayen, R., Sternberg, R.W., Kineke, G.C., and Tate, G.B. (1995) Measurements in the bottom boundary layer on the Amazon subaqueous delta. Mar. Geol. 125, 235–258.

Caddy, J.F. (1993) Towards a comparative evaluation of human impacts on fishery ecosystems of enclosed and semi-enclosed seas. Rev. Fish. Sci. 1, 57–95.

Cadée, G.C. (1978) Primary production and chlorophyll-a in the Zaire River estuary and plume. Neth. J. Sea Res. 12, 368–381.

Cadée, G.C., Dronkers, J., Heip, C., Martin, J.M., and Nolan, C. (eds.) (1994) ELOISE (European Land–Ocean Interaction Studies) Science Plan, Luxemborg: Off. Official Publ. Eur. Communities.

Caffrey, J.M. (2004) Factors controlling net ecosystem metabolism in U.S. estuaries. Estuaries 27, 90–101.

Caffrey, J.M., Cloern, J.E., and Grenz, C. (1998) Changes in production and respiration during a spring phytoplankton bloom in San Francisco Bay, California, U.S.A: implications for net ecosystem metabolism. Mar. Ecol. Prog. Ser. 172, 1–12.

Caffrey, J.M., and Kemp, W.M. (1990) Nitrogen cycling in sediments with estuarine populations of Potamogeton perfoliatus and Zostera marina. Mar. Ecol. Prog. Ser. 66, 147–160.

Caffrey, J.M., and Kemp, W.M. (1991) Seasonal and spatial patterns of oxygen production, respiration, and root-rhizome release in *Potamogeton perfoliatus* and *Zostera marina*. Aquat. Bot. 40: 109–128.

Cahoon, L.B. (1999) The role of benthic microalgae in neritic ecosystems. Oceanogr. Mar. Biol.: Ann. Rev. 37, 47–86.

Cai, W.J. (2003) Riverine inorganic carbon flux and rate of biological uptake in the Mississippi River plume. Geophys. Res. Lett. 30, 1032, doi10.1029/2002GL016312.

Cai, W.J., Pomeroy, L.R., Moran, M.A., and Wang, Y. (1999) Oxygen and carbon dioxide mass balance in the estuarine/intertidal marsh complex of five rivers in the southeastern U.S Limnol. Oceanogr. 44, 639–649.

Cai, W.J., and Wang, Y. (1998) The chemistry, fluxes and sources of carbon dioxide in the estuarine waters of the Satilla and Altamaha Rivers, Georgia. Limnol. Oceanogr. 43, 657–668.

Cai, W.J., Wang, Y., Krest, J., and Moore, W.S. (2003) The geochemistry of dissolved inorganic carbon in a surficial groundwater aquifer in North Inlet, South Carolina, and the carbon fluxes to the coastal ocean. Geochem. Cosmochim. Acta 67, 631–637.

Cai, W.J., Wang, Z.A., and Wang, Y. (2004) The role of marsh-dominated heterotrophic continental margins in transport of CO_2 between the atmosphere, the land–sea interface and the ocean. Limnol. Oceanogr. 49, 348–354.

Cai, W.J., Wiebe, W.J., Wang, Y., and Sheldon, J.E. (2000) Intertidal marsh as a source of dissolved inorganic carbon and a sink of nitrate in the Satilla River estuarine complex in the southeastern U.S. Limnol. Oceanogr. 45, 1743–1752.

Cai, W.J., Zhao, P., Theberge, S.M., Wang, Y., and Luther III, G. (2002) Porewater redox species, pH and pCO_2 in aquatic sediments—electrochemical sensor studies in Lake Champlain and Sapelo Island. *In* Environmental Electrochemistry: Analysis of Trace Element Biogeochemistry (Taillefert, M., and Rozan, T., eds.), pp. 188–209, American Chemical Society, Washington, DC.

Cairns, T., Doose, G.M., Froberg, J.E., Jacobson, R.A., and Siegmund, E.G. (1986) Analytical chemistry of PCBs. *In* PCBs and the Natural Environment (Ward, J.S., ed.), pp. 2–45, CRC Press, Boca Raton, FL.

Callender, E., and Hammond, D.E. (1982) Nutrient exchange across the sediment–water interface in the Potomac River estuary. Estuar. Coastal Shelf Sci. 15, 395–413.

Calvert, S.E., and Pedersen, T.F. (1992) Organic carbon accumulation and preservation in marine sediments: how important is anoxia? *In* Organic Matter (Whelan, J., and Farrington, J.W., eds.), pp. 231–263, Columbia University Press, New York.

Campbell, P.G.C. (1995) Interactions between trace metals and organisms: critique of the free-ion activity model. *In* Metal Speciation and Bioavailability in Aquatic Systems (Tessier, A., and Turner, D.R., eds.), pp. 45–102, John Wiley, Chichester, UK.

Canfield, D.E. (1989) Reactive iron in marine sediments. Geochim. Cosmochim. Acta 53, 619–632.

Canfield, D.E. (1993) Organic matter oxidation in marine sediments. *In* NATO–ARW Interactions of C, N, P and S Biogeochemical Cycles and Global Change (Wollast, R., Chou, L., and Mackenzie, F., eds.), pp. 333–365, Springer, New York.

Canfield, D.E. (1994) Factors influencing organic carbon preservation in marine sediments. Chem. Geol. 114, 315–329.

Canfield, D.E., and Berner, R.A. (1987) Dissolution and pyritization of magnetite in anoxic marine sediments. Geochim. Cosmochim. Acta 51, 645–659.

Canfield, D.E., and Des Marais, D.J. (1993) Biogeochemical cycles of carbon, sulfur, and free oxygen in a microbial mat. Geochim. Cosmochim. Acta 57, 3971–3984.

Canfield, D.E., Raiswell, R., and Bottrell, S. (1992) The reactivity of sedimentary iron minerals toward sulfide. Am. J. Sci. 292, 659–683.

Canfield, D.E., and Thamdrup, B. (1994) The production of ^{34}S-depleted sulfide during bacterial disproportionation of elemental sulfur. Science 266, 1973–1975.

Canfield, D.E., Thamdrup, B., and Hansen, J.W. (1993) The anaerobic degradation of organic matter in Danish coastal sediments: iron reduction, manganese reduction, and sulfate reduction. Geochim. Cosmochim. Acta 57, 3867–3883.

Canuel, E.A. (2001) Relations between river flow, primary production and fatty acid composition of particulate organic matter in San Francisco and Chesapeake Bays: a multivariate approach. Org. Geochem. 32, 563–583.

Canuel, E.A., Cloern, J.E., Ringelborg, D.B., Guckert, J.B., and Rau, G.H. (1995) Molecular and isotopic tracers used to examine sources or organic matter and its incorporation into food webs of San Francisco Bay. Limnol. Oceanogr. 40, 67–81.

Canuel, E.A., Freeman, K.H., and Wakeham, S.G. (1997) Isotopic compositions of lipid biomarker compounds in estuarine plants and surface sediments. Limnol. Oceanogr. 42, 1570–1583.

Canuel, E.A., and Martens, C.S. (1993) Seasonal variation in the sources and alteration of organic matter associated with recently-deposited sediments. Org. Geochem. 20, 563–577.

Canuel, E.A., and Martens, C.S. (1996) Reactivity of recently deposited organic matter: degradation of lipid compounds near the sediment–water interface. Geochim. Cosmochim. Acta 60, 1793–1806.

Canuel, E.A., Martens, C.S., and Benninger, L.K. (1990) Seasonal variations of ^{7}Be activity in the sediments of Cape Lookout Bight, North Carolina. Geochim. Cosmochim. Acta 54, 237–245.

Canuel, E.A., and Zimmerman, A.R. (1999) Composition of particulate organic matter in the Southern Chesapeake Bay: sources and reactivity. Estuaries 22, 980–994.

Capone, D.G. (1982) Nitrogen fixation (acetylene reduction) by rhizosphere sediments of the eelgrass Zostera marina. Mar. Ecol. Prog. Ser. 10, 67–75.

Capone, D.G. (1988) Benthic nitrogen fixation. In Nitrogen Cycling in Coastal Environments (Blackburn, T.H., and Sørensen, J., eds.), pp. 85–123, John Wiley, New York.

Capone, D.G. (1996) A biologically constrained estimate of oceanic N_2O flux. Ver. Der Inst. Verein. fur Theo. Angew. Limnol. 25, 105–113.

Capone, D.G. (2000) The marine nitrogen cycle. In Marine Microbial Ecology. (Kirchman, D., ed.), pp. 455–493, John Wiley, New York.

Capone, D.G., and Bautista, M.F. (1985) A groundwater source of nitrate in nearshore marine sediments. Nature 313, 214–216.

Capone, D.G., and Kiene, R.P. (1988) Comparison of microbial dynamics in marine and freshwater sediments: contrast in anaerobic carbon catabolism. Limnol. Oceanogr. 33, 725–749.

Capone, D.G., and Slater, J.M. (1990) Interannual patterns of water table height and groundwater derived nitrate in nearshore sediments. Biogeochemistry 10, 277–288.

Caraco, N.F. (1995) Influence of human population on P transfers to aquatic systems: a regional scale study using large rivers. In Phosphorus in the Global Environment (Tiessen, H., ed.), pp. 235–244, John Wiley, New York.

Caraco, N.F., and Cole, J.J. (2003) The importance of organic nitrogen production in aquatic systems: a landscape perspective. In Aquatic Ecosystems: Interactivity of Dissolved Organic Matter (Findlay, S.E.G., and Sinsabaugh, R.L., eds.), pp. 263–283, Academic Press, New York.

Caraco, N.F., Cole, J.J., and Likens, G.E. (1989) Evidence for sulphate-controlled phosphorus release from sediments of aquatic systems. Nature 341, 316–318.

Caraco, N.F., Cole, J.J., and Likens, G.E. (1990) A comparison of phosphorus immobilization in sediments of freshwater and coastal marine systems. Biogeochemistry 9, 277–290.

Caraco, N.F., Cole, J.J., and Likens, G.E. (1993) Sulfate control of phosphorus availability in lakes. Hydrobiologia 252, 275–280.

Caraco, N.F., Lapman, G., Cole, J.J., Limburg, K.E., Pace, M.L., and Fischer, D. (1998) Microbial assimilation of DIN in a nitrogen rich estuary: implications for food quality and isotope studies. Mar. Ecol. Prog. Ser. 167, 59–71.

Caraco, N.F., Tamse, A., Boutros, O., and Valiela, I. (1987) Nutrient limitation of phytoplankton growth in brackish coastal ponds. Can. J. Fish. Aquat. Sci. 44, 473–476.

Carder, K.L., Chen, F.R., Lee, Z.P., Hawes, S.K., and Kamykowski, D. (1999) Semi-analytic moderate-resolution imaging spectrometer algorithms for chlorophyll *a* and absorption with biooptical domains based on nitrate-depletion temperatures. J. Geophys. Res. 104, 5403–5421.

Cardoso, J.N., and Eglinton, G. (1983) The use of hydroxy acids as geochemical indicators. Geochem. Cosmochim. Acta 47, 723–730.

Carini, S., Orcutt, B.N., and Joye, S.B. (2003) Interactions between methane oxidation and nitrification in coastal sediments. Geomicrobiol. J. 20, 355–374.

Carini, S., Weston, N., Hopkinson, C., Tucker, J., Giblin, A., and Vallino, J. (1996) Gas exchange in the Parker estuary. Mar. Biol. Bull. 191, 333–334.

Carlson, P.R., and Forrest, J. (1982) Uptake of dissolved sulfide by *Spartina alterniflora*: evidence from natural sulfur isotope abundance ratios. Science 216, 633–635.

Carlsson, P., Graneli, E., and Segatto, Z. (1999) Cycling of biologically available nitrogen in riverine humic substances between marine bacteria, a heterotrophic nanoflagellate and a photosynthetic dinoflagellate. Aquat. Microbiol. Ecol. 18, 23–36.

Carlton, J.T., Thompson, J.K., Schemel, L.E., and Nichols, F.H. (1990) The remarkable invasion of San Francisco Bay (California, USA) by the Asian clam *Potamocorbula amurensis*. I, Introduction and dispersal. Mar. Ecol. Prog. Ser. 66, 81–94.

Carman, K.R., Bianchi, T.S, and Kloep, F. (2000) The influence of grazing and nitrogen on benthic algal blooms in diesel-contaminated saltmarsh sediments. Environ. Sci. Technol. 34, 107–111.

Carman, K.R., Fleeger, J.W., and Pomarico, S.M. (1997) Responses of a benthic food web to hydrocarbon contamination. Limnol. Oceanogr. 42, 561–571.

Carman, K.R., Means, J.C., and Pomarico, S.C. (1996) Response to sedimentary bacteria in a Louisiana salt marsh to contamination by diesel fuel. Aquat. Microb. Ecol. 10, 231–241.

Carman, K.R., and Todaro, M.A. (1996) Influence of polycyclic aromatic hydrocarbons on the meiobenthic-copepod community of a Louisiana salt marsh. J. Exp. Mar. Biol. Ecol. 198, 37–54.

Carman, R., and Jonsson, P. (1991) Distribution patterns of different forms of phosphorus in some surficial sediments of the Baltic Sea. Chem. Geol. 90, 91–106.

Caron, D.A. (1987) Grazing of attached bacteria by heterotrophic microflagellates. Microb. Ecol. 13, 203–218.

Carpenter, E.J., and Capone, D.G. (eds.) (1981) Nitrogen in the Marine Environment. Academic Press, New York.

Carpenter, E.J., van Raalte, C.D., and Valiela, I. (1978) Nitrogen fixation by algae in a Massachusetts salt marsh. Limnol. Oceanogr. 23, 318–327.

Carpenter, P.D., and Smith, J.D. (1984) Effect of pH, iron and humic acid on the estuarine behavior of phosphate. Environ. Sci. Technol. Lett. 6, 65–72.

Carritt, D.E., and Goodgal, S. (1954) Sorption reactions and some ecological implications. Deep-Sea Res. 1, 224–243.

Carter, M.W., and Moghissi, A.A. (1977) The decades of nuclear testing. Health Phys. 33, 55–71.

Carvalho, R.A., Benfield, M.C., and Santschi, P.H. (1999) Comparative bioaccumulation studies of colloidally complexed and free-ionic heavy metals in juvenile brown shrimp *Penaeus aztacus* (Crustacea: Decapoda:Penaeidae). Limnol. Oceanogr. 44, 403–414.

Castaing, P., and Guilcher, A. (1995) Geomorphology and sedimentology of rias. *In* Geomorphology and Sedimentology of Estuaries. Developments in Sedimentology 53 (Perillo, G.M.E., ed.), pp. 69–111, Elsevier Science, New York.

Castro, M.S., Driscoll, C.T., Jordan, T.E., Reay, W.G., and Boynton, W.R. (2003) Sources of nitrogen to estuaries in the United States. Estuaries 26, 803–814.

Caughey, M.E. (1982) A study of the dissolved organic matter in pore waters of carbonate-rich sediment cores from Florida Bay. M.S. Thesis, University of Texas at Dallas.

Cauwet, G. (2002) DOM in the coastal zone. *In* Biogeochemistry of Marine Dissolved Organic Matter (Hansell, D.A., and Carlson, C.A., eds.), pp. 579–602, Academic Press, New York.

Cebrian, J., Duarte, C.M., Marba, N., and Enriquez, S. (1997) Magnitude and fate of the production of four co-occurring Western Mediterranean seagrass species. Mar. Ecol. Prog. Ser. 155, 29–44.

Cebrian, J., Pedersen, M.F., Kroeger, K.D., and Valiela, I. (2000) Fate of production of seagrass *Cymodocea nodosa* in different stages of meadow formation. Mar. Ecol. Prog. Ser. 204, 119–130.

Cederwall, H., and Elmgren, R. (1980) Biomass increase of benthic macrofauna demonstrates eutrophication of the Baltic Sea. Ophelia 1, 287–304.

Cerco, C.F., and Seitzinger, S.P. (1997) Measured and modeled effects of benthic algae on eutrophication in Indian River–Rehoboth Bay, Delaware. Estuaries 20, 231–248.

Cerqueira, M.A., and Pio, C.A. (1999) Production and release of dimethylsulphide from an estuary in Portugal. Atmos. Environ. 33, 3355–3366.

Chacko, T., Cole, D.R., and Horita, J. (2001) Equilibrium oxygen, hydrogen and carbon isotope fractionation factors applicable to geologic systems. *In* Reviews in Mineralogy and Geochemistry: Stable Isotope Geochemistry (Valley, J.W., and Cole, D.R., eds.), Vol. 43, pp. 1–81, Mineralogical Society of America, Chantilly, VA.

Chambers, L.A., and Trudinger, P.A. (1978) Microbiological fractionation of stable sulfur isotopes: a review and critique. J. Geomicrobiol. 1, 249–295.

Chambers, L.A., Trudinger, P.A., Smith, J.W., and Burns, M.S. (1975) Fractionation of sulfur isotopes by continuous cultures of *Desulfovibrio desulfuricans*. Can. J. Microbiol. 21, 1602–1607.

Chambers, R.M., Osgood, D.T., Bart, D.J., and Montalto, F. (2003) *Phragmites australis* invasion and expansion in tidal wetlands: interactions among salinity, sulfide, and hydrology. Estuaries 26, 398–406.

Chang, C.C.Y., Kendall, C., Silva, S.R., Battaglin, W.A., and Campbell, D.H. (2002) Nitrate stable isotopes: tools for determining nitrate sources among different land uses in the Mississippi River basin. Can. J. Fish. Sci. 59, 1874–1885.

Chang, H.M., and Allen, G.G. (1971) Oxidation. *In* Lignins (Sarkanen, K.V., and Ludwig, C.H., eds.), pp. 433–485, Wiley Interscience, New York.

Chanton, J.P., Crill, P.M., Baertlett, K.B., and Martens, C.S. (1989b) Amazon Capims (floating grassmats): a source of ^{13}C-enriched methane to the troposphere. Geophys. Res. Lett. 16, 799–802.

Chanton, J.P., and Dacey, J.W.H. (1991) Effects of vegetation on methane flux, and carbon isotopic composition. *In* Trace Gas Emissions by Plants (Mooney, H.A., and Sharkey, T. eds.), pp. 65–92, Academic Press, San Diego, CA.

Chanton, J.P., and Lewis, F.G. (1999) Plankton and dissolved inorganic carbon isotopic composition in a river-dominated estuary: Apalachicola Bay, Florida. Limnol. Oceanogr. 22, 575–583.

Chanton, J.P., and Lewis, F.G. (2002) Examination of coupling between primary and secondary production in a river-dominated estuary: Apalachicola Bay, Florida. Limnol. Oceanogr. 47, 683–697.

Chanton, J.P., and Martens, C.S. (1987a) Biogeochemical cycling in an organic-rich coastal marine basin. 7. Sulfur mass balance, oxygen uptake and sulfide retention. Geochim. Cosmochim. Acta 51, 1187–1199.

Chanton, J.P., and Martens, C.S. (1987b) Biogeochemical cycling in an organic-rich coastal marine basin. 8. A sulfur isotopic budget balanced by differential diffusion across the sediment–water interface. Geochim. Cosmochim. Acta 51, 1201–1208.

Chanton, J.P., and Martens, C.S. (1988) Seasonal variations in ebullitive flux and carbon isotopic composition of methane in a tidal freshwater estuary. Global Biogeochem. Cycles 2, 289–298.

Chanton, J.P., Martens, C.S., and Kelley, C.A. (1989a) Gas transport from methane-saturated, tidal freshwater and wetland sediments. Limnol. Oceanogr. 34, 807–819.

Chanton, J.P., Smith, C.J., and Patrick, W. (1993) Methane release from Gulf coast wetlands. Tellus 35, 8–15.

Chanton, J.P., and Whiting, G.J. (1996) Methane stable isotopic distributions as indicators of gas transport mechanisms in emergent aquatic plants. Aquat. Bot. 54, 227–236.

Chanton, J.P., Whiting, G.L. Showers, W.J., and Crill, P.M. (1992) Methane flux from *Petulandra virginica* stable isotope tracing and chamber effects. Global Biogeochem. Cycles 6, 15–31.

Chapman, V. J. (1960) Salt Marshes and Salt Deserts of the World. Plant Science Monographs, Leonard Hill, London.

Chapman, V.J. (1966) Three new carotenoids isolated from algae. Phytochemistry 5, 1331–1333.

Characklis, W.G., and Marshall, K.C. (1989) Biofilms. John Wiley, New York.

Charette, M., and Buessler, K.O. (2004) Submarine groundwater discharge of nutrients and copper to an urban subestuary of Chesapeake Bay (Elizabeth River). Limnol. Oceanogr. 49, 376–385.

Charette, M., Buesseler, K.O., and Andrews, J.E. (2001) Utility of radium isotopes for evaluating the input and transport of groundwater-derived nitrogen to a Cape Cod estuary. Limnol. Oceanogr. 46, 465–470.

Charlson, R.J. (2000) The atmosphere. In Earth System Science, from Biogeochemical Cycles to Global Change (Jacobson, M.C., Charlson, R.J., Rodhe, H, and Orians, G.H., eds.), pp. 132–158, International Geophysics Series, Academic Press, New York.

Charlson, R.J., Lovelock, J.E., Andreae, M.O., and Warren, S.G. (1987) Oceanic phytoplankton atmospheric sulphur, cloud albedo and climate. Nature 326, 655–661.

Chauvaud, L., Jean, F., Ragueneau, O., and Thouzeau, G. (2000) Long-term variation of the Bay of Brest ecosystem: pelagic–benthic coupling revisited. Mar. Ecol. Prog. Ser. 200, 35–48.

Cheevaporn, V., Jacinto, G.S., and San Diego-McGlone, M.L. (1995) Heavy metal fluxes in Bang Pakong River estuary, Thailand: sedimentary versus diffusive fluxes. Mar. Pollut. Bull. 31, 290–294.

Chefetz, B., Chen, Y., Clapp, C.E., and Hatcher, P.G. (2000) Characterization of organic matter in soils by thermochemolysis using tetramethylammonium hydroxide (TMAH). Soil Sci. Am. J. 64, 583–589.

Chen, J.H., Lawrence, R.E., and Wasserburg, G.J. (1986) ^{238}U, ^{234}U, and ^{232}Th in seawater. Earth Planet. Sci. Lett. 80, 241–251.

Chen, N., Bianchi, T.S., and Bland, J.M. (2003a) Novel decomposition products of chlorophyll-*a* in continental shelf (Louisiana shelf) sediments: Formation and transformation of carotenol chlorine esters. Geochim. Cosmochim. Acta 67, 2027–2042.

Chen, N., Bianchi, T.S., McKee, B.A., and Bland, J. (2001) Historical trends of hypoxia on the Louisiana shelf: application of pigments as biomarkers. Org. Geochem. 32, 543–561.

Chen, N., Bianchi, T.S., McKee, B.A., and Bland, J. (2003b) Fate of chlorophyll-*a* in the lower Mississippi River and Louisiana shelf: Implications for pre- versus post-depositional decay. Mar. Chem. 83, 37–55.

Cherrier, J., Bauer, J.E., Druffel, E.R.M., Coffin, R.B., and Chanton, J.C. (1999) Radiocarbon in marine bacteria: evidence for the ages of assimilated carbon. Limnol. Oceanogr. 44, 730–736.

Chesney, Jr., E.J. (1989) Estimating the food requirements of striped bass larvae *Morone saxatilis*: Effects of light, turbidity and turbulence. Mar. Ecol. Prog. Ser. 53, 191–200.

Chesney, Jr., E.J., and Baltz, D.M. (2001) The effects of hypoxia on the northern Gulf of Mexico coastal ecosystem: a fisheries perspective. *In* Coastal hypoxia: Consequences for Living Resources and Ecosystems (Rabalais, N.N., and Turner, R.E., eds.), pp. 321–354, Coastal and Estuarine Studies 58, American Geophysical Union, Washington, DC.

Chester, R. (1990) Marine Geochemistry. Unwin Hyman, London.

Chester, R. (2003) Marine Geochemistry. Blackwell, London.

Childers, D.L., Davis, S.E., Twilley, R., and Rivera-Monroy, V.H. (1999) Wetland-water column interactions and the biogeochemistry of estuary–watershed coupling around the Gulf of Mexico. *In* Biogeochemistry of Gulf of Mexico Estuaries (Bianchi, T.S., Pennock, J.R., and Twilley, R.R., eds.), pp. 211–235, John Wiley, New York.

Childers, D.L., and Day, J.W. (1990) Marsh–water column interactions in two Louisiana estuaries. I. Sediment dynamics. Estuaries 13, 393–403.

Childers, D.L., Day, J.W., and McKellar, H.N. (2000) Twenty more years of marsh and estuarine studies: revisiting Nixon (1980). *In* Concepts and Controversies in Tidal Marsh Ecology (Weinstein, M.P., and Kreeger, D.Q., eds.), pp. 385–414, Kluwer Academic, New York.

Childers, D.L., McKellar, H.N., Dame, R.F., Sklar, F.H., and Blood, E.R. (1993) A dynamic nutrient budget of subsystem interactions in a salt marsh estuary. Estuar. Coastal Shelf Sci. 36, 105–131.

Chin, Y.P., Aiken, G.R., and Danielsen, K.M. (1997) Binding of pyrene to aquatic and commercial humic substances: the role of molecular weight and aromaticity. Environ. Sci. Technol. 31, 1630–1635.

Chin, Y.P., and Gschwend, P.M. (1991) The abundance, distribution, and configuration of porewater organic colloids in recent sediments. Geochim. Cosmochim. Acta 55, 1309–1317.

Chin, W.C., Orellana, M.V., and Verdugo, P. (1998) Spontaneous assembly of marine dissolved organic matter into polymer gels. Nature 391, 568–572.

Chin-Leo, G., and Benner, R. (1992) Enhanced bacterioplankton production at intermediate salinities in the Mississippi River plume. Mar. Ecol. Prog. Ser. 87, 87–103.

Cho, B.C., Park, M.G., Shim, J.H., and Azam, F. (1996) Significance of bacteria in urea dynamics in coastal waters. Mar. Ecol. Prog. Ser. 142, 19–26.

Choe, K.Y., Gill, G.A., and Lehman, R. (2003) Distribution of particulate, colloidal, and dissolved mercury in San Francisco Bay estuary. 1. Total mercury. Limnol. Oceanogr. 48, 1535–1546.

Christensen, D., and Blackburn, T.H. (1980) Turnover of tracer (^{14}C, ^{3}H labeled) alanine in inshore marine sediments. Mar. Biol. 58, 97–103.

Christensen, J.P., Smethie, W.M., and Devol, A.H. (1987) Benthic nutrient regeneration and denitrification on the Washington continental shelf. Deep Sea Res. 34, 1027–1047.

Christensen, S., and Tiedje, J.M. (1988) Sub-parts-per-billion nitrate method: Use of an N_2O producing denitrifier to convert NO_3^- or $^{15}NO_3^-$ to N_2O. Appl. Environ. Microbiol. 54, 1409–1413.

Christian, R.R., and Thomas, C.R. (2003) Network analysis of nitrogen inputs and cycling in the Neuse River estuary, North Carolina, USA. Estuaries 26, 815–828.

Chrzastowski, M.J., Kraft, J.C., and Stedman, S.M. (1987) Coastal Delaware sea-level rise based on marsh mud accumulation rates and ^{210}Pb dating. Geol. Soc. Am. (Abstracts and Programs) 9, 8.

Chuang, W.S., and Wiseman, W.J. (1983) Coastal sea level response to frontal passages on the Louisiana–Texas coast. J. Geophys. Res. 88, 2615–2620.

Church, J.A., Gregory, J.M., Huybrechts, P., Kuhn, M., Lambeck, K., Nhuan, M.T., Qin, D., and Woodworth, P.L. (2001) Changes in sea level. In Climate Change 2001. The Scientific Basis (Houghton, J.T., Ding, Y., Griggs, D.J., Noguer, M., van der Linden, P.J., Dai, X., Maskell, K., and Johnson, C.A., eds.), pp. 639–694. Cambridge University Press, Cambridge, UK.

Church, T.M., Lord, C.J., and Somayajula, L.K. (1981) Uranium, thorium, and lead nuclides in a Delaware salt marsh sediment. Estuar. Coastal Shelf Sci. 13, 267–275.

Church, T.M., Tramontano, J.M., and Murray, S. (1986) Trace metal fluxes through the Delaware Bat estuary. Rapp. P.V. Reun. Cons. Inl. Explor. Mer. 186, 271–276.

Cicerone, R.J., and Oremland, R.S. (1988) Biogeochemical aspects of atmospheric methane. Global Biogeochem. Cycles 2, 299–327.

Cifuentes, L.A., and Salata, G.G. (2001) Significance of carbon isotope discrimination between bulk carbon and extracted phospholipid fatty acids in selected terrestrial and marine environments. Org. Geochem. 32, 613–621.

Cifuentes, L.A., Sharp, J.H., and Fogel, M.L. (1988) Stable carbon and nitrogen isotope biogeochemistry in the Delaware estuary. Limnol. Oceanogr. 33, 1102–1115.

Clark, J.E., Henricks, M., Timmermans, M., Struck, C., and Hilbrunda, K.J. (1994) Glacial isostatic deformation of the Great Lakes region. Geol. Soc. Am. 106, 19–31.

Clark, J.F., Schlosser, P., Stute, M., and Simpson, H.J. (1996) SF_6^3–He tracer release experiment: a new method of determining longitudinal dispersion coefficients in large rivers. Environ. Sci. Technol. 30, 1527–1532.

Clark, J.F., Simpson, H.J., Bopp, R.F., and Deck, B. (1992) Geochemistry and loading history of phosphate and silica in the Hudson estuary. Estuar. Coastal Shelf Sci. 34, 213–233.

Clark, L.L., Ingall, E.D., and Benner, R. (1998) Marine phosphorus is selectively remineralized. Nature 393, 426.

Clauzon, G. (1973) The eustatic hypotheses and the pre-Pliocene cutting of the Rhone valley. Init. Rep. DSDP 13, 1251–1256.

Clegg, S.L., and Whitfield, M. (1993) Applications of generalized scavenging model to times series ^{234}Th and particle data obtained during the JGOFS North Bloom Experiment. Deep Sea Res. 40, 152–1545.

Cleven, R.F.M.J., and van Leeuwen, H.P. (1986) Electrochemical analysis of the heavy metal/humic acid interaction. J. Environ. Anal. Chem. 27, 11–28.

Clifford, D.J., Carson, D.M., McKinney, D.E., Bortiatynski, J.M., and Hatcher, P.G. (1995) A new rapid technique for the characterization of lignin in vascular plants: thermochemolysis with tetramethylammonium hydroxide (TMAH). Org. Geochem. 23, 169–175.

Cline, J.D., and Kaplan, I.R. (1975) Isotopic fractionation of dissolved nitrate during denitrification in the eastern tropical North Pacific Ocean. Mar. Chem. 3: 271–299.

Cloern, J.E. (1996) Phytoplankton bloom dynamics in coastal ecosystems: a review with some general lessons from sustained investigation of San Francisco Bay, California. Rev. Geophys. 34, 127–168.

Cloern, J.E. (2001) Our evolving conceptual model of the coastal eutrophication problem. Mar. Ecol. Prog. Ser. 210, 223–253.

Cloern, J.E., Canuel, E.A., and Harris, D. (2002) Stable carbon and nitrogen isotopic composition of aquatic and terrestrial plants in the San Francisco Bay estuarine system. Limnol. Oceanogr. 47, 713–729.

Clough, B. (1998) Mangrove forest productivity and biomass accumulation in Hinchbrook Channel, Australia. Mangroves Salt Marshes 77, 171–182.

Coale, K.H., and Bruland, K.W. (1987) Oceanic stratified euphotic zone as elucidated by ^{234}Th:^{238}U disequilibria. Limnol. Oceanogr. 32, 189–200.

Coale, K.H., and Bruland, K.W. (1988) Copper complexation in Northeast Pacific. Limnol. Oceanogr. 33, 1084–1101.

Coble, P.G. (1996) Characterization of marine and terrestrial DOM in seawater using excitation–emission matrix spectroscopy. Mar. Chem. 51, 325–346.

Coble, P.G., Del Castillo, C.E., and Avril, B. (1998) Distribution and optical properties of CDOM in the Arabian Sea during the 1995 Southwest Monsoon. Deep-Sea Res. Part II. 45, 2195–2223.

Coble, P.G., Green, S.A., Blough, N.V., and Gagosian, R.B. (1990) Characterization of dissolved organic matter in the Black Sea by fluorescence spectroscopy. Nature 348, 432–435.

Cochran, J.K. (1980) The flux of Ra-226 from deep-sea sediments. Earth Planet. Sci. Lett. 49, 381–392.

Cochran, J.K. (1992) The oceanic chemistry of the U- and Th-series nuclides. In Uranium Series Disequilibrium: Applications to Environmental Sciences (Ivanovich, M., and Harmon, R.S., eds.), pp. 334–395, Clarendon Press, Oxford, UK.

Cochran, J.K., and Aller, .C. (1979) Particle reworking in sediments from the New York Bight: evidence from ^{234}Th/^{238}U disequilibrium. Estuar. Coastal Shelf Sci. 9, 739–747.

Cochran, J.K., Carey, A., Sholkovitz, E.R., and Surprenant, L.D. (1986) The geochemistry of uranium and thorium in coastal sediments and sediment pore waters. Geochim. Cosmochim. Acta 50, 663–680.

Cochran, J.K., and Hirschberg, D. (1991) ^{234}Th as an indicator of biological reworking and particle transport. In Long Island Sound Study: Sediment Geochemistry and Biology (Cochran, J.K., Aller, R.C., Aller, J.Y., Hirschberg, D.J., and Mackin, J.E. eds.), EPA Final Report, CE 002870026.

Codispoti, L. A., Brandes, J. A., Christensen, J. P., Devol, A. H., Naqvi, S. W. A., Paerl, H.W., and Yoshinari, T. (2001) The oceanic fixed nitrogen and nitrous oxide budgets: Moving targets as we enter the anthropocene? Sci. Mar. 65 (Suppl. 2), 85–105.

Coelho, J.P., Flindt, M.R., Jensen, H.S., Lillebo, A.I., and Pardal, M.A. (2004) Phosphorus speciation and availability in intertidal sediments of a temperate estuary: relation to eutrophication and annual P-fluxes. Estuar. Coastal Shelf Sci. 61, 583–590.

Coffin, R.B. (1989) Bacterial uptake of dissolved free and combined amino acids in estuarine waters. Limnol. Oceanogr. 34, 531–542.

Coffin, R.B., and Cifuentes, L.A. (1999) Stable isotope analysis of carbon cycling in the Perdido estuary, Florida. Estuaries 22, 917–926.

Cole, J.J., and Caraco, N.F. (2001) Carbon in catchments: connecting terrestrial carbon losses with aquatic metabolism. Mar. Freshwat. Res. 52, 101–110.

Cole, J.J., Findlay, S., and Pace, M.L. (1988) Bacterial production in fresh and saltwater ecosystems: a cross-system overview. Mar. Ecol. Prog. Ser. 43, 1–10.

Cole, J.J., Peierls, B.L., Caraco, N.F., and Pace, M.L. (1993) Nitrogen loading of rivers as a human-driven process. In Humans as Components of Ecosystems (McDonnell, M.J., and Pickett, S.T.A., eds.), pp. 141–157, Springer-Verlag, New York.

Coleman, J.M. (1969) Brahmaputra River: channel processes and sedimentation. Sediment. Geol. 3, 131–239.

Coleman, J.M., and Gagliano, S.M. (1964) Cyclic sedimentation in the Mississippi River deltaic plain. Trans. Gulf Coast Assoc. Geol. Soc. 14, 67–80.

Coleman, J.D., and Wright, L.D. (1975) Modern river deltas: variability of processes and sand bodies. *In* Deltas, Models for Exploration (Broussard, M.L., ed.), pp. 99–149, Houston Geological Society, Houston, TX.

Colijn, F., and Dijekma, K.S. (1981) Species composition of the benthic diatoms and distribution of chlorophyll-*a* on an intertidal flat in the Dutch Wadden Sea. Mar. Ecol. Prog. Ser. 4, 9–21.

Collier, A., and Hedgepeth, J. (1950) An introduction to the hydrography of tidal waters of Texas. Contrib. Mar. Sci. 1, 121.

Collos, Y. (1989) A linear model of external interactions during uptake of different forms of inorganic nitrogen by microalgae. J. Plank. Res. 11, 521–533.

Colombo, J.C., Silverberg, N., and Gearing, J.N. (1996a) Biogeochemistry or organic matter in the Laurentian Trough. I. Composition and vertical fluxes of rapidly settling particles. Mar. Chem. 51, 277–293.

Colombo, J.C., Silverberg, N., and Gearing, J.N. (1996b) Biogeochemistry of organic matter in the Laurentian Trough. II. Bulk composition of the sediments and the relative reactivity of major components during early diagenesis. Mar. Chem. 51, 295–314.

Colombo, J.C., Silverberg, N., and Gearing, J.N. (1998) Amino acid biogeochemistry in the Laurentian Trough: vertical fluxes and individual reactivity during early diagenesis. Org. Geochem. 29, 933–945.

Colquhoun, D.J. (1968) Coastal plains. *In* The Encyclopedia of Geomorphology (Fairbridge, R.W., ed.), pp. 144–150, Reinhold Book Corp. New York.

Comet, P.A., and Eglinton, G. (1987) The use of lipids as facies indicators. *In* Marine Petroleum Source Rocks. Geological Society Special Publication 26, 99–117.

Compiano, A.M., Romano, J.C., Garabetian, F., Laborde, P., and Giraudiere, I. (1993) Monosaccharide composition of particulate hydrolysable sugar fraction in surface microlayers from brackish and marine waters. Mar. Chem. 42, 237–251.

Compton, J., Mallinson, D., Glenn, C.R., Filippelli, G., Follmi, K., Shields, G., and Zanin, Y. (2000) Variations in the global phosphorus cycle. *In* Authigenesis: from Global to Microbial to Microbial (Glenn, C.R., Prevot-Lucas, L, and Lucas, J., eds.), pp. 21–33, Spec. Publ-SEPM, Vol. 66, Tulsa, OK.

Condron, L.M., Goh, K.M., and Newman, R.H. (1985) Nature and distribution of soil phosphorus as revealed by a sequential extraction method followed by ^{31}P nuclear magnetic resonance analysis. J. Soil Sci. 36, 199–207.

Conley, D. (1988) Biogenic silica as an estimate of siliceous microfossil abundance in Great Lake sediments. Biogeochemistry 6, 161–179.

Conley, D.J. (1997) Riverine contribution of biogenic silica to the oceanic silica budget. Limnol. Oceanogr. 42, 774–777.

Conley, D.J. (1999) Biogeochemical nutrient cycles and nutrient management strategies. Hydrobiology 410, 87–96.

Conley, D.J. (2000) Biogeochemical nutrient cycles and nutrient management strategies. Hydrobiologia 410, 87–96.

Conley, D.J. (2002) Terrestrial ecosystems and the global biogeochemical silica cycle. Global Biogeochem. Cycle 16, 774–777.

Conley, D.J., Humborg, C., Rahm, L., Savchuk, O.P., and Wulff, F. (2002) Hypoxia in the Baltic Sea and basin-scale changes in phosphorus biogeochemistry. Environ. Sci. Technol. 36, 5315–5320.

Conley, D.J., and Johnstone, R.W. (1995) Biogeochemistry of N, P, and Si in Baltic Sea sediments: response to a simulated deposition of a spring diatom bloom. Mar. Ecol. Prog. Ser. 122, 265–276.

Conley, D.J., Kaas, H., Mohlenberg, F., Rasmussen, B., and Windole, J. (2000) Characteristics of Danish estuaries. Estuaries 23, 820–837.

Conley, D.J., Kilham, S.S., and Theriot, E. (1989) Differences in silica content between marine and freshwater diatoms. Limnol. Oceanogr. 34, 205–212.

Conley, D.J., and Malone, T.C. (1992) Annual cycle of dissolved silicate in Chesapeake Bay: implications for the production and fate of phytoplankton biomass. Mar. Ecol. Prog. Ser. 81, 121–128.

Conley, D.J., Quigley, M.A., and Schelske, C.L. (1988) Silica and phosphorus flux from sediments: Importance of internal recycling in Lake Michigan. Can. J. Fish. Aquat. Sci. 45, 1030–1035.

Conley, D.J., Schleske, C.L., and Stoermer, E.F. (1993) Modification of the biogeochemical cycle of silica with eutrophication. Mar. Ecol. Prog. Ser. 101, 179–192.

Conley, D.J., Smith, W.M., Cornwell, J.C., and Fisher, T.R. (1995) Transformation of particle-bound phosphorus at the land–sea interface. Estuar. Coastal Shelf Sci. 40, 161–176.

Costanza, R., Low, B., Ostrom, E., and Wilson, J. (eds.) (2001) Institutions, Ecosystems, and Sustainablilty. Lewis/CRC Press, Boca Raton, FL.

Constantz, B., and Weiner, S. (1988) Acidic macromolecules associated with the mineral phase of Scleractinian coral skeletons. J. Exp. Zool. 248, 253–258.

Cooper, S.R., and Brush, G.S. (1991) Long-term history of Chesapeake Bay anoxia. Science 254, 992–996.

Cooper, S.R., and Brush, G.S. (1993) A 2,500-year history of anoxia and eutrophication in Chesapeake Bay. Estuaries 16, 627–626.

Corbett, D.R., Chanton, J., Burnett, W., Dillon, K., and Rutkowski, C. (1999) Patterns of groundwater discharge into Florida Bay. Limnol Oceanogr. 44, 1045–1055.

Corbett, D.R., Dillon, K., Burnett, W., and Chanton, J. (2000) Estimating the groundwater contribution into Florida Bay via natural tracers, ^{222}Rn and CH_4. Limnol. Oceanogr. 45, 1546–1557.

Corbett, D.R., McKee, B.A., and Duncan, D. (2004) An evaluation of mobile mud dynamics in the Mississippi River deltaic region. Mar. Geol. 209, 91–112.

Cornwell, J.C., Conley, D.J., Owens, M., and Stevenson, J.C. (1996) A sediment chronology of the eutrophication of Chesapeake Bay. Estuaries 19, 488–499.

Cornwell, J.C., Kemp, W.M., and Kana, T.M. (1999) Denitrification in coastal ecosystems: methods, environmental controls and ecosystem level controls, a review. Aquat. Ecol. 33, 41–54.

Correll, D.L., and Ford, D. (1982) Comparison of precipitation and land runoff as sources of estuarine nitrogen. Estuar. Coastal Shelf Sci. 15, 45–56.

Correll, D.L., Jordan, T.E., and Weller, D.E. (1992) Nutrient flux in a landscape: effects of coastal land use and terrestrial community mosaic on nutrient transport to coastal waters. Estuaries 15, 431–442.

Costanza, R., d'Arge, R., de Groot, R., Farber, S., Grasso, M., Hanon, B., Limburg, K., Naeem, S., and van den Belt, M. (1997) The value of the world's ecosystem services and natural capital. Nature 387, 253–260.

Costanza, R., Sklar, F.H., and White, M.L. (1990) Modeling coastal landscape dynamics. BioScience 40, 91–107.

Costanza, R., and Voinov, A. (2000) Integrated ecological economic regional modeling: linking consensus: Building and analysis for synthesis and adaptive management. *In* Estuarine Science: A Synthetic Approach to Research and Practice (Hobbie, J.E., ed.), pp. 461–506, Island Press, Washington, DC.

Costanzo, S.D., O'Donohue, M.J., Dennison, W.C., Lonerargan, R.R., and Thomas, M. (2001) A new approach for detecting and mapping sewage inputs. Mar. Pollut. Bull. 42, 149–156.

Cotner, J.B., Suplee, M.W., Chen, N.W., and Shormann, D.E. (2004) Nutrient, sulfur and carbon dynamics in a hypersaline lagoon. Estuar. Coastal Shelf Sci. 59: 639–652.

Cottrell, M.T., and Kirchman, D.L. (2003) Contribution of major bacterial groups to bacterial biomass production (thymidine and leucine incorporation) in the Delaware estuary. Limnol. Oceanogr. 48, 168–178.

Countway, R.E., Dickut, R.M., and Canuel, E.A. (2003) Polycyclic aromatic hydrocarbons (PAH) distributions and associations with organic matter in surface waters of the York River, VA estuary. Org. Geochem. 34, 209–224.

Cowan, J.L., and Boynton, W.R. (1996) Sediment water oxygen and nutrient exchanges along the longitudinal axis of Chesapeake Bay: seasonal patterns, controlling factors and ecological significance. Estuaries 9, 562–580.

Cowan, J.L., Pennock, J.R., and Boynton, W.R. (1996) Seasonal and interannual patterns of sediment–water nutrient and oxygen fluxes in Mobile Bay, Alabama (USA): regulating factors and ecological significance. Mar. Ecol. Prog. Ser. 141, 229–245.

Cowie, G.L., and Hedges, J.I. (1984a) Carbohydrate sources in a coastal marine environment. Geochim. Cosmochim. Acta 48, 2075–2087.

Cowie, G.L., and Hedges, J.I. (1984b) Determination of neutral sugars in plankton, sediments and wood by capillary gas chromatography of equilibrated isomeric mixtures. Anal. Chem. 56, 497–504.

Cowie, G.L., and Hedges, J.I. (1992) The role of anoxia in organic matter preservation in coastal sediments: relative stabilities of the major biochemical's under oxic and anoxic depositional conditions. Org. Geochem. 19, 229–234.

Cowie, G.L., and Hedges, J.I. (1994) Biochemical indicators of diagenetic alteration in natural organic matter mixtures. Nature 369, 304–307.

Cowie, G.L., Hedges, J.I., and Calvert, S.E. (1992) Sources and relative reactivities of amino acids, neutral sugars, and lignin in an intermittently anoxic marine environment. Geochim. Cosmochim. Acta 56, 1963–1978.

Cowie, G.L., Hedges, J.I., Prahl, F.G., and de Lange, G.L. (1995) Elemental and biochemical changes across an oxidation front in a relict turbidite: an oxygen effect. Geochim. Cosmochim. Acta 59, 33–46.

Cox, R.A., Culkin, E., and Riley, J.P. (1967) The electrical conductivity/chlorinity relationship in natural seawater. Deep Sea res. 14, 203–220.

Craig, H. (1961) Isotopic variations in meteoric waters. Science 133, 1702–1703.

Cranwell, P.A. (1973) Chain-length distribution of n-alkanes from lake sediments in relation to post-glacial environmental change. Freshwat. Biol. 3, 259–265.

Cranwell, P.A. (1981) Diagenesis of free and bound lipids in terrestrial detritus in a lacustrine sediment. Org. Geochem. 3, 79–89.

Cranwell, P.A. (1982) Lipids of aquatic sediments and sedimenting particles. Prog. Lipid Res. 21, 271–308.

Cranwell. P.A. (1984) Lipid geochemistry of sediments from Upton Broad, a small productive lake. Org. Geochem. 7, 25–37.

Cranwell, P.A., and Volkman, J.K. (1985) Alkyl and steryl esters in a recent lacustrine sediment. J. Chem. Geol. 32, 29–43.

Craig, H. (1954) Carbon-13 in plants and the relationships between carbon-13 and carbon-14 variations in nature. J. Geol. 62, 115–149.

Crawford, C.C., Hobbie, J.E., and Webb, K.L. (1974) The utilization of dissolved free amino acids by estuarine microorganisms. Ecology 55, 551–563.

Crawford, R.L. (1981) Lignin Biodegradation and Transformation. John Wiley, New York.

Crill., P.M., and Martens, C.S. (1987) Biogeochemical cycling in an organic-rich coastal marine basin. 6. Temporal and spatial variation in sulfate reduction rates. Geochim. Cosmochim. Acta 51, 1175–1186.

Crusius, J., and Wanninkhof, R. (2003) Gas transfer velocities measured at low wind speed over a lake. Limnol. Oceanogr. 48, 1010–1017.

Crutzen, P.J., and Stoermer, E.F. (2000) The "Anthropocene." IGBP Newslett. 41, 17–18.

Culkin, F., and Cox, R.A. (1966) Sodium, potassium, magnesium, calcium and strontium in sea water. Deep-Sea Res. 13, 789–804.

Curćo, A., Ibanez, C., Day, J.W., and Prat, N. (2002) Net primary production and decomposition of salt marshes of the Ebre delta (Catalonia, Spain). Estuaries 25, 309–324.

Curray, J.R. (1965) Late Quaternary history. Continental shelves of the USA. *In* Quaternary of the United States (Wright, N.E., and Frey, D.G., eds.), pp. 725, Princeton University Press, NJ.

Currin, C.A., Newell, S.Y., and Paerl, H.W. (1995) The role of standing dead *Spartina alterniflora* and benthic microalgae in salt marsh food webs: considerations based on multiple stable isotope analysis. Mar. Ecol. Prog. Ser. 121, 99–116.

Currin, C.A., and Paerl, H.W. (1998) Epiphytic nitrogen fixation associated with standing dead shoots of smooth cordgrass, *Spartina alterniflora*. Estuaries 21, 108–117.

Cutter, G.A., and Cutter, L.S. (2004) Selenium biogeochemistry in the San Francisco Bay estuary: changes in water column behavior. Estuar. Coastal Shelf Sci. 61, 463–476.

Czerny, A.B., and Dunton, KH. (1995) The effects of in situ light reduction on the growth of two subtropical seagrasses, *Thalassia testudinum* and *Halodule wrightii*. Estuaries 18, 418–427.

Dacey, J.W.H. (1981) How aquatic plants ventilate. Oceanus 24, 43–51.

Dacey, J.W.H. (1987) Knudsen-transitional flow and gas pressurization in leaves of *Nelumbo*. Plant Physiol. 85, 199–203.

Dacey, J.W.H., and Howes, B.L. (1984) Water uptake by roots control water table movement and sediment oxidation in a short *Spartina* marsh. Science 224, 487–489.

Dacey, J.W.H., King, G.M., and Wakeham, S.G. (1987) Factors controlling emission of dimethylsulfide from salt marshes. Nature 330, 643–645.

Dacey, J.W.H., and Wakeham, S.G. (1986) Oceanic dimethylsulfide: Production during zooplankton grazing on phytoplankton. Science 233, 1314–1316.

Dagg, M., Benner, R., Lohrenz, S., and Lawrence, D. (2004) Transformation of dissolved and particulate materials on continental shelves influenced by large rivers: plume processes. Cont. Shelf Res. 24, 833–858.

Dahle, S., Savinov, V.M., Matishov, G.G., Evenset, A., and Naes, K. (2003) Polycyclic aromatic hydrocarbons (PAHs) in bottom sediments of the Kara Sea shelf, Gulf of Ob and Yenisei Bay. Sci. Total Environ. 306, 57–71.

Dai, M.H., Martin, J.M., and Cauwet, G. (1995) The significant role of colloids in the transport and transformation of organic carbon and associated trace metals (Cd, Cu, and Ni) in the Rhone delta (France). Mar. Chem. 51, 159–175.

Dale, A.W., and Prego, R. (2002) Physico-biogeochemical controls on benthic–pelagic coupling of nutrient fluxes and recycling in a coastal inlet affected by upwelling. Mar. Ecol. Prog. Ser. 235, 15–28.

Daley, R.J. (1973) Experimental characterization of lacustrine chlorophyll diagenesis. II. Bacterial, viral, and herbivore grazing effects. Arch. Hydrobiol. 72, 409–439.

Daley, R.J., and Brown, S.R. (1973) Experimental characterization of lacustrine chlorophyll diagenesis. 1. Physiological and environmental effects. Arch. Hydrobiol. 72, 277–304.

Dalrymple, R.W., Zaitlin, B.A., and Boyd, R. (1992) A conceptual model of estuarine sedimentation. J. Sed. Petrol. 62, 1130–1146.

Dalsgaard, J., St. John, M., Kattner, G., Muller-Navarra, D., and Hagen, W. (2003) Fatty acid trophic markers in the pelagic marine environment. Adv. Mar. Biol. 46, 225–340.

Dalsgaard, T. (2003) Benthic primary production and nutrient cycling in sediments with benthic microalgae and transient accumulation of macroalgae. Limnol. Oceanogr. 48, 2138–2150.

Dalsgaard, T., Canfield, D.E., Petersen, J., Thamdrup, B., and Acuna-Gonzalez, J. (2003) N_2 production by ammonex reaction in the anoxic water column of Golfo Dulce, Costa Rica. Nature. 422, 606–608.

Dame, R.F. (1989) The importance of *Spartina alterniflora* to Atlantic coast estuaries. Crit. Rev. Aquat. Sci. 1, 639–660.

Dame, R.F., Alber, M., Allen, D., Mallin, M., Montague, C., Lewitus, A., Chalmers, A., Gardner, R., Gilman, C., Kjerfve, B., Pinckey, J., and Smith, N. (2000) Estuaries of the South Atlantic Coast of North America: their geographical signatures. Estuaries 23, 793–819.

Dame, R.F., Dankers, N., Prins, T., Jongsma, H., and Smaal, A. (1991) The influence of mussel beds on nutrients in the western Wadden Sea and the eastern Scheldt estuaries. Estuaries 14, 130–138.

Danishefsky, I., Whistler, R.L. and Bettleheim, F.A. (1970) Introduction to polysaccharide chemistry. *In* The Carbohydrates: Chemistry and Biochemistry, 2nd edn. (Pigman, W., and Horton, D., eds.), pp. 375–412, Academic Press, New York.

Darnell, R.M. (1967) The organic detritus problem in estuaries. American Association for the Advancement of Science, Publication No. 83, 374–375.

Dauwe, B., and Middelburg, J.J. (1998) Amino acid and hexosamines as indicators of organic matter degradation state in North Sea sediments. Limnol. Oceanogr. 43, 782–798.

Dauwe, B., Middelburg, J.J., and Herman, P.M.J. (2001) The effect of oxygen on the degradability of organic matter in subtidal and intertidal sediments of the North Sea. Mar. Ecol. Prog. Ser. 215, 13–22.

Dauwe, B., Middelburg, J.J., van Rijswijk, Sinke, J., Herman, P.M.J., and Heip, C.H.R. (1999) Enzymatically hydrolysable amino acids in the North Sea sediments and their possible implication for sediment nutritional values. J. Mar. Res. 57, 109–134.

D'Avanzo, C., Kremer, J.N., and Wainright, S.C. (1996) Ecosystem production and respiration in response to eutrophication in shallow temperate estuaries. Mar. Ecol. Prog. Ser. 141, 263–274.

Davis, R.A. (1996) Coasts. Prentice Hall, Englewood cliffs, NJ.

Davis, R.A. (1985) Coastal Sedimentary Environments. Springer-Verlag, New York.

Davis III, S.E., Childers, D.L., Day, J.W., Rudnick, D.T., and Sklar, F.H. (2001) Wetland–water column exchange of carbon, nitrogen, and phosphorus in a southern everglades dwarf mangrove. Estuaries 24, 610–622.

Dawson, R., and Gocke, K. (1978) Heterotrophic activity in comparison to the free amino acid concentration in Baltic Sea water samples. Oceanol. Acta 1, 45–54.

Day, J., Hall, C.S., Kemp, W.M., and Yanez-Arancibia, A. (1989) Estuarine Ecology. John Wiley, New York.

Day, J.W., Madden, C., Twilley, R., Shaw, R., McKee, B.A., Dagg, M., Childers, D., Raynie, R., and Rouse, L. (1995) The influence of the Atchafalaya River discharge on Fourleague Bay, LA. *In* Changes in Fluxes in Estuaries (Dyer K., and Orth, R., eds.), pp. 151–160, Olsen and Olsen, Copenhagen, Denmark.

Day, J.W., Smith, W.G., Wagner, P.R., and Stowe, W.C. (1973) Community structure and carbon budget of a salt marsh and shallow bar estuarine system in Louisiana.

Office of Sea Grant Development, Center for Wetland Resources, Louisiana State University, Baton Rouge.

De Angelis, M.A., and Lilley, M.D. (1987) Methane in surface waters of Oregon estuaries and rivers. Limnol. Oceanogr. 32, 299–327.

De Angelis, M.A., and Scranton, M.D. (1993) Fate of methane in the Hudson River and estuary. Global Biogeochem. Cycles 7, 509–523.

de Beer, D., and Kühl, M. (2001) Interfacial Microbial Mates and Biofilms. *In* The Benthic Boundary Layer (Boudreau, B.P., and Jørgensen, B.B., eds.), pp. 374–394, Oxford University Press, New York.

Decho, A.W., and Herndl, G.J. (1995) Microbial activities and the transformation of organic matter within mucilaginous material. Sci. Total Environ. 165, 33–42.

Decho, A.W., and Lopez, G.R. (1993) Exopolymer microenvironments of microbial flora: multiple and interactive effects on trophic relationships. Limnol. Oceanogr. 38, 1633–1645.

Deeds, J.R., and Klerks, P.L. (1999) Metallothionein-like proteins in the freshwater oligochaete *Limnodrilis udekemianus* and their role as a homeostatic mechanism against cadmium toxicity. Environ. Pollut. 106, 381–389.

Deegan, C.E., and Garritt, R.H. (1997) Evidence for spatial variability in estuarine food webs. Mar. Ecol. Prog. Ser. 147, 31–47.

Deegan, C.E., Kirby, R., and Rae, I. (1973) The superficial deposits of the Firth of Clyde and its sea lochs. Inst. Geol. Sci. Rep. 73/9, 135 pp.

Deegan, L.A. (1993) Nutrient and energy transport between estuaries and coastal marine ecosystems by fish migration. Can. J. Fish. Aquat. Sci. 50, 74–79.

Deflandre, B., Mucci, A., Gagne, J., Guignard, C., and Sundby, B. (2002) Early diagenetic processes in coastal marine sediments disturbed by catastrophic sedimentation event. Geochim. Cosmochim. Acta 66, 2547–2558.

Deflandre, B., Sundby, B., Gremare, a., Lefrançois, L., and Gagné, J.P. (2000) Effects of sedimentary microenvironments on the vertical distributions of oxygen and DOC in coastal marine sediments: scales of variability. EOS: Trans Am. Geophys. Union 80, 115.

Degens, E.T. (1977) Molecular nature of nitrogenous compounds in seawater and recent marine sediments. *In* Organic Matter in Natural Waters (Hood, D.W., ed.), pp. 77–106, University of Alaska Press, Fairbanks, AK.

Degens, E.T. (1989) Perspectives on Biogeochemistry. Springer-Verlag, New York.

Degens, E.T., Guillard, R.R.L., Sackett, W.M., and Hellebust, J.A. (1968) Metabolic fractionation of carbon isotopes in marine plankton. I. Temperature and respiration experiments. Deep Sea Res. 15, 1–9.

Degens, E.T., Kempe, S., and Richey, J.E. (eds.) (1991) Biogeochemistry of Major Rivers. John Wiley, New York.

Degens, E.T., Kempe, S., and Spitzy, A. (1984) Carbon dioxide: a biogeochemical portrait. *In* The Handbook of Environmental Chemistry (Hutzinger, O., ed.), Vol. 1 (Part C), Springer-Verlag, New York.

Degobbis, D., Gilantin, M., and Relevante, N. (1986) An annotated nitrogen budget calculation for the North Adriatic Sea. Mar. Chem. 20, 159–177.

de Groot, C. (1990) Some remarks on the presence of organic phosphates in sediments. Hydrobiologia 207, 303–309.

Deines, P. (1980) The isotopic composition of reduced carbon. *In* Handbook of Environmental Isotope Geochemistry. Vol. I. The Terrestrial Environment (Fitz, P., and Fontes, J.C., eds.), pp. 329–406, Elsevier, Amsterdam.

de Jonge, V.N. (1985) The occurrence of "episammic" diatom populations a result of interaction between physical sorting of sediments and certain properties of diatom species. Estuar. Coastal Shelf Sci. 21, 607–622.

de Jonge, V.N., and Colijn, F. (1994) Dynamics of microphytobenthos in the Ems estuary measured as chlorophyll-a and carbon. Mar. Ecol. Prog. Ser. 104, 185–196.

de Jonge, V.N., Elliott, M., and Orive, E. (2002) Causes, historical development, effects and future challenges of a common environmental problem: eutrophication. Hydrobiology 475/476, 1–19.

de Jonge, V.N., and Villerius, L.A. (1989) Possible role of carbonate dissolution in estuarine phosphate dynamics. Limnol. Oceanogr. 34, 332–340.

DeKanel, J., and Morse, J.W. (1978) The chemistry of orthophosphate uptake from seawater onto calcite and aragonite. Geochim. Cosmochim. Acta 42, 1335–1340.

Delaney, M.L. (1998) Phosphorus accumulation in marine sediments and the oceanic phosphorus cycle. Global Biogeochem. Cycles 12, 563–572.

Delaune, R.D., Patrick, W.H., and Buresh, R.J. (1978) Sedimentation rates determined by ^{137}Cs dating in a rapidly accreting salt marsh. Nature 275, 532–533.

Delaune, M.L., Reddy, C.N., and Patrick, W.H. (1981) Accumulation of plant nutrients and heavy metals through sedimentation processes and accretion in a Louisiana salt marsh. Estuaries 4, 328–334.

Delaune, R., Smith, C.J., Patrick, W.H. (1983) Methane release from Gulf coast wetlands. Tellus 35, 8–15.

Delaune, R., Smith, C.J., and Patrick, W.H., and Roberts, H.H. (1987) Rejuvenated marsh and bay-bottom accretion on the rapidly subsiding coastal plain of the U.S. Gulf Coast: A second-order effect of the emerging Atchafalaya delta. Estuar. Coastal Shelf Sci. 25, 381–389.

Del Castillo, C.E., Coble, P.E., Morell, P.E., Lopez, J.M., and Corredor, J.E. (1999) Analysis of the optical properties of the Orinoco River plume by absorption and fluorescence spectroscopy. Mar. Chem. 66, 35–51.

Del Castillo, C.E., Gilbes, F., Coble, P.G., and Muller-Karger, F.E. (2000) On the dispersal of riverine colored dissolved organic matter over the West Florida Shelf. Limnol. Oceanogr. 45, 1425–1432.

de Leeuw, J.W., and Largeau, C. (1993) A review of macromolecular organic compounds that comprise living organisms and their role in kerogen, coal, and petroleum formation. In Organic Geochemistry—Principle and Applications (Engel, M.H., and Macko, S.A., eds.), pp. 23–72, Plenum Press, New York.

de Leeuw, J.W., Simoneit, B.R., Boon, J.J., Rijpstra, W.I.C., de Lange, F., van der Leedee, J.C.W., Correia, V.A., Burlingame, A.L., and Schenck, P.A. (1977) Phytol compounds in the geosphere. In Advances in Organic Geochemistry (Campos, R., and Goni, J., eds.), pp. 61–79, Enadisma, Madrid.

De Leiva Moreno, J.I., Agnostini, V.N., Caddy, J.F., and Carocci, F. (2000) Is the pelagic–demersal ratio from fishery landings a useful proxy for nutrient availability? A preliminary data exploration for the semi-enclosed seas around Europe, ICES. J. Mar. Sci. 57, 1091–1102.

D'Elia, C.F., Boynton, W.R., and Sanders, J.G. (2003) A watershed perspective on nutrient enrichment, science, and policy in the Patuxent River, Maryland: 1960–2000. Estuaries 26, 171–185.

D'Elia, C.F., Harding, L.W., Leffler, M., and Mackiernan, G.B. (1992) The role and control of nutrients in Chesapeake Bay. Wat. Sci. Technol. 26, 2635–2644.

D'Elia, C.F., Nelson, D.M., and Boynton, W.R. (1983) Chesapeake Bay nutrient and plankton dynamics III. The annual cycle of dissolved silicon. Geochim. Cosmochim. Acta 47, 1945–1955.

D'Elia, C.F., Sanders, J.G., and Boynton, W.R. (1986) Nutrient enrichment studies in a coastal plain estuary: phytoplankton growth in large-scale, continuous cultures. Can. J. Fish. Aquat. Sci. 43, 397–406.

Dellapenna, T.H., Kuehl, S.A., and Pitts, L. (2001) Transient, longitudinal, sedimentary furrows in the York River subestuary, Chesapeake Bay: furrow evolution and effects on seabed mixing and sediment transport. Estuaries 24, 215–227.

Dellapenna, T.M., Kuehl, S.A., and Schaffner, L.C. (1998) Sea-bed mixing and particle residence times in biologically and physically dominated estuarine systems: a comparison of lower Chesapeake Bay and York River subestuary. Estuar. Coastal Shelf Sci. 46, 777–795.

Delwiche, C.C. (1981) Atmospheric chemistry of nitrous oxide. In Denitrification, Nitrification, and Atmospheric Nitrous Oxide pp. 17–44. John Wiley, New York.

DeMaster, D.J. (1981) The supply and accumulation of silica in the marine environment. Geochim. Cosmochim. Acta 64, 2467–2477.

DeMaster, D.J. (2002) The accumulation and cycling of biogenic silica in the Southern Ocean: revisiting the marine silica cycle. Deep-Sea Res. II, 49, 3155–3167.

DeMaster, D.J., and Cochran, J.K. (1982) Particle mixing rates in deep-sea sediments determined from excess ^{210}Pb and ^{32}Si profiles. Earth Planet. Sci. Lett. 61, 257–271.

DeMaster, D.J., Kuehl, S.A., and Nittrover, C.A. (1986) Effects of suspended Sediments and geochemical processes near the mouth of the Amazon River: examination of biological silica uptake and the fate of particle-reactive elements. Cont. Shelf Res. 6, 107–125.

DeMaster, D.J., McKee, B.A., Nittrouer, C.A., Jiangchu, Q., and Quodong, C. (1985) Rates of sediment accumulation and particle reworking based on radiochemical measurements from the continental shelf deposits in the East China Sea. Cont. Shelf Res. 4, 143–158.

DeMaster, D.J., and Pope, R.H. (1996) Nutrient dynamics in Amazon shelf waters: results from AMASSEDS. Cont. Shelf Res. 16, 263–289.

DeMaster, D.J., Smith, W.O., Nelson, D.M., and Aller, J.Y. (1996) Biogeochemical processes in Amazon shelf waters: chemical distributions and uptake rates of silicon, carbon, and nitrogen. Cont. Shelf Res. 16, 617–643.

DeMaster, D.J., Thomas, C.J., Blair, N.E., Fornes, W.L., Plaia, G., and Levin, L.A. (2002) Deposition of bomb ^{14}C in continental slope sediments of the Mid-Atlantic Bight: assessing organic matter sources and burial rates. Deep-Sea Res. II. 49, 4667–4685.

De Mello, W.Z. , Cooper, D.J., Cooper, W.J., Saltzman, E.S., Zika, R.G, Savoie, D.L., and Prospero, J.M. (1987) Spatial and diel variability in the emissions of some biogenic sulfur compounds from Florida Spartina alterniflora coastal zone. Atmos. Environ. 21, 987–990.

Deming, J.W., and Baross, J.A. (1993) The early diagenesis of organic matter: bacterial activity. In Organic Geochemistry (Engel, M.H., and Macko, S.A., eds.), pp. 119–144, Plenum Press, New York.

den Hartog, C. (1970) The Seagrasses of the World. North-Holland, Amsterdam.

DeNiro, M.J., and Epstein, S. (1978) Influence of diet on the distribution of carbon isotopes in animals. Geochim. Cosmochim. Acta 42, 495–506.

DeNiro, M., and Epstein, S. (1981) Influence of diet on the distribution of nitrogen isotopes in animals. Geochim. Cosmochim. Acta 45, 341–351.

Denissenko, M.K., Pao, A., Tang, M., and Pfeifer, G.P. (1996) Preferential formation of benzo[a]pyrene adducts at lung cancer mutational hotspots in P53. Science 274, 430–432.

Dennison, W.C., and Alberte, R.A. (1982) Photosynthetic responses of Zostera marina L. (eelgrass) to in situ manipulations of light intensity. Oecologia 55, 137–144.

Dennison, W.C., Orth, R.J., Moore, K.A., Stevenson, J.C., Careter, V., Kollar, S., Bergstrom, P.W., and Batiuk, R.A. (1993) Assessing water quality with submerged aquatic vegetation. Bioscience 43, 86–94.

de Resseguier, A. (2000) A new type of horizontal in-situ water and fluid mud sampler. Mar. Geol. 163, 409–411.

De Stefano, C., Foti, C., Gianguzza, A., and Sammartano, S. (2000) The interaction of amino acids with the major constituents of natural waters at different ionic strengths. Mar. Chem. 72, 61–76.

Desvilettes, C., Bourdier, G., Amblard, C., and Barth, B. (1997) Use of fatty acids for the assessment of zooplankton grazing on bacteria, protozoans, and microalgae. Freshwat. Biol. 38, 629–637.

Detmers, J., Bruchert, V., Habicht, K.S., and Kuever, J. (2001) Diversity of sulfur isotope fractionations by sulfate-reducing Prokaryotes. Appl. Environ. Microbiol. 67, 888–894.

Dettmann, E.H. (2001) Effect of water residence time on annual export and denitrification of nitrogen in estuaries: a model analysis. Estuaries 24, 481–490.

Devol, A.H. (1991) Direct measurement of nitrogen gas fluxes from continental shelf sediments. Nature. 349: 319–321.

Devol, A.H., Anderson, J.J., Kuivila, K., and Murray, J.W. (1984) A model for coupled sulfate reduction and methane oxidation in the sediments of Saanich Inlet. Geochim. Cosmochim. Acta 48, 993–1004.

Devol, A.H., Quay, P.D., Richey, J.E., and Martinelli, L.A. (1987) The role of gas exchange in the inorganic carbon, oxygen, and ^{222}Rn budgets of the Amazon River. Limnol. Oceanogr. 32, 235–248.

de Wilde, H.P.J., and De Bie, M.J.M. (2000) Nitrous oxide in the Scheldt estuary: production by nitrification and emission to the atmosphere. Mar. Chem. 69, 203–216.

Dhakar, S.P., and Burdige, D.J. (1996) A coupled, non-linear, steady state model for early diagenetic processes in pelagic sediments. Am. J. Sci. 296, 296–330.

Diaz, F., and Raimbault, P. (2000) Nitrogen regeneration and dissolved organic nitrogen release during spring in a NW Mediterranean coastal zone (Gulf of Lions); implications for the estimation of new production. Mar. Ecol. Prog. Ser. 197, 51–65.

Diaz, R.J., and Rosenberg, R. (1995) Marine benthic hypoxia: a review of its ecological effects and the behavioral responses of benthic macrofauna. Oceanogr. Mar. Biol. Ann. Rev. 33, 245–303.

Dibb, J.E. (1989) Atmospheric deposition of beryllium-7 in Chesapeake Bay region. J. Geophys. Res. 94, 2261–2265.

Dibb, J.E., and Rice, D.L. (1989a) Temporal and spatial distribution of beryllium-7 in the sediments of Chesapeake Bay. Estuar. Coastal Shelf Sci. 28, 395–406.

Dibb, J.E., and Rice, D.L. (1989b) The geochemistry of beryllium-7 in Chesapeake Bay. Estuar. Coastal Shelf Sci. 28, 379–394.

Dickhut, R.M., Canuel, E.A., Gustafson, K.E., Liu, K., Arzayus, K.M., Walker, S.E., Edgecombe, G., Gaylor, M.O., and Macdonald, E.H. (2000) Automotive sources of carcinogenic polycyclic aromatic hydrocarbons associated with particulate matter in the Chesapeake bay region. Environ. Sci. Technol. 34, 4635–4640.

Dickinson, R.E., and Cicerone, R.J. (1986) Future global warming from atmospheric trace gases. Nature 319, 109–115.

Dickson, A.G. (1992) Thermodynamics of the dissociation of boric acid in synthetic seawater from 273.15 to 318.15 K. Deep-Sea Res. 37, 755–766.

Dickson, A.G. (1993) The measurement of pH in seawater. Mar. Chem. 44, 131–142.

Didyk, B.M., Simoneit, B.R.T., Brassell, S.C., and Eglinton, G. (1978) Organic geochemical indicators of paleoenvironmental conditions of sedimentation. Nature 272, 216–222.

Di Toro, D.M., Allen, H.E., Bergman, H.L., Meyer, J.S., Paquin, P.R., and Santore, R.C. (2001) Biotic ligand model of the acute toxicity of metals. 1. Technical basis. Environ. Toxicol. Chem. 20, 2383–2396.

Di Toro, D.M., Hallden, J.A., and Plafkin, J.L. (1991) Modeling ceriodaphnia toxicity in the Naugatuck river. 2. copper, hardness, and effluent interactions. Environ. Toxicol. Chem. 10, 261–274.

Dittmar, T. (2004) Evidence for terrigenous dissolved organic nitrogen in the Arctic deep sea. Limnol. Oceanogr. 49, 149–156.

Dittmar, T., Fitznar, H.P., and Kattner, G. (2001) Origin and biogeochemical cycling of organic nitrogen in the eastern Arctic Ocean as evident from D- and L-amino acids. Geochim. Cosmochim. Acta 65, 4103–4114.

Dittmar, T., and Lara, R.J. (2001) Driving forces behind nutrient and organic matter dynamics in a mangrove tidal creek in North Brazil. Estuar. Coastal Shelf Sci. 52(2), 249–259.

Doblin, M.A., Blackburn, S.I., and Hallegraeff, G.M. (1999) Growth and biomass stimulation of the toxic dinoflagellate Gymnodinium by dissolved organic substances. J. Exp. Mar. Biol. Ecol. 236, 33–47.

Dodson, J.J., Dauvin, J.C., Ingram, R.G., and D'Anglejan, B. (1989) Abundance of larval rainbow smelt (Osmerus mordax) in relation to the maximum turbidity zone and associated macrozooplanktonic fauna of the middle St. Lawrence estuary. Estuaries 12, 66–81.

Dollhopf, M.E., Nealson, K.H., Simon, D.M., and Luther III, G.W. (2000) Kinetics of Fe (III) and Mn (IV) reduction by the Black Sea strain of Shewanella putrifaciens using solid state voltammetric Au/Hg electrodes. Mar. Chem. 70, 171–180.

Donali, E., Olli, K., Heiskanen, A.S., and Andersen, T. (1999) Carbon flow patterns in the planktonic food web of the Gulf of Riga, the Baltic Sea: a reconstruction by the inverse method. J. Mar. Syst. 23, 251–268.

Donat, J.R., and Bruland, K.W. (1995) Trace elements in the oceans. In Trace Metals in Natural Waters (Salbu, B., and Steinnes, E., eds.), pp. 247–281, CRC Press, Boca Raton, FL.

Donat, J.R., Lao, K.A., and Bruland, K.W. (1994) Speciation of dissolved copper and nickel in South San Francisco Bay; a multi-method approach. Anal. Chim. Acta 284, 547–571.

Donat, J.R., and van den Berg, C.M.G. (1992) A new cathodic stripping voltammetric method for determining organic copper complexation in seawater. Mar. Chem. 38, 69–90.

Dong, L.F., Thornton, D.C.O., Nedwell, D.B., and Underwood, G.J.C. (2000) Denitrification in sediments of the River Colne estuary. Mar. Ecol. Prog. Ser. 203, 109–122.

Dortch, Q., Parsons, M.L., Doucette, G.J., Fryxell, G.A., Maier, A., Thessen, A., Powell, C.L., and Soniat, T.M. (2000) Pseudo-nitzschia spp. in the northern Gulf of Mexico: Overview and response to increasing eutrophication. In Symposium on Harmful Algae in the U.S., pp. 27, Dec. 4–9, Marine Biological Laboratory, Woods Hole, MA.

Dortch, Q., Robichaux, R., Pool, S., Milsted, D., Mire, G., Rabalais, N.N., Soniat, T.M., Fryxell, G.A., Turner, R.E., and Parsons, M.L. (1997) Abundance and vertical flux of Pseudo-nitzschia in the northern Gulf of Mexico. Mar. Ecol. Prog. Ser. 146, 249–264.

Dortch, Q., and Whitledge, T.E. (1992) Does nitrogen or silicon limit phytoplankton production in the Mississippi River plume and nearby regions? Cont. Shelf Res. 12, 1293–1309.

Downing, J.A. (1997) Marine nitrogen:phosphorus stoichiometry and the global N:P cycle. Biogeochemistry 37, 237–252.

Drake, L.A., Dobbs, F.C., and Zimmerman, R.C. (2003) Effects of epiphyte load on optical properties and photosynthetic potential of the seagrasses Thalassia testudinum Banks ex Konig and Zostera marina L. Limnol. Oceanogr. 48, 456–463.

Drenzek, N.J., Eglinton, T.I., Wirsen, C.O., May, H.D., Wu, Q., Sowers, K.R., and Reddy, C.M. (2001) The absence and application of stable carbon isotopic fractionation during reductive dechlorination of polychlorinated biphenyls. Environ. Sci. Technol. 35, 3310–3313.

Drier, C.A. (1982) Trace metal accumulations in Delaware salt marshes. M.S. Thesis, Univ. of Delaware, Newark, DE.

Druffel, E.R.M., Williams, P.M., Bauer, J.E., and Ertel, J. (1992) Cycling of dissolved and particulate organic matter in the open ocean. J. Geophys. Res. 97, 15639–15659.

Duarte, C.M. (1995) Submerged aquatic vegetation in relation to different nutrient regimes. Ophelia 41, 87–112.

Duce, R.A., Liss, P.S., Merill J.T., et al. (1991) The atmospheric input of trace species to the world ocean. Global Biogeochem. Cycle 5, 193–259.

Ducklow, H.W., and Kirchman, D.L. (1983) Bacterial dynamics and distribution during a spring diatom bloom in the Hudson River plume, USA. J. Plankton Res. 5, 333–355.

Dugdale, R.C., and Goering, J.J. (1967) Uptake of new and regenerated forms of nitrogen in primary productivity. Limnol. Oceanogr. 12, 196–206.

Dugdale, R.C., Wilkerson, F.P., and Minas, H.J. (1995) The role of a silicate pump in driving new production. Deep-Sea Res. I. 42, 697–719.

Duinker, J. C. (1980). Suspended matter in estuaries. In Chemistry and Biogeochemistry of Estuaries (Olausson, E., and Cato, I., eds.), pp. 121–152, John Wiley, Chichester, UK.

Duinker, J.C. (1983) Effects of particle size and density on the transport of metals to the ocean. In Trace Metals in Sea Water (Wong, C.S., Boyle, E., Bruland, K., Burton, J.D., and Goldberg, E.D., eds.), pp. 209–226, Plenum Press, New York.

Dunne, T., and Leopold, L.B. (1978) Water in Environmental Planning. W.H. Freeman, San Francisco.

Dunstan, G.A., Volkman, J.K., Barrett, S.M., Leroi, J.M., and Jeffrey, S.W. (1994) Essential polyunsaturated fatty acids from 14 different species of diatom (Bacillariophyceae). Phytochemistry 35, 155–161.

Dyer, K.R. (1973) Estuaries: A Physical Introduction. John Wiley, New York.

Dyer, K.R. (ed.) (1979) Estuaries and estuarine sedimentation. In Estuarine Hydrography and Sedimentation—A Handbook pp. 1–18, Cambridge University Press, Cambridge, UK.

Dyer, K.R. (1986) Coastal and Estuarine Sediment Dynamics. John Wiley, Chichester, UK.

Dyer, K.R., and Manning, A.J. (1999) Observation of the size, settling velocity and effective density of flocs, and their fractal dimensions. Neth. J. Sea Res. 41, 87–95.

Dyer, K.R., and Taylor, P.A. (1973) A simple segmented prism model of tidal mixing in well-mixed estuaries. Estuar. Coastal Shelf Sci. 1, 411–418.

Dyke, P.P.G. (2001) Coastal and Shelf Sea Modeling. Kluwer International Series: Topics in Environmental Fluid Mechanics. Kluwer Academic, London.

Dyrssen, D., and Sillén, L.G. (1967) Alkalinity and total carbonate in sea water. A plea for the P–T-independent data. Tellus 19, 113–120.

Dzombak, D.A., Fish, W., and Morel, F.M.M. (1986) Metal–humate interactions. 1. Discrete ligand and continuous distribution models. Environ. Sci. Technol. 20, 669–675.

Dzombak, D.A., and Morel, F.M.M. (1990) Surface Complexation Modeling. Hydrous Ferric Oxide. Wiley-Interscience, New York.

Eadie, B.J., McKee, B.A., Lansing, M.B., Robbins, J.A., Metz, S., and Trefrey, J.H. (1994) Records of nutrient-enhanced coastal ocean productivity in sediments from the Louisiana continental shelf. Estuaries 17, 754–765.

Eckert, J.M., and Sholkovitz, E.R. (1976) The flocculation of iron, aluminum and humates from river water by electrolytes. Geochim. Cosmochim. Acta 40, 847–848.

Eckman, J.E., Nowell, A.R.M., and Jumars, P.A. (1981) Sediment destabilization by animal tubes. J. Mar. Res. 39, 361–374.

Edzwald, J.K., and O'Melia, C.R. (1975) Clay distributions in recent marine sediments. Clay Clay Miner. 23, 39–44.

Edzwald, J.K., Upchurch, J.B., and O'Melia, C.R. (1974) Coagulation in estuaries. Environ. Sci. Technol. 8, 58–63.

Egeland, E.S., and Liaaen-Jensen, S. (1992) Eight new carotenoids from a chemosystematic evaluation of Prasinophyceae. Proceedings of 7th International Symposium on Marine Natural Products, Capri.

Egeland, E.S., and Liaaen-Jensen, S. (1993) New carotenoids and chemosystematics in the Prasinophyceae. Porceedings of 10th Symposium on Carotenoids, Tron II heim Norway.

Egge, J.K., and Aksnes, D.L. (1992) Silicate as regulating nutrient in phytoplankton competition. Mar. Ecol. Prog. Ser. 83, 281–289.

Eglinton, G., and Calvin, M. (1967) Chemical fossils. Sci. Am. 216, 32–43.

Eglinton, G., and Hamilton, R.J. (1963) The distribution of alkanes. *In* Chemical Plant Taxonomy (Swain, T., ed.), pp. 187–217, Academic Press, New York.

Eglinton, G., and Hamilton, R.J. (1967) Leaf epicuticular waxes. Science 156, 1322–1335.

Eglinton, G., Hunneman, D.H., and Douraghi-Zadeh, K. (1968) Gas chromatographic-mass spectrometric studies of long chain hydroxyl acids—II. The hydroxy acids and fatty acids of a 5000 year-old lacustrine sediment. Tetrahedron 24, 5929–5941.

Eglinton, T.I., Aluwihare, L., Bauer, J.E., Druffel, E.R.M., and McNichol, A.P. (1996) Gas chromatographic isolation of individual compounds from complex matrices for radiocarbon dating. Anal. Chem. 68, 904–912.

Eglinton, T.I. Benitez-Nelson, B., McNichol, A., Bauer, J.E., and Druffel, E.R.M. (1997) Variability in radiocarbon ages of individual organic compounds from marine sediments. Science 277, 796–799.

Ehleringer, J., Phillips, S.L., Schuster, W.S.F., and Sandquist, D.R. (1991) Differential utilization of summer rains by desert plants. Oecologia 88, 430–434.

Ehrenhauss, S., and Huettel, M. (2004) Advective transport and decomposition of chain-forming planktonic diatoms in permeable sediments. J. Sea Res. 52, 179–198.

Ehrenhauss, S., Witte, U., Janssen, F., and Huettel, M. (2004) Decomposition of diatoms and nutrient dynamics in permeable North Sea sediments. Cont. Shelf Res. 24, 721–737.

Eilers, H., Pernthaler, J., Glockner, F.O., and Amann, R. (2000) Culturability and in situ abundance of pelagic bacteria from the North Sea. Appl. Environ. Microbiol. 66, 3044–3051.

Einsele, W. (1936) Uber die Beziehungen des Eisenkreislaufs zum Phosphatkreislauf im eutrophen See. Arch. Hydrobiol. 29, 664–686.

Eisma, D. (1986) Flocculation and de-flocculation of suspended matter in estuaries. Neth. J. Sea Res. 20, 183–199.

Eisma, D., and Li, A. (1993) Changes in suspended matter floc size during the tidal cycle in the Dollard estuary. Neth. J. Sea. Res. 31, 107–117.

Elderfield, H., and Hepworth, A. (1975) Diagenesis, metals, and pollution in estuaries. Mar. Pollut. Bull. 6, 85–87.

Elderfield, H., Luedtke, N., McCaffrey, R.J., and Bender, M.L. (1981a) Benthic flux studies in Narragansett Bay. Am. J. Sci. 281, 768–787.

Elderfield, H., McCaffrey, R.J., Luedtke, N., Bender, M., and Truesdale, V.W. (1981b) Chemical diagenesis in Narragansett Bay sediments. Am. J. Sci. 281, 1021–1055.

Elliot, M., and McLusky, D.S. (2002) The need for definitions in understanding estuaries. Estuar. Coastal Shelf Sci. 55, 815–827.

Elliott, T. (1978) Clastic shorelines. *In* Sedimentary Environments and Facies (Reading, H.G., ed.), pp. 143–175, Elsevier, New York.

Elmgren, R. (1984) Trophic dynamics in the enclosed, brackish Baltic Sea. Rapp. Reun. Cons. Intl. Explor. Mer. 183, 152–169.

Elmgren, R. (2001) Understanding human impact on the Baltic ecosystem: changing views in recent decades. Ambio 30, 222–231.

Elmgren, R., Hansson, S., Larsson, U., Sundelin, B., and Boehm, P.D. (1983) The "Tsesis" oil spill: acute and long-term impacts on the benthos. Mar. Biol. 73, 51–65.

Elmgren, R., and Larrson, U. (2001) Eutrophication in the Baltic Sea area: integrated coastal management issues. *In* Science and Integrated Coastal Management (von Bodungen, B., and Turner, R.K., eds.), pp. 15–35, Dahlem University Press, Berlin.

Elmore, D., and Phillips, F.M. (1987) Accelerator mass spectrometry for measurements of long-lived radioisotopes. Science 236, 543–550.

Elsinger, R.T. and Moore, W.S. (1983) ^{224}Ra, ^{228}Ra, and ^{226}Ra in Winyah Bay and Delaware Bay. Earth Planet. Sci. Lett. 64, 430–436.

Elton, C.S. (1958) The Ecology of Invasions by Animals and Plants. Methuen, London.

Embelton, C., and King, C.A.M. (1970) Glacial and Periglacial Geomorphology. Macmillan of Canada, Toronto.

Emerson, S. (1985) Organic carbon preservation in marine sediments. *In* The Carbon Cycle and Atmospheric CO_2: Natural Variations Archean to Present (Sundquist, E.T., and Broecker, W.S., eds.), pp. 78–87, American Geophysical Union, Washington, DC.

Emerson, S., Jahnke, R., and Heggie, D. (1984) Sediment water exchange in shallow water estuarine sediments. J. Mar. Res. 4, 709–730.

Emery, K.O., and Aubrey, D.G. (1991) Sea Levels, Land Levels and Tide Gauges. Springer-Verlag, New York.

Emery, K.O., and Uchupi, E. (1972) Western Atlantic Ocean: Topography, Rocks, Structure, Water, Life, and Sediments. American Association of Petroleum Geology Memoires 17, Washington, DC.

Emmett, R., Llanso, R., Newton, J., Thom, R., Hornberger, M., Morgan, C., Levings, C., Copping, A., and Fishman, P. (2000) Geographical signatures of North America West Coast Estuaries. Estuaries 23, 765–792.

Engel, D.W., and Brouwer, M. (1993) Crustaceans as models for metal metabolism. I. Effects of the molt cycle on blue crab metal metabolism and metallothionein. Mar. Environ. Res. 35, 1–12.

Engel, M.H., and Macko, S.A. (1993) Organic Geochemistry—Principles and Applications. Plenum Press, New York.

Engelhaupt, E., and Bianchi, T.S. (2001) Sources and composition of high-molecular-weight dissolved organic carbon in a southern Louisiana tidal stream (Bayou Trepagnier). Limnol. Oceanogr. 46, 917–926.

Engelhaupt, E., Bianchi, T.S., Wetzel, R.G., and Tarr, M.A. (2002) Photochemical transformations and bacterial utilization of high-molecular-weight dissolved organic carbon in southern Louisiana tidal stream (Bayou Trepagnier). Biogeochemistry 62, 39–58.

Eppley, R.W. (1972) Temperature and phytoplankton growth in the sea. Fishery Bull. 70, 1063–1085.

Ernissee, J.J., and Abbott, W.H. (1975) Binding of mineral grains by a species of *Thallassiosira*. Nova. Hedwigia. Beth. 53, 241–252.

Ertel, J.R., and Hedges, J.I. (1984) Sources of sedimentary humic substances: vascular plant debris. Geochim. Cosmochim. Acta 48, 2065–2074.

Ertel, J.R., Hedges, J.I., Devol, A.H., and Richey, J.E. (1986) Dissolved humic substances of the Amazon River system. Limnol. Oceanogr. 31, 739–754.

Estep, M.F., and Dabrowski, H. (1980) Tracing food webs with stable hydrogen isotopes. Science 209, 1537–1538.

Estep, M.F., and Hoering, T.C. (1980) Biogeochemistry of the stable hydrogen isotopes. Geochim. Cosmochim. Acta 44, 1197–1206.

Etemad-Shahidi, A., and Imberger, J. (2002) Anatomy of turbulence in a narrow and weakly stratified estuary. Mar. Freshwater Res. 53, 757–768.

European Union (2000) Parliament and Council Directive 2000/60/EC of the 23rd October 2000, Establishing a framework for community action in the field of water policy. Official Journal PE-CONS 3639/1/00 REV 1, Brussels.

Eyre, B.D. (1995) A first-order nutrient budget for the tropical Moresby estuary and catchment of North Queensland, Aust. J. Coast. Res. 11, 717–732.

Eyre, B.D. (2000) A regional evaluation of nutrient transformation and phytoplankton growth in nine river dominated sub-tropical East Australian estuaries. Mar. Ecol. Prog. Ser. 205, 61–83.

Eyre, B.D., and McKee, L.J. (2002) Carbon, nitrogen, and phosphorus budgets for a shallow subtropical coastal embayment (Moreton Bay, Australia). Limnol. Oceanogr. 47, 1043–1055.

Eyre, B.D., Pepperell, P., and Davies, P. (1999) Budgets for Australian estuarine systems: Tropical systems. *In* Australian Estuarine Systems: Carbon, Nitrogen, and Phosphorus fluxes (Smith, S.V., and Crossland, C.J., eds.), pp. 9–11, LOICZ Reports and Studies 12.

Eyre, B.D., and Twigg, C. (1997) Nutrient behavior during post-flood recovery of the Richmond River estuary, northern NSW, Australia. Estuar. Coastal Shelf Sci. 44, 311–326.

Eyrolle, F., Benedetti, M.F., Benaim, J.Y., and Fevrier, D. (1996) The distributions of colloidal and dissolved organic carbon, major elements, and trace elements in small tropical catchments. Geochim. Cosmochim. Acta 60, 3643–3656.

Fabbri, D., Chiavari, G., and Galletti, G.C. (1996) Characterization of soil humin by pyrolysis (methylation)—gas chromatography/mass spectrometry: structural relationships with humic acids. J. Anal. Appl. Pyrol. 37, 161–172.

Facca, C., Sfriso, A., and Socal, G. (2002) Temporal and spatial distribution of diatoms in the surface sediments of the Venice Lagoon. Bot. Mar. 452, 170–183.

Fager, E.W. (1964) Marine sediments: effects of a tube-building polychaete. Science 143, 356–359.

Fain, A.M., Jay, D.A., Wilson, D., Orton, P.M., and Baptista, A.M. (2001) Seasonal and tidal monthly patterns of particulate matter dynamics in the Columbia River estuary. Estuaries 24, 770–786.

Fairbridge, R.W. (1961) Eustatic changes of sea level. Phys. Chem. Earth 4, 99–185.

Fairbridge, R.W. (1980) The estuary: its definition and geologic cycle. *In* Chemistry and Biogeochemistry of Estuaries (Olausson, E., and Cato, I., eds.), pp. 1–36, Wiley-Interscience, New York.

Falkowski, P.G. (1975) Nitrate uptake in marine phytoplankton: (Nitrate, chloride)-activated adenosine triphosphate from *Skeletonema costatum* (Bacillariophyceae). J. Phycol. 11, 323–326.

Falkowski P.G. (2000) Rationalizing elemental ratios in unicellular algae. J. Phycol. 36, 3–6.

Falkowski, P.G., and Rivkin, R.B. (1976) The role of glutamine synthetase in the incorporation of ammonium in *Skeletonema costatum* (Bacillariophyceae) J. Phycol. 12, 448–450.

Farley, K.J., and Morel, F.M.M. (1986) Role of coagulation in the kinetics of sedimentation. Environ. Sci. Technol. 20, 187–195.

Farquhar, G.D. (1983) On the nature of carbon isotope discrimination in C_4 species. Aust. J. Plant Physiol. 10, 205–226.

Farquhar, G.D., Hubrick, K.T., Condun, A.G., and Richard, R.A. (1989) Carbon isotope fractionation and plant water-use efficiency. *In* Stable Isotopes in Ecological Research (Rundel, P.W., Ehleringer, J.R., and Nagy, K.A., eds.), pp. 21–41, Springer Verlag, Berlin.

Farquhar, G.D., O'Leary, M.H., and Berry, J.A. (1982) On the relationship between carbon dioxide discrimination and the intracellular carbon dioxide concentration in leaves. Aust. J. Plant Physiol. 9, 121–137.

Faure, G. (1986) Principles of Isotope Geology. John Wiley, New York.

Fawley, M.N., and Lee, C.M. (1990) Pigment composition of the scaly green flagellate *Mesostigma virde* (Micromonadophyceae) is similar to that of the siphonous green alga *Bryopsis plumose* (Ulvophyceae). J. Phycol. 26, 666–670.

Fazio, S.A., Uhlinger, D.L., Parker, J.H., and White, D.C. (1982) Estimations of uronic acids as quantitative measures of extracellular and cell wall polysaccharide polymers from environmental samples. Appl. Environ. Microbiol. 43, 1151–1159.

Fell, J. W., Cefalu, R. C., Master, I. M., and Tallman, A. S. (1975) Microbial activities in the mangrove (*Rhizophora mangle*) detrital system. *In* Proceedings of the International Symposium on the Biology and Management of Mangroves (Walsh, G.E., Snedaker, S.C., and Teas, H.J., eds.), pp. 661–679, University Press, University of Florida, Gainesville.

Fenchel, T.M., and Blackburn, T.H. (1979) Bacteria and Mineral Cycling. Academic Press, New York.

Fenchel, T.M., and Jørgensen, B.B. (1977) Detritus food chains of aquatic ecosystems: the role of bacteria. Adv. Microb. Ecol., 1, 1–57.

Fenchel, T.M., King, G.M., and Blackburn, T.H. (1998) Bacterial Biogeochemistry: the Ecophysiology of Mineral Cycling. Academic Press, New York.

Fenchel, T.M., and Riedl, R.J. (1970) The sulfide system: a new biotic community underneath the oxidized layer of marine sand bottoms. Mar. Biol. 7, 255–268.

Feng, H., Cochran, J.K., and Hirschberg, D.J. (1999) [234]Th and [7]Be as tracers for the transport and dynamics of suspended particles in partially mixed estuary. Geochim. Cosmochim. Acta 63, 2487–2505.

Fenton, G.E., and Ritz, D.A. (1988) Changes in carbon and hydrogen stable isotope ratios of macroalgae and seagrass during decomposition. Estuar. Coastal Shelf Sci. 26, 429–436.

Ferguson, A., Eyre, B., and Gay, J. (2004) Nutrient cycling in the sub-tropical Brunswick estuary, Australia. Estuaries 27, 1–17.

Festa, J.F., and Hansen, D.V. (1978) Turbidity maxima in partially mixed estuaries: A two-dimensional numerical model. Estuar. Coastal Shelf Sci. 7, 347–359.

Fetter, C.W. (1988) Applied Hydrogeology. Prentice Hall, Englewood Cliffs, NJ.

Fetter, C.W. (2001) Applied Hydrology. Prentice Hall, Englewood Cliff, NJ.

Fettweis, M., Sas, M., and Monbaliu, J. (1998) Seasonal neap-spring and tidal variations of cohesive sediment concentration in the Scheldt estuary, Belgium. Estuar. Coastal Shelf Sci. 47, 21–36.

Ficken, K.J., Li, B., Swain, D.L., and Eglinton, G. (2000) An *n*-alkane proxy for the sedimentary input of submergent/floating freshwater aquatic macrophytes. Org. Geochem. 31, 745–749.

Filella, M., and Buffle, J. (1993) Factors controlling the stability of sub-micron colloids in natural waters. Colloids Surf. 73, 255–273.

Filip, Z., Newman, R.H., and Alberts, J.J. (1991) Carbon-13 nuclear magnetic resonance characterization of humic substances associated with salt marsh environments. Sci. Total Environ. 101, 191–199.

Filius, J.D., Lumsdon, D.G., Meeussen, J.C., Hiemstra, T., and van Riemsdijk, W.H. (2000) Adsorption of fulvic acid on goethite. Geochim. Cosmochim. Acta 64, 51–60.

Filley, T.R. (2003) Assessment of fungal wood decay by lignin analysis using tetra-methylammonium hydroxide (TMAH) and [13]-labelled TMAH thermochemolysis. *In* Wood Deterioration and Preservation: Advances in Our Changing World (Goodell, T., Nicholas, A., and Schultz, C., eds.), pp. 119–139, ACSociety, Series 845, American chemical Society, Washington, DC.

Filley, T.R., Freeman, K.H., Bianchi, T.S., Baskaran, M., Colarusso, L.A., and Hatcher, P.G. (2001) An isotopic biogeochemical assessment of shifts in organic matter input to Holocene sediments from Mud lake, Florida. Org. Geochem. 32, 1153–1167.

Filley, T.R., Minard, R.D., and Hatcher, P.G. (1999) Tetramethylammonium hydroxide (TMAH) thermochemolysis: proposed mechanisms based upon the application pf ^{13}C-labeled TMAH to a synthetic model lignin dimer. Org. Geochem. 30, 607–621.

Findlay, S.E.G. (2003) Bacterial response to variation in dissolved organic matter. *In* Aquatic Ecosystems: Interactivity of Dissolved Organic Matter (Findlay, S.E.G., and Sinsabaugh, R.L., eds.), pp. 363–379, Academic Press, New York.

Findlay, S.E.G., Pace, M.L., Lints, D., Cole, J.J., Caraco, N.F., and Peierls, B. (1991) Weak coupling of bacterial and algal production in a heterotrophic ecosystem: the Hudson River estuary. Limnol. Oceanogr. 36, 268–278,

Findlay, S.E.G., and Sinsabaugh, R.L. (eds.) (2003) Aquatic Ecosystems—Interactivity of Dissolved Organic Matter. Academic Press, New York.

Findlay S.E.G., and Tenore K.R. (1982) Effect of a free-living marine nematode (*Diplolaimella chitwoodi*) on detrital carbon mineralization. Mar. Ecol. Prog. Ser. 8, 161–166.

Finni, T., Laurila, S., and Laakonen, S. (2001) The history of eutrophication in the sea of Helsinki in the 20th century. Long-term analysis of plankton assemblages. Ambio 30, 264–271.

Fischer, J.M., Klug, J.L., Reed-Andersen, T., and Chalmers, A.G. (2000) Spatial pattern of localized disturbance along a southeastern salt marsh tidal creek. Estuaries 23, 565–571.

Fisher, D., and Oppenheimer, M. (1991) Atmospheric nitrogen deposition and the Chesapeake Bay estuary. Ambio 20, 102–108.

Fisher, H.B. (1972) Mass transport mechanisms in partially stratified estuaries. J. Fluid Mech. 53, 671–687.

Fisher, H.B. (1976) Mixing and dispersion in estuaries. Ann. Rev. Fluid. Mech. 8, 107–133.

Fisher, J.B., Lick, W.J., McCall, P.L., and Robbins, J.A. (1980) Vertical mixing of lake sediments by tubificid oligochaetes. J. Geophys. Res. 85, 3997–4006.

Fisher, N.S., Burns, K.A., Cherry, R.D., and Heyraud, M. (1983) Accumulation and cellular distribution of [241]Am, ^{210}Pb and [210]Po in two marine algae. Mar. Ecol. Prog. Ser. 11, 233–237.

Fisher, N.S., Teyssie, J.L., Krishnaswami, S., and Baskaran, M. (1987) Accumulation of Th, Pb, U, and Ra in marine phytoplankton and its geochemical significance. Limnol. Oceanogr. 32, 131–142.

Fisher, R.R., Carlson, P.R., and Barber, R.T. (1982) sediment nutrient regeneration in three North Carolina estuaries. Estuar. Coastal Shelf Sci. 14, 101–116.

Fisher, T.R., Gustaffson, A.B., Sellner, K., Lacouture, R., Haas, I.W., Wetzel, R.L., Magnien, R., Everitt, D., Michaels, B., and Karrh, R. (1999) Spatial and temporal variation of resource limitation in Chesapeake Bay. Mar. Biol. 133, 763–778.

Fisher, T.R., Harding, L., Stanley, D.W., and Ward, L.G. (1988) Phytoplankton, nutrients, and turbidity in the Chesapeake, Delaware, and Hudson estuaries. Estuary. Coastal Shelf Sci. 27, 61–93.

Fisk, H.N. (1944) Geological Investigation of the Alluvial Valley of the lower Mississippi River. Mississippi River Commission, Vicksburg, MS.

Fisk, H.N. (1955) Sand facies of Recent Mississippi Delta Deposits. World Petroleum Congress, Rome.

Fisk, H.N. (1960) Recent Mississippi River sedimentation and peat accumulation. Compte Rendu 4th Congrès 1 Avancement des Études de Stratigraphie et de Geologie du Carbonifère, Heerlen 1958, 1:187–199.

Fitzgerald, W.F., and Lamborg, C.H. (2003) Geochemistry of mercury in the environment. *In* Treatise in Geochemistry, Vol. 9 (Sherwood-Lollar, B, ed.), pp. 107–148, Elsevier, New York.

Flessa, K.W., Constantine, K.J., and Cushman, M.K. (1977) Sedimentation rates in a coastal marsh determined from historical records. Ches. Bay Sci. 18, 172–176.

Flindt, M.R., Kamp-Nielsen, L., Marques, J.C., Pardal, S.E., Bocci, M., Bendoricho, G., Nielsen, S.N., and Jørgensen, S.E. (1997) Description of the three shallow estuaries: Mondego River (Portugal), Rosklide Fjord (Denmark) and the Lagoon of Venice (Italy). Ecol. Model. 102, 17–31.

Florek, R.J., and Rowe, G.T. (1983) Oxygen consumption and dissolved inorganic nutrient production in marine coastal and shelf sediments of the Middle Atlantic Bight. Intl. Rev. Gesmaten. Hydrobiol. 68, 73–112.

Fogel, M.L., and Cifuentes, L.A. (1993) Isotope fractionation during primary production. *In* Organic Geochemistry—Principle and Applications (Engel M.H., and Macko, S.A., eds.), pp. 73–98, Plenum Press, New York.

Fogel, M.L., Cifuentes, L.A., Velinsky, D.J., and Sharp, J.H. (1992) Relationships of carbon availability in estuarine phytoplankton to isotopic composition. Mar. Ecol. Prog. Ser. 82, 291–300.

Fogel, M.L., Velinsky, D.J., Cifuentes, L.A., Pennock, J.R., and Sharp, J.H. (1988) Biogeochemical processes affecting the stable carbon isotopic composition of particulate carbon in the Delaware Estuary. Carnegie Inst. Wash. Annu. Rep. Director, 107–113.

Fonseca, M.S., and Kenworthy, W.J. (1987) Effects of current on photosynthesis and distribution of seagrasses. Aquat. Bot. 27, 59–78.

Fonselius, S.H. (1972) Marine pollution and sea Life. Mar. Pollut. Bull. 26, 64–67.

Foote, A. L., and Reynolds, K.A. (1997) Salt meadow cordgrass (*Spartina patens*) decomposition and its importance in Louisiana coastal marshes. Estuaries 20, 579–588.

Foreman, C.M., and Covert, J.S. (2003) Linkages between dissolved organic matter composition and bacterial community. *In* Aquatic Ecosystems: Interactivity of Dissolved Organic Matter (Findlay, S.E.G, and Sinsabaugh, R.L., eds.), pp. 343–359, Academic Press, New York.

Forja, J.M., Blasco, J., and Gomez-Parra, A. (1994) Spatial and seasonal variation of in situ benthic fluxes in the Bay of Cadiz (south-west Spain). Estuar. Coastal Shelf Sci. 39, 127–141.

Fortes, M.D. (1992) Comparative study of structure and productivity of seagrass communities in the ASEAN region. *In* Third ASEAN Science and Technology Conference Proceedings, Vol. 6, Marine Science: Living Coastal Resources, Department of Zoology, National University of Singapore and National Science Technology Board, Singapore.

Foss, P., Levin, R.A., and Liaaen-Jensen, S. (1987) Carotenoids of *Prochloron* sp. (Prochlorophyta). Phycologia 26, 142–144.

Fox, L.E. (1983) The removal of dissolved humic acid during estuarine mixing. Estuar. Coastal Shelf Sci. 16, 413–440.

Fox, L.E. (1984) The relationship between dissolved humic acids and soluble iron in estuaries. Geochim. Cosmochim. Acta 48, 879–884.

Fox, L.E. (1989) A model for inorganic control of phosphate concentrations in river waters. Geochim. Cosmochim. Acta 53, 417–428.

Francois, F., Poggiale, J., Durbec, J., and Stora, G. (2001). A new model of bioturbation for a functional approach to sediment reworking resulting from macrobenthic communities. In Organism–Sediment Interactions (Aller, J.Y., Woodin, S.A., and Aller, R.C., eds.), pp. 73–86, University of South Carolina Press, Columbia.

Frank, H.S., and Wen, W.Y. (1957) Structural aspects of ion–solvent interaction in aqueous solutions: a suggested picture of water structure. Disc. Faraday Soc. 24, 133–140.

Frankignoulle, M., Abril., G., Borges, A., Bourge, I., Canon, C., Delille, B., Libert, E., and Theate, J.M. (1998) Carbon dioxide emission from European estuaries. Science 282, 434–436.

Frankignoulle, M., and Bourge, I. (2001) European continental shelf as a significant sink for atmospheric CO_2. Global Biogeochem. Cycles 15, 569–576.

Frankignoulle, M., Bourge, I., and Wollast, D. (1996) Atmospheric CO_2 fluxes in a highly polluted estuary (the Scheldt). Limnol. Oceanogr. 41, 365–369.

Frankignoulle, M., and Middelburg, J. J. (2002) Biogases in tidal European estuaries: the BIOGEST project. Biogeochemistry 59, 1–4.

Frazier, D.E. (1967) Recent deltaic deposits of the Mississippi River: their development and chronology: Gulf Coast Assoc. Geol. Soc. Trans. 17, 287–315.

Freeman, K.H. (2001) Isotopic biogeochemistry of marine organic carbon (stable isotope geochemistry). In Reviews in Mineralogy and Geochemistry (Valley, J., and Colemen, D., eds.), pp. 579–606, Mineralogical Society of America, Washington, DC.

Freeman, K.H., Hayes, J.M., Trendel, J.M., and Albrecht, P. (1990) Evidence from carbon isotope measurements for diverse origins of sedimentary hydrocarbons. Nature 343, 254–256.

Freeze, R.A., and Cherry, J.A. (1979) Groundwater. Prentice Hall, Englewood Cliffs, NJ.

Freidig, A.P., Garciano, E.A., Busser, F.J.M., and Hermens, J.L.P. (1998) Estimating impact of humic acid on bioavailability and bioaccumulation of hydrophobic chemical in guppies using kinetic solid-phase extraction. Environ. Toxicol. Chem. 17, 998–1004.

Freyer, H.D., and Aly, A.I.M. (1974) Nitrogen-15 variations in fertilizer nitrogen. J. Environ. Qual. 3, 405–406.

Friedrich, J., Dinkel, C., Friedl, G., Pimenov, N., Wijsman, J., Gomoiu, M.-T., Cociasu, A., Popa, L., and Wehrli, B. (2002) Benthic nutrient cycling and diagenetic pathways in the north-western Black Sea. Estuar. Coastal Shelf Sci. 54,369–383.

Froelich, P.N. (1988) Kinetic control of dissolved phosphate in natural rivers and estuaries: a primer on the phosphate buffer mechanism. Limnol. Oceanogr. 33, 649–668.

Froelich, P.N., Bender, M.L., and Luedtke, N.A. (1982) The marine phosphorus cycle. Am. J. Sci. 282, 474–511.

Froelich, P.N., Klinkhammer, G.P., Bender, M.L., Luedtke, N.A., Heath, G.R., Cullen, D., Dauphin, P., Hammond, D, Hartman, B., and Maynard, V. (1979) Early oxidation of organic matter in pelagic sediments of the eastern equatorial Atlantic: suboxic diagenesis. Geochim. Cosmochim. Acta 43, 1075–1091.

Fry, B. (1986) Sources of carbon and sulfur nutrition for consumers in three meromictic lakes of New York State. Limnol. Oceanogr. 31, 79–88.

Fry, B. (1999) Using stable isotopes to monitor watershed influences on aquatic trophodynamics. Can. J. Fish. Aquat. Sci. 56, 2167–2171.

Fry, B. (2002) Conservative mixing of stable isotopes across estuarine salinity gradients: a conceptual framework for monitoring watershed influences on down stream fisheries production. Estuaries 25, 264–271.

Fry, B., Bern, A.L., Ross, M.S., and Meeder, J.F. (2000) $\delta^{15}N$ studies of nitrogen use by the red mangrove, *Rhizophora mangle* L. in south Florida. Estuar. Coastal Shelf Sci. 50, 291–296.

Fry, B., Gace, A., and McClelland, J.W. (2003) Chemical indicators of anthropogenic nitrogen loading in four Pacific estuaries. Pacific Sci. 57, 77–101.

Fry, B., Gest, H., and Hayes, J.M. (1988) $^{34}S/^{32}S$ fractionation in sulfur cycles catalyzed by anaerobic bacteria. Appl. Environ. Microbiol. 54, 250–256.

Fry, B., and Parker, P.L. (1979) Animal diets in Texas seagrass meadows: ^{13}C evidence for the importance of benthic plants. Estuar. Coastal Shelf Sci. 8, 499–509.

Fry, B., and Parker, P.L. (1984) ^{13}C enrichment and oceanic food web structure in the northwestern Gulf of Mexico. Contrib. Mar. Sci. 27, 49–63.

Fry, B., Scalan, R.S., Winters, J.K., and Parker, P.L. (1977) Stable carbon isotope evidence for two sources of organic matter in coastal sediments: seagrasses and plankton. Geochim. Cosmochim. Acta 41, 1876–1877.

Fry, B., Scalan, R.S., Winters, J.K., and Parker, P.L. (1982) Sulfur uptake by salt grasses, mangroves, and seagrasses in anaerobic sediments. Geochim. Cosmochim. Acta 46, 1121–1124.

Fry, B., and Sherr, E.B. (1984) $\delta^{13}C$ measurements as indicators of carbon flow in marine and freshwater ecosystems. Contrib. Mar. Sci. 27, 13–47.

Fry, B., and Smith III, T.J. (2002) Stable isotope studies of red mangrove and filter feeders from the shark river estuary, Florida. Bull. Mar. Sci. 70, 870–890.

Fry, B., and Wainwright, S.C. (1991) Diatom sources of ^{13}C-rich carbon in marine food webs. Mar. Ecol. Prog. Ser. 76, 149–157.

Fuhrman, J. (1990) Dissolved free amino acid cycling in an estuarine outflow plume. Mar. Ecol. Prog. Ser. 66, 197–203.

Fuhrman, J., and Azam, F. (1982) Thymidine incorporation as a measure of heterotrophic bacterioplankton production in marine surface waters—evaluation of field results. Mar. Biol. 66, 109–120.

Fukushima, T., Ishibashi, T., and Imai, A. (2001) Chemical characterization of dissolved organic matter in Hiroshima Bay, Japan. Estuar. Coastal Shelf Sci. 53, 51–62.

Furlong, E.T., and Carpenter, R. (1988) Pigment preservation and remineralization in oxic coastal marine sediments. Geochim. Cosmochim. Acta 52, 87–99.

Furukawa, K., and Wolanski, E. (1996) Sedimentation in mangrove forests. *In* Mangroves and Salt marshes, Vol. I., pp. 3–10, SPB Academic Publishing, Amsterdam.

GACGC (2000) World in transition. Strategies for managing global environmental risks. German Advisory Council on Global Change, Annual Report 1998, Springer, Berlin.

Gächter, R., and Meyer, J.S. (1993) The role of microorganisms in mobilization and fixation of phosphorus in sediments. Hydrobiologia 253, 103–121.

Gagnon, C., Micci, A., and Pelletier, E. (1995) Anomalous accumulation of acid-volatile sulphides (AVS) in a coastal marine sediment, Sagueny Fjord, Canada. Geochim. Cosmochim. Acta 59, 2663–2675.

Gagosian, R.B., Nigrelli, G.E., and Volkman, J.K. (1983) Vertical transport and transformation of biogenic organic compounds from sediment trap experiment off the coast of Peru. *In* Coastal Upwelling: Its Sediment Record. Part A. Response of the Sedimentary Regime to Present Coastal Upwelling (Suess, E., and Thiede, J., eds.), pp. 241–272, Plenum Press, New York.

Galimov, E.M. (1974) Organic geochemistry of carbon isotopes. *In* Advances in Organic Geochemistry 1973 (Tissot, B., and Bienner, F., eds.), pp. 439–452, Editions Technip, Paris.

Gallenne, B. (1974) Study of fine material in suspension in the estuary of the Loire and its dynamic grading. Estuar. Coastal Shelf Sci. 2, 261–272.

Galler, J. J., Bianchi, T. S., Allison, M. A., Campanella, R., and Wysocki, L. A. (2003) Biogeochemical implications of levee confinement in the lower-most Mississippi River. EOS. Vol. 84, No.44, 469, 475–476.

Galloway, J.N. (1985) The deposition of sulfur and nitrogen from the remote atmosphere. *In* The Biogeochemical Cycling of Sulfur and Nitrogen in the Remote Atmosphere (Galloway, J., Charlson, M., Andreae, O., and Rodhe, H., eds.), pp. 97–106, Reidel, Dordrecht.

Galloway, J.N. (1998) The global nitrogen cycle: changes and consequences. *In* Proceedings of the First International Nitrogen Conference, pp. 15–24, Elsevier Science, New York.

Galloway, J.N., Dentener, F.J., Capone, D.G., Boyer, E.W., Howarth, R.W., Seitzinger, S.P., Asner, G.P., Clevland, C., Green, P., Holland, E., Karl, D.M., Michaels, A.F., Porter, J.H., Townsend, A., and Vörösmarty, C. (2004) Nitrogen cycles: past, present and future. Biogeochemistry 70, 153–226.

Galloway, J.N., Levy, H., and Kasibhatia, P. (1994) Consequences of population growth and development on deposition of oxidized nitrogen. Ambio 23, 120–123.

Galloway, J.N., Schlesinger, W.H., Levy, II, V., Michaels, A., and Schnoor, J.L. (1995) Nitrogen fixation: anthropogenic enhancement—environmental response. Global Biogeochem. Cycles 9: 235–252.

Galloway, W.E. (1975) Process framework for describing the morphologic and stratigraphic evolution of deltaic depositional systems. *In* Deltas, Models for Exploration (Broussard, M.L., ed.), pp. 87–98, Houston Geological Society, Houston, TX.

Gao, H.Z., and Zepp, R.G. (1998) Factors influencing photoreactions of dissolved organic matter in a coastal river of the southeastern United States. Environ. Sci. Technol. 32, 2940–2946.

Gardner, L.R. (1973) The effect of hydrologic factors on the pore water chemistry of intertidal marsh sediments. Southeast. Geol. 15, 17–28.

Gardner, L.R., Sharma, P., and Moore, W.S. (1987) A regeneration model for the effect of bioturbation by fiddler crabs on ^{210}Pb profiles in salt marsh sediments. J. Environ. Radioact. 5, 25–36.

Gardner, L.R., Wolaver, T.G., and Mitchell, M. (1988) Spatial variations in the sulfur chemistry of salt marsh sediments at North Inlet, South Carolina. J. Mar. Res. 46, 815–836.

Gardner, W.S., Benner, R., Amon, R., Cotner, J., Cavaletto, J., and Johnson, J. (1996) Effects of high molecular weight dissolved organic matter on the nitrogen dynamics on the Mississippi River plume. Mar. Ecol. Prog. Ser. 133, 287–297.

Gardner, W.S., Cavaletto, J.F., Bootsma, H.A., Lavrentyev, P.J., and Tanvone, F. (1998) Nitrogen cycling rates and light effects in tropical Lake Maracaibo, Venezuela. Limnol. Oceanogr. 43, 1814–1825.

Gardner, W.S., Cavaletto, J.F., Cotner, J.B., and Johnson, J.R. (1997) Effects of natural light on nitrogen cycling rates in the Mississippi River plume. 42, 273–281.

Gardner, W.S., Escobar-Broines, E., Cruz-Kaegi, E., and Rowe, G.T. (1993) Ammonium excretion by benthic invertebrates and sediment-water nitrogen flux in the Gulf of Mexico near the Mississippi River outflow. Estuaries 16, 799–808.

Gardner, W.S., and Hanson, R.B. (1979) Dissolved free amino acids in interstitial waters of Georgia salt marsh soils. Estuaries 2, 113–118.

Gardner, W.S., and Lee, G.F. (1975) The role of amino acids in the nitrogen cycle of Lake Mendota. Limnol. Oceanogr. 20, 379–388.

Gardner, W.S. and Menzel, D.W. (1974) Phenolic aldehydes as indicators of terrestrially derived organic matter in the sea. Geochim. Cosmochim. Acta 38, 813–822.

Gardner, W.S., Nalepa, T.F., and Malczyk, J.M. (1987) Nitrogen mineralization and denitrification in lake Michigan sediments. Limnol. Oceanogr. 32, 1226–1238.

Gardner, W.S., Seitzinger, S.P., and Malczyk, J.M. (1991) The effects of sea salts on the forms of nitrogen released from estuarine and freshwater sediments: does ion pairing affect ammonium flux? Estuaries 14, 157–166.

Garnier, J., Servais, P., Billen, G., Akopian, M., and Brion, N. (2001) Lower Seine River and estuary (France) carbon and oxygen budgets during low flow. Estuaries 24, 964–976.

Garrido, J.L., Otero, J., Maestro, M.A., and Zapata, M. (2000) The main nonpolar chlorophyll-c from *Emiliana huxleyi* (Prymnesiophyceae) is a chlorophyllc$_2$-monogalactosyldiacylglyceride ester: A mass spectrometry study. J. Phycol. 36, 497–505.

Gassmann, G. (1994) Phosphine in the fluvial and marine hydrosphere. Mar. Chem. 45, 197–205.

Gastrich, M.D., Anderson, O.R., and Cosper, E.M. (2002) Viral-like particles (VLP) in the alga, *Aurecoccus anophagefferens* (Pelagophyceae). during 1999–2000 brown tide blooms in Little Egg Harbor, New Jersey. Estuaries 25, 938–943.

Gätcher, R., and Meyer, J.S. (1993) The role of microorganisms in mobilization and fixation of phosphorus in sediments. Hydrobiologia 253, 103–121.

Gattuso, J.P., Frankignoulle, M., and Wollast, R. (1998) Carbon and carbonate metabolism in coastal aquatic ecosystems. Annu. Rev. Ecol. Syst. 29, 405–434.

Gaudette, H.E., and Lyons, W.B. (1980) Phosphate geochemistry in nearshore carbonate sediments: suggestion of apatite formation. Soc. Econ. Paleon. Min. Spec. Publ. 29, 215–225.

Gearing, P.J., Gearing, J.N., Pruell, R.J., Wade, T.S., and Quinn, J.G. (1980) Partitioning of No. 2 fuel oil in controlled estuarine ecosystems. Sediments and suspended particulate matter. Environ. Sci. Technol. 14, 1129–1136.

Gelboin, H. (1980) Benzo[*a*]pyrene metabolism, activation, and carcinogenesis: Role and regulation of mixed functional oxidases and related enzymes. Physiol. Rev. 60, 1107–1166.

Geyer, W.R. (1993) The importance of suppression of turbulence by stratification on the estuarine turbidity maximum. Estuaries 16, 113–125.

Geyer, W.R., and Beardsley, R.C. (1995) Introduction to a special session on physical oceanography of the Amazon shelf. J. Geophys. Res. 100, 2281–2282.

Geyer, W.R., Morris, J.T., Prahl, F.G., and Jay, D.A. (2000) Interaction between physical processes and ecosystem structure: a comparative approach. *In* Estuarine Science: A Synthetic Approach to Research and Practice (Hobbie, J.E., ed.), pp. 177–206, Island Press, Washington, DC.

Geyer, W.R., and Nepf, H. (1996) Tidal pumping of salt in a moderately stratified estuary. Coast. Estuar. Stud. 53, 213–226.

Geyer, W.R., and Signell, R.P. (1992) A reassessment of the role of tidal dispersion in estuaries and bays. Estuaries 15, 97–108.

Geyer, W.R., Signell, R., and Kineke, G. (1998) Lateral trapping of sediment in a partially mixed estuary. *In* Physics of Estuaries and Coastal Seas: Proceedings of the 8th International Biennial Conference on Physics of Estuaries and Coastal Seas (Dronkers, J., and Sheffers, M., eds.), pp. 115–126, Rotterdam, The Netherlands.

Geyer, W.R., Woodruf, J.D., and Traykovski, P. (2001) Sediment transport and trapping in the Hudson River Estuary. Estuaries 24, 670–679.

Gibbs, R.J. (1967) The geochemistry of the Amazon River System: Part I. The factors that control the salinity and the composition and concentration of the suspended solids. Geol. Soc. Am. Bull. 78, 1203–1232.

Gibbs, R.J. (1970) Mechanisms controlling world water chemistry. Science 170, 1088–1090.

Giblin, A.E. (1988) Pyrite formation in marshes during early diagenesis. Geomicrobiol. J. 6, 77–97.

Giblin, A.E., Hopkinson, C.S., and Tucker, J. (1997) Benthic metabolism and nutrient cycling in Boston Harbor, Massachusetts. Estuaries. 20, 346–364.

Giblin, A.E., and Howarth, R.W. (1984) Porewater evidence for dynamics sedimentary iron cycle in salt marshes. Limnol. Oceanogr. 29, 47–63.

Gieskes, W.W.C., Kraay, G.W., Nontji, W., Setiapermana, A., and Sutomo, D. (1988) Monsoonal alternation of a mixed and layered structure in the phytoplankton of the euphotic zone of the Banda Sea (Indonesia): a mathematical analysis of algal pigment fingerprints. Neth. J. Sea Res. 22, 123–137.

Gill, G.A., Bloom, N.S., Cappellino, S., Driscoll, C.T., Mason, R., and Rudd, J.W.M. (1999) Sediment–water fluxes of mercury in Lavaca Bay, Texas. Environ. Sci. Technol. 33, 663–669.

Gill, G.A., Guentzel, J.J., Landing, W.M., and Pollman, C.D. (1995) Total gaseous mercury measurements in Florida: The FAMS Project (1992–1994). Wat. Air Soil Pollut. 80, 235–244.

Gillan, F.T., and Johns, R.B. (1986) Chemical biomarkers for marine bacteria: fatty acids and pigments. In Biological Markers in the Sediment Record (Johns, R.B., ed.), pp. 291–309, Elsevier, New York.

Gilmour, C.C., Henry, E.A., and Mitchell, R. (1992) Sulfate stimulation of mercury methylation in freshwater sediments. Environ. Sci. Technol. 26, 2281–2288.

Gilmour, C.C., Reidel, G.S., Ederington, M.C., Bell, J.T., Benoit, G.A., Gill, G.A., and Stordal, M.C. (1998) Methylmercury concentrations and production rates across a trophic gradient in the northern Everglades. Biogeochemistry 40, 327–345.

Giovanelli, J. (1987) Sulfur amino acids of plants: An overview. Methods Enzymol. 143, 419–426.

Gladden, J.B., Cantelmo, F.R., Croom, J.M., and Shabot, R. (1988) Evaluation of the Hudson River ecosystem in relation to the dynamics of fish populations. Amer. Fish. Soc. Monogr. 4, 37–52.

Glasby, G.P. (ed.) (1978) Sedimentation and sediment geochemistry of Caswell, Nancy and Milford Sounds, New Zealand. Mem. N.Z. Oceaogr. Inst. 79, 7–9.

Glasgow, H.B., and Burkholder, J.M. (2000) Water quality trends and management implications from a five-year study of a poorly flushed, eutrophic estuary. Ecol. Appl. 10, 1024–1046.

Glasgow, H.B., and Burkholder, J.M., Mallin, M.A., Deamer-Melia, N.J., and Reed, R.E. (2001) Field ecology of toxic Pfiesteria complex species, and conservative analysis of their role in estuarine fish kills. Environ. Health Perspec. 109, 715–730.

Gleason, P.J., and Spackman, W., Jr. (1974) Calcareous periphyton and water chemistry in the Everglades. In Environments of South Florida: Past and Present (Gleason, P.J., ed.), pp. 146–181, Miami Geological Society, Coral Gables, FL.

Glibert, P.M., Conley, D.J., Fisher, T.R., Harding, L.W., and Malone, T.C. (1995) Dynamics of the 1990 winter/spring bloom in Chesapeake Bay. Mar. Ecol. Prog. Ser. 122, 27–43.

Glibert, P.M., Lipshultz, F., McCarthy, J.J., and Altabet, M.A. (1982) Isotope dilution models and remineralization of ammonium by marine plankton. Limnol. Oceanogr. 27, 639–650.

Glindemann, D., Stottmeister, U., and Bergmann, A. (1996) Free phosphine from the anaerobic biosphere. Environ. Sci. Pollut. Res. 3, 17–19.

Gloe, A., Pfenning, N., Brockmann, H., and Trowitzsch, W. (1975) A new bacteriochlorophyll from brown-colored Chlorobiaceae. Arch. Microbiol. 102, 103–109.

Goericke, R., and Repeta, D.J. (1992) The pigments of Prochlorococcus marinus: the presence of divinyl chlorophyll a and b in a marine prokaryote. Limnol. Oceanogr. 37, 425–433.

Goericke, R., Strom, S.L., and Bell, M.A. (2000) Distribution and sources of cyclic pheophorbides in the marine environment. Limnol. Oceanogr. 45, 200–211.

Goering, J., Alexander, V., and Haubenstock, N. (1990) Seasonal variability of stable carbon and nitrogen isotope ratios of organisms in a North Pacific Bay. Estuar. Coastal Shelf Sci. 30, 239–260.

Goldberg, A.B., Maroulis, P.J., Wilner, L.A., and Brandy, A.R. (1981) Study of H_2S emissions in a salt water marsh. Atmos. Environ. 15, 11–18

Goldberg, E.D. (1985) Black Carbon in the Environment: Properties and Distribution. John Wiley, New York.

Goldberg, E.D., and Bruland, K. (1974) Determination of marine chronologies using natural radionuclides. In The Sea (Goldberg, E.D., ed.), pp. 34–48, Wiley-Interscience, New York.

Goldberg, E.D., Griffin, J.J., Hodge, V., Koide, M., and Windom, H. (1979) Pollution history of the Savannah River Estuary. Environ. Sci. Technol. 3, 588–594.

Goldberg, E.D., and Koide, M. (1962) Geochronological studies of deep sea sediments by ionium/thorium method. Geochim. Cosmochim. Acta 26, 417–450.

Goldberg, E.D., Koide, M., Hodge, M., Flegal, A.R., and Martin, J. (1983) U.S. mussel watch: 1977–1978 results on trace metals and radionuclides. Estuar. Coastal Shelf Sci. 16, 69–93.

Goldfinch, A.C., and Carman, K.R. (2000) Chironomid grazing on benthic microalgae in a Louisiana salt marsh. Estuaries 23, 536–547.

Goldhaber, M. B., Aller, R. C., Cochran, J. K., Rosenfeld, J. K., Martens, C. S., and Berner, R. A. (1977) Sulphate reduction, diffusion, and bioturbation in Long Island Sound sediments. Report of the FOAM group. Am. J. Sci. 277, 193–237.

Goldhaber, M.B., and Kaplan, I.R. (1974) The sulfur cycle. In The Sea, Vol. 5 (Goldberg, E.D., ed.), pp. 569–655, Chichester, UK.

Goldschmidt, V.M. (1954) Geochemistry. Oxford University Press, Fairlawn, NJ.

Gomez-Belinchon, J.I., Llop, R., Grimalt, J.O., and Albaiges, J. (1988) The decoupling of hydrocarbons and fatty acids in the dissolved and particulate water phases of a deltaic environment. Mar. Chem. 25, 325–348.

Gong, C., and Hollander, D.J. (1997) Differential contribution of bacteria to sedimentary organic matter in oxic and anoxic environments, Santa Monica basin, California. Org. Geochem. 26, 545–563.

Goñi, M.A., Gardner, R.L. (2003) Seasonal dynamics in dissolved organic carbon concentrations in a coastal water-table aquifer at the forest–marsh interface. Aquat. Geochem. 9, 209–232.

Goñi, M.A., and Hedges, J.I. (1990a) Potential applications of cutin-derived CuO reaction products for discriminating vascular plant sources in natural environments. Geochim. Cosmochim. Acta 54, 3073–3083.

Goñi, M.A., and Hedges, J.I. (1990b) Cutin derived CuO reaction products from purified cuticles and tree leaves. Geochim. Cosmochim. Acta 54, 3065–3072.

Goñi, M.A., and Hedges, J.I. (1990c) The diagenetic behavior of cutin acids in buried conifer needles and sediments from a coastal marine environment. Geochim. Cosmochim. Acta 54, 3083–3093.

Goñi, M.A., and Hedges, J.I. (1992) Lignin dimers: Structures, distribution, and potential geochemical applications. Geochim. Cosmochim. Acta 56, 4025–4043.

Goñi, M.A., Ruttenberg, K.C., and Eglinton, T.I. (1997) Sources and contribution of terrigenous organic carbon to surface sediments in the Gulf of Mexico. Nature 389, 275–278.

Goñi, M.A., Ruttenberg, K.C., and Eglinton, T.I. (1998) A reassessment of the sources and importance of land-derived organic matter in surface sediments from the Gulf of Mexico. Geochim. Cosmochim. Acta 62, 3055–3075.

Goñi, M.A., and Thomas, K.A. (2000) Sources and transformations of organic matter in surface soils and sediments from a tidal estuary (North Inlet, South Carolina, U.S.A). Estuaries 23, 548–564.

Gontier, G., Gerino, M., Stora, G., and Melquiond, J. (1991) A new tracer technique for in situ experimental study of bioturbation processes. In Radionuclide in the Study of Marine Processes (Kershaw, P.J., and Woodhead, D.S., eds.), pp. 198–196, Elsevier Science, New York.

Gonzalez-Davila, M., Santana-Casiano, J.M., Perez-Pena, J., and Millero, F.J. (1995) Binding of Cu(II) to the surface and exudates of the alga Dunaliella tertiolecta in seawater. Environ. Sci. Technol. 29, 289–301.

Goodberlet, M.A., Swift, C.T., Kiley, K.P., Miller, J.L., and Zaitzeff, J.B. (1997) Microwave remote sensing of coastal zone salinity. J. Coastal Res. 13, 363–372.

Goodfriend, G.A., and Rollins, H.B. (1998) Recent barrier beach retreat in Georgia: Dating exhumed salt marshes by aspartic acid racemization and post-bomb radiocarbon. J. Coast. Res. 14, 960–969.

Goodwin, T.W. (1980) The Biochemistry of the Carotenoids, 2nd edn., Vol. 1. Chapman and Hall, London.

Goodwin, T.W., and Mercer, E.I. (1972) Introduction to Plant Biochemistry. Pergamon Press, Oxford, UK..

Goolsby, D.A. (2000) Mississippi basin nitrogen flux believed to cause Gulf hypoxia. EOS Trans. 2000, 321.

Gordeev, V.V., Martin, J.M., Sidorov, I.S., and Sidorova, M.V. (1996) A reassessment of the Eurasian River input of water sediment, major elements, and nutrients to the Arctic Ocean. Am. J. Sci. 296, 664–691.

Gordon, A.S., Donat, J.R., Kango, R.A., Dyer, B.J., and Stuart, L.M. (2000) Dissolved copper-complexing ligands in cultures of marine bacteria and estuarine water. Mar. Chem. 70, 149–160.

Gordon, E.S., and Goñi, M.A. (2003) Sources and distribution of terrigenous organic matter delivered by the Atchafalaya River to sediments in the northern Gulf of Mexico. Geochim. Cosmochim. Acta 67, 2359–2375.

Gordon, E.S., Goñi, M.A., Roberts, Q.N., Kineke, G.C., and Allison, M.A. (2001) Organic matter distribution and accumulation on the inner Louisiana shelf west of the Atchafalaya River. Cont. Shelf Res. 21, 1691–1721.

Gossauer, A., and Engel, N. (1996) Chlorophyll catabolism structures, mechanisms, and conversions. J. Photochem. Photobiol. 32, 141–151.

Gosselink, J.G., and Turner, R.E. (1978) The role of hydrology in freshwater wetland ecosystems. In Freshwater Wetlands: Ecological Processes and Management Potential (Good, R.E., Whigham, D.F., and Simpson, R.L., eds.), pp. 63–78, Academic Press, New York.

Gould, D., and Gallagher, E. (1990) Field measurements of specific growth rate, biomass, and primary production of benthic diatoms of Savin Hill Cove, Boston. Limnol. Oceanogr. 35, 1757–1770.

Graneli, E. (1987) Nutrient limitation of phytoplankton biomass in a brackish water bay highly influenced by river discharge. Estuar. Coastal Shelf Sci. 25, 555–565.

Graneli, W., and Sundbäck, K. (1986) Can microphytobenthic photosynthesis influence below-halocline oxygen conditions in the Kattegat? Ophelia 26, 195–206.

Granéli, E., Wallstrom, K., Larsson, U., Graneli, W., and Elmgren, R. (1990) Nutrient limitation of primary production in the Baltic Sea. Ambio 19, 142–151.

Granger, J., Sigman, D.M., Needoba, J.A., and Harrison, P.J. (2004) Coupled nitrogen and oxygen isotopic fractionation of nitrate during assimilation by cultures of marine phytoplankton. Limnol. Oceanogr. 49, 1763–1773.

Grant, J., Bathmann, U.V., and Mills, E.L. (1986) The interaction between benthic diatom films and sediment transport. Estuar. Coastal Shelf Sci. 23, 225–238.

Green, E.P., and Short, F.T. (2003) World Atlas of Seagrass. California University Press, Berkeley, CA.

Green, F., and Edmisten, J. (1974) Seasonality of nitrogen fixation in Gulf Coast salt marshes. *In* Phenology and Seasonality Modeling (Lieth, H, ed.), pp. 113–126, Springer-Verlag, New York.

Griffin, C., Kaufman, A., and Broeker, W.S. (1963) J. Geophys. Res. 68, 1749–1757.

Griffith, P., Shiah, F.K., Gloersen, K., Ducklow, H.W., and Fletcher, M. (1994) Activity and distribution of attached bacteria in Chesapeake Bay. Mar. Ecol. Prog. Ser. 108, 1–10.

Griffith, T.W., and Anderson, J.B. (1989) Climatic controls on sedimentation in bays and fjords of the northern Antarctic Peninsula. Mar. Geol., 85, 181–204.

Grill, E., Winnacker, E.L., and Zenk, M.H. (1985) Phytochelatins: the principal heavy-metal complexing peptides of higher plants. Science 230, 674–676.

Grimalt, J., Albaiges, J., Al-Saad, H.T., and Douabul, A.A.Z. (1985) *n*-Alkane distributions in surface sediments from the Arabian Gulf. Naturwissenschaften 72, 35–37.

Gruebel, T.F., and Martens, C.S. (1984) Radon-222 tracing of sediment–water chemical transport in estuarine sediment. Limnol. Oceanogr. 29, 587–597.

Gschwend, P., and Hites, R.A. (1981) Fluxes of the polycyclic aromatic compounds to marine and lacustrine sediments in the northeastern United States. Geochim. Cosmochim. Acta 45, 2359–2367.

Gschwend, P., and Wu, S.C. (1985) On the constancy of sediment–water partitioning coefficients of hydrophobic organic pollutants. Environ. Sci. Technol. 19, 90–96.

Gu, B., Schmitt, J., Chen, Z., Liang, L., and McCarthy, J.F. (1995) Adsorption and desorption of different organic matter fractions on iron oxide. Geochim. Cosmochim. Acta 59, 219–229.

Guarini, J., Blanchard, G.F., Gros, P., Gouleau, D., and Bacher, C. (2000) Dynamic model of the short-term variability of microphytobenthic biomass on temperate intertidal mudflats. Mar. Ecol. Prog. Ser. 195, 291–303.

Guarini, J., Cloern, J.E., Edmunds, J., and Gros, P. (2002) Microphytobenthic potential productivity estimated in three tidal embayments of the San Francisco Bay: a comparative study. Estuaries 25, 409–417.

Guentzel, J.L., Landing, W.M., Gill, G.A., and Pollman, C.D. (1998) Mercury and major ions in rainfall: throughfall and foliage from the Florida Everglades. Sci. Total Environ. 213, 43–51.

Guentzel, J.L., Landing, W.M., Gill, G.A., and Pollman, C.D. (2001) Processes influencing rainfall deposition of mercury in Florida: The FAMS Project (1992–1996), Environ. Sci. Technol. 35, 863–873.

Guentzel, J.L., Powell, R.T., Landing, W.M., and Mason, R.P. (1996) Mercury associated with colloidal material in an estuarine and an open-ocean environment. Mar. Chem. 55, 177–188.

Guillard, R.R.L., Murphy, L.S., Foss, P., and Liaaen-Jensen, S. (1985) *Synechcoccus* spp. as a likely zeaxanthin-dominant ultraphytoplankton in the north Atlantic. Limnol. Oceanogr. 30, 412–414.

Guinasso, N.L., and Schink, D.R. (1975) Quantitative estimates of biological mixing rates in abyssal sediments. J. Geophys. Res. 80, 3032–3043.

Gunnars, A., and Blomqvist, S. (1997) Phosphate exchange across the sediment–water interface when shifting from anoxic to oxic conditions: an experimental comparison of freshwater and brackish-marine systems. Biogeochemistry 37, 203–226.

Gunnars, A., Blomqvist, S., and Martinsson, C. (2004) Inorganic formation of apatite in brackish seawater from the Baltic Sea: an experimental approach. Mar. Chem. 91, 15–26.

Gunnarsson, J.S., Grandberg, M.E., Nilsson, H.C., Rosenberg, R., and Hellman, B. (1999) Influence of sediment–organic matter quality on growth and polychlorobiphenyl bioavailability in Echniodermata (*Amphiura filiformis*). Environ. Toxicol. Chem. 18, 1534–1543.

Gunnarsson, L.A.H., and Ronnow, P.H. (1982) Inter-relationships between sulfate reducing and methane producing bacteria in coastal sediments with intense sulfide production. Mar. Biol. 69, 121–128.

Guo, L., Coleman, C.H., and Santschi, P.H. (1994) The distribution of colloidal and dissolved organic carbon in the Gulf of Mexico. Mar. Chem. 45, 105–119.

Guo, L., Hunt, B.J., Santschi, P.H., and Ray, S.M. (2001) Effect of dissolved organic matter on the uptake of colloidal organic carbon in seawater. Mar. Chem. 55, 113–127.

Guo, L., and Santschi, P.H. (1996) A critical evaluation of the cross-flow ultrafiltration techniques for sampling colloidal organic carbon in seawater. Mar. Chem. 55, 113–127.

Guo, L., and Santschi, P.H. (1997) Composition and cycling of colloids in marine environments. Rev. Geophys. 35, 17–40.

Guo, L., and Santschi, P.H. (2000) Sedimentary sources of old high molecular weight dissolved organic carbon from the ocean margin benthic nepheloid layer. Geochim. Cosmochim. Acta 64, 651–660.

Guo, L., Santschi, P.H., and Baskaran, M. (1997) Interaction of thorium isotopes with colloidal organic matter in oceanic environments. Colloids Surf. A, Physiochem. Eng. Aspect 120, 255–271.

Guo, L., Santschi, P.H., and Bianchi, T.S. (1999) Dissolved organic matter in estuaries of the Gulf of Mexico. *In* Biogeochemistry of Gulf of Mexico Estuaries (Bianchi, T.S., Pennock, J., and Twilley, R.R., eds.), pp. 269–299, John Wiley, New York.

Guo, L., Santschi, P.H., Cifuentes, L.A., Trumbore, S.E., and Southon, J. (1996) Cycling of high molecular-weight dissolved organic mater in the Middle Atlantic Bight as revealed by carbon isotopic (^{13}C and ^{14}C) signatures. Limnol. Oceanogr. 41, 1242–1252.

Guo, L., Santschi, P.H., and Ray, S.M. (2002) Metal partitioning between colloidal and dissolved phases and its relation with bioavailability to American oysters. Mar. Environ. Res. 54, 49–64.

Guo, L., Santschi, P.H., and Warnken, K.W. (1995) Dynamics of dissolved organic carbon (DOC) in oceanic environments. Limnol. Oceanogr. 40, 1392–1403.

Guo, L., Santschi, P.H., and Warnken, K.W. (2000a) Trace metal composition of colloidal organic matter in marine environments. Mar. Chem. 70, 257–275.

Guo, L., Wen, L., Tang, D., and Santschi, P.H. (2000b) Re-examination of cross-flow ultrafiltration for sampling aquatic colloids: evidence from molecular probes. Mar. Chem. 69, 75–90.

Gustafson, K.E., and Dickhut, R.M. (1997a) Gaseous exchange of polycyclic aromatic hydrocarbons across the air–water interface of southern Chesapeake Bay. Environ. Sci. Technol. 31, 1623–1629.

Gustafson, K.E., and Dickhut, R.M. (1997b) Particle/gas concentrations and distribution of PAHs in the atmosphere of southern Chesapeake Bay. Environ. Sci. Technol. 31, 140–147.

Gustafson, K.E., and Dickhut, R.M. (1997c) Distribution of polycyclic aromatic hydro-carbons in southern Chesapeake Bay surface waters: evaluation of three methods for determining freely dissolved water concentrations. Environ. Sci. Technol. 16, 452–461.

Gustafsson, O., Bucheli, T.D., Kukulska, Z., Andersson, M., Largeau, C., Rouzaud, J.N., Reddy, C.M., and Eglinton, T.I. (2001) Evaluation of a protocol for the quantification of black carbon in sediments, soils and aquatic particles. Global Biogeochem. Cycles 15, 881–890.

Gustafsson, O., and Gschwend, P.M. (1997) Aquatic colloids: concepts, definitions, and current challenges. Limnol. Oceanogr. 42, 519–528.

Guy, R.D., Fogel, M.L., Berry, J.A., and Hoering, T.C. (1987) Isotope fractionation during oxygen production and consumption by plants. Prog. Photosyn. Res. III. 9, 597–600.

Guy, R.D., Reid, D.M., and Krouse, H.R.(1986) Factors affecting $^{13}C/^{12}C$ ratios of inland halophytes. I. Controlled studies on growth and isotopic composition of *Puccinella nuttalliana*. Can. J. Bot. 64, 2693–2699.

Guyoneaud, R., Borrego, C.M., Martinez-Planells, A., Buitenhuis, E.T., and Garcia-Gil, L.J. (2001) Light responses in the green sulfur bacterium *Prosthecochloris aestuarii*: changes in prosthecae length, ultrastructure, and antenna pigment composition. Arch. Microbiol. 176, 278–284.

Habicht, K.S., and Canfield, D.E. (1997) Sulfur isotope fractionation during bacterial sulfate reduction in organic sediments. Geochim. Cosmochim. Acta 24, 5351–5361.

Hackney, C. T., and de la Cruz, A. A. (1980) In situ decomposition of roots and rhizomes of two tidal marsh plants. Ecology 61, 226–231.

Haddad, R.I., Newell, S.Y., Martens, C.S., and Fallon, R.D. (1992) Early diagenesis of lignin-associated phenolics in the salt marsh grass *Spartina alterniflora*. Geochim. Cosmochim. Acta 56, 3751–3764.

Hadjichristophorou, M., Argyrou, M., Demetropoulos, A., and Bianchi. T.S. (1997) A species list of the sublittoral soft-bottom macrobenthos of Cyprus. Acta Adriatica 38, 3–31.

Hager, A. (1980) The reversible, light-induced conversions of xanthophylls in the chloroplast. *In* Pigments in Plants, 2nd edn. (Czygan, F.C., ed.), pp. 57–79, Fischer, Stuttgart.

Hahn, J., and Crutzen, P.J. (1982) The role of fixed nitrogen in atmosphere photochemistry. Phil. Trans. R. Soc. Lond. 296, 521–541.

Haines, E.B. (1977) The origins of detritus in Georgia salt marsh estuaries. Oikos 29, 254–260.

Haines, E.B., and Montague, C.L. (1979) Food sources of estuarine invertebrates analyzed using $^{13}C/^{12}C$ ratios. Ecology 60, 48–56.

Haith, D.A., and Shoemaker, L.I. (1987) Generalized watershed loading functions for stream flow nutrients. Wat. Res. Bull. 23, 471–478.

Haitzer, M., Abbt-Braun, G., Traunspurger, W., and Steinberg, C.E.W. (1999) Effects of humic substances on the bioconcentration of polycyclic aromatic hydrocarbons: correlations with spectroscopic and chemical properties of humic substances. Environ. Sci. Technol. 18, 2782–2788.

Haitzer, M., Hoss, S., Traunspurger, W., and Steinberg, C. (1998) Effects of dissolved organic matter (DOM) on the biogeochemistry of organic chemicals in aquatic organisms, a review. Chemosphere 37, 1335–1362.

Hall, P. O. J., Hulth, S., and Hulthe, G. (1996) Benthic nutrient fluxes on a basin-wide scale in the Skagerrak (north-eastern North Sea). J. Sea Res. 35, 123–137.

Hallberg, R.O., and Larsson, C. (1999) Biochelates as a cause of metal cycling across the redoxcline. Aquat. Geochem. 5, 269–280.

Hallegraeff, G.M. (1988) Three estuarine Australian dinoflagellates that can produce paralytic shellfish toxins. J. Plank. Res. 10, 533–541.

Hallegraeff, G.M., and Jeffrey, S.W. (1985) Description of new chlorophyll *a* alteration products in marine phytoplankton. Deep-Sea Res. 32, 697–705.

Hamelink, J.L., Landrum, P.F., Bergman, H.L., and Benson, W.H. (1994) Bioavailability, Physical, Chemical, and Biological Interactions. Lewis Publications, Boca Raton, FL.

Hamilton, S.E., and Hedges, J.I. (1988) The comparative geochemistries of lignins and carbohydrates in an anoxic fjord. Geochim. Cosmochim. Acta 52, 129–142.

Hammerschmidt, C.R., Fitzgerald, W.F., Lamborg, C.H., Balcom, P.H., and Visscher, P.T. (2004) Biogeochemistry of methyl mercury in sediments of Long Island Sound. Mar. Chem. 90, 31–52.

Hammond, D.E., and Fuller, C. (1979) The use of radon-222 to estimate benthic exchange and atmospheric rates in San Francisco Bay. *In* Investigation into the Natural History of San Francisco Bay and Delta with Reference to the Influence of Man (Conomos, J.J., ed.), pp. 31–43, Pacific Division of the American Association for the Advancement of Science, San Francisco, CA.

Hammond, D.E., Fuller, C., Harmon, D., Hartman, B., Korosec, M., Miller, L.G., Rea, R.L., Warren, S., Berelson, W., and Hager, S. (1985) Benthic fluxes in San Francisco Bay. Hydrobiologia 129, 69–90.

Hammond, D.E., Simpson, H.J., and Mathieu, G. (1977) Radon-222 distribution and transport across the sediment–water interface in the Hudson River estuary. J. Geophys. Res. 82, 3913–3920.

Handbook of Environmental Chemistry, (1986) (Hutzinger, O., ed.), pp. 1–58, Springer-Verlag, Berlin.

Hansell, D.A., and Carlson, C.A. (eds.) (2002) Biogeochemistry of Marine Dissolved Organic Matter. Academic Press, New York.

Hansen, D.V., and Rattray, M. (1965) Gravitational circulation in straits and estuaries. J. Mar. Res. 23, 104–122.

Hansen, D.V., and Rattray, M. (1966) New dimensions in estuary classification. Limnol. Oceanogr. 11, 319–326.

Hansen, R.P. (1980) Phytol: its metabolic products and their distribution. A review. NZ J. Sci. 23, 259–275.

Hanson, R.B. (1983) Nitrogen fixation activity (acetylene reduction) in the rhizosphere of a salt marsh angiosperm, Georgia, USA. Bot. Mar. 26, 49–59.

Hanson, R.B., and Gardner, W.S. (1978) Uptake and metabolism of two amino acids by anaerobic microorganisms in four diverse salt marsh soils. Mar. Biol. 46, 101–107.

Hanson, R.S., and Hanson, T.E. (1996) Methanotrophic bacteria. Microb. Rev. 60, 439–471.

Harada, K., Burnett, W.C., LaRock, P.A., and Cowart, J.B. (1989) Polonium in Florida groundwater and its possible relationship to the sulfur cycle and bacteria. Geochim. Cosmochim. Acta 53, 143–150.

Harada, K., and Tsunogai, S. (1988) Is lead soluble at the surface of sediments in biologically productive seas? Cont. Shelf Res. 8, 387–396.

Hardy, E.P., Krey, P.W., Volchok, H.L. (1973) Global inventory and distribution of fallout plutonium. Nature 241, 444–445.

Hargrave, B.T. (1969) Similarity of oxygen uptake by benthic communities. Limnol. Oceanogr. 14, 801–805.

Harnett, H.E., Keil, R.G., Hedges, J.I., and Devol, A. (1998) Influence of oxygen exposure time on organic carbon preservation in continental margin sediments, Nature 391, 572–574.

Harradine, P.J., Harris, P.G., Head, R.N., Harris, R.P., and Maxwell, J.R. (1996) Steryl chlorine esters are formed by zooplankton herbivory. Geochim. Cosmochim. Acta 60, 2265–2270.

Harris, G.P. (1999) Comparison of the biogeochemistry of lakes and estuaries: ecosystem processes, functional groups, hysteresis effects and interactions between macro- and microbiology. Mar. Freshwater Res. 50, 791–811.

Harris, P.G., Carter, J.F., Head, R.N., Harris, R.P., Eglinton, G., and Maxwell, J.R. (1995) Identification of chlorophyll transformation products in zooplankton faecal pellets and marine sediment extracts by liquid chromatography/mass spectrometry atmospheric pressure chemical ionization. Rapid Commun. Mass Spectrom. 9, 1177–1183.

Harris, P.T., Baker, E.K., Cole, A.R., and Short, S.A. (1993) A preliminary study of sedimentation in the tidally dominated Fly River Delta, Gulf of Papua. Cont. Shelf Res. 13, 441–472.

Harris, P.T., and Collins, M.B. (1985) Bedform distributions and sediment transport paths in the Bristol Channel and Severn Estuary, UK. Mar. Geol. 62, 153–166.

Harrison, E.Z., and Bloom, A.L. (1977) Sedimentation rates on tidal salt marshes in Connecticut. J. Sed. Petrol. 47, 1484–1490.

Harrison, P.G. (1982) Control of microbial growth and of amphipod grazing by water soluble compounds from leaves of Zostera marina. Mar. Biol. 67, 225–230.

Harrison, P.G., and Mann, K.H. (1975) Detritus formation from eelgrass (Zostera marina): the relative effects of fragmentation, leaching, and decay. Limnol. Oceanogr. 20, 924–934.

Hart, B.A., and Scaife, B.D. (1997) Toxicity and bioaccumulation of cadmium in Chlorella pyrenoidosa. Environ. Res. 14, 401–417.

Hart, B.S. (1995) Delta front estuaries. In Geomorphology and Sedimentology of Estuaries. Developments in Sedimentology 53. (Perillo, G.M.E., ed.), pp. 207–224, Elsevier Science, New York.

Hart, B.S., Prior, D.B., Barrie, J.V., Currie, R.A., and Lutenauer, J.L. (1992) A river mouth submarine landslide and channel complex, Fraser Delta, Canada. Sed. Geol. 81, 73–87.

Hartman, B., and Hammond, D.E. (1984) Gas exchange rates across the sediment–water and air–water interfaces in the South San Francisco Bay. J. Geophys. Res. 89, 3593–3603.

Hartnett, H.E., Keil, R.G., Hedges, J.I., and Devol, A.H. (1998) Influence of oxygen exposure time on organic carbon preservation in continental margin sediments. Nature 391, 572–574.

Harvey, R.H., and Macko, S.A. (1997) Kinetics of phytoplankton decay during simulated sedimentation: changes in lipids under oxic and anoxic conditions. Org. Geochem. 27, 129–140.

Harvey, R.H., and Mannino, A. (2001) The chemical composition and cycling of particulate and macromolecular dissolved organic matter in temperate estuaries as revealed by molecular organic tracers. Org. Geochem. 32, 527–542.

Harvey, R.H., Tuttle, J.H., and Bell, J.T. (1995) Kinetics of phytoplankton decay during simulated sedimentation: changes in biochemical composition and microbial activity under oxic and anoxic conditions. Geochim. Cosmochim. Acta 59, 3367–3377.

Harwood, J.L. and Russell, N.J. (1984) Lipids in Plants and Microbes. George Allen and Unwin, London.

Hassellov, M., Lyven, V., Haraldsson, C., and Sirinawin, W. (1999) determination of continuous size and trace element distribution of colloidal material in natural water by on-line coupling of flow field fractionation with ICPMS. Anal. Chem.. 71, 3497–3502.

Hatcher, P.H. (1987) Chemical structural studies of natural lignin by dipolar dephasing solid state ^{13}C nuclear magnetic resonance. Org. Geochem. 11, 31–39.

Hatcher, P.H., Dria, K.J., Kim, S., and Frazier, S.W. (2001) Modern analytical studies of humic substances. Soil Sci. 166, 770–794.

Hatcher, P.G., and Minard, R.D. (1996) Comparison of dehydrogenase polymer (DHP) lignin with native lignin from gymnosperm wood by thermochemolysis using tetramethylammonium hydroxide (TMAH). Org. Geochem. 24, 593–600.

Hatcher, P.G., Nanny, M.A., Minard, R.D., Dible, S.C., and Carson, D.M. (1995) Comparisons of two thermochemolytic methods for the analysis of lignin in decomposing wood: The CuO oxidation method and the method of thermochemolysis with TMAH. Org. Geochem. 23, 881–888.

Hatton, R.S., Delaune, R.D., and Patrick, W.H. (1983) Sedimentation, accretion, and subsidence in marshes of Barataria Basin, Louisiana. Limnol. Oceanogr. 28, 494–502.

Haugen, J.E., and Lichtentaler, R. (1991) Amino acid diagenesis, organic carbon and nitrogen mineralization in surface sediments from the inner Oslofjord, Norway. Geochim. Cosmochim. Acta 55, 1649–1661.

Hauxwell J., Cebrian J., Herrera-Silveira J. A., Ramirez J., Zaldivar A., Gomez N., and Aranda-Cirerol N. (2001) Measuring production of Halodule wrightii Ascherson: additional evidence suggests clipping underestimates growth rate. Aquat. Bot. 69, 41–54.

Hawkes, G.E., Powlson, D.S., Randall, E.W., and Tate, K.R. (1984) A ^{31}P nuclear magnetic resonance study of the phosphorus species in alkali extracts from long-term field experiments. J. Soil Sci. 35, 35–45.

Hawkins, A.J.S., Bayne, B.L., Mantoura, R.F.C., and Llewellyn, C.A. (1986) Chlorophyll degradation and absorption through the digestive system of the blue mussel *Mytilus edulis*. J. Exp. Mar. Biol. Ecol. 96, 213–223.

Hayes, J.M. (1983) Geochemical evidence bearing on the origin of aerobiosis, a speculative interpretation. *In* The Earth's Earliest Biosphere: Its Origin and Evolution (Schopf, J.W., ed.), pp. 291–301, Princeton University Press, Princeton, NJ.

Hayes, J.M. (1993) Factors controlling the ^{13}C contents of sedimentary organic compounds: Principles and evidence. Mar. Geol. 113, 111–125.

Hayes, J.M. (2004) Isotopic order, biogeochemical processes, and Earth history. Geochim. Cosmochim. Acta 68, 1691–1700.

Hayes, M.O. (1975) Morphology and sand accumulations in estuaries. *In* Estuarine Research (Cronin, L.E., ed.), pp. 3–22, Academic Press, New York.

Head, E.J.H., Hargrave, B.T., and Subba Rao, D.V. (1994) Accumulation of a phaeophorbide *a*-like pigment in sediment traps during late stages of a spring bloom: a product of dying algae? Limnol. Oceanogr. 39, 176–181.

Head, E.J.H., and Harris, L.R. (1994) Feeding selectivity by copepods grazing on natural mixtures of phytoplankton determined by HPLC analysis of pigments. Mar. Ecol. Prog. Ser. 110, 75–83.

Head, E.J.H., and Harris, L.R. (1996) Chlorophyll destruction by *Calanus* grazing on phytoplankton: kinetic effects of ingestion rate and feeding history, and a mechanistic interpretation. Mar. Ecol. Prog. Ser. 135, 223–235.

Heaton, T.H.E. (1986) Isotopic studies of nitrogen pollution in the hydrosphere and atmosphere: a review. Chem. Geol. 59, 87–102.

Heck, K.L., Able, K.W., Roman, C.T., and Fahay, M.P. (1995). Composition, abundance, biomass, and production of macrofauna in a New England estuary: comparisons among eelgrass meadows and other nursery habitats. Estuaries 18, 379–389.

Hecky, R.E., and Kilham, P. (1988) Nutrient limitation of phytoplankton in freshwater and marine environments: a review of recent evidence on the effects of enrichment. Limnol. Oceanogr. 33, 796–822.

Hedges, J.I. (1978) The formation and clay mineral reactions of melanoidins. Geochim. Cosmochim. Acta 42, 69–76.

Hedges, J.I. (1988) Polymerization of humic substances in natural environments. *In* Humic Substances and their Role in the Environment (Frimmel, F.H., and Christman R.F., eds.), pp. 45–58, John Wiley, New York.

Hedges, J.I. (1992) Global biogeochemical cycles: progress and problems. Mar. Chem. 39, 67–93.

Hedges, J.I. (2002) Why dissolved organic matter? *In* Biogeochemistry of Marine Dissolved Organic Matter (Hansell, D.A., and Carlson, C.A., eds.), pp. 1–27, Academic Press, New York.

Hedges, J.I., Baldock, J.A., Gelinas, Y., Lee, C., Peterson, M.L., and Wakeham, S.G. (2002) The biochemical and elemental compositions of marine plankton: a NMR perspective. Mar. Chem. 78, 47–63

Hedges, J.I., Clark, W.A., and Cowie, G.L. (1988) Organic matter sources to the water column and surficial sediments of a marine bay. Limnol. Oceanogr. 33, 1116–1136.

Hedges, J.I., Cowie, G.L., Richey, J.E., and Quay, P. (1994) Origins and processing of organic matter in the Amazon River as indicated by carbohydrates and amino acids. Limnol. Oceanogr. 39, 743–761.

Hedges, J.I. and Ertel, J.R. (1982) Characterization of lignin by gas capillary chromatography of cupric oxide oxidation products. Anal. Chem. 54, 174–178.

Hedges, J.I., Ertel, J.R., Quay, P.D., Grootes, P.M., Richey, J.E., Devol, A.H., Farwell, G.W., Schmidt, F.W, and Salati, E. (1986) Organic carbon-14 in the Amazon River system. Science 231, 1129–1131.

Hedges, J.I., and Hare, P.E. (1987) Amino acid adsorption by clay minerals in distilled water. Geochim. Cosmochim. Acta 51, 255–259.

Hedges, J.I., Hatcher, P.H., Ertel, J.R., and Meyers-Schulte. (1992) A comparison of dissolved humic substances from seawater with Amazon River counterparts by ^{13}C-NMR spectrometry. Geochim. Cosmochim. Acta 56, 1753–1757.

Hedges, J.I., Hu, F.S., Devol, A.H., Hartnett, H.E., Tsamakis, E., and Keil, R.G. (1999) Sedimentary organic matter preservation: a test from selective degradation under oxic conditions. Am. J. Sci. 299, 525–555.

Hedges, J.I., and Keil, R. (1995) Sedimentary organic matter preservation; an assessment and speculative synthesis. Mar. Chem. 49, 81–115.

Hedges, J.I., Keil, R., and Benner, R. (1997) What happens to terrestrially-derived organic matter in the ocean? Org. Geochem. 27, 195–212.

Hedges, J.I. and Mann, D.C. (1979) The characterization of plant tissues by their lignin oxidation products. Geochim. Cosmochim. Acta 43, 1809–1818.

Hedges, J.I., Mayorga, E., Tsamakis, E., McClain, M.E., Aufdenkampe, A., Quay, P., Richey, J.E., Benner, R., Opsahl, S., Black, B., Pimental, T., Quintanilla, J., and Maurice, L. (2000) Organic matter in Bolivian tributaries of the Amazon River: A comparison to the lower mainstream. Limnol. Oceanogr. 45, 1449–1466.

Hedges, J.I. and Parker, P.L. (1976) Land-derived organic matter in the surface sediments from the Gulf of Mexico. Geochim. Cosmochim. Acta 40, 1019–1029.

Heidelberg, J.F., Heidelberg, K.B., and Colwell, R.R. (2002) Seasonality of Chesapeake bay bacterioplankton species. Appl. Environ. Microbiol. 68, 5488–5497.

Heijerick, D.G., De Schamphelaere, A.C., and Janssen, C.R. (2002) Predicting acute toxicity for Daphnia magna as a function of key chemistry characteristics: development and validation of a biotic ligand model. Environ. Toxicol. Chem. 21, 1309–1315.

Heip, C.H.R., Goosen, N.K., Herman, P.M., Kromkamp, J., Middelburg, J.J., and Soetaert, K. (1995) Production and consumption of biological particles in temperate tidal estuaries. Oceanogr. Mar. Biol. Rev. 33, 1.

HELCOM (1993) Second Baltic Sea Pollution Load Compilation. Baltic Sea Environment Proceedings No. 45, Helsinki.

HELCOM (1996) Third Periodic Assessment of the State of the Marine Environment of the Baltic Sea, 1989–1993. Helsinki Commission, Baltic Sea Environment Proceedings 69, Helsinki.

HELCOM (1998) Third Baltic Sea Pollution Load Compilation. Baltic Sea Environment Proceedings No. 70, Helsinki.

HELCOM (2001) Fourth Periodic Assessment of the State of the Baltic Marine Area, 1994–1998; Baltic Sea Environment Proceedings No. 82, Helsinki.

Hellou, J., Steller, S., and Albaiges, J. (2002) Alkanes, gerpanes, and aromatic hydro-carbons in superficial sediments of Halifax Harbor. Intl. J. Polycycl. Aromat. Compounds. 22, 631–642.

Heltz, G.R., Setlock, G.H., Cantillano, and Moore, W.S. (1985) Processes controlling the regional distribution of ^{210}Pb, ^{226}Ra and anthropogenic zinc in estuarine sediments. Earth Planet. Sci. Lett. 76, 23–34.

Hemminga, M.A., and Duarte, C.M. (2000) Seagrass Ecology. Cambridge University Press, Cambridge, UK.

Henderson, P. 1984. General geochemical properties and abundances of the rare Earth elements. In Rare Earth Element Geochemistry (Henderson, P., ed.), pp. 1–29, Elsevier, New York.

Henderson-Sellers, A. (1996) Soil moisture simulation: achievements of the RICE and PILPS intercomparison workshop and future directions. Global Planet. Change 13, 99–116.

Hendry, C.D., and Brezonik, P.L. (1980) Chemistry of precipitation at Gainesville, Florida. Environ. Sci. Technol. 14, 843–849.

Henrichs, S.M., and Farrington, J.W. (1979) Amino acids in interstitial waters of marine sediments. Nature 279, 319–322.

Henrichs, S.M., and Farrington, J.W. (1987) Early diagenesis of amino acids and organic matter in two coastal marine sediments. Geochim. Cosmochim. Acta 51, 1–15.

Henrichs, S.M., Farrington, J.W., and Lee, C. (1984) Peru upwelling region sediments near 15° S. 2. Dissolved free and total hydrolyzable amino acids. Limnol. Oceanogr. 29, 20–34.

Henrichs, S.M., and Reeburgh, W.S. (1987) Anaerobic mineralization of marine sediment organic matter: rates and the role of anaerobic processes in the oceanic carbon economy. Geomicrobiol. J. 5, 191–237.

Henrichs, S.M., and Sugai, S.F. (1993) Adsorption of amino acids and glucose by sediments of Resurrection Bay, Alaska, USA: Functional group effects. Geochim. Cosmochim. Acta 57, 823–835.

Henriksen, K., Hansen, J.I., and Blackburn, T.H. (1980) The influence of benthic infauna on exchange rates of inorganic nitrogen between sediment and water. Ophelia 1, 249–256.

Henriksen, K., Hansen, J.I., and Blackburn, T.H. (1981) Rates of nitrification, distribution of nitrifying bacteria, and nitrate fluxes in different types of sediment from Danish waters. Mar. Biol. 61, 299–304.

Henriksen, K., and Kemp, W.M. (1988) Nitrification in estuarine and coastal marine sed-iments. In Nitrogen Cycling in Coastal Marine Environments. SCOPE (Blackburn, T.H., and Sørensen, J., eds.), pp. 207–249, John Wiley, New York.

Henshaw, P.C., Charlson, R.J., and Burges, S.J. (2000) Water and the hydrosphere. In Earth System Science—From Biogeochemical Cycles to Global Change (Jacobson, M.C., Charlson, R.J., Rodhe, H., and Orians, G.H., eds.), pp. 109–131, Academic Press, New York.

Herczeg, A.C., and Hesslein, R.H. (1984) Determination of hydrogen ion concentration in softwater lakes using carbon dioxide equilibria. Geochim. Cosmochim. Acta 48, 837–845.

Hering, J., and Morel, F.M. (1989) Slow coordination reactions in seawater. Geochim. Cosmochim. Acta 53, 611–618.

Herman, P.M., and Heip, C.H.P. (1999) Biogeochemistry of the MAximum TURbidity zone of Estuaries (MATURE): some conclusions. J. Mar. Syst. 22, 89–104.

Hermes, J.D., Weiss, P.M., and Cleland, W.W. (1985) Use of nitrogen-15 and deu-terium isotopes effects to determine the chemical mechanism of phenylalanine ammonia-lyase. Biochem. 24, 2959–2967.

Hernandez, M.E., Mead, R.N., Peralba, M.C., and Jaffé, R. (2001) Linear *n*-alkane-2-ones as potential biomarkers for seagrass-derived organic matter in coastal environments. Org. Geochem. 32, 21–32.

Hernes, P.J., Benner, R., Cowie, G.L., Goni, M.A., Bergamaschi, B.A., and Hedges, J.I. (2001) Tannin diagenesis in mangrove leaves from a tropical estuary: a novel molecular approach. Geochim. Cosmochim. Acta 65, 3109–3122.

Hernes, P.J., Hedges, J.I., Peterson, L., Wakeham, S.G., and Lee, C. (1996) Neutral carbohydrate geochemistry on particulate material in the central equatorial Pacific. Deep sea res. II. 43, 1181–1204.

Herrera-Silveira, J.A., (1994) Nutrients from underground water discharges in a coastal lagoon (Celestun, Yucatan, Mexico). Verh. Intl. Ver. Limnol. 25, 1398–1401.

Herrera-Silveira, J.A., Medina-Gomez, I., and Colli, R. (2002) Trophic status based on nutrient concentration scales and primary producers community of tropical coastal lagoons influenced by groundwater discharges. Hydrobiologia 475/476, 91–98.

Herrera-Silveira, J.A., and Remirez-Remirez, J. (1996) Effects of natural phenolic material (tannin) on phytoplankton growth. Limnol. Oceanogr. 41, 1018–1023.

Hickey, B.M., and Banas, N. (2003) Oceanography of the U.S. Pacific Northwest coastal ocean and estuaries with application to coastal ecology. Estuaries 26, 1010–1031.

Hicks, R.E., Lee, C., and Marinucci, A.C. (1991) Loss and recycling of amino acids and protein from smooth cordgrass (*Spartina alterniflora*) litter. Estuaries 14, 430–439.

Hicks, S.D., Debaugh, H.A., Hickman, J.E. (1983) Sea level variations for the United States 1855–1980. U.S. Dept. of Commerce, NOAA, Rockville, MD.

Hill, P.S., Milligan, T.G., and Geyer, W.R. (2000) controls on effective settling velocity of suspended sediment in the Eel River flood plume. Cont. Shelf Res. 20, 2095–2111.

Hillman, K., Walker, D.I., Larkum, A.W.D. and McComb, A.J. (1989) Productivity and nutrient limitation. *In* A Treatise on the Biology of Seagrasses with Special Reference to the Australian Region. Aquatic Plant Studies 2A. (Larkum, W.D., McComb, A.J., and Shepherd, S.A. eds.), pp. 635–685, Elsevier, Amsterdam.

Hines, M.E. (1991) The role of certain infauna and vascular plants in the mediation of redox reactions in marine sediments. *In* Diversity of Environmental Biogeochemistry (Berthelin, J., ed.), pp. 275–286, Elsevier, Amsterdam.

Hines, M.E., Knollmeyer, S.L., and Tugel, J.B. (1989) Sulfate reduction and other sedimentary biogeochemistry in a northern New England salt marsh. Limnol. Oceanogr. 34, 578–590.

Hinga, K.R., Pilson, M.E.Q., Lee, R.F., Farrington, J.W., Tjessem, K., and Davis, A.C. (1980) Biogeochemistry of benzanthracene in an enclosed marine ecosystem. Environ. Sci. Technol. 14, 1136–1143.

Hinga, K.R., Sieburth, J., and Heath, G.R. (1979) The supply and use of organic material at the deep-sea floor. J. Mar. Res. 37, 557–579.

Hitchcock, D. (1975) Dimethyl sulfide emissions to the global atmosphere. Chemosphere 3, 137–138.

Hitchcock, G.L., Olson, D.B., Cavendish, S.L., and Kanitz, E.C. (1996) A tracked surface drifter with cellular telemetry capabilities. Mar. Tech. Soc. J. 30, 44–49.

Hobbie, J.E. (ed.) (2000) Estuarine science: the key to progress in coastal ecological research. *In* Estuarine Science: A Synthetic Approach to Research and Practice, pp. 1–11, Island Press, Washington, DC.

Hobbie, J.E., Copeland, B.J., and Harrison, W.G. (1975) Sources and fate of nutrients of the Pamlico River estuary, N.C. *In* Estuarine Research. Vol. 1. Chemistry, Biology, and the Estuarine System (Cronin, L.E., ed.), pp. 287–302, Academic Press, New York.

Hobbie, J. E., and Lee, C. (1980) Microbial production of extracellular material: importance in benthic ecology. *In* Marine Benthic Dynamics (Tenore, K., and Coull, B.,
eds.), pp. 341–346. Belle W. Baruch Institute for Marine Biology, University of
South Carolina Press, Columbia.

Hoch, M.P., and Kirchman, D.L. (1995) Ammonium uptake by heterotrophic bacteria
in the Delaware estuary and adjacent coastal waters. Limnol. Oceanogr. 40,
886–897.

Hodell, D.A., and Schelske, C.L. (1998) Production, sedimentation and isotopic
composition of organic matter in Lake Ontario. Limnol. Oceanogr. 43, 200–214.

Hodson, R.E, Christian, R.R., and Maccubbin, A.E. (1983) Lignocellulose and lignin
in the salt marsh grass *Spartina alterniflora*: initial concentration and short-term,
post-depositional changes in detrital matter. Mar. Biol. 81, 1–7.

Hoefs, J. (1980) Stable Isotope Geochemistry. Springer-Verlag, Heidelberg, Germany.

Hoegberg, P., and Johannisson (1993) ^{15}N abundance of forests is correlated with losses
of nitrogen. Plant Soil 157, 147–150.

Holden, M. (1976) Chlorophylls. *In* Chemistry and Biochemistry of Plant Pigments, 2nd
edn. (Goodwin, T.W., ed.), pp. 2–37, Academic Press, London.

Holland, H.D. (1978) The Chemistry of the Atmosphere and Oceans. John Wiley,
New York.

Holland, H.D. (1994) The Chemical Evolution of the Atmosphere and the Oceans.
Princeton University Press, Princeton, NJ.

Hollibaugh, J.T., Buddemeier, R., and Smith, S.V. (1991) Contributions of colloidal and
high molecular weight dissolved material to alkalinity and nutrient concentrations
in shallow marine and estuarine systems. Mar. Chem. 34, 1–27.

Holligan, P.M. (1992) Do marine phytoplankton influence global climate? *In* Primary
Productivity and Biogeochemical Cycles in the Sea (Falkowski, P.G., and Woodhead,
A.D., eds.), pp. 487–501.

Holloway, P.J. (1973) Cutins of *Malus pumila* fruits and leaves. Phytochemistry 12,
2913–2920.

Holmén, K. (2000) The global carbon cycle. *In* Earth System Science—from Biogeo-
chemical Cycles to Global Change (Jacobson, M.C., Charlson, R.J., Rodhe, H, and
Orians, G.H., eds.), pp. 282–321, Academic Press, International Geophysics Series,
New York.

Holmes, A.J., Costello, A., Lidstrom, M.E., and Murrell, J.C. (1995) Evidence that particulate methane monooxygenase and ammonia monooxygenase may be evolutionary
related. FEMS Microbial. Lett. 132, 203–208.

Holmes, R.M., McClelland, J.W., Sigman, D.M., Fry, B., and Peterson, B.J. (1998)
Measuring ^{15}N-NH_4^+ in marine, estuarine and fresh waters: an adaptation of the
ammonia diffusion method for samples with low ammonium concentrations. Mar.
Chem. 60, 235–243.

Holmes, R.M., Peterson, B.J., Gordeev, V.V., Zhulidov, A.V., Meybeck, M., Lammers,
R.B., and Vorosmarty, C.J. (2000) Flux of nutrients from Russian rivers to the Arctic
Ocean: can we establish a baseline against which to judge future changes? Wat.
Resour. Res. 36, 2309–2320.

Honeyman, B.D., and Santschi, P.H. (1988) Critical review: metals in aquatic systems.
Predicting their scavenging residence times from laboratory data remains a challenge.
Environ. Sci. Technol. 22, 862–871.

Honeyman, B.D., and Santschi, P.H. (1989) A Brownian-pumping model for oceanic
trace metal scavenging: evidence from Th isotopes. J. Mar. Res. 47, 951–992.

Honeyman, B.D., and Santschi, P.H. (1991) Coupling of trace metal adsorption and
particle aggregation: kinetics and equilibrium studies using ^{59}Fe-labelled hematite.
Environ. Sci. Technol. 25, 1739–1747.

Hopkinson, C.S., Buffam, I., Hobbie, J., Vallino, J., Perdue, M., Eversmeyer, B., Prahl, F., Covert, J., Hodson, R., Moran, M.A., Smith, E., Baross, J., Crump, B., Findlay, S., and Foreman, K. (1998) Terrestrial inputs or organic matter to coastal ecosystems: an intercomparison of chemical characteristics and bioavailability. Biogeochemistry 43, 211–234.

Hopkinson, C. S., Giblin, A.E., Tucker, J., and Garritt, H. (2001) Benthic metabolism and nutrient regeneration on the continental shelf off eastern Massachusetts, USA. Mar. Ecol. Prog. Ser. 224, 1–19.

Hopner, T., and Wonneberger, K. (1985) Examination of the connection between the patchiness of benthic nutrient efflux and epiphytobenthos patchiness on intertidal flats. Neth. J. Sea. Res. 11, 14–23.

Hoppe, H.G. (1991) Microbial extracellular enzyme activity: a new key parameter in aquatic ecology. In Microbial Enzymes in Aquatic Environments (Chrost, R.J., ed.), pp. 60–83, Springer-Verlag, New York.

Hoppema, J.M.J. (1991) The seasonal behaviour of carbon dioxide and oxygen in the coastal North Sea along the Netherlands. Neth. J. Sea Res. 28, 167–179.

Hoppema, J.M.J., and Goeyens, L. (1999) Redfield behavior of carbon, nitrogen, and phosphorus depletions in Antarctic surface waters. Limnol. Oceanogr. 44, 220–224.

Hori, T., Horiguchi, M., and Hayashi, A. (1984) Biogeochemistry of Natural C–P Compounds. Maruzen, Shiga, Japan.

Horne, R.A. (1969) Marine Chemistry. The Structure of Water and the Chemistry of the Hydrosphere. Wiley-Interscience, New York.

Horrigan, S.G., Montoya, J.P., Nevins, J.L., and McCarthy, J.J. (1990) Natural isotopic composition of dissolved inorganic nitrogen in the Chesapeake Bay. Estuar. Coastal Shelf Sci. 30, 393–410.

Horton, R.E. (1933) The role of infiltration in the hydrologic cycle. Trans. Am. Geophys. Union 14, 446–460.

Horton, R.E. (1940) An approach toward a physical interpretation of infiltration capacity. Soil Sci. Soc. Am. 4, 399–417.

Hostettler, F.D., Rapp, J.B., Kvenvolden, K.A., and Luoma, S.N. (1989) Organic markers as source discriminants and sediment transport indicators in south San Francisco Bay, California. Geochim. Cosmochim. Acta 53, 1563–1576.

Houde, E.D., and Rutherford, E.S. (1993) Recent trends in estuarine fisheries: Predictions of fish production and yield. Estuaries 16, 161–176.

Houghton, J.T., Meiro-Filho, L.G., Bruce, J., Hoesung, L., Callander, B.A., Haites, E., Harris, N., and Maskell, K. (1995) Climate change 1994, radiative forcing of climate change and an evaluation of the IPCC IS92 emission scenarios: reports of working groups I and II of the international panel on climate change. Cambridge University Press, New York.

Howard, J.D., Remmer, G.H., and Jewitt, J.L. (1975) Hydrography and sediments of the Duplin River, Sapelo Island, Georgia. Senckenberg. Mar. 7, 237–256.

Howarth, R.W. (1979) Pyrite: Its rapid formation in a salt marsh and its importance in ecosystem metabolism. Science 203, 49–51.

Howarth, R.W. (1984) The ecological significance of sulfur in the energy dynamics of salt marsh and coastal sediments. Biogeochemistry 1, 5–27.

Howarth, R.W. (1993) Microbial processes in salt-marsh sediments. In An Ecological Approach (Ford, T.E., ed.), pp. 239–259, Blackwell Publishers, Cambridge, MA.

Howarth, R.W. (1998) An assessment of human influences on inputs of nitrogen to the estuaries and continental shelves of the North Atlantic Oceans. Nutr. Cycl. Agroecosyst. 52, 213–223.

Howarth, R.W., Billen, G., Swaney, D., Townsend, A., Jawarski, N., Lajtha, K., Downing, J.A., Elmgren, R., Caraco, N., Jordon, T., Berendse, F., Freney, J., Kudeyarov, V., Murdoch, P., and Zhao-ling, Z. (1996) Regional nitrogen budgets and riverine inputs of N & P for the drainages to the North Atlantic Ocean: natural and human influences. Biogeochemistry 35, 75–139.

Howarth, R.W., Chan, F., and Marino, R. (1999) Do top-down and bottom-up controls interact to exclude nitrogen-fixing cyanobacteria from the plankton of estuaries?: explorations with a simulation model. Biogeochemistry 46, 203–231.

Howarth, R.W., and Cole, J.J. (1985) Molybdenum availability, nitrogen limitation, and phytoplankton growth in natural waters. Science 229: 653–655.

Howarth, R.W., Fruci, J.R., and Sherman, D. (1991) Inputs of sediment and carbon to an estuarine ecosystem: influence of land use. Ecol. Appl. 1, 27–39.

Howarth, R.W., and Giblin, A.E. (1983) Sulfate reduction in the salt marshes at Sapelo Island, Georgia. Limnol. Oceanogr. 28, 70–82.

Howarth, R.W., Jaworski, N., Swaney, D., Townsend, A., and Billen, G. (2000) Some approaches for assessing human influences on fluxes of nitrogen and organic carbon to estuaries. In Estuarine Science: A Synthetic Approach to Research and Practice (Hobbie, J.E., ed.), pp. 17–42, Island Press, Washington, DC.

Howarth, R.W., Marino, R., and Cole, J.J. (1988b) Nitrogen fixation in freshwater, estuarine, and marine ecosystems. 2. Biogeochemical controls. Limnol. Oceanogr. 33, 688–701.

Howarth, R.W., Marino, R., Lane, R., and Cole, J.J. (1988a) Nitrogen fixation in freshwater, estuarine, and marine ecosystems. 1. Rates and importance. Limnol. Oceanogr. 33, 669–687.

Howarth, R.W., Sharpley, A., and Walker, D. (2002) Sources of nutrient pollution to coastal waters in the United States: Implications for achieving coastal water quality goals. Estuaries 25, 656–676.

Howarth, R.W., and Teal, J.M. (1979) Sulfate reduction in a New England salt marsh. Limnol. Oceanogr. 24, 999–1013.

Howes, B.L., Dacey, J.W.H., and King, G.M. (1984) Carbon flow through oxygen and sulfate reduction pathways in salt marsh sediments. Limnol. Oceanogr. 29, 1037–1051.

Howes, B.L., Dacey, J.W.H., and Wakeham, S.G. (1985) Effects of sampling technique on measurements of porewater constituents in salt marsh sediments. Limnol. Oceanogr. 30, 221–227.

Howes B. L., Howarth R. W., Teal J. M., and Valiela I. (1981) Oxidation–reduction potentials in a salt marsh. Spatial patterns and interactions with primary production. Limnol. Oceanogr. 26, 350–360.

Hsieh, Y., and Yang, C. (1997) Pyrite accumulation and sulfate depletion as affected by root distribution in a Juncus (needlerush) salt marsh. Estuaries 20, 640–645.

Huang, W.Y., and Meinschein, W.G. (1979) Sterols as ecological indicators. Geochim. Cosmochim. Acta 43, 739–745.

Huerta-Diaz, M.A., and Morse, J.W. (1990) A quantitative method for determination of trace metal concentrations in sedimentary pyrite. Mar. Chem. 29, 119–144.

Huettel, M., Forster, S., Kloser, S., and Fossing, H. (1996) Vertical migration in the sediment-dwelling sulfur bacteria Thiploca spp. in overcoming diffusion limitations. Appl. Environ. Microbiol. 62, 1863–1872.

Huettel, M., and Gust, G. (1992) Solute release mechanisms from confined sediment cores in stirred benthic chambers and flume flows. Mar. Ecol. Prog. Ser. 82, 187–197.

Hughes, E.H., and Sherr, E.B. (1983) Subtidal food webs in a Georgia estuary: δ^{13}C analysis. J. Exp. Mar. Biol. Ecol. 67, 227–242.

Hughes, F.W., and Rattray, M. (1980) Salt flux and mixing in the Columbia River estuary. Estuar. Coast. Mar. Sci. 10, 479–493.

Hughes, J.E., Deegan, L.A., Peterson, B.J., Holmes, R.M., and Fry, B. (2000) Nitrogen flow through the food web in the oligohaline zone of a New England estuary. Ecology 81, 433–452.

Huheey, J. E., (1983) Inorganic Chemistry, 3rd edn., Harper and Row, New York.

Hulth, S., Aller, R.C., and Gilbert, F. (1999) Coupled anoxic nitrification/manganese reduction in marine sediments. Geochim. Cosmochim. Acta 63, 49–66.

Hulthe, G., Hulth, S., and Hall, P.O.J. (1998) Effect of oxygen on degradation rate of refractory and labile organic matter in continental margin sediments. Geochim. Cosmochim. Acta 62, 1319–1328.

Humborg, C., Blomqvist, S., Avsan, E., Bergensund, Y., Smedberg, E., Brink, J., and Morth, C.M. (2002) Hydrological alterations with river damming in northern Sweden: implications for weathering and river biogeochemistry. Global Biogeochem. Cycles 16, 1039.

Humborg, C., Conley, D.J., Rahm, L., Wulff, F., Cociasu, A., and Ittekkot, V. (2000) Silica retention in river basins: far-reaching effects on biogeochemistry and aquatic food webs in coastal marine environments. Ambio 29, 45–50.

Humborg, C., Ittekot, V., Cociasu, A., and von Bodungen, B. (1997) Effect of Danube river on Black Sea biogeochemistry and ecosystem structure. Nature 386, 385–388.

Humborg, ,C., Smedberg, E., Blomqvist, S, Mörth, C., Brink, J., Rahm, L., Danielsson, A., and Sahlberg, J. (2004) Nutrient variations in boreal and subarctic Swedish Rivers: Landscape control of land–sea fluxes. Limnol. Oceanogr. 49, 1871–1883.

Hume, T.H., and Herdendorf, C.H. (1988) A geomorphic classification of estuaries and its application to coastal resource management—a New Zealand example. Ocean Shoreline Manag. 11, 249–274.

Hung, C., Guo, L., Santschi, P.H., Alvarado-Quiroz, N., and Haye, J.M. (2003a) Distributions of carbohydrate species in the Gulf of Mexico. Mar. Chem. 81, 119–135.

Hung, C., Guo, L.,Schultz, G.E., Pinckney, J.L., and Santschi, P.H. (2003b) Production and flux of carbohydrate species in the Gulf of Mexico. Global Biogeochem. Cycles 17, 24–1.

Hung, C., Tang, D., Warnken, K.W., and Santschi, P.H. (2001) Distributions of carbohydrates, including uronic acids, in estuarine waters of Galveston Bay. Mar. Chem. 73, 305–318.

Hupfer, M., and Gachter, R., and Ruegger, H. (1995) Polyphosphate in lake sediments: ^{31}P NMR spectroscopy as a tool for its identification. Limnol. Oceanogr. 40, 610–617.

Hupfer, M., Rube, B., and Schmieder, P. (2004) Origin and diagenesis of polyphosphate in lake sediments: a^{31}P-NMR study. Limnol. Oceanogr. 49, 1–10.

Hurd, D.C. (1983) Physical and chemical properties of the siliceous skeletons. In Silicon Geochemistry and Biogeochemistry (Aston, S.R., ed.), pp. 187–244, Academic Press, New York.

Hutchinson, G.E. (1938) On the relation between the oxygen deficit and the productivity and typology of lakes. Intl. Rev. Gesmaten Hydrobiol. 36, 336–355.

Hutchinson, G.E. (1957) A Treatise on Limnology. 1. Geography, Physics, and Chemistry. John Wiley, New York.

Incze, L.S., Mayer, L.M., Sherr, E.B., and Macko, S.A. (1982) Carbon inputs to bivalve mollusks: A comparison of two estuaries. Can. J. Fish. Aquat. Sci. 39, 1348–1352.

Ingall, E.D., and Jahnke, R. (1994) Evidence for enhanced phosphate regeneration from marine sediments overlain by oxygen depleted waters. Geochim. Cosmochim. Acta 58, 2571–2575.

Ingall, E.D., and Jahnke, R. (1997) Influence of water column anoxia on the elemental fractionation of carbon and phosphorus during sediment diagenesis. Mar. Geol. 139, 219–229.

Ingall, E.D., Schroeder, P.A., and Berner, R.A. (1990) The nature of organic phosphorus in marine sediments: new insights from ^{31}P NMR. Geochim. Cosmochim. Acta 54, 2617–2620.

Ingall, E.D., and van Cappellen, P. (1990) Relation between sedimentation rate and burial of organic phosphorus and organic carbon in marine sediments. Geochim. Cosmochim. Acta 54, 373–386.

Ingalls, A.E., Lee, C., Wakeham, S.G., and Hedges, J.I. (2003) The role of biominerals in the sinking flux and preservation of amino acids in the Southern Ocean along 170° W. Deep-Sea Res. II 50, 713–738.

Ingram, R.G., and El-Sabh, M.I. (1990) Fronts and mesoscale features in the St. Lawrence Estuary. *In* Oceanography of a Large-Scale Estuarine System, The St. Lawrence (El-Sabh, M.I., and Silverberg, N, eds.), pp. 71–93, Springer-Verlag, New York.

Isla, F.I. (1995) Coastal lagoons. *In* Geomorphology and Sedimentology of Estuaries. Developments in Sedimentology 53 (Perillo, G.M.E., ed.), pp. 241–272, Elsevier Science, New York.

Ittekkot, V., Degens, E.T., and Brockmann, U. (1982) Monosaccharide composition of acid-hydrolyzable carbohydrates in particulate matter during a plankton bloom. Limnol. Oceanogr. 27, 770–776.

Ittekkot, V., Humborg, C., and Schaefer, P. (2000) Hydrlogical alterations and marine biochemistry: a silicate issue? Bioscience 50, 776–782.

Iverson, N., and Jørgensen, B.B. (1993) Diffusion coefficients of sulfate and methane in marine sediments: Influence of porosity. Geochim. Cosmochim. Acta 57, 571–578.

Iverson, R.L., Nearhof, F.L., and Andreae, M.O. (1989) Production of dimethylsulfonium proprionate and dimethylsulfide by phytoplankton in estuarine and coastal waters. Limnol. Oceanogr. 34, 53–67.

Jaakkola, T., Tolonen, K., Huttunen, P., and Leskinen, S. (1983) The use of fallout ^{137}Cs and ^{239,240}Pu for dating of lake sediments. Hydrobiologia 109, 15–19.

Jackson, G.A., and Williams, P.W. (1985) Importance of dissolved organic nitrogen and phosphorus to biological nutrient cycling. Deep-Sea Res. 32, 223–235.

Jacobsen, B.S., Smith, B.N., Epstein, S., and Laties, G.G. (1970) The prevalence of carbon-13 in respiratory carbon dioxide as an indicator of the type of endogenous substrate. J. Gen. Physiol. 55, 1–17.

Jacobsen, R., and Postma, D. (1999) Redox zoning, rates of sulfate reduction and interactions with Fe-reduction and methanogenesis in a shallow sandy aquifer, Romo, Denmark. Geochim. Cosmochim. Acta 63, 137–151.

Jaeger, J.M., and Nittrouer, C.A. (1995) Tidal controls on the formation of fine-scale sedimentary strata near the Amazon River mouth. Mar. Geol. 125, 259–282.

Jaffe, D.A. (1992) The nitrogen cycle. *In* Global Biogeochemical Cycles (Butcher, S.S., Charlson, R.J., Orians, G.H., and Wolfe, G.U., eds.), pp. 263–284, Academic Press, New York.

Jaffe, D.A. (2000) The nitrogen cycle. *In* Earth System Science—from Biogeochemical Cycles to Global Change (Jacobson, M.C., Charlson, R.J., Rodhe, H., and Orians, G.H., eds.), pp. 322–342, Academic Press, New York.

Jaffé, R., Boyer, J.N., Lu, X., Maie, N., Yang, C., Scully, N.M., and Mock, S. (2004) Source characterization of dissolved organic matter in a subtropical mangrove-dominated estuary by fluorescence analysis. Mar. Chem. 84, 195–210.

Jaffé, R., Mead, R., Hernandez, M.E., Peralba, M.C., and DiGuida, O.A. (2001) Origin and transport of sedimentary organic matter in two subtropical estuaries: a comparative, biomarker-based study. Org. Geochem. 32, 507–526.

Jaffé, R., Wolff, G.A., Cabrera, A.C., and Carvajal-Chitty, H. (1995) The biogeochemistry of lipids in rivers from the Orinoco basin. Geochim. Cosmochim. Acta 59, 4507–4522.

Jahnke, R.A. (2000) The phosphorus cycle. *In* Earth System Science—from Biogeochemical Cycles to Global Change (Jacobson, M.C., Charlson, R.J., Rodhe, H., and Orians, G.H., eds.), pp. 360–376, Academic Press, New York.

Jahnke, R.A., Emerson, S.R., Roe, K.K., and Burnett, W.C. (1983) The present day formation of apatite in Mexican continental margin sediments. Geochim. Cosmochim. Acta 47, 259–266.

Jahnke, R.A., Heggie, D., Emerson, S., and Graham, D. (1982) Pore waters of the central Pacific Ocean: nutrient results. Earth Planet. Sci. Lett. 61, 233–256.

Jahnke, R.A., Nelson, J.R., Marinelli, R.L., and Eckman, J.E. (2000) Benthic flux of biogenic elements on the southeastern U.S. continental shelf: influence of porewater advective transport and benthic microalgae. Cont. Shelf Res. 20, 109–127.

Jakobsen, R., and Postma, D. (1999) Redox zoning, rates of sulfate reduction and interactions with Fe-reduction and methanogenesis in a shallow sandy aquifer, Romo, Denmark. Geochim. Cosmochim. Acta 63, 137–151.

Jannasch, H.W. (1960) Denitrification as influenced by photosynthetic oxygen production. J. Gen. Microbiol. 23, 55–63.

Jansson, M. (1998) Degradation of dissolved organic matter in humic waters by bacteria. *In* Aquatic Humic Substances Ecology and Biogeochemistry (Hessen, D.O., and Tranvik, L.J., eds.), pp. 177–196, Springer-Verlag, Berlin.

Janus, L.L., and Vollenweider, R.A. (1984) Phosphorus residence time in relation to trophic conditions in lakes. Verh. Intl. Verein. Theor. Angew. Limnol. 22, 179–184.

Janzen, D.H. (1974) Tropical blackwater rivers, animals, and mast fruiting by the Dipterocarpaceae. Biotropica 6, 69–103.

Jarman, W.M., Hilkert, A., Bacon, C.E., Collister, J.W., Ballschmiter, K., and Risebrough, R.W. (1998) Compound-specific carbon isotopic analysis of Aroclors, Clophens, Kaneclors, and Phenoclors. Environ. Sci. Technol. 32, 833–836.

Jasper, J.P., and Hayes, J.M. (1990) A carbon-isotope record of CO_2 levels during the late Quaternary. Nature 347, 462–464.

Jay, D.A., Rockwell, W.R., and Montgomery, D.R. (2000) An ecological perspective on estuarine classification. *In* Estuarine Science: A Synthetic Approach to Research and Practice (Hobbie, J.E., ed.), pp. 149–176, Island Press, Washington, DC.

Jay, D. A., and Smith, J.D. (1988) Circulation in and classification of shallow, stratified estuaries. *In* Physical Processes in Estuaries (Dronkers, J., and van Leussen, W.), pp. 21–41, Springer-Verlag, Berlin.

Jay, D.A., Uncles, R.J., Largier, J., Geyer, W.R., Vallino, J., and Boynton, W.R. (1997) A review of recent developments in estuarine scalar flux estimation. Estuaries 20, 262–280.

Jeffrey, S.W. (1974) Profiles of photosynthetic pigments in the ocean using thin-layer chromatography. Mar. Biol. 26, 101–110.

Jeffrey, S.W. (1976a) The occurrence of chlorophyll c_1 and c_2 in algae. J. Phycol. 12, 349–354.

Jeffrey, S.W. (1976b) A report on green algal pigments in the central North Pacific Ocean. Mar. Biol. 37, 33–37.

Jeffrey, S.W. (1989) Chlorophyll c pigments and their distribution in the chromophyte algae. *In* The Chromophyte Algae: Problems and Perspectives (Green J.C., Leadbeater, B.S.C., and Diver, W.L., eds.), pp. 13–36, Clarendon Press, Oxford, UK.

Jeffrey, S.W. (1997) Application of pigment methods to oceanography. *In* Phytoplankton Pigments in Oceanography (Jeffrey, S.W., Mantoura, R.F.C., and Wright, S.W., eds.), pp. 127–178, UNESCO Publishing, Paris.

Jeffrey, S.W., and Hallegraeff, G.M. (1987) Phytoplankton pigments, species and photosynthetic pigments in a warm-core eddy of the East Australian Current. I. Summer populations. Mar. Ecol. Prog. Ser. 3, 285–294.

Jeffrey, S.W., and Mantoura, R.F.C. (1997) Development of pigment methods for oceanography: SCOR-supported working groups and objectives. *In* Phytoplankton Pigments in Oceanography (Jeffrey, S.W., Mantoura, R.F.C., and Wright, S.W., eds.), pp. 19–13, UNESCO Publishing, Paris.

Jeffrey, S.W., Sielicki, M., and Haxo, F.T. (1975) Chloroplast pigment patterns in dinoflagellates. J. Phycol. 11, 374–384.

Jeffrey, S.W., and Wright, S.W. (1987) A new spectrally distinct component in preparations of chlorophyll *c* from the microalga *Emiliana huxleyi* (Prymnesiophyceae). Biochim. Biophys. Acta 894, 180–188.

Jeffrey, S.W., and Wright, S.W. (1994) Photosynthetic pigments in the Haptophyta. *In* The Haptophyte Algae (Green, J.C., and Leadbeater, B.S.C., eds.), pp. 111–132, Clarendon Press, Oxford, UK.

Jenkins, M.C., and Kemp, W.M. (1984) The coupling of nitrification and denitrification in two estuarine sediments. Limnol. Oceanogr. 29: 609–619.

Jensen, H.S., Mortensen, P.B., Andersen, F.O., Rasmussen, E., and Jensen, A. (1995) Phosphorus cycling in a coastal marine sediment, Aarhus Bay, Denmark. Limnol. Oceanogr. 40, 908–917.

Jensen, H.S., and Thamdrup, B. (1993) Iron-bound phosphorus in marine sediments as measured by bicarbonate–dithionite extraction. Hydrobiologia 252, 47–59.

Jensen, K., Sloth, N.P., Rysgaard-Petersen, N., Rysgaard, S., and Revsbech, N.P. (1994) Estimation of nitrification and denitrification from microprofiles of oxygen and nitrate in model sediments. Appl. Environ. Microbiol. 60, 2094–2100.

Jensen, S., Johnels, A.G., Olsson, M., and Otterlind, G. (1969) DDT and PCB in marine animals from Swedish waters. Nature 224, 247–250.

Johannes, R. (1980) The ecological significance of the submarine discharge of ground water. Mar. Ecol. Prog. Ser. 3, 365–373.

Johansen, J.E., Svec, W.A., Liaaen-Jensen, S., and Haxo, F.T. (1974) Carotenoids of the Dinophyceae. Phytochemistry 13, 2261–2271.

Johnsen, G., and Sakshaug, E. (1993) Bio-optical characteristics and photoadaptive responses in the toxic and bloom-forming dinoflagellates *Gymnodinium aureolum, G. galatheanum,* and two strains of *Prorocentrum minimum.* J. Phycol. 29, 627–642.

Johnstone, J. (1908) Nitrate flux into the euphotic zone near Bermuda. Nature 331, 521–523.

Jones, J.B., and Mulholland, P.J. (1998) Influence of drainage basin topography and elevation on carbon dioxide and methane supersaturation of stream water. Biogeochemistry 40, 57–72.

Jones, R.D., and Amador, J.A. (1993) Methane and carbon monoxide production, oxidation and turnover times in the Caribbean Sea as influenced by the Orinoco river. J. Geophys. Res. 98, 2353–2359.

Jonsson, P. (2000) Sediment burial of PCBs in the offshore Baltic Sea. Ambio 29, 260–267.

Jonsson, P. Carman, R., and Wulff, F. (1990) Laminated sediments in the Baltic—a tool for evaluating nutrient mass balances. Ambio 19, 152–158.

Jordan, T.E., Correll, D.L., and Weller, D.E. (1997) Relating nutrient discharges from watersheds to land use and stream flow variability. Wat. Resource. Res. 33, 2579–2590.

Jørgensen, B.B. (1977) The sulfur cycle of coastal marine sediment (Limfjorden, Denmark). Limnol. Oceanogr. 28, 814–822.

Jørgensen, B.B. (1978) A comparison of methods for the quantification of bacterial sulfate reduction in coastal marine sediments. 1. Measurements with radiotracer techniques. Geomicrobiol. J. 1, 11–27.

Jørgensen, B.B. (1979) A theoretical model of the stable sulfur isotope distribution in marine sediments. Geochim. Cosmochim. Acta 43, 363–374.

Jørgensen, B.B. (1982) Mineralization of organic matter in the sea—the role of sulfate reduction. Nature 296, 643–645.

Jørgensen, B.B. (1989) Biogeochemistry of chemoautotrophic bacteria. In Autotrophic Bacteria (Shlegel, H.G., and Bowien, B., eds.), pp. 117–146, Science Technical Publishers, Madison, W1ø and Springer-Verlag, New York.

Jørgensen, B.B. (1996) Material flux in the sediments. In Eutrophication in Coastal Marine Ecosystems (Jørgensen, B.B., and Richardson, K., eds.), pp. 115–135, American Geophysical Union, Washington, DC.

Jørgensen, B.B. (2000) Bacteria and marine biogeochemistry. In Marine Geochemistry, (Schulz, H.D., and Zabel, M., eds.), pp. 173–207, Springer-Verlag, Berlin.

Jørgensen, B.B., Bang, M., and Blackburn, T.H. (1990) Anaerobic mineralization in marine sediments from the Baltic Sea–North sea transition. Mar. Ecol. Prog. Ser. 59, 55–61.

Jørgensen, B.B., and Boudreau, B.P. (2001) Diagenesis and sediment–water exchange. In The Benthic Boundary Layer (Boudreau, B.P., and Jørgensen, B.B., eds.), pp. 211–244, Oxford University Press, New York.

Jørgensen, B.B., and Des Marais, D.J. (1986) Competition for sulfide among colorless and purple sulfur bacteria in cyanobacterial mats. FEMS Microbiol. Ecol. 38, 179–186.

Jørgensen, B.B., and Okholm-Hansen, B. (1985) Emissions of biogenic sulfur gases from a Danish estuary. Atmos. Environ. 19, 1737–1749.

Jørgensen, B.B., and Revsbech, N.P. (1983) Colorless sulfur bacteria, Beggiatoa spp. and Thiovulum spp., on O_2 and H_2S microgradients. Appl. Environ. Microbiol. 45, 1261–1270.

Jørgensen, B.B., and Revsbech, N.P. (1985) Diffusive boundary layer and the oxygen uptake of sediments and detritus. Limnol. Oceanogr. 30, 111–122.

Jørgensen, B.B., and Richardson, K. (1996) Eutrophication in Coastal Marine Systems. American Geophysical Union, Washington, DC.

Jørgensen, K.S. (1989) Annual pattern of denitrification and nitrate ammonification in estuarine sediment. Appl. Environ. Microbiol. 55, 1841–1847.

Jørgensen, N.O.G. (1984) Microbial activity in the water–sediment interface: assimilation and production of dissolved free amino acids. Oceanus 10, 347–365.

Jørgensen, N.O.G. (1987) Free amino acids in lakes: Concentrations and assimilation rates in relation to phytoplankton and bacterial production. Limnol. Oceanogr. 32, 97–111.

Jørgensen, N.O.G., Kroer, N., Coffin, R.B., and Hoch, M.P. (1999) Relations between bacterial nitrogen metabolism and growth efficiency in an estuarine and an open water ecosystem. Aquat. Microb. Ecol. 18, 247–261.

Jørgensen, N.O.G., Kroer, N., Coffin, R.B., Yang, X.H., and Lee, C. (1993) Dissolved free amino acids, combined amino acids, and DNA as sources of carbon and nitrogen to marine bacteria. Mar. Ecol. Prog. Ser. 98, 135–148.

Jørgensen, N.O.G., Tranvik, L.J., Edling, H., Graneli, W., and Lindell, M. (1998) Effects of sunlight on occurrence and bacterial turnover of specific carbon and nitrogen compounds in lake water. FEMS Microb. Ecol. 25, 217–227.

Josefson, A.B., and Rasmussen, B. (2000) Nutrient retention by benthic macrofaunal biomass of Danish estuaries: importance of nutrient load and residence time. Estuar. Coastal Shelf Sci. 50, 205–216.

Joye, S.B., and Hollibaugh, J.T (1995) Sulfide inhibition of nitrification influences nitrogen regeneration in sediments. Science 270, 623–625.

Joye, S.B., and Paerl, H.W. (1993) Contemporaneous nitrogen fixation and denitrification in marine microbial mats: rapid response to runoff events. Mar. Ecol. Prog. Ser. 94, 267–274.

Joye, S.B., and Paerl, H.W. (1994) Nitrogen cycling in marine microbial mats: rates and patterns of denitrification nitrogen fixation. Mar. Biol. 119, 285–295.

Juday, C., Birge, E.A., Kemmerer, G.I., and Robinson, R.J. (1927) Phosphorus content of lake waters in northwestern Wisconsin. Trans. Wis. Acad. Arts. Lett. 23, 233–248.

Jumars, P.A., Mayer, L.M., Deming, J.W., Baross, J.A., and Wheatcroft, R.A. (1990) Deep-sea deposit-feeding strategies suggested by environmental and feeding constraints. Trans. Royal Soc. London 331, 85–102.

Junge, C.E. (1974) Residence time and variablility of troposheric trace gases. Tellus 26, 477–488.

Kadlec, R.H. (1990) Overload flow in wetlands: vegetation resistance. J. Hydraul. Eng. 116, 691–706.

Kahru, M. (1997) Using satellites to monitor large-scale environment change: a case study of cynaobacterial bloom in the Baltic Sea. In Monitoring Algal Bloom: New Techniques for Detecting Large-Scale Environmental Change (Kahru, M., and Brown, C.W., eds.), pp. 43–61, Landes Bioscience, Georgetown, TX.

Kahru, M., Horstmann, U., and Rud, O. (1994) Satellite detection of increased cyanobacteria in the Baltic Sea: natural fluctuation or ecosystem change? Ambio 23, 469–472.

Kahru, M., and Mitchell, B.G. (2001) Seasonal and nonseasonal variability of satellite-derived chlorophyll and dissolved organic matter concentration in the California Current. J. Geophys. Res. 106, 2517–2529.

Kaldy, J.F., Onuf, C.P., Eldridge, P.M., and Cifuentes, L.A. (2002) Carbon budget for a subtropical dominated coastal lagoon: how important are seagrasses to total ecosystem net primary production? Estuaries 25, 528–539.

Kamatani, A. (1982) Dissolution rates of silica from diatoms decomposing at various temperatures. Mar. Biol. 68, 91–96.

Kamen, M.D. (1963) Early history of carbon-14. Science 140, 584–590.

Kamer, K., Boyle, K.A., and Fong, P. (2001) Macroalgal bloom dynamics in a highly eutrophic southern California estuary. Estuaries 24, 623–635.

Kaneda, T. (1991) Iso- and anteiso-fatty acids in bacteria: biosynthesis, function, and taxonomic significance. Microbiol. Rev. 288–302.

Kanneworff, E., and Christensen, H. (1986) Benthic community respiration in relation to sedimentation of phytoplankton in the Oresund. Ophelia 26, 269–284.

Kaplan, I.R., Emer, K.O., and Rittenberg, S.C. (1963) The distribution and isotopic abundance of sulfur in recent marine sediments of Southern California. Geochim. Cosmochim. Acta 27, 297–331.

Kaplan, I.R., and Rittenberg, S.C. (1964) Microbiological fractionation of sulfur isotopes. J. Microbiol. 34, 195–212.

Kappenberg, J., and Grabemann, I. (2001) Variability of the mixing zones and estuarine turbidity maxima in the Elbe and Weser Estuaries. Estuaries 24, 699–706.

Kapralek, F., Jechova, E., and Otavova, M. (1982) Two sites of oxygen control in induced synthesis of respiratory nitrate reductase in Escherichia coli. J. Bacteriol. 149, 1142–1145.

Karickhoff, S.W. (1984) Organic sorption in aquatic systems. J. Hydraulic Eng. 110, 707–735.

Karickhoff, S.W., Brown, B.S., and Scott, T.A. (1979) Sorption of hydrophobic pollutants on natural sediments. Wat. Res. 13, 241–248.

Karl, D.M., Tien, G., Dore, J., and Winn, C.D. (1993) Total dissolved nitrogen and phosphorus concentrations at United States-JGOFS station ALOHA—Redefield reconciliation. Mar. Chem. 41, 203–208.

Kaufman, A., Li, Y.H., and Turekian, K.K. (1981) The removal rates of ^{234}Th and ^{228}Th from waters of the New York Bight. Earth Planet. Sci. Lett. 54, 385–392.

Kawamura, K., Ishiwatari, R., and Ogura, K. (1987) Early diagenesis of organic matter in the water column and sediments: microbial degradation and resynthesis of lipids in Lake Haruna. Org. Geochem. 11, 251–264.

Keefe, C.W. (1994) The contribution of inorganic compounds to the particulate carbon, nitrogen, and phosphorus in suspended matter and surface sediments of Chesapeake Bay. Estuaries 17, 122–130.

Keeling, C.D. (1973) Industrial production of carbon dioxide from fossil fuels and limestone. Tellus 25, 174–198.

Keely, B.J., and Maxwell, J.R. (1991) Structural characterization of the major chlorins in recent sediments. Org. Geochem. 17, 663–669.

Keil, R., and Kirchman, D. (1991a) Contribution of dissolved free amino acids and ammonium to the nitrogen requirements of heterotrophic bacterioplankton. Mar. Ecol. Prog. Ser. 72, 1–10.

Keil, R., and Kirchman, D. (1991b) Dissolved combined amino acids in marine waters as determined by vapor-phase hydrolysis method. Mar. Chem. 33, 243–259.

Keil, R.G., and Kirchman, D. (1993) Dissolved combined amino acids: chemical form and utilization by marine bacteria. Limnol. Oceanogr. 38, 1256–1270.

Keil, R.G., Mayer, L.M., Quay, P.D., Richey, J.E., and Hedges, J.I. (1997) Loss of organic matter from riverine particles in deltas. Geochim. Cosmochim. Acta 61, 1507–1511.

Keil, R.G., Montlucon, D.B., Prahl, F.G., and Hedges, J.I. (1994a) Sorptive preservation of labile organic matter in marine sediments. Nature 370, 549–552.

Keil, R.G., Tsamakis, E, Fuh, C.B., Giddings, J.C., and Hedges, J.I. (1994b) Mineralogical and textural controls on the organic composition of coastal marine sediments: hydrodynamic separation using SPLITT-fractionation. Geochim. Cosmochim. Acta 58, 879–893.

Keil, R.G., Tsamakis, E., Giddings, J.C. and Hedges, J.I. (1998) Biochemical distributions (amino acids, neutral sugars, and lignin phenols) among size-classes of modern marine sediments from the Washington coast. Geochim. Cosmochim. Acta 62, 1347–1364.

Keil, R.G., Tsamakis, E., Hedges, J.I. (2000) Early diagenesis or particulate amino acids in marine sediments. In Perspectives in Amino Acid and Protein Chemistry (Goodfriend, G.A., Collins, M.J., Fogel., M.L., Macko, S.A., and Wehmiller, J.F., eds.), pp. 69–82, Oxford University Press, New York.

Keith, D.J., Yoder, J.A., and Freeman, S.A. (2002) Spatial and temporal distribution of colored dissolved organic matter (CDOM) in Narragansett Bay, Rhode Island: Implications for phytoplankton in coastal waters. Estuar. Coastal Shelf Sci. 55, 705–717.

Keith, S.C., and Arnosti, C. (2001) Extracellular enzyme activity in a river-bay-shelf transect: variations in polysaccharide hydrolysis rates with substrate and size class. Aquat. Microb. Ecol. 24, 243–253.

Kelley, C.A., Martens, C.S., and Chanton, J.P. (1990) Variations in sedimentary carbon remineralization rates in the White Oak River estuary, N.C. Limnol. Oceanogr. 35, 372–383.

Kelley, C.A., Marten, C.S., and Ussler III, W. (1995) Methane dynamics across a tidally flooded riverbank margin. Limnol. Oceanogr. 40, 1112–1129.

Kelly, J.R., and Nixon, S.W. (1984) Experimental studies of the effect of organic deposition on the metabolism of a coastal marine bottom community. Mar. Ecol. Prog. Ser. 17, 157–169.

Kelly, J.R., and Nowicki, B.L. (1992) Sediment denitrification in Boston harbor. MWRA Technical Report 92–2.

Kelly, R.P., and Moran, S.B. (2002) Seasonal changes in groundwater input to a well-mixed estuary estimated using radium isotopes and implications for coastal nutrient budgets. Limnol Oceanogr. 47, 1796–1807.

Kemp, G.P. (1986) Mud deposition at the shoreface: wave and sediment dynamics in the chenier plain of Louisiana. Ph.D Dissertation, Louisiana State University, Baton Rouge, pp. 147.

Kemp, W.M., and Boynton, W.R. (1980) Influence of biological and physical processes on dissolved oxygen dynamics in an estuarine system: implications for measurement of community metabolism. Estuar. Coastal Shelf Sci. 11, 407–431.

Kemp, W.M., and Boynton, W.R. (1981) External and internal factors regulating metabolic rates of an estuarine benthic community. Oecologia 51, 19–27.

Kemp, W.M., and Boynton, W.R. (1984) Spatial and temporal coupling of nutrient inputs to estuarine primary production: the role of particulate transport and decomposition. Bull. Mar. Sci. 35, 522–535.

Kemp, W.M., Sampou, P., Caffrey, J., Mayer, M., Henriksen, K., and Boynton, W.R. (1990) Ammonium recycling versus denitrification in Chesapeake Bay sediments. Limnol. Oceanogr. 35, 1545–1563.

Kemp, W.M., Sampou, P., Garber, J., Tuttle, J., and Boynton, W.R. (1992) Seasonal depletion of oxygen from bottom waters of Chesapeake Bay—roles of benthic and planktonic respiration and physical exchange processes. Mar. Ecol. Prog. Ser. 85, 137–152.

Kemp, W.M., Smith, E.M., Marvin-DiPasquale, M., and Boynton, W.R. (1997) Organic carbon balance and NEM in Chesapeake Bay. Mar. Ecol. Prog. Ser. 150, 229–248.

Kemp, W.M., Wetzel, R., Boynton, W., D'Elia, C., and Stevenson, J. (1982) Nitrogen cycling and estuarine interfaces: some current research directions. In Estuarine Interactions (Kennedy, V., ed.), pp. 209–230, Academic Press, New York.

Kempe, S. (1982) Valdivia cruise, October 1981: carbonate equilibria in the estuaries of Elbe, Weser, Ems, and in the southern German Bight. Mitt. Geol. Paleont. Inst. Univ. Hamburg, SCOPE/UNEP, Sonderband 52, 719–742.

Kempe, S. (1984) Sinks of the anthropogenically enhanced carbon cycle in surface fresh waters. J. Geophys. Res. 89, 4657–4676.

Kempe, S. (1990) Alkalinity: the link between anaerobic basins and shallow water carbonates? Naturwiaaenschaffer 7, 426–427.

Kempe, S., Pettine, M., and Cauwet, G. (1991) Biogeochemistry of Europe rivers. In Biogeochemistry of Major World Rivers (Degens, E.T., Kempe, S., and Richey, J.E., eds.), pp. 169–211, John Wiley, New York.

Kendall, C. (1998) Tracing nitrogen sources and cycling in catchments. In Isotope tracers in catchment hydrology (Kendall, C., and McDonnell, J.J., eds.), pp. 534–569, Elsevier, New York.

Kendall, C., and Coplen, T.B. (2001) Distribution of oxygen-18 and deuterium in river waters across the USA. In Water Quality of Large U.S. Rivers: Results from the U.S. Geological Survey's National Stream Quality Accounting Network (Hooper, R.P., and Kelly, V.P., eds.), pp. 1361–1393, John Wiley, New York.

Kenne, L., and Linberg, B. (1983) Bacterial polysaccharides. In The Polysaccharides (Aspinall, G.O., ed.), pp. 287–363, Academic Press, New York.

Kennedy, H.A., and Elderfield, H. (1987) Iodine diagenesis in pelagic deep-sea sediments. Geochim. Cosmochim. Acta 51, 2489–2504.

Kennicutt II, M.C., Bidigare, R.R., Macko, S.A., and Keeney-Kennicutt, W.L. (1992) The stable isotopic composition of photosynthetic pigments and related biochemicals. Chem. Geol. 101, 235–245.

Kennicutt II, M.C., and Comet, P.A. (1992) research of sediment hydrocarbon sources: multiparameter approaches. *In* Organic Matter: Productivity, Accumulation, and Preservation in Recent and Ancient Sediments (Whelan, J.K., and Farrington, J.K., eds.), pp. 308–338, Columbia University Press, New York.

Kennish, M.J. (1986) Ecology of Estuaries. Vol. 1: Physical and Chemical Aspects. CRC Press, Boca Raton, FL.

Kennish, M.J. (1992) Ecology of Estuaries: Anthropogenic Effects. CRC Press, Boca Raton, FL.

Kennish, M.J. (1997) Practical handbook of Estuarine and Marine Pollution. CRC Press, Boca Raton, FL.

Kerner, M., and Spitzy, A. (2001) Nitrate regeneration coupled to degradation of different size fractions of DON by the picoplankton in the Elbe estuary. Microb. Ecol. 41, 69–81.

Kester, D.R. (1975) Dissolved gases other than CO_2. *In* Chemical Oceanography, 2nd edn. (Riley, J.P., and Skirrow, G., eds.), pp. 497–556, Academic Press, New York.

Ketchum, B.H. (ed.) (1983) Estuarine characteristics. *In* Estuaries and Enclosed Seas, pp. 1–13, Elsevier, New York.

Key, R., Stallard, R.F., Moore, W.S., and Sarmiento, J.L. (1985) Distribution and flux of [226]Ra and [228]Ra in the Amazon River estuary. J. Geophys. Res. 90, 6995–7004.

Khalil, M.A., and Rasmussen, R.A. (1992) The global sources of nitrous oxide. J. Geophys. Res. 97, 14651–14660.

Khalili, A., Huettel, M., and Merzkirch, W. (2001) Fine-scale flow measurements in the benthic boundary layer. *In* The Benthic Boundary Layer: Transport Processes and Biogeochemistry (Boudreau, B.P., and Jørgensen, B.B., eds.), pp. 44–77, Oxford University Press, New York.

Kieber, R.J., Jiao, J., Kiene, R.P., and Bates, T.S. (1996) Impact of dimethylsulfide photochemistry on methyl sulfur cycling in the Equatorial Pacific Ocean. J. Geophys. Res. 101, 3715–3722.

Kieber, R.J., Li, A., and Seaton, P.J. (1999) Production of nitrite from the photodegradation of dissolved organic matter in natural waters. Environ. Sci. Technol. 33, 993–998.

Kieber, R.J., Zhou, X., and Mopper, K. (1990) formation of carbonyl compounds from UV-induced photodegradation of humic substances in natural waters: fate of riverine carbon in the sea. Limnol. Oceanogr. 35, 1503–1515.

Kiene, R.P. (1990) Dimethyl sulfide production from dimethylsulfoniopropionate in coastal seawater samples and bacterial cultures. Appl. Environ. Microbiol. 56, 3292–3297.

Kiene, R.P., and Linn, L. (2000) The fate dissolved dimethylsulfoniopropionate (DMSP) in seawater: Tracer studies using ^{35}S-DMSP. Geochim. Cosmochim. Acta 64, 2797–2810.

Kiene, R.P., Linn, L.J., and Bruton, J.A. (2000) New and important roles for DMSP in marine microbial communities. J. Sea Res. 43, 209–224.

Kiene, R.P., and Linn, L.J. (2000) The fate of dissolved dimethylsulfopropionate (DMSP) in seawater: tracer studies using ^{35}S-DMSP. Geochim. Cosmochim. Acta 64, 2797–2810.

Kikuchi, Y., Mochida, Y., Miyagi, T., Fujimoto, K:, and Tsuda, S. (1999) Mangrove forests supported by peaty habitats on several islands in the Western Pacific. Tropics 8, 197–205.

Kilham, P. (1971) A hypothesis concerning silica and freshwater planktonic diatoms. Limnol. Oceanogr. 16, 10–18.

Kilham, S.S., and Kilham, P. (1984) The importance of resource supply rates in determining phytoplankton community structure. *In* Trophic Interactions Within Aquatic Systems (Meyers, D.G., and Strickler, J.R., eds.), pp. 7–28, Westview Press, Boulder, CO.

Kim, K. H., and Andreae, M. O. (1987) Carbon disulfide in seawater and the marine atmosphere over the North Atlantic. J. Geophys. Res. 92, 14733–14738.

Kineke, G.C., and Sternberg, R.W. (1989) The effects of particle settling velocity on computed suspended-sediment concentration profiles. Mar. Geol. 90, 159–174.

Kineke, G.C., and Sternberg, R.W. (1995) Distribution of fluid muds on the Amazon continental shelf. Mar. Geol. 125, 193–233.

Kineke, G.C., Sternberg, R.W., Trowbridge, J.H., and Geyer, W.R. (1996) Fluid-mud processes on the Amazon continental shelf. Cont. Shelf Res. 16, 667–696.

King, G.M. (1983) Sulfate reduction in Georgia salt marsh soils: an evaluation of pyrite formation by use of ^{35}S and ^{55}Fe tracers. Limnol. Oceanogr. 28, 987–995.

King, G.M. (1988) Patterns of sulfate reduction and the sulfur cycle in a South Carolina salt marsh. Limnol. Oceanogr. 33, 376–390.

King, G.M. (1990) Regulation by light of methane emissions from a wetland. Nature 345, 513–515.

King, G.M., Howes, B.L., and Dacey, J.W.H. (1985) Short-term endproducts of sulfate reduction in a salt marsh: the significance of acid volatile sulfide, elemental sulfur, and pyrite. Geochim. Cosmochim. Acta 49, 1561–1566.

King, G.M., Klug, M.J., Wiegert, R.G., and Chalmers, A.G. (1982) Relation of soil water movement and sulfide concentration to *Spartina alterniflora* production. Science 218, 61–63.

King, J., Kostka, J., Frischer, M., and Saunders, F. (2000) Sulfate-reducing bacteria methylate mercury at variable rates in pure cultures and in marine sediments. Appl. Environ. Microbiol. 66, 2430–2437.

King, K., and Hare, P.E. (1972) Amino acid composition of planktonic foraminifera: A paleobiochemical approach to evolution. Science 175, 1461–1463.

King, L.L., and Repeta, D.J. (1994) Novel pyropheophorbide steryl esters in Black Sea sediments. Geochim. Cosmochim. Acta 55, 2067–2074.

Kipphut, G.W., and Martens, C.S. (1982) Biogeochemical cycling in an organic-rich coastal marine basin, 3, Dissolved gas transport in methane saturated sediments. Geochim. Cosmochim. Acta 46, 2049–2060.

Kirby, C.J., and Gosselink, J.G. (1976) Primary production in a Louisiana Gulf Coast *Spartina alterniflora* marsh. Ecology 57,1072–1059.

Kirchman, D.L. (1994) The uptake of inorganic nutrients by heterotrophic bacteria. Microbiol. Ecol. 28, 255–271.

Kirchman, D.L. (2003) The contribution of monomers and other low-molecular weight compounds to the flux of dissolved organic material in aquatic ecosystems. *In* Aquatic Ecosystems: Interactivity of Dissolved Organic Matter (Findlay, S.E.G., and Sinsabaugh, R.L., eds.), pp. 218–237, Academic Press, New York.

Kirchman, D.L., and Borch, N.H. (2003) Fluxes of dissolved combined neutral sugars (polysaccharides) in the Delaware estuary. Estuaries 26, 894–904.

Kirchman, D.L., K'Nees, E., and Hodson, R. (1985) Leucine incorporation and its potential as a measure of protein synthesis by bacteria in natural aquatic systems. Appl. Environ. Microbiol. 49, 599–607.

Kistner, D.A., and Pettigrew, N.R. (2001) A variable turbidity maximum in the Kennebec Estuary, Maine. Estuaries 24, 680–687.

Kitto, M.E., Anderson, D.L., Gordon, G.E., and Olmez, I. (1992) Rare Earth distribution in catalysts and airborne particles. Environ. Sci. Technol. 267, 1368–1375.

Kjerfve, B.J. (ed.) (1994) Coastal lagoons. *In* Coastal lagoon Processes, pp. 1–7, Elsevier Oceanography Series, New York.

Kjerfve, B.J., Greer, J.E., and Crout, R.L. (1978) Low-frequency response of estuarine sea level to non-local forcing. *In* Estuarine Interactions (Wiley, M.L., ed.), pp. 497–513, Academic Press, New York.

Kjerfve, B.J., and Magill, K.E. (1989) Geographic and hydrodynamic characteristics of shallow coastal lagoons. Mar. Geol. 88, 187–199.

Kleeberg, A. (2002) Phosphorus sedimentation in seasonal anoxic Lake Scharmutzel, N.E. Germany. Hydrobiologia 472, 53–65.

Kleiner, D. (1985) Bacterial ammonium transport. FEMS Microbiol. Rev. 32, 87–100.

Klerks, P.L., and Lentz, S.A. (1998) Resistance to lead and zinc in the western mosquito fish *Gambusia affinis* inhabiting contaminated Bayou Trepagnier. Ecotoxicology 7, 11–17.

Klerks, P.L., and Levinton, J.S. (1989) Rapid evolution of metal resistance in a benthic oligochaete inhabiting a metal-polluted site. Biol. Bull. 176, 135–141.

Klinkhammer, G.P., and Bender, M.L. (1981) Trace metal distributions in the Hudson River estuary. Estuar. Coastal Shelf Sci. 12, 629–643.

Klinkhammer, G.P., and McManus, J. (2001) Dissolved manganese in the Columbia River estuary: production in the water column. Geochim. Cosmochim. Acta 65, 2835–2841.

Klinkhammer, G.P., and Palmer, M.R. (1991) Uranium in the oceans: where it goes and why. Geochim. Cosmochim. Acta 55, 1799–1806.

Klok, J., Cox, H., Baas, M., Schuyl, P.J.W., de Leeuw, J.W., and Schenck, P.A. (1984a) Carbohydrates in recent marine sediments. I. Origin and significance of deoxy and *O*-methyl sugars. Org. Geochem. 7, 73–84.

Klok, J., Cox, H., Baas, M., Schuyl, P.J.W., de Leeuw, J.W., and Schenck, P.A. (1984b) Carbohydrates in recent marine sediments. II. Occurrence and fate of carbohydrates in a recent stromatolitic environment. Org. Geochem. 7, 101–109.

Knicker, H. (2000) Solid-state 2D double cross polarization magic angle spinning ^{15}N ^{13}C NMR spectroscopy on degraded algal residues. Org. Geochem. 31, 337–340.

Knicker, H., and Hatcher, P.G. (1997) Survival of protein in an organic-rich sediment. Possible protection by encapsulation in organic matter. Naturwissenschaften 84, 231–234.

Knicker, H., and Ludemann, H.D. (1995) N-15 and C-13 CPMAS and solution NMR studies of N-15 enriched plant material during 600 days of microbial degradation. Org. Geochem. 23, 329–341.

Knobloch, K. (1966) Photosynthetische Sulfid-Oxidation gruner Pflanzen I. Mitteilung. Planta (Berl.) 70, 73–86.

Knudsen, M. (1902) Berichte uber die Konstantenbestimmungen zur Aufstellung der hydrographischen Tabellen. Kon Danske Videnskab. Selsk. Skrifter, 6 Raekke, Naturvidensk. Mathemat. Vol. XII, pp. 1–151.

Ko, F.C., and Baker, J.E. (1995) Partitioning of hydrophobic organic contaminants to resuspended sediments and plankton in the mesohaline Chesapeake Bay. Mar. Chem. 49, 171–188.

Koelmans, A.A., Gillissen, F., Makatita, W., and van den Berg, M. (1997) Organic carbon normalisation of PCB, PAH, and pesticide concentrations in suspended solids. Wat. Res. 31, 461–470.

Kofoed, L.H. (1975) The feeding biology of *Hydrobia ventrosa* Montagu. II. Allocation of the carbon budget and the significance of the secretion of dissolved organic material. J. Exp. Mar. Biol. Ecol. 19, 233–241.

Kogel-Knabner, I., Hatcher, P.G., and Zech, W. (1991) Chemical structural studies of forest soil humic acids: aromatic carbon fraction. Soil Sci. Soc. Am. J. 55, 241–247.

Köhler, H., Meon, B., Gordeev, V.V., Spitzy, A., and Amon, R.M.W. (2003) Dissolved organic matter (DOM) in the estuaries of Ob and Yenisei and the adjacent Kara sea, Russia. *In* Siberian River Run-off in the Kara Sea (Stein, R., Fahl, K., Futterer, D.K., Galimov, E.M., and Stepanets, O., eds.), pp. 281–308, Elsevier Science, New York.

Kohnen, M.E.L., Schouten, S., Sinninghe-Damste, J.S., de Leeuw, J.W., Merritt, D.A., and Hayes, J.M. (1992) Recognition of paleobiochemicals by a combined molecular sulfur and isotopic geochemical approach. Science 256, 358–362.

Kohring, L.L., Ringelberg, D.B., Devereux, R., Stahl, D.A., Mittelmann, M.M., and White, D.C. (1994) Comparison of phylogenetic relationships based on phospholipids' fatty acid profiles and ribosomal RNA sequence similarities among dissimilatory sulfate-reducing bacteria. FEMS Microbiol. Lett. 119, 303–308.

Koike, I., and Hattori, A. (1978) Denitrification and ammonia formation in aerobic coastal sediments. Appl. Environ Microbiol. 35, 278–282.

Koike, I., and Sørensen, J. (1988) Nitrate reduction and denitrification in marine sediments. In Nitrogen Cycling in Coastal Marine Environments. SCOPE 33 (Blackburn, T.H., and Sørensen, J., eds.), pp. 251–274, John Wiley, New York.

Koike, I., and Terauchi, K. (1996) Fine scale distribution of nitrous oxide in marine sediments. Mar. Chem. 52, 185–193.

Kolowith, L.C., Ingall, E.D., and Benner, R. (2001) Composition and cycling of marine phosphorus. Limnol. Oceanogr. 46, 309–320.

Komada, T., and Reimers, C.E. (2001) Resuspension-induced partitioning of organic carbon between solid and solution phases from a river–ocean transition. Mar. Chem. 76, 155–174.

Kononen, K. (1992) Dynamics of the toxic cyanobacterial blooms in the Baltic Sea. Finn. Mar. Res. 261, 1–36.

Koopmans, D. J., and Bronk, D. A. (2002) Photochemical production of inorganic nitrogen from dissolved organic nitrogen in waters of two estuaries and adjacent surficial groundwaters. Aquat. Microb. Ecol. 26, 295–304.

Koretsky, C.A., Moore, C.M., Lowe, K.L., Meile, C., Dichristina, T.J., and van Capellen, P. (2003) Seasonal oscillation of microbial iron and sulfate reduction in saltmarsh sediments (Sapelo Island, GA, USA.). Biogeochemistry 64, 179–203.

Kornitnig, S. (1978) Phosphorus in Handbook of Geochemistry, Vol. 2 (Wedephol, K.H., ed.), pp. 15E1–15E9, Springer-Verlag, New York.

Kostka, J.E., Gribsholt, B., Petrie, E., Dalton, D., Skelton, H., and Kristensen, E. (2002b) The rates and pathways of carbon oxidation in bioturbated saltmarsh sediments. Limnol. Oceanogr. 47, 230–240.

Kostka, J.E., and Luther III, G.W. (1994) Partitioning and speciation of solid phase iron in saltmarsh sediments. Geochim. Cosmochim. Acta 58, 1701–1710.

Kostka, J.E., and Luther III, G.W. (1995) Seasonal cycling of reactive Fe in salt-marsh sediments. Biogeochemistry 29, 159–181.

Kostka, J.E., Roychoudhury, A., and van Capellen, P. (2002a) Rates and controls of anaerobic microbial respiration across spatial and temporal gradients in saltmarsh sediments. Biogeochemistry 60, 49–76.

Kozelka, P.B., and Bruland, K.W. (1998) Chemical speciation of dissolved Cu, Zn, Cd, Pb in Narragansett Bay, Rhode Island. Mar. Chem. 60, 267–282.

Kozelka, P.B., Sanudo-Wilhelmy, S., Flegal, A.R., and Bruland, K.W. (1997) Physico-chemical speciation of lead in south San Francisco Bay. Estuar. Coastal Shelf Sci. 44, 649–658.

Kraft, J.C., Allen, E.A., Belknap, D.F., John, C.J., and Maurmeyer, E.M. (1979) Processes and morphologic evolution of an estuarine and coastal barrier system. In Barrier Islands from the Gulf of St. Lawrence to the Gulf of Mexico (Leatherman, S.P., ed.), pp. 149–183, Academic Press, New York.

Krajewski, K.P., van Cappellen, P., Trichet, J., Kuhn, O., Lucas, J., Martin-Algarra, A., Prevot, L., Tewari, V.C., Gaspar, L., Knight, R.I., and Lamboy, M. (1994) Biological processes and apatite formation in sedimentary environments. Ecol. Geol. Helv. 87, 701–745.

Krauss, K.W., Allen, J.A., and Cahoon, D.R. (2003) Differential rates of vertical accretion and elevation change among aerial root types in Micronesian mangrove forests. Estuar. Coastal Shelf Sci. 56, 251–259.

Kremer, J.N., Kemp, W.M., Giblin, A., Valiela, I., Seitzinger, S.P., and Hofmann, E.E. (2000) Linking biogeochemical processes to higher trophic levels. *In* Estuarine Science: A Synthetic Approach to Research and Practice (Hobbie, J.E., ed.), pp. 299–341, Island Press, Washington, DC.

Kremer, J.N., Nixon, S.W., Buckley, B., and Roques, P (2003b) Technical note: conditions for using the floating chamber method to estimate air–water gas exchange. Estuaries 26, 985–990.

Kremer, J.N., Reischauer, A., and D'Avanzo, C. (2003a) Estuary-specific variation in the air–water gas exchange coefficient for oxygen. Estuaries 26, 829–836.

Krest, J.M., and Harvey, J.W. (2003) Using natural distributions of short-lived radium isotopes to quantify groundwater discharge and recharge. Limnol. Oceanogr. 48, 290–298.

Krest, J.M., Moore, W.S., Gardner, L.R., and Morris, J.T. (2000) Marsh nutrient export supported by groundwater discharge: evidence from radium isotope measurements. Global Biogeochem. Cycles 14, 167–176.

Krest, J.M., Moore, W.S., and Rama, J. (1999) ^{226}Ra and ^{228}Ra in the mixing zones of the Mississippi and Atchafalaya Rivers: indicators of groundwater input. Mar. Chem. 64, 129–152.

Krezel, A., and Bal, W. (1999) Coordination chemistry of glutathione. Acta Biochim. 46, 567–580.

Krishnaswami, S., Benninger, L.K., Aller, R.C., and Von Damm, K.L. (1980) Atmospherically-derived radionuclides as tracers of sediment mixing and accumulation in near-shore marine and lake sediments: evidence from ^{7}Be, ^{210}Pb, and ^{239,240}Pu. Earth Planet. Sci. Lett. 47, 307–318.

Krishnaswami, S., and Lal, D. (1978) Radionuclide limnolochronology. *In* Lakes, Chemistry, Geology, Physics (Lerman, A., ed.), pp. 153–177, Springer-Verlag, New York.

Krishnaswami, S., Lal, D., Martin, J., and Meybeck, M. (1971) Geochronology of lake sediments. Earth Planet. Sci. Lett. 11, 407.

Krishnaswami, S., Monaghan, M.C., Westrich, J.T, Bennett, J.T., and Turekian, K.K. (1984) Chronologies of sedimentary processes of the FOAM site, Long Island Sound, Connecticut. Am. J. Sci. 284, 706–733.

Kristensen, E. (1988) Benthic fauna and biogeochemical processes in marine sediments: microbial activities fluxes. *In* Nitrogen Cycling in Coastal Marine Environments. SCOPE (Blackburn, T.H., and Sørensen, J., eds.), pp. 275–299, Scope, Chichester, UK.

Kristensen, E., Ahmed, S.I., and Devol, A.H. (1995) Aerobic and anaerobic decomposition of organic matter in marine sediment: Which is fastest? Limnol. Oceanogr. 40, 1430–1437.

Kristensen, E., and Blackburn, H. (1987) The fate of organic carbon and nitrogen in experimental marine sediment systems: influence of bioturbation and anoxia. J. Mar. Res. 45, 231–257.

Kristensen, E., and Hansen, K. (1995) Decay of plant detritus in organic-poor marine sediment: production rates and stoichiometry of dissolved C and N compounds. J. Mar. Res. 53, 675–702.

Kristensen, E., and Holmer, M. (2001) Decomposition of plant materials in marine sediment exposed to different electron acceptors (O_2, NO_3^-, and SO_4^{2-}), with emphasis on substrate origin, degradation kinetics, and the role of bioturbation. Geochim. Cosmochim. Acta 65, 419–433.

Kroer, N., Jørgensen, N.O.G, and Coffin, R.B. (1994) Utilization of dissolved nitrogen by heterotrophic bacterioplankton: a comparison of three ecosystems. Appl. Environ. Microbiol. 60, 4116–4123.

Kroeze, C., and Seitzinger, S.P. (1998) Nitrogen inputs to rivers, estuaries and continental shelves and related nitrous oxide emissions in 1990 and 2050: a global model. Nutrient Cycl. Agroecosyst. 52: 195–212.

Kröger, N., Deutzmann, R., and Sumper, M. (1999) Polycationic peptides from diatom biosilica that direct silica nanosphere formation. Science 286, 1129–1132.

Krom, M.D., and Berner, R.A. (1980a) The diffusion coefficients of sulfate, ammonium, and phosphate ions in anoxic marine sediments. Limnol. Oceanogr. 25, 327–337.

Krom, M.D., and Berner, R.A. (1980b) Adsorption of phosphate in anoxic marine sediments. Limnol. Oceanogr. 25, 797–806.

Krom, M.D., and Berner, R.A. (1981) The diagenesis of phosphorus in a near shore marine sediment. Geochim. Cosmochim. Acta 45, 207–216.

Krom, M.D., Brenner, S., Kress, N., Neori, A., and Gordon, L.I. (1992) Nutrient dynamics and new production in a warm-core eddy from the eastern Mediterranean Sea. Deep-Sea Res. 39, 467–480.

Krom, M.D., Kress, N., Brenner, S., and Gordon, L.I. (1991) Phosphorus limitation of primary productivity in the eastern Mediterranean Sea. Limnol. Oceanogr. 36, 424–432.

Krone, R.B. (1962) Flume studies of the transport of sediment in estuarial shoaling processes. Univ. of California, Hydraulics and Engineering Lab. and Sanitarian Engineering Research Lab. Berkeley, pp. 110.

Krumbein, W.C., and Sloss, L.L. (1963) Stratigraphy and Sedimentation, 2nd ed. p. 660, W.H. Freeman, San Francisco.

Kuehl, S.A., DeMaster, D.J., and Nittrourer, C.A.(1986) Nature of sediment accumulation on the Amazon continental shelf. Cont. Shelf Res. 6, 209–225.

Kuehl, S.A., Nittrouer, C.A., and DeMaster, D.J. (1982) Modern sediment accumulation and strata formation on the Amazon continental shelf. Mar. Geol. 49, 279–300.

Kuelegan, G.H. (1949) Interfacial instability and mixing in stratified flows. J. Res. Nat. Bur. Stds. 43, 487–500.

Kuhlbusch, T.A.J. (1998) Black carbon and the carbon cycle. Science 280, 1903–1904.

Kukkonen, J., and Oikari, A. (1991) Bioavailability of organic pollutants in boreal waters with varying levels of dissolved organic material. Wat. Res. 25, 455–463.

Kuo, A.Y., and Park, K. (1995) A framework of coupling shoals and shallow embayments with main channels in numerical modeling of coastal plain estuaries. Estuaries 18, 341–350.

Kuo, A.Y., Park, K., and Moustafa, M.Z. (1991) Spatial and temporal variability of hypoxia in the Rappahannock River, Virginia. Estuar. Coastal Shelf Sci. 14, 113–121.

Kuparinen, J., Leonardsson, K., Mattila, J., and Wilkner, J. (1996) Food web structure and function in the Gulf of Bothnia, the Baltic Sea. Ambio 8, 12–20.

Kure, L.K., and Forbes, T.L. (1997) Impact of bioturbation by Arenicola marina on the fate of particle-bound fluoranthene. Mar. Ecol. Prog. Ser. 156, 157–166.

Kurie, F.N.D. (1934) A new mode of disintegrations induced by neutrons. Phys. Rev. 45, 904–905.

Kuypers, M.M., Slickers, A.O., Lavik, G, Schmid, M., Jørgensen, B.B., Kuenen, J.G., Sinninghe-Damste, J.S., Strous, M., and Jetten, M.S. (2003) Anaerobic ammonium oxidation by ammonox bacteria in the Black Sea. Nature 422, 608–611.

Kwak, T.J., and Zedler, J. (1997) Food web analysis of southern California coastal wetlands using multiple stable isotopes. Oecologia 110, 262–277.

Laamanen, M.J. (1997) Environmental forms affecting the occurrence of different morphological forms of cyanprokaryotes in northern Baltic Sea. J. Plankton Res. 19, 1385–1403.

Laegreid, M., Alstad, J., Klaveness, D., and Seip, H.M. (1983) Seasonal variation of cadmium toxicity toward the algae *Selenatrum capricornutum* Printz in two lakes with different humus content. Environ. Sci. Technol. 17, 357–359.

Laflamme, R.E., and Hites, R.A. (1978) Tetra- and pentacyclic, naturally-occurring aromatic hydrocarbons in recent sediments. Geochim. Cosmochim. Acta 43, 1687–1691.

Lajtha, K., and Michener, R.H. (eds.) (1994) Stable Isotopes in Ecology and Environmental Science. Blackwell Scientific, Oxford.

Lal, D., and Lee, T. (1988) Cosmogenic ^{32}P and ^{33}P used as tracers to study phosphorus recycling in the upper ocean. Nature 333, 752–754.

Lal, D., Malhorta, P.K., and Peters, B. (1958) On the production of radioisotopes in the atmosphere by cosmic radiation and their application to meteorology. J. Atmos. Terr. Phys. 12, 306–328.

Lal, D., and Peters, B (1967) Cosmic ray produced activity on the Earth. Hanbuch Phys. 46, 551–612.

LaMontagne, M.G., Astorga, V., Giblin, A.F., and Valiela, I. (2002) Denitrification and stoichiometry of nutrient regeneration in Waquoit Bay, Massachusetts. Estuaries 25, 272–281.

LaMontagne, M.G., and Valiela, I. (1995) Dentrification measured by direct N_2 flux method in sediments of Waquoit Bay, MA. Biogeochemistry 31, 63–83.

Lamontagne, R.A., Swinnerton, J.W., Linnenbom, V.J., and Smith, W.D. (1973) Methane concentrations in various marine environments. J. Geophys. Res. 78, 5317–5323.

Lancelot, C., Billen, G., Sournia, A., Weisse, T., Colijn, F., Veldhuis, M.J.W., Davies, A., and Wassman, P. (1987) *Phaeocystis* blooms and nutrient enrichment in the continental coastal zones of the North Sea. Ambio 16, 38–46.

Landen, A., and Hall, P.O.J. (2000) Benthic fluxes and pore water distributions of dissolved free amino acids in the open Skagerrak. Mar. Chem. 71, 53–68.

Landing, W.M., Guentzel, J.L., Perry, J.J., and Pollman, C.D. (1998) Methods for measuring mercury and other trace species in rainfall and aerosols in Florida. Atmos. Environ. 32, 909–918.

Landrum, P.F., Reinhol, M.D., Nihart, S.R., and Eadie, B.J. (1985) Predicting the bioavailability of organic xenobiotics to *Pontoporeia hoyi* in the presence of humic and fulvic materials and natural dissolved organic matter. Environ. Toxicol. Chem. 4, 459–467.

Langhorne, D.N. (1977) Consideration of meteorological conditions when determining the navigational water depth over a sand wave field. Intl. Hydrogr. Rev. LIV, 17–30.

Lapointe, B.E., Littler, M.M., and Littler, D.S. (1992) Nutrient availability to macroalgae in siliciclastic versus carbonate-rich coastal waters. Estuaries 15, 75–82.

Lapointe, B.E., and Matzie, W.R. (1996) Effects of stormwater nutrient discharges on eutrophication processes in nearshore waters of the Florida Keys. Estuaries 19, 422–435.

Lapointe, B.E., O'Connell, J.D., and Garrett, G.S. (1990) Nutrient couplings between on-site disposal systems, groundwaters, and nearshore surface waters of the Florida Keys. Biogeochemistry 10, 289–307.

LaRoche, J., Nuzzi, R., Waters, R., Wyman, K., Falkowski, P.G., and Wallace, D.W.R. (1997) Brown tide blooms in Long Island's coastal waters linked to interannual variability in groundwater flow. Global Change Biol. 3, 397–410.

Larsen, I.L., and Cuttshall, N.H. (1981) Direct determination of ^{7}Be in sediments. Earth Planet. Sci. Lett. 54, 379–384.

Larson, T.E., and Buswell, A.M. (1942) Calcium carbonate saturation index and alkalinity interpretations. J. Am. Wat. Works Assoc. 34, 1664.

Larsson, P. Andersson, A., Broman, D., Nordback, J., and Lundberg, E. (2000) Persistent organic pollutants (POPs) in pelagic systems. Ambio 29, 202–209.

Larsson, U., Elmgren, R., and Wulff, F. (1985) Eutrophication and the Baltic Sea: causes and consequences. Ambio 14, 9–14.

Larsson, U., Hadju, S., Walve, J., and Elmgren, R. (2001) Baltic Sea nitrogen fixation estimated from summer increase in upper mixed layer total nitrogen. Limnol. Oceanogr. 46, 811–820.

Lasagna, A.C., and Holland, H.D. (1976) Mathematical aspects of non-steady state diagenesis. Geochim. Cosmochim. Acta 40, 257–266.

Laursen, A.E., and Seitzinger, S.P. (2002) Measurement of denitrification in rivers: an integrated, whole reach approach. Hydrobiologia 485, 67–81.

Laws, E.A., Popp, B.N., Bidigare, R.R., Kennicutt, M.C., and Macko, S.A. (1995) Dependence of phytoplankton carbon isotopic composition on growth rate and $[CO_2aq]$: theoretical considerations and experimental results. Geochem. Cosmochim. Acta 59, 1131–1138.

Leal, M.F.C., Vasconcelos, M.T.S.D., and van den Berg, C.M.G. (1999) Copper induced release of complexing ligands similar to thiols by *Emiliana huxleyi* in seawater cultures. Limnol Oceanogr. 44, 567–580.

Leaney, F.W., Osmond, C.B., Allison, G.B., and Ziegler, H. (1985) Hydrogen-isotope composition of leaf water in C_3 and C_4 plants: its relationship to the hydrogen-isotope composition of dry matter. Planta 164, 215–220.

Leavitt, P.R. (1993) A review of factors that regulate carotenoids and chlorophyll deposition and fossil pigment abundance. J. Paleolimnol. 1, 201–214.

Leavitt, P.R., and Carpenter, S.R. (1990) Aphotic pigment degradation in the hypolimnion—Implications for sedimentation studies and paleolimnology. Limnol. Oceanogr. 35, 520–534.

Leavitt, P.R., and Hodgson, D.A. (2001) Sedimentary pigments. *In* Tracking Environmental Changes Using Lake Sediments (Smol, J.P., Birks, H.J.B., and Last, W.M., eds.), pp. 2–21, Kluwer, New York.

Lebo, M.E. (1991) Particle-bound phosphorus along an urbanized coastal plain estuary. Mar. Chem. 34, 225–246.

Le Borgne, R. (1986) The release of soluble end products of metabolism. *In* The Biological Chemistry of Marine Copepods, (Corner, D.S., and O'Hara, S.C.M., eds.), pp. 109–164. Oxford University Press, Oxford, UK.

Leck, C., Larsson, U., Bagander, L.E., Johansson, S., and Hajdu, S. (1990) Dimethyl sulfide in the Baltic Sea: annual variability in relation to biological activity. J. Geophys. Res. 95, 3353–3364.

Lee, C. (1992) Controls on organic carbon preservation: the use of stratified water bodies to compare intrinsic rates of decomposition in oxic and anoxic systems. Geochim. Cosmochim. Acta 56, 3323–3335.

Lee, C., and Bada, J.L. (1977) Dissolved amino acids in the equatorial Pacific, the Sargasso Sea and Biscayne Bay. Limnol. Oceanogr. 22, 502–510.

Lee, K.K., Holst, R.W., Watanabe, I., and App, A. (1981) Gas transport through rice. Soil Sci. Plant Nutr. 27, 151–158.

Lee, R.F. (1980) Phycology. Cambridge University Press, Cambridge, UK.

Lee, R.F., and Loeblich, A.R. (1971) Distribution of 21:6 hydrocarbon and its relationship to 21:6 fatty acid in algae. Phytochemistry 10, 593–602.

Leeder, M. (1982) Sedimentology: Process and Product. George Allen and Unwin, London.

Lehmann, M.F., Reichert, P., Bernasconi, S.M., Barbieri, A., and McKenzie, J.A. (2003) Modeling nitrogen and oxygen isotope fractionation during denitrification in a lacustrine redox-transition zone. Geochim. Cosmochim. Acta 67, 2529–2542.

Lehmann, M.F., Sigman, D.M., and Berelson, W.M. (2004) Coupling the $^{15}N/^{14}N$ and $^{18}O/^{16}O$ of nitrate as a constraint on benthic nitrogen cycling. Mar. Chem. 88: 1–20.

Lemaire, E.A., Abril, G., de Wit, R., and Etcheber, H. (2002) Distribution of phytoplankton pigments in nine European estuaries and implications for an estuarine typology. Biogeochemistry 59, 5–23.

Lenanton, R.C., Longeragan, N.R., and Potter, I. (1985) Blue–green algal blooms and the commercial fishery of a large Australian estuary. Mar. Pollut. Bull. 16, 477–482.

Leonard, L.A., and Luther, M.E. (1995) Flow hydrodynamics in tidal marsh canopies. Limnol. Oceanogr. 40, 1474–1484.

Leppard, G.G, Flannigan, D.T., Mavrocordatos, D., Marvin, C.H., Bryant, D.W., and McCarry, B.E. (1998) Binding of polycyclic aromatic hydrocarbons by size classes of particulate matter in Hamilton harbor water. Environ. Sci. Technol. 32, 3633–3639.

Lerman, A. (1979) Geochemical Processes: Water and Sediment Environments. Wiley-Interscience, New York.

Levinton, J.S., and Bianchi, T.S. (1981) Nutrition and food limitation of deposit-feeders. I. The role of microbes in the growth of mud snails (Hydrobiidae). J. Mar. Res. 39, 531–545.

Levinton, J.S., Bianchi, T.S., and Stewart, S. (1984) What is the role of particulate organic matter in benthic invertebrate nutrition? Bull. Mar. Sci. 35, 270–282.

Levinton, J.S., Suatoni, E., Wallace, W., Junkins, R., Kelaher, B., and Allen, B.J. (2003) Rapid loss of genetically based resistance to metals after clean-up of a Superfund site. Proc. Natl. Acad. Sci. USA. 100, 9889–9891.

Levitt, M.H. (2001) Spin Dynamics: Basics of Nuclear Magnetic Resonance. John Wiley, New York.

Lewis, E. (1978) The Practical Salinity Scale 1978 and its antecedents. J. Ocean. Eng. 5, 3–8.

Lewitus, A.J., Koepfler, E.T., and Pigg, R.J. (2000) Use of dissolved organic nitrogen by a salt marsh phytoplankton bloom community. Arch. Hydrobiol. Spec. Issues Adv. Limnol. 55, 441–456.

Lewitus, A.J., Willis, B.M., Hayes, K.C., Burkholder, J.M., Glasgow, J.M., Glibert, P.M., and Burke, M.K. (1999) Mixotrophy and nitrogen uptake by *Pfiesteria piscicda* (Dinophyceae). J. Phycol. 35, 1430–1437.

Liaaen-Jensen, S. (1978) Marine carotenoids. *In* Marine Natural Products: Chemical and Biological Perspectives, Vol. 2 (Scheuer, P.J., ed.), pp. 1–73, Academic Press, New York.

Li, C., Chen, G., Yao, M., and Wang, P. (1991) The influence of suspended load on the sedimentation in the coastal zones and continental shelves of China. Mar. Geol. 96, 341–352.

Li, Y.H., and Chan, L.H. (1979) Desorption of barium and ^{226}Ra from river-borne sediments in the Hudson Estuary. Earth Planet. Sci. Lett. 43, 343–350.

Li, Y.H., Mathieu, G., Biscaye, P., and Simpson, H.J. (1977) The flux of Ra-226 from estuarine and continental shelf sediments. Earth Planet. Sci. Lett. 37, 237–241.

Li, Y.H., Santschi, P.H., Kaufman, A., Benninger, L.K., and Feely, H.W. (1981) Natural radionuclides in waters of the New York Bight. Earth Planet. Sci. Lett. 55, 217–228.

Libby, W.F. (1982) Nuclear dating. *In* An Historical Perspective (Currie, L.A, ed.), pp. 516, Nuclear and Chemical Dating Techniques. American Chemical Society Symposium Series, Washington, DC.

Libby, W.L. (1952) Radiocarbon Dating. University of Chicago Press, Chicago.

Libes, S.M. (1992) An Introduction to Marine Biogeochemistry. John Wiley, New York.

Lijklema, L. (1977) The role of iron in the exchange of phosphorus between water and sediments. *In* Interactions Between Sediments and Freshwater (Golterman, H.L., ed.), pp. 313–317, Dr. W. Junk B.V., The Hague.

Likens, G., Borman, F., and Johnson, M. (1974) Acid rain. Environment 14, 33–40.

Lillebo, A.I., Neto, J.M., Flindt, M.R., Marques, J.C., and Pardal, M.A. (2004) Phosphorus dynamics in a temperate intertidal estuary. Estuar. Coastal Shelf Sci. 61, 101–109.

Lin, J., and Kuo, A.Y. (2001) Secondary turbidity maximum in a partially mixed microtidal estuary. Estuaries 24, 707–720.

Lindström, G. (1855) Birdag till Kännedomen om Östersjöns invertebratfauna. Stockholm, Öfversigt af Knogi. Ventenskaps Akad. Förhandlingar 12, 49–73.

Linsalata, P., Wrenn, M.E., Cohen, N., and Singh, N.P. (1980) ^{239,240}Pu and ^{238}Pu in sediments of the Hudson River estuary. Environ. Sci. Technol.12, 1519.

Lipiatouk, E., and Saliot, A. (1991) Fluxes and transport of anthropogenic and natural polycyclic aromatic hydrocarbons in the western Mediterranean Sea. Mar. Chem. 32, 51–71.

Lipschultz, F. (1981) Methane release from a brackish intertidal salt-marsh embayment of Chesapeake Bay, Maryland. Estuaries 4, 143–145.

Lipschultz, F., Wofsy, S.C., and Fox, L.E. (1986) Nitrogen metabolism of the eutrophic Delaware River ecosystem. Limnol. Oceanogr. 31, 701–716.

Lirman, P.S., and Cropper, W.P. (2003) The influence of salinity on seagrass growth, survivorship, and distribution within Biscayne Bay, Florida: field, experimental, and modeling studies. Estuaries 26, 131–141.

Lisitzin, A.P. (1995) The marginal filter of the ocean. Oceanol. 34, 671–682.

Liss, P.S. (1976) Conservative and non-conservative behavior of dissolved constituents during estuarine mixing. In Estuarine Chemistry (Burton, J.D., and Liss, P.S., eds.), pp. 93–130, Academic Press, London.

Little, D.I. (1987) The physical fate of weathered crude and emulsified fuel oils as a function of intertidal sedimentology. In Fate and Effects of Oil in Marine Ecosystems (Kuiper, J., and van den Brink, W.J., eds.), pp. 3–18, Martinus Nijhoff, Boston, MA.

Liu, K.K., Atkinson, L., Chen, C.T., Gao, S., Hall, J., Macdonald, R.W., McManus, L.T., and Quinones, R. (2000) Exploring continental margin carbon fluxes on a global scale. EOS 81, 641–642.

Liu, Q., Parrish, C.C., and Helleur, R. (1998) Lipid class and carbohydrate concentrations in marine colloids. Mar. Chem. 60, 177–188.

Livingstone, D.A. (1963) Chemical composition of rivers and lakes. Prof. Pap. U.S. Geol. Surv. 440.

Llobet-Brossa, E., Rossello, R., and Amann, R. (1998) Microbial community composition of Wadden Sea sediments as revealed by fluorescence in situ hybridization. Appl. Environ. Microbiol. 64, 2691–2696.

Lobartini, J.C., Tan, K.H., Asmussen, L.E., Leonard, R.A., Himmelsbach, D., and Gingle, A.R. (1991) Chemical and spectral differences in humic matter from swamps, streams and soils in the southeastern United States. Geoderma 49, 241–254.

Loder, T.C., and Liss, P.S. (1985) Control by organic coatings of the surface-charge of estuarine suspended particles. Limnol. Oceanogr. 30, 418–421.

Lofty, M.F., and Frihy, O.E. (1993) Sediment balance in the nearshore zone of the Nile Delta coast, Egypt. J. Coastal Res. 9, 654–662.

Lohnis, F. (1926) Nitrogen availability of green manure. Soil Sci. 22, 253–290.

Lohrenz, S.E., Dagg, M.J., and Whitledge, T.E. (1990) Enhanced primary production at the plume oceanic interface of the Mississippi River. Cont. Shelf Res. 10, 639–664.

Lohrenz, S.E., Fahnenstiel, G.L., and Redalje, D.G. (1994) Spatial and temporal variations of photosynthetic parameters in relation to environmental conditions in coastal waters of the northern Gulf of Mexico. Estuaries 17, 779–795.

Lohrenz, S.E., Fahnenstiel, G.L., Redalje, D.G., Lang, G.A., Chen, X.G., and Dagg, M.J. (1997) Variations in primary production of northern Gulf of Mexico continental shelf waters linked to nutrient inputs from the Mississippi River. Mar. Ecol. Prog. Ser. 155 45–54.

Lohrenz, S.E., Fahnenstiel, G.L., Redalje, D.G., Lang, G.A., Dagg, M.J., Whitledge, T.E., and Dortch, Q. (1999) Nutrients, irradiance, and mixing as factors regulating primary production in coastal waters impacted by the Mississippi River plume. Cont. Shelf Res. 19, 1113–1141.

Loizeau, U., Abarnou, A., Cugier, P., Jaouen-Madoulet, A., Le Guellec, A.M., and Menesguen, A. (2001) A model of PCB bioaccumaulation in the sea bass food web from Seine estuary (eastern English Channel). Mar. Pollut. Bull. 43, 242–255.

Lomas, M.W., Glibert, P.M., Berg, G.M., and Burford, M. (1996) Characterization of nitrogen uptake by natural populations of *Aurecoccus anaphagefferens* (Chrysophyceae) as a function of incubation duration, substrate concentrations, light, and temperature. J. Phycol. 32, 907–916.

Lomas, M.W., Trice, T.M., Glibert, P.M., Bronk, D.A., and McCarthy, J.J. (2002) Temporal and spatial dynamics of urea uptake and regeneration rates and concentrations in Chesapeake Bay. Estuaries 25, 469–482.

Lomstein, B.A., Blackburn, T.H., and Henriksen, K. (1989) Aspects of nitrogen and carbon cycling in the Northern Bering shelf sediment. I. The significance of urea turnover in the mineralization of NH_4^+. Mar. Ecol. Prog. Ser. 57, 237–247.

Lomstein, B.A., Jensen, A.G.U., Hansen, J.W., Andreasen, J.B., Hansen, L.S., Berntsen, J., and Kunzendorf, H. (1998) Budgets of sediment nitrogen and carbon cycling in the shallow water of Knebel Vig. Denmark. Aquat. Microb. Ecol. 14, 69–80.

Long, E.R. (1992) Ranges in chemical concentrations in sediments associated with adverse biological effects. Mar. Pollut. Bull. 24, 38–45.

Lord, C.J., III., and Church, T.M. (1983) The geochemistry of salt marshes: sedimentary iron diffusion. Sulfate reduction, and pyritization. Geochem. Cosmochim. Acta 47, 1381–1391.

Lores, E.M., Patrick, J.M., and Summers, J.K. (1993) Humic acid effects on uptake of hexachlorobenzene and hexachlorobiphenyl by sheepshead minnows in static sediment/water systems. Environ. Toxicol. Chem. 12, 541–550.

Louchouaran, P., Lucotte, M., Canuel, R., Gagne, J.P., and Richard, L.F. (1997) Sources and early diagenesis of lignin and bulk organic matter in the sediments of the lower St. Lawrence estuary and the Saguenay Fjord. Mar. Chem. 58, 3–26.

Louchouaran, P., Opsahl, S., and Benner, R. (2000) Isolation and quantification of dissolved lignin from natural waters using solid-phase extraction (SPE) and GC/MS SIM. Anal. Chem. 72, 2780–2787.

Louda, W.J., Liu, L., and Baker, E.W. (2002) Senescence- and death-related alteration of chlorophylls and carotenoids in marine phytoplankton. Org. Geochem. 33, 1635–1653.

Louda, J.W., Loitz, J.W., Rudnick, D.T., and Baker, E.W. (2000) Early diagenetic alteration of chlorophyll-*a* and bacteriochlorophyll-*a* in a contemporaneous marl ecosystem: Florida Bay. Org. Geochem. 31, 1561–1580.

Lovley, D. (1991) Dissimilatory Fe(III) and Mn(IV) reduction. Microbiol. Rev. 55, 259–287.

Lovley, D., and Klug, M.J. (1983) Sulfate reducers can outcompete methanogens at freshwater sulfate concentrations. Appl. Environ. Microbiol. 45, 187–192.

Lovley, D., and Phillips, E.J.P. (1988) Novel mode of microbial energy metabolism: organic carbon oxidation coupled to dissimilatory reduction of iron or manganese. Appl. Environ. Mocrobiol. 54, 1472–1480.

Lovley, D., Phillips, E.J.P., and Lonergan, D.J. (1989) Hydrogen and formate oxidation coupled to dissimilatory reduction of iron and manganese by *Alteromonas putrefaciens*. Appl. Environ. Microbiol. 55, 700–706.

Lovley, D., Phillips, E.J.P., and Lonergan, D.J. (1991) Enzymatic versus non enzymatic mechanisms for Fe(III) reduction in aquatic sediments. Environ. Sci. Technol. 25, 1062–1067.

Lovley, D., Roden, E.E., Phillips, E.J.P., and Woodward, J.C. (1993) Enzymatic iron and uranium reduction by sulfate-reducing bacteria. Mar. Geol. 113, 41–53.

Lovley, D., Stolz, J.F., Nord, G.L., and Phillips, E.J.P. (1987) Anaerobic production of magnetite by a dissimilatory iron-reducing microorganism. Nature 330, 252–254.

Lucas, C.H., Widdows, J., and Wall, L. (2003) Relating spatial and temporal variability in sediment chlorophyll *a* and carbohydrate distribution with erodability of a tidal flat. Estuaries 26, 885–893.

Lucas, L.V., and Cloern, J.E. (2002) Effects of tidal shallowing and deepening on phytoplankton production dynamics: modeling study. Estuaries 25, 497–507.

Lucas, W.J., and Berry, J.A. (1985) Inorganic Carbon Uptake by Aquatic Photosynthetic Organisms. American Society of Plant Physiology, Rockville, MD.

Lucotte, M., and d'Angleian, B. (1993) Forms of phosphorus and phosphorus–iron relationships in the suspended matter of the St. Lawrence estuary. Can. J. Fish. Aquat. Sci. 20, 1880–1890.

Lugo, A.E., and Snedaker, S.C. (1974) The ecology of mangroves. Ann. Rev. Ecol. and Syst. 5, 39–64.

Luoma, S. (1989) Can we determine the biological availability of sediment-bound trace elements? Hydrobiology 176/177, 379–396.

Luoma, S.N., Johns, C., Fisher, N.S., Steinberg, N.S., Oremland, R.S., and Reinfelder, J.R. (1992) Determination of selenium bioavailability to a benthic bivalve from particulate and solute pathways. Environ. Sci. Technol. 26, 485–492.

Luther III, G.W. (1991) Pyrite synthesis via polysulfide compounds. Geochim. Cosmochim. Acta 55, 2839–2849.

Luther III, G.W., and Church, T.M. (1988) Seasonal cycling of sulfur and iron in porewaters of a Delaware salt marsh. Mar. Chem. 23, 295–309.

Luther III, G.W., Church, T.M., Scudlark, J.R., and Cosman, M. (1986) Inorganic and organic sulfur cycling in salt-marsh pore waters. Science 232, 746–779.

Luther III, G.W., Giblin, A., Howarth, R.W., and Ryans, R.A. (1982) Pyrite and oxidized iron mineral phases formed from pyrite oxidation in salt marsh and estuarine sediments. Geochim. Cosmochim. Acta 46, 2665–2669.

Luther III, G.W., Ma, S., Trouwborst, R., Glazer, B., Blickley, M., Scarborough, R.W., and Mensinger, M.G. (2004) The roles of anoxia, H_2S and storm events in fish kills of dead-end canals of Delaware inland bays. Estuaries 27, 551–560.

Luther III, G.W., Sundby, B., Lewis, B.L., Brendel, P.J., and Silverberg, N. (1997) Interactions of manganese with nitrogen cycle: alternative pathways to dinitrogen. Geochim. Cosmochim. Acta 61, 4043–4052.

Lyman, J., and Flemming, R. (1940) Composition of seawater. J. Mar. Res. 3, 134.

Lynch, J.C., Meriwether, J.R., McKee, B.A., Vera-Herrera, F., and Twilley, R.R. (1989) Recent accretion in mangrove ecosystems based on ^{137}Cs and [210]Pb. Estuaries 12, 284–299.

Lyons, W.B., and Gaudette, H.E. (1979) Sulfate reduction and the nature of organic matter in estuarine sediments. Org. Geochem. 1, 151–155.

Lyons, W.B., Gaudette, H.E., and Hewitt, A.D. (1979) Dissolved organic matter in pore waters of carbonate sediments from Bermuda. Geochim. Cosmochim. Acta 43, 433–437.

Ma, L., and Dolphin, D. (1996) Stereoselective synthesis of new chlorophyll a related antioxidants isolated from margin organisms. J. Org. Chem. 61, 2501–2510.

Maccubbin, A.E., and Hodson, R.E. (1980) Mineralization of detrital lignocelluloses by salt marsh sediment microflora. Appl. Environ. Microbiol. 40, 735–740.

MacGill, J.T. (1958) Map of coastal landforms of the world. Geogr. Rev. 48, 402–405.

Mackey, D.J., and Zirino, A. (1994) Comments on trace metal speciation in seawater or do "onions" grow in the sea? Anal. Chim. Acta 284, 635–647.

Mackey, M., Mackey, D., Higgins, H., and Wright, S. (1996) CHEMTAX—a program for estimating class abundances from chemical markers: application to HPLC measurements of phytoplankton. Mar. Ecol. Prog. Ser. 144, 265–283.

Mackin, J.E., and Aller, R.C. (1984) Ammonium adsorption in marine sediments. Limnol. Oceanogr. 29, 250–257.

Mackin, J.E., and Aller, R.C. (1986) The effects of clay mineral reactions on dissolved Al distributions in sediments and waters of the Amazon continental shelf. Cont. Shelf Res. 6, 245–262.

Mackin, J.E., and Swider, K.T. (1989) Organic matter decomposition pathways and oxygen consumption in coastal marine sediments. J. Mar. Res. 47, 681–716.

Macko, S.A., Estep, M.L., and Lee, W.Y. (1983) Stable hydrogen isotope analysis of food webs on laboratory and field populations of marine amphipods. J. Exp. Mar. Biol. Ecol. 72, 243–249.

Macko, S.A., Fogel, M.L., Hare, P.E., and Hoering, T.C. (1987) Isotopic fractionation of nitrogen and carbon in the synthesis of amino acids by microorganisms. Chem. Geol. 65, 79–92.

Macko, S.A., Helleur, R., Hartley, G., and Jackman, P. (1989) Diagenesis of organic matter—a study using stable isotopes of individual carbohydrates. Adv. Org. Geochem. 16, 1129–1137.

Madden, C.J., and Kemp, W.M. (1996) Ecosystem model of an estuarine submersed plant community: calibration and simulation of eutrophication responses. Estuaries 19, 457–474.

Maeda, M., and Windom, H.L. (1982) Behavior of uranium in two estuaries of the southeastern United States. Mar. Chem. 11, 427–436.

Maestrini, S.Y., Balode, M., Bechemin, C., and Purina, I. (1999) Nitrogenous organic substances as potential nitrogen sources, for summer phytoplankton in the Gulf of Riga, eastern Baltic Sea. Plankton Biol. Ecol. 46, 8–17.

Magenheimer, J.F., T.R. Moore, T.R., Chmura, G.L., and Daoust, R.J. (1996) Methane and carbon dioxide flux from a macrotidal salt marsh. Bay of Fundy, New Brunswick. Estuaries 19, 139–145.

Maguer, J., Wafer, M., Madec, C., Morin, P., and Denn, E. (2004) Nitrogen, and phosphorus requirements of an *Alexandrium minutum* bloom in the Penzé Estuary, France. Limnol. Oceanogr. 49, 1108–1114.

Maillacheruvu, K.Y., and Parkin, G.F. (1996) Kinetics of growth, substrate utilization and sulfide toxicity for proprionate, acetate, and hydrogen utilizers in anaerobic systems. Water Environ. Res. 68, 1099–1106.

Malcolm, R.I. (1990) The uniqueness of humic substances in each of soil, stream, and marine environments. Anal. Chim. Acta 232, 19–30.

Malcolm, R.I., and Durum, W.H. (1976) Organic carbon and nitrogen concentrations and annual organic carbon load of six selected rivers of the U.S. Geol. Surv. Water-Supply Paper 1817-F, Reston, VA.

Malin, G., Wilson, W.H., Bratbak, G., Liss, P.S., and Mann, N.H. (1998) Elevated production of dimethylsulfide resulting from viral infection of cultures of *Phaeocystis pouchetii*. Limnol. Oceanogr. 43, 1389–1393.

Malone, T.C., Boynton, W., Horton, T., and Stevenson, C. (1993) Nutrient loadings to surface waters: Chesapeake Bay case study. *In* Keeping Pace with Science and Engineering (Uman, M.F., ed.), pp. 8–38, National Academy Press. Washington, DC.

Malone, T.C., Conley, D.J., Fisher, T.R., Glibert, P.M., Harding, I.W., and Sellner, K.G. (1996) Scales of nutrient-limited phytoplankton productivity in Chesapeake Bay. Estuaries 19, 371–385.

Malone, T.C., Crocker, L.H., Pike, and Wendler, B.W. (1988) Influences of river flow on the dynamics of phytoplankton production in a partially stratified estuary. Mar. Ecol. Prog. Ser. 48, 235–249.

Malone, T.C., Ducklow, H.W., Peele, E.R., and Pike, S.E. (1991) Picoplankton carbon flux in Chesapeake Bay. Mar. Ecol. Prog. Ser. 78, 11–22.

Malone, T.C., Kemp, W.M., Ducklow, H.W., Boynton,, W.R., Tuttle, J.H., and Jonas, R.B. (1986) Lateral variation in the production and fate of phytoplankton in a partially stratified estuary. Mar. Ecol. Prog. Ser. 32, 149–160.

Mancuso, C.A., Franzmann, P.D., Pourtan, H.A., and Nichols, P.D. (1990) Microbial community structure and biomass estimates of a methanogenic Antarctic lake ecosystem as determined by phospholipid analyses. Microb. Ecol. 19, 73–95.

Mann, K.H. (1982) Ecology of Coastal Waters, A System Approach. University of California Press, Berkeley.

Mann, K.H., and Lazier, J.R.N. (1991) Dynamics of Marine Ecosystems—Biological–Physical Interactions in the Oceans. Blackwell Scientific Publications, Boston, MA.

Mannino, A., and Harvey, H.R. (1999) Lipid composition in particulate and dissolved organic matter in the Delaware Estuary: sources and diagenetic patterns. Geochim. Cosmochim. Acta 63, 2219–2235.

Mannino, A., and Harvey, H.R. (2000) Biochemical composition of particles and dissolved organic matter along an estuarine gradient: sources and implications for DOM reactivity. Limnol. Oceanogr. 45, 775–788.

Manny, B.A., and Wetzel, R.G. (1973) Diurnal changes in dissolved organic and inorganic carbon and nitrogen in a hard-water stream. Freshwater Biol. 3, 31–43.

Mantoura, R.F.C., Dickson, A., and Riley, J.P. (1978) The complexation of metals with humic materials in natural waters. Estuar. Coastal Shelf Sci. 6, 387–408.

Mantoura, R.F.C., Martin, J.M., and Wollast, R. (eds.) (1991) Ocean Margin Processes in Global Change. John Wiley, Chichester, UK.

Mantoura, R.F.C., and Woodward, E. (1983) Conservative behavior of riverine dissolved organic matter in the Severn estuary. Geochim. Cosmochim. Acta 47, 1293–1309.

Marchand, C., Baltzer, F., Lallier-Verges, E., and Alberic, P. (2004) Interstitial water chemistry in mangrove sediments in relationship to species composition and development stage (French Guiana). Mar. Geol. 208, 361–381.

Marinelli, R.L., Jahnke, R.A., Craven, D.B., Nelson, J.R., and Eckman, J.E. (1998) Sediment nutrient dynamics on the South Atlantic Bight continental shelf. Limnol. Oceanogr. 43, 1305–1320.

Marino, R., and Howarth, R.W. (1993) Atmospheric oxygen exchange in the Hudson River: dome measurements and comparison with other natural waters. Estuaries 16, 433–445.

Marino, R., Howarth, R.W., Shamess, J., and Prepas, E.E. (1990) Molybdenum and sulfate as controls on the abundance of nitrogen-fixing cyanobacteria in saline lakes in Alberta. Limnol. Oceanogr. 35, 245–259.

Mariotti, A., Germon, J.C., Hubert, P., Kaiser, P., Letolle, R., Tardieux, A., and Tardieux, P. (1981) Experimental determination of nitrogen kinetic isotope fractionation, some principles; illustration for the denitrification and nitrification principles. Plant Soil 62, 413–430.

Mariotti, A., Lancelot, C., and Billen, G. (1984) Natural isotopic composition of nitrogen as a tracer of origin for suspended matter in the Scheldt Estuary. Geochim. Cosmochim. Acta 48, 549–555.

Mariotti, A., Mariotti, F., Champigny, M.L., Amarger, N., and Moyse, A. (1982) Nitrogen isotope fractionation with nitrate reductase activity and uptake of NO_3^- by pearl millet. Plant Physiol. 69, 880–884.

Markaki, Z., Oikonomou, K., Kocak, M., Kouvarakis, G., Chaniotaki, A., Kubilay, N., and Mihalopoulos, N. (2003) Atmospheric deposition of inorganic phosphorus in the Levantine Basin, eastern Mediterranean: Spatial and temporal variability and its role in seawater productivity. Limnol. Oceanogr. 48, 1557–1568.

Marmorino, G.O., and Trump, C.L. (2000) Shore-based acoustic Doppler measurement of near-surface currents across a small embayment. J. Coastal Res. 16, 864–869.

Marsh, A.G. and K.R. Tenore (1990) The role of nutrition in regulating the population dynamics of opportunistic, surface deposit feeders in a mesohaline community. Limnol. Oceanogr. 35, 710–724.

Marsho, T.V., Burchard, R.P., and Fleming, R. (1975) Nitrogen fixation in the Rhode River estuary of Chesapeake Bay. Can. J. Microbiol. 21, 1348–1356.

Martens, C.S., and Berner, R.A. (1974) Methane production in the interstitial waters of sulfate depleted sediments. Science 185, 1067–1069.

Martens, C.S., and Chanton, J.P. (1989) Radon as a tracer of biogenic gas equilibration and transport from methane-saturated sediments. J. Geophys. Res. 94, 3451–3459.

Martens, C.S., Haddad, R.I., and Chanton, J.P. (1992) Organic matter accumulation, remineralization and burial in an anoxic marine sediment. In Productivity, Accumulation, and Preservation of Organic Matter in Recent and Ancient Sediments (Whelan, J.K., and Farrington, J.W., eds.), pp. 82–98, Columbia University Press, New York.

Martens, C.S., Kipphut, G.W., and Klump, J.V. (1980) Coastal sediment–water chemical exchange traced by in situ ^{222}Rn flux measurements. Science 208, 285–288.

Martens, C.S., and Klump, J.V. (1984) Biogeochemical cycling in an organic-rich coastal marine basin. 4. An organic carbon budget for sediments dominated by sulfate reduction and methanogenesis. Geochim. Cosmochim. Acta 48, 1987–2004.

Martin, F., Gonzalez-Vila, F.J., del Rio, J.C., and Verdejo, T. (1994) Pyrolysis derivatization of humic substances: I. Pyrolysis of fulvic acids in the presence of tetramethylammonium hydroxide. J. Anal. Appl. Pyrolysis 28, 71–80.

Martin, F., Gonzalez-Vila, F.J., del Rio, J.C., and Verdejo, T. (1995) Pyrolysis derivatization of humic substances: II. Pyrolysis of soil humic acids in the presence of tetramethylammonium hydroxide. J. Anal. Appl. Pyrolysis 31, 75–83.

Martin, J.B., Cable, J.E., Swarzenski, P.W., and Lindenberg, M.K. (2004) Mixing of ground and estuary waters: influences on ground water discharge and contaminant transport. Ground Water 42, 1000–1010.

Martin, J.M., Dai, M.H., and Cauwet, G. (1995) Significance of colloids in the biogeochemical cycling of organic carbon and trace metals in a coastal environment—example of the Venice Lagoon (Italy). Limnol. Oceanogr. 40, 119–131.

Martin, J.M., and Meybeck, M. (1979) Elemental mass-balance of material carried by major world rivers. Mar. Chem. 7, 173–206.

Martin, J.M., Meybeck, M., and Pusset, M. (1978b) Uranium behavior in the Zaire estuary. Netherlands J. Sea Res. 12, 338–344.

Martin, J.M., Mouchel, J.L., and Thomas, A.J. (1986) Time concepts in hydrodynamic systems with an application to ^{7}Be in the Gironde estuary. Mar. Chem. 18, 369–392.

Martin, J.M., Nijampurkar, V.M., and Salvadori, F. (1978a) Uranium and thorium isotope behavior in estuarine systems. In Biogeochemistry of Estuarine Sediments (Goldberg, E.D., ed.), pp. 111–127, UNESCO, Paris.

Martin, J.M., and Whitfield, M. (1981) The significance of river input of chemical elements to the ocean. In Trace Metals in the Sea (Wong, C.S., Boyle, E., Bruland, K.W., Burton, J.D., and Goldberg, E.D., eds.), pp. 265–296, Plenum Press, New York.

Martin, J.P. and Haider, K. (1986) Influence of mineral colloids on turnover rates of soil organic matter. *In* Interactions of Soil Minerals with Natural Organics and Microbes, (Huang, P.M., and Schnitzer, M., eds.), pp. 283–304, Soil Science Society of American Special Publication, Madison.

Martin, J.T., and Juniper, B.E. (1970) The Cuticles of Plants. Edward Arnold, London.

Maruya, K.A., and Lee, R.F. (1998) Aroclor 1268 and toxaphene in fish from a southeastern U.S. estuary. Environ. Sci. Technol. 32, 1069–1075.

Maruya, K.A., Loganathan, B.G., Kannan, K., McCumber-Kahn, S., and Lee, R.F. (1997) Organic and organometallic compounds in estuarine sediments from the Gulf of Mexico (1993–1994). Estuaries 20, 700–709.

Marvin-DiPasquale, M.C., Boynton, W.R., and Capone, D.G. (2003) Benthic sulfate reduction along the Chesapeake Bay central channel. II. Temporal controls. Mar. Ecol. Prog. Ser. 260, 55–70.

Marvin-DiPasquale, M.C., and Capone, D.G. (1998) Benthic sulfate reduction along the Chesapeake Bay central channel. I. Spatial trends and controls. Mar. Ecol. Prog. Ser. 168, 213–228.

Marvin-DiPasquale, M.C., and Oremland, R.S. (1998) Bacterial methylmercury degradation in Florida Everglades peat sediment. Environ. Sci. Technol. 32, 2556–2563.

Masiello, C.A., and Druffel, E.R.M. (1998) Balck carbon in deep-sea sediments. Science 280, 1911–1913.

Mason, R.P., Lawson, N.M., Lawrence, A.L., Leaner, J.J., Lee, J.G., and Sheu, G.R. (1999) Mercury in the Chesapeake Bay. Mar. Chem. 65, 77–96.

Mason, R.P., Reinfelder, J.R., and Morel, F.M.M. (1996) Uptake, toxicity, and trophic transfer of mercury in a coastal diatom. Environ. Sci. Technol. 30, 1835–1845.

Massé, A., Pringault, O., and de Wit, R. (2002) Experimental study of interactions between purple and green sulfur bacteria in sandy sediments exposed to illumination deprived of near-infrared wavelengths. Appl. Environ. Microbiol. 68, 2972–2981.

Matciak, M., Urbanski, J., Piekarek-Jankowska, H., and Szymelfenig, M. (2001) Presumable groundwater seepage influence on the upwelling events along the Hel Peninsula. Oceanol. Stud. 30, 125–132.

Mateo, M.A., Lizaso-Sanchez, J.L., and Romero, J. (2003) *Poisidonia oceania* 'banquettes': a preliminary assessment of the relevance for meadow carbon and nutrients budget. Estuar. Coastal Shelf Sci. 56, 85–90.

Matisoff, G. (1982) Mathematical models of bioturbation. *In* Animal–Sediment Relations (McCall, P.L., and Tevesz, M.J.S., eds.), pp. 289–330, Plenum Press, New York.

Matrai, P.A., and Vetter, R.D. (1988) Particulate thiols in coastal waters: The effect of light and nutrients on their planktonic production. Mar. Chem. 33, 624–631.

Mayer, L.M. (1982) Retention of riverine iron in estuaries. Geochim. Cosmochim. Acta 46, 1003–1009.

Mayer, L.M. (1994a) Surface area control of organic carbon accumulation on continental shelf sediments. Geochim. Cosmochim. Acta 58, 1271–1284.

Mayer, L.M. (1994b) Relationships between mineral surfaces and organic carbon concentrations in soils and sediments. Chem. Geol. 114, 347–363.

Mayer, L.M. (1999) Extent of coverage of mineral surfaces by organic matter in marine sediments. Geochim. Cosmochim. Acta 63, 207–215.

Mayer, L.M., Keil, R.G., Macko, S.A., Joye, S.B., Ruttenburg, K.C., and Aller, R.C. (1998) Importance of suspended particulates in riverine delivery of bioavailable nitrogen to coastal zones. Global Biogeochem. Cycles 12: 573–579.

Mayer, L.M., Schick, L.S., Hardy, K.R., Wagai, R., and McCarthy, J. (2004) Organic matter in small mesopores in sediments and soils. Geochim. Cosmochim. Acta 68, 3863–3872.

Mayer, L.M., Schick, L.S., Sawyer, T., Plante, C.J., Jumars, P.A., and Self, R.L. (1995) Bioavailable amino acids in sediments: A biomimetic, kinetic-based approach. Limnol. Oceanogr. 40, 511–520.

Mayer, L.M., Schick, L.S., and Setchell, F.S. (1986) Measurement of protein in nearshore marine sediments. Mar. Ecol. Prog. Ser. 30, 159–165.

Mazeas, L., and Budzinski, H. (2001) Polycyclic aromatic hydrocarbon $^{13}C/^{12}C$ ratio measurement in petroleum and marine sediments: application to standard reference material and a sediment suspected of contamination from Erika oil spill. J. Chrom. 923, 165–176.

Mazurek, M. A., and Simoneit, B.R.T. (1984) Characterization of biogenic and petroleum-derived organic matter in aerosols over remote, rural and urban areas. *In* Identification and Analysis of Organic Pollutants in Air (Keith, L.H., ed.), pp. 353–370, Ann Arbor Science/Butterworth, Boston, MA.

McCaffrey, R.J., and Thomson, J. (1980) A record of the accumulation of sediments and trace metals in a Connecticut salt marsh. *In* Advances in Geophysics, Estuarine Physics and Chemistry: Studies in Long Island Sound (Saltzman, B., ed.), pp. 165–236, Academic Press, New York.

McCall, P.L., and Fisher, J.B. (1980) Effects of tubificid oligochaetes on physical and chemical properties of Lake Erie sediments. *In* Aquatic Oligochaetes Biology (Brinkhurst, K.O., and Cook, D.G., eds.), pp. 253–318, Plenum Press, New York.

McCall, P.L., and Tevesz, M.J.S. (1982) The effects of benthos on physical properties of freshwater sediments. *In* Animal–Sediment Relations: The Biogenic Alteration of Sediments, Topics in Geobiology, Vol. 2 (McCall, P.L., and Tevesz, M.J., eds.), pp. 105–176, Plenum Press, New York.

McCallister, S.L., Bauer, J.E., Cherrier, J.E., and Ducklow, H.W. (2004) Assessing sources and ages of organic matter supporting river and estuarine bacterial production: A multiple-isotope ($\Delta^{14}C$, $\delta^{13}C$, and $\delta^{15}N$) approach. Limnol. Oceanogr. 49, 1687–1702.

McCarthy, J.F., Roberson, L.E., and Burrus, L.W. (1989) Association of benzo[a] pyrene with dissolved organic matter: prediction of K_{dom} from structural and chemical properties of the organic matter. Chemosphere 19, 1911–1920.

McCarthy, M.D., Hedges, J.I., and Benner, R. (1998) Major bacterial contribution to marine dissolved organic nitrogen. Science 281, 231–234.

McCave, I.N., (ed.) (1976) The Benthic Boundary Layer. Plenum Press, New York.

McClelland, J.W., Valiela, I. (1998) Linking nitrogen in estuarine producers to land-derived sources. Limnol. Oceanogr. 43, 577–585.

McClelland, J.W., Valiela, I., and Michener, R.H. (1997) Nitrogen-stable isotope signatures in estuarine food webs: A record of increasing urbanization in coastal watersheds. Limnol. Oceanogr. 42, 930–937.

McCready, R.G.L., Gould, W.D., and Barendregt, R.W. (1983) Nitrogen isotope fractionation during the reduction of NO_3^- to NH_4^+ by *Desulvovibrio* sp. Can. J. Microbiol. 29: 231–234.

McDonnell, J., and Kendall, C. (1994) Isotope Tracers in Catchment Hydrology. Elsevier, Amsterdam.

McFarlan, E. (1961) Radiocarbon dating of Late Quaternary deposits, South Louisiana. Geol. Soc. Am. Bull. 72, 129–158.

McGlathery, K.J., Krause-Jensen, D., Rysgaard, S., and Christensen, P.B. (1997) Patterns of ammonium uptake within dense mats of the filamentous macroalga *Chaetomorpha linum*. Aquat. Bot. 59, 99–115.

McGowen, J.H., and Scott, A.J. (1975) Hurricanes as geologic agents on the Texas Coast. *In* Estuarine Research, Vol. 2. Geology and Engineering (Cronin, L.E., ed.), pp. 23–46, Academic Press, New York.

McGroddy, S.E., and Farrington, J.W. (1995) Sediment porewater partitioning of ploy-cyclic aromatic hydrocarbons in three cores from Boston Harbor, Massachusetts. Environ. Sci. Technol. 29, 1542–1550.

McGroddy, S.E., Farrington, J.W., and Gschwend, P.M. (1996) Comparison of in situ and desorption sediment–water partitioning of polycyclic aromatic hydrocarbons and polychlorinated biphenyls. Environ. Sci. Technol. 30, 172–177.

McKee, B. (1972) Cascadia: The Geological Evolution of the Pacific Northwest. McGraw-Hill, New York.

McKee, B.A., Aller, R.C., Allison, M.A., Bianchi, T.S., and Kineke, G.C. (2004) Transport and transformation of dissolved and particulate materials on continental margins by major rivers: benthic boundary layer and seabed processes. Cont. Shelf Res. 24, 899–926.

McKee, B.A., and Baskaran, M. (1999) Sedimentary processes of the Gulf of Mexico. In Biogeochemistry of Gulf of Mexico Estuaries (Bianchi, T.S., Pennock, R., and Twilley, R.R., eds.), pp. 63–81, John Wiley, New York.

McKee, B.A., DeMaster, D.J., and Nittrouer, C.A. (1984) The use of ^{234}Th/^{238}U disequilibrium to examine the fate of particle-reactive species on the Yangtze continental shelf. Earth Planet. Sci. Lett. 68, 431–442.

McKee, B.A., DeMaster, D.J., and Nittrouer, C.A. (1986) Temporal variability in the partitioning of thorium between dissolved and particulate phases on the Amazon shelf: implications for the scavenging of particle-reactive species. Cont. Shelf Res. 6, 87–106.

McKee, B.A., DeMaster, D.J., and Nittrouer, C.A. (1987) Uranium geochemistry on the Amazon Shelf: evidence for uranium release from bottom sediments. Geochim. Cosmochim. Acta 51, 2779–2786.

McKee, B.A., Nittrouer, C.A., and DeMaster, D.J. (1983) Concepts of sediment deposition and accumulation applied to the continental shelf near the mouth of the Yangtze River. Geology 11, 631–633.

McKee, B.A., and Skei, J. (1999) Framvaren Fjord as a natural laboratory for examining biogeochemical processes in anoxic environments. Mar. Chem. 67, 147–148.

McKee, B.A., and Todd, J.F. (1993) Uranium behavior in a permanently anoxic fjord: microbial control? Limnol. Oceanogr. 38, 408–414.

McKee, B.A., Wiseman, W., and Inoue, M. (1995) Salt water intrusion and sediment dynamics in a bar-built estuary: Terrebonne Bay, LA. In Changes in Fluxes in Estuaries, pp. 13–16, Olsen and Olsen, Copenhagen.

McKee, K.L., Mendelssohn, I.A., and Hester, M.W. (1988) Reexamination of pore water sulfide concentrations and redox potentials near the aerial roots of Rhizophora mangle and Avicenna germinans. Am. J. Bot. 75: 1352–1359.

McKee, L.J., Eyre, B.D., and Hossan, S. (2000) Transport and retention of nitrogen and phosphorus in the sub-tropical Richmond River estuary, Australia. Biogeochemistry 50, 241–278.

McKelvie, I.D., Peat, D.M., and Worsfold, P.J. (1995) Techniques for the quantification and speciation of phosphorus in natural waters. Anal. Proc. Incl. Anal. Comm. 32, 437–445.

McKenna, T.E., and Martin, J.B. (2004) Ground water discharge to estuarine and coastal ocean environments. Ground Water 42, 1–5.

McKenzie, L., and Campbell, S. (2003) Seagrass resources of the Booral Wetlands and the Great Sandy Straight. Queensland Department of Primary Industries Information Series, No Q103016, QDPI, Brisbane.

McKinney, D.E., Carson, D.M., Clifford, D.J., Minard, R.D., and Hatcher, P.G. (1995) Off-line thermochemolysis versus flash pyrolysis for the in situ methylation of lignin: is pyrolysis necessary? J. Anal. Appl. Pyrol. 34, 41–46.

McKnight, D.M., and Aiken, G.R. (1998) Sources and age of aquatic humus. *In* Aquatic Humic Substances: Ecology and Biogeochemistry (Hessen, D.O., and Tranvik, L.J., eds.), pp. 9–39, Springer-Verlag, Berlin.

McKnight, D.M., Boyer, E.W., Westerhoff, P.K., Doran, P.T., Kulbe, T., and Andersen, D.T. (2001) Spectrofluormetric characterization of dissolved organic matter for indication of precursor organic material and aromaticity. Limnol. Oceanogr. 46, 38–48.

McKnight, D.M., Hood, E., and Klapper, L. (2003) Trace organic moieties of dissolved organic material in natural waters. *In* Aquatic Ecosystems: Interactivity of Dissolved Organic Matter (Findlay, S.E.G, and Sinsabaugh, R.L., eds.), pp. 71–93, Academic Press, New York.

McManus, J. (2002) Deltaic responses to changes in river regimes. Mar. Chem. 79, 155–170.

McManus, J., Berelson, W.M., Coale, K.H., Johnson, K.S., and Kilgore, T.E. (1997) Phosphorus regeneration in continental margin sediments. Geochim. Cosmochim. Acta 61, 2891–2902.

McManus, J., Hammond, D.E., Berelson, W.M., Kilgore, T.E., DeMaster, D.J., Ragueneau, O.G, and Collier, R.W. (1995) Early diagenesis of biogenic opal: dissolution rates, kinetics and paleoceanographic implication. Deep-Sea Res. II 38, 1481–1516.

McNichol, A.P., Ertel, J.R., and Eglinton, T.I. (2000) The radiocarbon content of individual lignin-derived phenols: technique and initial results. Radiocarbon 42, 219–227.

McPherson, B.F., and Miller, R.L. (1987) The vertical attenuation of light in Charlotte Harbor, a shallow, subtropical estuary, southwestern Florida. Estuar. Coastal Shelf Sci. 25, 721–737.

McVeety, B.D., and Hites, R.A. (1988) Atmospheric deposition of polycyclic aromatic hydrocarbons to water surfaces: a mass balance approach. Atmos. Environ. 22, 511–536.

Meade, R.H. (1969) Landward transport of bottom sediments in estuaries of the Atlantic coastal plain. J. Sed. Petrol. 39, 222–234.

Meade, R.H. (1996) River-sediment inputs to major deltas. *In* Sea Level Rise and Coastal Subsidence (J. Milliman J.D., and Haq, B.U., eds.), pp. 63–85, Kluwer Academic, Dordrecht, The Netherlands.

Meade, R.H., Dunne, T., Richey, J.E., Santos, U., and Salati, E. (1985) Storage and remobilization of suspended sediment in the lower Amazon River of Brazil. Science 228, 488–490.

Meade, R.H., and Parker, R.S. (1985) Sediment in rivers of the United States. *In* National Water Summary 1984—Hydrologic Events, Selected Water Quality Trends, and Groundwater Resources. U.S. Geol. Survey Water Supply Paper no. 2275, 1–467.

Means, J.C. (1995) Influence of salinity upon sediment–water partitioning of aromatic hydrocarbons. Mar. Chem. 51, 3–16.

Means, J.C., and Wijayaratne, R. (1982) Role of natural colloids in transport of hydrophobic pollutants. Science 215, 968–970.

Means, J.C., Wood, S.G, Hassett, J.J., and Banwart, W.L. (1980) Sorption properties of polynuclear aromatic hydrocarbons by sediments and soils. Environ. Sci. Technol. 14, 1524–1528.

Meentemeyer, V. (1978) Macroclimate and lignin control of litter decomposition rates. Ecology 59, 465–472.

Megens, L., van der Plicht, de Leuw, J.W., and Smedes, F. (2002) Stable carbon and radiocarbon isotope composition of particle size fractions to determine origins of sedimentary organic matter in an estuary. Org. Geochem. 33, 945–952.

Mei, M.L., and Danovaro, R. (2004) Virus production and life strategies in aquatic sediments. Limnol. Oceanogr. 49, 459–470.

Meister, A., and Anderson, M.E. (1983) Glutathione. Ann. Rev. Biochem. 52, 711–760.

Mendelssohn, I.A., McKee, K.L., and Patrick, W.H. (1981) Oxygen deficiency in *Spartina alterniflora* roots: metabolic adaptation to anoxia. Science 439–441.

Merriam-Webster (1979) Webster's New Collegiate Dictionary, G and C Merriam Co., Springfield, MA.

Meybeck, M. (1979) Concentration des eaux fluviales en elements majeurs et apports en solution aux oceans. Rev. Geol. Dyam. Geogr. Phys. 21, 215–246.

Meybeck, M. (1982) Carbon, nitrogen, and phosphorus transport by world rivers. Am. J. Sci. 282, 401–450.

Meybeck, M. (1983) Atmospheric inputs and river transport of dissolved substances in dissolved loads of rivers and surface water quantity/quality relationships. Intl. Union of Geodesy and Geophysics, Hamburg, Germany. IAHS Publication, pp. 173–192.

Meybeck, M., (1993) C, N, P and S in rivers: from sources to global inputs. *In* Interaction of C, N, phosphorus and S Biogeochemical Cycles and global Change (Wollast, R., Mackenzie, F.T., and Chou, L., eds.), pp. 163–193, NATO ASI Series I, Vol. 4, Springer-Verlag, Berlin.

Meybeck, M. (1998) Man and river interface: multiple impacts on water quality illustrated by the River Seine. Hydrobiologia 373/374, 1–20.

Meybeck, M. (2002) Riverine quality at the Anthropocene: propositions for global space and timer analysis, illustrated by the Seine River. Aquat. Sci, 64, 376–393.

Meybeck, M. (2003) Global analysis of river systems: from Earth system controls to Anthropocene syndromes. Phil. Trans. R. Soc. Lond. 358, 1935–1955.

Meybeck, M., and Vörösmarty, C. (2004) Fluvial filtering of land-to-ocean fluxes: from natural Holocene variations to Anthropocene. C.R. Geoscience 337, 107–123.

Meyer-Harms, B., and von Bodungen, B. (1997) Taxon-specific ingestion rates of natural phytoplankton by calanoid copepods in an estuarine environment (Pomeranian Bight, Baltic Sea) determined by cell counts and HPLC analyses of marker pigments. Mar. Ecol. Prog. Ser. 153, 181–190.

Meyers, P.A. (1994) Preservation of elemental and isotopic identification of sedimentary organic matter. Chem. Geol. 144, 289–302.

Meyers, P.A. (1997) Organic geochemical proxies of paleoceanographic, paleolimnologic, and paleoclimatic processes. Org. Geochem. 27, 213–250.

Meyers, P.A. (2003) Applications of organic geochemistry to paleolimnological reconstructions: a summary of examples from the Laurentian Great Lakes. Org. Geochem. 34, 261–290.

Meyers, P.A., and Eadie, B.J. (1993) Sources, degradation, and resynthesis of the organic matter on sinking particles in Lake Michigan. Org. Geochem. 20, 47–56.

Meyers, P.A., and Ishiwatari, R. (1993) Lacustrine organic geochemistry—an overview of indicators of organic matter sources and diagenesis in lake sediments. Org. Geochem. 20, 867–900.

Meyers, P.A., and Quinn, J.G. (1973) Factors affecting the association of fatty acids with mineral particles in sea water. Geochim. Cosmochim. Acta 37, 1745–1759.

Meyers, P.A., and Takeuchi, N. (1981) Environmental changes in Saginaw Bay, Lake Huron, recorded by geolipid contents of sediments deposited since 1800. Environ. Geol. 3, 257–266.

Meyerson, L.A., Saltonstall, K., Windham, L., Kiviat, E., and Findlay, S.E.G. (2000) A comparison of *Phragmites australis* in freshwater and brackish marsh environments in North America. Wetlands Ecol. Manag. 8, 89–103.

Meyers-Schulte, K.J., and Hedges, J.I. (1986) Molecular evidence for terrestrial component of organic matter dissolved in ocean water. Nature 321, 61–63.

Meziane, T., Bodineau, L., Retiere, C., and Thoumelin, G. (1997) The use of lipid markers to define sources of organic matter in sediment and food web of the intertidal salt-marsh-flat ecosystem of Mont-Saint-Michel Bay, France. J. Sea Res. 38, 47–58.

Michalopoulos, P., and Aller, R.C. (1995) Rapid clay mineral formation in Amazon delta sediments: reverse weathering and oceanic elemental cycles. Science 270, 614–617.

Michalopoulos, P., and Aller, R.C. (2004) Early diagenesis of biogenic silica in the Amazon Delta: alteration, authigenic clay formation, and storage. Geochim. Cosmochim. Acta 68, 1061–1085.

Michalopoulos, P., Aller, R.C., and Reeder, R. (2000) Conversion of diatoms to clay minerals during early diagenesis in tropical, continental shelf muds. Geology 28, 1095–1098.

Michener, R.H., and Schell, D.M. (1994) Stable isotope ratios as tracers in marine aquatic food webs. In Stable Isotopes in Ecology and Environmental Science (Lajtha, K., and Michener, R. eds.), pp. 138–157, Blackwell Scientific, Oxford.

Middelboe, M., Borch, N.H., and Kirchman, D.L. (1995) Bacterial utilization of dissolved free amino acids, dissolved combined amino acids and ammonium in the Delaware Bay estuary: effects of carbon and nitrogen limitation. Mar. Ecol. Prog. Ser. 128, 109–120.

Middelburg, J.J. (1989) A simple rate model for organic matter decomposition in marine sediments. Geochim. Cosmochim. Acta 53, 1577–1581.

Middelburg, J.J., Klaver, G., Nieuwenhuize, J., Markusse, R.M., Vlug, T., and van der Nat, J.W.A. (1995) Nitrous oxide emissions from estuarine intertidal sediments. Hydrobiologia 311, 43–55.

Middelburg, J.J., Klaver, G, Nieuwenhuize, J., Wielemaker, A., de Haas, W., and van der Nat, J.F.W.A (1996) Organic matter mineralization in intertidal sediments along an estuarine gradient. Mar. Ecol. Prog. Ser. 132, 157–168.

Middelburg, J.J., and Nieuwenhuize, J. (2000) Nitrogen uptake by heterotrophic bacteria and phytoplankton in the nitrate-rich Thames River. Mar. Ecol. Prog. Ser. 203, 13–21.

Middelburg, J.J., and Nieuwenhuize, J. (2001) Nitrogen isotope tracing of dissolved inorganic nitrogen behavior in tidal estuaries. Estuar. Coastal Shelf Sci. 53, 385–391.

Middelburg, J.J., Nieuwenhuize, J., Iverson, N., Hogh, N., De Wilde, H., Helder, W., Seifert, R., and Christof, O. (2002) Methane distribution in European tidal estuaries. Biogeochemistry 59, 95–119.

Middelburg, J.J., and Soetaert, K. (2003) The role of sediments in shelf ecosystem dynamics. In The Sea, Chap. 11, Vol. 13 (Robinson, A.R., McCarthy, J., and Rothsthild, B.J., eds.), pp. 353–373, The President and Fellows of Harvard College, Boston, MA.

Middelburg, J.J., Soetaert, K., and Herman, P.M.J. (1997) Empirical relationships for use in global diagenetic models. Deep Sea Res. 44, 327–344.

Middleton, G.V., and Southward, J.B. (1984) The Mechanics of Sediment Movement. Society for Sedimentary Geology, Short Course No. 3, RI. Society for Sedimentary Geology, Tulsa, OK.

Migniot, C. (1968) Tassement et rheologie des vases. La Houille Blanche 1, 11–111.

Mihalopoulos, N., Nguyen, B.C., and Belviso, S. (1992) The oceanic source of carbonyl sulfide (COS). Atmos. Environ. 26, 1383–1394.

Milan, C.S., Swenson, E.M., Turner, R.E., and Lee, J.M. (1995) Assessment of estimating sediment accumulation rates: Louisiana saltmarshes. J. Coastal Res. 11, 296–307.

Miley, G.A., and Kiene, R.P. (2004) Sulfate reduction and porewater chemistry in a Gulf coast Juncus roemerianus (Needlerush) marsh. Estuaries 27, 472–481.

Miller, R.L., and McPherson, B.F. (1991) Estimating estuarine flushing and residence times in Charlotte Harbor, Florida, via salt balance and a box model. Limnol. Oceanogr. 36, 602–612.

Miller, W.L., and Moran, M.A. (1997) Interaction of photochemical and microbial processes in the degradation of refractory dissolved organic matter from a coastal marine environment. Limnol. Oceanogr. 42, 1317–1324.

Miller, W.L., and Zepp, R.G. (1995) Photochemical production of dissolved inorganic carbon from terrestrial organic matter: significance to the oceanic organic carbon cycle. Geophys. Res. Lett. 22(4), 417–420.

Millero, F.J. (1982) The effect of pressure on the solubility of minerals in water and seawater. Geochim. Cosmochim. Acta 46, 11–22.

Millero, F.J. (1985) The effect of ionic interactions in the oxidation of metals in natural waters. Geochim. Cosmochim. Acta 49, 547–553.

Millero, F.J. (1995) Thermodynamics of the carbon dioxide system in the ocean. Geochim. Cosmochim. Acta 59, 661–677.

Millero, F.J. (1996) Chemical Oceanography, 2nd edn. CRC Press, Boca Raton, FL.

Millero, F.J., and Hawke, D.J. (1992) Ionic interactions of divalent metals in natural waters. Mar. Chem. 40, 19–48.

Miller-Way, T., Boland, G.S., Rowe, G.T., and Twilley, R.R. (1994) Sediment oxygen consumption and benthic nutrient fluxes on the Louisiana continental shelf: a methodological comparison. Estuaries 17, 809–815.

Millie, D.F., Paerl, H.W., and Hurley, J.P. (1993) Microalgal pigment assessments using high-performance liquid chromatography: a synopsis of organismal and ecological applications. Can. J. Fish. Aquat. Sci. 50, 2513–2527.

Milligan, T.G., Kineke, G.C., Blake, A.C., Alexander, C.R., and Hill, P.S. (2001) Flocculation and sedimentation in the ACE Basin, South Carolina. Estuaries 24, 734–744.

Milliman, J.D. (1980) Sedimentation in the Fraser River and its estuary, southwestern British Columbia (Canada). Estuar. Coastal Shelf Sci. 10, 609–633.

Milliman, J.D., and Haq, B.U. (eds.) (1996) Sea-level rise and coastal subsidence: towards meaningful strategies. In Sea-level Rise and Coastal Subsidence pp. 1–9. Kluwer Academic Dordrecht, The Netherlands.

Milliman, J.D., and Meade, R.H. (1983) World-wide delivery of river sediment to the oceans. J. Geol. 91, 1–21.

Milliman J.D., Qin,Y.S., Ren, M.E., and Saito, Y. (1987) Man's influence on the erosion and transport of sediment by Asian rivers: the Yellow River (Huanghe) example. J. Geol. 95, 751–762.

Milliman, J.D., and Syvitski, J.P.M. (1992) Geomorphic tectonic control of sediment discharge to the ocean—the importance of small mountainous rivers. J. Geol. 100, 525–554.

Millward, G.E., and Turner, A. (1995) Trace metals in estuaries. In Trace Elements in Natural Waters (Salbu, B., and Steinnes, E., eds.), pp. 223–245, CRC Press, Boca Raton, FL.

Miltner, A., and Zech, W. (1998) Beech leaf litter lignin degradation and transformation as influenced by mineral phases. Org. Geochem. 28, 457–463.

Minagawa, M., and Tsunogai, S. (1980) Removal of ^{234}Th from a coastal sea: Funka Bay, Japan. Earth Planet. Sci. Lett. 47, 51–64.

Minagawa, M., and Wada, E. (1984) Stepwise enrichment of ^{15}N along food chains: further evidence and the relation between δ^{15}N and animal age. Geochim. Cosmochim. Acta 48, 1135–1140.

Minor, E.C., Boon, J.J., Harvey, H.R., and Mannino, A. (2001) Estuarine organic matter composition as probed by direct temperature-resolved mass spectrometry and traditional geochemical techniques. Geochim. Cosmochim. Acta 65, 2819–2834.

Minor, E.C., Simjouw, J.P., Boon, J.J., Kerkhoff, A.E., and van der Horst, J. (2002) Estuarine/marine UDOM as characterized by size-exclusion chromatography and organic mass spectrometry. Mar. Chem. 78, 75–102.

Mistri, M. (2002) Ecological characteristics of the invasive Asian Date mussel, *Musculita senhousia*, in the Sacca de Goro (Adriatic Sea, Italy). Estuaries 25, 431–440.

Mitchell, J.G., Okubo, A., and Fuhrman, J.A. (1985) Microzones surrounding phytoplankton form the basis for a stratified marine microbial ecosystem. Nature 316, 58–59.

Mitra, S., Bianchi, T.S., Guo, L., and Santschi, P.H. (2000a) Terrestrially-derived dissolved organic matter in Chesapeake Bay and the Middle Atlantic Bight. Geochim. Cosmochim. Acta 64, 3547–3557.

Mitra, S., Bianchi, T.S., McKee, B.A., and Sutula, M. (2002) Black carbon from the Mississippi River: quantities, sources, and potential implications for the global carbon cycle. Environ. Sci. Technol. 36, 2296–2302.

Mitra, S., Dellapenna, T.M., and Dickhut, R.M. (1999a) Polycyclic aromatic hydrocarbon distribution within lower Hudson River estuarine sediments: physical mixing vs. sediment geochemistry. Estuar. Coastal Shelf Sci. 49, 311–326.

Mitra, S., and Dickhut, R.M. (1999) Three-phase modeling of polycyclic aromatic hydrocarbon association with pore-water-dissolved organic carbon. Environ. Toxicol. Chem. 18, 1144–1148.

Mitra, S., Dickhut, R.M., Kuehl, S.A., and Kimbrough, K.L. (1999b) Polycyclic aromatic hydrocarbon (PAH) source, sediments deposition patterns, and particle geochemistry as factors influencing PAH distribution coefficients in sediments of the Elizabeth River, VA, USA. Mar. Chem. 66, 113–127.

Mitra, S., Klerks, P.L., Bianchi, T.S., Means, J., and Carman, K.R. (2000b) Effects of estuarine organic matter biogeochemistry on the bioaccumulation of PAHs by two epibenthic species. Estuaries 23, 864–876.

Miyake, Y., and Wada, E. (1971) The isotope effect on the nitrogen in biochemical, oxidation–reduction reactions. Rec. Oceanogr. Works Jpn. 11, 1–6.

Mobed, J.J., Hemmingsen, S.L., Autry, J.L., and McGowan, L.B. (1996) Fluorescence characterization of IHSS humic substances: total luminescence spectra with absorbance correction. Environ. Sci. Technol. 30, 3061–3065.

Moers, M.E.C., and Larter, S.R. (1993) Neutral monosaccharides from a hypersaline tropical environment: applications to the characterization of modern and ancient ecosystems. Geochim. Cosmochim. Acta 57, 3063–3071.

Moffett, J.W., and Brand, L.E. (1996) Production of strong, extracellular Cu chelators by marine cyanobacteria in response to Cu stress. Limnol. Oceanogr. 41, 388–395.

Moffett, J.W., Brand, L.E., Croot, P.L., and Barbeau, K.A. (1997) Cu speciation and cyanobacterial distribution in harbors subject to anthropogenic Cu inputs. Limnol. Oceanogr. 42, 789–799.

Moisander, P.H., and Paerl, H.W. (2000) growth, primary productivity, and nitrogen fixation potential of Nodularia spp. (Cyanophyceae) in water from a subtropical estuary in the United States. J. Phycol. 36, 645–658.

Monbet, Y. (1992) Control of phytoplankton biomass in estuaries: a comparative analysis of microtidal and macrotidal estuaries. Estuaries 15, 563–571.

Montagna, P.A. (1989) Nitrogen process studies (NIPS): the effects of freshwater inflow on benthos communities and dynamics. Final report to the Texas Water Development Board, Austin, TX. UT Marine Science Institute Technical Report No. TR/89–011.

Montagna, P.A., Blanchard, G.F., and Dinet, A. (1995) Effect of production and biomass of intertidal microphytobenthos on meiofaunal grazing rates. J. Exp. Mar. Biol. Ecol. 185,149–165.

Montgomery, D.R., Zabowski, D., Ugolini, F.C., Hallerg, R.O., and Spaltenstein, H. (2000) Soils, watershed processes, and marine sediments. *In* Earth System Science, from Biogeochemical Cycles to Global Change (Jacobson, M.C., Charlson, R.J., Rodhe, H, and Orians, G.H., eds.), pp. 159–194, International Geophysics Series, Academic Press, New York.

Montoya, J.P. (1994) Nitrogen isotope fractionation in the modern ocean: implications for the sedimentary record. *In* Carbon Cycling in the Glacial Ocean: Constraints on the Ocean's Role in Global Change (Zahn, R., Pedersen, T.F., Kaminski, M.A., and Labeyrie, L., eds.), pp. 259–280, Springe, Berlin.

Mook, J.G., and Tan, F.C., (1991) Stable carbon isotopes in rivers and estuaries. *In* Biogeochemistry of Major World Rivers. SCOPE, 245–264.

Moore, D.G., and Scott, M.R. (1986) Behavior of ^{226}Ra in the Mississippi River mixing zone. J. Geophys. Res. 91, 14317–14329.

Moore, H.E., Poet, S.E., and Martell, E.A. (1973) ^{210}Bi and ^{210}Po profiles and aerosol Residence times versus altitude. J. Geophys. Res. 78, 7065–7075.

Moore, R.M., Burton, J.D., Willimas, P.L., and Young, M.L. (1979) The behavior of dissolved organic material, iron, and manganese in estuarine mixing. Geochim. Cosmochim. Acta 43, 919–926.

Moore, W.S. (1967) Amazon and Mississippi river concentrations of uranium, thorium and radium isotopes. Earth Planet. Sci. Lett. 2, 21–234.

Moore, W.S., (1992) Radionuclides of the uranium and thorium decay series in the estuarine environment. *In* Uranium-Series Disequilibrium, 2nd edn. (Ivanovich, M., and Harmon, R.S., eds.), pp. 396–422, Clarendon Press, New York.

Moore, W.S. (1996) Large groundwater inputs to coastal waters revealed by ^{226}Ra enrichments. Nature 380, 612–614.

Moore, W.S. (1999) The subterranean estuary: a reaction zone of groundwater and sea water. Mar. Chem. 65, 111–125.

Moore, W.S. (2003) Sources and fluxes of submarine groundwater discharge delineated by radium isotopes. Biogeochemistry 66, 75–93.

Moore, W.S., and Edmond, J.M. (1984) Radium and barium in the Amazon River system. J. Geophys. Res. 89, 2061–2065.

Moore, W.S., and Krest, J. (2004) Distribution of ^{223}Ra and ^{224}Ra in the plumes of the Mississippi and Atchafalaya Rivers and the Gulf of Mexico. Mar. Chem. 86, 105–119.

Moore, W.S., and Todd, J.F. (1993) Radium isotopes in the Orinoco estuary and eastern Caribbean Sea. J. Geophys. Res. 98, 2233–2244.

Mopper, K. (1977) Sugars and uronic acids in sediment and water from the Black Sea and North Sea with emphasis on analytical techniques. Mar. Chem. 5, 585–603.

Mopper, K., and Larsson, K. (1978) Uronic and other organic acids in Baltic Sea and Black Sea sediments. Geochim. Cosmochim. Acta 42, 153–163.

Mopper, K., and Lindroth, P. (1982) Diel and depth variations in dissolved free amino acids and ammonium in the Baltic Sea determined by shipboard HPLC analysis. Limnol. Oceanogr. 27, 336–347.

Mopper, K., Zhou, X., Kieber, R.J., Kieber, D.J., Sikorski, R.J., and Jones, R.D. (1991) Photochemical degradation of dissolved organic carbon and its impact on the ocean carbon cycle. Nature 353, 60–62.

Moran, M.A., and Covert, J.S. (2003) Photochemically mediated linkages between dissolved organic matter and bacterioplankton. *In* Aquatic Ecosystems: Interactivity of Dissolved Organic Matter (Findlay, S.E.G., and Sinsabaugh, R.L., eds.), pp. 244–259, Academic Press, New York.

Moran, M.A., and Hodson, R.E. (1989a) Formation and bacterial utilization of dissolved organic carbon derived from detrital lignocellulose. Limnol. Oceanogr. 34, 1034–1037.

Moran, M.A., and Hodson, R.E. (1989b) Bacterial secondary production on vascular plant detritus: relationships to detritus composition and degradation rate. Appl. Environ. Microbiol. 55, 2178–2189.

Moran, M.A., and Hodson, R.E. (1994) Dissolved humic substances of vascular plant origin in a coastal marine environment. Limnol. Oceanogr. 39, 762–771.

Moran, M.A., Sheldon, W.M., and Zepp, R.G. (2000) Carbon loss and optical property changes during long-term photochemical and biological degradation of estuarine dissolved organic matter. Limnol. Oceanogr. 45, 1254–1264.

Moran, M.A., Wicks, R.J., and Hodson, R.E. (1991) Export of dissolved organic matter from a mangrove swamp ecosystem: evidence from natural fluorescence, dissolved lignin phenols, and bacterial secondary production. Mar. Ecol. Prog. Ser. 76, 175–184.

Moran, M.A., and Zepp, R.G. (1997) Role of photoreactions in the formation of biologically labile compounds from dissolved organic matter. Limnol. Oceanogr. 42, 1307–1316.

Morel, F.M. (1983) Principles of Aquatic Chemistry. John Wiley, New York.

Morel, F.M.M., Dzombak, D.A., and Price, N.M. (1991) Heterogeneous reactions in coastal waters. In Ocean Margin Processes in Global Change (Mantoura. R.F.C., Martin, J.M., and Wollast, R., eds.), pp. 165–180, John Wiley, New York.

Morel, F.M.M., and Hering, J.G. (1993) Principles and Applications of Aquatic Chemistry. John Wiley, New York.

Morel, F.M.M., and Hudson, R.J.M. (1985) The geobiological cycle of trace elements in aquatic sediments: Redfield revisited. In Chemical Processes in Lakes (Stumm, W., ed.), pp. 251–281, John Wiley, New York.

Morris, A.W., Bale, A.J., Howland, R.J., Loring, D.H., and Rantala, R.T.T. (1987) Controls on the chemical composition of particle compositions in a macrotidal estuary (Tamar estuary, UK). Cont. Shelf Res. 7, 1351–1355.

Morris, A.W., and Riley, J.P. (1966) The bromide/chlorinity and sulphate/chlorinity ratio in sea water. Deep-Sea Res. 13, 699–705.

Morse, J.W., and Arakaki, T. (1993) Adsorption and coprecipitation of divalent metals with mackinawite (FeS). Geochim. Cosmochim. Acta 57, 3635–3640.

Morse, J.W., and Cook, N. (1978) The distribution and form of phosphorus in North Atlantic Ocean deep-sea and continental slope sediments. Limnol. Oceanogr. 23, 825–830.

Morse, J.W., and Cornwell, J.C. (1987) Analysis and distribution of iron sulfide minerals in recent anoxic marine sediments. Mar. Chem. 22, 55–69.

Morse, J.W., Cornwell, J.C., Arakaki, T., Lin, S., and Huerta-Diaz, M.A. (1992) Iron sulfide and carbonate mineral diagenesis in Baffin Bay, Texas. J. Sed. Petrol. 62, 671–680.

Morse, J.W., Presley, B.J., Taylor, R.J., and Santschi, P.H. (1993) Trace metal chemistry of Galveston Bay: water, sediments and biota. Mar. Environ. Res. 36, 1–37.

Morse, J.W., and Rowe, G.T. (1999) Benthic biogeochemistry beneath the Mississippi River plume. Estuaries 22, 206–214.

Morse, J.W., and Wang, Q. (1997) Pyrite formation under conditions approximating those in anoxic sediments: II. Influence of precursor iron minerals and organic matter. Mar. Chem. 57, 187–193.

Mortazavi, B., Iverson, R.L., Huang, W., Lewis, F.G., and Cafrey, J.M. (2000) Nitrogen budget of Apalachicola Bay, a bar-built estuary in the northeastern Gulf of Mexico. Mar. Ecol. Prog. Ser. 195, 1–14.

Mortimer, C.H. (1941) The exchange of dissolved substances between mud and water in lakes. J. Ecol. 29, 280–320.

Mortimer, R.J., Krom, M.D., Watson, P.G., Frickers, P.E., Davey, J.T., and Clifton, R.J. (1998) Sediment–water exchange of nutrients in the intertidal zone of the Humber Estuary, UK. Mar. Pollut. Bull. 37, 261–279.

Mountfort, D.O., Asher, R.A., Mays, E.L., and Tiedje, J.M. (1980) Carbon and electron flow in mud and sandflat intertidal sediments at Delaware Inlet, Nelson, New Zealand. Appl. Environ. Microbiol. 39, 686–694.

Mukhopadhyay, S.K., Biswas, H., De, T.K., Sen, S., and Jana, T.K. (2002) Seasonal effects on the air–water carbon dioxide exchange in the Hooghly estuary, NE coast of Bay of Bengal, Indian J. Environ. Monit. 4, 549–552.

Mulder, A., van de Graaf, A.A., Robinson, L.A., and Kuenen, J.G. (1995) Anaerobic ammonium oxidation discovered in denitrifying fluidized bed reactor. FEMS Microbiol. Ecol. 16, 177–184.

Mulholland, M.R., Glibert, P.M., Berg, G.M., van Heukelem, L., Pantoja, S., and Lee, C. (1998) Extraclualar amino acid oxidation by microplankton: A cross-ecosystem comparison. Aquat. Microb. Ecol. 15, 141–152.

Mulholland, M.R., Gobler, C.J., and Lee, C. (2002) Peptide hydrolysis, amino acid oxidation, and nitrogen uptake in communities seasonally dominated by *Aureococcus anophagefferens*. Limnol. Oceanogr. 47, 1094–1108.

Muller, F.L.L. (1998) Colloid/solution partitioning of metal-selective organic ligands, and its relevance to Cu, Pb, and Cd cycling in the Firth of Clyde. Estuar. Coastal Shelf Sci. 46, 419–437.

Muller, F.L.L. (1999) Evaluation of the effects of natural dissolved and colloidal organic ligands on the electrochemical lability of Cu, Pb, and Cd in the Arran Deep, Scotland. Mar. Chem. 67, 43–60.

Müller, P.J. (1977) C/N ratios in Pacific deep-sea sediments: effects of inorganic ammonium and organic nitrogen compounds sorbed to clays. Geochim. Cosmochim. Acta 41, 765–776.

Müller, P.J., and Schneider, R. (1993) An automated leaching method for the determination of opal in sediments and particulate matter. Deep-Sea Res. 40, 425–444.

Murphy, R.C., and Kremer, J.N. (1983) Community metabolism of Clipperton Lagoon, a coral atoll in the eastern Pacific. Bull. Mar. Sci. 33, 152–164.

Murray, J.W. (2000) The oceans. *In* Earth System Science, from Biogeochemical Cycles to Global Change (Jacobson, M.C., Charlson, R.J., Rodhe, H, and Orians, G.H., eds.), pp. 230–278, International Geophysics Series, Academic Press, New York.

Murray, J.W., Grundmanis, V., and Smethie, W.J. (1978) Interstitial water chemistry in the sediments of Saanich Inlet. Geochim. Cosmochim. Acta 42, 1011–1026.

Murrell, M.C., and Hollibaugh, J.T. (2000) Distribution and composition of dissolved and particulate organic carbon in northern San Francisco Bay during low flow conditions. Estuar. Coastal Shelf Sci. 51, 75–90.

Mutchler, T., Sullivan, M.J., and Fry, B. (2004) Potential of ^{15}N isotope enrichment to resolve ambiguities in coastal trophic relationships. Mar. Ecol. Prog. Ser. 266, 27–33.

Najdek, M., Deborris, D. Miokovic, D., and Ivancic, I. (2002) Fatty acid and phytoplankton composition of different types of mucilaginous aggregates in the northern Adriatic. J. Plankton Res. 24, 429–441.

Nakai, N., and Jensen, M.L. (1964) The kinetic isotope effect in the bacterial reduction and oxidation of sulfur. Geochim. Cosmochim. Acta 28, 1893–1912.

Nanny, M.A., and Minear, R.A. (1997) Characterization of soluble unreactive phosphorus using ^{31}P nuclear magnetic resonance spectroscopy. Mar. Geol. 139, 77–94.

Napolitano, G.E., Pollero, R.J., Gayoso, A.M., MacDonald, B.A., and Thompson, R.J. (1997) Fatty acids as trophic markers of phytoplankton blooms in the Bahia Blanca estuary (Buenos Aires, Argentina) and in Trinity Bay (Newfoundland, Canada). Biochem. Syst. Ecol. 25, 739–755.

National Research Council (1985) Oil in the Sea: Inputs, Fates, and Effects. National Academy Press, Washington, DC.

National Research Council (2000) Clean Coastal Waters: Understanding and Reducing the Effects of Nutrient Pollution. National Academy Press, Washington, DC.

Neale, P.J., and Kieber, D.J. (2000) Assessing biological and chemical effects of UV in the marine environment:spectral weighting functions. *In* Causes and Environmental Implications of Increased UV-B Radiation (Hester, R.E., and Harrison, R.M., eds.), pp. 61–83, The Royal Society of Chemistry, Cambridge, UK.

Nealson, K.H., and Myers, C.R. (1992) Microbial reduction of manganese and iron: new approaches to carbon cycling. Appl. Environ. Microbiol. 58, 439–443.

Nealson, K.H., and Saffarini, D. (1994) Iron and manganese in anaerobic respiration: environmental significance, physiology and regulation. Annu. Rev. Microbiol. 48, 311–341.

Neckles, H.A., and C. Neill (1994) Hydrologic control of litter decomposition in seasonally flooded prairie marshes. Hydrobiology 286, 155–165.

Nedwell, D.B., and Abram, J.W. (1979) Relative influence of temperature and electron donor and electron acceptor concentrations on bacterial sulfate reduction in saltmarsh sediment. Microbial. Ecol. 5, 67–72.

Neilson, A.H., and Lewin, R.A. (1974) The uptake and utilization of organic carbon by algae: an essay in comparative biochemistry. Phycologia 13, 227–264.

Nelson, D.C., and Castenholz, R.W. (1981) Organic nutrition of *Beggiatoa* sp. J. Bacteriol. 147, 236–247.

Nelson, D.M., and Dirtch, Q. (1996) Silicic acid depletion and silicon limitation in the plume of the Mississippi River: evidence from kinetic studies in spring and summer. Mar. Ecol. Prog. Ser. 136, 163–178.

Nelson, D.M., Goering, J.J., and Boisseau, D.W. (1981) Consumption and regeneration of silicic acid in three coastal upwelling systems. *In* Coastal Upwelling (Richards, F.A., ed.), pp. 242–256, American Geophysical Union, Washington, DC.

Nelson, D.M., Treguer, P., Brzezinski, M.A., Leynaert, A., and Queguiner, B. (1995) Production and dissolution of biogenic silica in the ocean: revised global estimates, comparison with regional data and relationship to biogenic sedimentation. Global Biogeochem. Cycles 9, 359–372.

Nelson, J.R., Eckman, J.E., Robertson, C.Y., Marinelli, R.L., and Jahnke, R.A. (1999) Benthic microalgal biomass and irradiance at the sea floor on the continental shelf of the South Atlantic Bight: spatial and temporal variability and storm effects. Cont. Shelf Res. 19, 477–505.

Nelson, J.R., and Wakeham, S.G. (1989) A phytol-substituted chlorophyll *c* from *Emiliana huxleyi* (Prymnesiophyceae). J. Phycol. 25, 761–766.

Neubauer, S.C., and Anderson, I.C. (2003) Transport of dissolved inorganic carbon from a tidal freshwater marsh to the York River estuary. Limnol. Oceanogr. 48, 299–307.

Neunlist, S., Bisseret, P., and Rohmer, M. (1988) The hopanoids of the purple non-sulfur bacteria *Rhodopseudomonas palustris* and *Rhodopseudomonas acidophila* and the absolute configuration of bacteriohopanetetrol. Eur. J. Biochem. 171, 245–252.

Nevissi, A. (1982) Atmospheric flux of ^{210}Pb in the northwestern United States. *In* Natural Environment (Vohra, K.G., Mishra, U.C., Pillai, K.C., and Sadasivan, S., eds.), pp. 614–620, John Wiley, New York.

Newcombe, C.L., Horne, W.A., and Shepard, B.B. (1938) Oxygen-poor waters of the Chesapeake Bay. Science 88, 80–81.

Newell, R.C. (1965) The role of detritus in the nutrition of two marine deposit-feeders, the Prosobranch *Hydrobia ulvae* and the bivalve *Macoma balthica*. Proc. Zool. Soc. Lond. 4, 25–45.

Newell, S.Y., Hopkinson, C.S., and Scott, L.A., (1992) Patterns of nitrogenase activity (acetylene reduction) associated with standing, decaying shoots of *Spartina alterniflora*. Estuar. Coastal Shelf Sci. 35, 127–140.

Nguyen, R.T., and Harvey, H.R. (1994) A rapid micro-scale method for the extraction and analysis of protein in marine samples. Mar. Chem. 45, 1–14.

Nguyen, R.T., and Harvey, H.R. (1997) Protein and amino acid cycling during phytoplankton decomposition in oxic and anoxic waters. Org. Geochem. 27, 115–128.

Nguyen, R.T., and Harvey, H.R. (2001) Protein preservation in marine systems: hydrophobic and other non-covalent associations as major stabilizing forces. Geochim. Cosmochim. Acta 65, 1467–1480.

Nguyen, R.T., Harvey, H.R., Zang, X., van Heemst, J.D.H., Hetenyi, M., and Hatcher, P.G. (2003) Preservation of algaenan and proteinaceous material during the oxic decay of *Botryococcus braunii* as revealed by pyrolysis–gas chromatography/mass spectrometry and ^{13}C NMR spectroscopy. Org. Geochem. 34, 483–498.

Nicholls, M., Johnson, G.H., and Peebles, P.C. (1991) Modern sediments and facies model for a microtidal coastal plain estuary, The James Estuary, Virginia. J. Sed. Petrol. 61, 883–899.

Nichols, M.N. (1974) Development of the turbidity maximum in the Rappahannock estuary, Summary. Mem. Inst. Geol. Bassin d'Aquitaine. 7, 19–25.

Nichols, M.N. (1984) Fluid mud accumulation processes in an estuary. Geo-Mar. Lett. 4, 171–176.

Nichols, M.N., and Biggs, R.B. (1985) Estuaries. *In* Coastal Sedimentary Environments (Davis, R.A., ed.), pp. 77–186, Springer-Verlag, New York.

Nichols, P.D., and Johns, R.B. (1985). Lipids of the tropical seagrass *Thallassia hemprichii*. Phytochemistry 24, 81–84.

Nichols, P.D., Palmisano, A.C., Rayner, M.S., Smith, G.A., and White, D.C. (1990) Occurrence of novel C_{20} sterols in Antarctic sea ice diatom communities during spring bloom. Org. Geochem. 15, 503–508.

Niell, F.X. (1977) Rocky intertidal benthic systems in temperate seas: a synthesis of their functional performances. Heigo Wiss. Meersunters 30, 315.

Nielsen, K., Nielsen, L.P., and Rasmussen, P. (1995) Estuarine nitrogen retention independently estimated by the denitrification rate and mass balance methods: a study of Norsminde Fjord, Denmark. Mar. Ecol. Prog. Ser. 119, 275–283.

Nielsen, L.P. (1992) Denitrification in sediment determined from nitrogen isotope pairing. FEMS Microbiol. Ecol. 86, 357–362.

Nielsen, S.L., Sand-Jensen, K., Borum, J., and Geertz-Hansen, O. (2002) Phytoplankton, nutrients, and transparency in Danish coastal waters. Estuaries 25, 930–937.

Niemi, A. (1979) Blue–green algal blooms and N:P ratios in the Baltic Sea. Acta Bot. Fenn. 110, 57–61.

Nier, A.O. (1947) A mass spectrometer for isotope and gas analysis. Rev. Sci. Instrum. 18, 398–411.

Nikaido, H., and Vaara, M. (1985) Molecular basis of bacterial outer membrane permeability. Microb. Rev. 49, 1–32.

Nishimura, M., and Koyama, T. (1977) The occurrence of stanols in various living organisms and the behavior of sterols in contemporary sediments. Geochim. Cosmochim. Acta 41, 379–385.

Nishio, T., Koike, I., and Hattori, A. (1982) Denitrification, nitrate reduction, and oxygen consumption in coastal and estuarine sediments. Appl. Environ. Microbiol. 43, 648–653.

Nissenbaum, A., and Kaplan, I.R. (1972) Chemical and isotopic evidence for the in situ origin of marine humic substances. Limnol. Oceanogr. 19, 570–582.

Nittrouer, C.A., DeMaster, D.J., Figueiredo, A.G., and Rine, J.M. (1991) AMASSEDS: An interdisciplinary investigation of a complex coastal environment, Oceanography 4, 3–7.

Nittrouer, C. A., DeMaster, D.J., Kuehl, S.A., McKee, B.A., and Thorbjarnarson, K.W. (1985) Some questions and answers about the accumulation of fine-grained sediment in continental margin environments. Geo-Mar. Lett. 4, 211–213.

Nittrouer, C.A., DeMaster, D.J., McKee, B.A., Cutshall, N.H., and Larsen, I.L. (1984) The effect of sediment mixing on Pb-210 accumulation rates for the Washington continental shelf. Mar. Geol. 54, 210–221.

Nittrouer, C.A., Kuehl, S.A., and DeMaster, D.L. (1986) The deltaic nature of Amazon shelf sedimentation. Geol. Soc. Am. Bull. 97, 444–458.

Nittrouer, C.A., Kuehl, S.A., Sternberg, R.W., Figueiredo, A.G., and Faria, L.E.C. (1995) An introduction to the geological significance of sediment transport and accumulation on the Amazon continental-shelf. Mar. Geol. 125, 177–192.

Nixon, S.W. (1980) Between coastal marshes and coastal waters—a review of twenty years of speculation and research on the role of salt marshes in estuarine productivity and water chemistry. In Estuarine Wetland Processes (Hamilton, P., and MacDonald, K.B., eds.), pp. 487–525, Plenum Press, New York.

Nixon, S.W. (1981) Remineralization and nutrient cycling in coastal marine ecosystems. In Estuaries and Nutrients, (Neilson, B.J., and Cronin, L.E., eds.), pp. 111–138, Humana, New York.

Nixon, S.W. (1982) Nutrient dynamics, primary production, and fisheries yields of lagoons. Oceanologic Acta Special edition: Proceedings, International Symposium on Coastal Lagoons, pp. 357–371, University of Rhode Island, Narragansett, RI.

Nixon, S.W. (1986) Nutrient dynamics and productivity of marine coastal waters In Coastal Eutrophication (Clayton, B., and Behbehani, M., eds.), pp. 97–115, The Alden Press, Oxford, UK.

Nixon, S.W. (1987) Chesapeake Bay nutrient budgets—a reassessment. Biogeochemistry 4, 77–90.

Nixon, S.W. (1988) Physical energy inputs and the comparative ecology of lake and marine ecosystems. Limnol. Oceanogr. 33, 1005–1025.

Nixon, S.W. (1992) Quantifying the relationship between nitrogen input and the productivity of marine ecosystems. Proc. Adv. Mar. Tech. Conf. 5, 57–83.

Nixon, S.W. (1995) Coastal marine eutrophication: a definition, social causes, and future concerns. Ophelia 4, 199–219.

Nixon, S.W (1997) Prehistoric nutrient inputs and productivity in Narragansett Bay. Estuaries 20, 253–261.

Nixon, S.W., Ammerman, J.W., Atkinson, L.P., Berounsky, V.M., Billen, G., Boicourt, W.C., Boynton, W.R., Church, T.M., Ditoro, D.M., Elmgren, R., Garber, J.H., Giblin, A.E., Jahnke, R.A., Owens, N.J.P., Pilson, M.E.Q., and Seitzinger, S.P. (1996) The fate of nitrogen and phosphorus at the land–sea margin of the north Atlantic Ocean. Biogeochemistry 35, 141–180.

Nixon, S.W., Furnas, B.N., Lee, V., Marshall, N., Ong, J.E., Wong, C.H., Gong, W.K., and Sasekumar, A. (1984) The role of mangroves in the carbon and nutrient dynamics of Malaysia estuaries. In Proceedings of the Asian Symposium on Mangrove Environment: Research and Management (Soepadmo, E., ed.), pp. 535–544, University of Malaya, Kuala Lumpur.

Nixon, S.W., Granger, S.L., and Nowicki, B.L. (1995) An assessment of the annual mass balance of carbon, nitrogen, and phosphorus in Narragansett Bay. Biogeochemistry 31, 15–61.

Nixon, S.W., and Oviatt, C.A. (1973) Ecology of a New England salt marsh. Ecol. Monogr. 43, 463–498.

Nixon, S.W., Oviatt, C.A., Fristhen, J., and Sullivan, B. (1986) Nutrients and the productivity of estuarine and coastal marine ecosystems. J. Limnol. Soc. South Afr. 12, 43–71.

Nixon, S.W., Oviatt, C.A., and Hale, S.S. (1976) Nitrogen regeneration and the metabolism of coastal marine bottom communities. *In* The Role of Terrestrial and Aquatic Organisms in Decomposition Processes (Anderson, J.M., and MacFadyen, A., eds.), pp. 269–283, Blackwell, London.

Nixon, S.W., and Pilson, M.Q. (1983) Nitrogen in estuarine and coastal marine ecosystems. *In* Nitrogen in the Marine Environment (Carpenter, E.J., and Capone, D.G., eds.), pp. 565–648, Academic Press, New York.

Nobel, M., and Butman, B. (1979) Low frequency wind-induced sea level oscillations along the east coast of North America. J. Geophys. Res., 84, 3227–3236.

North, E.W. and Houde, E.D. (2001) Retention of white perch and striped bass larvae: biological–physical interactions in Chesapeake Bay estuarine turbidity maximum. Estuaries 24, 756–769.

Novelli, P.C., Michelson, A.R., Scranton, M.I., Banta, G.T., Hobbie, J.E., and Howarth, R.W. (1988) Hydrogen and acetate in two sulfate-reducing sediments: Buzzards Bay and Town Cove, Mass. Geochim. Cosmochim. Acta 52, 2477–2486.

Nowell, A.R.M., and Jumars, P.A. (1987) Flumes: theoretical and experimental considerations for simulation of benthic environments. Oceanogr. Mar. Biol. Annu. Rev. 25, 91–112.

Nowicki, B.L. (1994) The effect of temperature, oxygen, salinity, and nutrient enrichment on estuarine denitrification rates measured with a modified nitrogen gas flux technique. Estuar. Coastal Shelf Sci. 38, 137–156.

Nowicki, B.L., Kelly, J.R., Requintina, E., and van Keuren, D. (1997) Nitrogen losses through sediment denitrification in Boston Harbor and Massachusetts Bay. Estuaries 20, 626–639.

Nowicki, B.L., and Nixon, S.W. (1985) Benthic nutrient remineralization in a coastal lagoon ecosystem. Estuaries 8: 182–190.

Nowicki, B.L., Reguintina, E., van Kevren, D., and Portnoy, J. (1999) The role of sediment denitrification in reducing groundwater-derived nitrate inputs to Nauset Marsh estuary, Cape Cod, Massachusetts. Estuaries 22, 245–259.

Nozaki, Y. (1991) The systematics and kinetics of U/Th decay series nuclides in ocean water. Rev. Aquat. Sci. 4, 75–105.

Nozaki, Y., Cochran, J.K., Turekian, K.K., and Keller, G. (1977) Radiocarbon and ^{210}Pb distribution in submersible-taken deep-sea cores from project FAMOUS. Earth Planet. Sci. Lett. 34, 167–173.

Nunes, R.A., and Simpson, J.H. (1985) Axial convergence in a well-mixed estuary. Estuar. Coastal Shelf Sci. 20, 637–649.

Nyman, J.A. (1999) Effect of crude oil and chemical additives on metabolic activity of mixed microbial populations in fresh marsh soils. Microb. Ecol. 37, 152–162.

Oades, J.M. (1989) An introduction to organic matter in mineral soils. *In* Mineral in Soil Environments (Dixon, J.B., and Weed, S.B., eds.), pp. 89–159, Soil Science Society of America, Madison, WI.

O'Brien, B.J. (1986) The use of natural and anthropogenic ^{14}C to investigate the dynamics of soil organic carbon. Radiocarbon 28, 358–362.

Odd, N.U.M., Bentley, M.A., and Waters, C.B. (1993) Observations and analysis of the movement of fluid muds in an estuary. *In* Nearshore and Estuarine Studies (Mehta, A.J., ed.), Vol. 42, pp. 430–446, American Geophysical Union, Washington, DC.

O'Donnell, J. (1993) Surface fronts in estuaries: a review. Estuaries 16, 12–39.

O'Donnell, T.H., Macko, S.A., Chou, J., Davis-Hartten, K.L., and Wehmiller, J.F. (2003) Analysis of δ^{13}C, δ^{15}N, and δ^{34}S in organic matter from the biominerals of modern and fossil *Mercenaria* spp. Org. Geochem. 34, 165–184.

O'Donnell, J., Marmorino, G.O., and Trump, C.L. (1998) Convergence and downwelling at a river plume front. J. Phys. Oceanogr. 28, 1481–1495.

O'Donohue, M.J.H., and Dennison, W.C. (1997) Phytoplankton productivity response to nutrient concentrations, light availability, and temperature along an Australian estuarine gradient. Estuaries 20, 521–533.

Odum, E.P. (1968). A research challenge: evaluating the productivity of coastal and estuarine water. Proceedings of 2nd Sea Grant Conference, Graduate School of Oceanograply, University of Rhode Island, Narragansett, RI.

Odum, E. P., and de la Cruz, A.A. (1967) Particulate organic detritus in a Georgia salt marsh–estuarine ecosystem. In Estuaries (Lauff, G.H., ed.), pp. 383–388, American Association for the Advancement of Science, Washington, DC.

Odum, W.E., Zieman, J.C., and Heald, E.J. (1973) The importance of vascular plant detritus to estuaries. In Coastal Marsh and Estuary Symposium (Chabreck, R.H., ed.), pp. 91–135, LSU, Baton Rouge, LA.

Oenema, O. (1990a) Sulfate reduction in fine-grained sediments in the Eastern Scheldt, southwest Netherlands. Biogeochemistry 9, 53–74.

Oenema, O. (1990b) Pyrite accumulation in salt marshes in the Eastern Scheldt, southwest Netherlands. Biogeochemistry 9, 75–98.

Officer, C.B. (1976) Physical Oceanography of Estuaries (and Associated Coastal Waters). John Wiley, New York.

Officer, C.B. (1979) Discussion of the behavior of non-conservative dissolved constituents in estuaries. Estuar. Coastal Shelf Sci. 9, 91–94.

Officer, C.B. (1980) Box models revisited. In Estuarine and Wetland Processes with Emphasis on Modeling (Hamilton, P., and MacDonald, K.B., eds.), pp. 65–114, Plenum Press, New York.

Officer, C.B., Biggs, R.B., Taft, J.L., Cronin, M.A., and Boynton, W.R. (1984) Chesapeake Bay anoxia: origin, development, and significance. Science 223, 22–26.

Officer, C.B., and Lynch, D.R. (1981) Dynamics of mixing in estuaries. Estuar. Coastal Shelf Sci. 12, 525–534.

Officer, C.B., and Ryther, J.H. (1980) The possible importance of silicon in marine eutrophication. Mar. Ecol. Prog. Ser. 9, 91–94.

Ogan, M.T. (1990) The nodulation and nitrogen activity of natural stands of mangrove legumes in a nitrogen swamp. Plant Soil Sci. 123, 125–129.

Ogawa, N., Koitabashi, T., Oda, H., Nakamura, T., Ohkouchi, N., and Wada, E. (2001) Fluctuations of nitrogen isotope ratios of gobiid fish (IIsaza) specimens and sediments in lake Biwa, Japan, during the 20th century. Limnol. Oceanogr. 46, 1228–1236.

Ogram, A., Sayler, G.S., Gustin, D., and Lewis, R.J. (1978) DNA adsorption to soils and sediments. Environ. Sci. Technol. 22, 982–984.

Ohmoto, H. (1992) Biogeochemistry of sulfur and the mechanisms of sulfide-sulfae mineralization in Archean oceans. In Early Organic Evolution: Implications for Mineral and Energy Resources (Schidlowski, M., Golubic, S., Kimberley, M.M., Mckirdy, D.M., and Trudinger, P.A., eds.), pp. 378–397, Springer-Verlag, Berlin.

Okubo, A. (1973) Effect of shoreline irregularities on streamwise dispersion in estuaries and other embayments. Neth. J. Sea Res. 6, 213–224.

Olah, J. (1972) Leaching, colonization, and stabilization during detritus formation. Mem. Inst. Ital. Idrobiol. 29, 105–127.

O'Leary, M.H. (1981) Carbon isotope fractionation in plants. Phytochemistry 20, 553–567.

O'Leary, M.H. (1988) Carbon isotopes in photosynthesis. Bioscience 38, 328–336.

Oliveira, A., and Baptista, A.M. (1997) Diagnostic modeling of residence times in estuaries. Wat. Resources. Res. 33, 1935–1946.

Olmez, I., and Gordon, G.E. (1985) Rare Earths: atmospheric signatures for oil fired power plants and refineries. Science 229, 966–968.

Olmez, I., Sholkovitz, E.R., Hermann, D., and Eganhouse, R.P. (1991) Rare Earth elements sediments off Southern California: a new anthropogenic indicator. Environ. Sci. Technol. 25, 310–316.

Olsen, C.R., Larsen, I.L., Lowry, P.D., and Cutshall, N.H. (1986) Geochemistry and deposition of ^{7}Be in river–estuarine and coastal waters. J. Geophys. Res. 91, 896–908.

Olsen, C.R., Larsen, I.L., Lowry, P.D., Cutshall, N.H., Todd, J.F., Wong, G.T.F., and Casey, W.H. (1985) Atmospheric fluxes and marsh-soil inventories of ^{7}Be and ^{210}Pb. J. Geophys. Res. 90, 10487–10495.

Olsen, C.R., Larsen, I.L., Mulholland, P.J., Von Damm, K.L., Grebmeier, J.M., Schaffner, L.C., Diaz, R.J., and Nichols, M.M. (1993) The concept of an equilibrium surface applied to particle sources and contaminant distributions in estuarine sediments. Estuaries 16, 683–696.

Olsen, C.R., Simpson, H.J., Bopp, R.F., Williams, S.C., Peng, T.H., and Deck, B.L. (1978) A geochemical analysis of the sediments and sedimentation in the Hudson Estuary. J. Sed. Petrol. 48, 410–418.

Olsen, C.R., Simpson, H.J., Peng, T.H., Bopp, F., and Trier, R.M. (1981) Sediment mixing and accumulation rate effects on radionuclide depth profiles in Hudson Estuary sediments. J. Geophys. Res. 86, 11020–11028.

Olsson, M., Bignert, A., Eckhell, J., and Jonsson, P. (2000) Comparison of temporal trends (1940s–1990s) of DDT and PCB in Baltic sediment and biota in relation to eutrophication. Ambio 29, 195–201.

O'Malley, V.P., Abrajano, T.A., and Hellou, J. (1994) Determination of ^{13}C/^{12}C ratios of individual PAH from environmental samples: can PAH sources be source apportioned? Org. Geochem. 21, 809–822.

Onstad, G.D., Canfield, D.E., Quay, P.D., and Hedges, J.I. (2000) Sources of particulate organic matter from rivers from the continental USA: lignin phenol and stable carbon isotope compositions. Geochim. Cosmochim. Acta 64, 3539–3546.

Opsahl, S., and Benner, R. (1995) Early diagenesis of vascular plant tissues: lignin and cutin decomposition and biogeochemical implications. Geochim. Cosmochim. Acta 59, 4889–4904.

Opsahl, S., and Benner, R. (1997) Distribution and cycling of terrigenous dissolved organic matter in the ocean. Nature 386, 480–482.

Opsahl, S., and Benner, R. (1998) Photochemical reactivity of dissolved lignin in river and ocean waters. Limnol. Oceanogr. 43, 1297–1304.

Opsahl, S., and Benner, R. (1999) Characterization of carbohydrates during early diagenesis of five vascular plant tissues. Org. Geochem. 30, 83–94.

Opsahl, S., Benner, R., and Amon, R.M.W. (1999) Major flux of terrigenous dissolved organic matter through the Arctic Ocean. Limnol. Oceanogr. 44, 2017–2023.

Orem, W.H., and Hatcher, P.G. (1987) Solid-state ^{13}C NMR studies of dissolved organic matter in pore waters from different depositional environments. Org. Geochem. 11, 73–82.

Orem, W.H., Hatcher, P.G., and Spiker, E.C. (1986) Dissolved organic matter in anoxic pore waters from Mangrove Lake, Bermuda. Geochim. Cosmochim. Acta 50, 609–618.

Ormaza-Gonzalez, F.I., and Statham, P.J. (1991) The occurrence and behavior of different forms of phosphorus in the waters of four English estuaries. In Estuaries and Coasts: Spatial and Temporal Intercomparisons (Elliott, M., and Ducrotoy, J.P., eds.), pp. 77–83, Olsen and Olsen, Copenhagen.

Osterman, L.E., Poore, R.Z., Swarzenski, P.W., and Turner, R.E. (2005) Reconstructing a 180 yr record of natural and anthropogenic induced low-oxygen conditions from Louisiana continental shelf sediments. J. Geol. 33, 329–332.

Ostrom, N.E., Macko, S.A., Deibel, D., and Thompson, R.J. (1997) Seasonal variation in the stable carbon and nitrogen isotope biogeochemistry of a coastal cold ocean environment. Geochim. Cosmochim. Acta 61, 2929–2942.

Ourisson, G., Rohmer, M., and Poralla, K. (1987) Prokaryotic hopanoids and other polyterpenoid sterol surrogates. Annu. Rev. Microbiol. 41, 310–333.

Ouverney, C.C., and Fuhrman, J.A. (2000) Marine planktonic Archaea take up amino acids. Appl. Env. Microbiol. 66, 4829–4833.

Overmann, J., Cypionka, H., and Pfennig, N. (1992) An extremely low-light adapted phototrophic sulfur bacterium from the Black Sea. Limnol. Oceanogr. 37, 150–155.

Overnell, J. (2002) Manganese and iron profiles during early diagenesis in Loch Etive, Scotland. Application of two diagenetic models. Estuar. Coastal Shelf Sci. 54, 33–44.

Oviatt, C.A., Lane, P., French, F., and Donaghay, P. (1989) Phytoplankton species and abundance in response to eutrophication in coastal marine ecosystems. J. Plankton Res. 11, 1223–1244.

Owens, N.J.P. (1985) Variations in the natural abundance of ^{15}N in estuarine suspended particulate matter: a specific indicator of biological processing. Estuar. Coastal Shelf Sci. 20, 505–510.

Packard, T.T. (1979) Half-saturation constants for nitrate reductase and nitrate translocation in marine phytoplankton. Deep Sea Res. 26, 321–326.

Paerl, H.W. (1985) Enhancement of marine primary production by nitrogen-enriched acid rain. Nature 315, 747–749.

Paerl, H.W. (1991) Ecophysiological and trophic implications of light-stimulated amino acid utilization in marine picoplankton. Appl. Environ. Microbiol. 57, 473–479.

Paerl, H.W. (1995) Coastal eutrophication in relation to atmospheric nitrogen deposition: current perspectives. Ophelia 41, 237–259.

Paerl, H.W. (1996) Microscale physiological and ecological studies of aquatic cyanobacteria: macroscale implications. Micros. Res. Techn. 33, 47–72.

Paerl, H.W. (1997) Coastal eutrophication and harmful algal blooms: Importance of atmospheric deposition and groundwater as "new" nitrogen and other nutrient sources. Limnol. Oceanogr. 42, 1154–1165.

Paerl, H.W. 2002. Connecting atmospheric deposition to coastal eutrophication. Environ. Sci. Technol. 36, No.15:323A–326A.

Paerl, H.W., Crocker, K.M., and Prufert, L.E. (1987) Limitation of N_2 fixation in coastal marine waters: relative importance of molybdenum, iron, phosphorus, and organic matter availability. Limnol. Oceanogr. 32, 525–536.

Paerl, H.W., Dennis, R.L., and Whitall, D.R. (2002) Atmospheric deposition of nitrogen: implications for nutrient over-enrichment of coastal waters. Estuaries 25, 677–693.

Paerl, H.W., and Fogel, M.L. (1994) Isotopic characterization of atmospheric nitrogen inputs as sources of enhanced primary production in coastal Atlantic Ocean waters. Mar. Biol. 119, 635–645.

Paerl, H.W., Rudek, J., and Mallin, M.A. (1990) Stimulation of phytoplankton production in coastal waters by natural rainfall inputs: nutritional and trophic implications. Mar. Biol. 107, 247–254.

Paez-Osuna, F., and Mandelli, E.F. (1985) ^{210}Pb in a tropical coastal lagoon sediment core. Estuar. Coastal Shelf Sci. 20, 367–374.

Page, H.M. (1995) Variation in the natural abundance ^{15}N in the halophyte, *Salicornia virginica*, associated with groundwater subsidies of nitrogen in a southern California salt-marsh. Oecologia 104, 181–188.

Painter, T.J. (1983) Algal polysaccharides. *In* The Polysaccharides (Aspinall, G.O., ed.), pp. 195–285, Academic Press, New York.

Pakulski, J.D., and Benner, R. (1994) Abundance and distribution of carbohydrates in the ocean. Limnol. Oceanogr. 39, 930–940.

Pakulski, J.D., Benner, R., Whitledge, T., Amon, R, Eadie, B., Cifuentes, L., Ammerman, J., and Stockwell, D. (2000) Microbial metabolism and nutrient cycling in the Mississippi and Atchafalaya River plumes. Estuar. Coastal Shelf Sci. 50, 173–184.

Palenik, B., and Morel, F.M. (1991) Amine oxidases of marine phytoplankton. Appl. Environ. Microb. 57, 2440–2443.

Pankow, J.F. (1991) Aquatic Chemistry Concepts. Lewis Publishers, Boca Raton, FL.

Pantoja, S. and Lee, C. (1994) Cell-surface oxidation of amino acids in seawater. Limnol. Oceanogr. 39, 1718–1726

Pantoja, S., and Lee, C. (1999) Peptide decomposition by extracellular hydrolysis in coastal seawater and salt marsh sediment. Mar. Chem. 63, 272–291.

Park, P.K., Gordon, L.I., Hager, S.W., and Cissel, M.C. (1969) Carbon dioxide partial pressure in the Columbia River. Science 166, 867–868.

Park, R., and Epstein, S. (1960) Carbon isotope fractionation during photosynthesis. Geochim. Cosmochim. Acta 21, 110–126.

Parker, C.A., and O'Reilly, J.E. (1991) Oxygen depletion in Long Island Sound: a historical perspective. Estuaries 14, 248–264.

Parker, P.L., and Leo, R.F. (1965) fatty acids in blue-green algal mat communities. Science 148, 373–374.

Parkes, R.J., and Buckingham, W.J. (1986) The flow of organic carbon through aerobic respiration and sulfate reduction. In Proceedings of 4th International Symposium on Microbial Ecology (Megusar, F., and Gantar, M., eds.), pp. 617–624, Ljubljana, Slovenia.

Parkes, R.J., and Taylor, J. (1983) The relationship between fatty acid distributions and bacterial respiratory types in contemporary marine sediments. Estuar. Coastal Shelf Sci. 16, 173–189.

Parrish, C.C., Abrajano, T.A., Budge, S.M., Helleur, R.J., Hudson, E.D., Pulchan, K., and Ramos, C. (2000) Lipid and phenolic biomarkers in marine ecosystems: analysis and applications. In The Handbook of Environmental Chemistry, Vol. 5, Part D, Marine Chemistry (Wangersky, P., ed.), pp. 193–223, Springer-Verlag, Berlin.

Parrish, C.C., Eadie, B.J., Gardner, W.S., and Cavaletto, J.F. (1992) Lipid class and alkane distribution in settling particles of the upper Laurentian Great Lakes. Org. Geochem. 18, 33–40.

Parson, J.W. (1988) Isolation of humic substances from soils and sediments. In Humic Substances and their Role in the Environment (Frimmel, F.H., and Christman, R.F. eds.), pp. 3–14, John Wiley, New York.

Parsons, T.R., and Lee Chen, Y.L. (1995) The comparative ecology of a subarctic and tropical estuarine ecosystem as measured with carbon and nitrogen isotopes. Estuar. Coastal Shelf Sci. 41, 215–224.

Parsons, T.R., Takahashi, M., and Hargrave, B. (1984) Biological Oceanographic Processes. Pergamon Press, New York.

Part, P., Svanberg, O., and Kiessling, A. (1985) The availability of cadmium to perfused rainbow trout gills in different water qualities. Wat. Res. 19, 427–429.

Passow, U. (2002) Production of transparent exopolymer particles (TEP) by phyto- and bacterioplankton. Mar. Ecol. Prog. Ser. 236, 1–12.

Patrick, O., Slawayk, G., Garcia, N., and Bonin, P. (1996) Evidence of denitrification and nitrate ammonification in the river Rhone plume (northwest Mediterranean Sea). Mar. Ecol. Prog. Ser. 141, 275–281.

Payne, J.W. (1976) Reduction of nitrogenous oxides by microorganisms. Bact. Rev. 37, 409–452.

Payne, J.W. (ed.) (1980) Transport and utilization of peptides by bacteria. *In* Microorganisms and Nitrogen Sources, pp. 211–256, John Wiley, New York.

Peckol, P., DeMeo-Anderson, B., Rivers, J., Valiela, I., Maldonado, M., and Yates, J. (1994). Growth, nutrient uptake capacities and tissue constituents of the macroalgae, *Cladophora vagabunda* and *Gracilaria tikvahiae*, related to site-specific nitrogen loading rates. Mar. Biol. 121, 175–185.

Peierls, B.L., Caraco, N.F., Pace, M.L., and Cole, J.J. (1991) Human influence on river nitrogen. Nature 350, 386–387.

Peierls, B.L., and Paerl, H.W. (1997) Bioavailability of atmospheric organic nitrogen deposition to coastal phytoplankton. Limnol. Oceanogr. 42, 1819–1823.

Peng, T.H., Broecker, W.S., Mathieu, G.G., and Li, Y.H. (1979) Radon evasion rates in the Atlantic and Pacific oceans as determined during the GEOSECS program. J. Geophys. Res. 84, 2471–2486.

Penland, S., and Ramsey, K.E. (1990) Relative seal-level rise in Louisiana and the Gulf of Mexico: 1908–1988. J. Coastal Res. 6, 323–342.

Pennington, F.C., Haxo, F.T., Borch, G., and Liaaen-Jensen, S. (1985) Carotenoids of Cryptophyceae. Biochem. Syst. Ecol. 13, 215–219.

Pennock, J.R., Boyer, J.N., Herrera-Silveira, J.A., Iverson, R.I., Whitledge, T.E., Mortazavi, B., and Comin, F.A. (1999) Nutrient behavior and phytoplankton production in Gulf of Mexico estuaries. *In* Biogeochemistry of Gulf of Mexico Estuaries (Bianchi, T.S., Pennock, J.R., and Twilley, R.R., eds.), pp. 109–162, John Wiley, New York.

Pennock, J.R., Sharp, J.H., Ludlam, J.H., Velinsky, D.J., and Fogel, M.L. (1988) Isotopic fractionation of nitrogen during uptake of NH_4^+ and NO_3^- by *Skeletonema costatum*. EOS 69, 1098.

Pennock, J.R., Velinsky, D.J., Ludlum, and Sharp, J.H. (1996) Isotopic fractionation of ammonium and nitrate during uptake by *Skeletonema costatum*: implications for $\delta^{15}N$ dynamics under bloom condition. Limnol. Oceanogr. 41, 451–459.

Percival, E. (1970) Algal carbohydrates. *In* The Carbohydrates: Chemistry and Biochemistry, 2nd edn. (Pigman, W., and Horton, D., eds), pp. 537–568, Academic Press, New York.

Perdue, E.M., and Lytle, C.R. (1983) Distribution model for binding of protons and metal ions by humic substances. Environ. Sci. Technol. 17, 564–660.

Perez, M.T., Pausz, C., and Herndl, G.J. (2003) Major shift in bacterioplankton utilization of enantiomeric amino acids between surface waters and the ocean's interior. Limnol. Oceanogr. 48, 755–763.

Periera, W., and Rostad, C.E. (1990) Occurrence, distribution and transport of herbicides and their degradation products in the lower Mississippi River and its tributaries. Environ. Sci. Technol. 24, 1400–1408.

Perillo, G.M.E. (ed.) (1995) Definitions and geomorphic classifications of estuaries. *In* Geomorphology and Sedimentology of Estuaries. Developments in Sedimentology 53, pp. 17–47, Elsevier Science, New York.

Pernetta, J.C., and Millman, J.D. (eds.) (1995) Land–Ocean Interactions in the Coastal Zone: Implementation Plan. IGBP Rep. 33, 1–215.

Perret, D., Gaillard, J.F., Dominik, J., and Atteia, O. (2000) The diversity of natural hydrous iron oxides. Environ. Sci. Technol. 34, 3540–3546.

Perry, G.J., Volkman, J.K., Johns, R.B., and Bavor, H.J. (1979) Fatty acids of bacterial origin in contemporary marine sediments. Geochim. Cosmochim. Acta 43, 1715–1725.

Peteet, D.M., and Wong, J.K. (2000) Late Holocene environmental changes from NY–NJ estuaries. Proc. Geol. Soc. Am. Northeastern Section Meeting, p. A-65.

Peters, H. (1999) Spatial and temporal variability of turbulent mixing in an estuary. J. Mar. Res. 57, 805–845.

Peters, H. (2003) Broadly distributed and locally enhanced turbulent mixing in a tidal estuary. J. Phys. Oceanog. 33, 1967–1977.

Peters, H., and Bokhorst, R. (2000) Microstructure observations of turbulent mixing in a partially mixed estuary. Part I: Dissipation rate. J. Phys. Oceanogr. 30, 1232–1244.

Peters, K.E., and Moldowan, J.M. (1993) The Biomarker Guide: Interpreting Molecular Fossils in Petroleum and Ancient Sediments. Prentice-Hall, Englewood Cliffs, NJ.

Peterson, B.J., and Fry, B. (1989) Stable isotopes in ecosystems studies. Annu. Rev. Ecol. Syst. 18, 293–320.

Peterson, B.J., and Howarth, R.W. (1987) Sulfur, carbon, and nitrogen isotopes used to trace organic matter flow in the salt marsh estuaries of Sapelo Island, Georgia. Limnol. Oceanogr. 32, 1195–1213.

Peterson, B.J., Howarth, R.W., and Garritt, R.H. (1985) Multiple stable isotopes used to trace the flow of organic matter in estuarine food webs. Science 227, 1361–1363.

Peterson, B.J., Howarth, R.W., and Garritt, R.H. (1986) Sulfur and carbon isotopes as tracers of salt-marsh organic matter flow. Ecology 67, 865–874.

Petsch, S.T. (2000) A study on the weathering of organic matter in black shales and implications for the geochemical cycles of carbon and oxygen. Ph.D dissertation, Yale University.

Petsch, S., Eglinton, T.I., and Edwards, K.J. (2001) ^{14}C-dead living biomass: evidence for microbial assimilation of ancient organic carbon during shale weathering. Science 292, 1127–1131.

Pfennig, N. (1989) Ecology of phototrophic purple and green sulfur bacteria. In Autotrophic Bacteria (Schlegel, H.G., and Bowien, B., eds.), pp. 97–116, Springer, Berlin.

Phillips, R.C., and McRoy, C.P. (1980) Handbook of Seagrass Biology: An Ecosystem Perspective. Garland STPM Press, New York.

Phinney, J.T., and Bruland, K.W. (1994) Uptake of lipophilic organic Cu, Cd, and Pb complexes in the coastal diatom *Thalassiosira weissflogii*. Environ. Sci. Technol. 28, 1781–1790.

Pierard, C., Budzinski, H., and Garrigues, P. (1996) Grain-size distribution of polychlorobiphenyls in coastal sediment. Environ. Sci. Technol. 30, 2776–2783.

Pinckney, J.L., Millie, D.F., Howe, K.E., Paerl, H.W., and Hurley, J.P. (1996) Flow scintillation counting of ^{14}C-labeled microalgal photosynthetic pigments. J. Plankton Res. 18, 1867–1880.

Pinckney, J.L., Paerl, H.W., Harrington, M.B., and Howe, K.E. (1998) Annual cycles of phytoplankton community-structure and bloom dynamics in the Neuse River estuary, North Carolina. Mar. Biol. 131, 371–381.

Pinckney, J.L., Richardson, T.L., Millie, D.F., and Paerl, H.W. (2001) Application of photopigment biomarkers for quantifying microalgal community composition and in situ growth rates. Org. Geochem. 32, 585–595.

Pinckney, J.L., and Zingmark, R.A. (1993) Modeling the annual production of intertidal benthic microalgae in estuarine ecosystems J. Phycol. 2, 396–407.

Pirc, H. (1985) Growth dynamics in *Posidonia oceanica* (L.) Delile. I. Seasonal changes of soluble carbohydrates, starch, free amino acids, nitrogen and organic anions in different parts of the plant. Publ. Staz. Zool. Napoli I Mar. Ecol. 6, 141–165.

Plante-Cuny, M.R., Salen-Picard, C., Grenz, C., Plante, R., Alliot, E., and Barranguet, C. (1993) Experimental field study of the effects of crude oil, drill cuttings and natural biodeposits on microphyto- and microzoobenthic communities in a Mediterranean area. Mar. Biol. 117, 355–366.

Pohl, C., and Hennings, U. (1999) The effect of redox processes on the partitioning of Cd, Pb, Cu, and Mn between dissolved and particulate phases in the Baltic Sea. Mar. Chem. 65, 41–53.

Pohl, C., Loffler, A., and Hennings, U. (2004) A sediment trap flux for trace metals under seasonal aspects in the stratified Baltic Sea (Gotland Basin: 57° 19.20′N; 20°03.00′E). Mar. Chem. 84, 143–160.

Pollak, M.J. (1960) Wind set-up and shear-stress coefficient in Chesapeake Bay. J. Geophys. Res., 65, 3383–3389.

Pollard, P.C., and Moriarty, D.J.W. (1991) Organic carbon decomposition, primary and bacterial productivity, and sulphate reduction, in tropical seagrass beds of the Gulf of Carpentaria, Australia. Mar. Ecol. Prog. Ser. 69, 149.

Pollman, C.D., Landing, W.M., Perry, J.J., and Fitzpatrick, T. (2000) Wet deposition of phosphorus in Florida. Atmos. Environ. 36, 2309–2318.

Pomeroy, L.R. (1974) The ocean's food web: a changing paradigm. Bioscience 24, 499–504.

Pomeroy, L.R., Smith, E.E., and Grant, C.M. (1965) The exchange of phosphate between estuarine water and sediments. Limnol. Oceanogr. 10, 167–172.

Pomeroy, L.R., and R.G. Wiegert (eds.). 1981. The Ecology of a Salt Marsh. Springer Verlag, New York.

Porra, R.J., Pfundel, E.E., and Engel, N. (1997) Metabolism and function of photosynthetic pigments. In Phytoplankton Pigments in Oceanography (Jeffrey, S.W., Mantoura, R.F.C., and Wright, S.W., eds.), pp. 85–126, UNESCO Publishing, Paris.

Postgate, J.R. (1984) The Sulphate-Reducing Bacteria, 2nd edn., Cambridge Press, Cambridge, UK.

Postma, H. (1980) Sediment transport and sedimentation. In Chemistry and Biogeochemistry of Estuaries (Olausson, E., and Cato, I., eds.), pp. 153–183, John Wiley, New York.

Poutanen, E.L., and Nikkila, K. (2001) Carotenoid pigments as tracers of cyanobacterial blooms in recent and post-glacial sediments of the Baltic Sea. Ambio 30, 179–183.

Powell, G.V.N., Kenworthy, W.J., and Fourqurean, J.F. (1989) Experimental evidence for nutrient limitation of seagrass growth in a tropical estuary with restricted circulation. Bull. Mar. Sci. 44, 324–340.

Powell, R.D. (1990) Glaciomarine processes at grounding-line fans and their growth to ice-contact deltas. In Glaciomarine Environments: Processes and Sediments (Dowdeswell, J.A., and Scorse, J.D., eds.), pp. 53–73, Geological Society of London, Special Publication.

Powell, R.T., and Donat, J.R. (2001) Organic complexation and speciation of iron in the South and Equatorial Atlantic. Deep-Sea Res. II 48, 2877–2893.

Powell, R.T., Landing, W.M., and Bauer, J.E. (1996) Colloidal trace metals, organic carbon, and nitrogen in southeastern U.S. estuary. Mar. Chem. 55, 165–176.

Powell, R.T., and Wilson-Finelli, A. (2003) Importance of organic Fe complexing ligands in the Mississippi River plume. Estuar. Coastal Shelf Sci. 58, 757–763.

Prahl, F.G. (1985) Chemical evidence of differential particle dispersal in the southern Washington coastal environment. Geochim. Cosmochim. Acta 49, 2533–2539.

Prahl, F.G., and Carpenter, R. (1979) The role of zooplankton fecal pellets in the sedimentation of polycyclic aromatic hydrocarbons in Dabob Bay, Washington. Geochim. Cosmochim. Acta 44, 1967–1976.

Prahl, F.G., and Carpenter, R. (1983) Polycyclic aromatic hydrocarbons (PAH)-phase associations in Washington coastal sediments. Geochim. Cosmochim. Acta 47, 1013–1023.

Prahl, F.G., and Coble, P.G. (1994) Input and behavior of dissolved organic carbon in the Columbia River estuary. *In* Changes in Fluxes in Estuaries: Implications from Science and Management (ECSA22/ERF Symp., Plymouth, England) (Dyer, K.R., and Orth, R.J., eds.), pp. 451–457, Olsen and Olsen, Copenhagen.

Prahl F.G., Eglinton G., Corner E.D.S. and O'Hara S.C.M. (1984) Copepod faecal pellets as a source of dihydrophytol in marine sediments. Science 224, 1235–1237.

Prahl, F.G., Ertel, J.R., Goni, M.A., Sparrow, M.A., and Eversmeyer, B. (1994) Terrestrial organic carbon contributions in sediments on the Washington margin. Geochim. Cosmochim. Acta. 58, 3035–3048.

Prahl, F.G., Haynes, J.M., and Xie, T.M. (1992) Diploptene: an indicator of terrigenous organic carbon in Washington coastal sediments. Limnol. Oceanogr. 37, 1290–1300.

Prahl, F.G., and Muelhausen, L.A. (1989) Lipid biomarkers as geochemical tools for paleoceanographic study. *In* Productivity of the Ocean: Present and Past (Berger, W.H., Smetacek, V.S., and Wefer, G., eds.), pp. 271–289, John Wiley, New York.

Precht, E., and Huettel, M. (2003) Advective porewater exchange driven by surface gravity waves and its ecological implications. Limnol. Oceanogr. 48, 1674–1684.

Prego, R. (2002) Nitrogen fluxes and budget seasonality in the Rio Vigo (NW Iberian Peninsula). Hydrobiologia 475/476, 161–171.

Prego, R., and Bao, R. (1997) Upwelling influence on the Galician coast: silicate in shelf water and underlying surface sediments. Cont. Shelf Res. 17, 307–318.

Prego, R., and Fraga, F. (1992) A simple model to calculate the residual flows in a Spanish ria. Hydrographic consequences in the ria of Vigo. Estuar. Coastal Shelf Sci. 34, 603–615.

Presley, B.J., Brooks, R.R., and Kappel, H.M. (1967) A simple squeezer for removal of interstitial water from ocean sediments. J. Mar. Res. 25, 355–357.

Presley, B.J., Kolodny, Y., Nissenbaum, A., and Kaplan, I.R. (1972) Early diagenesis in a reducing fjord, Saanich Inlet, British Columbia—II. Trace element distribution in interstitial water and sediment. Geochim. Cosmochim. Acta 36, 1073–1090.

Presley, B.J., Refrey, J.H., and Shokes, R.F. (1980) Heavy metal inputs to Mississippi Delta sediments. Wat. Air Soil Pollut. 13, 481–494.

Presley, B.J., and Trefrey, J.H. (1980) Sediment–water interactions and the geochemistry of interstitial waters. *In* Chemistry and Biogeochemistry of Estuaries (Olausson, E., and Cato, I., eds.), pp. 187–222, John Wiley, New York.

Preston, R.L. (1987) D-Alanine transport and metabolism by coelomocytes of the bloodworm, *Glycera dibranchiate* (Polychaeta). Comp. Biochem. Physiol. 87, 63–71.

Preston, R.L., McQuade, H., Oladokun, O., and Sharp, J. (1997) Racemization of amino acids by invertebrates. Bull. Mount Desert Island Biol. Lab. 36, 86.

Prins, T. C., and Smaal, A. C. (1994) The role of the blue mussel *Mytilus edulis* in the cycling of nutrients in the Oosterschelde estuary (The Netherlands). *In* The Oosterschelde Estuary: a Case Study of a Changing Ecosystem (Nienhuis, P.H., and A. C. Smaal, A.C., eds.), pp. 413–429, Kluwer, Dordrecht.

Pritchard, D.W. (1952) Salinity distribution and circulation in the Chesapeake Bay estuaries system. J. Mar. Res. 11, 106–123.

Pritchard, D.W. (1954) A study of the salt balance in a coastal plain estuary. J. Mar. Res. 13, 133–144.

Pritchard, D.W. (1955) Estuarine circulation patterns. Proc. Am. Soc. Civil Eng. 81, No. 717.

Pritchard, D.W. (1956) The dynamic structure of a coastal plain estuary. J. Mar. Res. 15, 33–42.

Pritchard, D.W. (1967) Observations of circulation in coastal plain estuaries. *In* Estuaries (Lauff, G.H., ed.), pp. 3–5, American Association for the Advancement of Science, Publ. 83, Washington, DC.

Pritchard, D.W. (1989) Estuarine classification—a help or hindrance. *In* Estuarine Circulation (Nielson, B.J., Kuo, A., and Brubaker, J., eds.), pp. 1–38, Humana Press, Clifton, NJ.

Probst, J.L., Mortatti, J., and Tardy, Y. (1994) Carbon river fluxes and weathering CO_2 consumption in the Congo and Amazon river basins. Appl. Geochem. 9, 1–13.

Pulchan, K., Abrajano, T.A., and Helleur, R.J. (1997) Characterization of tetramethylammonium hydroxide thermochemolysis products of near-shore marine sediments using gas chromatography/mass spectrometry and gas chromatography/combustion/isotope ratio mass spectrometry. J. Anal. Appl. Pyrolysis 42, 135–150.

Pulchan, K.J., Helleur, R., and Abrajano, T.A. (2003) TMAH thermochemolysis characterization of marine sedimentary organic matter in a Newfoundland fjord. Org. Geochem. 34, 305–317.

Pulliam, W.M. (1993) Carbon dioxide and methane exports from a southeastern floodplain swamp. Ecol. Monograph. 63, 29–53.

Pulliam-Holoman, T.R., Elberson, M.A., Cutter, L.A., May, H.D., and Sowers, K.R. (1998) Characterization of a defined 2,3,5,6-tetracholobiphenyl-ortho-dechlorinating microbial community by comparative sequence analysis of genes coding for 16S rRNA. Appl. Environ. Microbiol. 64, 3359–3367.

Pyzik, A.J., and Sommer, S.E. (1981) Sedimentary iron monosulfides: kinetics and mechanism of formation. Geochim. Cosmochim. Acta 45, 687–698.

Qian, Y., Kennicutt, M.C., Svalberg, J., Macko, S.A., Bidigare, R.R., and Walker, J. (1996) Suspended particulate organic matter (SPOM) in Gulf of Mexico estuaries: compound-specific isotope analysis and plant pigment compositions. Org. Geochem. 24, 875–888.

Qualls, R.G., and Haines, B.L. (1990) the influence of humic substances on the aerobic decomposition of submerged leaf litter. Hydrobiology 206, 133–138.

Quemeneur, M., and Marty, Y. (1992) Sewage influence in a macrotidal estuary: fatty acid and sterol distributions. Estuar. Coastal Shelf Sci. 34, 347–363.

Quigley, M.S., Honeyman, B.D., and Santschi, P.H. (1996) Thorium sorption in the marine environment: equilibrium partitioning at the hematite–water interface, sorption/desorption kinetics and particle tracing. Aquat. Geochem. 1, 277–301.

Quigley, M.S., Santschi, P.H., Guo, L., and Honeyman, B.D. (2001) Sorption irreversibility and coagulation behavior of ^{234}Th with marine organic matter. Mar. Chem. 76, 27–45.

Quigley, M.S., Santschi, P.H., Hung, C.C., Guo, L., and Honeyman, B.D. (2002) Importance of acid polysaccharides for ^{234}Th complexation to marine organic matter. Limnol. Oceanogr. 47, 367–377.

Quin, L.D. (1967) The natural occurrence of compounds with the carbon–phosphorus bond. *In* Topics in Phosphorus Chemistry (Grayson, M., and Griffith, E.J., eds.), Vol. 4, pp. 23–48, John Wiley, New York.

Quinn, P.K., Charlson, R.J., and Bates, T.S. (1988) Simultaneous measurements of ammonia in the atmosphere and ocean. Nature 335, 336–338.

Quivira, M.P. (1995) Structural estuaries. *In* Geomorphology and Sedimentology of Estuaries. Developments in Sedimentology 53 (Perillo, G.M.E., ed.), pp. 227–239, Elsevier Science, New York.

Rabalais, N.N., and Nixon, S.W. (2002) Preface: nutrient over-enrichment of the coastal zone. Estuaries 25, 639.

Rabalais, N. N. and Turner, R.E. (eds.) (2001) Coastal Hypoxia: Consequences for Living Resources and Ecosystems. Coastal and Estuarine Studies 58, American Geophysical Union, Washington, DC.

Rabalais, N.N., Turner, R.E., Justic, D., Dortch, Q., Wiseman, W.J., and Sen Gupta, B.K. (1996) Nutrient changes in the Mississippi River and system responses on the adjacent continental shelf. Estuaries 19, 386–407.

Ragueneau, O., Conley, D.J., Longphuirt, S., Slomp, C.P., and Leynaert, A. (2005a) A review of the Si biogeochemical cycle in coastal waters, I: diatoms in coastal food webs and the coastal Si cycle. *In* Land–Ocean Nutrient Fluxes: Silica Cycle, (Ittekot, V., Humborg, C., and Garnier, L., eds.), SCOPE, Linköping, Sweden.

Ragueneau, O., Conley, D.J., Longphuirt, S., Slomp, C.P., and Leynaert, A. (2005b) A review of the Si biogeochemical cycle in coastal waters, II: anthropogenic perturbation of the Si cycle and responses of coastal ecosystems. *In* Land–Ocean Nutrient Fluxes: Silica Cycle, (Ittekkot, V., Humborg, C., and Garnier, L., eds.), SCOPE, Linköping, Sweden.

Ragueneau, O., Deblasvarela, E., Treguer, P., Queguiner, B, and Elamo, Y.D. (1994) Phytoplankton dynamics in relation to the biogeochemical cycle of silicon in a coastal ecosystem of western Europe. Mar. Ecol. Prog. Ser. 106: 157–172.

Ragueneau, O., Lancelot, C., Egorov, V., Vervlimmeren, J., Cociasu, A., Deliat, G., Krastev, A., Daoud, N.; Rousseau, V., Popovitchev, V., Brion, N., Popa, L., and Cauwet, G. (2002). Biogeochemical transformations of inorganic nutrients in the mixing zone between the Danube River and the north-western Black Sea. Estuar. Coastal. Shelf Sci. 54, 321–336.

Ragueneau, O., Queguiner, B., and Treguer, P. (1996) Contrast in biological responses to tidally-induced vertical mixing for two macrotidal ecosystems of Western Europe. Estuar. Coastal Shelf Sci. 42, 645–665.

Rainbow, P.S., and Phillips, J.H. (1993) Cosmopolitan biomonitors of trace metals. Mar. Pollut. Bull. 26, 593–601.

Raiswell, R. (1982) Pyrite texture, isotopic composition and the availability of iron. Am. J. Sci. 282, 1244–1263.

Raiswell, R., and Berner, R.A. (1985) Pyrite formation in euxinic and semi-euxinic sediments. Am. J. Sci. 285, 710–724.

Raiswell, R., and Canfield, D.E. (1996) Rates of reaction between silicate iron and dissolved sulfide in Peru margin sediments. Geochim. Cosmochim. Acta 60, 2777–2787.

Ralph, E.K. (1971) Carbon-14 dating. *In* Dating Techniques for the Archeologist (Michael, H.N., and Ralph, E.K., eds), pp. 1–48, MIT Press, Cambridge, MA.

Rama, and Moore, W.S. (1996) Using radium quartet for evaluating groundwater input and water exchange in salt marshes. Geochim. Cosmochim. Acta 60, 4645–4652.

Rama, M., Koide, M., and Goldberg, E.D. (1961) Lead-210 in natural waters. Science 134, 98–99.

Ramirez, A.J., and Rose, A.W. (1992) Analytical geochemistry of organic phosphorus and its correlation with organic marine and fluvial sediments and soils. Am. J. Sci. 292, 421–454.

Ramos, C.S., Parrish, C.C., Quibuyen, T.A.O., and Abrajano, T.A. (2003) Molecular and carbon isotopic variations in lipids in rapidly settling particles during a spring phytoplankton bloom. Org. Geochem. 34, 195–207.

Ransom, B., Benett, R.H., Baerwald, R., and Shea, K. (1997) TEM study of in situ organic matter on continental margins: occurrence and the "monolayer" hypothesis. Mar. Geol. 138, 1–9.

Ransom, B., Kim, D., Kastner, M., and Wainwright, S. (1998) Organic matter preservation on continental slopes: importance of mineralogy and surface area. Geochim. Cosmochim. Acta 62, 1329–1345.

Rao, G.G., and Dhar, N.R. (1934) Photolysis of amino acids in sunlight. J. Indian Chem. Soc. 11, 617–622.

Rasmussen, B., and Josefson, A.B. (2002) Consistent estimates for the residence time of micro-tidal estuaries. Estuar. Coastal Shelf Sci. 54, 65–73.

Rasmussen, M.B., Henricksen, K., and Jensen, A. (1983) Possible causes of temporal fluctuations in primary production of microphytobenthos in the Dutch Wadden Sea. Mar. Biol. 73, 109–114.

Rattray, M., and Hansen, D.V. (1962) A similarity solution for the circulation in an estuary. J. Mar. Res. 20, 121–133.

Rau, G.H., Riebesell, U., and Wolf-Gladrow, D. (1997) CO_2 aq-dependent photosynthetic ^{13}C fractionation in the ocean: a model versus measurements. Global Biogeochem. Cycles 11, 267–278.

Rau, G.H., Takahashi, T., and Desmarais, D.J. (1989) Latitudinal variations in plankton $\delta^{13}C$. Implications for CO_2 and productivity in past oceans. Nature 341, 516–518.

Rau, G.H., Takahashi, T., Desmarais, D.J., Repeta, D.J., and Martin, J. (1992) The relationship between organic matter $\delta^{13}C$ and $[CO_2(aq)]$ in ocean surface water: data from a JGOFS site in the northeast Atlantic Ocean and model. Geochim. Cosmochim. Acta 56, 1413–1419.

Ravenschlag, K., Sahm, K., Knoblauch, C., Jørgensen, B., and Amann, R. (2000) Community structure, cellular rRNA content and activity of sulfate-reducing bacteria in marine Arctic sediments. Appl. Environ. Microbiol. 66, 3592–3602.

Ravichandran, M. (1996) Distribution of rare Earth elements in sediment cores of Sabine-Neches estuary. Mar. Pollut. Bull. 32, 719–726.

Ravichandran, M., Baskaran, M., Santschi, P.H., and Bianchi, T.S. (1995a) Geochronology of sediments of Sabine–Neches estuary, Texas. Chem. Geol. 125, 291–306.

Ravichandran, M., Baskaran, M., Santschi, P.H., and Bianchi, T.S. (1995b) History of trace metal pollution in Sabine–Neches estuary, Beaumont, TX. Environ. Sci. Technol. 29, 1495–1503.

Ravikumar, S., Kathiresan, K., Ignatiammal, S., Selvum, M., and Shanthy, S. (2004) Nitrogen-fixing azobacters from mangrove habitat and their utility as marine biofertilizers. J. Exp. Mar. Biol. Ecol. 312, 5–17.

Ravit, B., Ehrenfield, J.G., and Haggblom, M.M. (2003) A comparison of sediment microbial communities associated with *Phragmites australis* and Spartina alterniflora in two brackish wetlands of New Jersey. Estuaries 26, 465–474.

Raymond, P.A., and Bauer, J.E. (2001a) Use of ^{14}C and ^{13}C natural abundances for evaluating riverine, estuarine and coastal DOC and POC sources and cycling: a review and synthesis. Org. Geochem. 32, 469–485.

Raymond, P.A., and Bauer, J.E. (2001b) Riverine export of aged terrestrial organic matter to the North Atlantic Ocean. Nature 409, 497–500.

Raymond, P.A., and Bauer, J.E. (2001c) DOC cycling in a temperate estuary: a mass balance approach using natural ^{14}C and ^{13}C. Limnol. Oceanogr. 46, 655–667.

Raymond, P.A, Bauer, J.E., and Cole, J.J. (2000) Atmospheric CO_2 evasion, dissolved inorganic carbon production, and net heterotrophy in the York River estuary. Limnol. Oceanogr. 45, 1707–1717.

Raymond, P.A., Caraco, N.F., and Cole, J.J. (1997) Carbon dioxide concentration and atmospheric flux in the Hudson River. Estuaries 20, 381–390.

Raymond, P.A., and Cole, J.J. (2001) Gas exchange in rivers and estuaries: choosing a gas transfer velocity. Estuaries 24, 312–317.

Raymond, P.A., and Cole, J.J. (2003) Increase in the export of alkalinity from North America's largest river. Science 301, 88–91.

Reckhow, K.H., and Gray, J. (2000) Neuse River Estuary modeling and monitoring project stage 1: stage 1 executive summary and long-term modeling recommendations. Report no. 325-A of the Water Resources Research Institute, University of North Carolina, Chapel Hill, N.C.

Redalje, D.G. (1993) The labeled chlorophyll *a* technique for determining photoautotrophic carbon specific growth rates and biomass. *In* Handbook of Methods in Aqautic Microbial Ecology (Kemp, P.F., ed.), pp. 563–572, Lewis Publishers, Boca Raton, FL.

Redalje, D.G., Lohrenz, S.E., and Fanenstiel, G.L. (1994) The relationship between primary production and the vertical export of particulate organic matter in a river-impacted coastal ecosystem. Estuaries 17, 829–838.

Redfield, A.C. (1958) The biological control of chemical factors in the environment. Am. J. Sci. 46, 205–221.

Redfield, A.C. (1972) Development of a New England salt marsh. Ecol. Monogr. 42, 201–237.

Redfield, A.C., Ketchum, B.H., and Richards, F.A. (1963) The influence of organisms on the composition of seawater. *In* The Sea, Vol. 2 (Hill, M.N., ed.), pp. 26–87, Interscience, New York.

Reeburgh, W.S., Whalen, S.C., and Alperin, M.J. (1993) The role of methylotrophy in the global methane budget. *In* Microbial Growth on CI compounds (Murrell, J.C., and Kelly, D.P., eds.), pp. 1–14, Intercept, Andover UK.

Regnier, P., Mouchet, A., Wollast, R., and Ronday, F. (1998) A discussion of methods for estimating residual fluxes in strong tidal estuaries. Cont. Shelf Res. 18, 1543–1571.

Regnier, P., Wollast, R., and Steefel, C.I. (1997) Long term fluxes of reactive species in macrotidal estuaries: Estuaries from a fully transient, multi-component reaction transport model. Mar. Chem. 58, 127–145.

Rehder, G., Keir, R.S., Suess, E. and Pohlmann, T. (1998) The multiple sources and patterns of methane in the North Sea waters. Aquat. Geochem. 4, 403–427.

Reimer, A., Brasse, S., Doerffer, R., Durselen, C.D., Kempe, S., Michaelis, W., Rick, H.J., and Siefert, R. (1999) Carbon cycling in the German Bight: an estimate of transformation processes and transport. Dt. Hydrogr. Z. 51, 311–327.

Rejmankova, E., Komarkova, J., and Rejmanek, M. (2004) $\delta^{15}N$ as an indicator of N_2-fixation by cynaobacterial mates in tropical marshes. Biogeochemistry 67, 353–368.

Relexens, J.C., Meybeck, M., Billen, G., Brugeaille, M., Etcheber, H., and Somville, M. (1988) Algal and microbial processes involved in particulate organic matter dynamics in the Loire estuary. Estuar. Coastal Shelf Sci. 27, 625–644.

Repeta, D.J. (1993) A high resolution historical record of Holocene anoxygenic primary production in the Black Sea. Geochim. Cosmochim. Acta 57, 4337–4342.

Repeta, D.J., and Gagosian, R.B. (1987) Carotenoid diagenesis in recent marine sediments. I. The Peru continental shelf (15°S, 75°W). Geochim. Cosmochim. Acta 51, 1001–1009.

Repeta, D.J., Quan, T.M., Aluwihare, L.I., and Accardi, A. (2002) Chemical characterization of high molecular weight dissolved organic matter in fresh and marine waters. Geochim. Cosmochim. Acta 66, 955–962.

Repeta, D.J., and Simpson, D.J. (1991) The distribution and cycling of chlorophyll, bacteriochlorophyll and carotenoids in the Black Sea. Deep-Sea Res. 38, 969–984.

Repeta, D.J., Simpson, D.J., Jørgensen, B.B., and Jannasch, H.W. (1989) Evidence for anoxygenic photosynthesis from the distribution of bacteriochlorophylls in the Black Sea. Nature 342, 69–72.

Reuss, N., Conley, D.J., and Bianchi. T.S. (2005) Sediment pigments as a proxy for long-term changes in plankton community structure. Mar. Chem. 95, 283–302.

Reuss, N., and Poulsen, L.K. (2002) Evaluation of fatty acids as biomarkers for a natural plankton community. A field study of a spring bloom and a post-bloom off West Greenland. Mar. Biol. 141, 423–434.

Reynolds, C.S. (1995) River plankton: the paradigm regained. *In* The Ecological Basis for River Management (Harper, D.M., and Ferguson, J.D., eds.), pp. 161–173, John Wiley, New York.

Rhoads, D.C. (1974) Organism–sediment relations on the muddy sea floor. Oceanogr. Mar. Biol. Annu. Rev. 12, 263–300.

Rhoads, D.C., and Boyer, L.F. (1982) Effects of marine benthos on physical properties of sediments. A successional perspective. it In Animal–Sediment Relations (McCall, P.L., and Tevesz, M.J.S., eds.), pp. 3–51, Plenum Press, New York.

Rhoads, D.C., McCall., P.L., and Yingst, J.Y. (1978) Disturbance and production on the estuarine seafloor. Am. J. Sci. 66, 577–586.

Rice, D., Rooth, J., and Stevenson, J.C. (2000) Colonization and expansion of *Phragmites australis* in upper Chesapeake Bay tidal marshes. Wetlands 20, 280–299.

Rice, D.L. (1982) The detritus nitrogen problem. New observations and perspectives from organic geochemistry. Mar. Ecol. Prog. Ser. 9, 153–162.

Rice, D.L. (1986) Early diagenesis in bioadvective sediments: relationships between the diagenesis of beryllium-7, sediment reworking rates, and the abundance of conveyor-belt deposit-feeders. J. Mar. Res. 44, 149–184.

Rice, D.L., Bianchi, T.S., and Roper, E. (1986) Experimental studies of sediment reworking and growth of *Scoloplos* spp (Orbiniidae:Polychaeta). Mar. Ecol. Prog. Ser. 30, 9–19.

Rice, D.L., and Hanson, R.B. (1984) A kinetic model for detritus nitrogen: role of the associated bacteria in nitrogen accumulation. Bull. Mar. Sci. 35, 326–340.

Rice, D.L., and Rhoads, D.C. (1989) Early diagenesis of organic matter and the nutritional value of sediment. *In* Ecology of Marine Deposit-Feeders (Lopez, G., Taghon, G., and Levinton, J.S., eds.), pp. 59–97, Springer-Verlag, New York.

Rice, D.L., and Tenore, K.R. (1981) Dynamics of carbon and nitrogen during the decomposition of detritus derived from estuarine macrophytes. Estuar. Coastal Shelf Sci. 13, 681–690.

Rich, H.W., and Morel, F.M.M. (1990) Availability of well-defined colloids to the marine diatom *Thalassiosira weissflogii*. Limnol. Oceanogr. 35, 652–662.

Rich, J.H., Ducklow, H.W., and Kirchman, D.L. (1996) Concentrations and uptakes of neutral monosaccharides along 140°W in the equatorial Pacific: contribution of glucose to heterotrophic bacterial activity and the DOM flux. Limnol. Oceanogr. 41, 595–604.

Richard, G.A. (1978) Seasonal and environmental variations in sediment accretion in a Long island salt marsh. Estuaries 1, 29–35.

Richards, F.A. (1965) Anoxic basins and fjords. *In* Chemical Oceanography Vol. 1. (Riley, J.P., and Skirrow, G., eds.), pp. 611–645, Academic Press, New York.

Richardson, T.I. (1997) Harmful or exceptional phytoplankton blooms in the marine ecosystem. Adv. Mar. Biol. 31, 302–385.

Richardson, T.I., Pinckney, J.L., and Paerl, H.W. (2001) Responses of estuarine phytoplankton communities to nitrogen form and mixing using microcosm bioassays. Estuaries 24: 828–839.

Richey, J.E., Devol, A.H., Wofsy, S.C., Victoria, R., and Riberio, M.N.G. (1988) Biogenic gases and the oxidation of carbon in Amazon river and floodplain waters. Limnol. Oceanogr. 33, 551–561.

Richey, J.E., Melack, J.M., Aufdenkampe, A.K., Ballester, V.M., and Hess, L.L. (2002) Out-gassing from Amazonian rivers and wetlands as a large tropical source of atmospheric CO_2. Nature 416, 617–620.

Richter, D.D., Markewitz, D., Trumbore, S.E., and Wells, C.G. (1999) Rapid accumulation and turnover of soil carbon in a re-establishing forest. Nature 400, 56–58.

Rickard, D.T. (1975) Kinetics and mechanisms of pyrite formation at low temperatures. Am. J. Sci. 275, 636–652.

Rickard, D.T. (1997) Kinetics of pyrite formation by the H_2S oxidation of iron (II) monosulfide in aqueous solutions between 25 and 125deg C: the rate equation. Geochim. Cosmochim. Acta 61, 115–134.

Rickard, D.T., and Luther III, G.W. (1997) Kinetics of pyrite formation by the H_2S oxidation of iron (II) monosulfide in aqueous solutions between 25 and 125°C; the mechanism. Geochim. Cosmochim. Acta 61, 135–147.

Ricketts, T.R. (1966) Magnesium-2,4-divinylphaeoporphyrin-a-5-monomethyl ester, a protochlorophyll-like pigment present in some unicellular flagellates. Phytochem. 5, 223–229.

Ridal, J.J., and Moore, R.M. (1990) A re-examination of the measurement of dissolved organic phosphorus in seawater. Mar. Chem. 29, 19–31.

Riedel, G.F. (1984) The influence of salinity and sulfate on the toxicity of chromium (VI) to the estuarine diatom *Thalassiosira pseudonana*. J. Phycol. 20, 496–500.

Riedel, G.F. (1985) The relationship between chromium (VI) uptake, sulfate uptake and chromium (VI) toxicity to the estuarine diatom *Thalassiosira pseudonana*. Aquat. Toxicol. 7, 191–204.

Riedel, G.F., Sanders, J.G., and Breitburg, D.L. (2003) Seasonal variability in response of estuarine phytoplankton to stress: linkages between toxic trace elements and nutrient enrichment. Estuaries 26, 323–338.

Rieley, G., Collier, R.J., Jones, D.M., and Eglinton, G. (1991) The biogeochemistry of Ellesmere Lake, UK—I. Source correlation of leaf wax inputs to the sedimentary record. Org. Geochem. 17, 901–912.

Rietsma, C.S., Valiela, I., and Sylvester-Serianni, A. (1982) Food preferences of dominant salt marsh herbivores and detritivores. Mar. Ecol. 3, 179–189.

Rijstenbil, J.W., and Wijnholds, J.A. (1996) HPLC analysis of nonprotein thiols in planktonic diatoms: pool size, redox state and response to copper and cadmium exposure. Mar. Biol. 127, 45–54.

Riley, J.P., and Tongadai, M. (1967) The major cation/chlorinity ratios in seawater. Chem. Geol. 2, 263–269.

Risatti, J.B., Rowland, S.J., Yon, D., and Maxwell, J.R. (1984) Sterochemical studies of acyclic isoprenoids—XII. Lipids of methanogenic bacteria and possible contributions to sediments. *In* Advances in Organic Geochemistry (Schenck, P.A., and de Leeuw, J.W., eds.), pp. 93–103, Pergamon Press, Oxford, UK.

Risgaard-Petersen, N. (2003) Coupled nitrification–denitrification in autotrophic and heterotrophic estuarine sediments: on the influence of benthic algae. Limnol. Oceanogr. 48, 93–105.

Ritchie, J.C., Spraberry, J.A., and McHenry, J.R. (1974) Estimating soil erosion from the redistribution of fallout Cs-137. Soil Sci., Soc. Am. Proc. 38, 137–139.

Rivera-Monroy, V.H., Day, J.D., Twilley, R.R., Vera-Herrera, F., and Coronado-Molina, C. (1995) Flux of nitrogen and sediments in a fringe mangrove forest in terminus Lagoon, Mexico. Estuar. Coastal Shelf Sci. 40, 139–160.

Rivera-Monroy, V.H., and Twilley, R. (1996) The relative role of denitrification and immobilization in the fate of inorganic nitrogen in mangrove sediments (Terminos Lagoon, Mexico). Limnol. Oceanogr. 41, 284–296.

Rizzo, W.M. , Lackey, G.J., and Christian, R.R. (1992) Significance of euphotic, subtidal sediments to oxygen and nutrient cycling in a temperate estuary. Mar. Ecol. Prog. Ser. 86, 51–61.

Robbins, J.A. (1986) A model for particle-selective transport of tracers in sediments with conveyor-belt deposit feeders. J. Geophys. Res. 91, 8542–8558.

Robbins, J.A., Krezoski, J.R., and Mozley, S.C. (1977) Radioactivity in sediments of the Great Lakes: post-depositional redistribution by deposit-feeding organisms. Earth Planet. Sci. Lett. 36, 325–333.

Robbins, J.A., McCall, P.L., Fisher, J.B., and Krezoski, J.R. (1979) Effect of deposit-feeders on migration of Cs-137 in lake sediments. Earth Planet. Sci. Lett. 42, 277–287.

Roberts, H.H. (1997) Dynamic changes of the Holocene Mississippi River Delta Plain: the delta cycle. J. Coastal Res. 13, 605–627.

Robertson, A.I., and Alongi, D.M. (eds.) (1992) Tropical Mangrove Ecosystems. American Geophysical Union Press, Washington, DC.

Robertson, L. A., and Kuenen, J. G. (1984). Aerobic denitrification: a controversy revived. Arch. Microbiol. 139, 351–354.

Robinson, N., Cranwell, P.A., and Eglinton, G. (1987) Sources of the lipids in the bottom sediments of an English oligomesotrophic lake. Freshwater Biol. 17, 15–33.

Robinson, N., Cranwell, P.A., Finlay, B.J., and Eglinton, G. (1984) Lipids of aquatic organisms as potential contributors to lacustrine sediments. Org. Geochem. 6, 143–152.

Rod, S.R., Ayres, R.U., and Small, M. (1989) Reconstruction of historical loadings of heavy metals and chlorinated hydrocarbon pesticides in the Hudson–Raritan Basin, 1880–1980. Report to the Hudson River Foundation, New York.

Roden, E.E., and Edmonds, J.W. (1997) Phosphate mobilization in iron-rich anaerobic sediments: microbial Fe(III) oxide reduction versus iron-sulfide formation. Arch. Hydrobiol. 139, 347–378.

Roden, E.E., and Lovley, D.R. (1993) Dissimilatory Fe (III) reduction by the marine microorganism, Desulfuromonas acetoxidans. Appl. Environ. Microbiol. 59, 734–742.

Roden, E.E., and Tuttle, J.H. (1992) Sulfide release from estuarine sediments underlying anoxic bottom water. Limnol. Oceanogr. 37, 725–738.

Roden, E.E., and Tuttle, J.H. (1993a) Inorganic sulfur cycling in mid- and lower Chesapeake Bay sediments. Mar. Ecol. Prog. Ser. 93, 101–118.

Roden, E.E., and Tuttle, J.H. (1993b) Inorganic sulfur turnover in oligohaline estuarine sediments. Biogeochemistry 22, 81–105.

Roden, E.E., and Tuttle, J.H. (1996) Carbon cycling in the mesohaline Chesapeake Bay. 2: kinetics of particulate and dissolved organic carbon turnover. J. Mar. Res. 54, 343–383.

Roden, E.E., Tuttle, J.H., Boynton, W.R., and Kemp, W.M. (1995) Carbon cycling in mesohaline Chesapeake bay sediments. 1: POC deposition rates and mineralization pathways. J. Mar. Res. 53, 799–819.

Roden, E.E., and Zachara, J.M. (1996) Microbial reduction of crystalline iron (III) oxides: influences of oxide surface area and potential for cell growth. Environ. Sci. Technol. 30, 1618–1628.

Roditi, H.A., Caraco, N.F., Cole, J.J., and Strayer, D.L. (1996) Filtration of Hudson River water by the zebra mussel (Dreissena polymorpha). Estuaries 19, 824–832.

Roditi, H.A., and Fisher, N.S. (1999) Rates and routes of trace element uptake in zebra mussels. Limnol. Oceanogr. 44, 1730–1749.

Roditi, H.A., Fisher, N.S., and Sanudo-Wilhelmy, S.A. (2000) Uptake of dissolved organic carbon and trace metals by zebra mussels. Nature 407, 78–80.

Roesijadi, G. (1994) Behavior of metallothionein-bound metals in a natural population of an estuarine mollusk. Mar. Environ. Res. 38, 147–152.

Rolff, C. (2000) Seasonal variation in $\delta^{13}C$ and $\delta^{15}N$ of size-fractionated plankton at a coastal station in the northern Baltic proper. Mar. Ecol. Prog. Ser. 203, 47–65.

Rolinski, S. (1999) On the dynamics of suspended matter transport in the tidal river Elbe: Description and results of a Lagrangian model. J. Geophys. Res. 104, 26043–26057.

Roman, C.T., and Able, K.W. (1988) Production ecology of eelgrass (Zostera marina L.) in a Cape Cod salt marsh estuarine system, Massachusetts. Aquat. Bot. 32, 353–363.

Roman, C.T., Able, K.W., Lazzari, M.A., and Heck, K.L. (1990) Primary productivity of angiosperm and macroalgae dominated habitats in a New England salt marsh: a comparison analysis. Estuar. Coastal Shelf Sci. 30, 35–46.

Roman, C.T., Jaworski, N., Short, F.T., Findlay, S., and Warren, R.S. (2000) Estuaries of the Northeastern United States: habitat and land use signatures. Estuaries 23, 743–764.

Romankevich, E.A. (1984) Geochemistry of Organic Matter in the Ocean. Springer-Verlag, New York.

Romero, O.E., Dupont, L., Wyputta, U., Jahns, S., and Wefer, G. (2003) Temporal variability of fluxes of eolian-transported freshwater diatoms, phytoliths, and pollen grains off Cape Blanc as reflection of land–atmosphere–ocean interface in northwest Africa. J. Geophys. Res. 108, 3153.

Rooney-Varga, J.N., Devereux, R., Evans, R.S., and Hines, M.E. (1997) Seasonal changes in the relative abundance of uncultivated sulfate-reducing bacteria in salt marsh sediments and in the rhizosphere of *Spartina alterniflora*. Appl. Environ. Microbiol. 63, 3895–3901.

Rooth, J.E., Stevenson, J.C., and Cornwell, J.C. (2003) Increased sediment accretion rates following invasion by *Phragmites australis*: the role of litter. Estuaries 26, 475–483.

Roques, P.F. (1985) Rates and stoichiometry of nutrient remineralization in an anoxic estuary, the Pettaquamscutt River. Ph.D. dissertation, University of Rhode Island, Narragansett, RI.

Rosemond, A.D., Pringle, C.M., Ramirez, A., Paul, M.J., and Meyer, J.L. (2002) Landscape variation in phosphorus concentration and effects on detritus-based tropical streams. Limnol. Oceanogr. 47, 278–289.

Rosenberg, G., and Ramus, J. (1984) Uptake and inorganic nitrogen seaweed surface area:volume ratios. Aquat. Bot. 19, 65–72.

Rosenberg, R., Nilsson, H.C., and Diaz, R.J. (2001) Response of benthic fauna and changing sediment redox profiles over a hypoxic gradient. Estuar. Coastal Shelf Sci. 53, 343–350.

Rosenfield, J.K. (1979) Amino acid diagenesis and adsorption in nearshore anoxic sediments. Limnol. Oceanogr. 24, 1014–1021.

Rowan, K.S. (1989) Photosynthetic Pigments of Algae. Cambridge University Press, Cambridge, UK.

Rowe, G.T., Boland, G.S., Phoel, W.C., Anderson, R.F., and Biscayne, P.E. (1992) Deep-sea floor respiration as an indication of lateral input of biogenic detritus from continental margins. Cont. Shelf Res. 24, 132–139.

Rowe, G.T., Clifford, C.H., Smith, K.L., and Hamilton, P.L. (1975) Benthic nutrient regeneration and its coupling to primary productivity in coastal waters. Nature 255, 215–217.

Rowland, S., and Robson, J.N. (1990) The widespread occurrences of highly branched acyclic C_{20}, C_{25}, and C_{30} hydrocarbons in recent sediments and biota—a review. Mar. Environ. Res. 30, 191–216.

Roy, R.N., Roy, L.N., Vogel, K.M., Porter-Moore, C., Pearson, T., Good, C.E., Millero, F.J., and Campbell, D.M. (1993) The dissociation constants of carbonic acid in seawater at salinities 5 to 45 and temperatures 0 to 45°C. Mar. Chem. 44, 249–267.

Roy, S., Chanut, J.P., Gosselin, M., and Sime-Ngando, T. (1996) Characterization of phytoplankton communities in the lower St. Lawrence estuary using HPLC-detected pigments and cell microscopy. Mar. Ecol. Prog. Ser. 142, 55–73.

Rozan, T.F., Taillefert, M., Trouwborst, R.E., Glazer, B.T., Ma, S., Herszage, J., Valdes, L.M., Price, K.S., and Luther III., G.W. (2002) Iron–sulfur–phosphorus cycling in the sediments of a shallow coastal bay: implications for sediment nutrient release and benthic macroalgal blooms. Limnol. Oceanogr. 47, 1346–1354.

Ruch, P., Mirmand, M., Jouanneau, J.M., and Latouch, C. (1993) Sediment budget and transfer of suspended sediment from the Gironde estuary to Cape Ferret Canyon. Mar. Geol. 114, 37–57.

Rue, E.L., and Bruland, K.W. (1995) Complexation of iron (III) by natural organic ligands in the Central North Pacific as determined by a new competitive ligand equilibration/adsorptive cathodic stripping voltammetric method. Mar. Chem. 50, 117–138.

Russell, M.J., and Hall, A.J. (1997) The emergence of life from iron monosulfide bubbles at a submarine hydrothermal redox and pH front. J. Geol. Soc. London 154, 377–402.

Russell, R.J. (1936) Physiography of the lower Mississippi River delta. In Reports on the Geology of Plaquemines and St. Bernard Parishes (Geol. Bull. 8), pp. 1–199, Louisiana Dept. of Conservation, Baton Rouge, LA.

Rustenbil, J.W., and Wijnholds, J.A. (1996) HPLC analysis of nonprotein thiols in planktonic diatoms: Pool size, redox state and response to copper and cadmium exposure. Mar. Biol. 127, 45–54.

Rutherford, D.W., Chiou, C.T., and Kile, D. (1992) Influence of soil organic matter composition on the partitioning of organic compounds. Environ. Sci. Technol. 26, 336–340.

Rutkowski, C., Burnett, W., Iverson, R., and Chanton, J. (1999) The effect of ground water seepage on nutrient delivery and seagrass distribution in the northeastern Gulf of Mexico. Estuaries 22, 1033–1040.

Ruttenberg, K.C. (1992) Development of a sequential extraction method for different forms of phosphorus in marine sediments. Limnol. Oceanogr. 37, 1460–1482.

Ruttenberg, K.C. (1993) Reassessment of the oceanic residence time of phosphorus. Chem. Geol. 107, 405–409.

Ruttenberg, K.C., and Berner, R.A. (1993) Authigenic apatite formation and burial in sediments from non-upwelling continental margin environments. Geochim. Cosmochim. Acta 57, 991–1007.

Rysgaard, S., Christensen, P.B., and Nielsen, L.P. (1995) Seasonal variation in nitrification and denitrification in estuarine sediment colonized by benthic microalgae and bioturbating infauna. Mar. Ecol. Prog. Ser. 126, 111–121.

Rysgaard, S., and Glud, R.N. (2004) Anaerobic N_2 production in Arctic sea ice. Limnol. Oceanogr. 49, 86–94.

Rysgaard, S., Glud, R.N., Risgaard-Petersen, N., and Dalsgaard, T. (2004) Denitrification and anammox activity in Arctic marine sediments. Limnol. Oceanogr. 49, 1493–1502.

Rysgaard, S., Risgaard-Petersen, N., Nielsen, L.P., and Revsbech, N.P. (1993) Nitrification and denitrification in lake and estuarine sediments measured by the ^{15}N dilution technique and isotope pairing. Appl. Environ. Microbiol. 59: 2093–2098.

Rysgaard, S., Risgaard-Petersen, Sloth, N.P., Jensen, K., and Nielsen, L.P. (1994) Oxygen regulation of nitrification and denitrification in sediments. Limnol. Oceanogr. 39, 1634–1652.

Rysgaard, S., Thastum, P., Dalsgaard, T., Christensen, P.B., and Sloth, N.P. (1999) Effects of salinity on NH_4^+ adsorption capacity, nitrification, and denitrification in Danish estuarine sediments. Estuaries 22, 21–30.

Sachs, J.P., Repeta, D.J., and Goericke, R. (1999) Nitrogen and carbon isotopic ratios of chlorophyll from marine phytoplankton. Geochim. Cosmochim. Acta 63, 1431–1441.

Sackett, W.M., Mo, T., Spalding, R.E., and Exner, M.E. (1973) A reevaluation of the marine geochemistry of uranium, in Radioactive Contamination of the Marine Environment (IAEA), Vienna.

Saenger, P.E. (1994) Mangroves and salt marshes. In Marine Biology (Hammond, L., and Synnot, R.N., eds.), Longman Cheshire, Melbourne.

Sahm, K.C., Knoblauch, C., and Amann, R. (1999) Phylogenetic affiliation and quantification of psychrophilic sulfate-reducing isolates in marine Arctic sediments. Appl. Environ. Microbiol. 65, 3976–3981.

Sakugawa, H., and Handa, N. (1985) Isolation and chemical characterization of dissolved and particulate polysaccharides in Mikawa Bay. Geochim. Cosmochim. Acta 49, 1185–1193.

Sampou, P., and Oviatt, C.A. (1991) Seasonal patterns of sedimentary carbon and anaerobic respiration along a simulated eutrophication gradient. Mar. Ecol. Prog. Ser. 72, 271–282.

Sandberg, J., Andersson, A., Johansson, S., and Wikner, J. (2004) Pelagic food web structure and carbon budget in the northern Baltic Sea: potential importance of terrigenous carbon. Mar. Ecol. Prog. Ser. 268, 13–29.

Sandberg, J., Elmgren, R., and Wulff, F. (2000) Caron flows in Baltic Sea food webs—a re-evaluation using a mass balance approach. J. Mar. Syst. 25, 249–260.

Sanders, J.G., Cibik, S.J., D'Elia, C.F., and Boynton, W.R. (1987) Nutrient enrichment studies in a coastal plain-estuary: changes in phytoplankton species composition. Can. J. Fish. Aquat. Sci. 44, 83–90.

Sanford, L.P., Suttles, S.E., and Halka, J.P. (2001) Reconsidering the physics of the Chesapeake Bay estuarine turbidity maximum. 24, 655–669.

Sanger, J.E., and Gorham, E. (1970) The diversity of pigments in lake sediments and its ecological significance. Limnol. Oceanogr. 15, 59–69.

Sansone, F.J., Holmes, M.E., and Popp, B.N. (1999) Methane stable isotope ratios and concentrations as indicators of methane dynamics in estuaries. Global Biogeochem. Cycles 13, 463–474.

Sansone, F.J., Rust, T.R., and Smith, S.V. (1998) Methane distribution and cycling in Tomales Bay. Estuaries 21, 66–77.

Santore, R.C., Di Toro, D.M., Paquin, P.R., Allen, H.E., and Meyer, J.S. (2001) Biotic ligand model of the acute toxicity of metals. 2. Application to acute copper toxicity in freshwater fish and $Daphnia$. Environ. Toxicol. Chem. 20, 2397–2402.

Santschi, P.H. (1995) Seasonality in nutrient concentrations in Galveston Bay. Mar. Environ. Res. 40, 337–362.

Santschi, P.H., Adler, D., Amdurer, M., Li, Y.H., and Bell, J.J. (1980) Thorium isotopes as analogues for "particle-reactive" pollutants in coastal marine environments. Earth Planet. Sci. Lett. 47, 327–335.

Santschi, P.H., Balnois, E., Wilkinson, K.J., Zhang, J., and Buffle, J. (1998) Fibrillar polysaccharides in marine macromolecular organic matter as imaged by atomic force microscopy and transition electron microscopy. Limnol. Oceanogr. 43, 896–908.

Santschi, P.H., Guo, L., Baskaran, M., Trumbore, S., Southon, J., Bianchi, T.S., Honeyman, B., and Cifuentes, L. (1995) Isotopic evidence for the contemporary origin of high-molecular weight organic matter in oceanic environments. Geochim. Cosmochim. Acta 59, 625–631.

Santschi, P.H., Guo, L., Means, J.C., and Ravichandran, M. (1999) Natural organic matter binding of trace metals and trace organic contaminants in estuaries. In Biogeochemistry of Gulf of Mexico Estuaries (Bianchi, T.S., Pennock, J.R., and Twilley, R.R., eds.), pp. 347–380, John Wiley, New York.

Santschi, P.H., Hohener, P., Benoit, G., and Buchholtzen, M. (1990) Chemical processes at the sediment–water interface. Mar. Chem. 30, 269–315.

Santschi, P.H., Lenhart, J.J., and Honeyman, B.D. (1997) Heterogeneous processes affecting trace contaminant distribution in estuaries: the role of natural organic matter. Mar. Chem. 58, 99–125.

Santschi, P.H., Li, Y.H., and Bell, J.J. (1979) Natural radionuclides in Narragansett Bay. Earth Planet. Sci. Lett. 47, 210–213.

Santschi, P.H., Li, Y.H., Bell, J.J., Adler, D., Amdurer, M., and Nyffeler, U.P. (1983) The relative mobility of natural (Th, Pb, Po) and fallout (Pu, As, Am) radionuclides in the coastal marine environment: results from model ecosystems (MERL) and Narragansett Bay studies. Geochim. Cosmochim. Acta 47, 201–210.

Santschi, P.H., Nixon, S., Pilson, M., and Hunt, C. (1984) Accumulation rate of sediments, trace metals and total hydrocarbons in Narragansett Bay, Rhode Island. Estuar. Coastal Shelf Sci. 19, 427–449.

Sargent, J.R., Bell, M.V., Bell, J.G., Henderson, R.J., and Tocher, D.R. (1995) Requirement criteria for essential fatty acids. J. Appl. Ichthyol. 11, 183–198.

Sargent, J.R., and Falk-Petersen, S. (1988) The lipid biochemistry of calanoid copepods. Hydrobiologia 167/168, 101–114.

Sargent, J.R., Gatten, R.R., and McIntosh, R. (1977) Wax esters in the marine environment—their occurrence, formation, transformation and ultimate fates. Mar. Chem. 5, 573–584.

Sarin, M.M., Krishnaswami, S., Dilli, K., Somayjulu, B.L., and Moore, W.S. (1989) Major ion chemistry in the Ganges–Bramaputra River system: weathering processes and fluxes to the Bay of Bengal. Geochim. Cosmochim. Acta 53, 997–1009.

Sarkanen, K.V., and Ludwig, C.H. (1971) Lignins: Occurrence, Formation, Structure, and Reactions. Wiley-Interscience, New York.

Sarma, V.V.S.S., Kumar, M.D., and Manerikar, M. (2001) Emission of carbon dioxide from a tropical estuarine system, Goa, India. Geophys. Res. Lett. 28, 1239–1242.

Sass, R.L., Fisher, F.M., Turner, F.T., and Jund, M.F. (1991) Mitigation of methane emissions from rice fields: possible adverse effects of incorporated rice straw. Global Biogeochem. Cycles 5, 275–282.

Saucier, R.E. (1963) Recent geomorphic history of the Pontchartrain Basin: L.S.U. Studies, Coastal Studies Series 9, p. 114.

Savchuk, O.P. (2000) Studies of the assimilation capacity and effects of nutrient load reductions in the eastern Gulf of Finland with a biogeochemical model. Boreal Env. Res. 5, 147–163.

Savchuk, O.P. (2002) Nutrient biogeochemical cycles in the Gulf of Riga: scaling up field studies with a mathematical model. J. Mar. Syst. 32, 253–280.

Savchuk, O.P., and Wulff, F. (2001) A model of the biogeochemical cycles of nitrogen and phosphorus in the Baltic. In Ecological Studies, A System Analysis of the Baltic Sea (Wulff, F., ed.), pp. 374–415, Springer-Verlag, Berlin.

Savela, K. (1983) Nitrogen fixation by the blue–green alga Calothrix scopulorum in coastal waters of the Baltic. Ann. Bot. Fenn. 20, 399–405.

Savenko, V.S., and Zakharova, E.A. (1995) Phosphorus in riverine runoff. Dokl. Ross. Akad. Nauk 345, 682–685.

Savidge, G., and Hutley, H.T. (1977) Rates of remineralization and assimilation of urea by fractionated plankton populations in coastal waters. J. Exp. Mar. Biol. Ecol. 28, 1–16.

Saxby, J.D. (1969) Metal-organic chemistry of the geochemical cycle. Rev. Pure Appl. Chem. 19, 131–150.

Scarton, F., Day, J.W., Rismondo, A., Cecconi, G., and Ave, D. (2000) Effects of an intertidal sediment fence on sediment elevation and vegetation distribution in a Venice (Italy) lagoon salt marsh. Ecol. Eng. 16, 223–233.

Schafer, J., Blanc, G., Lapaquellerie, Y., Maillet, N., Maneaux, E., and Etcheber, H. (2002) Ten-year observation of the Gironde tributary fluvial system: fluxes of suspended matter, particulate organic carbon and cadmium. Mar. Chem. 79, 229–242.

Schaffner, L.C. (1990) Small-scale organism distributions and patterns of species diversity: evidence for positive interactions in an estuarine benthic community. Mar. Ecol. Prog. Ser. 61, 107–117.

Schaffner, L. C., Dellapenna, T. M., Hinchey, E. K., Friedrics, C. T., Neubauer, M. T., Smith, M. E., and Kuehl, S. A. (2001). Physical energy regimes, seabed dynamics and organism–sediment interactions along an estuarine gradient. *In* Organism–Sediment Interactions (Aller, J. Y., Woodin, S. A., and Aller, R. C., eds.), pp. 161–182, University of South Carolina Press, Columbia.

Schaffner, L.C., Dickhut, R.M., Mitra, S., Lay, P.W., and Brouwer-Riel, C. (1997) Effects of physical chemistry and bioturbation by estuarine macrofauna on the transport of hydrophobic organic contaminants in the benthos. Environ. Sci. Technol. 31, 3120–3125.

Schedel, M., and Truper, H. (1980) Anaerobic oxidation of thiosulfate and elemental sulfur in *Thiobacillus denitrificans*. Arch. Microbiol. 124, 205–210.

Schelske, C.L., and Stoermer, E.F. (1971) Eutrophication, silica depletion, and predicted changes in algal quality in Lake Michigan. Science 173, 423–424.

Schiel, D.R. (1994) Kelp communities. *In* Marine Biology (Hammond, L.S., and Synnot, R.N., eds.), pp. 23–45, Longman Cheshire, Melbourne.

Schiff, S.L., Aravena, R., Trumbore, S.E., and Dillon, P.J. (1990) Dissolved organic carbon cycling in forested watersheds: a carbon isotopic approach. Wat. Res. 26, 2949–2957.

Schimel, D.S., Enting, I.G., Heimann, M., Wigley, T.M.L., Raynaud, D., Alves, D., and Siegenthaler, U. (1995) CO_2 and the carbon cycle. *In* Climate Change 1994: Radiative Forcing of Climate Change and an Evaluation of the IPCC IS92 Emission Scenarios (Houghton, J.T., ed.), pp. 35–71, Cambridge University Press, Cambridge, UK.

Schindler, D.W. (1997) Evolution of phosphorus limitation in lakes. Science 195, 260–262.

Schindler, D.W. (1987) Detecting ecosystem responses to anthropogenic stress. Can. J. Fish. Aquat. Sci. 44, 6–25.

Schmid, H., Bauer, F., and Stich, H.B. (1998) Determination of algal biomass with HPLC pigment analysis from lakes of different trophic state in comparison to microscopically measured biomass. J. Plankton Res. 20, 1651–1661.

Schmidt, J.E., and Ahring, B.K. (1994) Extracellular polymers in granular sludge from different upflow anaerobic sludge blanket (UASB) reactors. Appl. Microbiol. Biotechnol. 42, 457–462.

Schnitzer, M., and Khan, S.U. (1972) Humic Substances in the Environment. Marcel Dekker, New York.

Schnitzer, M., and Preston, C.M. (1986) Analysis of humic acids by solution and solid-state carbon-13 nuclear magnetic resonance. Soil Sci. Soc. Am. J. 50, 326–331.

Schoell, M., McCaffrey, M.A., Fago, F.J., and Moldowan, J.M. (1992) Carbon isotopic compositions of 28,30-bisnorhopanes and other biological markers in a Monterey crude oil. Geochim. Cosmochim. Acta 56, 1391–1399.

Schoeninger, M.J., and DeNiro, M.J. (1984) Nitrogen and carbon isotope composition of bone collagen from marine and terrestrial animals. Geochim. Cosmochim. Acta 46, 625–639.

Scholln, C.A., and 25 others. (2000) Mortality of sea lions along the central California coast linked to a toxic diatom bloom. Nature 403, 80–84.

Schouten, S., Klien, W.C.M., Breteler, K., Blokker, P., Schogt, N., Irene, W., Rupstra, I.C., Grice, K., Bass, M., and Damste J.S.S. (1998) Biosynthetic effects on the stable carbon isotopic compositions of algal lipids: Implications for deciphering the carbon isotopic biomarker record. Geochim. Cosmochim. Acta 62, 1397–1406.

Schroeder, W.W., and Wiseman, W.J. (1986) Low-frequency shelf–estuarine exchange processes in Mobile Bay and other estuarine systems on the northern Gulf of Mexico. *In* Estuarine Variability (Wolfe, D., ed.), pp. 365–367, Academic Press, New York.

Schroeder, W.W., and Wiseman, W.J. (1999) Geology and hydrodynamics of Gulf of Mexico estuaries. *In* Biogeochemistry of Gulf of Mexico Estuaries (Bianchi, T.S., Pennock, J.R., and Twilley, R.R., eds.), pp. 3–28, John Wiley, New York.

Schubauer, J. P., and Hopkinson, C.S. (1984) Above- and below ground production dynamics of *Spartina alterniflora* and *Spartina cynosuroides*. Limnol. Oceanogr. 29, 1052–1065.

Schubel, J.R. (1968) Turbidity maximum of the northern Chesapeake Bay. Science 161, 1013–1015.

Schubel, J.R. (1971) Tidal variation of the size distribution of suspended sediment at a station in the Chesapeake Bay turbidity maximum. Neth. J. Sea Res. 5, 252–266.

Schubel, J.R. (ed.) (1972) Classification according to mode of basin formation. *In* The Estuarine Environment: Estuaries and Estuarine Sedimentation, pp. 2–8, American Geological Institute, Washington, DC.

Schubel, J.R., and Biggs, R.B. (1969) Distribution of seston in upper Chesapeake Bay. Ches. Sci. 10, 18–23.

Schubel, J.R., and Hirschberg, D.J. (1978) Estuarine graveyard and climatic change. *In* Estuarine Processes (Wiley, M., ed.), pp. 285–303, Academic Press, New York.

Schubel, J.R., and Kana, T.W. (1972) Agglomeration of fine-grained suspended sediment in northern Chesapeake Bay. Power Technol. 6, 9–16.

Schubel, J.R., and Meade, R.H. (1977) Man's impact on estuarine sedimentation. *In* Estuarine Pollution Control and Assessment, Proceedings of Conference, Vol. 1, U.S. Government Printing Office, pp. 193–209, Washington, DC.

Schultz, D.M., and Quinn, J.G. (1977) Suspended material in Narragansett Bay: fatty acid and hydrocarbons composition. Org. Geochem. 1, 27–36.

Schultz, H.D., Dahmke, A., Schinzel, U., Wallmann, K., and Zabel, M. (1994) Early diagenetic processes, fluxes, and reaction rates in sediments of the South Atlantic. Geochim. Cosmochim. Acta 58, 2041–2060.

Schulz, H.N., Brinkhoff, T., Ferdelman, T.G., Hernandez, M.M., Teske, A., and Jørgensen, B.B. (1999) Dense population of a giant sulfur bacterium in Namibian shelf sediments. Science 284, 493–495.

Schutte, H.R. (1983) Secondary plant substances. Aspects of carotenoid biosynthesis. Prog. Bot. 45, 120–135.

Schutz, H., Schroder, P., and Rennenberg, H. (1991) Role of plants in regulating the methane flux to the atmosphere. *In* Trace Gas Emissions from Plants (Sharkey, T., ed.), pp. 29–64, Academic Press, San Diego, CA.

Schwartz, M. (2003) Significant groundwater input to a coastal plain estuary: assessment from excess radon. Estuar. Coastal Shelf Sci. 56, 31–42.

Schwarzenbach, R.P., Gschwend, P.M., and Imboden, D.M. (1993) Environmental Organic Chemistry. John Wiley, New York.

Sciare, J., Mihalopoulos, N., and Nguyen, B.C. (2002) Spatial and temporal variability of dissolved sulfur compounds in European estuaries. Biogeochemistry 59, 121–141.

Scranton, M.I., and McShane, K. (1991) Methane fluxes in the southern North Sea: the role of European rivers. Cont. Shelf Res. 11, 37–52.

Scudlark, J.R., and Church, T.M. (1994) Atmospheric input of nitrogen to Delaware Bay. Estuaries 16, 747–759.

Sebilo, M., Billen, G., Grably, M., and Mariotti, A. (2003) Isotopic composition of nitrate-nitrogen as a marker of riparian and benthic denitrification at the scale of the whole Seine River system. Biogeochemistry 63: 35–51.

Seim, H.E., and Gregg, M.C. (1997) The importance of aspiration and channel curvature in producing strong mixing over a sill. J. Geophys. Res. 102, 3451–3471.

Seitzinger, S.P. (1987) Nitrogen biogeochemistry in an unpolluted estuary: The importance of benthic denitrification. Mar. Ecol. Prog. Ser. 41, 177–186.

Seitzinger, S.P. (1988) Denitrification in freshwater and coastal marine ecosystems: ecological and geochemical significance. Limnol. Oceanogr. 33: 702–724.

Seitzinger, S.P. (1990) Denitrification in aquatic sediments. *In* Denitrification in Soil and Sediment (Revsbech, N.P., and Sørensen, J., eds.), pp. 301–322, Plenum Press, New York.

Seitzinger, S.P. (1998) An analysis of processes controlling N:P ratios in coastal marine ecosystems. *In* Effects of Nitrogen in the Aquatic Environment, pp. 65–83, Swedish Royal Academy of Sciences, Stockholm.

Seitzinger, S.P. (2000) Scaling up: Site-specific measurements to global estimates of denitrification. *In* Estuarine Science: A Synthetic Approach to Research and Practice (Hobbie, J.E., ed.), pp. 211–240, Island Press, Washington, DC.

Seitzinger, S.P., Gardner, W.S., and Spratt, A.K. (1991) The effect of salinity on ammonium sorption in aquatic sediments: implications for benthic nutrient recycling. Estuaries 14, 167–174.

Seitzinger, S.P., and Giblin, A.E. (1996) Estimating denitrification in North Atlantic continental shelf sediments. Biogeochemistry 35, 235–259.

Seitzinger, S.P., and Kroeze, C. (1998) Global distribution of nitrous oxide production and N inputs in freshwater and coastal marine ecosystems. Global Biogeochem. Cycles 12, 93–113.

Seitzinger, S.P., and Kroeze, C., Bouman, A.F., Caraco, N., Dentener, F., and Styles, R.V. (2002a). Global patterns of dissolved inorganic and particulate nitrogen inputs to coastal systems: recent conditions and future projections. Estuaries 25, 640–655.

Seitzinger, S.P., Kroeze, C., and Styles, R.V. (2000) Global distribution of N_2O emissions from aquatic systems: natural emissions and anthropogenic effects. Chemosphere: Global Change Science 2, 267–279.

Seitzinger, S.P., and Nixon, S.W. (1985) Eutrophication and the rate of denitrification and N_2O production in coastal marine sediments. Limnol. Oceanogr. 30, 1332–1339.

Seitzinger, S.P., Nixon, S.W., and Pilson, M.E.Q. (1984) Denitrification and nitrous oxide production in a coastal marine ecosystem. Limnol. Oceanogr. 29, 73–83.

Seitzinger, S.P., and Nixon, S.W., Pilson, M.E.Q., and Burke, S. (1980) Denitrification and N_2O production in near-shore marine sediments. Geochim. Cosmochim. Acta 44, 1853–1860.

Seitzinger, S.P., Pilson, M.E.O., and Watson, S.W. (1983) Nitrous oxide production in nearshore marine sediments. Science 222, 1244–1245.

Seitzinger, S.P., and Sanders, R.W. (1997) Contribution of dissolved organic nitrogen from rivers to estuarine eutrophication. Mar. Ecol. Prog. Ser. 159, 1–12.

Seitzinger, S.P., and Sanders, R.W. (1999) Atmospheric inputs of dissolved organic nitrogen stimulate estuarine bacteria and phytoplankton. Limnol. Oceanogr. 44, 721–730.

Seitzinger, S.P., Sanders, R.W., and Styles, R. (2002c) Bioavailability of DON from natural and anthropogenic sources to estuarine plankton. Limnol. Oceanogr. 47, 353–366.

Seitzinger, S.P., Styles, R.V., Boyer, E.W., Alexander, R.B., Billen, G., Howarth, R.W., Mayer, B., and Breemen, N.V. (2002b) Nitrogen retention in rivers: model development and application to watersheds in the northeastern USA. Biogeochemistry 57/58, 199–237.

Seliger, H.H., Boggs, J.A., and Biggley, W.H. (1985) Catastrophic anoxia in the Chesapeake Bay in 1984. Science 228, 70–73.

Selmer, J.S. (1988) Ammonium regeneration in eutrophicated coastal waters of Sweden. Mar. Ecol. Prog. Ser. 44, 265–273.

Sempere, R., and Cauwet, G. (1995) Occurrence of organic colloids in the stratified estuary of the Krka River, Croatia. Estuar. Coastal Shelf Sci. 40, 105–114.

Sen Gupta, B.K., Lee, R.F., May, M.S. (1981) Upwelling and an unusual assemblage of benthic foraminifera on the northern Florida continental slope. J. Paleontol. 55, 853–857.

Sen Gupta, B.K., and Machain-Castillo, M.L. (1993) Benthic foraminifera in oxygen-poor habitats. Mar. Micropaleontol. 20, 183–210.

Sen Gupta, B.K., Turner, R.E., and Rabalais, N.N. (1996) Seasonal oxygen depletion in continental-shelf waters of Louisiana: Historical record of benthic foraminifera. Mar. Geol. 24, 227–230.

Senior, W., and Chevolot, L. (1991) Studies of dissolved carbohydrates (or carbohydrate-like substances) in an estuarine environment. Mar. Chem. 32, 19–35.

Serodiø, J.J., Da Silva, M., and Catarino, F. (1998) Non-destructive tracing of migratory rhythms of intertidal benthic microalgae using in vivo chlorophyll-a fluorescence. J. Phycol. 33, 542–553.

Shaffer, G., and Rönner, U. (1984) Denitrification in the Baltic proper deep water. Deep Sea Res. 31, 197–220.

Shank, G.C., Skrabal, S.A., Whitehead, R.F., and Kieber, R.J. (2004a) Strong copper complexation in an organic-rich estuary: the importance of allochthonous dissolved organic matter. Mar. Chem. 88, 21–39.

Shank, G.C., Skrabal, S.A., Whitehead, R.F., and Kieber, R.J. (2004b) Fluxes of strong Cu-complexing ligands from sediments of an organic-rich estuary. Estuar. Coastal Shelf Sci. 60, 349–358.

Shannon, L.V., Cherry, R.D., and Orren M.J. (1970) Polonium-210 and lead-210 in the marine environment. Geochim. Cosmochim. Acta 34, 701–711.

Sharma, P., Gardner, L.R., Moore, W.S., and Bollinger, M.S. (1987) Sedimentation and bioturbation in a salt marsh as revealed by ^{210}Pb, ^{137}Cs, and ^{7}Be studies. Limnol. Oceanogr. 32, 313–326.

Sharp, J.H. (1973) Size classes of organic carbon in seawater. Limnol. Oceanogr. 18, 441–447.

Sharp, J.H. (1983) The distribution of inorganic nitrogen and dissolved and particulate organic nitrogen in the sea. In Nitrogen in the Marine Environment (Carpenter, E.J., and Capone, D.G., eds.), pp. 1–35, Academic Press, New York.

Shaw, P. M., and Johns, R.B. (1985) Organic geochemical studies of a recent Great Barrier Reef sediment. I. Assessment of input sources. Org. Geochem. 8, 147–156.

Shaw, T.J., Gieskes, J.M., and Jahnke, R.A. (1990) Early diagenesis in differing depositional environments: the response of transition metals in pore water. Geochim. Cosmochim. Acta 54, 1233–1246.

Sheng, Y.P., Lee, H.K., and Demas, C.E. (1993) Simulation of flushing in Indian River Lagoon using 1-D and 3-D models. In Estuarine and Coastal Modeling III (Spaulding, M.L., ed.), pp. 366–380, ASCE, Monterey, CA.

Shepard, F.P. (1973) Submarine Geology. Harper and Row, New York.

Sherman, R. (1952) The genesis and morphology of the alumina-rich laterite clays. Am. Inst. Min. Met. Eng. 154–161.

Shi, W., Sun, M.Y., Molina, M., and Hodson, R.E. (2001) Variability in the distribution of lipid biomarkers and their molecular isotopic composition in Altamaha estuarine sediments: implications for the relative contribution or organic matter from various sources. Org. Geochem. 32, 453–468.

Shiklomanov, I.A., and Sokolov, A.A. (1983) Methodological basis of world water balance investigation and computation, In New Approaches in Water Balance computations, International Association for Hydrological Sciences Publication No.148, Proceedings of the Hamburg Symposium.

Shiller, A.M. (1996) The effect of recycling traps and upwelling on estuarine chemical flux estimates. Geochim. Cosmochim. Acta 60, 4321–4330.

Shiller, A.M., and Boyle, E.A. (1987) Variability of dissolved trace metals in the Mississippi River. Geochim. Cosmochim. Acta 51, 3273–3277.

Shimeta, J., Starczak, V.R., Ashiru, O.M., and Zimmer, C.A. (2001) Influences of benthic-layer flow on feeding rates of ciliates and flagellates at the sediment–water interface. Limnol. Oceanogr. 46, 1709–1719.

Sholkovitz, E.R. (1976) Flocculation of dissolved organic and inorganic matter during the mixing of river water and seawater. Geochim. Cosmochim. Acta 40, 831–845.

Sholkovitz, E.R. (1978) The geochemistry of plutonium in fresh and marine water environments. Earth Sci. Rev. 64, 95–161.

Sholkovitz, E.R. (1983) The geochemistry of plutonium in fresh and marine environments. Earth Sci. Rev. 64, 95–161.

Sholkovitz, E.R. (1993) The geochemistry of rare Earth elements in the Amazon river estuary. Geochem. Cosmochim. Acta 57, 2181–2190.

Sholkovitz, E.R. (1995) The aquatic chemistry of rare Earth elements in rivers and estuaries. Aquat. Chem. 1, 1–34.

Sholkovitz, E.R., Boyle, E.A., and Price, N.B. (1978) The removal of dissolved humic acids and iron during estuarine mixing. Earth Planet. Sci. Lett. 40, 130–136.

Short, F.T. (1987) Effects of sediment nutrients on seagrasses: literature review and mesocosm experiment. Aquat. Bot. 27, 41–57.

Short, F.T., and Burdick, D.M. (1996) Quantifying eelgrass habitat loss in relation to housing development and nitrogen loading in Waquoit Bay, Massachusetts. Estuaries 19, 730–739.

Short, F.T., Davis, M.W., Gibson, R.A., and Zimmermann, C.F. (1985) Evidence for phosphorus limitation in carbonate sediments of the seagrass *Syringodium filiforme*. Estuar. Coastal Shelf Sci. 20, 419–430.

Short, F.T., Dennison, W.C., and Capone, D.G. (1990) Phosphorus-limited growth of the tropical seagrass *Syringodium filiforme* in carbonate sediments. Mar. Ecol. Prog. Ser. 62, 169–174.

Shum, K.T., and Sundby, B. (1996) Organic matter processing in continental shelf sediments—the subtidal pump revisited. Mar. Chem. 53, 81–87.

Shuman, F.R., and Lorenzen, C.J. (1975) Quantitative degradation of chlorophyll by a marine herbivore. Limnol. Oceanogr. 20, 580–586.

Siccama, T.G., and Porter, E. (1972) Lead in a Connecticut salt marsh. Bioscience 22, 232–234.

Siefert R.L., Pehkonen, S.O., Johansen, A.M. and Hoffmann, M.R. (1998) Trace metal (Fe, Cu, Mn, Cr) redox chemistry in fog and stratus clouds. J. Air Waste Manag. 48, 128–143.

Siegenthaler, U., and Sarmiento, J.L. (1993) Atmospheric carbon dioxide and the ocean. Nature, 365, 119–125.

Sigleo, A.C., and Macko, S.A. (1985) Stable isotope and amino acid composition of estuarine dissolved colloidal material. *In* Marine and Estuarine Geochemistry (Sigleo, A.C., and Hattori, A., eds.), pp. 29–46, Lewis Publishers, Boca Raton, FL.

Sigman, D.M., Altabet, M.A., Michener, R., McCorkle, D.C., Fry, B., and Holmes, R.M. (1997) Natural abundance-level measurement of the nitrogen isotopic composition of oceanic nitrate: An adaptation of the ammonia diffusion method. Mar. Chem. 57, 227–242.

Silliman, J.E., Meyers, P.A., and Eadie, B.J. (1998) Perylene: an indicator of alteration processes or precursor materials. Org. Geochem. 29, 1737–1744.

Silliman, J.E., and Schelske, C.L. (2003) Saturated hydrocarbons in the sediments of Lake Apopka, Florida. Org. Geochem. 34, 253–260.

Simenstad, C.A., and Wissmar, R.C. (1985) $\delta^{13}C$ evidence of the origins and fates of organic carbon in estuarine and nearshore food webs. Mar. Ecol. Prog. Ser. 22, 141–152.

Simkiss, K., and Taylor, M.G. (1989) Metal fluxes across the membranes of aquatic organisms. Rev. Aquat. Sci. 1, 173–188.

Simmons, G.M. (1992) Importance of submarine groundwater discharge (SGWD) and seawater cycling to material flux across the sediment–water interfaces in marine environments. Mar. Ecol. Prog. Ser. 84, 173–184.

Simo, R., Grimalt, J.O., and Albaiges, J. (1997) Dissolved dimethylsulfide, dimethyl-sulphonioproprionate and dimethyl-sulphoxide in western Mediterranean waters. Deep-Sea Res. II, 44, 929–950.

Simon, N.S., and Kennedy, M.M. (1987) The distribution of nitrogen species and adsorption of ammonium in sediments from the tidal Potomac River and Estuary. Estuar. Coastal Shelf Sci. 25, 11–26.

Simoneit, B.R.T. (1977) The Black Sea, a sink for terrigenous lipids. Deep-Sea Res. 24, 813–830.

Simoneit, B.R.T. (1978) Organic chemistry of marine sediments. In Chemical Oceanography, Vol. 7, 2nd edn. (Chester, J.P., ed.), pp. 233–311, Academic Press, London.

Simoneit, B.R.T. (1984) Organic matter of the troposphere—III. Characterization and sources of petroleum and pyrogenic residues in aerosols over the western United States. Atmos. Environ. 18, 51–67.

Simoneit, B.R.T., and Mazurek, M.A. (1982) Organic matter of the troposphere—II. Natural background of biogenic lipid matter in aerosols over the rural western United States. Atmos. Environ. 16, 2139–2159.

Simoneit, B.R.T., Sheng, G., Chen, X., Fu, J., Zhang, J., and Xu, Y. (1991) Molecular marker study of extractable organic matter in aerosols from urban areas of China. Atmos. Environ. 25A, 2111–2129.

Simpson, H.J., Olsen, C.R., Trier, R.M., and Willimas, S.C. (1996) Man-made radionuclides and sedimentation in the Hudson River estuary. Science 194, 179–183.

Sinninghe-Damste, Rijpstra, W.I.C., Schouten, S., Fuerst, J.A., Jetten, M.S.M., and Strous, M. (2004) The occurrence of hopanoids in planctomycetes: implications for the sedimentary biomarker record. Org. Chem. 35, 561–566.

Sinsabaugh, R.L., and Findlay, S.E.G. (2003) Dissolved organic matter: out of the black box into the mainstream. In Aqautic Ecosystems: Interactivity of Dissolved Organic Matter (Findlay, S.E.G., and Sinsabaugh, R.L., eds.), pp. 479–496, Academic Press, New York.

Sinsabaugh, R.L., and Foreman, C.M. (2003) Integrating dissolved organic matter metabolism and microbial diversity: an overview of conceptual models. In Aquatic Ecosystems: Interactivity of Dissolved Organic Matter (Findlay, S.E.G., and Sinsabaugh, R.L., eds.), pp. 426–449, Academic Press, New York.

Sjöström, E. (1981) Wood Chemistry, Fundamentals and Applications. Academic Press, New York.

Skei, J., Larsson, P., Rosenberg, R., Jonsson, P., Olsson, M., and Broman, D. (2000) Eutrophication and contaminants in aquatic ecosystems. Ambio 29, 184–194.

Skoog, A., and Benner, R. (1997) Aldoses in various size fractions of marine organic matter: implications for carbon cycling. Limnol. Oceanogr. 42, 1803–1813.

Skrabal, S.A. (1995) Distributions of dissolved titanium in Chesapeake Bay and the Amazon River Estuary. Geochim. Cosmochim. Acta 59, 2449–2458.

Skrabal, S.A., Donat, J.R., and Burdige, D.J. (1997) Fluxes of copper-complexing ligands from estuarine sediments. Limnol. Oceanogr. 42, 992–996.

Skrabal, S.A., Donat, J.R., and Burdige, D.J. (2000) Pore water distributions of dissolved copper and copper-complexing ligands in estuarine and coastal marine sediments. Geochim. Cosmochim. Acta 64, 1843–1857.

Skrabal, S.A., Ullman, W.J., and Luther III, G.W. (1992) Estuarine distributions of dissolved titanium. Mar. Chem. 37, 83–103.

Sleath, H. (1984) Sea Bed Mechanics. John Wiley, New York.

Slim, F.J., Hemminga, M.A., Ochieng, C., Jannink, N.T., Cocheret de la Morinière , E., and van der Velde, G. (1997) Leaf litter removal by the snail *Terebralia palustris* (Linnaeus) and sesarmid crabs in an East African mangrove forest (Gazi Bay, Kenya). J. Exp. Mar. Biol. Ecol. 215, 35–48.

Slomp, C.P., Epping, E.H., Helden, W., and Raaphorst, W.V. (1996) A key role for iron-bound phosphorus in authigenic apatite formation in North Atlantic continental platform sediments. J. Mar. Res. 54, 1179–1205.

Slomp, C.P., Malschaert, J.F.P., Lohse, L., and van Raaphorst, W. (1997) Iron and manganese cycling in different sedimentary environments on the North Sea continental margin. Cont. Shelf Res. 17, 1083–1117.

Sloth, N.P., Blackburn, H., Hansen, L.S., Risgaard-Petersen, N., Lomstein, B.A. (1995) Nitrogen cycling in sediments with different organic loading. Mar. Ecol. Prog. Ser. 116, 163–170.

Smayda, T.J. (1990) Novel and nuisance phytoplankton blooms in the sea: evidence for a global epidemic. *In* Toxic Marine Phytoplankton (Graneli, E., Sunderstrom, B., Elder, L., and Anderson, D.M., eds.), pp. 29–40, Elsevier, New York.

Smethie, W.J.J., Nittrouer, C.A., and Self, R.F.L. (1981) The use of radon-222 as a tracer of sediment irrigation and mixing on the Washington continental shelf. Mar. Geol. 42, 173–200.

Smith, B.N., and Epstein, S. (1970) Biogeochemistry of the stable isotopes of hydrogen and carbon in salt marsh biota. Plant Physiol. 46, 738–742.

Smith, B.N., and Epstein, S. (1971) Two categories of $^{13}C/^{12}C$ ratios for higher plants. Plant Physiol. 47, 380–384.

Smith, C.J., Wright, W.F., and Patrick, W.H. (1983) The effect of soil redox potential and pH on the reduction and production of nitrous oxide. J. Environ. Qual. 12, 186–188.

Smith, K.L., Jr. (1987) Food energy supply and demand: a discrepancy between particulate organic flux and sediment community oxygen consumption in the deep ocean. Limnol. Oceanogr. 32, 21–220.

Smith, L., Kruszynah, H., and Smith, R.P. (1977) The effect of metheglobin on the inhibition of cytochrome c oxidase by cyanide, sulfide or azide. Biochem. Pharmacol. 26, 2247–2250.

Smith, N.P. (1978) Long-period, estuarine-shelf exchange in response to meteorological forcing. *In* Hydrodynamics of Estuaries and Fjords (Nichoul, J.C.J., ed.), pp. 147–159, Elsevier, New York.

Smith, S.V. (1984) Phosphorus versus nitrogen limitation in the marine environment. Limnol. Oceanogr. 29, 1149–1160.

Smith, S.V. (1991) Stoichiometry of C:N:P fluxes in shallow-water marine ecosystems. *In* Comparative Analyses of Ecosystems—Patterns, Mechanisms, and Theories (Cole, J.J., Lovett, G., and Findlay, S.G., eds.), pp. 259–286, Springer-Verlag, Berlin.

Smith, S.V., and Atkinson, M.J. (1984) Phosphorus limitation of net production in a confined aquatic ecosystem. Nature 307, 626–627.

Smith, S.V., and Hollibaugh, J.T. (1993) Coastal metabolism and the oceanic carbon balance. Rev. Geophys. 31, 75–89.

Smith, S.V., and Hollibaugh, J.T. (1997) Annual cycle and interannual variability of ecosystem metabolism in a temperate climate embayment. Ecol. Monogr. 67, 509–533.

Smith, S.V., and Hollibaugh, J.T., Dollar, S.J., and Vink, S. (1991) Tomales Bay metabolism: C–N–P stoichiometry and ecosystem heterotrophy at the land–sea interface. Estuar. Coastal Shelf Sci. 33, 223–257.

Smoak, J.M., DeMaster, D.J., Kuehl, S.A., Pope, R.H., and McKee, B.A. (1996) The behavior of particle-reactive tracers in a high turbidity environment: [234]Th and ^{210}Pb on the Amazon continental shelf. Geochim. Cosmochim. Acta 60, 2123–2137.

Smullen, J.T., Taft, J.L., and Macknis, J. (1982) Nutrient and sediment loads to the tidal Chesapeake Bay system. In United States Environmental Protection Agency, Chesapeake Bay Program, Technical Studies: A Synthesis, pp. 147–258, Washington, DC.

Socha, S.B., and Carpenter, R. (1987) Factors affecting pore water hydrocarbon concentrations in Puget Sound sediments. Geochim. Cosmochim. Acta 51, 1273–1284.

Soetaert, K., and Herman, P.M.J. (1995) Nitrogen dynamics in the Westerschelde Estuary (S.W. Netherlands) estimated by means of the ecosystem model MOSES. Hydrobiologia 311, 225–246.

Solis, R.S, and Powell, G.L. (1999) Hydrography, mixing characteristics, and residence times of Gulf of Mexico estuaries. In Biogeochemistry of Gulf of Mexico (Bianchi, T.S., Pennock, J., and Twilley, R.R., eds.), pp. 29–61, John Wiley, New York.

Solomons, T.W.G. (1980) Organic Chemistry. John Wiley, New York.

Sommerfield, C.K., Nittrouer, C.A., and Alexander, C.R. (1999) [7]Be as a tracer of flood sedimentation on the northern California continental margin. Cont. Shelf Res. 19, 335–361.

Sørensen, J. (1978) Denitrification rates in a marine sediment measured by the acetylene inhibition technique. Appl. Environ. Microbiol. 36: 139–143.

Sørensen, J. (1982) Reduction of ferric iron in anaerobic, marine sediment and interaction with reduction if nitrate and sulfate. Appl. Environ. Microbiol. 43, 319–324.

Sørensen, J. (1987) Nitrate reduction in marine sediment: pathways and interactions with iron and sulfur cycling. Geomicrobiol. J. 5, 401–421.

Sørensen, J. (1988) Dimethylsulfide and methane thiol in sediment porewater of a Danish estuary. Biogeochemistry 6, 201–210.

Sørensen, J., and Jørgensen, B.B. (1987) Early diagenesis in sediments from Danish coastal waters: microbial activity and Mn–Fe–S geochemistry. Geochim. Cosmochim. Acta 51, 1583–1590.

Soudant, P., Marty, Y., Moal, J., and Samain, J.F. (1995) Separation of major polar lipids in Pecten maximus by high performance liquid chromatography and subsequent determination of their fatty acids using gas chromatography. J. Chromalogy 673, 15–26.

Spalding, M.D., Blasco, F., and Field, C.D. (eds.) (1997) World Mangrove Atlas. The International Society for Mangrove Ecosystems, Okinawa, Japan.

Spenceley, A.P. (1982) Sedimentation patterns in a mangal on Magnetic Island near Townsville, North Queensland, Australia. Singapore J. Trop. Geogr. 3, 100–107.

Spiker, E.C. (1980) The behavior of ^{14}C and ^{13}C in estuarine water: effects of in situ CO_2 production and atmospheric exchange. Radiocarbon 22, 647–654.

Spiker, E.C., and Rubin, M. (1975) Petroleum pollutants in surface and groundwater as indicated by carbon-14 activity of dissolved organic carbon. Science 187, 61–64.

Spiker, E.C., and Schemel, L.E. (1979) Distributions and stable isotope composition of carbon in San Francisco Bay. In San Francisco Bay: The Urbanized Estuary. Proceedings 58th Annual Meeting Pacific Division/American Assoc. Adv. Sci., pp. 192–212, California Academy of Science, San Francisco.

Spinner, G.P. (1969) Serial atlas of the marine environment. In The Wildlife Wetlands and Shellfish Areas of the Atlantic Coastal Zone, Vol. I, Folio 18, American Geographic Society, New York.

Squier, A.H., Hodgson, D.A., and Keely, B.J. (2002) Sedimentary pigments as markers for environmental change in an Antarctic lake. Org. Geochem. 33, 1655–1665.

Squier, A.H., Hodgson, D.A., and Keely, B.J. (2004) Identification of bacteriophaeophytin a esterified with geranylgeraniol in an Antarctic lake sediment. Org. Geochem. 35, 203–207.

Stacey, M.T., Burau, J.R., and Monismith, S.G. (2001) Creation of residual flows in a partially stratified estuary. J. Geophys. Res. 106, 17013–17037.

Stacey, M.T., Monismith, S.G., and Burau, J.R. (1999) Observations of turbulence in a partially stratified estuary. J. Phys. Oceanogr. 29, 1950–1970.

Stallard, R.F. (1980) Major element geochemistry of the Amazon river system. Ph.D Thesis, Massachusetts Institute of Technology, and Wood Hole Oceanographic Institute, Boston, MA.

Stallard, R.F., and Edmond, J.M. (1983) Geochemistry of the Amazon: II. The influence of geology and weathering environment on the dissolved load. J. Geophys. Res. 88, 9671–9688.

Stallard, R.F., and Edmond, J.M. (1986) Geochemistry of the Amazon: I. Precipitation chemistry and the marine contribution to the dissolved load at the time of peak discharge. J. Geophys. Res. 86, 9844–9852.

Stallard, R.F., and Edmond, J.M. (1987) Geochemistry of the Amazon: III. Weathering and limits to dissolved inputs. J. Geophys. Res. 92, 8293–8302.

Standley, L.J. (1997) Effect of sedimentary organic matter composition on the partitioning and bioavailability of dieldrin to the Oligochaete *Lumbriculus variegatus*. Environ. Sci. Technol., 31(9), 2577–2583.

Stanley, D.W., and Nixon, S.W. (1992) Stratification and bottom-water hypoxia in the Pamlico River estuary. Estuaries 15, 270–281.

Starik, I.E., and Kolyadnin, L.B. (1957) The occurrence of uranium in ocean water. Geochemistry 2, 245–256.

Stark, A., Abrajano, T., Hellou, J., and Metcalf-Smith, J.L. (2003) Molecular and isotopic characterization of polycyclic aromatic hydrocarbon distribution and sources at the international segment of the St. Lawrence River. Org. Geochem. 34, 225–237.

Stauber, J.L., and Jeffrey, S.W. (1988) Photosynthetic pigments in fifty-one species of marine diatoms. J. Phycol. 24, 158–172.

Staudinger, B., Peiffer, S., Avnimelech, Y., and Berman, T. (1990) Phosphorus mobility in interstitial waters in Lake Kinneret, Israel. Hydrobiologia 207, 167–177.

Stauffer, R.E. (1990) Electrode pH error, seasonal epilimnetic pCO_2, and the recent acidification of the Maine lakes. Wat. Air Soil Poll. 50, 123–148.

Stedman, D.H., and Shetter, R. (1983) The global budget of atmospheric nitrogen species. *In* Trace Atmospheric Constituents: properties, Transformations, and Fates (Schwartz, S.S., ed.), pp. 411–454, John Wiley, New York.

Steers, J.A. (1964) The Coastline of England and Wales. Cambridge University Press, Cambridge, UK.

Steever, E.Z., Warren, R.S., and Niering, W.A. (1976) Tidal energy subsidy and standing crop production of *Spartina alterniflora*. Estuar. Coastal Shelf Sci. 4, 473–490.

Steidinger, K.A., and Baden, D.G. (1984) Toxic marine dinoflagellates. *In* Dinoflagellates (Spector, D.L., ed.), pp. 201–261, Academic Press, New York.

Stepanauskas, R., Edling, H., and Tranvik, L.J. (1999) Differential dissolved organic nitrogen availability and bacterial aminopeptidase activity in limnic and marine waters. Microb. Ecol. 38, 264–272.

Stepanauskas, R., Laudon, H., and Jørgensen, N.O.G. (2000) High DON bioavailability in boreal streams during a spring flood. Limnol. Oceanogr. 45, 1298–1307.

Stephen, A.M. (1983) Other plant polysaccharides. *In* The Polysaccharides (Aspinall, G.O., ed.), pp. 97–193, Academic Press, New York.

Sternbeck, J., Sohlenius, G., and Hallberg, R.O. (2000) Sedimentary trace elements as proxies to depositional changes induced by a Holocene fresh–brackish water transition. Aquat. Geochem. 6, 325–345.

Sternberg, R.W., Berhane, I., and Ogston, A.S. (1999) Measurement of size and settling velocity of suspended aggregates on the northern California continental shelf. Mar. Geol. 154, 43–53.

Sterner, R.W., and Elser, J.J. (2002) Ecological Stoichiometry—The Biology of Elements from Molecules to the Biosphere. Princeton University Press, Princeton NJ.

Steudler, P.A., and Peterson, B.J. (1985) Annual cycle of gaseous sulfur emissions from a New England *Spartina alterniflora* marsh. Atmos. Environ. 19, 1411–1416.

Stevenson, J.C. (1988) Comparative ecology of submerged grass beds in freshwater estuarine, and marine environments. Limnol. Oceanogr. 33, 867.

Stevenson, R.J. (1990) Benthic algal community dynamics in lake Michigan and Lake Superior. Biogeochemistry 1, 197–218.

Stiller, M., and Nissenbaum, A. (1980) Variations of stable isotopes in plankton from a freshwater lake. Geochim. Cosmochim. Acta 44, 1099–1101.

Stirling, H.P., and Wormald, A.P. (1977) Phosphate/sediment interaction in toto and Long Harbors, Hong Kong, and its role in estuarine phosphorus availability. Estuar. Coastal Shelf Sci. 5, 631–642.

Stommel, H. (1953) Computation of pollution in a vertically mixed estuary. Sewage Ind. Wastes 25, 1065–1071.

Stommel, H., and Farmer, H.G. (1952) Abrupt change in width in two-layer open channel flow. J. Mar. Res. 11, 205–214.

Stommel, H., and Farmer, H.G. (1953) Control of salinity in an estuary by transition. J. Mar. Res. 11, 13–20.

Stordal, M.C., Gill, G.A., Wen, L.S., and Santschi, P.H. (1996) Mercury phase speciation in the surface waters of three Texas estuaries: importance of colloidal forms. Limnol. Oceanogr. 41, 52–61.

Strayer, D.L., and Smith, L.C. (1993) Distribution of the zebra mussel (*Dreissena polymorpha*) in estuaries and brackish waters. *In* Zebra Mussels: Biology, Impacts, and Control (Nalepa, T.F., and Scholoesser, eds.), pp. 715–727, Lewis Publishers, Ann Arbor, MI.

Stribling, J.M., and Cornwell, J.C. (1997) Identification of important primary producers in a Chesapeake Bay tidal creek system using stable isotopes of carbon and sulfur. Estuaries 20, 77–85.

Strickland, J.D.H., and Parsons, T.R. (1972) A Practical Handbook of Seawater Analysis. Fisheries Research Board of Canada, Ottawa.

Strom, S.L. (1991) Growth and grazing rates of an herbivorous dinoflagellate (*Gymnodinium* sp.) from the open subarctic Pacific Ocean. Mar. Ecol. Prog. Ser. 78, 103–113.

Strom, S.L. (1993) Production of phaeopigments by marine protozoa: results of laboratory experiments analyzed by HPLC. Deep-Sea Res. 40, 57–80.

Strom, S.L., and Strom, M.W. (1996) Microplankton growth, grazing, and community structure in the northern Gulf of Mexico. Mar. Ecol. Prog. Ser. 130, 229–240.

Stuiver, M. (1978) Atmospheric carbon dioxide and carbon reservoir changes. Science 199, 253–258.

Stuiver, M., and Polach, H.A. (1977) Discussion: reporting of ^{14}C data. Radiocarbon 19, 355–363.

Stuiver, M., and Quay, P.D. (1981) Atmospheric ^{14}C changes resulting from fossil fuel CO_2 release and cosmic ray flux variability. Earth Planet Sci. Lett. 53, 349–362.

Stumm, W., and Leckie, J.O. (1971) Phosphate exchange with sediments; its role in the productivity of surface waters. Proceedings 5th Intl. Wat. Pollut. Res. Conf. pp. 1–16 Pergamon Press, London.

Stumm, W., and Morgan, J.J. (1981) Aquatic Chemistry. An Introduction Emphasizing Chemical Equilibria in Natural Waters. John Wiley, New York.

Stumm, W., and Morgan, J.J. (1996) Aquatic Chemistry. Chemical Equilibria and Rates in Natural Waters (3rd edn.). John Wiley, New York.

Suberkropp, K., Godshalk, G., and Klug, M.J. (1976) Changes in the chemical composition of leaves during processing in a woodland stream. Ecology 57, 720–727.

Suess, E. (1906) The Face of the Earth. Clarendon Press, Oxford University Press, New York.

Suess, H.E. (1958) Radioactivity of the atmosphere and hydrosphere. Annu. Rev. Nucl. Sci. 8, 243–256.

Suess. H.E. (1968) Climatic changes, solar activity and the cosmic ray production rate of radiocarbon. Meteorol. Monogr. 8, 146–150.

Sugai, S.F., Alperin, M.J., and Reeburgh, W.S. (1994) Episodic deposition and ^{137}Cs immobility in Skan Bay sediments: a ten year ^{210}Pb and ^{137}Cs time series. Mar. Geol. 116, 351–372.

Sullivan, B.E., Prahl, F.G., Small, L.F., and Covert, P.A. (2001) Seasonality of phyto-plankton production in the Columbia River: a natural or anthropogenic pattern? Geochim. Cosmochim. Acta 65, 1125–1139.

Sullivan, M.J., and Moncreiff, C.A. (1990) Edaphic algae are an important component of salt marsh food-webs: evidence from multiple stable isotope analyses. Mar. Ecol. Prog. Ser. 62, 149–159.

Summers, J.K., Wade, T.L., and Engle, V.D. (1996) Normalization of metal concentrations in estuarine sediments from the Gulf of Mexico. Estuaries 19, 581–594.

Summons, R.E., Jahnke, L.L., and Roksandic, Z. (1994) Carbon isotopic fractionation in lipids from methanotrophic bacteria: relevance for interpretation of the geochemical record of biomarkers. Geochim. Cosmochim. Acta 58, 2853–2863.

Sun, M.Y., Aller, R.C., Lee, C. (1994) Spatial and temporal distributions of sedimentary chloropigments as indicators of benthic processes in Long Island Sound. J. Mar. Res. 52, 149–176.

Sun, M.Y., Aller, R.C., and Lee, C., and Wakeham, S.G. (2002) Effects of oxygen and redox oscillation on degradation of cell-associated lipids in surficial marine sediments. Geochim. Cosmochim. Acta 66, 2003–2012.

Sun, M.Y., and Wakeham, S.G. (1994) Molecular evidence for degradation and preserva-tion of organic matter in the anoxic Black Sea Basin. Geochim. Cosmochim. Acta 58, 3395–3406.

Sun, M.Y., and Wakeham, S.G. (1998) A study of oxic/anoxic effects on degrada-tion of sterols at the simulated sediment–water interface of coastal sediments. Org. Geochem. 28, 773–784.

Sun, M.Y., Wakeham, S.G., Aller, R.C., and Lee, C. (1998) Impact of seasonal hypoxia on diagenesis of phytol and its derivatives in Long Island Sound. Mar. Chem. 62, 157–173.

Sun, M.Y., Wakeham, S.G., and Lee, C. (1997) Rates and mechanisms of fatty acid degradation in oxic and anoxic coastal marine sediments. Geochim. Cosmochim. Acta 61, 341–355.

Sun, M.Y., Zou, L., Dai, J., Ding, H, Culp, R.A., and Scanton, M.I. (2004) Molecular carbon isotopic fractionation of algal lipids during decomposition in natural oxic and anoxic seawaters. Org. Geochem. 35, 895–908.

Sunda, W.G., and Ferguson, R.L. (1983) Sensitivity of natural bacterial communities to additions of copper and to cupric ion activity: A bioassay of copper complexation in seawater. In Trace Metals in Sea Water (Wong, C., ed.), pp. 871–891, Plenum Press, New York.

Sunda, W.G., and Guillard, R.R.L. (1976) The relationship between cupric ion activity and the toxicity of copper to phytoplankton. J. Mar. Res. 34, 511–529.

Sunda, W.G., Kieber, D.J., Kiene, R.P., and Huntsman, S. (2002) An antioxidant function for DMSP and DMS in marine algae. Nature 418, 317–320.

Sunda, W.G., and Lewis, J.A.M. (1978) Effects of complexation by natural organic matter on the toxicity of copper to a unicellular alga. Limnol. Oceanogr. 23, 870–876.

Sundbäck, K., Enoksson, V., Granéli, W., and Pettersson, K. (1991) Influence of sublittoral microphytobenthos on the oxygen and nutrient flux between sediment and water: a laboratory continuous-flow study. Mar. Ecol. Prog. Ser. 74: 263–279.

Sundbäck, K., and Graneli, W. (1988) Influence of microphytobenthos on the nutrient flux between sediment and water: a laboratory study. Mar. Ecol. Prog. Ser. 43, 63–69.

Sundbäck, K., Linares, F., Larson, F., and Wulff, A. (2004) Benthic nitrogen fluxes along a depth gradient in a microtidal fjord: the role of denitrification and microphytobenthos. Limnol. Oceanogr. 49, 1095–1107.

Sundbäck, K., and Miles, a. (2000) Balance between denitrification and microalgal incorporation of nitrogen in microtidal sediments, NE Kattegat. Aquat. Microb. Ecol. 22, 291–300.

Sundby, B., Gobeil, C., Silverburg, N., and Mucci, A. (1992) The phosphorus cycle in coastal marine sediments. Limnol. Oceanogr. 37, 1129–1145.

Sundby, B., Silverberg, N., and Chesselet, R. (1981) Pathways of manganese in an open estuarine system. Geochim. Cosmochim. Acta 45, 293–307.

Sunderland, E.M., and Chmura, G.L. (2000) An inventory of historical mercury pollution in Maritime Canada: implications for present and future contamination. Sci. Total Environ. 256, 39–57.

Sunderland, E.M., Gobas, F.A.P.C., Heyes, A., Brainfireun, B.A., Bayer, A.K., Cranston, R.E., and Parsons, M.B. (2004) Speciation and bioavailability of mercury in well-mixed estuarine sediments. Mar. Chem. 90, 91–105.

Sundquist, E.T. (1993) The global carbon dioxide budget. Science 259, 934–941.

Suttle, C.A. (1994) The significance of viruses to mortality in aquatic microbial communities. Microb. Ecol. 28, 237–243.

Sutula, M., Bianchi, T.S., and McKee, B. (2004) Effect of seasonal sediment storage in the lower Mississippi River on the flux of reactive particulate phosphorus to the Gulf of Mexico. Limnol. Oceanogr. 49, 2223–2235.

Swackhamer, D.L., and Skoglund, R.S. (1991) The role of phytoplankton in the partitioning of hydrophobic organic contaminants in water. *In* Organic Substrates and Sediments in Water (Baker, R.A., ed.), pp. 91–106, American Chemical Society, Washington, DC.

Swackhamer, D.L., and Skoglund, R.S. (1993) Bioaccumulation of PCBs by algae: kinetics versus equilibrium. Environ. Sci. Technol. 12, 831–838.

Swain, E. (1985) Measurement and interpretation of sedimentary pigments. Freshwat. Biol. 15, 53–75.

Swain, T. (1977) Secondary compounds as protective agents. Rev. Plant Physiol. 28, 479–501.

Swaney, D.P., Sherman, D., and Howarth, R.W. (1996) Modeling water, sediment, and organic carbon discharges in the Hudson/Mohawk Basin: coupling to terrestrial sources. Estuaries 19, 833–847.

Swanson, V.E., and Palacas, J.G. (1965) Humate in coastal sands of northwest Florida, U.S. U.S. Geol. Survey Bull. 1214-B, 1–29.

Swarzenski, P.W., and McKee, B.A. (1998) Seasonal uranium distributions in the coastal waters off the Amazon and Mississippi Rivers. Estuaries 21, 379–390.

Swarzenski, P.W., McKee, B.A., Sorenson, K., and Todd, J.F. (1999) ^{210}Pb and ^{210}Po, manganese and iron cycling across the O_2/H_2S interface of a permanently anoxic fjord: Framvaren, Norway. Mar. Chem. 67, 199–217.

Sweeney, R.E., Kalil, E.K., and Kaplan, I.R. (1980) Characterization of domestic and industrial sewage in southern California coastal sediments using nitrogen, carbon, sulfur and uranium tracers. Mar. Environ. Res. 3, 225–243.

Sweeney, R.E., and Kaplan, I.R. (1973) Pyrite framboid formation: laboratory synthesis and marine sediments. Econ. Geol. 68, 618–634.

Swenson, E.M., and Sasser, C.E. (1992) Water level fluctuations in the Atchafalaya Delta, Louisiana: tidal forcing versus river forcing. *In* Dynamics and Exchanges in Estuaries and the Coastal Zone, Coastal and Estuarine Studies 40 (Prandle, D., ed.), pp. 191–208, American Geophysical Union, Washington, DC.

Swinnerton, J.W., and Lamontagne, R.A. (1974) Oceanic distribution of low-molecular-weight hydrocarbons: baseline measurements. Environ. Sci. Technol. 8, 657–663.

Syvitski , J.P.M., Burrell, D.C., and Skei, J.M. (1987) Fjords: Processes and Products. Springer-Verlag, New York.

Syvitski, J.P.M., Morehead, M.D., Bahr, D., and Mulder, T. (2000) Estimating fluvial sediment transport: the rating parameters. Wat. Resour. Res. 36, 2747–2760.

Syvitski, J.P.M., and Shaw, J. (1995) Sedimentology and geomorphology of Fjords. *In* Geomorphology and Sedimentology of Estuaries. Developments in Sedimentology 53 (Perillo, G.M.E., ed.), pp. 113–178, Elsevier Science, New York.

Taft, J.L., and Taylor, W.R. (1976) Phosphorus dynamics in some coastal plain estuaries. *In* Estuarine Processes. I. Use, Stresses, and Adaptations to the Estuary (Wiley, M., ed.), pp. 79–89, Academic Press, New York.

Tahir, A., Fletcher, T.C., Houlihan, D.F., and Secombes, C.J. (1993) Effect of short-term exposure to oil-contaminated sediments on the immune response of dab, *Limanda limanda* (L.). Aquat. Toxicol. 27, 71–82.

Taillefert, M., Bono, A.B., and Luther III, G.W. (2000) Reactivity of freshly formed Fe(III) in synthetic solutions and porewaters: voltammetric evidence of an aging process. Environ. Sci. Technol. 34, 2169–2177.

Talbot, H.M., Head, R., Harris, R.P., and Maxwell, J.R. (1999) Distribution and stability of steryl chlorin esters in copepod faecal pellets from diatom grazing. Org. Geochem. 30, 1163–1174.

Talbot, M.M.B., Knoop, W.T., and Bate, G.C. (1990) The dynamic of estuarine macrophytes in relation to flood/situation cycles. Bot. Mar. 33, 159–164.

Tamminen, T., and Irmisch, A. (1996) Urea uptake kinetics of a midsummer planktonic community on the SW coast of Finland. Mar. Ecol. Prog. Ser. 130, 201–211.

Tang, D., Chin-Chang, H., Warnken, K.W., and Santschi, P.H. (2000) The distribution of biogenic thiols in surface waters of Galveston Bay. Limnol. Oceanogr. 45, 1289–1297.

Tang, D., Warnken, K.W., and Santschi, P.H. (2001) Organic complexation of copper in surface waters of Galveston Bay. Limnol. Oceanogr. 46, 321–330.

Tang, D., Warnken, K.W., and Santschi, P.H. (2002) Distribution and partitioning of trace metals (Cd, Cu, Ni, Pb, Zn) in Galveston Bay waters. Mar. Chem. 78, 29–45.

Tang, K., Damm, H., and Visscher, P.T. (1999) Dimethysulfoniopropionate (DMSP) in marine copepods and its relation with diets and salinity. Mar. Ecol. Prog. Ser. 179, 71–79.

Taniguchi, M., Burnett, W.C., Cable, J.E., and Turner, J.V. (2002) Investigation of submarine groundwater discharge. Hydrol. Process. 16, 2115–2129.

Tanoue, E., and Handa, N. (1979) Differential sorption of organic matter by various sized sediment particulates in recent sediment from the Bering Sea. J. Oceanogr. Soc. Jpn. 35, 199–208.

Tarr, M.A., Wang, W., Bianchi, T.S., and Engelhaupt, E. (2001) Mechanisms of ammonia and amino acid photoproduction from aquatic humic and colloidal matter. Wat. Res. 35, 3688–3696.

Taylor, G.T., Way, J., and Scranton, M.I. (2003) Planktonic carbon cycling and transport in surface waters of the highly urbanized Hudson River estuary. Limnol. Oceanogr. 48, 1779–1795.

Teague, K., Madden, C., and Day, J. (1988) Sediment oxygen uptake and net sediment–water nutrient fluxes in a river-dominated estuary. Estuaries 11, 1–9.

Teal, J.M. (1962) Energy flow in the salt marsh ecosystem of Georgia. Ecology 42, 614–624.

Teal, J.M. Valiela, I., and Berlo, D. (1979) Nitrogen fixation by rhizosphere and free-living bacteria in salt marsh sediments. Limnol. Oceanogr. 24, 126–132.

Telang, S.A., Puckington, R., Naidu, A.S., Romankevich, E.A., Gitelson, I.I., and Gladyshev, M.I. (1991) Carbon and mineral transport in major North American, Russian Arctic and Siberian Rivers, the St. Lawrence, the Mackenzie, the Arctic Alaskan rivers, the Arctic Basin rivers in the Soviet Union and the Yenisei. *In* Biogeochemistry of Major Rivers (Degens, E.T., Kempe, S., and Richey, J.E., eds.), pp. 77–104, John Wiley, New York.

Tenore, K.R. (1977) Growth of *Capitella capitata* cultured on various levels of detritus derived from different sources. Limnol. Oceanogr. 22, 936–941.

Tenore, K.R., Cammen, L., Findlay, S.E.G., and Phillips, N. (1982) Perspectives of research on detritus: do factors controlling the availability of detritus to macroconsumers depend on its source? J. Mar. Res. 40, 473–480.

Tenore, K.R., Hanson, R.B., McClain, J., Maccubbin, A.E., and Hobson, R.E. (1984) Changes in compositional nutritional value to a benthic deposit-feeder of decomposing detritus pools. Bull. Mar. Sci. 35, 299–311.

Tester, P.A., Geesey, M.E., Guo, C., Paerl, H.W., and Millie, D.F. (1995) Evaluating phytoplankton dynamics in the Newport River estuary (North Carolina, USA.) by HPLC-derived pigment profiles. Mar. Ecol. Prog. Ser. 124, 237–245.

Thamdrup, B. (2000) Bacterial manganese and iron reduction in aquatic sediments. *In* Advances in Microbial Ecology (Schink, B., ed.), vol. 16, pp. 86–103, Kluwer Academic, New York.

Thamdrup, B., and Dalsgaard, T. (2002) Production of N_2 through anaerobic ammonium oxidation coupled to nitrate reduction in marine sediments. Appl. Environ. Microbiol. 68, 1312–1318.

Thamdrup, B., Fossing, H., and Jørgensen, B.B. (1994) Manganese, iron, and sulfur cycling in a coastal marine sediment, Aarhus Bay, Denmark. Geochim. Cosmochim. Acta 58, 5115–5129.

Theberge, S.M., and Luther III, G.W. (1997) Determination of electrochemical properties of a soluble aqueous FeS species present in sulfidic solutions. Aquat. Geochem. 3, 191–211.

Thimsen, C.A., and Keil, R.G. (1998) Potential interactions between sedimentary dissolved organic matter and mineral surfaces. Mar. Chem. 62, 65–76.

Thode-Andersen, S. and Jørgensen, B. B. (1989) Sulphate reduction and the formation of ^{35}S-labeled FeS, FeS_2, and S^0 in coastal marine sediments. Limnol. Oceanogr. 34, 793–806.

Thomann, R.V., and Komlos, J. (1999) Model of biota-sediment accumulation factor for polycyclic aromatic hydrocarbons. Environ. Tox. Chem. 18, 1060–1068.

Thompson, P.A. (1998) Spatial and temporal patterns of factors influencing phytoplankton in a salt wedge estuary, the Swan River, Western Australia. Estuaries 21, 801–817.

Thurman, E.M. (1985) Organic Geochemistry of Natural Waters. Nijhoff/Junk, Boston, MA.

Thybo-Christensen, M., Rasmussen, M.B., and Blackburn, T.H. (1993) Nutrient fluxes and growth of *Cladophora sericea* in a shallow Danish Bay. Mar. Ecol. Prog. Ser. 100, 273.

Tilman, D. (1977) Resource competition between plankton algae: an experimental and theoretical approach. Ecology 58, 338–348.

Tilman, D., Fargione, J., Wolff, B., D'Antonio, C., Dobson, A., Howarth, R., Schindler, D., Schlesinger, W.H., Simberloff, D., and Swackhamer, D. (2001) Forecasting agriculturally driven global environmental change. Science 292, 281–284.

Timmons, M., and Price, K.S. (1996) The macroalgae and associated fauna of Rehoboth and Indian Bays in Delaware. Bot. Mar. 39, 231–238.

Timperley, M.H., Vigor-Brown, R.J., Kawashima, M., and Ishigamo, M. (1985) Organic nitrogen compounds in atmospheric precipitation: their chemistry and availability to phytoplankton. Can. J. Fish. Aquat. Sci. 42, 1171–1177.

Tipping, E. (1981) The adsorption of aquatic humic substances by iron oxides. Geochim. Cosmochim. Acta 45, 191–199.

Tipping, E. (1993) Modeling the competition between alkaline Earth cations and trace metal species for binding by humic substances. Environ. Sci. Technol. 27, 520–529.

Tipping, E. (1994) WHAM—A chemical equilibrium model and computer code for waters, sediments and soils incorporating a discrete-site/electrostatic model of ion-binding by humic substances. Comp. Geosci. 20, 973–1023.

Tipping, E., Berggren, D., Mulder, J., and Woof, C. (1995a) Modeling the solid-solution distribution of protons, aluminium, base cations, and humic substances in acid soils. Eur. J. Soil Sci. 46, 77–94.

Tipping, E., and Hurley, M.A. (1992) A unifying model of cation binding by humic substances. Geochim. Cosmochim. Acta 56, 3627–3641.

Tipping, E., Lofts, S., and Lawlor, A.J. (1998) Modeling the chemical speciation of trace metals in the surface waters of the Humber system. Sci. Total Environ. 210/211, 63–77.

Tipping, E., Woof, C., and Harley, M.A. (1991) Humic substances in acid surface waters; modelling aluminium binding, contribution to ionic charge-balance and control of pH. Wat. Res. 25, 425–435.

Tipping, E., Woof, C., Kelly, M., Bradshaw, K., and Rowe, J.E. (1995b) Solid-solution distributions of radionuclides in acid soils: applications of the WHAM chemical speciation model. Environ. Sci. Technol. 29, 1365–1372.

Tissot, B.P., and Welte, D.H. (1984) Petroleum Formation and Occurrence. Springer-Verlag, Berlin.

Tobias, C.R., Anderson, I.C., Canuel, A.C., and Macko, S.A. (2001) Nitrogen cycling through a fringing marsh-aquifer ecotone. Mar. Ecol. Prog. Ser. 210, 25–39.

Todd, J.F., Elsinger, R.J., and Moore, W.S. (1988) The distributions of uranium, radium, and thorium isotopes in two anoxic fjords: Framvaren Fjord (Norway) and Saanich Inlet (British Columbia). Mar. Chem. 23, 393–415.

Todd, J.F., Wong, G.T.F., Olsen, C.R., and Larsen, I.L. (1989) Atmospheric depositional characteristics of beryllium-7 and lead-210 along the southeastern Virginia coast. J. Geophys. Res. 94, 11,106–11,116.

Tomasko, D.A., and Dunton, K.H. (1995) Primary productivity in *Halodule wrightii*: a comparison of techniques based on daily carbon budgets. Estuaries 18, 271–278.

Tomasko, D.A., and Lapointe, B.E. (1991) Productivity and biomass of *Thalassia testudinum* as related to water column nutrient availability and epiphyte levels: field observations and experimental studies. Mar. Ecol. Prog. Ser. 75: 9–17.

Tomasky, G., and Valiela, I. (1995) Nutrient limitation of phytoplankton growth in Waquoit Bay. Biol. Bull. 189, 257–258.

Toole, J., Baxter, M.S., and Thomson, J. (1987) The behaviour of uranium isotopes with salinity change in three UK estuaries. Estuar. Coastal Shelf Sci. 25, 283–297.

Topp, E., and Hanson, R.S. (1991) Metabolism of a radiatively important trace gas by methane-oxidizing bacteria. *In* Microbial Production and Consumption of Greenhouse Gases (Rogers, J.E., and Whitman, W.B., eds.), pp. 71–90, ASM Press, Washington, DC.

Törnqvist, T.E., and Gonzalez, J.L. (2002) Reconstructing "background" rates of sea-level rise as a tool for forecasting coastal wetland loss, Mississippi Delta. EOS 83, 530–531.

Törnqvist, T.E., Kidder, T.R., Autin, W.J., van der Borg, K., de Jong, A.F.M., Klerks, C.J.W., Snijders, E.M.A., Stroms, J.E.A., van Dam, R.L., and Wiemann, M.C. (1996) A revised chronology for Mississippi River subdeltas. Science 273, 1693–1696.

Tovar-Sanchez, A., Sañudo-Wilhelmy, S.A., and Flegal, A.R. (2004) Temporal and spatial variations in the biogeochemical cycling of cobalt in two urban estuaries: Hudson River Estuary and San Francisco Bay. Estuar. Coastal Shelf Sci. 60, 717–728.

Tranvik, L.J. (1998) Degradation of dissolved organic matter in humic waters by bacteria. In Aquatic Humic Substances: Ecology and Biogeochemistry (Hessen, D.O., and Tranvik, L.J., eds.), pp. 259–278, Springer-Verlag, New York.

Tranvik, L.J., Sherr, E.B., and Sherr, B.F. (1993) Uptake and utilization of colloidal DOM by heterotrophic flagellates in seawater. Mar. Ecol. Prog. Ser. 92, 301–305.

Trefry, J.H., Metz, S., Nelsen, T.A., Trocine, T.P., and Eadie, B.A. (1994) Transport and fate of particulate organic carbon by the Mississippi River and its fate in the Gulf of Mexico. Estuaries 17, 839–849.

Tréguer, P., Nelson, D.M., van Bennekom, A.J., demister, D.J., Leynaert, A., and Quegiuner, B. (1995) The silica balance in the world ocean: A re-estimate. Science 268, 375–379.

Trowbridge, A.C. (1930) Building of the Mississippi delta. Bull. Am. Assoc. Petrol. Geol. 14, 867–901.

Trowbridge, J.H., Geyer, W.R., Butman, A., and Chapman, R.J. (1989) The 17-meter flume at the coastal research laboratory. Part II: Floe characteristics. Woods Hole Oceanographic Institution, Tech. Pap. WHOI-89–11.

Trowbridge, J.H., and Kineke, G.C. (1994) Structure and dynamics of fluid muds on the Amazon continental shelf. J. Geophys. Res. 99, 865–874.

Trumbore, S. (2000) Age of soil organic matter and soil respiration: radiocarbon constraints on belowground carbon dynamics. Ecol. Appl. 10, 399–411.

Trumbore, S., Vogel, J.S., and Southon, J. (1989) AMS ^{14}C measurements of fractionated soil organic matter: an approach to deciphering the soil carbon cycle. Radiocarbon 31, 644–654.

Trust, B.A., and Fry, B. (1992) Stable sulfur isotopes in plants: a review. Plant Cell. Environ. 15, 1105–1110.

Tseng, C.M., Amouroux, D., Abril, G., Tessier, E., Etcheber, H., and Donard, O.F.X. (2001) Speciation of mercury in a fluid mud profile of a highly turbid macrotidal estuary (Gironde, France). Environ. Sci. Technol. 35, 2627–2633.

Tsunogai, S., Watanabe, S., and Sato, T. (1999) Is there a "continental shelf pump" for the absorption of atmospheric CO_2? Tellus, Series B51, 701–712.

Tulloch, A.P. (1976) Chemistry of waxes of higher plants. In Chemistry and biochemistry of Natural Waxes (Kolattukudy P.E., ed.), pp. 235–287, Elsevier Amsterdam.

Tunnicliffe, V. (2000) A fine-scale record of 130 years of organic carbon deposition in an anoxic fjord, Saanich Inlet, British Columbia. Limnol. Oceanogr. 45, 1380–1387.

Turekian, K.K., Benninger, L.K., and Dion, E.P. (1983) ^{7}Be and ^{210}Pb total deposition fluxes at New Haven, Connecticut, and at Bermuda. J. Geophys. Res. 88, 5411–5415.

Turekian, K.K., and Cochran, J.K. (1978) Determination of marine chronologies using natural radionuclides. In Chemical Oceanography (Riley, J.P., ed.), pp. 313–360, Academic Press, New York.

Turekian, K.K., Tanaka, N., Turekian, V.C., Torgersen, T., and Deangelo, E.C. (1996) Transfer rates of dissolved tracers through estuaries based on ^{228}Ra: study of Long Island Sound. Cont. Shelf Res. 7, 863–873.

Turner, A., Martino, M., and Le Roux, S.M. (2002) Trace metal distribution coefficients in the Mersey Estuary, UK: evidence for salting out of metal complexes. Environ. Sci. Technol. 36, 4578–4584.

Turner, A., and Millward, G.E. (2002) Suspended particles: their role in estuarine biogeochemical cycles. Estuar. Coastal Shelf Sci. 55, 857–883.

Turner, A., Millward, G.E., and Le Roux, S.M. (2004) Significance of oxides and particulate organic matter in controlling trace metal partitioning in a contaminated estuary. Mar. Chem. 88, 179–192.

Turner, B.L., Clark, W.C., Kates, R., Richards, J.F., Mathews, J.T., and Meyer, W.B. (eds.) (1990) The Earth as Transformed by Human Action. Cambridge University Press, Cambridge, UK.

Turner, D.R. (1995) Problems in trace metal speciation modeling. In Metal Speciation and Bioavailability in Aquatic Systems (Tessier, A., and Turner, D.R., eds.), pp. 150–203, John Wiley, Chichester, UK.

Turner, D.R., Whitfield, M., and Dickson, A.G. (1981) The equilibrium speciation of dissolved components in freshwater and seawater at 25°C and 1 atm pressure. Geochim. Cosmochim. Acta 45, 855–881.

Turner, R.E. (1976) Geographic variations in salt marsh macrophyte production: a review. Contrib. Mar. Sci. 20, 47–69.

Turner, R.E. (1990) Landscape development and coastal wetland loss in the northern central Gulf of Mexico. Am. Zool. 30, 89–105.

Turner, R.E. (1991) Tide gauge records, water level rise, and subsidence in the northern Gulf of Mexico. Estuaries 14, 139–147.

Turner, R.E., and Rabalais, N.N. (1991) Changes in Mississippi River water quality this century: implications for coastal food webs. Bioscience 41, 140–147.

Turner, R.E., and Rabalais, N.N. (1994) Coastal eutrophication near the Mississippi river delta. Nature 368, 619–621.

Turner, R.E., Rabalais, N.N., Justic, D., and Dortch, Q. (2003) Global patterns of dissolved N, P, and Si in large rivers. Biogeochemistry 64, 297–317.

Turner, S.M., Malin, G., Liss, P.S., Harbor, D.S., and Holligan, P.M. (1988) The seasonal variation of dimethyl sulfide and dimethylsulfoniopropionate concentrations in nearshore waters. Limnol. Oceanogr. 33, 364–375.

Turner, S.M., Malin, G., Nightingale, P.D., and Lis, P.S. (1996) Photochemical production and air–sea exchange of OCS in the eastern Mediterranean Sea. Mar. Chem. 53, 25–39.

Tuttle, J.H., Jonas, R.B., and Malone, T.C. (1987) Origin, development, and significance of Chesapeake Bay anoxia. In Contaminant Problems and Management of Living Chesapeake Bay Resources (Majumdar, S.K., Hall, L.W., and Hebert, M.A., eds.), pp. 442–472, Pennsylvania Academy of Natural Sciences, Philadelphia.

Twilley, R.R (1985) The exchange of organic carbon in basin mangrove forest in a southwestern Florida estuary. Estuar. Coast. Shelf Sci. 20, 543–557.

Twilley, R.R., and Chen, R. (1998) A water budget and hydrology model of a basin mangrove forest in Rookery Bay, Florida. Mar. Freshwat. Res. 49, 309–323.

Twilley, R.R., Chen, R.H., and Hargis, T. (1992) Carbon sinks in mangroves and their implications to carbon budget of tropical coastal ecosystems. Wat. Air Soil Pollut. 64, 265–288.

Twilley, R.R., Cowan, J., Miller-Way, T., Montagna, P.A., and Mortazavi, B. (1999) Benthic nutrient fluxes in selected estuaries in the Gulf of Mexico. In Biogeochemistry of Gulf of Mexico Estuaries (Bianchi, T.S., Pennock, J.R., and Twilley, R.R., eds.), pp. 163–209, John Wiley, New York.

Twilley, R.R., Kemp, W.M., Staver, K.W., Stevenson, J.C., and Boynton, W.R. (1985) Nutrient enrichment of estuarine submersed vascular plant communities. I. Algal growth and effects on production of plants and associated communities. Mar. Ecol. Prog. Ser. 23, 179–191.

Twilley, R.R., and McKee, B.A. (1996) Ecosystem analysis of the Louisiana Bight and adjacent shelf environments. Vol. I. The fate of organic matter and nutrients in the sediments of the Louisiana Bight. OCS study/MMS No., U.S. Dept. of the Interior, Minerals Management Service, Gulf of Mexico OCS Regional Office, New Orleans.

Twilley, R.R., Pozo, M., Garcia, V.H., Rivera-Monroy, V.H., Zambrano, R., and Bodero, A. (1997) Litter dynamics in riverine mangrove forests in the Guayas River estuary, Ecuador. Oecologia 111, 109–122.

Tyler, A.C., McGlathery, K.J., and Anderson, I.C. (2003) Benthic algae control sediment–water column fluxes of organic and inorganic nitrogen compounds in a temperate lagoon. Limnol. Oceanogr. 48, 2125–2137.

Tyrell, T. (1999) The relative influences of nitrogen and phosphorus on oceanic primary production. Nature 400, 525–531.

Uhlinger, D.J., and White, D.C. (1983) Relationship between physiological status and formation of extracellular polysaccharideglycocalyx in *Pseudomonas atlantica*. Appl. Environ. Microbiol. 45, 64–70.

Ulanowicz, R.E. (1987) NETWRK4: A package of computer algorithms to analyze ecological flow networks. Chesapeake Biological Laboratory, University of Maryland, Solomons, MD.

Ulanowicz, R.E., and Wulff, F. (1991) comparing ecosystem structure: The Chesapeake Bay and the Baltic Sea. *In* Comparative Analyses of Ecosystems (Cole, J., Lovett, G., and Findlay, S.G., eds.), pp. 140–166, Springer-Verlag, New York.

Ullman, W.J., and Aller, R.C. (1982) Diffusion coefficients in nearshore marine sediments. Limnol. Oceanogr. 27, 552–556.

Ullman, W.J., and Aller, R.C. (1983) Rates of iodine remineralization in terrigenous near-shore sediments. Geochim. Cosmochim. Acta 47, 1423–1432.

Ullman, W.J., and Aller, R.C. (1985) The geochemistry of iodine in near-shore carbonate sediments. Geochim. Cosmochim. Acta 49, 967–978.

Ullman, W.J., and Aller, R.C. (1989) Nutrient release rates from the sediments of Saginaw Bay, Lake Huron. Hydrobiologia 171, 127–140.

Ulshofer, U.S., Uher, G., and Andreae, M.O. (1995) Evidence for a winter sink of atmospheric carbonyl sulfide in the northeast Atlantic Ocean. Geophys. Res. Lett. 22, 2601–2604.

U.S. Standard Atmosphere (1976) NOAA, NASA, U.S. Air Force. NOAA-SFR 76–1562, Washington, DC.

Uncles, C.M., Lavender, S.J., and Stephens, J.A. (2001) Remotely sensed observations of the turbidity maximum in the high turbid Humber Estuary, UK. Estuaries 24, 745–755.

Uncles, R.L. (2002) Estuarine physical processes research: some recent studies and progress. Estuar. Coastal Shelf Sci. 55, 829–856.

Uncles, R.L., Barton, M.L., and Stephens, J.A. (1994) Seasonal variability of fine-sediment concentrations in the turbidity maximum region of the Tamar Estuary. Estuar. Coastal Shelf Sci. 38, 19–39.

Uncles, R.J., Easton, A.E., Griffiths, M.L. Harris, C., Howland, R.J.M., King, R.S., Morris, A.W., and Plummer, D.H. (1998) Seasonality of the turbidity maximum in the Humber–Ouse estuary, UK. Mar. Poll. Bull. 37, 206–215.

Underwood, G.C., and Krompkamp, J. (1999) Primary production by phytoplankton and microphytobenthos in estuaries. Adv. Ecol. Res. 29, 93–153.

Uppström, L.R. (1974) The boron/chlorinity ratio of deep-sea water from the Pacific Ocean. Deep Sea Res. 21, 161–162.

Urien, C.M. (1972) Rio de la Plata estuary environments. Geol. Soc. Am. Mem. 133, 213–234.

Ulshofer, V.W., Flock, O.R., Uher, G., and Andreae, M.O. (1996) Photochemical production and air-sea exchange of carbonyl sulfide in the eastern Mediterranean Sea. Mar. Chem. 53, 25–39.

Usui, T., Koike, I., and Ogura, N. (2001) N_2O production, nitrification and denitrification in an estuarine sediment. Estuar. Coastal Shelf Sci. 52, 769–781.

Vachet, R.W., and Callaway, M.B. (2003) Characterization of Cu(II)-binding ligands from Chesapeake Bay using high-performance size-exclusion chromatography and mass spectrometry. Mar. Chem. 82, 31–45.

Valencia, J., Abalde, J., Bode, A., Cid, A., Fernandez, E., Gonzalez, N., Lorenzo, J., Teira, E., and Varela, M. (2003) Variations in planktonic bacterial biomass and production, and phytoplankton blooms off A Coruna (NW Spain). Sci. Mar. 67, 143–157.

Valiela, I. (1983) Nitrogen in salt marsh ecosystems. In Nitrogen in the Marine Environment (Carpenter, E.J., and Capone, D.G., eds.), pp. 649–678, Academic Press, New York.

Valiela, I. (1995) Marine Ecological Processes, 2nd edn. Springer, New York.

Valiela, I., Costa, J.E., Foreman, K., Teal., J.M., Howes, B., and Aubrey, D. (1990) Transport of groundwater-borne nutrients from watersheds and their effects on coastal waters. Biogeochemistry 10, 177–197.

Valiela, I., and D'Elia, C. (1990) Groundwater inputs to coastal waters. Special Issue. Biogeochemistry 10, 328.

Valiela, I., Foreman, K., LaMontagne, M., Hersh, D., Costa, J., Peckol, P., DeMeo-Anderson, B., D'Avanzo, C., Babione, M., Sham, C.-H., Brawley, J., and Lajtha., K. (1992) Coupling of watersheds and coastal waters: Sources and consequences of nutrient enrichment in Waquoit Bay, Massachusetts. Estuaries 15, 443–457.

Valiela, I., Koumjian, L., Swain, T., Teal., J.M., and Hobbie, J.E. (1979) Cinnamic acid inhibition of detritus feeding. Nature 280, 55–57.

Valiela, I., and Teal, J.M. (1976) production and dynamics of experimentally enriched salt marsh vegetation: belowground biomass. Limnol. Oceanogr. 21, 245–252.

Valiela, I., Teal., J.M., Allan, S.D., van Etten, R., Goehungel, D., and Volkman, S. (1985) Decomposition in salt marsh ecosystems: the phases and major factors affecting disappearance of above-ground organic matter. J. Exp. Mar. Biol. Ecol. 89, 29–54.

Valigura, R., Luke, W., Artz, R., and Hicks, B. (1996). Atmospheric nutrient inputs to coastal areas: reducing the uncertainty. U.S. National Oceanic and Atmospheric Administration Coastal Ocean Program Decision Analysis Series No. 9, Washington, DC.

Valle-Levinson, A., and O'Donnell, J. (1996) Tidal interaction with buoyancy-driven flow in a coastal plain estuary. Buoyancy effects on coastal and estuarine dynamics. Coast. Estuar. Stud. 53, 265–281.

Valle-Levinson, A., and Wilson, R.E. (1994a) Rotation and vertical mixing effects on volume exchange in eastern Long Island Sound. Estuar. Coast. Shelf Sci. 46, 573–585.

Valle-Levinson, A., and Wilson, R.E. (1994b) Effects of sill processes and tidal forcing on exchange in eastern Long Island Sound. J. Geophys. Res. 99, 12667–12681.

van Bennekom, A.J., and Solomons, W. (1981) Pathways of nutrients and organic matter from land to ocean through river. In River Inputs to Ocean Systems (Martin, J., Burton, J.D., and Eisma, D., eds.), pp. 33–51, UNEP, IOC, SCOR, United Nations, New York.

van Capellen, P., and Berner, R.A. (1988) A mathematical model for the early diagenesis of phosphorus and fluorine in marine sediments: apatite precipitation. Am. J. Sci. 288, 289–333.

van Capellen, P., and Berner, R.A. (1989) Marine apatite precipitation. *In* Water–Rock Interaction, Proceedings of 6th International Symposium (WRI-6), (Miles, D.I., ed.), pp. 707–710, A.A. Balkema, Rotterdam, The Netherlands.

van Capellen, P., Dixit, S., and van Buesekom, J. (2002) Biogenic silica dissolution in the oceans: reconciling experimental and field-based dissolution rates. Global Biogeochem. Cycles 16, 1075–1085.

van Capellen, P., and Ingall, E.D. (1996) Redox stabilization of the atmosphere and oceans by phosphorus-limited marine productivity. Science 271, 493–496.

van Capellen, P., and Wang, Y. (1996) Cycling of iron and manganese in surface sediments: a general theory for the coupled transport and reaction of carbon, oxygen, nitrogen, sulfur, iron, and manganese. Am. J. Sci. 296, 197–243.

van de Kreeke, J. (1988) Dispersion in shallow estuaries. *In* Hydrodynamics of Estuaries. Vol. 1, Estuarine Physics (Kjerfve, B., ed.), pp. 27–39, CRC Press, Boca Raton, FL.

van den Berg, C.M.G., Merks, A.G., and Duursma, E. (1987) Organic complexation and its control on the dissolved concentrations of copper and zinc in the Scheldt estuary. Estuar. Coastal Shelf Sci. 24, 785–797.

van Diggelen, J., Rozema, J., Dickson, D.M.J., and Broekman, R. (1986) β-3-Dimethylsulphoniopropionate, proline and quaternary ammonium compounds in *Spartina anglica* in relation to sodium chloride, nitrogen and sulphur. New Phytologist 103, 573–586.

van der Nat, F.J.W.A., and Middelburg, J.J. (2000) Methane emissions from tidal freshwater marshes. Biogeochemistry 49, 103–121.

van Heemst, J.D.H., del Rio, J.C., Hatcher, P.G., and deLeeuw, J.W. (2000) Characterization of estuarine and fluvial dissolved organic matter by thermochemolysis using tetramethylammonium hydroxide. Acta Hydrochim. Hydrobiol. 28, 69–76.

van Leussen, W. (1988) Aggregation of particles, settling velocity of mud flocs—a review. *In* Physical Processes in Estuaries (Dronkers, J., and van Leussen, W., eds.), pp. 345–403, Springer, Berlin.

van Loosdrecht, M.C.M., Norde, W., Lyklema, J., and Zehnder, A.J.B. (1990) Hydrophobic and electrostatic parameters in bacterial adhesion. Aquat. Sci. 52, 103–114.

van Mooy, B.A.S., Keil, R.G., and Devol. A.H. (2002) Impact of suboxia on sinking particulate organic carbon: Enhanced carbon flux and preferential degradation of amino acids via denitrification. Geochim. Cosmochim. Acta 66, 457–465.

Vance, D.E., and Vance, J.E. (1996) Biogeochemistry of Lipids, Lipoproteins, and Membranes. Elsevier Science, Amsterdam.

Vanderborght, J.P., Wollast, R., Loijens, C. M., and Regnier, P. (2002) Application of a transport-reaction model to the estimation of biogas fluxes in the Scheldt estuary. Biogeochemistry 59, 207–237.

Velinsky, D.J., Wade, T.L., and Wong, G.T.F. (1986) Atmospheric deposition of organic carbon to Chesapeake Bay. Atmos. Environ. 20, 941–947.

Verduin, J.J., Walker, D.I., and Kuo, J. (1996) In-situ submarine pollination in the seagrass *Amphibolis Antarctica*: research notes. Mar. Ecol. Prog. Ser. 133, 307–309.

Verity, P.G. (2002a) A decade of change in the Skidaway River estuary. I. Hydrography and nutrients. Estuaries 25, 944–960.

Verity, P.G. (2002b) A decade of change in the Skidaway River estuary. II. Particulate organic carbon, nitrogen, and chlorophyll *a*. Estuaries 25, 961–975.

Vernadski, V.I. (1926) The Biosphere (translated and annotated version, 1998). Copernicus and Springer, New York.

Vernet, M., and Lorenzen, C.J. (1987) The presence of chlorophyll *b* and the estimation of phaeopigments in marine phytoplankton. J. Plankton Res. 9, 255–265.

Vesk, M., and Jeffrey, S.W. (1987) Ultrastructure and pigments of two strains of the picoplanktonic alga *Pelagococcus subviridis* (Chrysophyceae). J. Phycol. 23, 322–336.

Viarengo, A. (1989) Heavy metals in marine invertebrates: mechanisms of regulation and toxicity at the cellular level. Rev. Aquat. Sci. 1, 295–298.

Vince, S., and Valiela, I. (1973) The effects of ammonium and phosphate enrichments on chlorophyll a pigment ratios and species composition of phytoplankton of Vineyard Sound. Mar. Biol. 19, 69–73.

Viso, A. C., and Marty, J.C. (1993) Fatty acids from 28 marine microalgae. Phytochemistry 34, 1521–1533.

Vitousek, P.M., Aber, J.D., Howarth, R.W., Likens, G.E., Matson, P.A., Schindler, D.W., Schlesinger, W.H., and Tilman, D.G. (1997) Human alteration of the global nitrogen cycle: sources and consequences. Ecol. Appl. 7, 737–750.

Vitousek, P.M., Cassman, K., Cleveland, C., Crews, T., Field, C.B., Grimm, N.B., Howarth, R.W., Marino, R., Martinelli, L., Rastetter, E.B., and Sprent, J.I. (2002) Towards an ecological understanding of biological nitrogen fixation. Biogeochemistry 57/58: 1–45.

Vitousek, P.M., Walker, L.R., Whiteaker, L.D., Mueller-Dombois, D., and Matson, P.A. (1988) Element interactions in forest ecosystems: succession, allometry and input–output budgets. Biogeochemistry 5, 7–34.

Vodacek, A., Blough, N.V., DeGrandpre, M.D., Peltzer, E.T., and Nelson, R.K. (1997) Seasonal variation of CDOM and DOC in the Middle Atlantic Bight: Terrestrial inputs and photooxidation. Limnol. Oceanogr. 42, 674–686.

Vodacek, A., Hoge, F., Swift, R.N., Yungei, J.K., Peltzer, E.T., and Blough, N.V. (1995) The use of in situ and airborne fluorescence measurements to determine UV absorption coefficients and DOC concentrations in surface waters. Limnol. Oceanogr. 40, 411–415.

Voet, D., and Voet, J.G. (2004) Biochemistry. John Wiley, New York.

Vogel, S. (1981) Life in Moving Fluids. Princeton University Press, Princeton NJ.

Vold, R.D., and Vold, M.J. (1983) Colloid and Interface Chemistry. Addison-Wesley, Reading, MA.

Volkman, J.K. (1986) A review of sterol markers for marine and terrigenous organic matter. Org. Geochem. 9, 83–99.

Volkman, J.K.(2003) Sterols in microorganisms. Appl. Microb. Biotechnol. 60, 495–506.

Volkman, J.K., Barrett, S.M., and Blackburn, S.I. (1999) Eustigmatophyte microalgae are potential sources of C_{29} sterols, n-C_{23}–n-C_{28} n-alkanols and C_{28}–C_{32} n-alkyl diols in freshwater environments. Org. Geochem. 30, 307–318.

Volkman, J.K., Barrett, S.M., Blackburn, S.I., Mansour, M.P., Sikes, E.L., and Gelin, F. (1998) Microalgal biomarkers: a review of recent research developments. Org. Geochem. 29, 1163–1179.

Volkman, J.K., Eglinton, G., and Corner, E.D.S. (1980a) Sterols and fatty acids of marine diatom *Biddulphia sinensis*. Phytochemistry 19, 1809–1813.

Volkman, J.K., Eglinton, G., Corner, E.D.S., and Forsberg, T.E.V. (1980b) Long-chain alkenes and alkenones in the marine coccolithophorid *Emiliania huxleyi*. Phytochemistry 19, 2619–2622.

Volkman, J.K., Farrington, J.W., and Gagosian, R.B. (1987) Marine and terrigenous lipids in coastal sediments from the Peru upwelling region at 15°S: sterols and triterpene alcohols. Org. Geochem. 6, 463–477.

Volkman, J.K., Farrington, J.W., Gagosian, R.B., and Wakeham, S.G. (1983) Lipid composition of coastal marine sediments from the Peru upwelling region *In* Advances in Organic Geochemistry (Bjoroy, M., ed.), pp. 228–240, John Wiley, Chichester, UK.

Volkman, J.K., and Hallegraeff, G.M. (1988) Lipids in marine diatoms of the genus *Thalassiosira*: predominance of 24-methylenecholesterol. Phytochemistry 27, 1389–1394.

Volkman, J.K., Jeffrey, S.W., Nichols, P.D., Rogers, G.I., and Garland, C.D. (1989) Fatty acid and lipid composition of 10 species of microalgae used in mariculture. J. Exp. Mar. Biol. Ecol. 128, 219–240.

Volkman, J.K., and Maxwell, J.R. (1984) Acyclic isoprenoids as biological markers. *In* Biological Markers in the Sedimentary Record (Johns, R.B., ed.), pp. 1–42, Elsevier, New York.

Volkman, J.K., Smith, D.J., Eglinton, G., Forsberg, T.E.V., and Corner, E.D.S. (1981) Sterol and fatty acid composition of four marine haptophycean algae. J. Mar. Biol. Assoc. UK 61, 509–527.

Vollenweider, R.A. (1968) Scientific fundamentals of the eutrophication of lakes and flowing waters, with particular reference to nitrogen and phosphate as factor in eutrophication. OECD, Paris, Tech. Report DAS/SCI/68.27.

Vollenweider, R.A. (1975) Input–output models, with special reference to the phosphorus loading concept in limnology. Schweiz. Z. Hydrobiol. 37, 53–82.

Vollenweider, R.A. (1976) Advances in defining critical loading levels of phosphorus in lake eutrophication. Mem Ist. Ital. Idrobiol. 33, 53–83.

von Gunten, U., and Furrer, G. (2000) Steady-state modeling of biogeochemical processes in columns with aquifer material: 2. Dynamics of iron–sulfur interactions. Chem. Geol. 167, 271–284.

Vörösmarty, C.J., and Peterson, B.J. (2000) Macro-scale models of water and nutrient flux to the coastal zone. *In* Estuarine Science, a Synthetic Approach to Research and Practice (Hobbie, J.E., ed.), pp. 43–79, Island Press, Washington, DC.

Vörösmarty, C.J., Sharma, K., Fekete, B., Copeland, A.H., Holden, J., Marble, J., and Lough, J.A. (1997) The storage and aging of continental runoff in large reservoir systems of the world. Ambio 26, 210–219.

Voss, M., Larsen, B., Leivuori, M., and Vallius, H. (2000) Stable isotope signals of eutrophication in Baltic Sea sediments. J. Mar. Syst. 25, 287–298.

Voss, M., and Struck, U. (1997) Stable nitrogen and carbon isotopes as indicators of eutrophication of the Oder River (Baltic Sea). Mar. Chem. 59, 35–49.

Vymazal, J., and Richardson, C.J. (1995) Species composition, biomass, and nutrient content of periphyton in the Florida Everglades. J. Phycol. 31, 343–354.

Wada, E. (1980) Nitrogen isotope fractionation and its significance in biogeochemical processes occurring in marine environments. *In* Isotope Marine Chemistry (Goldberg, E.D., Horibe, Y., and Saruhashi, K., eds.), pp. 375–398, Uchida Rokakudo, Tokyo.

Wada, E., and Hattori, A. (1978) Nitrogen isotope effects in the assimilation of inorganic nitrogenous compounds by marine diatoms. Geomicrobiology 1, 85–101.

Wada, E., Kabaya, Y., Tsuru, K., and Ishiwatari, R. (1990) ^{15}N abundance of sedimenting organic matter in estuarine area of Tokyo Bay, Japan. Mass. Spectr. 38, 307–318.

Wahby, S.D., and Bishara, N.F. (1979) The effect of the river Nile on Mediterranean water, before and after the construction of the High Dam at Aswan. *In* River Inputs to Ocean Systems (Martin, J.M., Burton, J.D., and Eisma, D., eds.), pp. 311–318, U.N. Environ. Prog. Intergov. Oceanogr. Comm. Sci. Comm. Ocean. Res., Rome.

Wakeham, S.G., and Canuel, E.A. (1988) Organic geochemistry of particulate matter in the eastern tropical North Pacific Ocean: implications for particle dynamics. J. Mar. Res. 46, 183–213.

Wakeham, S.G., and Farrington, J.W. (1980) Hydrocarbons in contemporary aquatic sediments. *In* Contaminants and Sediments (Baker, R.A., ed.), pp. 3–32, Ann Arbor Science, Ann Arbor, MI.

Wakeham, S.G., and Lee, C. (1989) Organic geochemistry of particulate matter in the ocean: the role of particles on oceanic sedimentary cycles. Org. Geochem. 14, 83–96.

Wakeman, S.G., and Lee, C. (1993) production, transport, and alteration of particulate organic matter in the marine water column. *In* Organic Geochemistry (Engel, M.H., and Macko, S.A., eds), pp. 145–169, Plenum Press, New York.

Wakeham, S.G., Schaffer, C., and Giger, W. (1980a) Polycyclic aromatic hydrocarbons in recent lake sediments—I. Compounds having anthropogenic origins. Geochim. Cosmochim. Acta 44, 403–413.

Wakeham, S.G., Schaffer, C., and Giger, W. (1980b) Polycyclic aromatic hydrocarbons in recent lake sediments—II. Compounds derived from biogenic precursors during early diagenesis. Geochim. Cosmochim. Acta 44, 415–429.

Waldichuk, M. (1989) The state of pollution in the marine environment. Mar. Pollut. Bull. 20, 598–601.

Walker, H.A., Latimer, J.S., and Dettmann, E.H. (2000) Assessing the effects of natural and anthropogenic stressors in the Potomac estuary: implications for long-term monitoring. Environ. Monit. Assess. 63, 237–251.

Walker, J.G.G. (1977) Evolution of the Atmosphere. Macmillan, New York.

Walsh, J.J. (1988) On the Nature of Continental Shelves. Academic Press, San Diego, CA.

Walsh, J.J. (1994) Particle export at Cape Hatteras. Deep Sea Res. II 41, 603–628.

Walsh, J.J., Biscaye, P.E., and Csanady, G.T. (1988) The 1983–1984 shelf-edge exchange processes (SEEP)—I. Experiment: hypothesis and highlights. Cont. Shelf Res. 8, 435–456.

Walsh, J.J., Premuzic, E.T., Gaffney, J.S., Rowe, G.T., Harbottle, G., Stoenner, R.W., Balsam, W.L., Betzer, P.R., and Macko, S.A. (1985) Organic storage of carbon dioxide on the continental slope off the Mid-Atlantic Bight, the southeastern Bering Sea, and the Peru coast. Deep-Sea Res. 32, 853–883.

Walters, R.A. (1997) A model study of tidal and residual flow in Delaware Bay and River. J. Geophys. Res. 102, 12689–12704.

Wang, D.P., and Elliott, A.J. (1978) Non-tidal variability in the Chesapeake Bay and Potomac River: evidence for non-local forcing. J. Phys. Oceanogr. 8, 225–232.

Wang, W., Tarr, M.A., Bianchi, T.S., and Engelhaupt, E. (2000) Ammonium photoproduction from aquatic humic and colloidal matter. Aquat. Geochem. 6, 275–292.

Wang, W.C., Yung, Y.L., Lacis, A.A., Mo, J., and Hansen, J.E. (1976) Greenhouse effects due to man-made perturbations of trace gases. Science 194, 685–690.

Wang, W.X., and Fisher, N.S. (1997) Modeling metal bioavailability for marine mussels. Rev. Environ. Contam. Toxicol. 151, 39–65.

Wang, X.C., Chen, R.F., and Gardner, G.B. (2004) Sources and transport of dissolved and particulate organic carbon in the Mississippi River estuary and adjacent coastal waters of the northern Gulf of Mexico. Mar. Chem. 89, 241–256.

Wang, X.C., Druffel, E.R.M., and Lee, C. (1996) Radiocarbon in organic compound classes in particulate organic matter and sediments in the deep northeast Pacific Ocean. Geophys. Res. Lett. 23, 3583–3586.

Wang, X.C., and Lee, C. (1990) The distribution and adsorption behavior of aliphatic amines in marine and lacustrine sediments. Geochim. Cosmochim. Acta 54, 2759–2774.

Wang, X.C., and Lee, C. (1993) Adsorption and desorption of aliphatic amines, amino acids and acetate by clay minerals and marine sediments. Mar. Chem. 44, 1–23.

Wang, X.C., and Lee, C. (1995) Decomposition of aliphatic amines and amino acids in anoxic marine sediment. Geochim. Cosmochim. Acta 59, 1787–1797.

Wang, X.W., and Guo, L. (2000) Bioavailability of colloid-bound Cd, Cr, and Zn to marine plankton. Mar. Ecol. Prog. Ser. 202, 41–49.

Wang, Y., and van Capellen, P. (1996) A multicomponent reactive transport model of early diagenesis: application to redox cycling in coastal marine sediments. Geochim. Cosmochim. Acta 60, 2993–3014.

Wang, Z.A., and Cai, W. (2004) Carbon dioxide degassing and inorganic carbon export from a marsh-dominated estuary (the Duplin River): A marsh CO_2 pump. Limnol. Oceanogr. 49, 341–354.

Wangersky, P. (1965) The organic chemistry of sea water. Amer. Sci. 53, 358–374.

Wanninkhof, R. (1992) Relationship between gas exchange and wind speed over the ocean. J. Geophys. Res. 97, 7373–7382.

Ward, L.G., Kearney, M.S., and Stevenson, J.C. (1986) Accretion rates and recent changes in sediment composition of estuarine marshes. Ches. Bay Sci. EOS 67, 998.

Ward, L.G., Kemp, W.M., and Boynton, W.R. (1984) The influence of waves and seagrass communities on suspended particles in an estuarine embayment. Mar. Geol. 59, 85–103.

Ward, T., Butler, E., and Hill, B. (1998) Environmental Indicators for Natural State of Environmental Reporting: Estuaries and the Sea, Australia: State of the Environment (Environmental Indicators Report). Department of the Environment, Canberra, Australia.

Warnken, K.W., Gill, G.A., Santschi, P.H., and Griffin, L.L. (2000) Benthic exchange of nutrients in Galveston Bay, Texas. Estuaries 23, 647–661.

Waser, N.A.D., Bacon, M.P., and Michaels, A.F. (1996) Natural activities of ^{32}P and ^{33}P and the $^{33}P/^{32}P$ ratio in suspended particulate matter and plankton in the Sargasso Sea. Deep Sea Res. II. 43, 421–436.

Wasmund, N. (1997) Occurrence of cynaobacterial blooms in the Baltic Sea in relation to environmental conditions. Intl. Rev. Ges. Hydrobiol. 82, 169–184.

Wassmann, R., Thein, U.G., Whiticar, M.J., Rennenberg, H., Seiler, W., and Junk, W.J. (1992) Methane emissions from the Amazon floodplain: characterization of production and transport. Global Biogeochem. Cycles 6, 3–13.

Watts, C.D., and Maxwell, J.R. (1977) Carotenoid diagenesis in a marine sediment. Geochim. Cosmochim. Acta 41, 493–497.

Watts, S.F. (2000) The mass budgets of carbonyl sulfide, dimethyl sulfide, carbon disulfide and hydrogen sulfide. Atmos. Environ. 34, 761–779.

Webster, I.T., Ford, P.W., and Hodgson, B. (2002) Microphytobenthos contribution to nutrient–phytoplankton dynamics in a shallow coastal lagoon. Estuaries 25, 540–551.

Webster, I.T., Hancock, G.J., and Murray, A.S. (1995) Modeling the effect of salinity on radium desorption from sediments. Geochim. Cosmochim. Acta 59, 2469–2476.

Webster, J. R., and Benfield, E.F. (1986) Vascular plant breakdown in freshwater ecosystems. Annu. Rev. Ecol. Syst. 17, 567–594.

Wei, C.L., and Murray, J.W. (1994) The behavior of scavenged isotopes in marine anoxic environments: ^{210}Pb and ^{210}Po in the water column of the Black Sea. Geochim. Cosmochim. Acta 58, 1795–1811.

Weigner, T.N., and Seitzinger, S.P. (2001) Photochemical and microbial degradation of external dissolved organic matter inputs to rivers. Aquat. Microb. Ecol. 24, 27–40.

Weiler, R.R., and Mills, A.A. (1965) Surface properties and pore structure of marine sediments. Deep-Sea Res. 12, 511–529.

Weisburg, R.H., and Sturges, W. (1976) Velocity observations in the west passage of Narragansett Bay: a partially mixed estuary. J. Phys. Oceangr. 6, 721–734.

Weisberg, R.H., and Zheng, L. (2003) How estuaries work: a Charlotte harbor example. J. Mar. Res. 61, 635–657.

Weiss, R.F. (1974) Carbon dioxide in water and seawater: the solubility of non-ideal gas. Mar. Chem. 2, 203–215.

Wells, J.T. (1995) Tide-dominated estuaries and tidal rivers. *In* Geomorphology and Sedimentology of Estuaries. Developments in Sedimentology 53 (Perillo, G.M.E., ed.), pp. 179–205, Elsevier Science, New York.

Wells, J.T. (1996) Subsidence, sea-level, and wetland loss in the lower Mississippi River Delta. *In* Sea-Level Rise and Coastal Subsidence (Milliman, J.D., and Haq, B.U., eds.), pp. 281–311, Kluwer Academic Dordrecht, The Netherlands.

Wells, M.L. (2002) Marine colloids and trace metals. *In* Biogeochemistry of Marine Dissolved Organic Matter (Hansell, D.A., and Carlson, C.A., eds.), pp. 367–397, Academic Press, New York.

Wells, M.L., and Goldberg, E.D. (1991) Occurrence of small colloids in seawater. Nature 353, 342–344.

Wells, M.L., Kozelka, P.B., and Bruland, K.W. (1998) The complexation of "dissolved" Cu, Zn, Cd, and Pb by soluble and colloidal organic matter in Narragansett Bay, RI. Mar. Chem. 62, 203–217.

Wells, M.L., Smith, G.J., and Bruland, K.W. (2000) The distribution of colloidal and particulate bioactive metals in Narragansett Bay, RI. Mar. Chem. 71, 143–163.

Wells, M.L., Zorkin, N.G., and Lewis, A.G. (1983) The role of colloidal chemistry in providing a source of iron to phytoplankton. J. Mar. Res. 41, 731–746.

Welschmeyer, N.A., and Lorenzen, C.J. (1985) Chlorophyll budgets: zooplankton grazing and phytoplankton growth in a temperate fjord and the Central Pacific Gyres. Limnol. Oceanogr. 30, 1–21.

Wen, L., Santschi, P.H., Gill, G.A., and Tang, D. (2002) Silver concentrations in Colorado, USA, watersheds using improved methodology. Environ. Toxicol. Chem. 21, 2040–2051.

Wen, L., Shiller, A., Santschi, P.H., and Gill, G. (1999) Trace element behavior in Gulf of Mexico estuaries. *In* Biogeochemistry of Gulf of Mexico Estuaries (Bianchi, T.S., Pennock, J.R., and Twilley, R.R., eds.), pp. 303–346, John Wiley, New York.

Wernecke, G., Floser, G., Korn, S., Weitkamp, C., and Michaelis, W. (1994) First measurements of the methane concentrations in the North Sea with a new in-situ device. Bull. Geol. Soc. Denmark 41, 5–11.

Westerhausen, L., Poynter, J., Eglinton, G., Erlenkeuser, H., and Sarnthein, M. (1993) Marine and terrigenous origin of organic matter in modern sediments of the equatorial East Atlantic: the molecular record. Deep Sea Res. I. 40, 1087–1121.

Westman, P., Borhendahl, J., Bianchi, T.S., and Chen, N. (2003) Probable causes for cyanobacterial expansion in the Baltic Sea: role of anoxia and phosphorus retention. Estuaries 26, 680–689.

Westrich, J.T., and Berner, R.A. (1984) The role of sedimentary organic matter in bacterial sulfate reduction: the G model tested. Limnol. Oceanogr. 29, 236–249.

Wetzel, R.G. (1990) Land–water interfaces: metabolic and limnological regulators. Verh. Intl. Veriinig. Theor. Angew. Limnol. 24, 6–24.

Wetzel, R.G. (1995) Death, detritus, and energy flow in aquatic ecosystems. Freshwat. Biol. 33: 83–89.

Wetzel, R.G. (1999) Organic phosphorus mineralization in soils and sediments. *In* Phosphorus Biogeochemistry of Subtropical Ecosystems (Reddy, K.R., O'Connor, G.A., and Schelske, C.L., eds.), pp. 225–245, CRC Press, Boca Raton, FL.

Wetzel, R.G. (2001) Limnology—Lake and River Ecosystems. Academic Press, New York.

Wetzel, R.G. (2003) Dissolved organic carbon: detrital energetics. Metabolic regulators, and drivers of ecosystem stability of aquatic ecosystems. *In* Aquatic Ecosystems: Interactivity of Dissolved Organic Matter (Findlay, S.E.G., and Sinsabaugh, R.L., eds.), pp. 455–475, Academic Press, New York.

Wetzel, R.G., Hatcher, P.G., and Bianchi, T.S. (1995) Natural photolysis by ultraviolet irradiance of recalcitrant dissolved organic matter to simple substrates for rapid bacterial metabolism. Limnol. Oceanogr. 40, 1369–1380.

Wetzel, R.L., and Penhale, P.A. (1983) Production ecology of seagrass communities in the lower Chesapeake Bay. Mar. Tech. Soc. J. 17, 22–31.

Wheatcroft, R.A., Jumars, P.A., Smith, C.R., and Nowell, A.R.M. (1991) A mechanistic view of the particulate biodiffusion coefficient: step lengths, rest periods and transport direction. J. Mar. Res. 48, 177–207.

Wheeler, P.A., Watkins, J.M., and Hansing, R.L. (1997) Nutrients, organic carbon, and organic nitrogen in the upper water column of the Arctic Ocean: implications for the sources of dissolved organic carbon. Deep Sea Res. 44, 1571–1592.

Whelan, J.K., and Emeis, K. (1992) Sedimentation and preservation of amino acid compounds and carbohydrates in marine sediments. In Productivity, Accumulation, and Preservation of Organic Matter: Recent and Ancient Sediments (Whelan, J.K., and Farrington, J.W., eds.), pp. 176–200, Columbia University Press, New York.

Whistler, R.L., and Richards, E.L. (1970) Hemicelluloses. In The Carbohydrates; Chemistry and Biochemistry, 2nd edn. (Pigman, W., and Horton, D., eds.), pp. 447–468, Academic Press, New York.

Whitall, D.R., and Paerl, H.W. (2001) Importance of atmospheric nitrogen deposition to the Neuse River estuary, North Carolina. J. Environ. Qual. 30, 1508–1515.

White, R.H. (1982) Analysis of dimethyl sulfonium compounds in marine algae. J. Mar. Res. 40, 529–536.

Whitehouse, B.G., Macdonald, R.W., Iseki, K., Yunker, M.B., and McLaughlin, F.A. (1989) Organic carbon and colloids in the Mackenzie River and Beaufort Sea. Mar. Chem. 26, 371–378.

Whiticar, M.J. (1999) Carbon and hydrogen isotope systematics of bacterial formation and oxidation of methane. Chem. Geol. 161, 291–314.

Whiting, G.J., and Chanton, J.P. (1992) Plant-dependent CH_4 emissions in a subarctic Canadian fen. Global Biogeochem. Cycles 6, 225–231.

Whiting, G.J., and Chanton, J.P. (1996) Control of the diurnal pattern of methane emission from emergent aquatic macrophytes by gas transport mechanisms. Aquat. Bot. 54, 237–253.

Widerlund, A. (1996) Early diagenetic remobilization of copper in near-shore marine sediments: A quantitative pore-water model. Mar. Chem. 54, 41–53.

Wiegert, R.G., and Freeman, B.J. (1990) Tidal salt marshes of the Southeastern Atlantic Coast: A community profile. U.S. Fish and Wildlife Biol. Rep. 85, Washington, DC.

Wiegner, T.N., and Seitzinger, S.P.(2001) Photochemical and microbial degradation of external dissolved organic matter inputs to rivers. Aquat. Microb. Geol. 24, 27–40.

Wijararatne, R. and Means, J.C. (1984a) Affinity of natural estuarine colloids for hydrophobic pollutants in aquatic environments. Environ. Sci. Technol. 18, 121–123.

Wijararatne, R. and Means, J.C. (1984b) Sorption of polycyclic aromatic hydrocarbons (PAHs) by natural colloids. Mar. Environ. Res. 11, 77–89.

Wilhelm, S.W., and Suttle, C.A. (1999) Viruses ant nutrient cycles in the sea. Bioscience 49, 781–788.

Williams, E.G. (1960) Marine and fresh water folliferous beds in the Pottsville and Allegheny Group of western Pennsylvania. J. Paleontol., 34, 905–922.

Williams, E.G., and Druffel, E.R.M. (1987) Radiocarbon in dissolved organic matter in the central north Pacific Ocean. Nature 330, 246–248.

Williams, G.P. (1971) Aids in designing laboratory flumes. Open file report, U.S. Geol. Survey, Washington, D.C.

Williams, P.M., and Gordon, L.I. (1970) Carbon-13: carbon-12 ratios in dissolved and particulate organic matter in the sea. Deep-Sea Res. 17, 17–27.

Williams, W.A., and May, R.J. (1997) Low-temperature microbial aerobic degradation of polychlorinated biphenyls in sediment. Environ. Sci. Technol. 31, 3491–3496.

Wilson, J.O., Buchsbaum, R., Valiela, I., and Swain, T. (1986) Decomposition in salt marsh ecosystems: phenolic dynamics during decay of litter of *Spartina alterniflora*. Mar. Ecol. Prog. Ser. 29, 177–187.

Wilson, J.O., Valiela, I., and Swain, T. (1985) Sources and concentrations of vascular plant material in sediments of Buzzards Bay, Massachusetts, USA. Mar. Biol. 90, 129–137.

Wilson, M.A., Airs, R.L., Atkinson, J.E., and Keely, B.J. (2004) Bacteriovirdins: novel sedimentary chlorines providing evidence for oxidative processes affecting paleobacterial communities. Org. Geochem. 35, 199–202.

Windom, H.I. (1992) Contamination of the marine environment from land-based sources. Mar. Pollut. Bull. 25, 32–36.

Windom, H.I., Byrd, J.T., Smith, R.G., and Huan, F. (1991) Inadequacy of nasquan data for assessing metal trends in the nation's rivers. Environ. Sci. Technol. 25, 1137–1142.

Windom, H.I., Smith, R., and Rawlinson, C. (1989) Particulate trace metal composition and flux across the southeastern U.S. continental shelf. Mar. Chem.. 27, 283–297.

Windom, H.I., Smith, R., Rawlinson, C., Hungspreugs, M., Dharmvanij, S., and Wattayakorn, J. (1988) Trace metal transport in a tropical estuary. Mar. Chem. 24, 293–305.

Wines, R.A. (1985) Fertilizer in America: from Waste Recycling to Resource Exploitation. Temple University Press, Philadelphia, PA.

Winger, P.V., and Lasier, P.J. (1994) Effects of salinity on striped bass eggs and larvae from the Savannah River, Georgia. Trans. Am. Fish. Soc. 123, 904–912.

Winner, W.E., Smith, C.L., Koch, G.W., Mooney, H.A., Bewley, J.D., and Krouse, H.R. (1981) Rates of emission of H_2S from plants and patterns of stable sulfur isotope fractionation. Nature 289, 672–673.

Winterwerp, J.C. (1998) A simple model for turbulence induced flocculation of cohesive sediments. J. Hydraul. Res. 36, 309–326.

Wolanski, E., Drew, E., Abel, K., and O'Brian, J. (1988) Tidal jets, nutrient upwelling and their influence on the production of the alga *Halimeda* in the Ribbon Reefs, Great Barrier Reef. Estuar. Coastal Shelf Sci. 26, 168–201.

Wollast, R. (1983) Interactions in estuaries and coastal waters. *In* The Major Biogeochemical Cycles and Their Interactions (Bolin, B., and Cook, R.B., eds.), pp. 385–409, Wiley-Interscience, New York.

Wollast, R. (1988) The Scheldt estuary. *In* Pollution of the North Sea: An assessment (Salomon, W., Bayne, R., Duursma, E.K., and Forstner, U., eds.), pp. 183–193, Springer-Verlag, Berlin.

Wollast, R. (1993) Interactions of carbon and nitrogen cycles in the coastal zone. *In* Interactions of Carbon, Nitrogen, Phosphorus, and Sulfur Biogeochemical cycles and Global Change (Wollast, R., Mackenzie, F.T., and Chou, L., eds.), pp. 195–210, Springer-Verlag, Berlin.

Wollast, R. (1998) Evaluation and comparison of the global carbon cycle in the coastal zone and in the open ocean. *In* The Sea (Brink, K.H., and Robinson, A.R., eds.), pp. 213–252, John Wiley, New York.

Wollast, R., and Mackenzie, F.T. (1983) The global cycle of silica. *In* Silicon Geochemistry and Biogeochemistry (Aston, S.R., ed.), pp. 39–76, Academic Press, San Diego, CA.

Wommack, K.E., and Colwell, R.R. (2000) Virioplankton: viruses in aquatic ecosystems. Microb. Molec. Biol. Rev. 64, 69–114.

Wong, K.C., and Moses-Hall, J.E. (1998) On the relative importance of the remote and local wind effects on the subtidal variability in a coastal plain estuary. J. Geophys. Res. 103, 18393–18404.

Wong, K.C., and Valle-Levinson, A. (2002) On the relative importance of the remote and local wind effects on the subtidal exchange at the entrance to the Chesapeake Bay. J. Mar. Res., 60, 477–498.

Woodroffe, E. (1995) Transport of sediment in mangrove swamps. Hydrobiologia 295, 31–42.

Wright, D.L. (1977) The effect of calcium on cadmium uptake by the shore crab *Carcinus maenas*. J. Exp. Biol. Ecol. 67, 163–165.

Wright, D.L., and Welbourn, P.M. (2002) Environmental Toxicology. Cambridge Press, Cambridge, UK.

Wright, L.D. (1977) Sediment transport and deposition at river mouths: a synthesis. Geol. Soc. Am. Bull. 88, 857–868.

Wright, L.D., Coleman, J.M., and Thom, B.G. (1975) Sediment transport and deposition in a macrotidal river channel, Ord River, Western Australia. *In* Estuarine Research, Vol. 2 (Cronin, L.E., ed.), pp. 309–322, Academic Press, New York.

Wright, L.D., Wiseman, W.J., Yang, Z., Bornhold, B.D., Keller, G.H., Prior, D.J., and Suhayda, J.N. (1990) Processes of marine dispersal and deposition of suspended silts off the modern mouth of the Huanghe (Yellow River). Cont. Shelf Res. 10, 1–40.

Wright, S., Thomas, D., Marchant, H., Higgins, H., Mackey, M., and Mackey, D. (1996) Analysis of phytoplankton of the Australian sector of the Southern Ocean: comparisons of microscopy and size frequency data with interpretations of pigment HPLC data using the CHEMTAX matrix factorization program. Mar. Ecol. Prog. Ser. 144, 285–298.

Wu, F., Midorikawa, T., and Tanoue, E. (2001) Fluorescence properties of organic ligands for copper (II) in Lake Biwa and its rivers. Geochim. J. 35, 333–346.

Wulff, F., and Stigebrandt, A. (1989) A time-dependent budget model for nutrients in the Baltic Sea. Global Biogeochem. Cycles 3, 63–78.

Wulff, F., and Ulanowicz, R. (1989) A comparative anatomy of the Baltic Sea and Chesapeake Bay ecosystems. *In* Flow analysis of Marine Ecosystems: Theory and Practice (Wulff, F., Field, J.G., and Mann, K.H., eds.), pp. 82–89, Springer-Verlag, New York.

Wyda, J.C., Deegan, L.A., Hughes, J.E., and Weaver, M.J. (2002) The response of fishes to submerged aquatic vegetation complexity in two ecoregions of the mid-Atlantic bight: Buzzards Bay and Chesapeake Bay. Estuaries 25, 86–100.

Wysocki, L.A., Bianchi, T.S., Powell, R.T., and Reuss, N. (2006) Spatial variability in the coupling of organic carbon, nutrients, and phytoplankton pigments in surface waters and sediments of the Mississippi River plume. Estuar. Coastal Shelf Sci. (in press).

Xue, C. (1993) Historical changes in the Yellow River delta, China. Mar. Geol. 113, 321–329.

Yamaoka, Y. (1983) Carbohydrates in humic and fulvic acids from Hiroshima Bay sediments. Mar. Chem. 13, 227–237.

Yamashita, Y., and Tanoue, E. (2003) Chemical characterization of protein-like fluorophores in DOM in relation to aromatic amino acids. Mar. Chem. 82, 255–271.

Yang, M., and Sanudo-Wilhelmy, S.A. (1998) Cadmium and manganese distributions in the Hudson River estuary: Interannual and seasonal variability. Earth Planet. Sci. Lett. 160, 403–418.

Yanik, P.J., O'Donnell. T.H., Macko, S.A., Qian, Y., and Kennicutt II, M.C. (2003) Source apportionment of polychlorinated biphenyls using compound specific isotope analysis. Org. Geochem. 34, 239–251.

Yao, W., and Millero, F.J. (1995) The chemistry of the anoxic waters in the Framvaren Fjord, Norway. Aquat. Geochem. 1, 53–88.

Yao, W., and Millero, F.J. (1996) Oxidation of hydrogen sulfide by hydrous Fe (III) oxides in seawater. Mar. Chem. 52, 1–16.

Yentsch, C., and Phinney, D. (1997) Yellow substances in the coastal waters of the Gulf of Maine: Implications for ocean color algorithms. *In* Ocean Optics XIII (Ackleson, S.G., ed.), pp. 120–131, SPIE Proceedings 2963.

Yingst, J.Y., and Rhoads, D.C. (1978) Sea floor stability in central Long Island Sound. Part II. Biological interactions and their potential importance for seafloor erodibility. *In* Estuarine Interactions (Wiley, M.A., ed.), pp. 245–260, Academic Press, New York.

Yingst, J. Y., and Rhoads, D.C. (1980) The role of bioturbation in the enhancement of bacterial growth rates in marine sediments. *In* Marine Benthic Dynamics (Tenore, K.R., and Coull, B.C., eds.), pp. 407–421, University of South Carolina Press, Columbia.

Yokokawa, T., Nagata, T., Cottrell, M.T., and Kirchman, D.L. (2004) Growth rate of the major phylogenetic bacterial groups in the Delaware estuary. Limnol. Oceanogr. 49, 1620–1629.

Yoon, W. B., and Benner, R. (1992) Denitrification and oxygen consumption in sediments of two south Texas estuaries. Mar. Ecol. Prog. Ser. 90, 157–167.

Yoshinari, T. (1976) Nitrous oxide in the sea. Mar. Chem. 4, 189–202.

Yunker, M.B., Cretney, W.J., Fowler, B.R., Macdonald, R.W., and Whitehouse, B.G. (1991) On the distribution of dissolved hydrocarbons in natural water. Org. Geochem. 17, 301–307.

Yunker, M.B., Macdonald, R.W., Cretney, W.J., Fowler, B.R., and McLaughlin, F.A. (1993) Alkane, terpene, and ploycyclic aromatic hydrocarbon geochemistry of the Mackenzie river and Mackenzie shelf: riverine contributions to Beaufort Sea coastal sediment. Geochim. Cosmochim. Acta 57, 3041–3061.

Yunker, M.B., Macdonald, R.W., Goyette, D., Paton, D.W., Fowler, B.R., Sullivan, D., and Boyd, J. (1999) Natural and anthropogenic inputs of hydrocarbons to the Strait of Georgia. Sci. Total Environ. 225, 181–209.

Yunker, M.B., Macdonald, R.W., Veltkamp, D.J., and Cretney, W.J. (1995) Terrestrial and marine biomarkers in a seasonally ice-covered Arctic estuary—integration of multivariate and biomarker approaches. Mar. Chem. 49, 1–50.

Yunker, M.B., Macdonald, R.W., Vingarzan, R., Mitchell, R.H., Goyotte, D., and Sylvestre, S. (2002) PAHs in the Fraser River basin: a critical appraisal of PAH ratios as indicators of PAH source and composition. Org. Geochem. 33, 489–515.

Zajac, R. N. (2001) Organism–sediment relations at multiple spatial scales: implications for community structure and successional dynamics. *In* Organsim–Sediment Interactions (Aller, J.Y., Woodin, S.A., and Aller, R.C., eds.), pp. 119–140, University of South Carolina Press, Columbia.

Zamuda, C.D., and Sunda, W.G. (1982) Bioavailability of dissolved copper to the American oyster *Crassostrea virginica*: Importance of chemical speciation. Mar. Biol. 66, 77–79.

Zang, X., Nguyen, R.T., Harvey, R., Knicker, H., and Hatcher, P.G. (2001) Preservation of proteinaceous material during the degradation of the green alga *Botryococcus braunii:* a solid-state 2D ^{15}N ^{13}C NMR spectroscopy study. Geochim. Cosmochim. Acta 65, 3299–3305.

Zappa, C.J., Raymond, P.A., Terray, E.A., and McGillis, W.R. (2003) Variation in surface turbulence and the gas transfer velocity over a tidal cycle in a macro-tidal estuary. Estuaries 26, 1401–1415.

Zepp, R.G., Callaghan, T.V., and Erickson, D.J. (1995) Effects of increased solar ultraviolet radiation on biogeochemical cycles. Ambio 24, 181–187.

Zhai, W., Dai, M., Cai, W.J., Wang, Y., and Wang, Z. (2005) High partial pressure of CO_2 and its maintaining mechanism in a subtropical estuary: the Pearl River estuary, China. Mar. Chem. 93, 21–32.

Zhang, J.G., Huang, W.W., and Wang, Q. (1990) Concentration and partitioning of particulate trace metals in the Changjiang. Wat. Air Soil Pollut. 52, 57–70.

Zhang, L., Walsh, R.S., and Cutter, G.A. (1998) Estuarine cycling of carbonyl sulfide: production and air–sea flux. Mar. Chem. 61, 127–142.

Zhang, L.J., Wang, B.Y., and Zhang, J. (1999) $pCO2$ in the surface water of the East China Sea in winter and summer. J. Ocean U. Qingdoa (suppl.), 149–153. (In Chinese.)

Zhang, Y., Zhu, L., Zeng, X., and Lin, Y. (2004) The biogeochemical cycling of phosphorus in the upper ocean of the East China Sea. Estuar. Coastal Shelf Sci. 60, 369–379.

Ziegler, R., Blaheta, A., Guha, N., and Schonegge, B. (1988) Enzymatic formation of pheophorbide and pyrophephorbide during chlorophyll degradation in a mutant of *Chlorella fusca* Shihira et Kraus. J. Plant Physiol. 132, 327–332.

Zieman, J.C., Fourqurean, J.W., and Iverson, R.L. (1989) Distribution, abundance, and productivity of seagrasses an macroalgae in Florida bay. Bull. Mar. Sci. 44, 292–311.

Zieman, J.C., and Wetzel, R.G. (1980) Productivity in seagrass methods and rates. *In* Handbook of Seagrass Biology: An Ecosytem Perspective (Phillips, R.C., and McRoy, C.P., eds), pp. 87–116, Garland STPM Press, New York.

Zieman, J.C., and Zieman, R.T. (1989) The ecology of the seagrass meadows of the west coast of Florida: a community profile. U.S. Fish and Wildlife Ser. Biol. Report 85(7.25).

Zimmerman, A.R., and Benner, R. (1994) Denitrification, nutrient regeneration and carbon mineralization in sediment of Galveston Bay, Texas, USA. Mar. Ecol. Prog. Ser. 114, 275–288.

Zimmerman, A.R., and Canuel, E.A. (2000) A geochemical record of eutrophication and anoxia in Chesapeake Bay sediments: anthropogenic influence on organic matter composition. Mar. Chem. 69, 117–137.

Zimmerman, J.T.F. (1988) Estuarine residence times. *In* Hydrodynamics of Estuaries. Vol. I. Estuarine Physics (Kjerfve, B., ed.), pp. 75–84, CRC Press, Boca Raton, FL.

Zimmerman, R.C. (2003) A biooptical model of irradiance distribution and photosynthesis in seagrass canopies. Limnol. Oceanogr. 48, 568–585.

Zlotnik, I., and Dubinsky, Z. (1989) The effect of light and temperature on DOC excretion by phytoplankton. Limnol. Oceanogr. 34, 831–839.

Zobell, C.E., and Feltham, C.B. (1935) The occurrence and activity of urea-splitting bacteria in the sea. Science 81, 234–236.

Zobrist, J., and Stumm, W. (1979) Chemical dynamics of the Rhine catchment area in Switzerland: extrapolation into the 'pristine' Rhine River input into the oceans. *In* Proceedings Review and Workshop on River Inputs to Ocean Systems, pp. 26–30, FAO, Rome.

Zou, L., X. Wang., Callahan, J., Culp, R.A., Chen, R.F., Altabet, M., and Sun, M. (2004) Bacterial roles in the formation of high-molecular-weight dissolved organic matter in estuarine and coastal waters: evidence from lipids and the compound-specific isotopic ratios. Limnol. Oceanogr. 49, 297–302.

Zubkov, M.V., Fuchs, B.M., Archer, S.D., Kiene, R.P., Amann, R., and Burkill, P.H. (2001) Linking the composition of bacterioplankton to rapid turnover of dissolved dimethylsulphoniopropionate in an algal bloom in the North Sea. Environ Microbiol. 3, 304–311.

Zucker, W.V. (1983) Tannins: does structure determine function? An ecological perspective. Am. Nat. 121, 335–365.

Zweifel, U.L. (1999) Factors controlling accumulation of labile dissolved organic carbon in the Gulf of Riga. Estuaries 48, 357–370.

Zwolsman, J.J.G. (1994) Seasonal variability and biogeochemistry of phosphorus in the Scheldt Estuary, South-West Netherlands. Estuar. Coastal Shelf Sci. 39, 227–248.

Index

Lightning Source UK Ltd.
Milton Keynes UK
UKHW020310190419
341300UK00012B/769/P